Fifteenth Scientific and Practical International Conference on International Transport Infrastructure, Industrial Centers and Corporate Logistics (NTI-UkrSURT 2019)

SHS Web of Conferences Volume 67 (2019)

Kharkiv, Ukraine
6 – 8 June 2019

Editors:

Sergly Panchenko
Glib Vatulia
Volodymyr Dykan

ISBN: 978-1-5108-9605-5

Printed from e-media with permission by:

Curran Associates, Inc.
57 Morehouse Lane
Red Hook, NY 12571

Some format issues inherent in the e-media version may also appear in this print version.

For additional information, please contact EDP Sciences – Web of Conferences
at the address below.

EDP Sciences – Web of Conferences
17, Avenue du Hoggar
Parc d'Activité de Courtabœuf
BP 112
F-91944 Les Ulis Cedex A
France

Phone: +33 (0) 1 69 18 75 75
Fax: +33 (0) 1 69 28 84 91

contact-edps@webofconferences.org

Additional copies of this publication are available from:

Curran Associates, Inc.
57 Morehouse Lane
Red Hook, NY 12571 USA
Phone: 845-758-0400
Fax: 845-758-2633
Email: curran@proceedings.com
Web: www.proceedings.com

TABLE OF CONTENTS

Statement of Peer review

In submitting conference proceedings to SHS Web of Conferences, I certify to the Publisher that I adhere to the **Policy on Publishing Integrity** of the journal in order to safeguard good scientific practice in publishing.

1. All articles have been subjected to peer review administered by the proceedings editors.
2. Reviews have been conducted by expert referees, who have been requested to provide unbiased and constructive comments aimed, whenever possible, at improving the work.
3. Proceedings editors have taken all reasonable steps to ensure the quality of the materials they publish and their decision to accept or reject a paper for publication has been based only on the merits of the work and the relevance to the journal.

Title, date and place of the conference

Fifteenth Scientific and Practical International Conference «INTERNATIONAL TRANSPORT INFRASTRUCTURE, INDUSTRIAL CENTERS AND CORPORATE LOGISTICS», June 6–8, 2019, Kharkiv, Ukraine

Proceedings editor(s): **Prof. Sergiy Panchenko, Prof. Glib Vatulia, Prof. Volodymyr Dykan**

26.09.2019

Date and editor's signature

17, avenue du Hoggar - PA de Courtabœuf – BP 112 - 91944 Les Ulis Cedex A (France)
Tél. : 33 (0)1 69 18 75 75 - Fax : 33(0)1 69 07 45 17 – www.edpsciences.org

Identification, Measurement and Reflection of Risks in Accounting

Natalia Chebanova[1,1], *Victoria* Orlova[1] , *Liliy* Revutska[1], and *M. Karpushenko*[2]

[1] Ukrainian State University of Railway Transport, Economy and Management of Industrial and Commerce Business Department, Feuerbach sq. 7, 61050 Kharkiv, Ukraine
[2]O.M.Beketow National University of Urban Economy in Kharkiv

Abstract. In the modern environment, the company constantly faces various types of risks in its business activities. Therefore, the problem of identifying and measuring risks is extremely relevant. The article proposes a scheme for managing financial risks, which includes identifying risk factors, determining the permissible risk level, analyzing individual transactions, developing risk mitigation measures. The article proposes to create the following funds, reserves and collateral: a bad debts reserve, provision for warranty service of clients, provision for social orientation, provision for restructuring, provision for burdensome contracts, fiscal (tax) reserves, commercial, industrial, informational risk reserves, future costs and payments reserve, legal provisions, provisions for impairment of assets, reserve fund. Risks should be taken only if the level of return on risky operations exceeds the level of risk. The issue of the choice of certain reserves, funds and provisions is regulated by the accounting policy of the enterprise, where their types and the order of their creation should be clearly defined. Such measures allow planning contingency expenses and informing users of financial statements of future risk events.

1 Introduction

The issue of identification and measurement of business risks is extremely important for modern business, as under all circumstances a situation can occur that will negatively affect the company and lead to unexpected financial consequences. All this requires the establishment of an effective risk management system that would identify, predict, and measure risks. It should be emphasized that it is accounting where there are certain rules and standards allowing pre-evaluating of the business risks. Therefore, accounting norms and principles of risk management should be combined to create such a system.

Business risks were analyzed in works by C. Drury [1], P. Atamas [2], L. Gnilitska [3], I. Deroun [4], I. Chibisova [5], B. Zasadny [6] and other scientists.

Having analyzed the study of risks, we can conclude that there is no single concept of risk management, since this problem has been considered from various aspects. For instance, I. Deroun [4] focused on approaches to understanding risks and developed a set of indicators to identify risks in the accounting system. L. Gnilitska [3] separated the risk-causation and risk-consequence and proposed a generalizing scale for assessing the probable consequences of the business risk, and analyzed the effect of risk on the value of

[1] Corresponding author: n_chebanova@ukr.net

accounting objects. Some scientists, namely P. Atamas [2], I. Chibisova [4] and others, focused on risk management. However, some important issues related to risk measurement and forecasting remain unclear and not sufficiently specified.

The concept of "risk" is considered by most researchers as the possibility of occurrence at an enterprise, which can cause loss of property, funds, expected profit deficiency, income, the introduction of additional costs as a result of adverse circumstances or events. But there is another interpretation. Thus, according to B. Zasadniy [5], "risk" should be interpreted as a combination of phenomena, events and factors that influence the making managerial decisions and can lead to both positive and negative result of the enterprise. According to C. Druri [1], if there are more than one possible outcomes, we can speak about risk. I.e., the "risk" is associated with an unexpected event – either negative or positive. In our opinion, the risk of business activity is associated with unexpected or anticipated future events that are likely to lead to expenditures. The risk may be caused by various factors that are related both to the internal circumstances associated with the production, financial and commercial activities of the enterprise, and with external factors such as inflation, political and economic situation, exchange rate changes, demographic situation, etc.

2 Presentation of the main results

The most common type of risk is financial risk, which can include the risk of insolvency, crime risk, loan risk, risk of financial stability, risk of inflation, tax risk, currency risk, deposit risk, investment risk, and others. In addition to the traditional types of risks that are already known for a sufficient amount of time, the use of IT technologies has brought new types of risks - risks associated with cyber attacks, viruses, Internet fraud, loss of reputation due to negative information and other risks. Information risks present the most serious threat to any enterprise today, especially if this company is actively working on the Internet. But of course, each enterprise has its own types of risks associated with the specifics of their activities.

The result of the occurrence of risk events - the losses that can be estimated in cash.

Recently, the influence of the accounting system on the adoption of managerial decisions and the assessment of any events is increasing. An enterprise should strive to reflect in the financial statements not only information about risks, but also about their impact on the financial position of the enterprise and the size of possible losses. The concept of risk is associated with the expectation of any events, business operations, the cost of which is not predetermined. You can identify many types of risks, but surely all risks should not be displayed. In order to confirm this position, it is necessary to refer to accounting norms.

The concept of materiality in the financial statements is determined by IAS 1 "Presentation of Financial Statements" [7].According to the standard, significant omissions or errors occur when they can, individually or collectively, affect the economic decisions taken by users on the basis of their financial statements. The boundary between significant events is determined by the specific value that each enterprise determines, depending on the size of the assets, income or profit in the accounting policy. Thus, it is necessary to compare the costs associated with risk events, with the limit of materiality. If these costs do not exceed the limit, they should not be displayed if their expected size is greater - they should be reflected to provide accurate reporting. An important role is played not only by the size of costs but also by the probability of a risk event occurring. For example, according to L. Gnilitska [3], it is expedient to reflect only such events, the probability of occurrence of which exceeds 50%. And unforeseen obligations that have a small probability of manifestation (10-50%), it is expedient to display off-balance sheet for further control of their offensive. Contingent liabilities, characterized by a minimal probability of their occurrence (0-10%), should be considered a real uncertainty, which is not reflected in the balance sheet.

Based on the fact that the risk is due to negative so positive events, they should be taken if the level of return on risky operations will exceed the size of expected costs. At the same time, the company's expenses for the neutralization of risk should not exceed the amount of losses on it. According to I. Chibisova [4], risk management is based on the application of the principles of risk-taking. P. Atamas [2] states that enterprise risk management involves substantiation of leading tasks and methods of risk neutralization, among which there are two main types of strategies: passive and active. Choosing a passive strategy, the company carries out risk avoidance operations, that is, it refuses or reduces its risk operations. Active strategy is aimed at carrying out risky operations, but commercial insurance and self-insurance of risks, their diversification and hedging prevent their negative impact from being reduced.

It is expedient to structure the process of risk management as a sequence of certain stages (fig. 1):

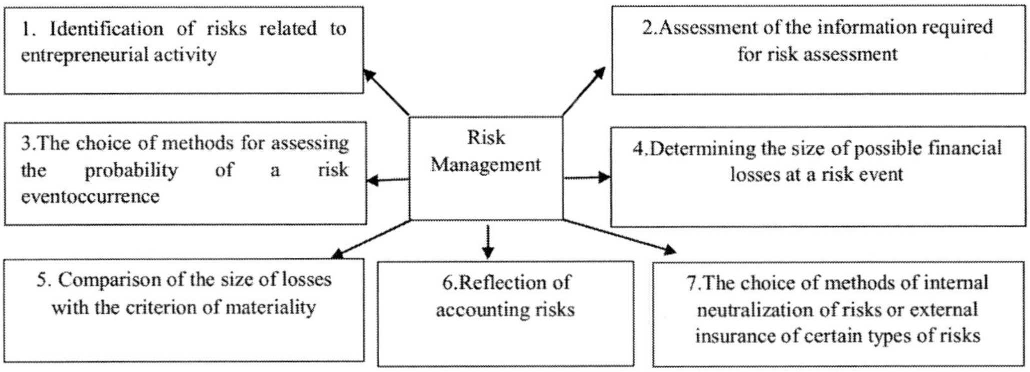

Fig. 1. Risk Management Process.

The most difficult step is to determine the scope of possible losses on risky events. To do this, you need to choose a particular method of evaluation. In practice, the following methods are most commonly used to assess risks:

— expert-based on a group of experts' survey, with further processing of the results of this survey;

— statistical - apply in the presence of statistical data for a number of previous periods;

— computational analytical - designed to calculate the precise quantitative expression of the level of risk based on the internal information of the enterprise itself;

— analogue - allow you to assess the level of risk in individual transactions based on comparison with similar, already multiple operations. In order to ensure the reliability of the comparison, both their own and external experience of such operations are used.

The use of a method depends on the specific conditions and types of risk. The choice of any of the methods is related to solving the question of the adequacy and reliability of the information that is present in the enterprise.

From the point of view of the possibility of influence and application of assessment methods, the following types of risk can be distinguished (Figure 2):

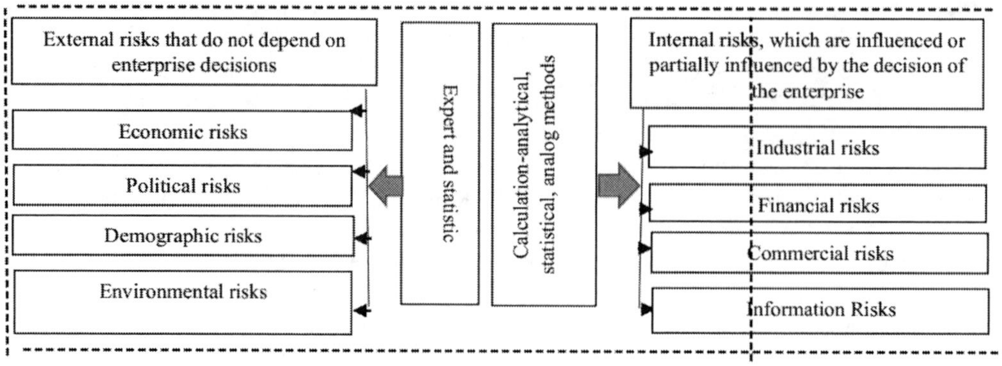

Fig. 2. Application of assessment methods for various types of risk.

Having determined the probable amount of losses from risk events and making sure that it is really significant and may affect the decision of the owners of the company, it must be reflected in the accounting. This can be done by one of the following approaches:

1) the creation of certain reserves, provisions, funds - the application of such an approach will lead to the enterprise recognizing in advance the costs that will be incurred if the risk event occurs. If there is a need for this, you can defer these funds to a deposit account and then use them in case if such an event will occur in the future;

2) reflection in the notes to the financial statements - the enterprise records the existing risk events that may affect the company and lead to significant changes in the financial statements.

Indeed, in accordance with IAS 1, "Presentation of Financial Statements" [7], the notes disclose non-financial information, in particular the sources of estimation uncertainty at the end of the reporting period that present a significant risk and may lead to a significant adjustment of the balance cost of assets and liabilities for the following fiscal year. In respect of such assets and liabilities, notes should contain detailed information about their nature and their balance cost at the end of the reporting period. It should be emphasized that all features of the creation of reserves, funds and provisions should be reflected in the accounting policies of the enterprise.

To create reserves, collateral and funds, it is the most optimal to use the computational and analytical methods, which are based on the indicators of a particular enterprise and take into account the norms of accounting standards.

Let's consider the features of creating reserves, collateral and funds. If we take into account the generalities of calculating the cost of risk events, it is advisable to classify them into three groups.

The first group is the reserves, funds and collateral that can be calculated on the basis of accounting data of previous periods with the use of calculation and analytical methods. This group includes the following reserves and collateral (Fig. 3):

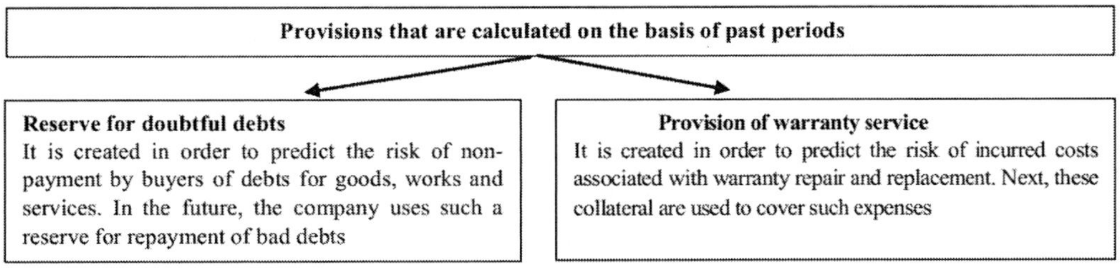

Fig. 3. Types of provisions that are calculated based on past periods.

To calculate these reserves, the accounting data of previous periods is used. I.e, it is assumed that the amount of expenses which the enterprise had in the event of a risk event in past periods will be approximately equal to the amount of costs in future periods. The more periods taken into account, the more accurate the results of the calculations. For example, to calculate the warranty service provision, it is necessary to gather information on the amount of costs actually incurred by the company in connection with the performance of warranty substitutes. For this purpose the following formula can be used:

$$S_{ri}= R_i\ x((\sum E_n /R_n)/n); \qquad (1)$$

S_{ri}- provision of warranty service of the i-th period;
R_i- revenues of the i-th period;
E_n- expenses for warranty service of past periods;
R_n – revenues of past periods
n- the number of previous periods.

The second group is the reserves, funds, provisions, which are based on reliable information of the company about future costs. This information can be estimated on the basis of available documents-calculations, estimates, accounts, etc. (Fig. 4).

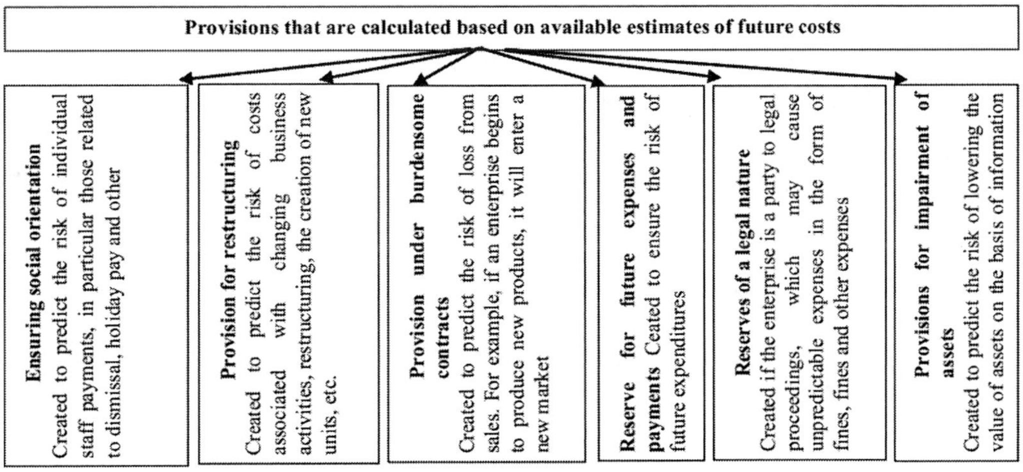

Fig. 4. Types of provisions that are calculated on the basis of future periods.

To determine the size of the cost of risk events for the given group evaluated certain documents, on the basis of which are made calculations. For example, in order to calculate employee benefits, information is required on the size of the wage bill and the amount of holiday pay planned for the year as well as the actual wage bill for the current period.

But there is such a provision, the size of which is not calculated, but is based on the data contained in the estimates, calculations and other documents. This, for example, is a provision for restructurings, a reserve for future payments and payments, etc. (Fig. 5):

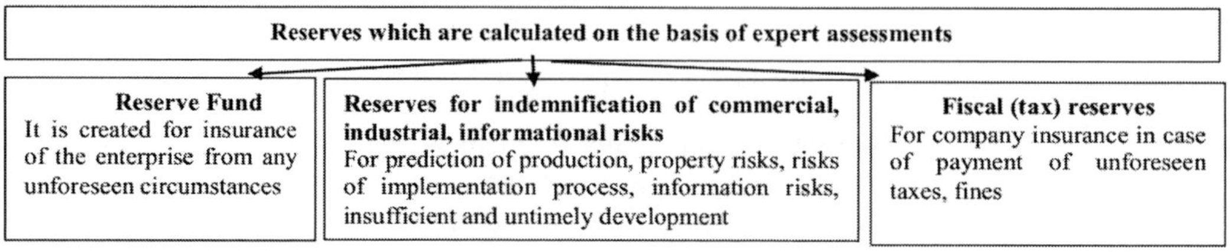

Fig. 5. Types of reserves that are calculated on the basis of expert assessments.

Indeed, the specified reserves can not be assessed based on past periods or available documents. For example, a reserve fund is formed from a net profit of an enterprise, the decision on its size is taken by the owner of the enterprise, taking into account a number of features of the operation of the enterprise.

Determining such reserves (provisions, funds) is a problem that should be solved by experts, the owner, the chief accountant, the financial director, the head of the company and other specialists, taking into account the events of future periods.

3 Conclusions

The problem of identification, measurement and risk assessment is extremely relevant, since today, enterprises may have situations that involve risks. Risk management is a rather complicated process that depends on its types and involves a procedure for certain actions and the application of methods for assessing possible losses. Particular attention is paid to accounting methods of risk insurance. These methods include the creation of reserves and provisions, certain funds that will be used in future to offset potential losses. Establishing reserves, funds, provisions for each enterprise is regulated in the accounting policy of the enterprise, where their types and procedure of creation should be clearly defined.

References

1. Druri, C. *Management and cost accounting*. Moscow: Uniti-Dana, (2005)
2. Atamas, P.J., Atamas, O.P., Kramarenko, G.O. Academic review, **1 (44)**, 60-69, (2016)
3. Gnilitska, L. V. Economic innovation, **57**, 88-100, (2014)
4. Derun, I.A. Economic Annals-XXI, **159(5–6)**, 97–100, (2016)
5. Chibisova, I.V., Levchyn B.V. The problems of e systemic approach in economic, **2(56)**, 113-116, (2016)
6. Zasadnyi, B. A. Nay Scientific herald of Uzhgorod National University, **14 (1)**, P. 111-115, (2017)
7. IAS 1 "Presentation of Financial Statements"- URL: https://www.ifrs.org/issued-standards/list-of-standards/ias-1-presentation-of-financial-statements/

Formation of an Innovative and Creative System of Managing the Development of Industrial Enterprises

Viktoriia Chobitok[1*], *Larisa* Chumak[1], *Tetiana* Demyanenko[1], *Yulia* Us[2]

[1] Ukrainian Engineering Pedagogical Academy (Kharkiv)
[2] Simon Kuznets Kharkiv National University of Economics (Kharkiv)

Abstract. The article deals with the issues of forming an innovative and creative system of managing the development of industrial enterprises. An analysis of the state of the development of domestic enterprises has been carried out. The essence of the directions of the development of domestic industrial enterprises has been defined. The analysis of the essence of the concept of "enterprise development" has been conducted, with the authors' own definition of the term being provided. Theoretical bases for the formation of the concept of "an innovative and creative system of managing the development of industrial enterprises" have been devised. Types of innovative and creative directions of managing the development of industrial enterprises have been singled out.

1 Introduction

The world economy at the present stage of its development is characterized by a significant increase in competition. The domestic industrial sector, as a budget-builder, is coming under considerable pressure from major foreign competitors. Effective functioning and constant development of domestic industrial enterprises actively affects the level of social and economic state of the country.

Successful functioning of domestic industrial enterprises, in the modern economic conditions, taking into account the permanent unfavorable influences of external and internal factors on their activity, is possible subject to formation and introduction of innovative and creative management of the development of all their structural elements.

2 Presenting main material

Some questions of the formation of an innovative and creative system of managing the development of industrial enterprises have been considered by such scientists as O. Kuzmin

*Corresponding author: vika_chobitok@ukr.net

[1], I. Lapina [3], O. Pashchenko [3], Yu. Pohorelov [4], V. Prokhorova [5], M. Sushko [6], V. Shylova [7] et al. But there still remain a number of debatable issues which concern the effective formation and implementation of an innovative and creative system of managing the development of industrial enterprises with regard to the permanent conditions of the environment, so the problem field under investigation needs further clarification.

The purpose of the research is to form an innovative and creative system of managing the development of industrial enterprises in modern economic conditions.

In the economy of any country a significant role is played by the level of industrial development. In Ukraine, the industry produces about 40% of the national output of goods and services. The domestic industry has about one third of the main means of production of the total number used in the country's economy. Although the domestic industry is in a difficult condition, it is one of the main sources of salaries for the employed, businessmen's profits, tax revenues, etc.

The indices of industrial production by types of activity for the period 2013-2018, according to the State Statistics Service of Ukraine [9], are given in Table 1.

Table 1. Indices of industrial production by types of activity for 2013-2018, %.

Industry types	Years					
	2013	2014	2015	2016	2017	2018
Industry including:	95.7	89.9	87.0	102.8	100.4	101.6
Mining and quarrying	100.8	86.3	85.8	99.8	94.3	102.4
Processing industry	92.7	90.7	87.4	104.3	104.8	101.1
Supply of electricity, gas, steam and air conditioning	98.9	93.4	88.0	102.5	93.5	102.8

The data provided in Table 1 were used to plot a histogram of the changes in indices of industrial production by types of activity for 2013-2018. The analysis of changes in indices of industrial production by types of activity for 2013-2018 shows that throughout almost the whole period under study, with the exception of 2018, the volumes of industrial production tended to decrease compared with the previous period, which is strongly related to the influence of internal factors (revolutions, unstable domestic political situation, decrease in the support from the state, constant increase in energy tariffs, raising unemployment, etc.) and external factors (decline in the level of investment in the domestic industry, decrease in the level of competitiveness, military conflicts, etc.). All this results in destabilization of the activity of domestic industry enterprises and negatively affects their level of development. An innovative and creative management system should be formed in order to increase the competitiveness of domestic industrial enterprises. The basis for achieving promising directions in the development of domestic industrial enterprises is the use of intellectual as well as scientific and technological potential in the organizational and economic support of production processes and provision of services. Justification of the priorities in the development of domestic industrial enterprises will help restore their significant role in the economic growth of the country and ensure its level of economy as a whole.

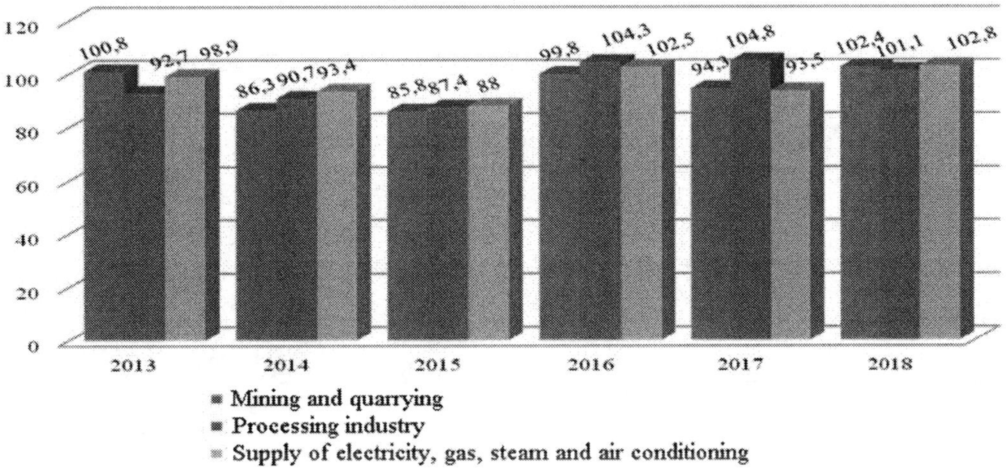

Fig. 1. Histogram of the changes in indices of industrial production by types of activity for 2013-2018.

The essence of the directions of the development of domestic industrial enterprises is shown in Fig. 2.

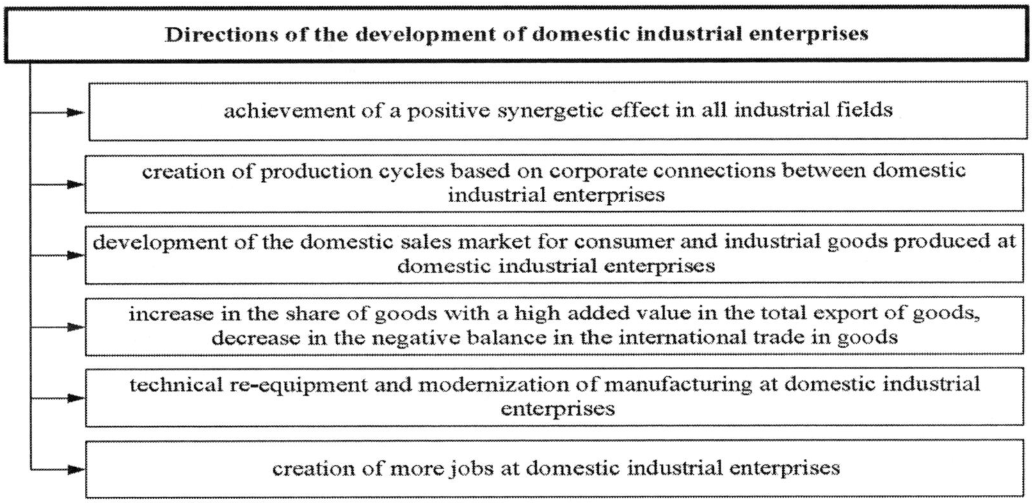

Fig. 2. Directions of the development of domestic industrial enterprises.

In modern economic literature, scientists give multiple different definitions of the essence of the concept of "enterprise development", some of which are presented in Table 2.

Table 2. Definitions of the concept of "enterprise development".

Author	Definition
V. Andriichuk	These are irreversible, purposeful and regular changes of the economic system, the sequence of its transition from one state to a qualitatively different one on the basis of the improvement of equipment, technology and labor organization as

Author	Definition
	well as by means of introducing innovations in management
R. Ackoff	Development is primarily characterized by qualitative changes such as improvement of technological equipment of manufacturing, professional development of employees, change of managerial principles and mechanisms, etc.
N. Afanasiev, V. Gohozhyn, V. Rudyka	The objective change only of the qualitative characteristics of the system is conditioned by both the fundamental laws of nature (unity and struggle of opposites, transfer of quantity into quality, spiral and upward development of society) and the laws of the functioning of specific systems (equipment aging, experience and knowledge acquisition by employees, depletion of natural resources), as a result of which some new properties of the system are formed
S. Dunda	A set of directed, intensive and qualitative changes of the economic nature occurring at the enterprise as a result of contradictions in the internal environment and influences of outer factors
Economic Encyclopedia	A natural process of quantitative and qualitative changes within the economic system, its transition to a new quality, a more perfect form
O. Kuzmin	The development of each organization which has a cyclic nature, i.e. passes through certain stages of the life cycle that determine the features of production, economic, investment, financial and other activities
M. Mescon, M. Albert, F. Khedouri	A long-term program for improving problem-solving capabilities and capabilities for renewal, especially by increasing the effectiveness of managing the culture of the organization
O. Raievnieva	A unique process of transformation of an open system in space and time, characterized by a permanent change in the global goals of its existence by means of forming a new dissipative structure, as well as by means of its transition to a new attractor of functioning (one of the alternative trajectories of enterprise development)

The analysis of the essence of the concept of "enterprise development" shows that the authors put forward their own definition of the notion, which they regard as the process of quantitative and qualitative changes taking place at the enterprise under the influence of internal and external factors and leading to an increase in the level of its operational, investment and financial activities through the formation of an innovative and creative management system.

Throughout the history of enterprise development, a system of managerial relations has been formed, with an opportunity to register events being discovered at the beginning of the fifth millennium BC, when the Sumerians invented the cuneiform characters. Since then, humanity has continuously worked on and implemented an innovative and creative management system which has undergone the following stages of development: religious and commercial revolution; secular and administrative revolution; production and construction revolution; industrial revolution; bureaucratic revolution and information revolution.

Theoretical bases for the development of the concept of "an innovative and creative system of managing the development of industrial enterprises" are shown in Fig. 3.

The innovative and creative system of managing the development of industrial enterprises consists in the formation and implementation of purposeful and logical changes in the economic system, the sequence of its transition from one state to a qualitatively different one on the basis of the improvement of equipment, technology, labor organization, etc., which is the foundation for increasing their competitiveness on domestic and foreign markets.

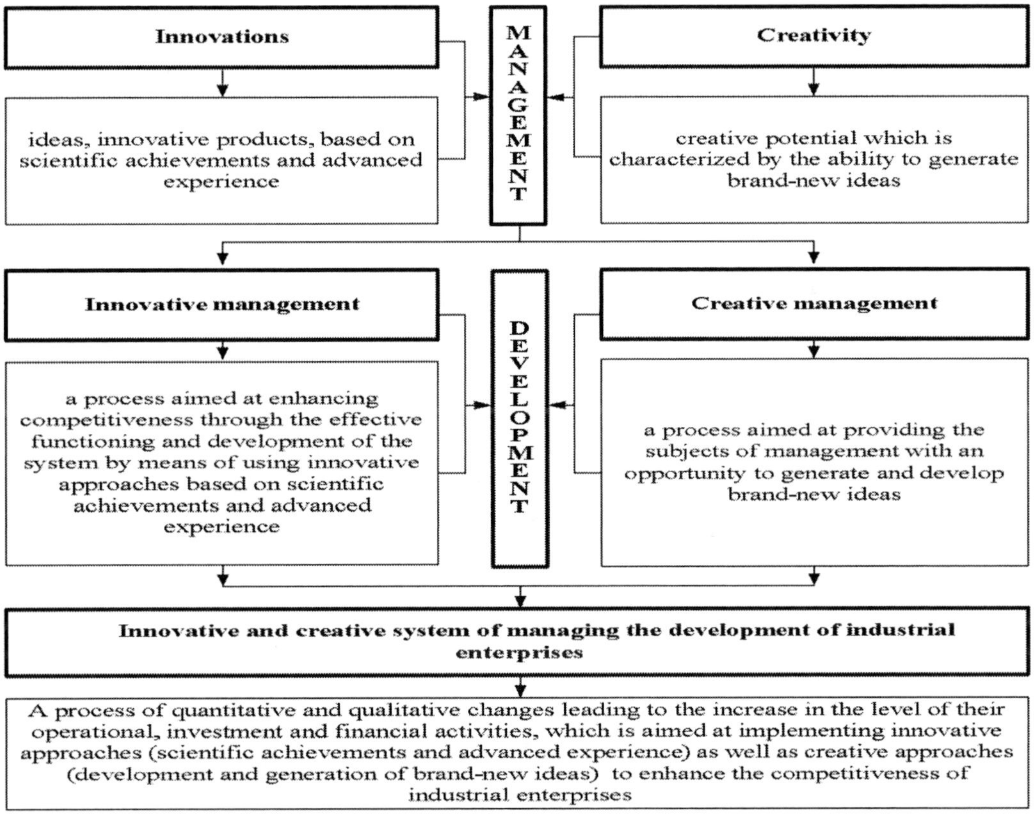

Fig. 3. Theoretical bases for the formation of the concept of "an innovative and creative system of managing the development of industrial enterprises".

Fig. 4 demonstrates the types of innovative and creative directions of managing the development of industrial enterprises which are most often considered.

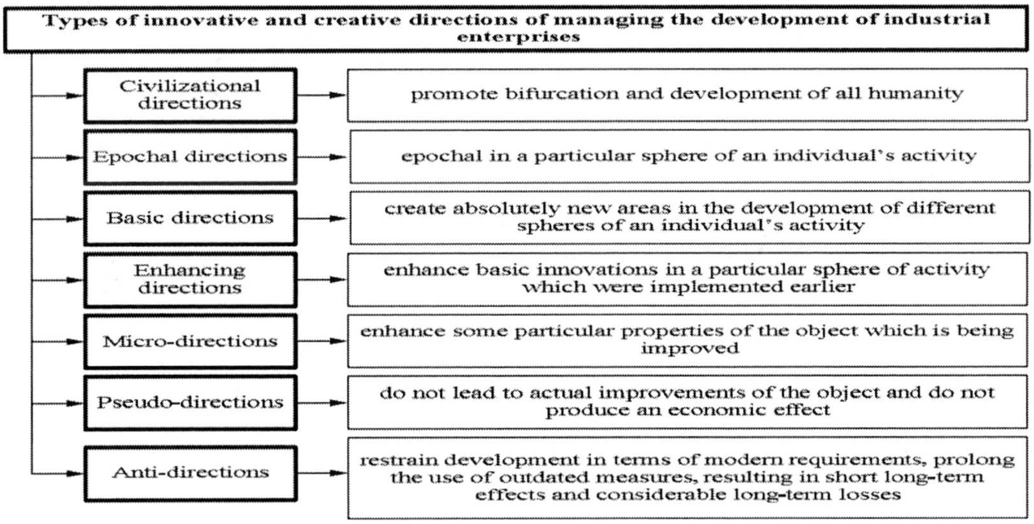

Fig. 4. Types of innovative and creative directions in the management of the development of industrial enterprises.

Certainly, today's growing demand is the introduction of basic and creative innovations into the system of management of industrial enterprises. This will help effectively organize the activity of industrial enterprises in the conditions of constant changes in market requirements. The possibility of introducing an innovative and creative system of managing the development of industrial enterprises becomes a reality through the use of advances in all areas of "high technology".

3 Conclusions from the conducted research.

According to the results of the research, it is determined that domestic industrial enterprises need to implement an innovative and creative system of managing the development of industrial enterprises, which will enable them to enhance their competitiveness. Only those managers who have innovative and creative thinking that differs in the originality of views and who are able to act flexibly and constructively in non-standard situations and form independent judgments can make effective managerial decisions, they achieve their goals, learn from mistakes and are ready to take risks. Leaders who have developed innovative and creative thinking make industrial enterprises brighter, more interesting, turning everything into something brand-new, effective and rational.

References

1. O. Ye.Kuzmin, D. K .Zinkevich, A. M. Chushak-Holoborodko Naukovyi visnyk of NLTU of Ukraine, **21.2**, 146-152 (2011)

2. I. M. Lapina Modern innovative technologies of management, Retrieved from http://klasnaocinka.com.ua/uk/article/suchasni--innovatsiini--tekhnologiyi--upravlinnya.html

3. O. P. Pashchenko. Global and national economic problems, Mykolaiv National University named after V. O. Sukhomlynskyi, **17**, 406- 410 (2017)

4. Yu. S.Pohorelov, K. A. Leiko. Economy and region, **6**, 58–63 (2015)

5. V. V. Prokhorova, V. M.,Protsenko, V. I.Chobitok *Formation of a competitive strategy of enterprises on the basis of innovation-directed investment*, Kharkiv, Ukrainian Engineering Pedagogics Academy, 291 (2015)

6. M. Yu.Sushko. Bulletin of economic science of Ukraine, **1,** 93–98 (2017)

7. V. V. Shylova Creative technologies in the system of managing foreign economic activity, Retrieved from http://conf.management.fmm.kpi.ua/proc/article/view/62558

8. L. F. Chumak, Bulletin of the University of Dnipropetrovsk, Series "Economics", **20.10/1**, 176–181 (2012)

9. Official website of the State Statistics Service of Ukraine, Retrieved from http://www.ukrstat.gov.ua/

10. I.V. Litvin. Scientific Bulletin of NLTU of Ukraine. **21.7,** 360-364 (2011)

Strategic Knowledge Management as Means of Ensuring the Competitiveness of Railway Transport Enterprises

Volodymyr Dykan [1], *Viktoria* Ovchynnikova [1*], *Valeria* Dykan[2], *Natalia* Kalicheva[1] and *Myroslava* Korin[1]

[1] Ukrainian State University of Railway Transport, Economy and Management of Industrial and Commerce Business Department, Feuerbach sq. 7, 61050 Kharkiv, Ukraine
[2]Karazin business school, management and administration Department, Mironositska st., 1, 61001 Kharkiv, Ukraine

Abstract. In article timeliness and expediency of application of strategic management of knowledge as effective means of ensuring competitiveness of the enterprises of railway transport is proved. Analysis of existing models of knowledge management has led to the conclusion that each of the existing models of knowledge management requires certain conditions regarding the possibility of its use. First of all, it is the presence of a stable organizational culture in the enterprise, a favorable moral climate, the initiative of employees, in fact, is a prerequisite for the creation of implicit knowledge, attention to the value of human and intellectual capital on the part of management, as well as the possibility of accumulating knowledge for their further analysis, use and improvement. This created the basis for improving the conceptual provisions of strategic knowledge management in railway transport, revealing management technologies in the following areas: the development of competencies of personnel focused on the implementation of strategic objectives; the creation of a comprehensive integrated information environment that improves the quality of business processes; the implementation of effective interaction of railway transport with stakeholders to improve its image; the formation of knowledge subsystem of economic security management.

1 Problem statement and its connection with scientific or practical tasks

The formation and development of market relations in our country is highly dependent on the conditions in which business entities are forced to conduct economic activity. The competition continues between enterprises, it is a feature inherent to competitive relations. In order to survive the competition and make a profit, railway transport enterprises must constantly ensure their competitiveness and maintain it at a sufficient level.

* Corresponding author: vikols75@ukr.net

Strengthening of crisis phenomena and transformation processes occurring in the world and domestic economic and transport systems, determine the need for new approaches and directions to improve the efficiency of railway transport. In the context of fierce competition in the domestic and world markets of transport services, one of the most effective tools to improve the efficiency of rail transport is to change the organizational and economic approaches and means of ensuring their competitiveness, of particular importance among which is knowledge management.

2 Analysis of recent research and publications

The problems of ensuring the competitiveness of rail transport, in particular through knowledge management, deals with a significant number of scientists and practitioners, among which are the works by V. Dykan, I. Tokmakova, V. Ovchynnikova, M. Porter, V. Shynkarenko, A. Bondarenko, N. Kalycheva, E. Sych, et al. [1-8]. During the research conceptual framework on this topic found that you can singled out three areas of knowledge management and "European", considering the knowledge management from the perspective of measurement (Karl Sveiby), the “American” approach is the direct “knowledge management” (Karl Wiig) and focuses on the “creation of knowledge” (by Ikujiro Nonaka and Hirotaka Takeuchi). The Japanese approach characterizes the four-phase model of knowledge management SECI, or “spiral of knowledge”, proposed by Japanese researchers I. Nonaka and H. Takeuchi [6], which is a system of knowledge transfer between the four phases and socialization, externalization, combination, internalization and through three levels of social aggregation, namely: individual, group, context.

The representatives of the European approach K. Sveiby, L. Edvinsson [7] model of knowledge management from the perspective of the intellectual capital of enterprises, which represents the interaction of four components: human capital, structural capital, business assets and intellectual property. Such models are purely theoretical in nature, and its practical use is complicated by the burden of managing individual business processes.

The American approach proposed by K. Wiing [8] reveals the model of knowledge management based on four main provisions: the creation and application of knowledge in production technologies and procedures, systematic knowledge research, knowledge assessment for each type of business and active knowledge management throughout the life cycle of goods and innovations. However, there are a number of limitations that do not allow effective use of this model in practice. In particular, it is the probability of problems in the process of assessing knowledge, methods and procedures of assessment, the controversy of their results on the basis of which, in fact, there is knowledge management.

Having analyzed the existing models of knowledge management, we state that each of the models of knowledge management requires certain conditions regarding the possibilities of its use. First of all, it is the presence of a stable organizational culture in the enterprise, a favorable moral climate, the initiative of employees, in fact, is a prerequisite for the creation of implicit knowledge, attention to the value of human and intellectual capital on the part of management, as well as the possibility of accumulating knowledge for their further analysis, use and improvement. This will create a basis for further improvement of the conceptual provisions of strategic knowledge management in railway transport, will contribute to the competitiveness of the industry.

3 Conceptual provisions of knowledge management in railway transport

The beginning of the new millennium is marked by the entry of human society into a phase of its development, when knowledge becomes the main source of competitive advantages of individuals and organizations. Information and knowledge produced on its basis allow responding quickly and adequately to changes in the external and internal environment of the organization. Awareness of knowledge as an economic resource, vital for the activities of companies, is a good reason for their inclusion in the sphere of management.

In the modern economy of knowledge is really becoming the main driving force of socio-economic development. The generation of new knowledge acquires the status of a priority strategic activity in the course of which a competitive product (work, services) is produced.

Knowledge management includes the following components [8, 9]: the stimulation of the growth of knowledge; selection and storage of meaningful information from external to the organization sources; the preservation, classification, transformation, ensuring the availability of knowledge; the dissemination and exchange of knowledge, including in the framework of the organization; knowledge in business processes, particularly in decision-making; the embodiment of knowledge in products, services, documents, databases, and software; evaluation, measurement, and use of intangible assets of the organization; protecting knowledge.

The basic principles of knowledge management include: the principle of science (involves the construction of a knowledge management system based on scientific recommendations, which requires the use of the full range of modern scientific achievements); the principle of consistency (involves taking into account the interaction and interdependence of all components of the knowledge management system in the enterprise); the principle of temporality (includes accounting for the value of the environment and feedback to the success of the enterprise, management decisions based on the study of the totality of situational factors); the principle of understanding and use of psychological factors (involves the creation of conditions for the most complete realization of the personal potential of each employee and the optimization of interpersonal relationships in the team); the principle of authority and responsibility (is that each employee must have sufficient authority to be responsible for the quality of the work); the principle of democratic centralism (provides for the granting of autonomy with the preservation of the leadership of the coordination functions); the principle of integration the effectiveness of knowledge management depends on the successful integration of skills, knowledge, experience, scientific potential, etc.; the principle of development of conditions and opportunities is that it is necessary to create such conditions under which participants of knowledge management possess all information, analyze and consider organizational culture, know all fundamental provisions; the principle of ensuring the novelty of knowledge – knowledge changes over time, become obsolete: knowledge has a natural life cycle, so some elements of knowledge must be constantly updated; the principle of remuneration for knowledge and the necessary monitoring and analysis of the results of the creation, application and development of knowledge; the principle of turning knowledge into action is to create new knowledge and use it for innovative learning and research projects.

4 Strategic knowledge management system as a means of ensuring the competitiveness of railway transport

The strategic approach to knowledge management in railway transport is implemented through ensuring the relationship of knowledge management with the strategy of Ukrzaliznytsia JSC and the needs of external stakeholders. In particular, it establishes causal relationships between knowledge and needs of external stakeholders, identifies key knowledge needed to implement the strategy and identifies a gap in key knowledge [10]. Given the above provisions, we present the author's approach to the formation of the concept of knowledge management in railway transport (Fig. 1).

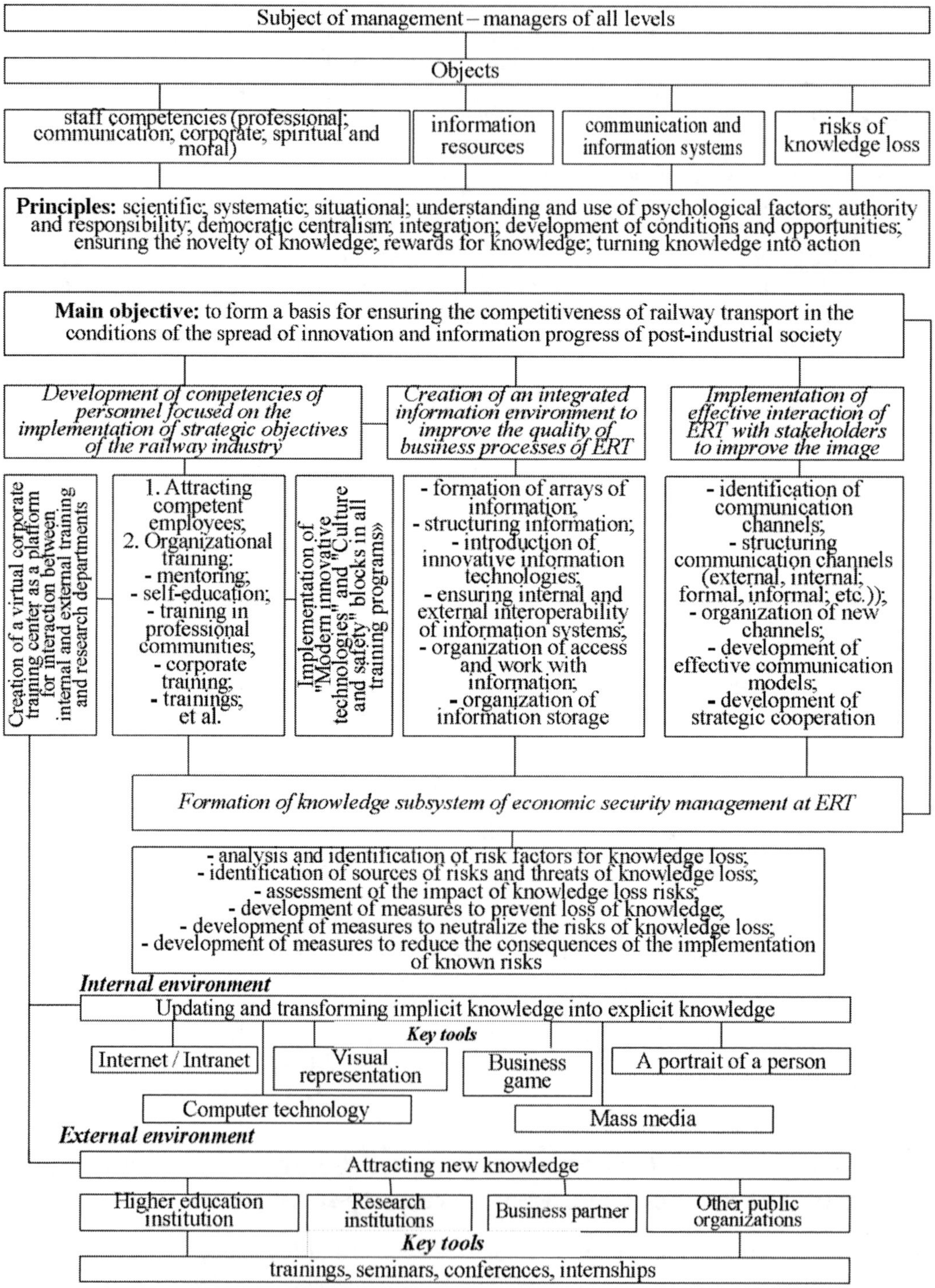

Fig. 1. Formation of strategic knowledge management system in railway transport

The subjects of knowledge management at railway transport enterprises are the top managers of all levels of the management hierarchy. The object of management identified the competence of personnel, information resources, information and communication systems and the risks of loss of knowledge of railway transport enterprises. The main goal of knowledge management should be the formation of a basis for ensuring the competitiveness of railway transport in the conditions of the spread of innovation and information progress of post-industrial society. The objectives of knowledge management in railway transport should be defined: development of competencies of personnel focused on the implementation of strategic objectives of the railway industry; creation of a comprehensive integrated information environment that improves the quality of business processes of railway transport enterprises; implementation of effective interaction of railway transport enterprises with stakeholders to improve the image of the railway industry; formation of knowledge subsystem of economic security management at railway transport enterprises.

A huge number of authors consider training to be the main source of knowledge for the enterprise. The relationship between knowledge management and the staff development function as an element of the personnel management system is not in doubt. To implement the concept of knowledge management based on the personnel management system, the key functions of the system are the functions of personnel development, evaluation, motivation and regulation of social and psychological processes. The formation of knowledge in the process of development of staff competencies occurs as a result of attracting competent employees, and as a result of organizational training.

In the light of modern approaches, organizational learning can be seen through the prism of comprehensive employee ownership of knowledge management through their comprehensive development and corporate learning.

It should be noted that in order to provide the industry with qualified specialists in the Ukrainian railway transport, an integrated system of training for the industry was created, an industry program for improving the material and technical base of educational institutions was developed, which is aimed at strengthening the direct connection between the enterprises of railway transport and educational institutions by transferring devices and equipment, normative and technical documentation, performing repair work on the reconstruction of educational buildings, dormitories.

The system of organizational training in the system of Ukrzaliznytsia JSC should cover mentoring, self-education, training in professional communities, corporate training, trainings, etc. Indicative activities in the field of organizational training:

1. Development of new knowledge, transformation of implicit knowledge into explicit: identification of gaps in knowledge; conducting research; synthesis of information, the results of various studies and the development of new knowledge; conducting research that is valuable for management processes and publication of their results; conducting specialized reviews of the results of research; providing participants with the results of research.

2. Dissemination of best practices: benchmarking: search for partners, concrete examples of best practices; documentation of best practices; compilation and dissemination of experience.

3. Meetings and seminars: implementation of training programs based on feedback from the audience; organization of video seminars; conferences, meetings with experts on topical issues.

An effective mechanism of formalization of corporate knowledge can be a virtual training center of Ukrzaliznytsia JSC. In the context of the development of the "knowledge economy" it is advisable to understand the training center as not just a training structural unit, but a platform for interaction with other economic agents, consulting companies, educational institutions, professional associations in the direction of joint design of the corporate training and development system. The basis for ensuring the competitiveness of railway transport enterprises and the formation of its strategic advantages in the long term is the constant updating of knowledge. It is possible to carry out in full the specified by means of the offered conceptual provisions of strategic management of knowledge at the enterprises of railway transport opening the administrative technologies in the directions: development of competences of the personnel focused on implementation of strategic tasks; creation of a complex integrated information environment, ensuring the quality of business processes; implementation of effective interaction of railway transport with stakeholders to improve its image; formation of knowledge subsystem of economic security management.

References

1. V.L. Dykan, I.V. Tokmakova, V.O. Ovchynnikova, M.V. Korin, A.V. Tolstova, *Osnovy biznesu*, Kharkiv, 291 (2018)
2. M. Porter, *Competitive strategy: techniques for analyzing industries and competitors* Alpina-Business-Books, 454 (2005)
3. V.G. Shynkarenko, A.S. Bondarenko, *Upravleniye konkurentosposobnostyu predpriyatiya*. Kharkiv, 186 (2003)
4. N.E. Kalycheva, Scientific Bulletin of Kherson State University, **27,** 139-142 (2017)
5. E.M. Sych, Bulletin of Kharkiv national University named after V. N. Karazin, **512**, 152-154 (2001)
6. V.L. Dykan, The bulletin of transport and industry economics, **43**, 11-20 (2013)
7. L.Edvinsson, M. Malone, Intellectual Capital: realizing your company's true value by fi nding its hidden brainpower, N.Y.: Harper Business, 230 (1997)
8. K. Wiig, Journal of Knowledge Management, **1**, 6-14 (1997)
9. L.A. Trofimova, V.V. Trofimov, Knowledge Management: textbook, SPb.,77 (2012)
10. V.O. Zadoya, Problems of transport Economics, **9**, 71-76 (2015)

Competitiveness Assessment for Transport Enterprises in the Context of Accelerated Development of the International Logistics Infrastructure

Olga Ievsieieva[1]

[1]Ukrainian State University of Railway Transport, Feierbakh Square 7, 61050, Kharkiv, Ukraine

Abstract. In the study, the authors have identified the dominant trends in the development of the international road transport services market. These include monopolization and protectionism, structural imbalances, namely, predominance of European transport companies in the international motor transport market, prevailing of private transport companies. Hence, the researchers determined the general vector of improvement of the motor transport services market in the modern stage of internationalization of global economic relations. The current trends in the development of the international road transport services market in Ukraine are substantiated. The authors propose a competitiveness assessment methodology for transport enterprises. It is proposed to quantify the following elements as part of the competitiveness coefficient of a transport enterprise: 1) the range of offered transportation services; 2) pricing policy; 3) advertising; 4) public relations; 5) sales system; 6) employee incentive system; 7) innovative activity of the enterprise. Using the proposed methodological approach will help making reasonable management decisions to improve the efficiency of transport enterprises, to expand the information base of economic research, and to improve the quality of analysis in the context of accelerated development of the international logistics infrastructure.

Searching for a solution of the problem how to increase the competitiveness of enterprises in the context of the development of the international logistics infrastructure is becoming increasingly important in view of the globalization of the world economy, transformation of the economic environment, and the growing needs of people and businesses in transport services. The competitiveness level is based on the assessment of the enterprise, which shows how successful this enterprise is in the market and how effective its activity is.

The transport system is one of the most significant components of the global infrastructure and economic environment of Ukraine. The transport system creates conditions for transportation of goods and services, as well as the mobility of the population and production factors, and thus serves as a link in the global economy, ensuring its integrity and unity. It was the development of the transport infrastructure that served as the initial prerequisite for the formation of a global economy. Today, the Ukrainian transport

[1] Corresponding author: polkya@meta.ua

infrastructure retains this function, allowing it to take advantage of the international division of labor, enhancing the internationalization of world economic relations.

The increasing degree of integration occurring within the world economic space in recent years confirms the imperative of the development of the Ukrainian transport infrastructure in order to ensure the effective functioning of these integration associations, mainly focused on the development of trade and transport enterprises. In this context, the modern international transport infrastructure of Ukraine needs objective analysis and evaluation for the needs determined by the imperatives of global development. The competitive analysis of Ukrainian transport companies is of paramount importance, since it would allow identifying the direction of their modernization and improving the competitiveness of enterprises.

In the study, we have identified dominant trends in the development of the international road transport services market (monopolization and protectionism; structural imbalances, namely, predominance of European transport companies in the international motor transport market; predominance of private transport companies), which allowed us to determine the general vector of its improvement in the current stage of internationalization of global economic relations.

The current trends in the development of the international transport infrastructure include the globalization of logistics, prevalence of non-monetary factors in the development of international transport infrastructure to increase the investment attractiveness of the international transport infrastructure. In our opinion, now when the global competition is formed in the world market of services of transport infrastructure enterprises, price competition is becoming an increasingly relevant area. Therefore, the systems of integrated logistics and intermodal transport technologies are sustainably developing, since they allow optimal combination of various types of transport.

Today, transport companies seek to minimize their risks and costs. Therefore, integrated associations of transport companies from different countries are becoming ever popular. The efficiency of transport services is increased with the integration. The integrated transportation scheme involves concluding an agreement with a major international transport company, which plots and organizes the optimal transportation route and concludes further transportation agreements with forwarding companies operating in the countries which the customer's freight is transported through.

The main problems in the development of international transport infrastructure include the lack of unified standards of the international customs system, poor accessibility of transport infrastructure, uneven involvement of various types of transport in the transport infrastructure, the problem of security and attraction of investments to upgrade the transport industry, the problem of developing regulatory support for the improvement of the transport system and transport services market, shortage of qualified professionals, territorial and structural inconsistencies in the operation of transport infrastructure, insufficient labor mobility in the transport infrastructure, poor quality of transport services, increased adverse impact of vehicle operation on the environment, insufficient use of integrated logistics systems, poor development of the intermodal transport technology, and low usage of information systems.

The conducted analysis revealed that the swift increase in demand for international motor transportation and the development of modern trade leads to a qualitative change in the international road transport services market associated with increased competition: the position of European transport companies has improved in recent years; in 2012, Western European countries controlled more than 20% of the market under investigation. Therefore, we concluded about a tendency of forming the so-called "European monopoly on international road transport".

In the course of the research, we revealed that the current directions of the development of the international road transport services market in Ukraine include: 1) overcoming the protectionist policies of most countries of the world with regard to Ukrainian road transport enterprises; reducing tariff and non-tariff restrictions in this area; 2) improvement of the international integration of road transport companies to accumulate large financial resources, which is a prerequisite to introduce innovations; providing transportation services of higher quality at a lower cost; 3) promotion of antimonopoly policy in the international market of services of motor transport enterprises and stimulation of fair competition; 4) introduction of unification, standardization and system formation in the international motor transport services market in Ukraine.

Improving the competitiveness of transport enterprises during the accelerated development of the international logistics infrastructure is an urgent task of economics. Today, the competitiveness assessment itself should be conducted using up-to-date economic and mathematical methods and a personal computer. In addition, the methodology should be understandable for practitioners of the analytical departments of enterprises and be based on accessible software.

Domestic and foreign scientific literature focuses on marketing management methods, which allow taking into account the influence of hardly predictable market factors on the activity of the enterprise and, thus, increasing its survival rate in the intense market competition [1]. A company needs to study, analyze and assess these features, i.e. the strength of the influence of an unstable environment on the activities of the transport enterprise, the conditions of competition and hardly predictable behavior of competitors, first of all, in order to: 1) identify their competitive advantages and disadvantages vs. competitors; 2) determine the degree of its adaptation to the impact of external and internal factors of the market environment; 3) assess their competitiveness level; 4) develop complexes of organizational, economic, technical and technological measures aimed at its improvement; 5) develop various development scenarios for the market environment and effective competitive strategies for its development in this environment.

Today, when the competitiveness of an enterprise is a criterion of sustainability and survival of a transport enterprise during the accelerated development of the international logistics infrastructure, its analysis, evaluation and forecasting becomes objectively necessary, because in modern competitive struggle, with its large scale, changes and intensity, they who analyzes and fights for its competitive position win. In the market economy, only resilient enterprises can function, which flexibly adjust and use new mechanisms in management [2]. Quantitative assessment of the competitiveness level of a transport enterprise, as well as the development of measures to manage it becomes one of the important tasks.

Competitiveness can be assessed in the general and local aspects [3]. For transport enterprises, competitiveness in the general aspect is described as the satisfaction of consumers' demand and the ratio of performance indices with the parameters of a business system. The need to solve both economic and social problems of the territory where the company is located, determines the differentiation of indices by the same characteristic [4]. Activities of transport enterprises by the territorial authorities should be monitored mainly for solving social problems. Economic indices are indicators of the state of this system and information for making strategic decisions regarding stimulating the development of the logistics sector associated with specific innovation and investment programs [5].

The local aspect of competitiveness of the enterprise includes the need to assess the efficiency of resource use [6]. This task is set within transport enterprises and is a function of business structures. We propose a method for calculating the competitiveness coefficient, which belongs to the group of efficiency indicators of resource use and most fully reflects the business activity of the management of a transport enterprise. Other things being equal, business activity determines the development and effective use of the materials and technical facilities of an enterprise, as well as labor and financial resources [7].

It is proposed to consider the following elements as part of the competitiveness coefficient of a transport enterprise: 1) the range of the offered transportation services; 2) pricing policy; 3) advertising; 4) public relations; 5) sales system; 6) employee incentive system; 7) innovative activity of the enterprise.

We used the prioritization method for the quantitative determination of the competitiveness coefficient of the enterprise. The use of the expert assessment method is preconditioned by the fact that such organizational and economic problems are solved with the absent or incomplete necessary initial information. Using this method quite acceptable results can be obtained, which contributes to a noticeable improvement in the marketing activity of the enterprise [8].

A quantitative competitiveness assessment of the enterprise was obtained on the basis of expert information gathered by making specialists of the transport enterprises involved in the study to fill in specially designed questionnaires. The expert assessment procedure was based on the use of the method of pairwise comparison, in which all indices are compared in pairs, each subsequent assessment not being related to the previous one; all pairwise estimates form a pairing preference matrix which is further processed to obtain weight coefficients to the elements of the marketing complex.

The algorithm for assessing the competitiveness of transport enterprises includes: 1) determining the most significant elements that form competitiveness; 2) compilation of comparison systems for the surveyed enterprises by elements of the competitiveness coefficient; 3) determining the significance (weight) of elements that form competitiveness and their assessment; 4) assessment of the competitiveness level of a particular transport enterprise: 4.1) analysis of competitive factors; 4.2) determining the regression equation; 5) development of ways to improve competitiveness.

Our study covered 20 transport companies. Comparison systems for enterprises by the main factors of competitiveness were compiled as a result of a survey of 79 analysts and managers of enterprises in the region. Such comparison systems were used as the basis to plot square adjacency matrices which are further used to calculate the values of relative priorities for each competitive factor. Assessment of the primary competitive factors is given in Table 1.

Table 1. Assessment of the competitive factors.

Competitive factors	Score
1. Range of offered transportation services	0.190
2. Pricing policy	0.132
3. Advertising	0.105
4. Public relations	0.156
5. Sales system	0.116
6. Employee incentive system	0.159
7. Innovative activity of the enterprise	0.141

To calculate the competitiveness coefficients, we have determined the significance of its elements, and then solved the problem of linear convolution of elements. Competitiveness scoring formed the basis for developing the competitiveness assessment methodology for the transport enterprises (Table 2).

Table 2. Competitive scores of enterprises.

Enterprise	Total competitive scores
1	0.052257
2	0.057002
3	0.072843
4	0.061758
5	0.070468
6	0.037213
7	0.046714
8	0.036421
9	0.048298
10	0.068092
11	0.053840
12	0.041964
13	0.062550
14	0.038797
15	0.041964
16	0.041964
17	0.041964
18	0.041964
19	0.041964
20	0.041964

We used the score assessment of such a quality factor as competitiveness of a transport enterprise (Figure 1) to calculate the operation performance using the correlation and regression analysis.

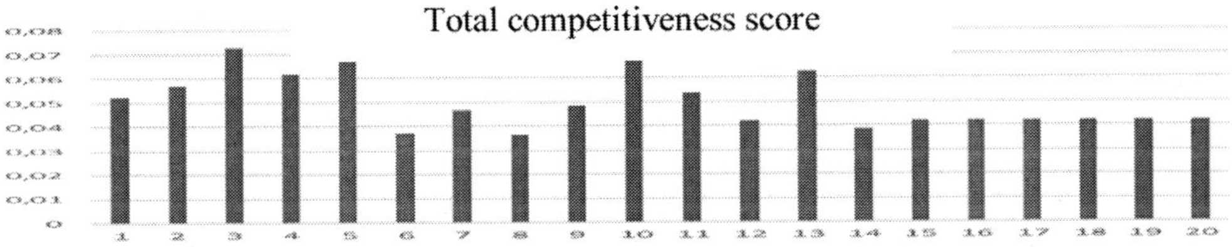

Fig. 1. Competitiveness assessment of transport enterprises in the region.

In our opinion, the advantage of the proposed methodology for determining the competitiveness of transport enterprises is that the equation for the amount of profit can be used to determine which specific competitive factor it is most significantly affected it. This will help make informed management decisions to improve the efficiency of economic activity in general, expand the information base of economic research, and improve the quality of analytical work.

We have also developed methodological tools for assessing the economic condition of a motor transport enterprise in the context of the accelerated development of the international transport infrastructure on the competitiveness of transport enterprises.

There is a relationship between the development of the international transport infrastructure and the competitiveness of Ukrainian road transport enterprises. We adjusted the competitiveness assessment methodology of enterprises on the basis of an assessment of their financial condition to the features of the transport market, and also related it to the international transport infrastructure.

The developed methodology involves a regression analysis to identify the relationship between the level of development of the international transport infrastructure and the competitiveness of transport enterprises, reveals the degree of correlation of these indices, and uses them as the basis for recommendations in improvement of the competitiveness of transport enterprises.

In our method of economic analysis, the vector of all indices is the same. In other words, higher values of indices are evidence of a better economic state of the organization, while lower ones, on the contrary, indicate a worse economic state. The main advantages of the developed methodology include the complexity and multidimensionality of the analysis of the economic standing of the organization; comparability of data and the opportunity to compare different organizations, since standardized financial (accounting) reports are used for economic analysis; ability to identify the changes in of the development of the economic state of the analyzed organization; based on the results of the analysis, it is possible to make more efficient management decisions by the organization; the computational algorithm is flexible, provides the implementation of mathematical modeling of the economic situation of the organization; economic assessment provides the identification of deviations of production and business indices of the organization and allows to take appropriate measures to improve the competitiveness of the enterprise.

Assessment of the company's finance and production, planning the company's activities and forecasting its development is possible based on the results of the economic analysis.

Conducting a comprehensive economic analysis using the selected indices provides a description of the economic condition of the organization and allows planning a strategy based on the obtained results to improve the competitiveness of the transport enterprise.

References

1. Vasiutina I.O. State and Regions. Series: Economics and business, **5**, 114–118 (2011)
2. Geleta I. V. *Ekonomika organizatsii (predpriyatiya)*, Moscow: Magistr, 140 (2014)
3. Nikolaev M.V. Bulletin of Nizhny Novgorod University named after N. I. Lobachevsky. Series "Economics and Finance", **1 (7)**, 392-397, (2005)
4. Yankovoi O. (ed.) *Konkurentospromozhnist pidpryiemstva: otsinka rivnia ta napriamy pidvyshchennia*, Odessa: Atlant, 514 (2013)
5. Lifits I.M. Marketing, **3(88)**, 25-29, (2014)
6. Mansurov R.E. Marketing in Russia and abroad, **2 (52)**, 94–102, (2011)
7. Rodionova L.N., Khakimova O.G. Marketing in Russia and abroad, **1**, 36-42, (2012)
8. Yatskovyi D. Visnyk sotsialno-ekonomichnykh doslidzhen, **4 (51)**, 183–188 (2013)

Methodological Principles of Assessing the Level of Technology Readiness for Transfer on the Basis of the Theory of Reliability of Hierarchically Branched Systems

Vasyl Kozyk[1], *Oleksandra* Mrykhina[1*], *Igor* Novakivskyi[1] and *Orest* Koleshchuk[1]

[1]Lviv Polytechnic National University, Institute of Economics and Management, Bandery Str., 12, 79013 Lviv, Ukraine

Abstract. The methodological principles of assessing the level of technology readiness for transfer from universities to the business environment are developed, in particular: a model and a set of tools for evaluating the readiness of technologies for transfer, a method for evaluating the level of technology readiness for transfer on the basis of the theory of reliability of hierarchically branched systems. The development is aimed at shortening the term of technology withdrawal into the market and its distribution on it; increasing the efficiency of technology commercialization; to become the basis for substantiating the investment indicators of technology projects. Methodical development was tested on the example of the technology of the means of personal passive optically stimulated luminescence (OSL) dosimetry of ionizing radiation (IR). The methodological principles for assessing the level of technology readiness for transfer are intended for use by specialists of technology transfer centers at higher education institutions or similar entities, auditors, technology appraisers, institutional level managers in the field of innovation development, etc.

Introduction

The model of open innovation has updated the need to review traditional approaches to innovation in the countries of the world. Universities are one of the main participants in the modern world ecosystem of innovation, which, besides educational and research roles, accumulate entrepreneurial activity, which is realized on the principles of technology transfer. The rethinking of society of the importance of universities as technology generators has focused attention on the problem of increasing the efficiency of their transfer to production. This necessitates the development of new approaches and tools for transferring technology from universities to the business environment.

* Corresponding author: oleksandramrykhina@gmail.com

Over the past decades, many technology scientists and practitioners have been discussing technology transfer issues, in particular [1-10]. Despite the significant number of developments in technology transfer pricing, there are no developments that would make it possible to set the level of technology readiness for the transfer and also indicate possible scenarios of technology development after its market launch. The unresolved problem of assessing the readiness of technologies for transfer in a methodical plan makes it impossible to substantiate further actions with these technologies.

Such a situation hinders the strategic development of universities in the system of interaction "University – Governmental Authority – Business". At the time, the development of methodological principles for assessing the readiness of technologies for transfer, which would facilitate the efficient and effective transfer of technologies from universities to the business environment on the one hand, on the other – would form a culture of university entrepreneurship.

1 Formation of a three-level hierarchical model of technology transfer assessment

Assessing the readiness of technologies for the purpose of their transfer often requires an analysis of interdisciplinary aspects, the consideration of components that significantly vary in meaning content and so on. This requires aggregation of the indicators included in the model. The international and domestic experience in evaluating technologies from the point of view of their transfer suggests that such an assessment should be based on an integrated approach that takes into account the value and cost parameters of the technology.

Based on the conducted researches of international practice on assessing the readiness of technologies for transfer from universities to the business environment, studying the national specifics of scientific and technical activities of universities, *a model for evaluating the level of readiness of technologies for transfer is proposed.* The model is based on an integrated approach, consisting of five units of evaluation: *the consumer value of technology; the competitiveness of technology; technological readiness of technology; technology cost; the riskiness of the technology.*

The preparation of technology for transfer is expedient to consider in the context of three conceptual levels that explain the content of this or that stage of technology readiness:

- I level: determination of technology transferability;
- II level: substantiation of the possibilities for transfer of the selected technology;
- III level: the choice of the technology transfer option.

2 Expert evaluation of parameters of the three-level hierarchical model of technology transfer estimation

The proposed model integrally covers five main blocks, each of which contains nine stages of technology readiness for the transfer. The need for the allocation of these units is confirmed by the statistical research of technological institutions of higher education of Ukraine during 2018-2019. The purpose of this statistical study was to determine the problems of technology transfer in the identified field in domestic developers. The research was conducted on the basis of a questionnaire.

In the process of statistical analysis, the opinion of representatives of research departments, marketing services of the Lviv Polytechnic National University and enterprises-manufacturers, as well as representatives of the consumer segment was studied. The essence of the survey was to assess the likelihood of a technology transfer ready according to the specified evaluation steps for the indicated blocks per 100 ball scale, which represents the probability scale from 0 to 1. The information base of the conducted research consists of the results of a survey of 21 participants in the development of technologies (Higher Education Institution) of various types of economic activity regarding the effectiveness of selecting and evaluating technologies at one or another stage of the model blocks.

It should be noted that the survey conducted is representative, since in Ukraine there are 277 institutions of higher education conducting scientific and technical activities. In all calculations the level of significance $\alpha = 0.05$; error is 1 point; the variance does not exceed 4. In such conditions, the required sample size is n = 12.06 representatives of the business entities involved in the survey.

We will conduct the initial study of the obtained sample by calculating relevant indicators. The value of the quantilet$_{tabl}$ according to the Student's table for the technology transfer assessment units was $t_{tabl}(n-1,\alpha/2) = t_{tabl}(48;\ 0.025) = 2.01063476$.

It is worth mentioning that the confidence intervals for all units of assessment of the availability of technologies for the transfer are largely high and close to each other. The largest spread has shown the assessment of the technology cost-absorption unit. This is due to the fact that subjects of the survey have different views on the importance of this or that technology transfer assessment unit.

The assertion that estimates for the evaluation units of this evaluation model were not accidental was checked on the hypothesis of the homogeneity of the mean values of the sampled samples obtained for each pair of blocks of the technology transfer readiness assessment model. Without reservations, we can assume that the sample mean has a normal distribution. In order to prove the homogeneity of technology transfer assessment units, it is enough to prove that a zero hypothesis is satisfied $H_0: \overline{Y}_i = \overline{Y}_j$. The competing hypothesis has the form $H_1: \overline{Y}_i \neq \overline{Y}_j$, so the critical area is two-sided. In this case, a random variable can be used to estimate the hypothesis:

$$Z(Y_i, Y_j) = |\overline{Y}_i - \overline{Y}_j| \Big/ \sqrt{\frac{D(Y_i)}{n_{Y_i}} + \frac{D(Y_j)}{n_{Y_j}}}.$$

If, then $Z(Y_i, Y_j) < Z_{crit}$, the null hypothesis is accepted; otherwise, it is rejected, where the critical value is $Z_{crit} = 1.645$ from the equation $W(Z_{crit}) = (1-2\alpha)/2 = 0.495$, where is $W(Z_{crit})$ the Laplace function. The calculated values $Z(Y_i, Y_j)$ of pair interactions between i^{th} and j^{th} technology transfer estimation blocks are given in Table 1.

Table 1. The calculated values of pair interactions are calculated $Z(Y_i, Y_j)$.

i \ j	Consumer value of technology	Competitivenes s of technology	Technological readiness of technology	The cost of technology	Risk of technology
Consumer value of Technology		18.210	25.738	47.835	97.910
Competitiveness of technology	18.210		7.341	32.563	85.340
Readiness of technology	25.738	7.341		26.967	81.596
The cost of technology	47.835	32.563	26.967		49.006

Risk of technology	97.910	85.340	81.596	49.006	

As can be seen from Table 1, for all pairwise comparisons, there is an inequality $Z(Y_i, Y_j) > Z_{crit}$, which means a clear separation of the respondents with the role of these blocks of technology transfer assessment.

3 Practical testing of the model

For the practical implementation of the proposed model for evaluating the readiness of technologies for the transfer, a special toolkit has been developed, one of the methods of which is a method for evaluating technologies based on the theory of reliability of hierarchically branched systems.

From the point of view of the theory of reliability of hierarchically branched systems, it is expedient to decompose the proposed model for assessing the readiness of technologies for the transfer (hereinafter referred to as the model) to three levels:

— *the zero level of the hierarchy* (P_0), where technology transfer centers (or other similar units of universities) receive technology and substantiate its transferability (reflecting a higher level of modeling);

— *the first level of the hierarchy* (P_1), which specifies the possibilities for the transfer of the selected technology (contains the main blocks of the technology transfer assessment model);

— *the second level of the hierarchy* (P_2), which is the multi-stage target assessment of the technology transfer.

In the given three-level hierarchical model on the branches $P_0 - P_1$ and $P_1 - P_2$ elements of the model are placed. At the highest level (roots) of the hierarchy there is a technology that is appreciated.

Given that the main five blocks of the evaluation model are placed on the first level of the hierarchically branched system, we will construct a generic function to determine the level of readiness of the technologies for the transfer. In the classical model, we assume that the evaluation blocks on the branches $P_0 - P_1$ are homogeneous and symmetric. Under these assumptions, the creature function can be written as follows:

$$S_1(z)_{\text{сим}} = P_0(P_1 z + q_1)^{a_1} + q_0, \qquad (1)$$

where P_0, P_1 – is the probability of passage of the zero stage and the first, respectively, levels of the model of evaluation of the readiness of technologies for transfer; q_0, q_1 – is the probability of non-passage of the phase of zero and the first, respectively, levels of the model of the evaluation of the readiness of the technologies for the transfer; ai is the number of branches at the ith level of assessment (a_1=5), z – is an artificial auxiliary indicator, in partial cases it can be interpreted as an indicator of the relative term of technology withdrawal into the market ($0 \leq z \leq 1$). Note that for the evaluated technology, the probability of its successful conclusion is unconditional, that is P_0=1. Obviously, the probability of turning off the transfer is impossible, but because q_0=0.

The creature function (1) should be used when determining the level of readiness of technologies in the first level of the model on the branches $P_0 - P_1$ also in the case of heterogeneous elements of the first level. Having made the corresponding mathematical transformations, by analogy obtained:

$$S_1(z) = P_0 \prod_{i=1}^{a_1}(P_{1i} z + q_i) + q_0, \qquad (2)$$

where P_1 – is the probability of passing the first stage, the level of the model for assessing the availability of technologies to the transfer by blocks: consumer value, competitiveness, technological readiness, cost and riskiness:

$$P_1(5) = P_0P_{11}P_{12}P_{13}P_{14}P_{15};\ P_1(4) = P_0(P_{12}P_{13}P_{14}P_{15}q_1 + P_{11}P_{13}P_{14}P_{15}q_2 + P_{11}P_{12}P_{14}P_{15}q_3 + P_{11}P_{12}P_{13}P_{15}q_4 + P_{11}P_{12}P_{13}P_{14}q_5);$$
$$P_1(3) = P_0(P_{11}P_{12}P_{13}q_4q_5 + P_{11}P_{12}P_{14}q_3q_5 + P_{11}P_{13}P_{14}q_2q_5 + P_{12}P_{13}P_{14}q_1q_5 + P_{11}P_{12}P_{15}q_3q_4 + P_{11}P_{13}P_{15}q_2q_4 + P_{12}P_{13}P_{15}q_1q_4 + P_{11}P_{14}P_{15}q_2q_3 + P_{12}P_{14}P_{15}q_1q_3 + P_{13}P_{14}P_{15}q_1q_2);\ P_1(2) = P_0(P_{11}P_{12}q_3q_4q_5 + P_{11}P_{13}q_2q_4q_5 + P_{11}P_{14}q_2q_3q_5 + P_{11}P_{15}q_2q_3q_4 + P_{12}P_{13}q_1q_4q_5 + P_{12}P_{14}q_1q_3q_5 + P_{12}P_{15}q_1q_3q_4 + P_{13}P_{14}q_1q_2q_5 + P_{13}P_{15}q_1q_2q_4 + P_{14}P_{15}q_1q_2q_3);\ P_1(1) = P_0(P_{11}q_2q_3q_4q_5 + P_{12}q_1q_3q_4q_5 + P_{13}q_1q_2q_4q_5 + P_{14}q_1q_2q_3q_5 + P_{15}q_1q_2q_3q_4);\ P_1(0) = P_0q_1q_2q_3q_4q_5 + q_0 \quad (3)$$

According to the probability theory, the sum of probable states at each level of the hierarchical model should be equal to 1. To do this, it is necessary to set the value of each of the elements (stages) on the branches $P_0 - P_1$ so that they total 1.

Accordingly, the mathematical expectation of meeting the requirements for estimating the number of output working elements in this system can be determined from the expression:

$$M(x_1) = 5\,P_1(5) + 4P_1(4) + 3P_1(3) + 2\,P_1(2) + P_1(1) \quad (4)$$

In assessing the transfer of technology of personal passive OSL dosimetry of IR, the given analysis shows the readiness indicators at the first level, that is, this technology is between the zero and the first levels of the hierarchical model of evaluation. In this case, the evaluation of each of the blocks of the hierarchical model is carried out, that is, they carry out work coordination of the transfer of technologies. By analogy, the process of technology evaluation is transferred to the next nine stages of the second level of the hierarchical model for evaluating the level of technology readiness for the transfer.

The proposed conceptual model for assessing the level of technology readiness for transfer is symmetric in its structure, branched to the second level (P_0, P_1, P_2). Thus, the number of branches in the model is five on the first ($a_1 = 5$) and nine on the second ($a_2 = 9$) levels. The constructed branched three-level hierarchical model can be represented as a generalized creature function:

$$S_2(z) = P_0 \prod_{i=1}^{a_1}(P_{1i} \prod_{j=1}^{a_2}(P_{2j}z + q_{2j}) + q_{1i}) + q_0 \quad (5)$$

The method of determining mathematical expectation is calculated in accordance with the law of discrete distribution on the theory of probability:

$$M(x_2) = \frac{\partial S_2(z)}{\partial z}|_{z=1} \quad (6)$$

In this work, the level of readiness of the technology for this model was calculated. It should be noted that due to the complexity of representing the formulas, the necessary calculations at the second level of the three-level hierarchical model for the technology of the means of personal passive OSL dosimetry of IR were conducted in the environment EXCEL. It is worth noting that in each individual case, the assessment of a particular stage is determined individually. Separate technologies do not require the obligatory passage of individual stages. In this case $P_{ij}=0$, since it does not affect the final result of the calculation.

4 Conclusion

The methodical principles of estimating the level of readiness of technologies for transfer were developed, in particular:

1) a model for assessing the level of technology readiness for transfer based on five technology assessment units (consumer value, competitiveness, technological readiness, cost, risk), each of which consists of nine stages of readiness assessment, which determine three levels readiness of technology.

The proposed method of estimating the level of technology readiness for transfer on the basis of the theory of reliability of hierarchically branched systems is based on the development: an extensive creature function for a system of a conceptual model for assessing the level of technology readiness for the transfer; recursive expression for calculating the distribution of probabilities of the number of output working elements of the conceptual model of the evaluation of the level of readiness of the technology for the transfer; mathematical expectation of the number of output working elements of the conceptual model of the evaluation of the level of readiness of the technology for the transfer;

2) testing of the developed toolkit on the example of the technology of personal passive OSL dosimetry of IR showed the following result of 79.60% of the readiness of this technology to the transfer.

In general, application of the theory of reliability of hierarchically branched systems in the framework of the proposed model for assessing the level of readiness of the technology for transfer allows: to increase the accuracy of obtaining the indicator of readiness of technology for the transfer; on the basis of recurrent expressions, to determine the nature of the interaction of elements among themselves, on the basis of which to draw conclusions about the level of development of technology and features of both its evaluation and transfer.

References

1. J. Bercovitz, M. Feldman,. The Journal of Technology Transfer, **31**, 1 (2006).
2. S. Bradley, C. Hayter, A. Link, Foundations and Trends in Entrepreneurship, **9(6)** (2013).
3. *Goldsmith Technology Commercialization Model.* Nebraska Business Development Center. College of Business Administration (2018). Available at: https://www.unomaha.edu/nebraska-business-development-center/technology-commercialization/goldsmith-technology/index.php
4. S. Hess, R. Y. Siegwart. Business and Management Research, **2** (2013)
5. V. Kozyk, O. Mrykhina, O. Koleshchuk, T. Mirkunova, Technology audit and production reserves, **3/4 (41)** (2018).
6. D. Ye. Marunchak, A. R. Sydor, *Nadiinist rozghaluzhenykh system [Reliability of branched systems]*, Lviv polytechnic publishing house (2007).
7. O. B. Mrykhina, Actual Problems of Economics, **8** (2014).
8. NASA *Technology Readiness Level* (2017). Available at: https://www.nasa.gov/directorates/heo/scan/engineering/technology/txt_accordion1.html
9. I. I. Novakivskyi, Bulletin of the Lviv Polytechnic National University. Series: Problems of Economics and Management, **754**, (2013)
10. D. S. Siegel, D. A.Waldman, L. E. Atwater, A. N. Link, Journal of Engineering and Technology Management, **21(1-2)** (2004).

Marketing Positions of Kremenchuk Industrial Center in International Markets

Oleh Kratt[1], *Andryi* Pochtovyuk[1], *Iryna* Trunina[1], *Kateryna* Pryakhina[1*]

[1]Kremenchuk Mykhailo Ostrohradskyi National University, Ukraine

Abstract. In the article, the authors analyze the role of Kremenchuk as an industrial center. It is determined that Kremenchuk is represented by such flagships machine building enterprises as PJSC "Kryukiv Rail Car Plant", PJSC "AvtoKrAZ", PJSC "Kredmash", PJSC "Kremenchuk Wheel Plant", PJSC "Kremenchuk Steel Works". Therefore, the author proposes to carry out the marketing positioning of the Kremenchuk industrial center on the international market, which includes powerful enterprises. The financial economic and marketing activity of machine-building enterprises - members of the Kremenchuk industrial center is analyzed. Through the dialectical connection of methods of analogy and deduction, the stages of development of marketing positioning for the center are proposed for expanding international sales markets for enterprises, marketing positioning is realized by developing a set of marketing tools. It is determined that the effective positioning of the Kremenchuk industrial center will lead to the formation of the brand.

1 Introduction

The competitiveness of the Ukrainian machine building complex at the international level is formed of competitiveness of a complex of regions. So, regions operate on the basis of the efficiency of economic entities operating in their territory. Machine building industry creating the most active part of the main production assets, significantly influences the pace and direction of scientific and technological progress, the growth of labor productivity and other economic indicators. The deterioration of fixed assets of the existing industrial enterprises, lack of working capital and their limited ability to attract investments to upgrade the existing equipment and to introduce innovative products create a threat to the preservation and increase of the competitiveness of the regional economy as a whole [1].

Therefore, the mission of the industrial center is to be the leader and instrument of realization of sustainable economic development of the region by its effective marketing positioning.

Kremenchuk is a Ukrainian city of regional subordination, an administrative and industrial center and has always been positioned as a major industrial center in the Poltava region. As there are 86 powerful industrial enterprises in the city, among which there are oil refining, chemical, machine-building, metallurgical, metalworking, food, light, and construction industries [2]. About 75 enterprises (86% of total number) export industrial

*Corresponding author: katerinapryakhina@gmail.com

goods [3]. It should be noted that the uniqueness of the center lies in its machine-building orientation, as evidenced by the presence of such flagships of the Ukrainian machine building, which are well known abroad as PJSC "Kryukiv Rail Car Plant", PJSC "AvtoKrAZ", PJSC "Kredmash", PJSC "Kremenchuk Wheel Plant", PJSC "Kremenchuk Steel Works". That is, Kremenchuk as an industrial center has competitive advantages in international markets; however, according to the data of the Main Department of Statistics in the Poltava region, the foreign trade in goods and services of Kremenchuk in recent years has been characterized by negative growth. Therefore, the question of developing a unified marketing positioning of the center for these machine-building enterprises arises. The choice of a single strategy is possible taking into account the specific features of enterprises within a single industry, and using the method of analogy to formulate a strategy of marketing positioning of one enterprise, it is possible to generalize it to the center as a whole.

2 Data and Methods

Indicators of the Ukrainian for Enterprise and Stock Market Infrastructure Development Agency (SMIDA) were used for the analysis.

The study has selected absolute indicators that characterize the financial and economic activities of the companies – "net revenue from sales of products (goods and services)" and marked with a symbol W_i; "net financial profit" marked with a symbol P_i and "net financial loss" is indicated $-P_i$. The indicator of the rate of increase (T_p^I) was chosen to justify the assessment of changes in the performance of enterprises. For analysis, such synthetic indicators are used: $T_p^I W$ – "rate of increase of net revenue from sales over the period", $T_p^I P$, –"rate of increase the net financial profit", where i is section of the studied period (calendar year), value of i is reduced to the last two digits of the year, so $i = \{16,17\}$.

Taking into account the similar financial instability of enterprises and identifying common risks at the exit and expansion of foreign markets, the creation of a unified marketing positioning of Kremenchug industrial center in the international market is proposed using analog and deduction methods (Fig. 1).

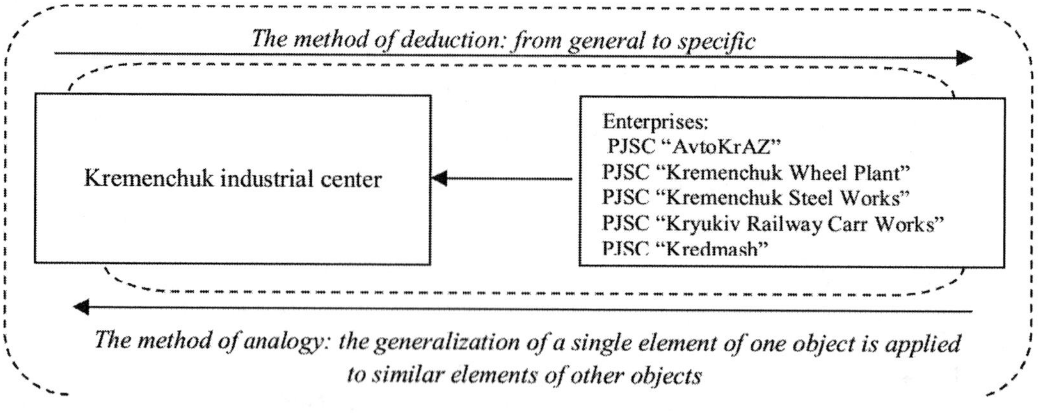

**author's development*

Fig. 2. Sales to international markets (share, %).

Analogy is a method of scientific research, in which from the similarity of objects of the class in one of the features a conclusion is made about their similarity in other features. The essence of the analogy can be expressed using the formula: A has signs abcd, B has signs abc, so, B, apparently, has the sign d. The method of analogy gives the most valuable results when an organic relationship is established not only between such signs, but also with the signs that is transferred to the object under investigation [4]. In this research, the method is in inseparable dialectic connection with the deduction

According to Fig. 1, the method of analogy with the financial and marketing activities of enterprises allowed to generalize the marketing positioning for the Kremenchuk industrial center. In the future, the method of deduction will allow to develop measures in general for the center and which will be universal for each enterprise individually [5].

3 Results

3.1 Financial activity of enterprises-members of the Kremenchuk industrial center

To determine the main directions of the marketing positioning of the Kremenchuk Center, it is necessary to consider the main financial indicators of the abovementioned enterprises (see Table 1).

Table 1. The analysis of indicators of financial and economic activity of enterprises.

Enterprise / Indicators	W_i, UAH			P_i /$-P_i$, UAH		
	2016	2017	$T_p^l W$, %	2016	2017	$T_p^l P$, %
PJSC “Kryukiv Rail Car Plant”	2,150,001	3,847,179	44.11	−11,2848	331,983	–
PJSC “AvtoKrAZ”	1,231,793	1,047,864	−17.55	−19,1815	−89,728	(113.77)
PJSC “Kredmash”	513,613	894,583	42.58	51,434	95,769	46.29
PJSC “Kremenchuk Wheel Plant”	526,217	687,392	23.44	26,473	21,951	20.60
PJSC “Kremenchuk Steel Works”	297,250	710,920	58.19	32,132	−9,673	–

* *author's development on the basis of [6]*

According to Table 1, it should be noted that PJSC “Kryukiv Rail Car Plant”, PJSC “Kredmash” and PJSC “Kremenchuk Wheel Plant” according to the results of 2017, earned profit 331,983 UAH, UAH 95,769 and UAH 21,951, respectively. In this case, the activity of PJSC “AvtoKrAZ” and PJSC “Steelmaking Works” was unprofitable.

The value of the indicator W_{17} of PJSC “Kryukiv Rail Car Plant” was 3,847,179 UAH, which indicates an increase the revenue for 44.11%. The value of the indicator $-P_{16}$ was −112,848 UAH, and P_i was UAH 33,183, it means that enterprise got the profit in 2017. “Kredmash”, during the reporting period, increases revenues and profits, as evidenced by the value of the indicator $T_p^l W$=42.58% and $T_p^l P$=46.29%. Similarly tendency was at “Kremenchuk Wheel Plant”, but its value is somewhat lower than that of Kredmash and was $T_p^l W$=23.44 % and $T_p^l P$=20.60 %.

As was noted during the 2016-2017 period, PJSC “AvtoKrAZ” worked with loss, but it decreased by 13% in 2017. At the same time the revenue decreased and the value of $T_p^l W$=−17.55%. Despite the fact that the “Kremenchuk Steel Works” in 2017 compared to 2016 revenue increased and value of $T_p^l W$ was 58.19%, received a loss of 9673 UAH. Thouse, the analysis of indicators of Kremenchuk enterprises showed instability as the enterprises had a loss or insignificant revenue during the reporting period.

3.2 Activity of the members of Kremenchuk industrial center in foreign markets and marketing tasks

However, the activity of enterprises in the international arena and foreign markets should be analyzed for further research (Fig. 2).

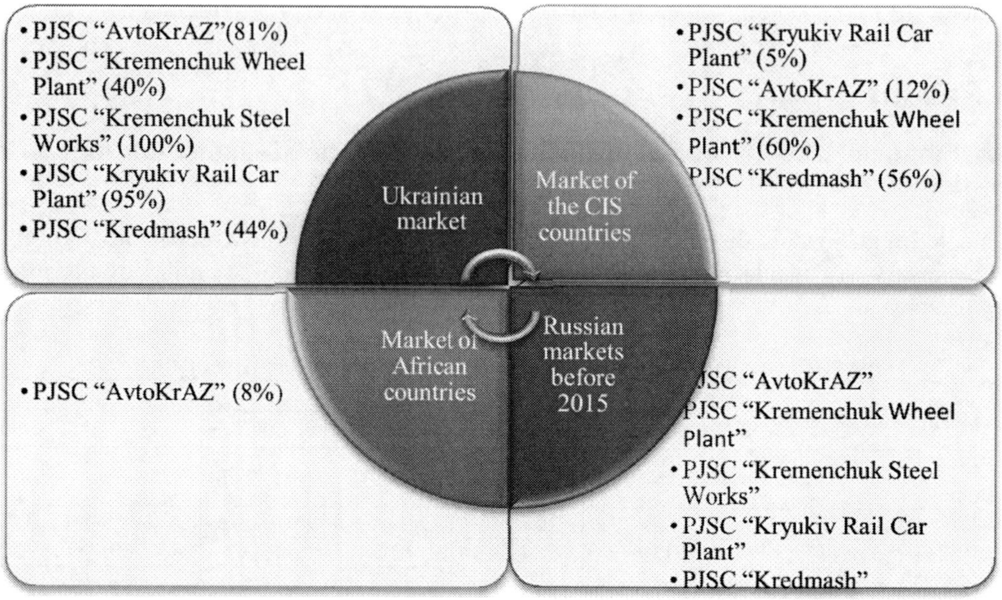

* *author's development on the basis of [6]*

Fig. 2. Sales to international markets (share, %).

In recent years, PJSC "Kremenchuk Steel Works" has not been present at international markets at all, being oriented only to the Ukrainian one, since the enterprise was previously working in the Russian market. The main consumers are car production and car repair enterprises. Products are sold on a contractual basis both to private enterprises, as well as to the public ones according to the procedure of trades on the Prozorro electronic platform. The marketing department is actively working towards creating an image of a trusted company, defining the goals and strategies of market activity on the domestic and foreign markets. The main risks affecting the activity of the issuer are the influence of the political situation in the state on the development of rail transport and the associated unstable demand for the carriage industry in the country. Also, the risks may include the possible expansion of foreign manufacturers; in particular Chinese with dumping prices for similar products, but with an uncertain level of quality and obscure warranty obligations. The methods of protecting their activities include the development of programs for the protection of domestic producers.

A similar situation has arisen at PJSC "AvtoKrAZ" regarding the impossibility of delivering products to the Russian market. The Russian market for a long period of time was the key to PJSC "AvtoKrAZ": by 2013, supplies to Russia accounted for more than 50% of the annual total sales volume of PJSC "AvtoKrAZ" products. During the last three years, supplies to the domestic market have increased significantly, deliveries have been made both to traditional foreign partners and to new consumer countries. The risks of company are the growth of exchange rates as the consequence is to increase the cost of finished products and reduce the profit from its realization. For example, in 2017, the growth of exchange rates caused a rise in materials by 31%, component parts – by 16%. To overcome this situation, the company applies a flexible pricing policy in order to maximize its position on a specific market. Further impossibility of deliveries of products to the Russian market through the inter-state conflict led to the loss of a large market of sales. Therefore, sales departments have intensified their efforts to find new international sales markets.

The main obstacles in entering the international markets of PJSC "Kryukiv Carriage Works" are as follows: the absence of real foreign investors, the lack of an effective mechanism for the promotion of Ukrainian products at the state level and lobbying the interests of domestic commodity producers in foreign markets, in the free trade zone between Ukraine and the EU. The main production risks are the transition of Ukrainian manufacturers to EU standards. At the same time, the enterprise continuously implements quality management systems of the international standard, integrates resources for the development of high-tech system engineering projects.

The main risks of in PJSC "Kremenchuk Wheel Plant" are the political aspects – the possibility of increasing taxes, problems with the customs clearance of goods, and the economic – related to changes in the exchange rate, which can lead to unplanned losses. The main direction of the company's marketing policy is to obtain maximum profit, achieve maximum customer satisfaction, and increase the competitiveness of products.

PJSC "Kredmash" is focused on expansion of markets at the expense of the Middle East, Africa, and Southeastern Asia. The risks that may affect the financial and business activities of PJSC "Kredmash" are political instability, recession, lowering the pace of economic development, inflation growth, the procedure for certification of products in importing countries, the development of new technologies and the orientation of consumers for technical innovation of a certain nature, a high degree of dependence of production on the supply of components and materials.

Thus, all of these enterprises lost a large Russian market of sales. Taking into account the political conflict and economic situation, marketing departments point out the need for reorientation and are constantly looking for new international markets. There are also the same risks for all enterprises: political instability, exchange rate change, lack of investment, reorientation and technical standards of the EU, etc. Taking into account the common challenges and problems of enterprises, this indicates the need to combine efforts to create a positive image of companies through marketing positioning as the only Kremenchuk industrial center.

3.3 Marketing positioning of the Kremenchuk industrial center

Today, the attitude of heads of domestic enterprises to unions in the centers has changed. They are considered as a tool for further development, a prerequisite for the introduction of business in a context of increased competition between national enterprises and between Ukrainian and foreign companies. The formation of such centers of the enterprise prompts a number of prerequisites of internal and external nature. In particular, the most important external prerequisites are general economic ones.

Marketing positioning is a complex of measures, which, in the consciousness of the target buyers, this product is relatively intrinsically comparable to other competing goods and a favorable place [7]. For effective positioning, the use of all components of the marketing mix is important, but promotion activities are one of the main roles, as they are the mechanism for reporting positioning to the target audience [8]. Significant moments of this issue are constant positioning and consistency in advancement. So, PR and advertising programs should be linear rather than spatial. In a spatial program, all its elements start simultaneously, but in different places. In the linear program, the elements, revealing for a certain time, have the ability to interact, reinforcing each other. In the formation of a stable position, a significant role is played by all elements of the promotion of the mix: ATL and

BTL instruments, in particular, merchandising plays an important role. In the final analysis, positioning should lead to the formation of the brand.

Taking into account the above mentioned risks, obstacles and problems of the enterprises, it is possible to propose the following algorithm of the market positioning of enterprises and the center as a whole on the international markets:

analyze competitors and evaluate their positioning;

analyze the product and compare it with the positioning of the product of competitors;

analyze the target audience;

determine the current position of enterprises;

create maps of desirable states and positions of enterprises;

develop positioning and implement it;

evaluate the effectiveness of the introduced positioning and will promote the "brand" of the center in the future.

Effective marketing positioning of machine-building enterprises will allow the Kremenchuk industrial center to create the image that will promote the attraction of investors and socio-economic development of the city as a whole and will further enable the development of marketing positioning from the standpoint of branding.

References

1. Buzko, O. Vartanova, I. Trunina, & I. Khovrak, SHS Web of Conferences. Innovative Economic Symposium 2018 Milestones and Trends of World Economy (IES2018), **61**, 01001, (2019).

2. Official website of "Kremenchuk City Council and Executive Committee". [online]. Available at: https://www.kremen.gov.ua/?view=main.

3. The State Statistics Service of Ukraine [online]. Available at: www.ukrstat.gov.ua/

4. A. Uemov, *Analogy in the practice of scientific research: From the history of physical and mathematical sciences.* Nauka Publishing House, 258. (1970).

5. K. Pryakhina, Effective economy: electronic scientific professional edition. Dniprovsky Agrarian and Economic University, **4** (2018). [online]. Available at: http://www.economy.nayka.com.ua/pdf/4_2018/153.pdf

6. Official site "Agency for the development of the infrastructure of the stock market of Ukraine" [online]. Available at: https://smida.gov.ua/.

7. N. Kudenko, Formation of a Market Economy: Coll. sciences works "Marketing: Theory and Practice". **1**, 393-403. (2007).

8. O. Maslak, P. Sokurenko, N. Grishko, M. Maslak, SHS Web of Conferences. Innovative Economic Symposium 2018 Milestones and Trends of World Economy (IES2018), 61, 01014. (2019).

Development of a Competitiveness Enhancement Mechanism for National Car Production Enterprises

Antonina Kuzub[1*1]

[1]Ukrainian State University of Railway Transport, Department of Economics and Management of Industrial and Commercial Business, 61050 Feerbach Square 7, Kharkov, Ukraine

Abstract. The article discusses the development and implementation of a competitiveness enhancement mechanism for national car production enterprises. It indicates the development stages of the competitiveness enhancement mechanism for car production enterprises. Four subsystems making the competitiveness enhancement mechanism for enterprises are distinguished: owners and investors, top management, lower level management and average executives. An algorithm of implementing the competitiveness enhancement mechanism for national car production enterprises has been developed.

1. Introduction

The development dynamics of the Ukrainian economy has been unstable over recent years showing different levels of industries readiness to cope with modern challenges, and being a consequence of a decrease in the investment and innovative activity, technological development and competitiveness of enterprises in the internal and external markets.

The Ukrainian economy can develop due to enhancement of car production enterprises' competitiveness. Enhancement of competitiveness of enterprise and products in the current economic context is a rather complex mechanism for the majority of car production enterprises due to the lack of effective methods covering the enterprise's business integrity.

The issue of car production enterprises' competitiveness enhancement can be resolved based on implementation of the competitiveness enhancement mechanism that is an effective tool to achieve national priorities and the basic principles of innovative and technological development of the Ukrainian economy.

The car production enterprises need well-developed and operating comprehensive tools so as to enhance their competitiveness that take into account branch specifics and current status of the industry, for further development of the Ukrainian economy. Therefore, functioning of the enhancement mechanism tools with a synergy effect is a precondition to reach the target competitiveness level.

* Corresponding author: njyz11@meta.ua

2. Analysis of the Latest Researches and Publications.

Works of many foreign and Ukrainian scientists are concerned with research of the competitiveness enhancement mechanism, including those by V.I. Blonska, M. I. Ivanova, I. Ya. Kachmaryk, Ya.D. Klymenko, V.O Kovalenko, P.O. Kutsyk, R.L. Lupak, I.M. Pushkarchuk, P. Yu. Yatsus. However, despite a considerable number of scientific publications in the field, the car production enterprises have no such mechanism, and therefore the issue is still relevant.

3. Statement of Purposes.

The main objective hereof is to develop a competitiveness enhancement mechanism for national car production enterprises.

4. Statement of Basic Materials.

The competitiveness enhancement mechanism for car production enterprises is a set of functions, tools, means and methods, using which a manager influences the whole internal capacity of a car production enterprise, as well as controlled external environment parameters, in order to identify and implement its competitive advantages.

The development of a competitiveness enhancement mechanism for car production enterprises involves the following stages:

1) making a decision on scope and areas of enhancement of the reached competitiveness indicator of the enterprise under study. In practice, top managers take such decision in coordination with the Director of the enterprise, but an approach when a decision is made by each employee involved in the performance of works on competitiveness enhancement along with planning the achievement of this purpose within each department is deemed logically correct.

2) Determining external (taxes, loan interest rate, exchange rate, etc.) and internal (planning, management system, quality control, incentives, etc.) factors influencing specific competitiveness indicators and the weighted average competitiveness indicator. Identification of dependent (variable) and independent (non-variable) factors. Whereas competitiveness is always relative in time or space, the analysis of influence at meso and macro levels and changes forecast for the period beyond completion of competitiveness enhancement mechanism implementation are an important stage in the development of such mechanism. At this stage, it is necessary to collect, predict and systematize opportunities and risks of the national car production enterprises.

Determination and forecast of external impact on the competitiveness enhancement mechanism implementation make it possible to create a corporate strategy with due account for possible market changes. The work on analysis and forecast determines the future of the enterprise, allowing it to hold competitive positions in the long term.

In this regard, in order to hold competitive positions based on response time, it is required to designate the following actions as important for market forecast: market (competitors, prices, goods, quality) analysis; economic and political analysis (forecast of inflation, exchange and interest rates growth); resources availability analysis (forecast of availability and growth in prices for resources used in the production); human resources availability analysis (forecast of availability of human resources as per principal trade jobs).

3) Selection and justification of competitiveness enhancement tools are based on assessment of the enterprise's competitiveness, determination of opportunities and threats, as well as specification of indicators that need improvement. Justification of competitiveness

enhancement tools is performed based on the relevance, need, opportunity and effectiveness of their use in order to improve competitiveness as per such area (indicator).

Identification of competitiveness enhancement tools makes it possible to list specific measures aimed at improvement of the car production enterprises' competitiveness. A set of competitiveness enhancement actions for car production enterprises based on changes in impact of tools ensuring the competitive position is a system of individual projects involving specific employees, budgets and plans.

4) Development of detailed plans for introduction of competitiveness enhancement actions by structural departments of the enterprise. A plan providing for detailed and specific functions, business processes, specific competitiveness enhancement resources is developed based on external and internal impact by each competitiveness enhancement area and its estimated indicator. As a result, the annual competitiveness enhancement plan represents a system of individual projects involving specific employees, budgets and plans. The specified "design department" (DD) exercises an important function in coordination of works within the enterprise.

5) Assessment of results of the competitiveness enhancement mechanism implementation in the short, medium and long terms. The key function of the DD at the final stage of the competitiveness enhancement involves assessment of cost efficiency of the competitiveness enhancement (to be performed upon introduction of each action under the general methodological principle of correlation between the result and the related expenses), as well as reassessment of the enterprise's competitive position with identification of profit and further implementation monitoring.

The competitiveness enhancement mechanism for a car production enterprise is represented in Figure 1.

The top managers and personnel should take decisions to increase the efficiency of the competitiveness enhancement mechanism by specific areas based on the logic of creation of such mechanism for national car production enterprises.

The competitiveness enhancement mechanism for car production enterprises consists of four subsystems: owners and investors, top management, lower level management and average executives.

The main task of the first subsystem involves formation of a strategic objective and desirable competitiveness level of an enterprise incorporated by its owners and investors. The competition strategy is developed and adopted at the top management level, using the strategic analysis tools and taking into account all possible impact factors. The lower level managers define tools for improvement of the enterprise's competitiveness by means of which the task and actions aimed to enhance the national car production enterprises' competitiveness are developed. At the final stage, the average executives take actions to enhance competitiveness of car production enterprises, approved by all above subsystems.

The author considers it necessary to take actions aimed at enhancement of the car production enterprises' competitiveness as follows: improvement of technical production standards; improvement of product competitiveness; decrease of expenses; improvement of management efficiency; improvement of marketing and sales management; diversification; innovations.

When developing actions aimed at competitiveness enhancement, it is essential to perform diagnostics of problems in the car production enterprise's management system. The quality of the enterprise's (or its divisions) management system at all levels is one of the major factors of the internal corporate environment influencing the competitive position, as well as understanding an accurate development strategy and tactics by employees, knowledge and observance of the existing business processes.

In order to ensure modern hi-tech infrastructure of the industrial complex, and in particular the car-production complex, and for development of the economy of Ukraine, in order to meet

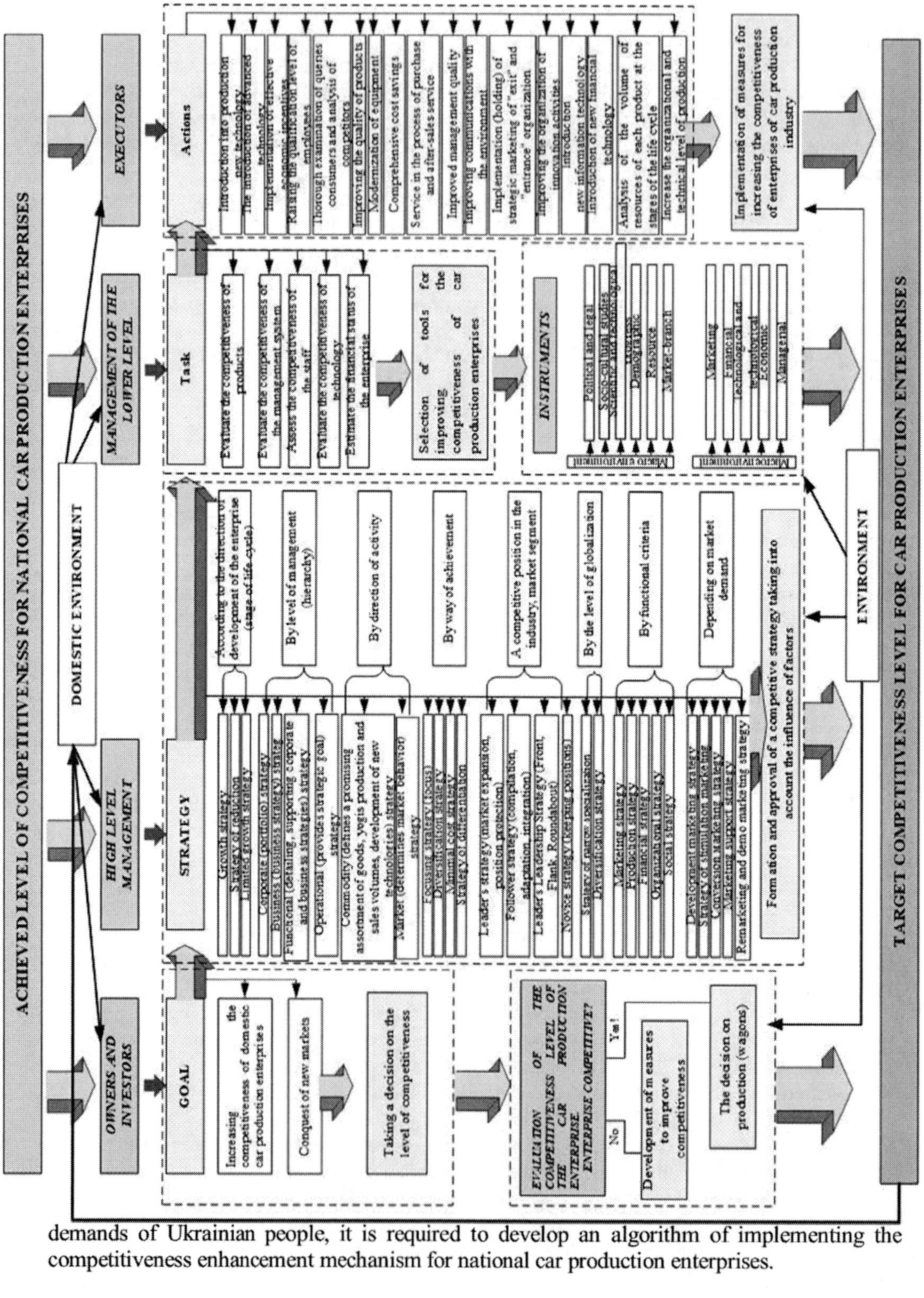

demands of Ukrainian people, it is required to develop an algorithm of implementing the competitiveness enhancement mechanism for national car production enterprises.

Fig. 1. Block diagram of the competitiveness enhancement mechanism for a car production enterprise.

The algorithm of implementing the competitiveness enhancement mechanism for national car production enterprises includes elements to be implemented in the following order:

1 - Assessment of the enterprise's competitive status in each segment of the occupied market (in the context of enterprises, products and services). Such assessment is carried out using the standard technique involving selection of a reference object and comparison of indicators of the enterprise under study (products or service) with the enterprise taken as the reference object;

2 - identification and assessment of a competitiveness parameter (indicator) that needs improvement. This stage involves comparison of competitiveness indicators system of the reference enterprise with the indicators of the enterprise under study to identify indicators reducing /enhancing competitive status of such enterprise within the occupied marked segments;

3 - analysis of the competitiveness enhancement tools, selection of tool by each indicator that needs improvement.

At this stage, it is important to select a tool improving competitiveness, which can be used effectively at such enterprise. In the course of such analysis, it is essential to consider resource, organizational and administrative opportunities of the use of each tool capable to improve competitiveness;

4 - development of actions aimed to improve competitiveness and ranging them as per importance degree, taking into account external impact and internal environment of the enterprise. Organizational and technical actions, the priority of which depends on the current competitive status in the specific market segment are developed for the revealed system of competitiveness parameters that need improvement;

5 - development of a program for competitiveness enhancement of the enterprise for a planning period (year, six months, quarter). The competitiveness enhancement program of the enterprise for the specific planning period is based on the list of actions ranged as per importance degree. The period of actions planning is determined by the emergency measured aimed at competitiveness enhancement and resource restrictions: financial, material and organizational, etc.;

6 - management of each action within the developed competitiveness enhancement mechanism. At this stage, it is required to perform the following: to define areas of responsibility of managers and personnel of the enterprise in regard to competitiveness enhancement as per each parameter (indicator and/or action), taking into account the content of measures to be taken and the term of performance; to develop internal standard and legal documents (order, regulation, instruction, statute, etc) ensuring introduction of actions aimed at enhancement of the enterprise's competitiveness;

7 - introduction of the competitiveness enhancement mechanism at the enterprise. At this stage, it is essential to consider the following: degree of personnel readiness to implement the competitiveness enhancement program, need for preparation and retraining of the personnel for performance of the planned actions aimed competitiveness enhancement; level of possible resistance of personnel at implementation of the competitiveness enhancement program;

8 - assessment of the implemented competitiveness enhancement mechanism efficiency.

Adaptation of the algorithm for the competitive status of the car production enterprise will ensure further successful implementation and functioning of the competitiveness enhancement mechanism at national car production enterprises, giving an opportunity for development of the economy of Ukraine in general in order to meet demands of the Ukrainian people.

Monitoring of the competitiveness enhancement mechanism in actual practice of a car production enterprise should be accompanied by analytical and information support of the relevant services and departments performed under accurate management of the enterprise and with a frequency in order to reduce risks of decrease in competitiveness and economic efficiency.

The efficiency of competitiveness enhancement mechanism implementation is reflected by indicators of the car production enterprise's economic activity, significantly improved as a result of competitiveness enhancement. Including: profitability of production and products; amount of profit; production cost; increase of the enterprise's market share.

5. Conclusion.

The technique of competitiveness enhancement mechanism development proposed herein by the author offers the possibility to resolve of the competitiveness issue systematically, providing the enterprise's management system with the tools for identification and realization of the enterprise's competitive advantages.

The competitiveness enhancement mechanism developed can be used at car production enterprises of any legal form or level as, due to correct selection of competitiveness enhancement tools, it is possible to consider specifics of any car production or industrial enterprise and develop an effective mechanism to enhance its competitiveness.

The technology of the competitiveness enhancement mechanism implementation can be applied in practice of national car-production enterprises in order to enhance their competitiveness and to foster the national economy development in general.

Therefore, in order to resolve the issue of the Ukrainian economy development, it is required to introduce the competitiveness enhancement mechanism for car-production enterprises in Ukraine using the author's algorithm of such mechanism implementation, which also have influence the development of the industrial complex and meeting the demand of the Ukrainian people.

References

1. Danko M.I., Dykan V.L., Kalinichenko L.L., *Pidvyshchennya investytsiyno-innovatsiynoho potentsialu promyslovykh pidpryyemstv zaliznychnoho transportuvannya intehratsiynykh protsesiv*. UkrSART, 167 (2010)
2. Panchenko S.V., Dykan V.L., Volovelska I.V., Maslova V.O., *Upravlinnya biznesom*. UkrSURT, 312 (2016)
3. Rail EXPO 2017, the strictly railroad-related international exhibition to be held for the first time in Ukraine [Electronic resource]. - URL: http://uprom.info/anons-podiy/vpershe-v-ukrayini-proyde-suto-zaliznichna-mizhnarodna-vistavka-rail-expo-2017/
4. International Specialized Exhibition Rail EXPO 2018 [Electronic resource]. - URL: https://railexpoua.com/
5. Kuzub A.V. Derzhava ta rehiony. Seriya: Ekonomika ta pidpryyemnytstvo, **3 (120),** 90-98 (2018)
6. Kuzub A.V. Prichornomorski ekonomichni studiyi, **26**, 10-15 (2018)
7. Kuzub A.V. The quality control system of domestic industrial companies of railway transport. Problems of systemic approach in the economy, **1 (63)**, 64-69 (2018)
8. Danko M.I., Dykan V.L., Yakimenko N.V., *Zabezpechennya konkurentospromozhnosti promyslovykh pidpriyemstv v uslovyyakh mizhnarodnoho transportnykh koridoriv*, UkrSART, 169 (2008)
9. Tolstova A.V. The bulletin of transport and industry economics, **30**, 97-102 (2010)
10. Dykan V.L., Zaiets A.P. The bulletin of transport and industry economics, **62 (SPECIAL ISSUE)**, 297-305 (2018)

Simulating Development of Science in a Country with the Use of the Cognitive Approach

Mykola Kyzym[1], *Olena* Reshetniak[2] and *Oleksij* Lelyuk[2]

[1]Research Centre for Industrial Problems of Development of NAS of Ukraine, Director, 1-a Inzhenernyi Ln., Kharkiv, 61166, Ukraine
[2]Research Centre for Industrial Problems of Development of NAS of Ukraine, Applicant, 1-a Inzhenernyi Ln., Kharkiv, 61166, Ukraine

Abstract. The study proposes a methodological approach to justifying the directions for reforming the scientific sphere of a country which is based on the cognitive approach. Using the proposed approach, the groups of the main problems in the development of science in Ukraine are determined and indicators for their evaluation are singled out; the cognitive model of the development of science in Ukraine is constructed. On the basis of the constructed cognitive model, impulse simulation of the influence of the main factors on development of science in a country is carried out and, according to the hypotheses on the nature of their influence, four scenarios for the development of the process under study are constructed. The analysis of the developed scenarios allowed to justify the most effective measures to stimulate the development of science in Ukraine. The practical evaluation of the proposed methodological approach has proved the feasibility of using cognitive models for analyzing and simulating development of science in a country, which allows to quickly and accurately assess a situation and identify promising measures to support it.

1 Introduction

The development of modern society is directly linked to globalization processes which imply the strengthening of integration ties among regions and countries of the world. These processes also transform the development of science which is influenced by many internal and external factors having complex interrelations and affecting each other with different strength. All this makes the problem of simulating the development of science in countries of the world extremely difficult to analyze and puts forward requirements for the choice of tools to use. Many scientists have proved the feasibility of applying the cognitive approach in analyzing and simulating complex semistructured systems and problems, including the one under study [1]. The methodological foundations of the cognitive approach were laid in the works of R. Axelrod, Vesa A. Niskanen, R. Atkin, F. Ritter, A. S. Huff and others. They were further developed in works of Z. Avdeev, S. Kovriga, V. Maximov, V. Khaustova, and others [1–7]. But insufficient attention has been paid to studying processes of scientific development with the use of the cognitive approach. At the same time, its application will contribute to a better understanding of problems in the development of domestic science, identification of contradictions and a high-quality analysis of the entire system of research activity in Ukraine. Therefore, the aim of the article is studying the current problems in the development of the Ukrainian science and simulating its development in the context of globalization challenges by means of the cognitive approach.

2 Methodology

Cognitive simulation is based on representation of a complex system (problem) in a simplified form as a specific model and allows, based on it, to consider possible scenarios for the development of the system (problem) at certain external and internal changes.

The analysis of the existing in theory and practice approaches to diagnosing problems of functioning and development of a system (based on methods of goal-setting, strategic management, management theory, etc.) has proved the feasibility of using functional cognitive maps to represent knowledge about the development of problem situations to carry out a formal analysis of structural and dynamic features of the development of socio-economic systems in solving practical problems of reforming an economic policy [1–7]. This made it possible to form a methodological approach to justifying the directions for reforming the scientific sphere in Ukraine, the block diagram of which is present in Fig. 1.

3 Results

In order to build a cognitive model for studying problems of science development in Ukraine, in accordance with the methodological approach to justifying the directions for reforming the country's scientific sphere (Fig. 1), at the first stage, directions for reforming the scientific sphere were determined based on forming a conceptual model of the research goals, structuring the goals, and defining the goal image. This process was implemented with the involvement of a group of experts, by analyzing the peculiarities, trends and problems in the development of science during the period of Ukraine's independence, as well as studying the experience in ensuring the development of science in the leading countries of the world. As a result of the study, there was justified the need to strengthen the export potential by increasing the share of high-tech products in the context of globalization, which can be achieved by enhancing the scientific and technical potential, improving the quality of scientific staff training, bringing the R & D funding to the level of the leading countries of the world. It is these indicators that have formed the goal image of reforming the scientific sphere in the country.

At the second stage of the study, factors affecting the processes of research activity and the development of a scientific product were analyzed, which has allowed to form the problem domain, identify the main problems that are most significant as well as the main indicators for their evaluation (Table 1).

Table 1. Groups of main problems of science development in Ukraine and indicators for their assessment.

Deno-tation	Problem	Indicator for assessing the problem	Assessment
X_1	Low demand for science in the economy	Decrease in the share of innovatively active enterprises	Statistical assessment, calculating the growth rate
X_2	Inefficient system of state management of science	Lack of a strategy for the development of science in the country	Expert assessment, linguistic rating, determining uncertain tendencies
X_3	Low level of communication (integration processes)	Level of integration of scientific, educational institutions and business to create a scientific product	Expert assessment, linguistic rating, determining uncertain tendencies
X_4	Lack of funding and other socio-economic incentives	R & D funding	Statistical assessment, calculating the growth rate
X_5	Low level of effectiveness	Share of exports of high-tech	Statistical assessment, calculating

	of scientific results, low export potential	products	the growth rate
X_6	Staffing problem	Decrease in the number of researchers	Statistical assessment, calculating the growth rate
X_7	Low level of authority of science in the society	Level of social status of scientists	Expert assessment, linguistic rating, determining uncertain tendencies

Source: developed by the authors

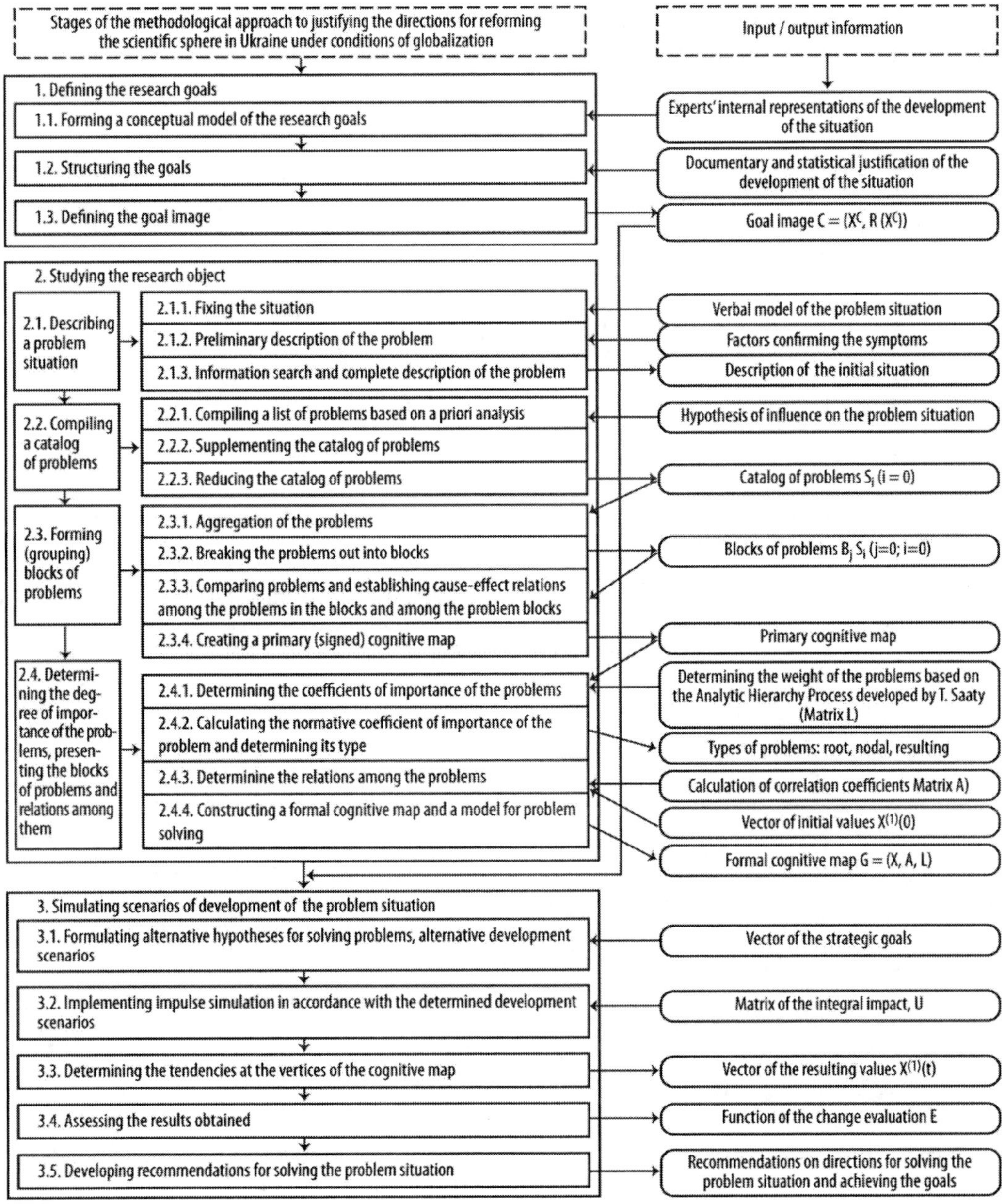

Fig. 1. Structural chart of the methodological approach to justifying the directions for reforming the scientific sphere in Ukraine.
Source: developed by the authors

After identifying the main problems of the problem domain and building the primary cognitive map, the importance of the problems and connections between them was assessed based on the Analytic Hierarchy Process (AHP) developed by T. Saaty [8]. Accordingly, for this purpose, matrices of pairwise comparisons were formed and the effect of each problem which is included into another problem (or causes its occurrence) is determined.

Similarly, the matrices of pairwise comparisons for other groups of problems were formed and the effect of each problem which causes the emergence of others was determined. Based on the calculation of the weighing coefficients for problems in a group of problems, the types of problems are identified. The determination of root and nodal problems and their effect on other problems are given in Table 2.

Table 2. Determination of types of problems in the primary cognitive model

Deno-tation	Group of problems	Weighting coefficient of the problem in the group	Type of problem
X_1	Low demand for science in the economy	0.509	nodal problem
X_2	Inefficient system of state management of science	3.922	root problem
X_3	Low level of communication (integration processes)	0.235	nodal problem
X_4	Lack of funding and other socio-economic incentives	0.456	nodal problem
X_5	Low level of effectiveness of scientific results, low export potential	0	resulting problem
X_6	Staffing problem	0.502	nodal problem
X_7	Low level of authority of science in the society	0.041	nodal problem

Source: developed by the authors

At the next stage of the analysis, there were determined the values of the matrix L which includes the value of the calculated coefficients of problem importance calculated based on the data of the State Statistics Service of Ukraine for the period 2000-2018 [9].

For some problems in the development of science, the impossibility to carry out a quantitative analysis of the indicators which characterize them was determined. Such problems are described by a linguistic variable and have an uncertain tendency in the development of indicators. To assess these problems, it was proposed to apply a rating scale using the approach which takes into account the Fibonacci series and the "golden proportion" based on expert assessment [10]. According to the proposed scale, the following problems were evaluated: the inefficient system of state management of science (insignificant increase – 1.119), a low level of communication (integration processes) – 0.942, a low level of authority of science in the life of the society – 0.618.

To determine the degree of the cause-effect relations of the problems in the development of science in Ukraine and the factors which characterize them, based on the matrices of statistical data describing their dynamics, the coefficients of correlation between the quantitative parameters of the influence of the factor i on the factor j were calculated. Measuring the causal factors and their weight, taking into account the correlation described by the matrix $A \times L$, is given in Table 3.

Based on these calculations, a cognitive model for studying the problems of the development of science in Ukraine was built (Fig. 2). Based on the cognitive model, impulse simulation of the influence of factors on the development of science in Ukraine was carried out and, according to the formed hypotheses on the nature of their influence, four scenarios for the development of the studied processes were built.

Scenario 1 is based on the hypothesis that the development of science in the country will be influenced by forming and implementing an effective strategy for the development of science. It was simulated by introducing an impulse (disturbance) into the vertex X_2, the value of $u_1 = +1$.

Scenario 2 is based on the hypothesis that the development of science in the country will be influenced by increasing the R & D funding. It was simulated by introducing an impulse (disturbance) into the vertex X_4 the value of $u_4 = +1$. The disturbance (impulse) $U = \{u_4 = 0, 0, 0, +1, 0, 0, 0\}$ simulates the hypothesis that at increasing R & D funding a gradual increase in the share of high-tech products in the GDP is expected.

Table 3. Determining the values of the matrix $A \times L$ of measuring the causal factors and significance of the factors, taking into account their correlation.

Denotation	Group of problems	X_1	X_2	X_3	X_4	X_5	X_6	X_7
X_1	Low demand for science in the economy	0.000	0.000	0.000	-0.146	0.132	0.000	0.000
X_2	Inefficient system of state management of science	0.504	0.000	0.618	0.543	0.226	0.257	0.379
X_3	Low level of communication (integration processes)	0.064	0.000	0.000	0.034	0.008	0.000	0.016
X_4	Lack of funding and other socio-economic incentives	0.000	0.000	0.000	0.000	0.272	-0.018	0.000
X_5	Low level of effectiveness of scientific results, low export potential	0.000	0.000	0.000	0.000	0.000	0.000	0.000
X_6	Staffing problem	-0.037	0.000	0.000	0.000	0.143	0.000	0.104
X_7	Low level of authority of science in the society	-0.013	0.000	0.000	0.000	0.000	0.000	0.000

Source: developed by the authors based on [9]

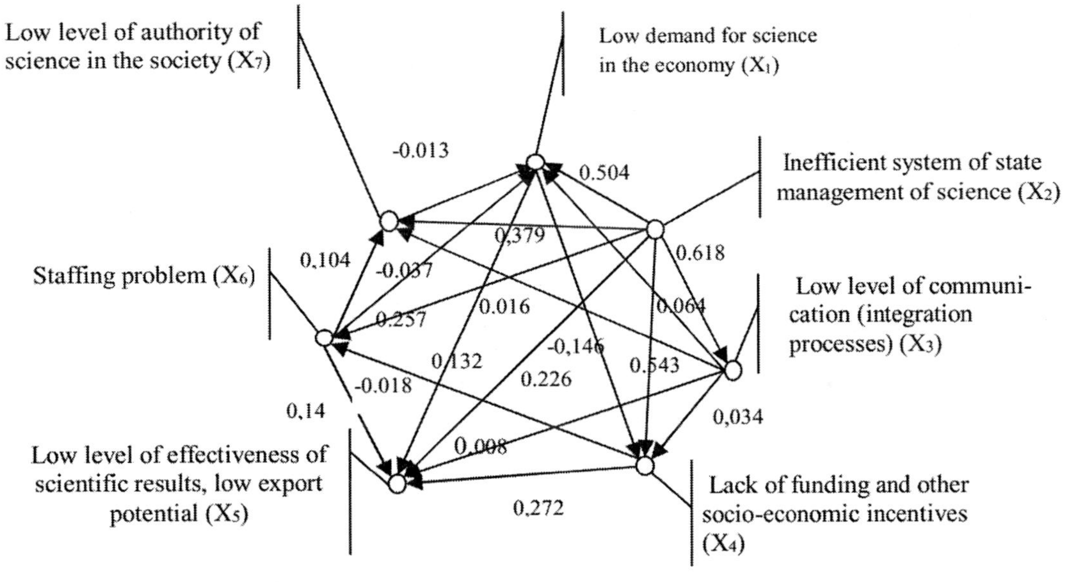

Fig. 2. Cognitive model for studying problems in the development of science in Ukraine
Source: developed by the authors.

Scenario 3 is based on the hypothesis that the development of science in the country will be influenced by enlarging the number of researchers due to improving the quality of scientific staff training. It was simulated by introducing an impulse (disturbance) into the vertex X_6, the value of $u_6 = +1$. The disturbance (impulse) $U = \{u_6 = 0, 0, 0, 0, 0, +1, 0\}$ simulates the hypothesis that at enlarging the number of researchers, due to improving of the quality of scientific staff training, a gradual increase in the share of high-tech products in the GDP is expected.

Scenario 4 is based on the hypothesis that the development of science in the country will be influenced by forming and implementing an effective strategy for the development of science along with increasing the R & D funding and enlarging the number of researchers due to improving the quality of scientific staff training. It was simulated by introducing an impulse (disturbance) into the vertex X_2, the value of $u_1 = +1$; introducing a disturbance into the vertex x_4, the value of $u_4 = +1$; introducing an impulse into the vertex x_6, the value of $u_6 = +1$. The disturbance (impulse) $U = \{0, +1, 0, +1, 0, +1, 0\}$ simulates the hypothesis that at formulating an adequate policy for the development of science and elaborating a strategy for the development of science in Ukraine, enlarging the number of researchers due to improving the quality of scientific staff training, increasing R & D funding a gradual increase in the share of high-tech products in the GDP is expected.

The analysis of the scenarios for the development of science in Ukraine, built on the basis of a cognitive model of the existing problems, has made it possible to determine that at present the most effective way, which will bring the most significant results (all parameters of the goal image increase), is forming and implementing an effective strategy for the development of science. Enlarging the number of researchers by improving the quality of scientific training and increasing R&D funding under conditions which have developed in the country will not yield significant results without paying attention to strategic priorities in the development of science and relevant government support.

4 Conclusions

Thus, the proposed methodological approach has allowed to justify the most effective measures to stimulate the development of science in Ukraine. The practical evaluation of the proposed methodological approach has proved the feasibility of using cognitive models for analyzing and simulating development of science in a country, which allows to quickly and accurately assess a situation and determine promising measures to support it.

References

1. Vesa A. Niskanen. Theoretical advances and applications of fuzzy logic and soft computing, **42**, 119–127 (2007).
2. Ritter F. E., Major N. P. Useful mechanisms for developing simulations for cognitive models. AI and Simulation of Behaviour Quarterly, **91**, 7–18 (1995).
3. Huff S. Mapping strategic thought. Chichester, UK: Wiley, 11–49 (1990).
4. Avdeeva Z., Kovriga S. The technology of the strategic goal-setting and monitoring of a system development on the basis of cognitive mapping. Procedia Computer Science, **122**, 977–984 (2017). https://www.sciencedirect.com/science/ article/pii/S1877050917327163
5. Kyzym M. O., Uzunov V. V. *Program-Targeted Approach to State Management of Social Tension in the Regions of the Country*. Kharkiv, PH "INZHEK", 224. (2008)
6. Khaustova V. Ye., Zinchenko V. A., Moschitskaya T. A. Business Inform, **12 (2)**, 200-203 (2009)
7. Tishchenko A. N., Khaustova V. Ye. International Economic Journal Economic Cybernetics, **1–2 (55–56)**, 54–57 (2009).

8. Saaty T., Sodenkamp M. Making decisions in hierarchic and network systems. Int. J. Applied Decision Sciences, **1**, 24–79 (2008).
9. The State Statistics Service of Ukraine (2019). http://www.ukrstat.gov.ua/
10. Davydov A. A., Churakov A. N. *Modular analysis and modeling of society*. Moscow, Institute of Sociology, 208. (2000).

Digital Transformation of Innovative Business Processes on Railway Transport

Viktoria Ovchynnikova [1*], *Anna* Kuzmenko[2], *Tetyana* Yusupova[1], *Vladislava* Toropova[3], and *Nadia* Gontar[1]

[1]Ukrainian State University of Railway Transport, Economy and Management of Industrial and Commerce Business Department, Feyerbakh sq. 7, 61050 Kharkiv, Ukraine
[2]Kharkiv Institute of Finance of Kyiv National University of Trade and Economics, Management Department, Pletnevsky prov. 5, 61003 Kharkiv, Ukraine
[3]Simon Kuznets Kharkiv National University of Economics, Management and Marketing Department, Nauky av. 9-a, 61166 Kharkiv, Ukraine

Abstract. The article finds that acceleration of the pace of global high-intellectual development causes the digital transformation of innovative business processes in the Ukrainian railway transport, and consequently, the continuous improvement of intellectual information systems and technologies of management of innovative processes. The study of the existing models of the innovation process enabled to distinguish its sixth type of “model of digital networks and artificial intelligence”. The main feature of this stage of the development of the innovation process is the formation of a single digital platform (innovation portal), which will allow to make transparent the basic business procedures and to form Big Data with the ability to optimize the main innovative business processes. It is proved that a new approach to innovative business processes should include the formation of an intellectual information platform for innovative management which should become a key link to support innovative business processes in the industry. An integrated data bank (Big Data) and situational control centers should also be created as tools for promoting the process of making innovative decisions. As the coordinating situation centers of innovation activity, it is advisable to distinguish between “Investments”, “Innovation”, “Personal”, “Transport Logistic”.

1 Problem statement and its connection with scientific or practical tasks

At present, one of the acute and urgent problems is the problem of effective organization of innovative activity of railway transport and efficient use of its production potential in the transition to the digital economy, because the development of the latter forms the main trends in the development of transport and logistics business worldwide, including in Ukraine. The most significant among them are: globalization of transport and logistics business; active development of transnational Internet resources; rapprochement of

* Corresponding author: vikols75@ukr.net

transport and logistics services providers and consumers; emergence of new distribution channels; change of consumption model: personification of the approach in the organization of transportation of goods and passengers, etc. Therefore, only knowledge-intensive, high-performance operational activities of railway transport can become the basis and the main condition for economic growth of the industry and the whole country in the digital economy.

2 Analysis of recent research and publications

The modern policy of management of innovative activity of railway transport should first of all be consider on transformation of the most innovative process. Regarding the current models of the innovation process, they do not meet the modern realities of the functioning of the world rail transport. They can be divided into five types:

- the first type of innovation process model (1950s - 1960s) is characterized as "a model that is pushed by technology (technology push-model), is a simple linear-sequential process with an emphasis on the crucial importance of research and development work and the attitude to the market only as a consumer of the results of technological activity of production" [1]. This type of model of the innovation process can meet the scientists: S. Ilyenkova [2], V. Grinev [3], V. Kardash [4], L. Vodachek and O. Vodachkova [5], T. Alimova [6];
- the second type of model of the innovation process (1960s and 1970s) scientific opinion defines this type of model as a "linear sequential model, but with an emphasis on the importance of the market, which needs to react to R&D (need pull model)" [151]. This type of model can be seen in the works of J. Bradbury [6], H. Holstein [7], P. Doyle [8], Donald J. Margus [6];
- the third type of model of the innovation process (1970s and 1980s) called the "coupled type of model (coupling model), which is a combination of the models of the first and second type, with focus on communication of technological possibilities and capabilities with the needs of the market". A striking representative of this type is the chain model (chain link model) S. Kline – N. Rosenberg [9];
- the fourth type of innovation process model (1980s and 2010s) the innovation process is defined as "simultaneous, but not joint work of several groups of specialists from different structural divisions of the enterprise on one innovation". An example of the fourth type of model is the process of development of a new product of Nissan [1], as well as the "Japanese" model of innovation process [10];
- the fifth type of innovation process model (early 2010s until the present) is characterized as "strategic networking model and model of innovation and strategic networks, integration and networking" [1].

3 Transformation of the innovation process in the digital economy

The acceleration of the global high-intellectual development determines the digital transformation of innovative business processes in the railway transport of Ukraine, respectively, the constant improvement of intellectual and information systems and technologies of management of innovative processes. Digital transformation is a continuous process that undoubtedly carries a huge potential for improving the efficiency and development of the industry, as well as provides for fundamental changes in the conceptual provisions of innovative business processes. This allowed us to identify the sixth type of innovation process model (Fig. 1), namely, a model of digital networks and artificial intelligence.

	Period of existence	Generation feature	Source of innovation	Marketing support of innovations	The link between the stages of the innovation process
I type of model	from the 1950s to the mid-1960s	The model, which is pushed by technology (technology push-model), is a simple linear-sequential process with a focus on the role of research and development	Innovations are based on scientific knowledge and capabilities of enterprises	The acceptability of innovations and market demand are either not studied at all or are studied after the launch of an innovation into production. Attitude to the market only as a consumer of the results of technological activity of production	One-way communication, which in fact means the inability to learn from their own mistakes, experience
II type of model	late 1960s - early 1970s	Linear-sequential model, but with an emphasis on the importance of the market (need pull model)	Innovations are based on scientific knowledge and capabilities of enterprises; market needs	The organization of the innovation process is carried out taking into account the requirements of the market at the early stages of innovation development	One-way communication that makes it impossible to learn from their own mistakes, experience
III type of model	early 1970s – mid 1980s	Linked model (coupling model) is a combination of I and II generation models with a focus on linking technological capabilities and capabilities to market needs	Sources of innovative ideas are: the scope of research and development, new market needs, training on their own experience and an array of existing external knowledge	The organization of the innovation process is carried out taking into account the requirements of the market at the early stages of innovation development	Feedback accounting, which allows you to take into account your own mistakes
IV type of model	the middle of 1980-ies - beginning of the 2010s	The attention is focused on the parallel activities of integrated groups and external horizontal and vertical links	Sources of innovative ideas are: the scope of research and development, new market needs, training on their own experience and an array of existing external knowledge	Marketing at all stages of the innovation process. The perspective demand is analyzed, and at the last stages of innovative process on the basis of the made forecast the market demand is formed	Presence of external, horizontal, vertical and feedbacks
V type of model	the beginning of the 2010s - now	Strategic networks, integration and networking model (Strategic networking model)	Sources of innovative ideas are: the needs of the market; external (existing in the world) and opening (research) knowledge in relation to the enterprise; research, obtained on their own experience	Marketing is carried out from basic research to after-sales service of finished products. The perspective demand is analyzed, and at the last stages of innovative process on the basis of the executed forecast the market demand is formed	Presence of external, horizontal, vertical and feedbacks
VI type of model	now	Model of digital networks and artificial intelligence (Model of digital networks and artificial intelligence)	The sources of innovative ideas are: the needs of the market; the needs of the production process; external (existing in the world) and opening (research) knowledge in relation to the enterprise; knowledge (scientific and non-scientific), obtained on their own experience	Smart marketing is carried out from basic research to after-sales service of finished products. Smart-marketing by analyzing Big Data and other achievements of modern technology affects the formation of needs, as well as develops and promotes in the market value propositions customized to the needs of each client	Two-way communication between innovative business processes through a single digital platform (innovation portal), which will make the main business procedures transparent, and form BigData with the ability to optimize the main innovative business processes

Fig. 1. The main characteristic features of the innovation process models.

The main feature of this stage of development of the innovation process is the formation of a single digital platform (innovation portal), which will make the main business procedures transparent, and form Big Data with the ability to optimize the main innovative business processes. At the same time, smart marketing is advisable to carry out from basic research to after-sales service of finished products, in addition, by analyzing Big Data and other achievements of modern technology to influence the formation of consumer needs, as well as to develop and promote in the market value propositions customized to the needs of each client.

4 Integrated intellectual and information platform for innovative changes as an imperative factor of innovation in the digitalization of the economy

A new approach to innovative business processes should provide for the formation of an integrated intellectual and information environment of interaction between branches of Ukrzaliznytsia JSC, because one of the important areas of strategic management of the enterprise is the process of making management decisions in modern conditions of formation of significant information flows and uncertainty of the environment is much more complicated. Therefore, now the management of railway transport should definitely form an intellectual and information platform for innovation management, which should become a key link in the future, which will support innovative business processes in the industry, because its feature is the concentration and reflection of information about the impact of each branch of Ukrzaliznytsia JSC on the processes of implementation of innovative changes in the railway industry.

The effectiveness of innovative processes in the railway industry largely depends on the quality of the formation of an integrated intellectual and information platform to ensure innovative changes in the railway industry [11-14], which, on the one hand, creates conditions for the introduction and use of leading intellectual and information systems and technologies for the development of branches of Ukrzaliznytsia JSC, and on the other □ provides their strategic interaction within the virtual space. In this context, the definition of intellectual and information systems and technologies that will become the basis for the formation of such an integrated intellectual and information platform is of strategic importance. The most common of them include the platform DSS (Decision Support System) – a class of automated information systems that help to collect the necessary information for decision-making and choose the best solution to achieve a specific goal. Structurally, the DSS system consists of [15]:

1. Data Warehouse – information platform DSS. Data warehouses allow you to integrate information that reflects different perspectives on the same subject area.

2. Business Intelligence Tools (business intelligence tools) – software that enables users to observe and use large amounts of complex data. The main components of the intelligent information system DSS is [16, 17]: OLAP (OnLine Analytical Processing); Data Mining; COMFAR; BEST (Business Environment Strategic Toolkit); FIT (Financial Improvement Toolkit), etc.

Given the above, it is also important to create an integrated data Bank (Big Data) and situational management centers as tools to facilitate the process of innovative decision-making based on a comprehensive analysis of the factors influencing the subjects of the relevant system on the process of implementing strategic initiatives of highly intelligent development of railway transport through the use of the necessary intellectual and information systems and technologies. As coordinating situation centers of innovation activity, it is advisable to distinguish the following: "Investments", "Innovation", "Personal", and "Transport Logistic".

Intellectual and information platform of innovative development of subjects of investment activity should be represented by intellectual and information systems and technologies that provide analytical processing of large volumes of information about financial subjects of innovative activity; automation of processes of financial and economic analysis of innovative projects; the use of information tools to support decision-making on the effectiveness of the implementation of priority innovative projects and the like. The processes of implementation of the relevant intellectual and information systems and technologies are controlled by the situation center of management "Investments".

Automation and improvement of intellectual and information support of processes of development of innovative products as the basis of innovative and technological renewal of railway infrastructure and rolling stock is implemented within the intellectual and information space of subjects of innovative activity, cooperate in the framework of the project. The introduction of intellectual and information technologies in this area will provide automation and mutual coordination of design and modeling of innovative products; automation of production and technological processes; management of supply processes of raw materials; automation of processes of adjustment and readjustment of equipment; improvement of relationships with suppliers and consumers; introduction of intelligent information systems to promote strategic management decisions in conditions of uncertainty and risk, and the like. Control of the processes of implementation and adaptation of intellectual and information systems and technologies of development of subjects of innovative activity takes place within the framework of the situation center of management "Innovation".

An important stage in the implementation of intelligent information systems is the training of modern specialists in the field of intelligent transport systems, which includes the study of system-wide features of intelligent transport systems, as well as the analysis and formal description of the processes of creating such automated systems that are "able to learn" (generate new knowledge in the form of mathematical models, control patterns), "classify" (automatically group objects, actions, control signals), adapt to changing conditions and parameters of the operating environment – "adapt". The solution of these problems is not possible without the formation of future specialists in the field of intelligent transport systems of system thinking, knowledge and skills of the system approach and system analysis to solve complex organizational, technical and technological problems, as well as knowledge bases of certain methods and examples of the implementation of complex projects of the transport complex on the basis of global integrated information and communication technologies. Thus, the development and implementation of intelligent information systems and technologies is closely related to the adaptability and flexibility of railway personnel to strategic changes, which makes it important to ensure its intellectual development. Among the main intellectual and information systems that provide distance and multimedia training, it is advisable to highlight: Webinar, Comdi, Acrobat Connect Pro, Mirapolis, Virtual Room, eLearning Server, iMind, iWebinar, Moodle. Some of them have already become widespread among the specialized higher educational institutions of the country, providing the opportunity to obtain the necessary knowledge and skills on the job through the use of these intellectual and information systems and technologies. Implementation of the processes of intellectual development of the personnel is planned within the limits of the situational center of management "Personal".

A key element in the formation of an integrated intellectual and information environment of interaction between the branches of Ukrzaliznytsia JSC should be intellectual and information systems and technologies for the development of transport and logistics activities. In turn, the unified information space of Ukrzaliznytsia JSC should be provided with a set of existing unified automated control systems for freight and passenger transportation, infrastructure, traction, production and service.

References

1. R. Rothwell, The Changing Nature of the Innovation Process, Technovation, (1993) http://technopark.al.ru/business/innovation.htm. Accessed 19 Apr 2019
2. S.D. Ilenkova, *Innovatsionnyy menedzhment*, Moscow, 327 (1997)
3. V.F. Grinev, *Innovatsionnyy menedzhment*, Kiev, 148 (2000)
4. V.Ya. Kardash, *Tovarna innovatsiyna polityka*, Kiev, 124 (1999)
5. L. Vodachec, O.Vodachkova, *Strategicheskoye upravleniye innovatsiyami na predpriyatii*, Moscow, 167 (1989)
6. V.N. Goncharov, E.V. Ivanova, Concept of strategic organization of innovative activity of enterprises, URL: www.maop.vorstu.ru/Gancharov.html. Accessed 07 May 2019
7. P. N. Zavlyn, A. K. Kazantsev, L. E. Mindeli, *Innovatsionnyy menedzhment*, Moscow, 173 (1998)
8. P. Doyle, Menedzhment: strategiya i taktika, St. Petersburg, 560 (1999)
9. E.O. Naumenko On the model of management of the innovative process of the enterprise in modern conditions, URL: http://ej.kubagro.ru/2006/04/pdf/03.pdf. Accessed 03 May 2019
10. F. Kodama, *Emerging patterns of innovation sources of Japan's technological edge, Harvard Business School*, 245 (1995)
11. V.L. Dykan, Bulletin of the Economy of Transport and Industry, **42**, 144 - 149 (2013)
12. V.O. Ovchynnikova, *Stratehichne upravlinnya rozvytkom zaliznychnoho transportu Ukrayiny*, Kharkiv, 427 (2017)
13. O.V. Dykan, *Orhanizatsiyno-ekonomichnyy instrumentariy zabezpechennya konkurentospromozhnosti promyslovykh pidpryyemstv zaliznychnoho transportu*, Kharkiv, 443 (2016)
14. I.V. Tokmakova, *Zabezpechennya harmoniynoho rozvytku zaliznychnoho transportu Ukrayiny*, Kharkiv, 403 (2015)
15. G.A. Mamaeva, The Age of Science, **13**, 81-86 (2018)
16. Methods and systems for supporting decision making, http://i-intellect.ru/articles/45/. Accessed 21 Apr 2019
17. Intelligent Information Systems in Management, https://rep.bntu.by/bitstream/handle/data/23986/%D0%A2..B8.pdf?sequence=11. Accessed 19 Apr 2019

Capitalization as a Tool for Managing the Development of Industry and Rail Transport

Igor Posokhov[1*], *Victoriia* Cherepanova [1], *Olha* Podrez[1]

[1] National Technical University "Kharkiv Polytechnical Institute", Chair of Innovative Entrepreneurship Management and International Economic Relations, Kirpichova str. 2. Kharkiv, 61002 Ukraine

Abstract. The Ukrainian real economy encounters similar problems that do not allow its rapid development – high level of wear of productive assets, lack of modern equipment, outdated technologies, inappropriate environmental measures, high rate of occupational injury, etc. All this requires designing of new tools to manage the development of such important sectors of economy as industry and rail transport. Therefore, the urgent issues at the current stage of development of these industries include the definition of conditions for fixed assets capitalization and the sources of its financing. The scientific novelty of the results is identification and justification of the capitalization main components, determining the sources of funding and the mechanism for their attraction. The tool for managing the productive and environmental protection assets capitalization has been designed which is optimized using a two-dimensional dynamic programming model. The results obtained are the basis for the practical solution of the problem, and for further scientific research. This approach allows solving the problem of rail transport and industrial enterprises capitalization in a comprehensive manner, which contributes to their sustainable development.

1 Introduction

Many countries of the world that are gradually overcoming the financial recession and whose objective is strong economic growth and enhancing quality of life, have same global problems: industry recovery, the introduction of innovative technologies in production and environmental protection, the provision of rail transport to supply raw materials, and final products marketing. Therefore, the issue of industrial facilities and rail transport development should be resolved in parallel. This is, especially, case of capitalization, which J. M. Keynes [1] describes as a "method of capital accumulation, one part of which is consumed, while the other is invested in long-term production and investment assets".

* Corresponding author: posokhov7@gmail.com

The importance of these two sectors of capitalization is confirmed by the fact that Ukrzaliznytsia has the old rolling stock fleet and infrastructure (51.7%) [2], while productive assets wear of industrial enterprises is over 60%, and in some cases, over 80%.

Taking into account the urgency of the problem, the main purpose of this study is to characterize the state of capital investments in the industrial enterprises fixed assets (for example, coking plants) and rail transport and to develop a tool for managing the enterprises development in capitalization conditions. To achieve these objectives, we have specified the following tasks:

–developing the conceptual construct of "capitalization" category;

–indicating general lines and financial sources of the enterprises capitalization;

–proposing the structure of managerial tool for the enterprises development in capitalization conditions.

2 Literature review

In the economic literature, the notion of "capitalization" has been used for a long time and differently. We propose to consider this term based on scientific approaches.

Karl Marx [3] wrote that capitalization is a "process of turning a part of profit (income) or total profit (income) into additional capital and factors of production (plant and machinery, material and supplies, manpower, etc.) and using it to expand production". Unlike K. Marx, S.V. Mocherny [4] complements his production approach with financial one – "the investment of part of the income into securities to receive profit from them in the form of interest". V.G. Zolotogorov [5] states that this term relates to the "estimation of an enterprise value", i.e., a cost approach is used, while R. Koch [6], M. Bertonesh and R. Knight [7], I.P. Buleev and M. A. Solod [8], S. Raza [9] and A. Singal [10] specify that capitalization is "the market value of a company whose shares are quoted on a stock exchange" [6], thus combining cost, market and financial approaches. I.P. Buleeva and N.E. Bryukhovetska [11] consider the definition of "capitalization" by economy levels (enterprise, branch of industry, region, national economy, international economy) and insist on production market approach to its conceptual framework. [11].

M.A. Kozoriz [12], V.M. Geits and A.A. Gritsenko [13] insist on a broader approach to the term "capitalization" which would cover different levels, spheres and the management mechanism with respect to production-financial approach.

Based on the generalization of the existing approaches to the concept of 'capitalization', we define this term as an economic category whose actions are aimed at obtaining additional capital to create productive assets in order to increase revenues from the sale of products and services in conditions of environmental safety of production.

The category of "capitalization" has a wide range of applications, such as production, financial, market-based, institutional; it is used at all levels of government – state (macro), industry branch-based (meso-) and entrepreneurial (micro level) and spheres of economy – processing industry, agriculture, transport, financial and credit system, etc.; it has its own management mechanism, which includes the appropriate tools. In this case, it relates to fixed assets, intangible assets, material resources.

3 Problem description and methodology

The problems of Ukrzaliznytsia and coking plants functioning are associated with outdated fixed assets and low environmental protection of the production process and services rendering. Thus, as of 01.01.19, the total number of coke batteries of the Ukrainian Research and Industrial Association (URIA) Ukrkoks [14] decreased by 17 pcs. vs. 1.01.14, the average age of the batteries is 29.07 years while the norm is 20 years, that is, 66.7% of the batteries are outdated, and there are 8 batteries with an operating life exceeding 35 years.

URIA Ukrkoks solves the coke batteries problem by investing its own funds into the KHP assets (Fig. 1). During the period from 2013 to 2018, capital investments of KHP increased twice from UAH 421,215 thousand in 2013 to UAH 958,940 thousand in 2018. Starting from 2015, the cost of major repairs exceeds the cost of acquiring and creating assets. The costs for equipment reconstruction and upgrading equal only to 24.56% of the total capital investment amount [14].

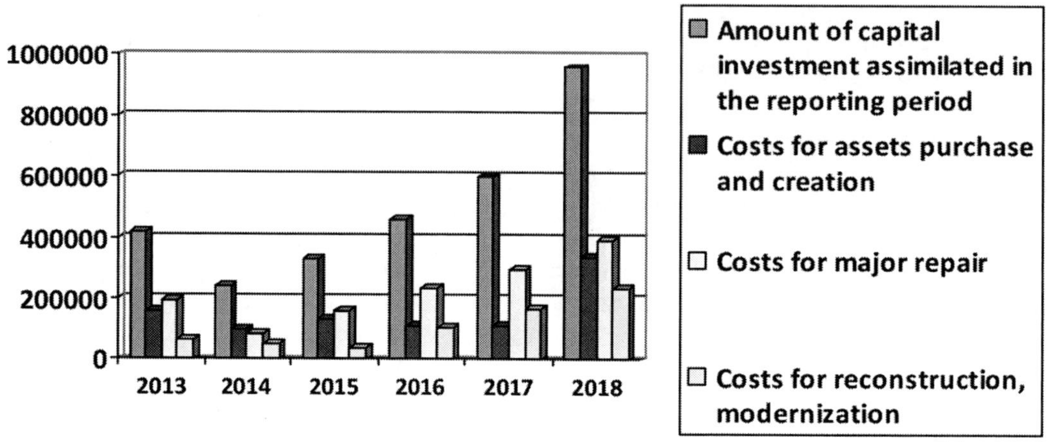

Fig. 1. Dynamics of KHP capital investments, thou. UAH [14].

All capital investments were made at the expense of the company's own funds.

A similar situation is observed in Ukrzaliznytsya. According to Ukrzaliznytsya, JSC financial reporting [15], the fixed assets depreciation is 60%, which is 0.9 of that in 2017. In the two last years, Ukrzaliznytsia's capital investments grew 1.55 times. Ukrzaliznytsya JSC pays the major attention to the construction of new objects of fixed assets, modernization and repair of the old ones. To renew the rolling stock, regional branches, branch offices allocated by 1.5 times more funds than in 2017, performed 4,282,7 million UAH worth capital construction works, acquired mechanisms and equipment, and other noncurrent tangible and intangible assets by 1.3 times more than in the previous year.

Thus, the by-product coking industry and Ukrzaliznytsya are upgrading, repairing and modernizing the main productive assets (MPA), but Ukrzaliznytsa has paid more attention to capital construction in these recent years, while KHP – to the capital repairs.

Issues of capitalization of production are directly related to the availability of financial sources (Table 1).

Increased capital investments of Ukrzaliznytsya JSC in 2018 were mainly funded with its own funds and attracted credit resources, which increased 2.73 times this year [15]. Ukrzaliznytsya and coking plants have another feature in common – the negative impact on the environment caused by their activities. Emission volumes depend on coke production and on measures implemented by KHP at their enterprises and capital investments allocated to environmental protection. Environmental expenditures of URIA Ukrkoks were allocated to capital investments, including major repairs, and current investments.

Table 1. Analysis of Ukrzaliznytsya JSC sources of financing of capital investments for the period from 2017 to 2018 [15].

Sources of financing	2017	2018	Absolute deviation	Accretion, %

Capital investment sources, total, including:	10,904,079	16,912,841	6008762	55.1
contracted loans	538,844	2,012,476	1473632	273
Budget funding	-	-	-	-
Own costs	10,337,225	14,860,811	4523586	43.76
Other sources	28,010	39,554	11544	41.21

In the cost structure, the largest share was borne by current expenditures in 2017-2018, which in 2018 decreased by 9%, and the cost of major repairs increased by 56.3%. [14].

A similar situation is observed in Ukrzaliznytsya JSC In the freight transportation performed in 2018, the largest share, by hazard class [15], belongs to class 3 (easy-combustible fluids) – 47%.

Thus, issues of capitalization of fixed assets are relevant for both the industry and Ukrzaliznytsia.

The methodological basis of this article is the design of a tool for managing production capitalization with respect to fixed assets, which consists of three management levels:

1. Production technical, through which the choice of fixed assets is made for production by type of improvement – purchase or new assets construction, major repairs, modernization (reconstruction);
2. Production environmental – the choice of the fixed environmental assets by type of improvement is made – purchase or construction of new assets, major repairs, current disposal measures and reverse water supply;
3. Financial investment – justification of the amount of capital investments is made by two levels of management and sources of their financing – own, contracted attracted loans, budget funds and other sources;
4. Organizational optimization – a model of dynamic programming is structured which takes into account: replacement of equipment (industrial and environmental), time of capital repairs according to schedule of planned production repairs and types of funds improvement (purchase or construction, major repairs, modernization (reconstruction) or current measures.

4 Findings

Fig. 2 provides the structure of fixed assets capitalization management of the industrial enterprise by management levels.

I.V. Kryvovyazyuk [16] emphasizes necessity of dynamic programming in the process of production facility capitalization. However, in our case, economic mathematical model of the dynamic issue is a two-dimensional investment resources allocation model [16] for productive and environmental assets.

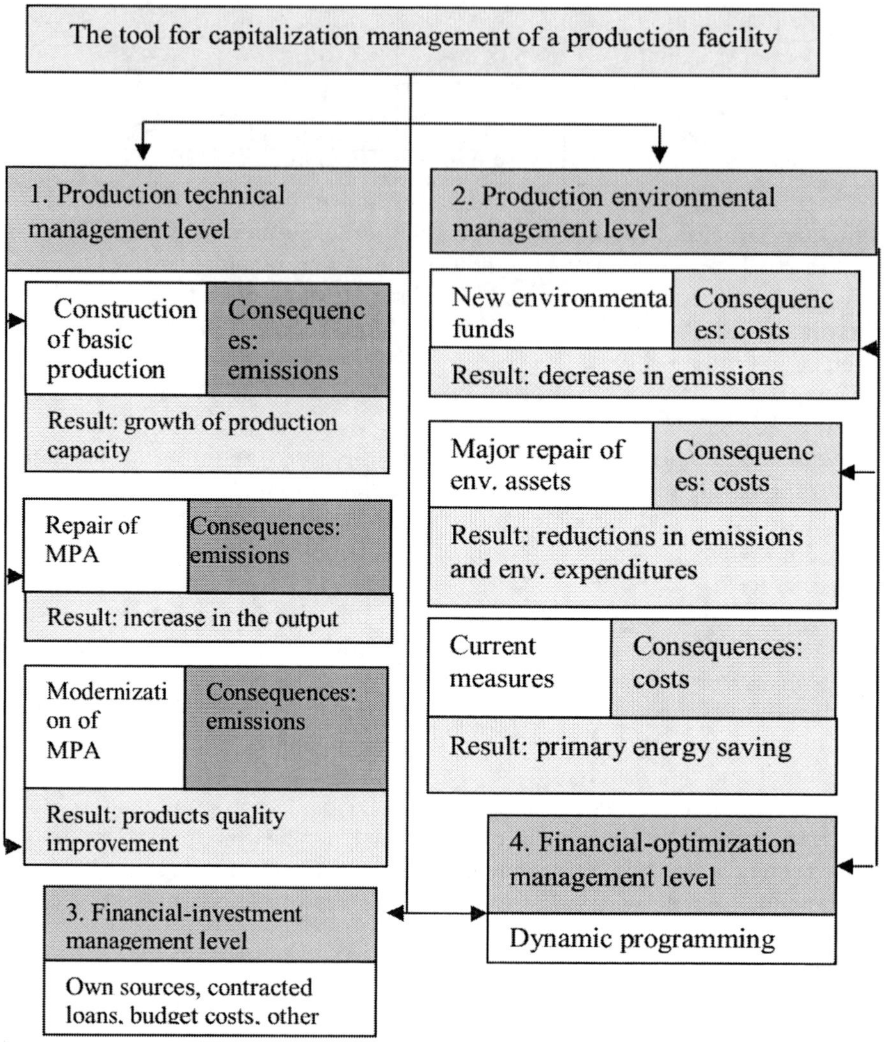

Fig. 2. The tool for managing production facility capitalization (developed by the authors).

5 Conclusions

Capitalization of production is important for the industry and Ukrzaliznytsia. It requires making managerial decisions regarding the improvement of productive and environmental fixed assets (purchase of new assets, major repair old assets, modernization, reconstruction and current measures), optimal allocation of investment resources in order to obtain the maximum income by the enterprise from the production and sale of products (services) and reduction in environmental payments due to emission reductions.

We propose to resolve this problem by constructing a managerial tool for capitalization of production in relation to fixed assets, which consists of four management levels: production and technical, production and environmental, financial and investment and organizational and optimization. With the help of the first one, the justification of the choice of the fixed production assets as to the lines of their improvement, the second one – the main means of environmental protection, the third one – the amounts of necessary investments and financial sources for its coverage, and the fourth one – the allocation of capital investments between the fixed production funds and the environmental protection assets. The optimization of funds allocation is proposed to be carried out by a two-dimensional dynamic programming model, which allows to maximize income of the enterprise due to replacement of fixed assets and to receive income from increased output and reduction in environmental expenditure by emissions reduction.

The research results can be used both by industrial enterprises and Ukrzaliznytsya.

References

1. J. M. Keynes, The general theory of employment, interest and money. Favorites, Moscow (2007).
2. Purposes of the Sustainable Development Goals: Ukraine. The Ministry of Ukraine's economic development and trade, Kiev (2017).
3. K. Marx, Capital: Critics of political economics. - T. 2. - Kn. II: Capital circuit process, Moscow (1974).
4. Economic encyclopaedia: in 3 vol. [edited by S. V. Mocherny]., Kiev (2000).
5. V.G. Zolotogorov Economics: encyclopaedic dictionary, Minsk (2004).
6. R. Koch, Management and Finance from A to Z., St. Petersburg (1999).
7. M. Bertonesh, R. Knight Money, Flow Management, (St. Petersburg) (2004).
8. I.P. Buleev, M.A Solod, Problems of capitalization of metallurgical enterprises In modern conditions. Economy of Industry, **1** (73), 73-82 (2016).
9. S.Raza, T. Jawaid, S. Afshan, M. Ka-rim, Is stock market sensitive to foreign capital inflows and economic growth? Evidence from Pakistan Journal of Chinese Economic and Foreign Trade Studies. **8** (3), 142-164 (2015) https://doi.org/10.1108/JCEFTS-03-2015-0012
10. A. Singal, A. Jain, Emerging market firms: measuring their success with strategic positioning maps. Journal of Business Strategy. **35** (1), 20-28 (2014). https://doi.org/10.1108/JBS-04-2013-0026
11. Capitalization of enterprises: theory and practice: monograph / edited by I.P. Buleeva, N.E. Bryukhovetska, NAN of Ukraine, Institute of industrial economy, Donetsk (2011).
12. M.A. Kozoriz, Capitalization role and function in ensuring economic development of economic entities. Regional economics, **2,** 42-48 (2007).
13. Capitalization of Ukrainian economy / edited by acad. of NAN of Ukraine V.M. Geits, A.A. Grizenko. Institute of Economics and Forecasting of NAN Ukraine, Kiev (2007).
14. Ukrainian Scientific-Industrial Association "Ukrkoks": Results of the work of coke-chemical enterprises and industries of Ukraine in 2018, Dnepr (2019).
15. Consolidated financial plan of Ukrzaliznytsia JSC, [online], Available at: https://www.uz.gov.ua/about/investors/financial_statements/kfp/
16. A.M. Romanovskaya, M.V. Mendziv, Dynamic programming: Handbook, Omsk (2010).

Technology Transfer as a Prerequisite of Innovative Development of Enterprises

Viktoriia Prokhorova[1]*, *Nadiia* Reznik[2], *Olena* Bozhanova[3], *Krystyna* Slastianykova[4]

[1] Ukrainian Engineering Pedagogics Academy, Department of Economics and Organization of the Activity of Business Entities, 61003, 16 Universitetskaya st., Kharkiv, Ukraine
[2]National University of Life and Environmental Science of Ukraine, Department of Labor Economics and Rural Development, 03041, 19 Generala Rodimtseva Str, Kyiv, Ukraine
[3]National Metallurgical Academy of Ukraine, Department of Finance, 49600, 4 Gagarina avenue, Dnipro, Ukraine
[4] Ukrainian Engineering Pedagogics Academy, Department of Economics and Organization of the Activity of Business Entities, 61003, 16 Universitetskaya st., Kharkiv, Ukraine

Abstract. Low success in the implementation of technology transfer projects in Ukraine is mainly caused by two groups of factors: the first one, directly related to the development of innovative processes, which hinders the country's participation in the international exchange of technologies, and the second one, caused by the shortcomings of domestic legislation on the definition and protection of intellectual property rights of participants in the innovation process. In order to develop the economy both at the enterprise level and at the national level, the following technology transfers should be developed: firstly, the classical one, which provides for the work of enterprises and research institutions in close relationship; secondly, it requires the development of transformational technology transfer, which ensures closer cooperation in science with other countries; thirdly, it is necessary to attract unique and advanced technologies to the regions. As a result of technology transfer, developments can be successfully commercialized for the benefit of their authors and, as a result, tax revenues to the country's budget can be increased.

1 Introduction

In the context of globalization and integration, increased competition in world markets, the level of economic development of the country determines its scientific and technological progress and the ability to commercialize intellectual capital. One of the most

*Corresponding author: vkprohkorova@gmail.com

important components of the innovation process and a promising mechanism of economic stabilization of the state is the transfer of technology, the essence of which is the transfer of know-how, new technologies, technological equipment and scientific and technical knowledge from the owner to the customer.

The issue of determining the essence of the mechanisms of technology transfer has been studied by many domestic and foreign authors, such as S. Arkhiyereyev [1], T. Tarasenko [1], I. Evgrafova [3], H. Pratnenko [4], N. Tkalenko [5], D. Makarov [6], N. Fonshtein [7], L. Balastryk [8], E. Dzhur [9], A. Sadekov [10] and others. Despite the existing scientific developments in this subject area, theoretical and applied issues of the organization and implementation of technology transfer require more in-depth study and improvement. The efficiency of technology transfer is an important factor in the formation of competitive advantages of any enterprise.

2 Presenting main material

The process of commercialization of own intellectual capital includes the transfer of new technologies, technological equipment and scientific and technical knowledge from the owner to the consumer. In the developed countries of the world, GDP increases by 75-80% due to the commercialization of intellectual capital.

In Ukraine, according to UNESCO, the relation between patent activity and GDP growth is the lowest among the countries of the international reference group (correlation index of the USA is 0.985, Germany – 0.916, UK – 0.895, Poland – 0.878, Ukraine – 0.330) [1,2].

Low success in the implementation of technology transfer projects in Ukraine is mainly due to two groups of factors: the first one, directly related to the development of innovative processes, which hinders the country's participation in the international exchange of technologies, and the second one, caused by the shortcomings of domestic legislation on the definition and protection of intellectual property rights of participants in the innovation process.

There is no general and free access to the international patent classification of intellectual property objects, which contains not only descriptions, but also characteristics and advantages. Therefore, researchers do not have the opportunity to learn about already developed and patented analogues and their technical specifications which prevents from focusing on the international technology market.

The issue of valuation of intellectual property objects and their reflection in accounting is not settled. Today, so the assessment is extremely difficult because of the use of each time a specially developed technique, which makes this procedure very long and costly. The absence of a unified method of determining the amount of material damage caused by the violation of intellectual property rights makes it impossible to punish violators.

The number of applications for inventions tends to diminish.

These facts confirm the existence of negative trends in the negative losses in the sphere of technological transfer in Ukraine.

Commercial sale of intellectual property is carried out through their actual sale-horizontal transfer-and use in their own production – vertical transfer. These two forms of realization of intellectual property objects are interrelated and constantly pass into each other during the life cycle of intellectual property objects. These processes can occur simultaneously or in parallel, with the use of intellectual property objects by different subjects.

Also, scientists decided to identify three main forms of technology transfer:

- internal transfer, when there is a transfer of technology between different organizations;

- introduce quasi-internal transfer, that is, the movement of technology within alliances, unions and associations of independent legal entities;

- external transfer, i.e. the process of technology diffusion involving independent technology developers and consumers, between different companies and countries [3].

Ukraine has established a national technology transfer network, which has a national scale and aims to promote the commercialization of high-tech technologies and attract the scientific potential of Ukraine in the global commercial turnover [4].

The analysis of the creation of advanced production technologies by industry showed that the dominant role is played by enterprises of mechanical engineering, metallurgical, chemical and petrochemical production, and food industry.

The economy of Ukraine is characterized by an unreasonably long period of use of technologies used in production and, consequently, a high degree of wear and tear of machinery and equipment. These problems can be solved as follows: to transfer technologies by acquiring the results of scientific research of domestic scientists and bringing them to industrial use; to transfer technologies from developed countries.

It should be noted that domestic technologies that are ready for industrial use are not enough, and when purchasing Research & development results at the early stages of the innovation cycle, the investor bears great risks, since from 100 ideas to the market reach from 5 to 10. In addition, additional investment is needed, which is about 90% of the cost of research. In the case of acquisition of foreign technologies, the company does not receive additional revenue from innovation activity, but the risks are minimal, as the products that are released according to the acquired technology has already been tested and received a positive assessment in the market. However, this will not allow the country to become a leader in the world market. Many countries of the world, especially Germany and Japan, paid special attention to the import of technologies, as a result, these countries are now among the world leaders in terms of scientific and technological development.

Today, society is trying to move to an innovative platform for building the economy, from the accumulation of material wealth as the basis of personal well-being – to the accumulation of information as the basis of social progress. The determining priority of the country's place in the globalizing world is the availability of information, intellectual potential and, as a consequence, the possession of material goods in the post-industrial paradigm.

According to N. V. Tkalenko, scientific-technological revolution of the first decades of the XXI century which includes new imperatives, corresponding to the essence of humanistic-noospheric post-industrial society [5].

In order to develop the economy both at the level of enterprises and at the national level, the following technology transfers should be developed. Firstly, the classic one, providing for the work of enterprises and research institutions in close cooperation. Such cooperation provides for making more active use of the results of domestic science and bringing it to industrial and market applications. Secondly, a transformational technology transfer should be developed, ensuring closer cooperation in science with other countries. It is very difficult for domestic enterprises to enter the free world market, as it has long been divided between specialized firms of the USA, Japan, Germany and other developed countries. Selling technologies gives the opportunity to enter the closed market of other countries, because the technology receives related goods and services. Thirdly, unique and advanced technologies to the region should be attracted.

As a result of technology transfer, developments can be successfully commercialized for the benefit of their authors and, as a consequence, tax revenues to the country's budget will be increased. However, developers and owners of new technologies – scientific organizations, small innovative firms, organizations of innovative infrastructure – have problems with finding buyers of their developments or partners to create production.

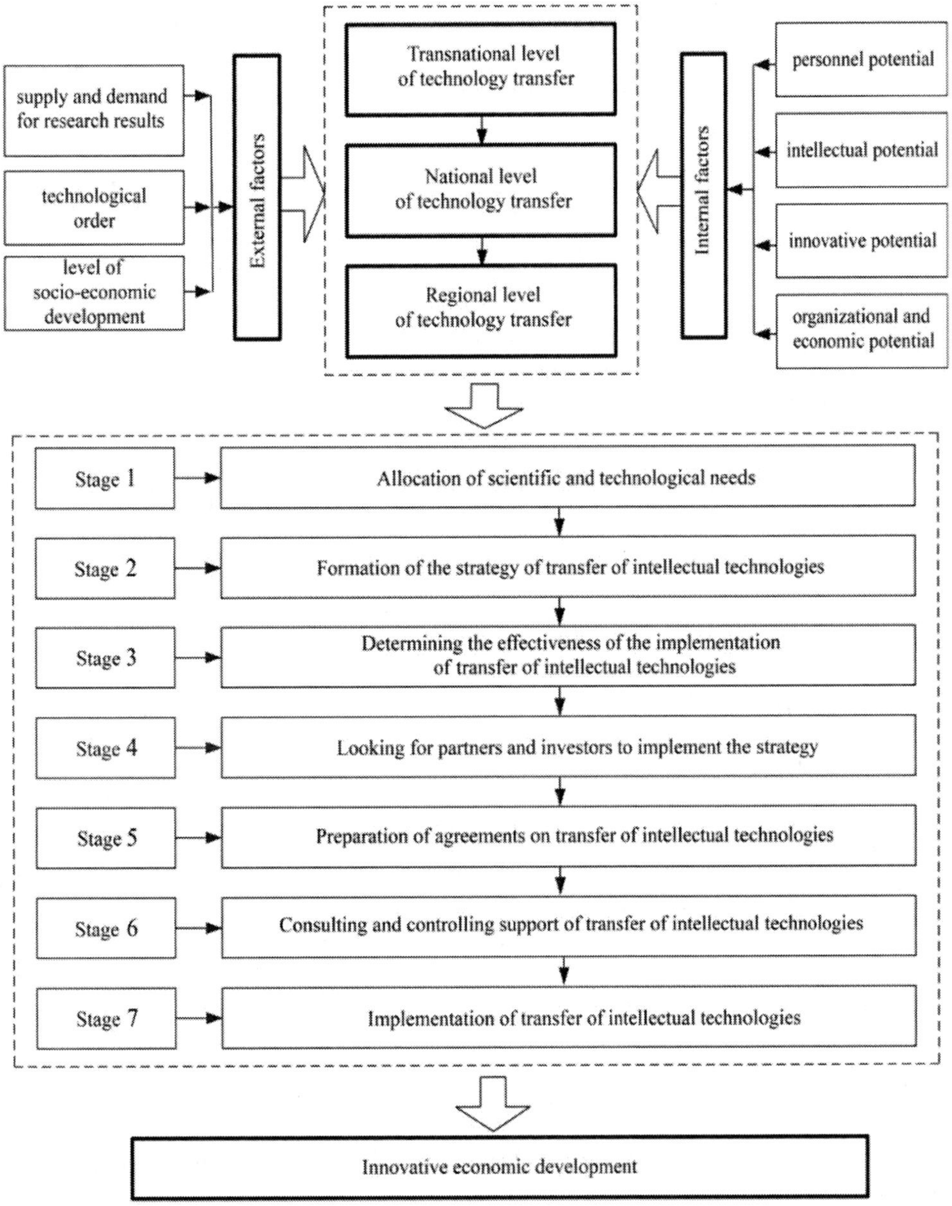

Fig. 1. Stages of implementation of the economic mechanism of technology transfer.

In addition, scientists usually do not have the skills to do business. There is another side to the problem. If the company plans to achieve competitive advantages by improving technologies, then the question will arise where to find information about technologies that can improve business efficiency.

To solve these problems, it is necessary to create institutions-intermediaries of the innovation market, which should provide participants of innovation processes with all necessary services for the development of their potential and innovative opportunities on the principle of "single window".

The efficient transfer of technology accelerates innovation processes at the regional, national and transnational levels. In order to create a mechanism for optimal interaction of organizations on a national and transnational scale, it is necessary to determine the parameters of modeling that allow the most effective way to implement the process of technology transfer (Fig. 1). The peculiarity of these parameters is that in determining them it is necessary to take into account the goals and objectives of each of the participants in the process of technology transfer.

Modeling allows obtaining a real system of economic relations with a certain accuracy, taking into account many different components [6]. As the main elements of modeling we can define the following: deepening of analytical approaches aimed at the study of significant economic problems, and the use of new methods in solving economic problems. The most effective mechanism for the implementation of the technology transfer process is a three-level model that allows you to activate the flow of scientific and technological information in short steps. Benefits from the implementation of the three-tier system of technological cooperation are: reduction of time of search of partners and investors; increase in the number of projects with scientific-technological cooperation; increasing the scope of the search for technological partners and investors. The implementation of a three-level technology transfer system will provide additional profit, increase innovation potential, and improve the competitiveness of the economy.

The number of factors influencing the choice of technology transfer mechanism is quite large. The effectiveness of technology transfer should be determined by comparing the costs to the financial result for a certain period of time:

$$D = \sum_{p=1}^{p}\left[\sum_{i=1}^{n}(Vip * Cip - Zip) + \sum_{j=1}^{m}(Fjp - Zjp)\right](1+k)^{-p} \qquad (1)$$

Where Vip, Cip - the volume and price of innovative products;
Fjp - royalties from the implementation of an innovative product;
k – normal reduction of costs (%);
p – point in time

Thus, the modeling of the technology transfer process allows us to explore trends and prospects of possible ways to spread innovation. The practical implementation of the three-level technology transfer system will allow to intensively increase innovation activity in the real sector of the economy.

3 Conclusions from the conducted research

For the development of the country's economy, it is necessary to increase the pace of transfer of unique and advanced technologies, attract financial resources for the development of innovation. At the same time, the introduction of innovations in the enterprise should be constant. In addition, during the development and commercialization of new products, it is necessary to focus on the needs of the market, investors, specific buyers of new developments, technologies, goods and services. The experience of developed countries shows that the process of technology transfer can be organized by creating organizations that provide a wide range of services in the field of technology transfer. A particularly important factor in improving the technological level of the country is the commercialization of the results of innovation, for which it is necessary to create tools for technology transfer.

References

1. Arkhiyereyev S., Tarasenko T. URL: *old.niss.gov.ua/Monitor/Oktober/7.htm*

2. Otsinka innovatsiynoho potentsialu Ukrayiny. – Tekhnichna dopovid' № 1 (kviten' 2013 r.)/ Proekt «Rozbudova spromozhnosti dlia nyz'kovuhletsevoho zrostannia v Ukrayini». – Berlin,(2013)
3. Evgrafova I. Innovatsionnyi menedzhment. URL:Rezhym dostupu: https://www.e-reading.club/chapter.php/103739/40/Evgrafova_-_Innovacionnyii_menedzhment._Shpargalka.html
4. Pratnenko H. URL: http://www.irbis-nbuv.gov.ua/cgi-bin/irbis_nbuv/cgiirbis_64.exe?C21COM=F&I21DBN=UJRN&P21DBN=UJRNsoc_gum/skhid/2009_7/13.pdf
5. Tkalenko N.V. Naukovyi visnyk CHDIEU, **4,** 92-97, (2014)
6. Makarov D.V. *Vestnik KRAUNTS. Fiziko-matematicheskiye nauki.* **1,** 67, (2014)
7. Fonshtein N.M. *Transfer tekhnologiy i effektivnaya realizatsiya innovatsiy: khrestomatiya*, M. ANKh pri Pravitel'stve RF, 296 (1999)
8. Balastryk L.O. Teoretychni ta prykladni pytannia ekonomiky. Zb. nauk. prats', **7,** 188-195, (2005)
9. Dzhur O.Ye. Visnyk ZhDTU. SERIYA: Ekonomichni nauk*y* **4,** 290-295, (2012)
10. Sadekov A.A. *Upravleniye predpriyatiyem v usloviyakh krizisa: monohrafiya.* – Donetsk: DonGUET, 178, (2006)

Criteria for Choosing the Priorities of Smart Development of Industrial Centers in the Context of Globalization

Volodymyr Rodchenko[1*], *Jiafei* Guo[2], *Yuliia* Prus[1], *Liu* Zeyu[2]

[1]V. N. Karazin Kharkiv National University, 4 Svobody Sq., Kharkiv, 61022, Ukraine
[2]Luoyang Normal University, Guangwen Rd, Jianxi Qu, Luoyang Shi, Henan Sheng, 471132, China

Abstract. The purpose of the study is to justify the criteria for choosing the priorities of a smart development of industrial centers in the context of globalization. By analyzing, systematizing and summarizing scientific works, it has been determined that the process of globalization has gone so far that even economically developed countries cannot independently solve production, scientific, technical, financial, trade and other problems effectively. The time of separate and independent economies is over. Industrial centers are the generators and re-translators of global change, the "backbone" of the world economy. In the course of the study, using a structural approach, the authors proposed five groups to assess the components of the potential and the prospects for the smart development of industrial centers in the context of globalization.

In the last decade, one of the most important factors in a world development has been the phenomenon of globalization, which covers not only the economy and sphere of entrepreneurial activity, but also the political and social sphere. The concept of "globalization" was introduced into a broad scientific revolution in the late 1980s and early 1990s. The first author who used this term was the American scientist T. Levit, who in 1983 in Harvard Business Review's article wrote about the merger of the markets of individual products produced by large multinational corporations [1].

To date, the process of globalization has gone so far that even economically developed countries cannot independently solve production, scientific, technical, financial, trade and other problems effectively. The time of separate and independent economies is over. Predicting the future, scientists point view to the inevitable formation of the economy of the planet, where national economies will become part of a certain whole system. Globalization is interpreted by science as objective data, which simply cannot be reckoned with. Objective means that isolation, exclusion and insularity are the direct path to the loss of economic benefits, backwardness and non-competitiveness.

World experience has shown that industrial centers are the generators and re-translators of global change, the "backbone" of the world economy. These are modern global industrial centers (which are the peaks of the global geopolitical and geo-economic landscape), and

* Corresponding author: rodchenko@karazin.ua

dozens of other industrial centers that are only approaching the upper levels of the hierarchy. At the same time, the industrial center is the recipient of global change with all the threats and benefits. Given the public significance of the industrial center as a widespread habitat (since 2007, more than half of the world's population live in industrial centers) and the potential problem of such intensive urbanization (by 2050, about 80% of the world's urban population will live in developing regions, especially in the cities of Africa and Asia), the relevance of research in this area is undoubted.

The development of industrial centers in a new global environment is studied by scientists from various fields of activity, since it is the industrial centers that serve as the main places of globalization. For example, F. Erixon considers the industrial centers as peculiar "gateways to the global world" and emphasizes the uneven spread of globalization [2]. P. Samimi analyzes the process of increasing competition between regions and industrial centers, defines approaches to the management of the development of industrial centers, trends and processes caused by globalization in her works [3]. In other works, industrial centers are considered in the context of sustainable development [4] and the level of globalization [5].

In the leading scientific schools, the tradition of studying global industrial centers has existed for several decades with dozens of major scientific programs and projects. Instead, the Ukrainian scientific schools of the regional economy have insufficient studies on the issue of patterns and trends shaping and development of the industrial centers of Ukraine under globalization. In connection with the foregoing, the purpose of the study is to justify the criteria for choosing the priorities for the smart development of industrial centers in a globalizing environment.

To date, a new hierarchy of industrial centers has been formed: that ones that have focus on new functions are growing faster. Some industrial centers, being components of global economic networks, receive the bulk of the rent from the spread of innovations. Globalization is an innovation process. Innovative impulses originate in the most advanced industrial centers, leading in the development and development of new technologies with financial resources and highly skilled personnel.

Determining the criteria for choosing the priorities for smart development is controversial. The idea of the impossibility of choosing a single criterion basis and the existence of an "own way" of choosing a strategy for smart development prevails for every industrial center. At the same time, the structural proportions considered in different sections can form the starting positions for choosing priorities.

In order to choose the criteria for the smart development of industrial centers in a globalized environment, it is appropriate to use the simplest approaches to assessing the level of concentration of business entities in a particular market (Herfindahl-Hirschman index, Hall-Taydman index, etc.).

For example, the US Department of Justice determines the degree of concentration of the market by the Herfindahl-Hirschman index, guided by the following scale [6]:

– index is below 0.1 – a slight concentration of the market;

– index is from 0.1 to 0.18 – an average market concentration;

– index is above the 0.18 – a high concentration of the market.

In the United States any association of industrial centers that increases the Herfindahl-Hirschman index by 0.1 points (or 100 points) in the average concentration market (and by 0.05/50 points in the high-concentrated market) automatically leads to application of antimonopoly legislation [6].

When substantiating the criteria for choosing the priorities for the smart development of industrial centers, the most informative indicator is gross added value. It is a direct source of the welfare of the industrial center. However, in the current statistical base, its calculation and selection at the level of industrial centers is extremely complex. Therefore, it is suggested to use the gross output indicator for the analysis, although the gross value added is more meaningful and indicative.

Structural indicators are fundamental in the methodology of the selection of smart strategy specialization by European Commission (RIS 3) [7]. The main criteria for choosing the priority sectors (types of economic activities) that can ensure the positive dynamics of the industrial centers development are: indicators of the relative share of output, number of employees, volume of exports, number of patents of a possible priority type of economic activity in the total volume of relevant indicators of the industrial center. The dominance of the specific weight of an economic activity by the above indicators allows us to determine the prospects of industry specialization.

In order to diagnose the priority areas of smart development, it is advisable to use a structural approach that allows to determine and compare different sections of the activity of industrial centers. The improvement of existing approaches and their adaptation to the specifics of the industrial centers development can be proposed by the following five groups of indicators for the assessment of potential components and prospects for smart development:

I. Indicators of the internal technological structure:

1. Structural proportions of the industrial centers gross output by types of economic activity. It is advisable to consider the types of economic activity with the most accessible level of detail. According to the authors, it is acceptable to detail at the level of the second mark of the Classification of Types of Economic Activity. In assessing this indicator, the absence of types of economic activity, whose share exceeds a third of the total volume of gross output, should be considered as positive. Also, a situation of high concentration, when three or less types of economic activity form more than half of the gross output of the industrial center, can be threatening. The assessment of this indicator will help to determine the level of sectorial structure and form the basis for concluding that there is a need to support new types of activities or existing production proportions. As examples, the scenario of the development of industrial centers with imperfect structural proportions of the output can include scenarios for the development of the Detroit city (USA), one-industry towns in eastern Ukraine, and others.

2. Structural proportions of output by types of economic activity, grouped by type of technological processes. The dominance in the total volume of production of certain technological processes allows us to determine the level of "technological age" of the industrial center, and, accordingly, to form conclusions about the urgency of the need for a transition to a higher technological level. This indicator will allow us to formulate judgments about the acceptability of an evolutionary, modernizing or hopping trajectory for the industrial center development. An in-depth analysis of this indicator, combined with an analysis of the main factors of the potential for the industrial center development, provides the basis for discussion on the choice of development path.

II Indicators of the export (import) specialization of the industrial center:

1. Structural proportions of the product export (import) of industrial centers by the economic activity types. The experience of developed industrial centers shows that the optimal conditions of effective foreign economic activity is importing raw materials and exporting finished products, providing more volume of added value.

For example, today, according to experts' estimates, more than 50% of Ukrainian exports are products with low value added. The annual volume of exports is about $50 billion. Only agrarian products with low added value in 2018 Ukraine were exported in the amount of $19 billion. Assuming that raw materials export – 50%, which gives a multiple increase in value added, the Ukrainian economy loses about $25 billion [8].

Importance in the development of industrial centers is inclusiveness in global value added chains. They are characterized by a network type of organization, by the complexity and diversity of development. Under such conditions, the indicators of structural proportions are not enough informative and do not give the desired result. To diagnose global value creation chains, it is appropriate to use big data. Being apart from the large array of statistical information (which access in Ukraine is limited), it is necessary to take into account the information that characterizes the inclusion of industrial centers into clusters, which parts of the value-adding chains bring the greatest profit, possible exchange of resources between traditional chains creation of value and those at the stage of formation.

III Indicators of the intellectual capital of the industrial center:

1. Age structure of the employed population in industrial centers by types of economic activity. This indicator shows the perception of the population by innovations. The smart industrial centers development underlies this. It has been scientifically proven that there is a reciprocal correlation relationship between the age of a person and their propensity to accept innovations. As a rule, young people are ready to experiment freely. They are more likely to learn something new and easily perceive innovations, while older generations give up the advantage of a well-established business model and, in most cases, are unwilling to accept innovation.

Confirming this, the president of the National Academy of Educational Sciences of Ukraine V.G. Kremen emphasizes that only by forming an innovative person capable of creating change and perceptions of change, we can become a competitive nation. Because the variability, transformation ceases to be an exception, but becomes the rule, an essential feature of the functioning of society and every member of the XXI century. In this regard, Ukrainian society should prepare a person for life in new conditions, to form an innovative person [9].

2. Structural proportions of patents' concentration for inventions in industrial centers by types of economic activity. In modern economic conditions, a patent portfolio is considered as an asset of an industrial center that can be used for various tasks.

For example, in 2012 the Ukrainian startup Viewdle was sold to Google for more than $ 30 million. At the time of sale, Viewdle has filed 11 patents in the United States. Another case is Looksery, which Snapchat bought for $ 150 million in 2015. Looksery had 4 patents for inventions and 15 patents applications in the US [10].

It should be noted that patenting is also beneficial to the state, since for the registration of each patent the inventor or the enterprise pay the State Service of Intellectual Property.

The concentration of patents is one of the competitive advantages of industrial centers; however, for their effective implementation, it is necessary to develop a developed infrastructure that will allow early inventions to be turned into finished products, as well as increase the security of business.

3. The ratio of structural proportions of the number of employed population and the amount of wage payments in the industrial centers by types of economic activity. It is extremely important in every industrial center to achieve a balance between these two indicators. Since in most cases most of the population of the country employs certain types of economic activity, while the largest amounts of wage payments are accumulated in other areas. This increases the gap between the individual strata of society and, accordingly, exacerbates the level of social tension.

For example, by 2017, the largest number of Ukrainians are employed in wholesale and retail trade (21.8%), agriculture (17.7%) and industry (15.1%), while the largest volume of wages is typical for industry (25.8%), education (14.9%) and public administration and defense (9.8%) [11].

IV. Indicators of the structure of investment flows of the industrial center:

1. Structural proportions of volumes of capital investments in the industrial centers by types of economic activity. The ratio in the structure of capital investments

in the development of active (machinery, equipment etc.) and passive (building, construction etc.) fixed assets is an indicator of the availability of additional funds in industrial centers to expand the scope of activity and the level of resistance to innovation. As a rule, at the stage of product output and growth of the industrial center, the share of investments, which is directed to the expansion of the active part of the funds due to the limited funds, increases. At the stage of maturity, the share of investments in the expansion of the passive part of funds increases.

2. Structural proportions of investments in industrial centers by sources of financing. For effective functioning of the industrial center it is important to determine the real need for financial resources, the optimal structure of the sources of their formation. The prevalence in the structure of industrial investment center's own funds is a testament to the closure of industries. This is typical for the crisis or high-risk environments, while the involvement of credit are for the introduction of intensive development model.

V. Indices of the structural dynamics of the industrial center:

1. Structural proportions of dynamics of output volumes in the industrial centers by economic activity types. To compare the obtained values as a benchmark criterion, it is appropriate to use indicators of the dynamics of the national economy. For the products that are in the introduction stages of the life cycle (launch to the market) or growth, the growth rate of production will be higher than the growth rate of the economy. For the products in the stage of maturity, the growth rates of output are stable and are in a range close to the rate of economic growth. For the products in the downturn phase, slowdowns are characteristic, which, respectively, are lower than the growth rate of the economy.

The proposed set of indicators forms a methodological basis for identifying the priority directions of the industrial centers development. The main target orientations are:

- destimulation of raw exportation and outsourcing service models and localization of raw material processing chains;
- diversification of the economy and transformation of rental models of the industrial center functioning;
- restriction of adsorption of obsolete technologies and prevention of technological lagging preservation. The main benchmark is the best available technology solutions;
- provision of optimal age structure of personnel capital, including through replacement of non-competitive workplaces with technological equipment;
- possibilities of increasing the effective human capital through mechanisms of remuneration of workers from economically unproductive spheres and from the outside;
- the replacement of the paternalistic outlook and the full support of entrepreneurial initiatives.

Crisis manifestations in the dynamics of the global situation determine the relevance of the decentralization of the management national models in the economic development processes in the context of targeting inclusion into the global value added chains. In such conditions, the implementation of an adequate approach for a smart strategic choice of the priorities of the industrial centers development becomes a special role. The program approach, which usually ensures the effectiveness of such a choice, is not sufficiently effective. This results to the transfer of the center from the national to the local level and the corresponding change in the methodological apparatus and the target guidelines for choosing the strategic development of the industrial centers.

References:

1. Levitt T., *Harvard Business Review,* Autumn (Fall), **92** (1983).
2. Erixon F., *European centre for international political economy,* **21** (2018).
3. Shangquan G., Economic & Social Affairs, **1**, 10 (2000).
4. Samimi P., *Journals PLoS One,* April (2014).
5. Lascurain Fernández, M. *Revista de Relaciones Internacionales, Estrategia y Seguridad,* **12(1)**, 23 (2016).
6. Barr W., *The United States Department of Justice* (2018).
7. Guide to Research and Innovation Strategies for Smart Specialisation (RIS3). URL : http://ec.europa.eu/regional_policy/sources/docgener/presenta /smart_specialisation/ smart_ris3_2012.pdf.
8. Amelin A., Ukrainian Institute for the Future. URL: https://www.uifuture.org/publications/news/24557-amelin-polovuny-ekonomiky-daruyemo-svitu.
9. Kremen V., *Education and development of gifted personality,* **8-9**, 7 (2013).
10. Mishchenko T., Snapchat bought for $150 million Ukrainian startup Looksery. URL: https://itc.ua/news/snapchat-kupil-za-150-mln-ukrainskiy-startap-looksery/.

Transport Service as a Component of the Tourism Industry Development of Ukraine

Hanna Brusiltseva[1], and *Olena* Akhmedova[1]

[1]Simon Kuznets Kharkiv National University of Economics, Nauky av. 9-a, 61166 Kharkiv, Ukraine

Abstract. The research on the impact of transport services on the development of the tourism industry has been carried out. For the selection of the most significant factors of the transport infrastructure quality and tourist demand, which can affect the effectiveness of development indicators of a tourist destination, the correlation analysis methods have been used. To determine the development level of countries according to the quality and demand indicators of the transport services, the methods of cluster analysis have been used, which confirm the established hypotheses concerning the connection between the formed groups of factors and the development of tourist destination.

1 Introduction

Tourism sector is connected with the activity of more than 50 industries, the development of which generally contributes to the employment generation, economy diversification, cultural and natural potential preservation and development. It also increases the level of innovativeness of the national economy, promotes harmonization of relations between different countries. Thus, according to the Tourism and Travel Annual Report, in 2017 the contribution of the tourism sector was about USD 7.6 trillion that constituted 10.2% of GDP; the growth of international tourist flows allowed creating 292 million jobs [1]. In Ukraine, in 2017, tourism and related industries gave the budget UAH 40 billion, which is only 1.5 % of GDP.

According to the results of the numerous studies, such a component as the transport service has a significant proportion in the structure of the tourist products cost. Thus, according to the Tourism Act of Ukraine, the tourism product includes a pre-designed tourist services complex that combines transportation services, accommodation services and other travel services that are not related to transportation and accommodation. As stated in various publications, the cost of the transport services can reach 60% or more of the tourist product value. The role of transport in the development of tourism is unconditional, as evidenced by the results of the Ukrainian and foreign scientists' researches and studies.

2 Theoretical development and hypotheses formulation

Transport infrastructure is very important to regional development and it is an important factor supporting economic activity. Transport becomes not only a means of transportation, ensuring accessibility of tourist resources, the factor of the industry development, but also acquires properties, functions of the tourist resource, attractor.

S. Page stated that transportation is one of the most important factors contributing to the success of the tourism industry [2].

* Corresponding author: anna.brusiltseva@hneu.net

According to B. Lamb and S. Davidson, transportation is crucial in the tourism industry, since it directly connects the supply (production) and the demand (market) [3]. Transportation network and infrastructure is a significant factor affecting the success of tourism development [4]. D. Pearce argues that tourism growth is closely linked to advances in transport technology, which gradually facilitates access to the markets and destinations around the world [5]. Government plays a significant role in developing the infrastructure for tourist transportation [6].

T. Litman identified 12 transport factors that affect the accessibility of the destination, in particular: 1) transportation modes – the quality of transportation options, such as speed, comfort and safety; 2) transport network – density, communication routes or links between destinations; 3) fare or accessibility; 4) mobility – speed and distance, traffic performance or traveling time; 5) integration of the connections and modes into the transport system; 6) transport demand; 7) a user of information – the availability of reliable information about the possibilities of mobility and accessibility; 8) mobility substitutes – telecommunications and physical travel replacement services; 9) transport management; 10) factors of the land use; 11) priority of the tourism development; 12) the level of inaccessibility or isolation [7].

The results of Kenichi Shimamoto' researches in this area allow to identify the factors influencing the competitiveness of tourist destinations: the cost of experiencing the tourist destination; the disutility/utility of the tourist destination; the level of the destination's branding; the transportation fee; the costs related to traveling time; and the transportation means disutility/utility [8]. Jameel Khadaroo and Boopen Seetanah stated that traveling income, distance, and relative prices are important ingredients of their own respect in the tourism demand equation [6].

The factor of transportation is also taken into account while calculating the Global Competitiveness Index (GCI), the Travel and Tourism Competitiveness Index (TTCI), which contains two groups of indicators of the transport infrastructure development, namely, the infrastructure of air and land transport. Thus, the paper objective is to determine the key transport factors that significantly affect the tourism industry of a certain destination. In the result of the study, two groups of factors have been distinguished: the factors of the transport infrastructure and the factors of the tourists' transport demand. Thus, the following hypotheses have been proposed:

Hypothesis 1. The development of a tourist destination depends on the quality of the transport infrastructure.

Hypothesis 2. The development of a tourist destination depends on tourist demand.

3 Results

Ukraine's transport infrastructure consists of different modes of transport. Each of them has its own specifics. In order to ensure the country's economy operation, all modes of transport should be interrelated and operate harmoniously. General transport network of Ukraine includes 21.0 thousand kilometres of railways, 159.4 thousand kilometres of highways with a hard surface, 1.6 thousand kilometres of operational river shipping routes with an access to the Azov and the Black Sea.

The research of the transport infrastructure factors should include the following indicators: the number of airports, the number of commercial aircrafts, the density of railways, the length of railways, the density of roads, the number of deaths on highways, etc.

While analysing the level of the transport infrastructure development according to the competitiveness index, it is necessary to note that in 2017, Ukraine took 79 place in terms of the development level of the aviation transport infrastructure and 81 place in terms of the development level of the land transport infrastructure. The main factors that negatively influenced the level of the transport infrastructure development are Quality of air transport infrastructure (102 position), Airport density (117), Quality of roads (132) [1].

Among the indicators, which characterize the demand for transport services, the number of passengers carried by the various modes of transport, revenues from the passenger transportation should be taken into account. In recent years, in Ukraine there has been a general tendency to reduce the passenger transportation market by almost all modes of transport that indicates a decrease in demand for transport services. Some positive changes are observed in the air transport sector. It is in place here to mention that the main mode of transport in Ukraine, which is in great demand among the consumers, is automotive. While there is a completely different situation in the world. Thus, according to UNWTO in 2017, slightly over half of all overnight visitors travelled to their destination by air (57%), while the remainder travelled by surface transport (45%) – whether by road (37%), rail (2%) or water (4%) [9]. The trend over time has been for air transport to grow at a somewhat faster pace than surface transport, thus the share of air transport is gradually increasing.

To study the influence factors of tourism infrastructure and tourist demand on the tourist destination development, the following algorithm has been proposed.

Stage 1. The choice of the indicator of effectiveness of the tourist destination development.

Stage 2. The collection of information on the main indicators of tourism infrastructure and tourist demand that can affect the tourist destination development.

Stage 3. The calculation of the pair correlation coefficients between the indicator of effectiveness of the tourist destination development and the factors of influence, the determination of significant correlation coefficients according to the Student's criterion and clarification of the relationship nature between the relevant factors.

Stage 4. Grouping of the countries according to the level of the tourist destination development and transportation availability with the cluster analysis methods.

During *the first stage*, the tourist flows of European countries were selected as an indicator of effectiveness of the tourist destination development.

During *the second and the third stages* the indicators that have significant correlation coefficients (Table 1) were selected. Significance was estimated with the help of the Student's coefficient.

Table 1. Factors influencing the effectiveness of the tourist destination development that have significant correlation coefficients.

Factors of transport infrastructure quality	The value of correlation coefficient	Factors of tourist demand	The value of correlation coefficient
The commercial aircraft fleet, number **Var 1**	0.58	Passenger transportation by rail, number, **Var 9**	0.74
The commercial aircraft fleet with the useful lifetime less than 5 years, number	0.45	Passenger transportation by air, number, **Var 10**	0.55
Commercial airports (more than 15,000 passengers), number **Var 3**	0.82	Revenues from passenger transportation, mln. USD, **Var 11**	0.65
Air transport, registered take-offs, number	0.23	Transport expenses, mln. USD, **Var 12**	0.57
The length of railway tracks, km **Var 5**	0.73		
Deaths on highways, number	0.33		

At *the fourth stage*, based on the desirable level of the study detalisation, three clusters of countries were identified: with the high, medium and low quality of transport infrastructure and factors of tourist demand.

Based on the iterative method of clustering k-means, the following cluster results have been obtained (Fig. 3, Fig. 4).

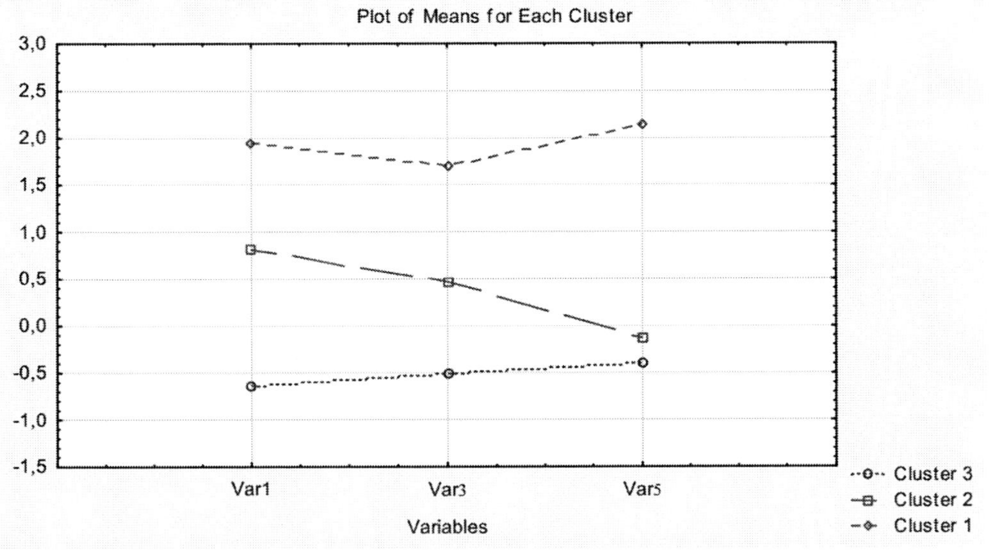

Fig. 3. Average means of transport infrastructure quality variables by clusters.
Source: Own research

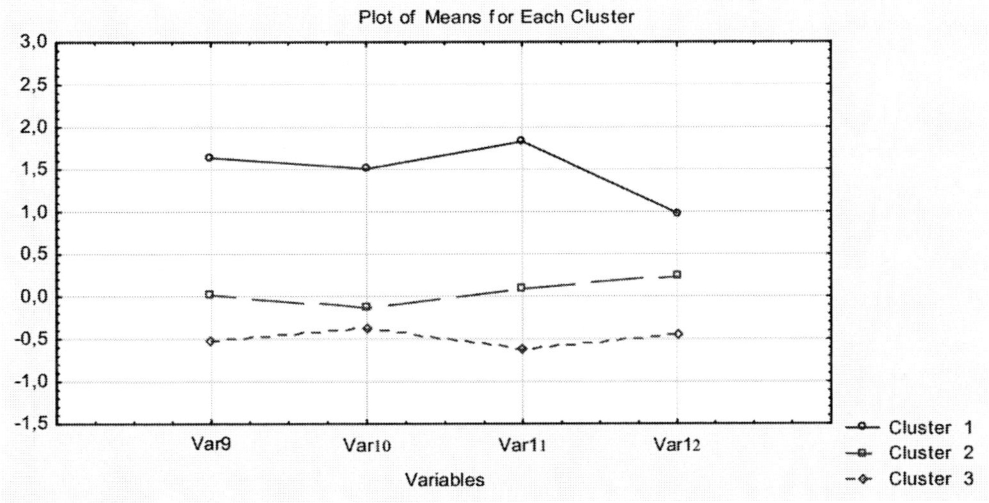

Fig. 4. Average means of tourist demand variables by clusters.
Source: Own research

The first cluster includes the countries with high indicators of the transport infrastructure quality and indicators that characterize tourist demand (Table 2). In addition, significant volumes of tourist flows characterize these countries.

Table 2. Members of Cluster 1 and Distances from Respective Cluster Centre.

Factors of transport infrastructure quality			Factors of tourist demand		
Case №	State	Distance	Case №	State	Distance
C_1	Germany	1.707896	C_1	Germany	0.845747
C_2	France	1.237087	C_2	France	0.930060
C_4	Spain	1.631217	C_4	Spain	0.668152
C_27	UK	1.376760	C_27	UK	0.843107
C_31	Turkey	0.672639	C_31	Turkey	0.906590

The second cluster includes the countries with the average indicators of the transport infrastructure quality and indicators that characterize tourist demand (Table 3).

Table 3. Members of Cluster 2 and Distances from Respective Cluster Center.

Factors of transport infrastructure quality			Factors of tourist demand		
Case №	State	Distance	Case №	State	Distance
C_3	Italy	0.783980	C_3	Italy	0.404520
C_5	Finland	0.311705	C_5	Finland	0.452581
C_6	Austria	0.326173	C_11	Greece	0.681613
C_9	Netherlands	0.244367	C_21	Denmark	0.287382
C_10	Portugal	0.493652	C_24	Poland	0.787039
C_12	Ireland	0.681203	C_26	Sweden	0.160468
C_24	Poland	1.070442	C_28	Switzerland	0.167171
C_21	Denmark	0.332759	C_24	Poland	0.787039
C_26	Sweden	1.257147			
C_28	Switzerland	0.224566			

The third cluster includes the countries with the lower indicators of the transport infrastructure quality and indicators that characterize tourist demand (Table 4).

Table 4. Members of Cluster 2 and Distances from Respective Cluster Centre.

Factors of transport infrastructure quality			Factors of tourist demand		
Case №	Distance	Distance	Case №	Distance	Distance
C_7	Slovakia	0.067171	C_7	Slovakia	0.067171
C_8	Belorus	0.164068	C_8	Belorus	0.164068
C_11	Greece	0.156566	C_9	Netherlands	0.156566
C_13	Lithuania	0.066646	C_10	Portugal	0.066646
C_14	Latvia	0.069404	C_12	Ireland	0.069404
C_15	Estonia	0.072214	C_13	Lithuania	0.072214
C_16	Slovenia	0.069191	C_14	Latvia	0.069191
C_17	Moldova	0.084050	C_15	Estonia	0.084050
C_19	Bulgaria	0.068108	C_16	Slovenia	0.068108
C_20	Czech Republic	0.188716	C_17	Moldova	0.188716
C_22	Croatia	0.063330	C_18	Ukraine	0.087582
C_23	Hungary	0.193069	C_19	Bulgaria	0.063330
C_18	Ukraine	0.287382	C_20	Czech Republic	0.193069
C_25	Romania	0.026193	C_22	Croatia	0.026193
C_29	Montenegro	0.097160	C_23	Hungary	0.097160
C_30	Macedonia	0.094974	C_25	Romania	0.094974
			C_29	Montenegro	0.164068
			C_30	Macedonia	0.156566

It should be noted that the structure of the second and the third clusters in the selected groups of factors is somewhat different. The reason for this is the close values of the indicators of both factors, namely, the length of the railways and the number of passengers transported by air transport. This is due to the relatively equal size of the territory of the countries included in the specified clusters. However, the overall results for Ukraine remain the same: in both cases, Ukraine is in the cluster with such countries as Moldova, Bulgaria, Hungary, Latvia, Lithuania, Belarus, Bulgaria, Estonia, Romania, etc.

4 Conclusion

Thus, the research on the impact of certain groups of transport infrastructure quality factors and tourist demand for transport services make it possible to claim their influence on the tourist destination development. In addition, the cluster groups obtained on the basis of these indicators confirm the hypotheses established in the research.

Nowadays the transport sector of Ukraine as a whole satisfies only the basic needs of the population and the economy in the volume of traffic, but not in the quality, which also creates significant obstacles to the development of domestic tourism. The current state of the transport industry does not fully meet the requirements of effective implementation of the European integration course of Ukraine and the integration of the national transport network into the Trans-European transport network.

According to the Association Agreement with the EU [10], it is necessary to increase the efficiency and competitiveness of the transport sector, primarily by improving the legal mechanism of public-private partnership, strengthening the interaction between the public and private sectors, state authorities and local self-government bodies, carrying out the necessary reforms through coordinated initiatives of the state politics. The implementation of the appropriate measures should provide a solid basis for the sustainable development of the transport industry and create a free and competitive market for tourist services.

References

1. The Travel & Tourism Competitiveness Report 2017. Available at: https://www.weforum.org/reports/the-travel-tourism-competitiveness-report-2017
2. Page S. Transport and Tourism. Global Perspectives. Harlow: Pearson Prentice Hall, 447 (2009)
3. Lamb B., Davidson S. Tourism and Transportation in Ontario, Canada: A Vital Link. In Practicing Responsible Tourism: International Case Studies in Tourism Planning, Policy and Development. (1996)
4. Prideaux B. The role of the transport system in destination development. Tourism Management, **21**(1), 53-63, (2000)
5. Lohmann G., Pearce D. Tourism and transport relationships: the suppliers' perspective in gateway destinations in New Zealand. Asia Pacific Journal of Tourism Research, **17**(1), 14-29, (2012)
6. Khadaroo J., Seetanah B. The role of transport infrastructure in international tourism development: A gravity model approach. Tour. Management, **29**, 831-840, (2008)
7. Litman T. "The New Transportation Planning Paradigm," ITE Journal (www.ite.org), **83**, June, pp. 20-28; 2013 Available at https://www.vtpi.org/paradigm.pdf.
8. Shimamoto K. The Factors that Influence the Boundary Between the Markets of Competing Tourist. Destinations Tourism and Hospitality Management, **24**(1), 1-12, (2018)
9. UNWTO Tourism Highlights: 2018 Edition. Available at: https://www.e-unwto.org/doi/pdf/10.18111/9789284419876
10. Association Agreement between the European Union and Ukraine. Available at: https://www.kmu.gov.ua/en/yevropejska-integraciya/ugoda-pro-asociacyu

Definition features of the efficiency in railways management

Oleksandr Deineka[1], *Liubov* Pozdniakova[2], *Olena* Sementsova[1*], and *Yuliia* Krykhtina[1]

[1] Ukrainian State University of Railway Transport, Management and Administration Department, Feuerbach sq., 7, Kharkiv, 61050, Ukraine

[2] Ukrainian State University of Railway Transport; Economics, Business and Personnel Management on Transport Department, Feuerbach sq., 7, Kharkiv, 61050, Ukraine

Abstract. The article deals with the determination of general effectiveness of managerial influences on the activity of the railway industry, taking into account such partial indicators as public, functional and economic efficiency. appropriate degree of satisfaction of requirements and expectations of consumers on the quality of transport services, as well as the degree of satisfaction of staff with the work conditions are used to assess the public effectiveness of management in railway transport. Functional efficiency of management in railway transport can be estimated by the conformity of managerial decisions to the functional goals and tasks of management, achievement of optimal results within the requirements of state regulation of the industry, etc. The economic efficiency of railway transport management has traditionally been evaluated by the degree of maximization of profitability of freight transport and minimization of losses from passenger transportation with rational use of industry resources.

1 Problem statement

It is well known that the railway industry at any time of historical development has had the status of a carrier of national strategic importance. It has always been due to the several factors. First, it is the crucial role of the railways in ensuring national security and defense; and, secondly, the exclusive role of rail transport in the connection of sectors of the national economy. In view of this, the railway industry has always experienced a special administrative influence of the government on the implementation of production and economic activity. Today, relations between the government and society are increasingly transformed from rigid centralized administration towards open, transparent, democratic governance. It happens due to the request of the railway industry reform in order to integrate it into international transport infrastructure, as well as the globalization of the national economy. In view of this, issues concerning the clarification and detailing of methodological tools for assessing the efficiency of railways management become more and more relevant.

*Corresponding author: sementsova.elen@gmail.com

2 Analysis of research and publications

By conducting an analytical study of global experience in evaluating the effectiveness of management in rail transport, it can be noted that the majority of researchers try to improve the evaluation of the effectiveness of using certain resources (such as labor, energy, etc.).

We [1] stressed the need to improve the assessment of the efficiency of management of labor resources of railway companies. They propose to improve the method of measuring labor productivity of personnel, which should be based on the assessment of the level of development of human potential. According to them, such an approach is generated by the need to adapt the performance management system and motivation staff to dynamically change the environmental factors.

The main purpose of [2] is to assess the efficiency of energy resources. The authors reveal the principles of assessing the energy efficiency of trains with electric traction, limiting existing systems, measuring and processing information on electricity consumption, "bottlenecks" of the traction power system, etc.

In [3], the authors substantiated the approach to assessing the economic efficiency of railway enterprises, which considers the natural environment as an active element of management. In assessing the economic efficiency of the enterprises, the authors are limited considering the environmental aspect of their functioning and suggest to take into account non-productive costs, as well as the formation of effective interaction of railway enterprises and society in solving environmental problems.

In contrast to the above-mentioned works, other authors [4] offer a comprehensive approach to determining the efficiency of railways. They emphasize the importance of analyzing the efficiency of management in logistics infrastructure.

They compare the Stochastic-DEA and Fuzzy-DEA models to assess the impact of randomness and fuzziness in the indicators of the volume of railway operations (inputs and outputs) for ranking the efficiency of railways functioning in the example of 5 countries in Asia: Japan, Thailand, Vietnam, Malaysia, and Indonesia.

A significant number of works are devoted to the aspects of determining the effectiveness of rail transport management in Ukraine and on individual directions of its work in particular [5-10].

The work by Yu. S. Barash [5] is generally concerned with improving the competitiveness of Ukrainian railways. It proposes an economic criterion for choosing the best rail transport management model. The study conducted by V. L. Dykan, M. V. Korin [6] focuses on the aspects of the impact of the globalization of economic systems on the functioning of the railway industry and its management.

O. M. Kirilenko [7] focuses on the determination of the efficiency of the operation of sites with low traffic intensity. P. O. Yanovskyi [8] is limited to the study of financial results to evaluate the efficiency of the operation of the suburban passenger complex in Ukraine. In [9], V. A. Zadoia approaches to assessing the efficiency of railways functioning are presented, based on indicators of economic efficiency. M.P. Sinhaievska [10] also studies the evaluation of the economic efficiency of railways functioning and provides analytical models for determining the influence of factors (extensive and intense) on the financial results of the industry.

The effectiveness of management on the railways of Ukraine cannot be assessed without analyzing the main volumes of operation of the industry, because they describe the results of the implementation of managerial influences. The analysis of public reporting on the results of the Ukrainian railway [11] indicates significant fluctuations in the scope of their work (Fig. 1).

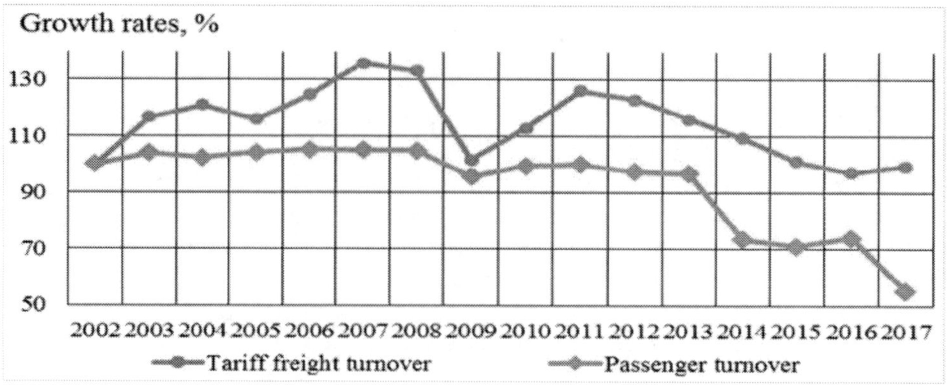

Fig. 1. Growth rate of tariff freight turnover and passenger turnover in the calculation before the first year (2002) of the analyzed period [*developed by authors by source 11*].

In general, during the analyzed period, the volumes of cargo and passenger transportation tended to decrease. This is especially true for passenger traffic which decreased almost twice during the period has. On the other hand, the cargo turnover has almost a stable position for growth during the first five years of the period, but the global economic crisis in 2008 had a significant negative impact on the Ukrainian economy, and the rail transport experienced a significant reduction of demand for freight during 2008-2009 (more than 35%). In the following two years, the revival of economic ties has made it possible to almost restore lost positions. But further, again there was a reduction in the number of services for the transportation of goods, due to the loss of competitive positions in the market of transport services. In the subsequent period (after 2014), this tendency was aggravated by a decrease in cargo flows due to military actions and loss of control over certain territories, which led to the loss of the image of Ukraine as a transit state.

Along with the decline in the total volumes of cargo and passenger transportation by rail, the contingent occupied by the main activity (Fig. 2) has a steady upward trend even in times of worsening of the recession, which led to a significant loss of traffic volumes. This fact proves the ineffective use of labor resources on the railways and worsens the overall efficiency of railways. However, the pace of change in the contingent of management is entirely consistent with changes in the volumes of transportation of railways.

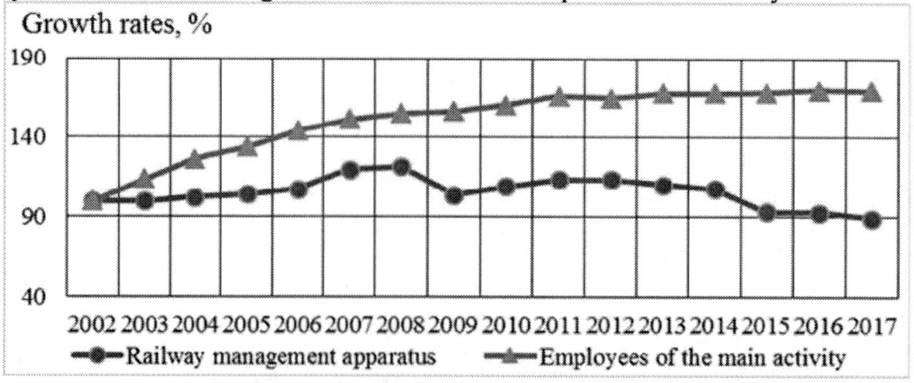

Fig. 2. The growth rate of the number of railway management apparatus and the average number of employees of the main activity in the calculation before the first year (2002) of the analyzed period [*developed by authors by source 11*].

Despite the recession in the national economy, the railways continue to increase the operational length of the track to improve the railway infrastructure to integrate into the European community and the globalization of the national economy. When investigating the efficiency of railways management, it is interesting to study the ratio of above-mentioned indicators (tariff freight turnover, passenger turnover, number of railway management apparatus and the average number of employees of the main activity) and operating length railroad (Fig. 3).

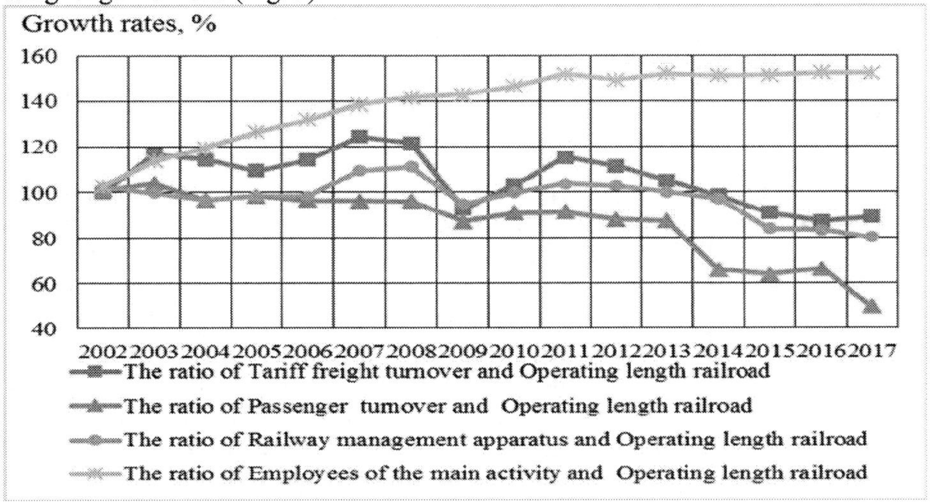

Fig. 3. Growth rate of the ratio of tariff freight turnover, passenger turnover, the number of the railway management apparatus and contingent of the main activity of the operating length railroad of Ukraine up to the first (2002) year of the analyzed period [*developed by authors by source 11*].

The figures in the graphs also illustrate the deterioration of the analyzed indicators, although at a slightly lower pace than in previous calculations. This situation proves the expediency of improving the management activity of the railway industry of Ukraine to increase efficiency.

3 Presentation of the main research

Summing up the conducted review of scientific publications on the assessment of the effectiveness of managerial impacts on the railway industry, it can be concluded that a limited number of studies on a comprehensive assessment of the efficiency of rail transport management. Analytical calculations testify to the necessity of improving the management activity in the railway industry to increase its efficiency.

General evaluation of the effectiveness of managerial influences on the railway industry activity is proposed to determine the aggregate of such most important partial indicators as public efficiency, functional efficiency, and economic efficiency.

The components of a generalizing indicator of the effectiveness of managerial influences on the railway industry activities are described in more detail in article [12]. The tools for assessing the effectiveness of management in the railways are shown in Fig. 4.

As partial criteria for assessing the public effectiveness of management in railway transport, it is advisable to distinguish:

- degree of satisfaction of requirements and expectations of consumers on the quality of transport services;
- the degree of staff satisfaction with the conditions of stay in the organization.

Estimating the degree of satisfaction of requirements and expectations of consumers about the quality of transport services should be carried out through satisfaction of passengers with quality of service, convenience of schedule, time of stay on a trip,

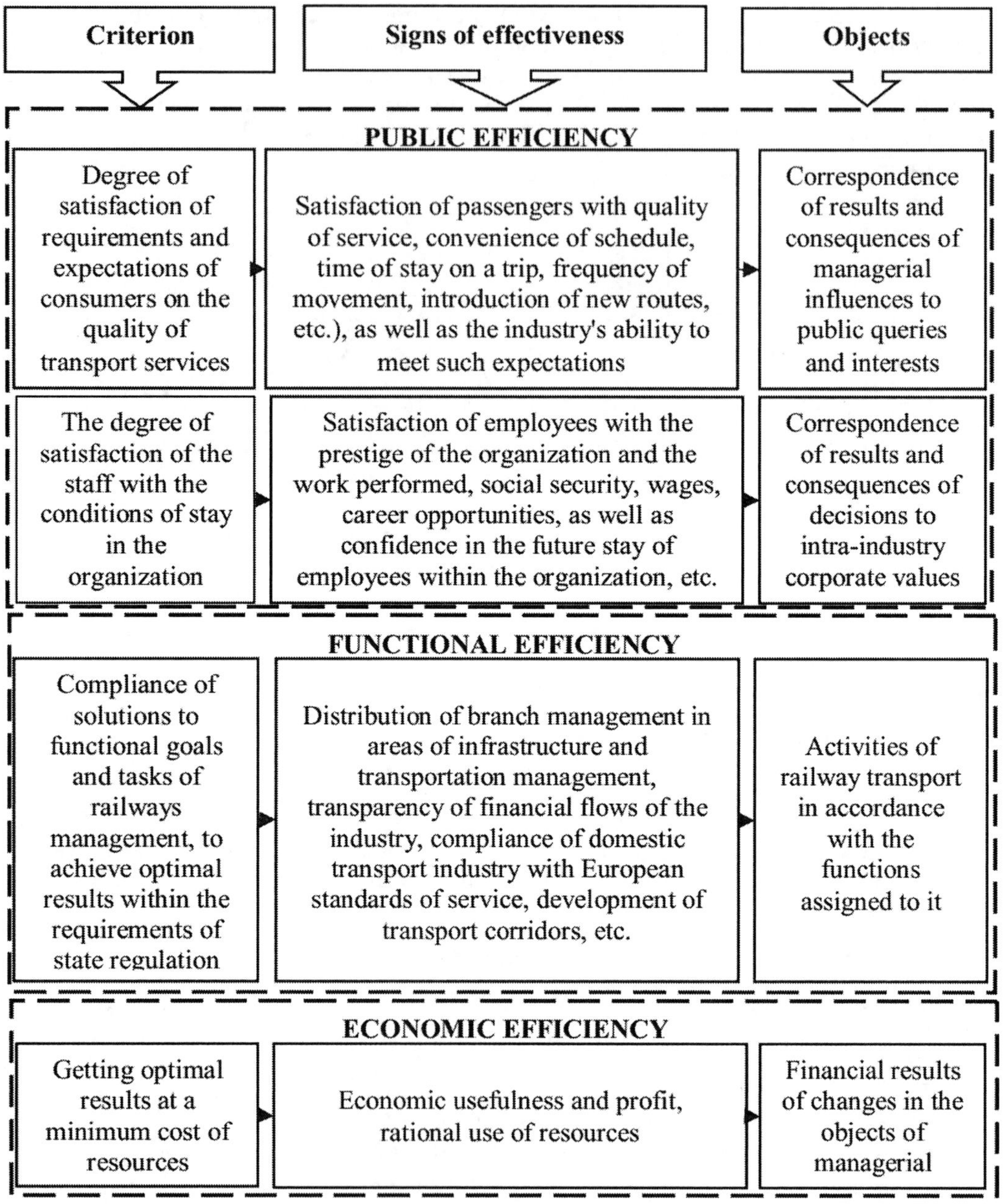

Fig. 4. Tools for assessing the effectiveness of management in the railway transport.

frequency of movement, introduction of new routes, etc.), as well as the industry's ability to meet such expectations.

Assessment of the degree of satisfaction of the organization's staff with the conditions of stay in the organization is carried out through satisfaction of the employees with the prestige of the organization and the work performed, social security, pay, career opportunity, as well as confidence in the future stay of employees within the organization, etc.

For measuring public effectiveness, expert appraisal methods are predominantly used. Unfortunately, the results of such an assessment have a certain subjectivity, which determines the probability and variability depending on the focus group of consumers and experts involved in the process of evaluation, the accuracy of the formulation of relevant issues for the railway industry, communication obstacles, etc.

The criteria for evaluating the functional efficiency of management in the railway transport can relevance of solutions to the functional objectives and tasks of management in the rail transport, achieving optimal results in accordance with the requirements of state regulation of the industry, the compliance of solutions with the expectations of competitors. The object of the evaluation of this criterion is the activity of the railway transport, respectively, the functions assigned to it. For the measurement of functional efficiency, the methods of expert evaluation are also used predominantly.

The criteria for assessing the economic efficiency of rail transport management traditionally include maximizing the profitability of freight traffic and minimizing losses from passenger transport when the industry's resources are used rationally. The objects of assessing under this criterion are the financial results of managerial influences that arise in the implementation of financial and economic activities. To measure economic efficiency, it is advisable to apply methods based on the formalized approach, which provides an opportunity to obtain an accurate, objective and reasoned assessment of the results of managerial influences.

4 Conclusion

The study of tools for assessing the effectiveness of management in rail transport creates conditions for conducting a qualitative, systematic and integrated analysis of managerial impacts on the functioning of railways, which will be the basis for making appropriate management decisions on the reform and integration of the national rail network to the international transport infrastructure.

References

1. A. Davydov, A. Dementev, V. Burovtsev, MATEC Web of Conferences, **239** (2018)
2. V. Cheremisin, M. Nikiforov, A. Kashtanov, S. Ushakov, Advances in Intelligent Systems and Computing, **692**, 83-90 (2018)
3. J. A. Tagiltseva, E. L. Kuzina, N. A. Drozdov, M. A. Vasilenko, International Conference "Quality Management, Transport and Information Security, Information Technologies", 163-168 (2018)
4. P. Wanke, M. A. Kalam Azad, Transportation Planning and Technology, **41**, 573-599 (2018)
5. Yu. S. Barash *Upravlinnia zaliznychnym transportom krainy* (Dnipropetr. nats. un-t zalizn. transp. im. akad. V. Lazariana, Dnipropetrovsk, (2006)
6. V. L. Dykan, M. V. Korin, The bulletin of transport and industry economics, **33**, 13-19 (2011)
7. O. M. Kyrylenko, *Zabezpechennia efektyvnosti funktsionuvannia malodiialnykh dilnyts PAT «Ukrzaliznytsia»* (Dnipropetr. nats. un-t zalizn. transp. im. akad. V. Lazariana, Dnipro, (2018)
8. P. O. Yanovskyi, Zaliznychnyi transport Ukrainy, **5**, 9 – 12 (2009)
9. V. O. Zadoia, Zbirnyk naukovykh prats Ukrainskoi derzhavnoi akademii zaliznychnoho transportu, **153**, 15-19 (2015)
10. M. P. Sinhaievska, Problemy informatyzatsii ta upravlinnia, **13**, 105-113 (2005)
11. *Dovidnyk osnovnykh pokaznykiv roboty rehionalnykh filii PAT «Ukrainska Zaliznytsia»* (Upravlinnia statystyky PAT «Ukrainska Zaliznytsia», Kyiv, (2018)
12. O. H. Deineka, O. V. Sementsova, O. L. Vasyliev, The bulletin of transport and industry economics, **59**, 204 - 207 (2017)

Study of Model for Selection of Priorities for Development of Cargo Transportation

Valerii Kuleshov [1*], *Ganna* Shapoval [1] , *Maksym* Kutsenko [1]

[1] Ukrainian State University of Railway Transport, Department of railway stations and junctions, Feuerbach sq. 7, 61050 Kharkiv, Ukraine

Abstract. The transport market of operator companies of various forms of ownership in Ukraine is constantly evolving, but the tasks associated with the selection and implementation of the priorities of the development of the transport market, the provision of railway services by the carriers - owners of rolling stock, are not fully resolved. There is no regulatory framework that determines the procedure for choosing the priorities of transport market development, their status is not defined, the relation of the selected priorities with the budget of Ukraine is not defined. Therefore, the task of improving the mechanism for selecting scientific and technological priorities for the development of these processes is relevant.

1 Introduction

Under current conditions of operation of the Ukrainian railway transport and operator companies (OC) that are owners of rolling stock, there are factors that are difficult to consider and control. The marketing subsystem can provide information support for making managerial decisions. Means of modern software and hardware provide an opportunity to solve the tasks of macroeconomic analysis of the operation of rail transport: to formulate tariff policies, to determine the return of individual areas on the network of railways, to calculate the yield on the type of rolling stock to determine the rent etc.

For the payback of the car OC need more than 30 years. If discounting and the value of the attracted investments are neglected, then the invested amount for a car can be returned in 8 years. But in fact, only captive OC can afford it – companies that export their own products or are part of large financial and industrial groups whose business guarantees the flow of goods.

The availability of a car model and an electronic card index of cars showing actual operations in real time gives the opportunity to receive all the established reporting, and also provides new opportunities in the management of car fleet.

2 Literature review

* Corresponding author: author@email.org

The papers [1] do not take into account advanced information technologies at transporting organisation which may provide the development of using the railway services by users in the transport organisation.

The research [2-4] does not study in the full scale the approaches of forming the interaction technology between railway administration and rolling stock owners in conditions of the sole system operated by the different owners of cargo car fleet on the basis of information systems in order to optimize the capacity of railway transport systems.

The books [5-6] do not reveal the issues of determination the optimal fleet of operators.

The study [7] in the full scale does not analyse the issues of route transportations organisation with the same requirements for different rolling stock owners, especially considering international transportation, the issues about organisation of route transportation with the account of incomplete information.

The papers [8-9] published abroad showed basic principles of transportation organisation, but organisation of heavy and long cargo trains is not considered there.

3 The proposed organisation-technological model

As of 04.04.2019, the total rolling stock fleet of the inventory cars of Ukrzaliznytsia JSC (UZ) with the sign CTL is 91737 cars, including by owners: branch of PVRZ 26%; branch of DVRZ 26%; branch CTL 12%; branch of SVRZ 12%; Branch RVK 9%; Branch "Centre of transport service Liski" 1%. The general CTL fleet is divided according to the type of rolling stock: open (gondola) cars 52%, other 33%, tanks 9%, covered 6%, refrigerators 0.3%. The operated fleet there makes 61% of the total fleet, and in the non-operated fleet in repair or waiting is 39%.

The share of transportation in the total volume of transport by gondola cars owned by UZ and Ukrainian enterprises: UZ - 39%; large domestic producers (captive OC) - 24%; other owners of cars whose volumes of transportation make up more than 1% - 24%; other owners of cars whose volumes of transportation is less than 1% - 8%; small domestic producers (captive OC) - 5%.

The share of large domestic producers that carry goods independently (MPS Corporation, Lemtrans LLC, Poltavsky GOK PJSC, Metinvest-shipping, KMMC "Kryvorizhstal OJSC) is 24%. The share of small domestic producers that carry goods independently is 5%. Thus, 29% is a non-competitive railroad market, which together with the share of UZ makes up 68%.

In 2014, the total fleet of Ukrzaliznytsya Joint Stock Company comprised 182.1 thousand cars. In 2015, in comparison with the previous year, the fleet decreased by 2.0%, in 2016 by 2.9%, in 2017 by 0.7%, and in 2018 there was an increase of 4.2%.

In 2014, the Ukrzaliznytsia's inventory fleet totalled 40.6 thousand cars. In 2015, the fleet decreased by 6.0%, in 2016 by 3.7%, in 2017 by 5.4%, in 2018 by 79.4%. In 2014, the Ukrzaliznytsia's own fleet numbered 141.5 thousand cars. In subsequent years there was a decrease of own fleet, respectively, in 2015 by 0.8%; in 2016 by 2.7%. Since 2017 there has been an increase of 0.6%, and in 2018 this number was 25.4%. The change of the total fleet of Ukrainian cargo cars for the period 2014-2018 is shown in Figure 1.

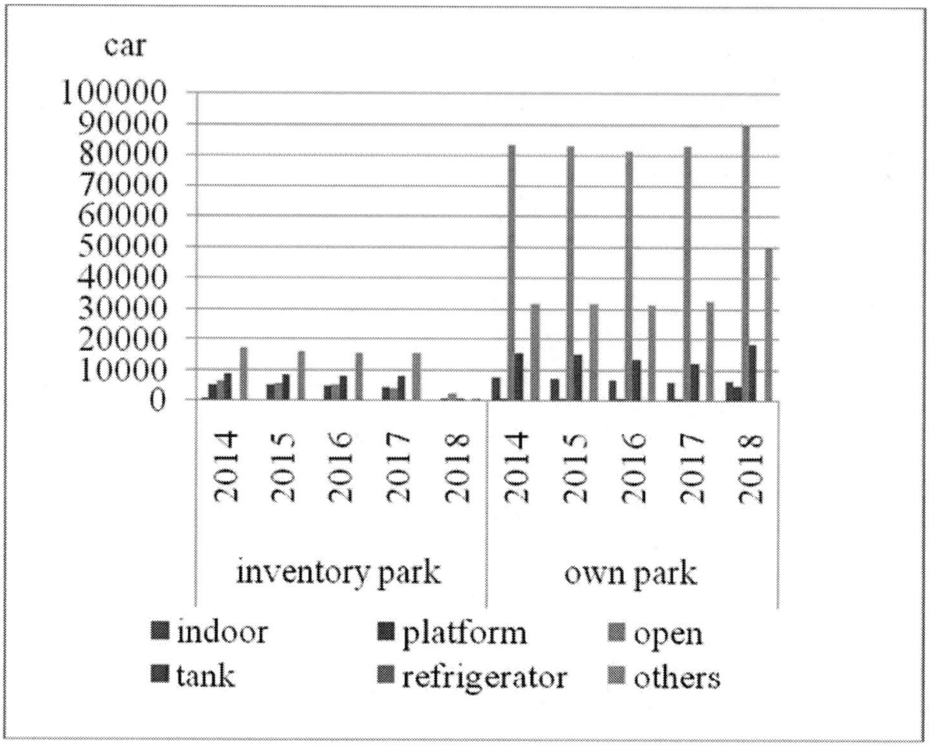

Fig. 1. Change of the total fleet of freight cars of Ukraine.

The aim of the research is to upgrade the route organisation and technological model organising from the different groups of cars of different owners using an interactive automated system map.

The organisation and formation of the route from the groups of cars of different owners may be provided according to the station technological processes by a transporter, operator - rolling stock owner, structural subdivisions of railways and access roads of senders or on station tracks, which are leased by the branch railway lines owners and their counterparties. The following routes of i-categories are taken in account in the routing: 1 — the main rail network direct routes of the operator; 2 — stepping routes under contracts for the operation of access to rail or for serving and cleaning cars taking into account the responsibility of the cargo sender; 3 — routes to the station for spraying or to spraying points for forwarding cars picked up at unloading stations, districts, cargo points and consignees within the limits established by the operator of the main railway network; 4 — routes to incoming or distribution stations receiving fuel cargoes agreed by the main rail network operator [2-4, 7]

The model of organization of the route from the groups of cars of different owners in the conditions of operation of the transport and logistics centre is a two-stage stochastic task with recursion and random right-hand side restrictions. The variables of the first stage are denoted n_j, and the recursive variables, respectively, s_i^l and x_{ik}^l. If the current state of the network $\{a_{ik}, y^0\}$ is given then the task is to minimize the costs of forming and following the departure route q_i^+, the expected network maintenance costs and the conditions for non-compliance fees, and other car flows on the voyage route.

$$f(E_i) = \sum_{j=1}^{r} c_j n_j t_j + \sum_{l=1}^{L} p_l \left[\sum_{j=1}^{r} q_j \left(\sum_{i=1}^{v} \sum_{k \in K(t)} a_{kj} c_{ik} x_{ik} \right) + \sum_{i=1}^{v} q_i^+ s_i^l \right] \rightarrow min \quad (1)$$

with restrictions

$$\left.\begin{aligned} \sum_{i=1}^{m}\sum_{k\in K(i)} a_{kj}x_{ik}^{l} \le n_j^0 + n_j, j = 1,\dots,r,\ \ l = 1,\dots,L, \\ \sum_{k\in K(i)} x_{ik}^{l} + s_i^l = d_i^l, i = 1,\dots,v,\ \ l = 1,\dots,L, \\ 0 \le n_j \le u_j,\ \ s_i^l \ge 0, x_{ik}^l \ge 0. \end{aligned}\right\} \quad (2)$$

where n – route of a train car; v – number of vertices of the network vector to be served; d – the random variable of maintenance requirements between pairs of vertices that are set using a list of implementations or probabilities; L –number of cars; p_i – probability of occurrence of a variant; d_i^l – i component vectors option; $K(i)$ – a set of paths that can be used to perform service between two points; $a_k \in Z^n$ – incident vector;

$$a_{kj} = \begin{cases} 1, j \in K \\ 0, j \notin K \end{cases}; \quad (3)$$

$y^0 \in R^n$ –existing number of technical routes that take the bandwidth of the network; s_i^l – number of requirements not fulfilled for the option r; q_i^+ – unit of penalties for failure to comply s_i^l ; x_{ik}^l – number of connections serving the couple i by κ for the option l.

When choosing the optimal mode of car traffic on the network of stations from the sender to the consignee at the stage of planning operating costs does not take into account the third structural part of the model $\sum_{i=1}^{v} q_i^+ s_i^l$, when unknown deviations from transportation requirements.

The first structural part of the model involves the combined costs of the initial-end operations of the transport process - in our case, this is loading and unloading. The second structural part of the model reflects the cost of running the carriage.

Thus, in the built-in dynamic model, the problem is reduced to determining the priorities of the delivery of the car under load to specific customers – users of rail services – in order to optimize the volume of traffic, increase profits with available resources (cars, shunting and train locomotives, track development).

3 Conclusion

In order to achieve breakeven performance of gondola cars, the minimum required total cargo turnover per month should be not less than 76.8 thousand t-km (calculation of breakeven: average costs per car per month are 5902 UAH, average income rate (car) is 76.84 kop/10 t-km).

As a result of replacing old cars with new ones, the use of car fleets improves: the average number of turns of a new car is 12% more than the old one; the average number of clutches for technical reasons is reduced by 10 times; the average turnover of a new car is 40% more than the old one.

There is a need to develop services for the route formation in accordance with the contract of payment for the infrastructure use, accumulation of the route, the actual duration of idle on the route and each route from the groups of cars of different owners, operators, finding the economic options of supply with significant changes in the number of cars in the groups of route formation.

The model of accumulation, transiting, unloading of the route allows taking into account the fluctuations of the car flow, the infrastructure capacity of these access tracks, first of all for mass transportation of senders and consignees of cargo.

References

1. National transport strategy of Ukraine for the period of 2030 year, Regulation of the cabinet of Ministers of Ukraine from May 30 2018. No. 430-p, available online: https://zakon.rada.gov.ua/laws/show/430-2018-p.
2. M.I. Danko, V.V.Kuleshov, Sc. Works URA, **57**, 121-128, (2004)
3. I.V. Berestov, V.V. Kuleshov, D.O. Kovalenko, A.O. Rekal, International economical-technical journal "Ukrainian railway", **5(59)**, 12-16, (2018)
4. A.V. Kuleshov, V.V.Kuleshov, East-European Journal of Advanced Technologies, **2/3 (68)**, 16-20, (2014)
5. E.V. Shikin, A.G. Chkhartishvili, *Mathematical methods and models in control*, M.: Case, 437, (2004)
6. T.V. Butko, D.V. Lomotko, A.V. Prokhorchenko, T.O. Oliynyk, Ukr. state acad. railroad trans. science materials, **111**, 23-30, (2009)
7. S. Panchenko, O. Ohar, V. Kuleshov, M. Kutsenko, A. Kuleshov Improvement of the organizational-technological model of the route from groups of wagons of different owners International Journal of Engineering & Technology, **7 (4.3)**, 266-269, (2018), Website: www.sciencepubco.com/index.php/IJET. https://www.sciencepubco.com/index.php/ijet/article/view/19801.
8. M. Newman, SIAM Review, **45**, 167-256, (2003).
9. B. Korte, J. Vygen, Combinatorial optimization, Springer, (2006).

Improvement of Characteristics of the Railway Direction Kharkiv-Dnipro with the Purpose of Introducing the high-speed Passenger Traffic

Maksym Kutsenko[1*], *Sergii* Zmii[1], *Oleksandr* Rozsokha[1], *Ganna* Shapoval[1], *and Olga* Semenova[2]

[1]Ukrainian State University of Railway Transport, 61050 Feierbakh Square 7, Kharkiv, Ukraine
[2] Shevchenkivska carriage station of Ukrzaliznytsya JSC, 20731 Nezalezhnosti street 1, Smila, Ukraine

Abstract. The high efficiency of high-speed rail transport contributes to its development both in the service sectors and in the construction of devices for servicing high-speed trains, in particular, track development. At the same time, it is necessary to take into account the obvious problems that accompany the process of reconstruction of railways during the transition to high-speed passenger traffic. Among them, the presence of small radii of circular curves, an insufficient length of straight inserts and transition curves should be noted. In this paper, it is proposed to consider small-radius curves located close to each other as one common module when reconstructing railways with the aim of introducing high-speed passenger traffic. The implementation of this approach will allow solving one of the main problems - straightening of the railway section. This will increase the speed of passenger trains and reduce travel time by reducing the length of the route. The article deals with the task of reconstruction of modules consisting of three adjacent curves, which occur when improving the characteristics of the plan of the Kharkiv-Dnipro railway direction.

1 Introduction

In [1] it is noted that the activity of the railway transport, as part of the Ukrainian unified transport system, contributes to the normal functioning of all sectors of the economy and the strengthening of the defense capabilities of the country, which are important factors in joining the European Union. Modern tendencies in the market of transport services for passenger transportation lead to a reduction of the role of railways and an increase in the importance of road and air transport. Therefore, certain measures should be taken in order to increase the attractiveness of rail passenger traffic by introducing high-speed highways. It is these measures that enable the railway industry to maintain and effectively use existing technical capacity for structural changes in technology and to maintain competitive advantages over other modes of transport.

* Corresponding author: **maksimus84@meta.ua**

At the same time, the transition to high-speed passenger traffic is associated with a number of problems, the main of which is the task of rebuilding the railway plan, and in the first place, increasing the radii of circular curves to 1200 and more meters, the extension of straight inserts and transitional curves. Domestic and foreign science of the railway industry paid enough attention to these issues, but now life is putting forward new and more stringent requirements.

2 Formulation of the problem

According to [2], the practice of designing and reconstructing a railway plan shows that rearrangements are for the most part single-curved small radius, while complex divisions consisting of two, three, and four adjacent curves remain in the same form as they do not exist. rather vivid and accessible methods of their reconstruction.

However, under very difficult plan conditions, the reconstruction of blocks containing three or more curves becomes complicated and extremely expensive, especially since the reconstruction works are carried out on the operating line and it becomes expedient to consider variants of radical straightening of the route at a sufficiently long distance, which allows to improve not only the parameters of the plan, but also the operational parameters of the site by reducing its length.

3 Recent research and publications analysis

Historically, a number of calculation methods have been developed when solving problems of reconstruction of a plan. The most perfect one is method of normal lines [1], which has found only a limited application by virtue of known difficulties. Approximate evolvent methods have been widespread.

With the introduction of computer technology the most widespread methods were obtained using rectangular coordinates, free from imperfections of evolvent methods. There are two classes of tasks for the reconstruction of the plan: bringing the plan to the correct geometric outline; reconstruction of the plan in order to improve its characteristics (increasing the radius of circular curves, increasing the lengths of transition curves, increasing the lengths of direct inserts between adjacent curves). The tasks for the reconstruction of the plan are considered in the work [3]. In total six types of tasks have been analyzed. The main disadvantage of this method is the complexity; in addition, the application of the method of angular diagrams deliberately brings a certain error to the result. Also, the disadvantages include the fact that only two adjacent curves have been considered, while in practice, often there are problems of reconstruction of a section containing more than two curves. It should also be noted that these objectives did not directly aim at increasing the radii of circular curves, which became relevant nowadays when the introduction of high-speed passenger trains became an urgent task. In the work [4] the nine design tasks, as well as the task of calculating landslides and project interchanges have been described. Relatively simple cases have been considered. These tasks are focused on the use of computers. Here the problem of increasing the radius of circular curves is not set. The reconstruction problem in relation to two adjacent curves is considered. Improving the characteristics of the plan requires consideration of tasks that include the reconstruction of complex plan districts, which consist of not only two, but also three or more adjacent curves. Therefore, there is a need to develop new methods of reconstruction of the existing position of the complex plan.

4 The selection of unresolved parts of the general problem

As shown by the analysis of some of the executed reconstruction projects, as a rule, designers are limited to the restructuring of single circular curves, rarely considering two, and almost do not touch cases with three or more adjacent curves.

5 The purpose of the article

The purpose of the article is to reconstruct modules consisting of three adjacent curves that take place while improving the characteristics of Kharkiv-Dnipro railway plan, which, according to the work, provides for the introduction of high-speed passenger traffic.

6 Statement of the main research material

The main disadvantages of the technical equipment of this section, which do not allow to increase the speed of passenger trains up to 160 km/h, are: curves of small radius, the presence of turnouts in curves, worn top structure of the track, emergency condition of a number of artificial structures, narrow interlayers and platforms on separate points, "sick" earthworks, large number of crossings, insufficient technical equipment of a number of stations. After eliminating the disadvantages in this direction, it is possible to significantly increase the speed of passenger trains (up to 160 km/h), and reduce the time of travel. The determination of the cost of reconstruction and construction of a new recaptured section at the comparative stage should be carried out on the following indicators: earthworks, upper structure of the track, electrification and man-made structures. Table 1 shows the cost indicators of the main components during the reconstruction of the site and the construction of a new track.

Table 1. The components of construction cost in the reconstruction and construction of a new site.

The components of the construction cost	Reconstruction of the site	The construction of a new site
Earth works	$L_r q_a^{I} \gamma$	$L_b q_a^{II}$
Upper structure of the track	$L_r a_{us}$	$L_{us} a_t$
Electrification	$L_r a_e$	$L_b a_e$
Artificial structures	$L_r q_a^{I} \delta$	$L_b q_a^{II} \delta$

The table shows the unit values, which depend on the length of the section $-a_{us}$, a_e, δ, and also γ – is the coefficient, which takes into account the increase in the cost of construction works on the reconstruction of the route with the continuous movement of trains. The presented expenses for the construction of a new decoiled area will amount to thousands UAH.

$$E_b = L_b a_b + L_b(\alpha Ne + k)\frac{1}{E}, \quad (1)$$

$L_{б}$ – length of the ordered section, km;

$a_{б}$ – cost of construction of one kilometer of the decoiled site, thousand UAH/km;

α – coefficient taking into account the uniform distribution of the traffic flow by directions;

N – the number of trains in the freight direction, thousand trains per year;

e – cost of one train-km, thousand UAH;
k – the cost of maintaining permanent devices, thousand UAH/km per year;
E – discount rate.

At the same time, at a known fraction of the length of the section to be reconstructed under high-speed traffic, the costs of reconstruction of the line, (thousand UAH), from the total length of the existing route (λ) are given as follows

$$E_r = \lambda L_e a_r + L_e(\alpha Ne + k)\frac{1}{E}, \quad (2)$$

L_e – length of the existing area, km;
a_r – cost of reconstruction of one kilometer, thousand UAH/km;

It is advisable to direct the route, provided that $E_r > E_b$ when the specific length of the reconstruction sites is equal to

$$\lambda \geq \frac{L_b a_b + c\Delta L}{L_e a_r}, \quad (3)$$

c – operating costs per kilometer, thousand UAH/km;
ΔL – shortening the length of the existing area when the route is adjusted, km. From formula (3) it is obvious that λ depends, first of all, on the lengths of the existing and new stations.

The key factor when introducing high-speed passenger traffic is the duration of the trip. When aligning the route, the gain in time (Δt) may be more significant than while reconstructing the line plan.

The gain in time can be found using the following formula, min.

$$\Delta t = 60 \cdot \left(\frac{L_e}{V_e} - \frac{L_e - \Delta L}{V_b}\right), \quad (4)$$

V_e, V_b – respectively the speed at the existing and the decoiled section, km/h;

The analysis of the main technical characteristics of the Kharkiv-Dnipro line showed that with an increase in the speed of passenger trains up to 160 km/h, it is necessary to reconstruct the plan in order to increase the radius and the lengths of straight inserts. For the indicated site, three variants were designed: East, West and Global. The main features of the options are presented in Table 2.

Table 2. The main characteristics of the station variants that decoil.

Characteristics	Options for sites that decoil		
	Global	Eastern	Western
Line plan: - radius, m - sum of angles rotation, degrees	 2000 75	 2000 156	 2000 135
Longitudinal profile: - length of sites - length of sections with guide slope, m	 0 11800	 1600 3600	 0 4000
The volume of earthworks, thousand m³	4553,94	652,72	301,62

The next step is to conduct a feasibility study of the options for rectification based on the above criteria. Substituting the corresponding values in the expression (1), and then in the formula (2) for all three variants, we obtain that according to the construction cost criterion and the difference in length it is expedient to align the route with the Western variant. The results of calculated costs are given in Table 3.

Table 3. The costs are for the reconstruction of the current plan and for the rectification options.

The improvement of the parameters of the current plan	Value of reduced costs, ths. UAH
Option of reconstruction	251852.34
Eastern version of rectification	152043.71
Western version of rectification	102691.26
Global version of rectification	325362.18

According to the results of Table 3, it can be said that the alignment of the route along the Eastern and Western variants requires less costs than the reconstruction of complex lines of the line plan for the introduction of high-speed trains. So, after the comparative analysis, we can say that from the considered variants of the alignment the most expedient to realization is the Western one. At the previous stage, it satisfies all the criteria. It was this variant of rectification that was used in the reconstruction of the block of curves on Merefa – Vodalaga running line. This running line requires the solution to the problem of the reconstruction of a block of three closely spaced curves with insufficient values of radii. In order to increase the speed of passenger trains, it is necessary to increase the radius (R_r) and the value of the direct insertion between the curves (d_r). Next, for finding the values of the normal, the analytical model of the plan was used.

To determine the values of the displacement between the project and the existing plan for the purpose of comparison, the method of the angular diagrams is used. To compare the results of calculating the values of normal lines obtained by the method of the angular diagrams and using the analytical model of the plan, a joint displacement graph was constructed. The graph has shown that the application of the method of the angular diagrams does not always allow accurate determination of the values of normal (Figure 1).

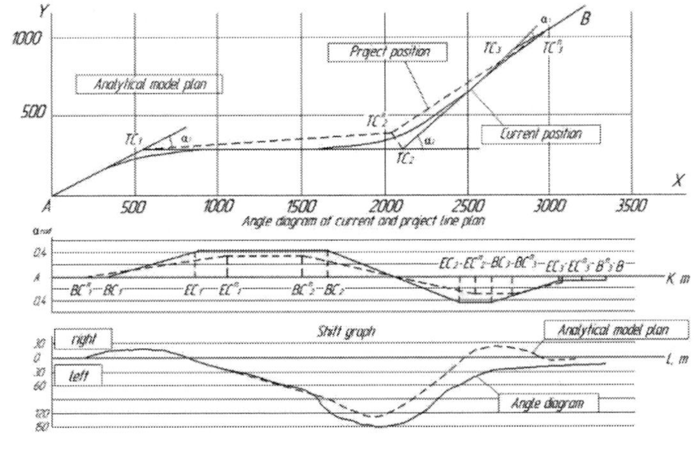

Fig. 1. The analytical plan model and angle chart method, shift graph.

The main component for determining capital investments is the cost of earthworks. The determination of the excavation volume on the basis of a digital model of relief was carried out on the pickets in widths from the condition of constant height of the mound. The calculation of the volume of the required soil itself was carried out using the Simpson formula. For the construction of the mound of the project line 23 thousand m³ will be required. Performance data includes: reduction of speed limit; line shortening; cutting of running time. In order to determine the performance, the following traction calculations for Merefa – Vodolaga running line for the freight and passenger traffic were made, taking into account the limit on this curve (85 km/h) and after lifting it. The results are presented in Table 4.

Table 4. The results of the traction calculations for Merefa – Vodolaga running line

Characteristics	Restricted				Without restriction			
	passenger		freight		passenger		freight	
	forth	back	forth	back	forth	back	forth	back
R_m, MJ	1953	2012	4583	4667	2072	2181	4751	4797
R_r, MJ	1965	2000	4629	4620	2085	2168	4800	4748
t, min	14.3	14.4	18.5	18.6	13.8	13.5	17.6	17.6
ΣR_m, MJ	3965		9250		4253		9548	
ΣR_r, MJ	3965		9249		4253		9548	
Σt, min	28.7		37.1		27.3		35.2	

The difference between the parameters of the mechanical work (R_m), the resistance forces (R_r) and the running time for passenger and freight traffic are given in Table 5.

Table 5. The difference between the traction calculations.

Traction calculation indices	Passenger service	Freight service
R_m, MJ	288	298
R_r, MJ	288	299
t, min	1.4	1.9

7 Conclusions

Consequently, the above studies convincingly show that when the length of the running line is reduced after the reconstruction of only one module on ΔL =79.09 m and the speed limit on the section of the module is eliminated, i.e. reconstructed, the traveling time of passenger trains will be reduced by 1.4 minutes, and freight by 1.9 minutes.

References

1. M. Y. Kutsenko, O. A. Dudin, G. V. Pakhar, O. M. Danilenko, O. M. Dubina. Information and control systems on the railway transport, 5, 55–59, (2018)
2. O.Ohar, O.Rozsokha, M. Kutsenko, Y. Smachilo, Eastern-European journal of enterprise technologies, **3-90**, 48-57, (2017.)
3. I. V. Turbin, A. V. Gavrilenkov, I. I. Kantor. *Izyskaniya i proyektirovaniye zheleznykh dorog*. Moscow: Transport, 479, (1989)
4. G. Z. Wertsman and A. P. Volodinю. *Proyektirovaniye vtorykh puteyю*. Moscow: Transport, 359, (1970)
5. P. Ferreira, R. Maciel, J. Estaire, M. Rodriguez-Plaza. Structure and Infrastructure Engineering, **15.1**,. 1 – 13, (2019)
6. Y.-Z. Zhou, L.-J.Chen, C. Gao. Bridge Construction. **48.5**, 11 – 15 (2018).

Technological and economic estimation of efficiency of a route choice for transportation of dangerous goods

Oleksandr Lavrukhin[1,*], *Anton Kovalov*[1] , and *Daria Kulova*[1]

[1] Ukrainian State University of Railway Transport, Department of Management of Freight and Commercial Work, Feuerbach sq., 7, Kharkiv, 61001, Ukraine

Abstract. One of the main conditions for the transportation of goods, especially dangerous, is minimizing all types of eventual cost in the event of an emergency. The cost value depends on multiples factors, which are significantly influenced by the location of the train in different places (such as running line, technical stations, residential places, etc.). The proposed approaches on the basis of the queuing theory will allow choosing traveling routes of cars with dangerous goods providing the minimum average cost reduced to one emergency situation.

1 Introduction

The transport industry needs accelerated development in the current competitive conditions during the transition to market relations. Improving the service quality is one of the key conditions that affect the level of demand for transport services. Railway transport has a leading position in the transport system and is the main carrier of goods. Transportation of dangerous goods (DG) takes an important place in railway transportation; thus, it should be not only prompt, but also safe, while complying with high international standards, since emergency situations with trains which include cars with DG always have more severe consequences, which in turn leads to heavier material damage.

2. Literature review

Leading scientists of the world have been engaged in raising the safety level and reducing costs caused by the onset of transport emergencies during transportation of DG. In [1], emergencies during transportation of dangerous goods were assessed. The analysis of the causes of emergencies demonstrated that they mostly include commercial failures, critical wear-out of fixed assets and violation of safety requirements, which leads to significant material damage; however, the specific place of an emergency is not taken into account.

In [2], a comprehensive approach was proposed to determine the safety level during the transportation by railway transport. Various impact factors are taken into account, such as: state of technical facilities, current operation technology, specific features of work with

* Corresponding author: tilavalval@gmail.com

personnel, and cost of ensuring an adequate level of traffic safety. Total of twelve factors with different distinctnesses are identified. An additive indicator was proposed as an evaluation criterion that determines the level of railway traffic safety, however, increased danger during transportation of DG is insufficiently accounted for in this indicator.

In [3], calculations of the main losses from the onset of emergencies during the transportation of DG are given. These include: direct loss of a transport company; accident localization and investigation costs, socio-economic losses; environmental damage; damaged caused by loss of labor resources due to the death or disability of people. However, the influence of the probability of cars with DG to be in different states on the cost value in the event of emergency is not taken into account.

In [4], a mathematical method for assessing the complex safety level in the railway transport is proposed. The traffic safety index was calculated, which is used as the basis of the simulation model "Railway Traffic Safety", which will allow railway staff to provide a high level of safety during the transportation and will reduce the cost of eliminating the consequences of transport accidents, but the transportation of DG needs special consideration in determining the safety level during the transportation of mass cargoes.

3. Main part of the study

In study [5], the amount of risk depends on the value of determined complex criterion U and the consequences of the onset of an adverse event

$$R = U \cdot \sum_{a=1}^{l} E(w_a), \quad (1)$$

where U – conditional confidence of occurrence of more significant consequences as a result of an emergency (complex criterion);
$E(w_a)$ – average cost reduced to one emergency situation;
l – number of components of average cost.

Average cost consists of:
– $E(w_1)$ – average reduced cost that relates to the damage of fixed assets (buildings, structures), cars, tracks, locomotives and third party's property;
– $E(w_2)$ – average reduced cost incurred as a result of environmental harm;
– $E(w_3)$ – average reduced cost for monetary compensation for causing adverse consequences to a person (death, injury, disability).

$$E(w_1) = A_1 + A_2 + A_3, \quad (2)$$

where A_1 – cost caused by the damage or total or partial destruction of the main productive and non-productive assets;
A_2 – cost caused by the damage or total or partial destruction of tracks, locomotives, cars, signals and interlocking devices, etc;
A_3 – cost caused by the damage or total or partial destruction of the third party's property.

In turn, A_1 can occur when a train is in all types of interstations, as well as during crossings and overtakes. A_2 can also occur when a train is in a running line and moves within residential places. A_3 can arise when a train moves within residential places.

$E(w_2)$ can occur when a train is in all types of interstations, as well as during crossings and overtakes, is in a running line and moves within residential places.

$$E(w_2) = B_1 + B_2 + B_3 + B_4 + B_5, \quad (3)$$

where B_1 – cost caused by air pollution;
B_2 – cost caused by water pollution;
B_3 – cost caused by contamination of the upper track structure;
B_4 – cost caused by the destruction of biological resources;
B_5 – other cost caused by adverse consequences of an emergency.

$E(w_3)$ will get maximum values when a train runs within residential places and when it is at technical stations, i.e.

$$E(w_3) = C_1 + C_2, \quad (4)$$

where C_1 – socio-economic cost for monetary compensation in case of death, disability or injury with permanent disability, injury without permanent disability;
C_2– cost caused by the loss of labor resources [3].

It is known that the process of delivery of goods from the sender to the consignee can be considered as a queuing system (QS) [6-10]. Functioning of any QS can be represented through its possible states and intensities of transitions from one state to another. Determining the probability of the train with DG to be in different states during the transportation is expedient for a more accurate and precise definition of all components of different costs:

S_1 –the train with cars loaded with DG is between the interstations points of all types (outside the residential places);
S_2 – the train with cars loaded with DG moves within residential places without stopping;
S_3 - the train with cars loaded with DG is under technical operations at the technical stations (change of a locomotive, locomotive crew, type of current, direction of movement, etc.);
S_4 – the train with cars loaded with DG is under crossing;
S_5 –the train with cars loaded with DG is under overtaking.

Then, the graph of states of the composition of the train loaded with DG will have the form (Fig. 1)

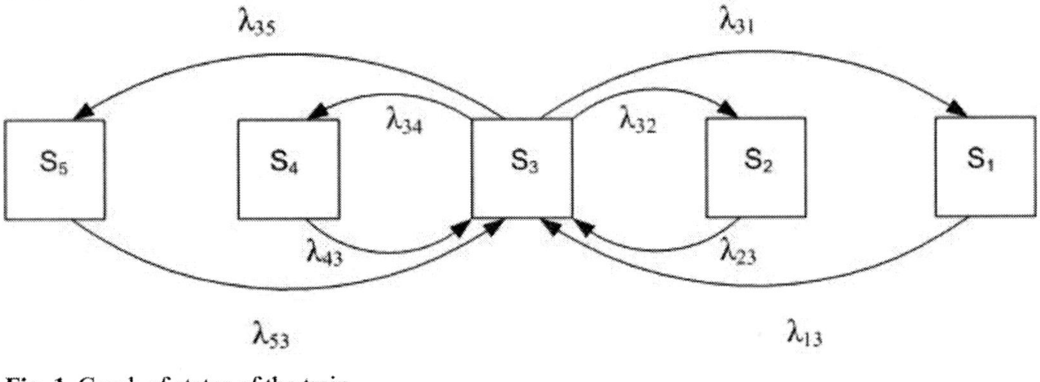

Fig. 1. Graph of states of the train.

Differential equation system:

$$\begin{cases} \frac{dp_1}{dt} = -(\lambda_{12} + \lambda_{13} + \lambda_{14} + \lambda_{15}) \cdot P_1 + \lambda_{21}P_2 + \lambda_{31}P_3 + \lambda_{41}P_4 + \lambda_{51}P_5 \\ \frac{dp_2}{dt} = -\lambda_{21}P_2 + \lambda_{12}P \\ \frac{dp_3}{dt} = -\lambda_{31}P_3 + \lambda_{13}P \\ \frac{dp_4}{dt} = -\lambda_{41}P_4 + \lambda_{14}P_1 \\ \frac{dp_5}{dt} = -\lambda_{51}P_5 + \lambda_{15}P \end{cases} \quad (5)$$

Normalizing condition:

$$P_1 + P_2 + P_3 + P_4 + P_5 = 1 \quad (6)$$

After calculating the probabilities of states of the train with cars loaded with DG, the value of average cost reduced to one emergency situation E(w$_a$) will be calculated as follows:

$$E(w_a) = E(w_1) + E(w_2) + E(w_3) = P_1A_2 + (P_3 + P_4 + P_5)\cdot(A_1 + A_2) + P_2(A_2 + A_3) + \\ + E(w_2) + (P_2 + P_3)E(w_3) \quad (7)$$

Thus, the results obtained in (7) can help more accurately predict the possible costs value in the event of an emergency situation with DG when choosing a route of the train.
The calculation of a conditional DG transportation site (Fig. 2) is shown as an example, when train passes, consequently:
1. two large residential places, one technical station (replacement locomotive) and one crossing;
2. three residential places and two technical stations with corresponding intensities of transfer of the train from one state to another.

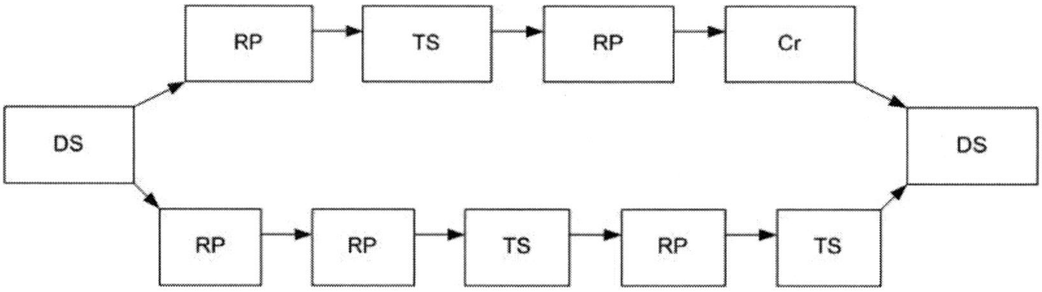

Fig. 2. Conditional DG transportation site.

Conventions:
DS – departure station; RP – residential place; TS – technical station; Cr – crossing; DS – destination station.

To calculate the first transportation route, a typical graph (Fig. 1), as well as the known principles of constructing a differential equation system in the Matlab environment are used, and the following results are obtained (Fig. 3).

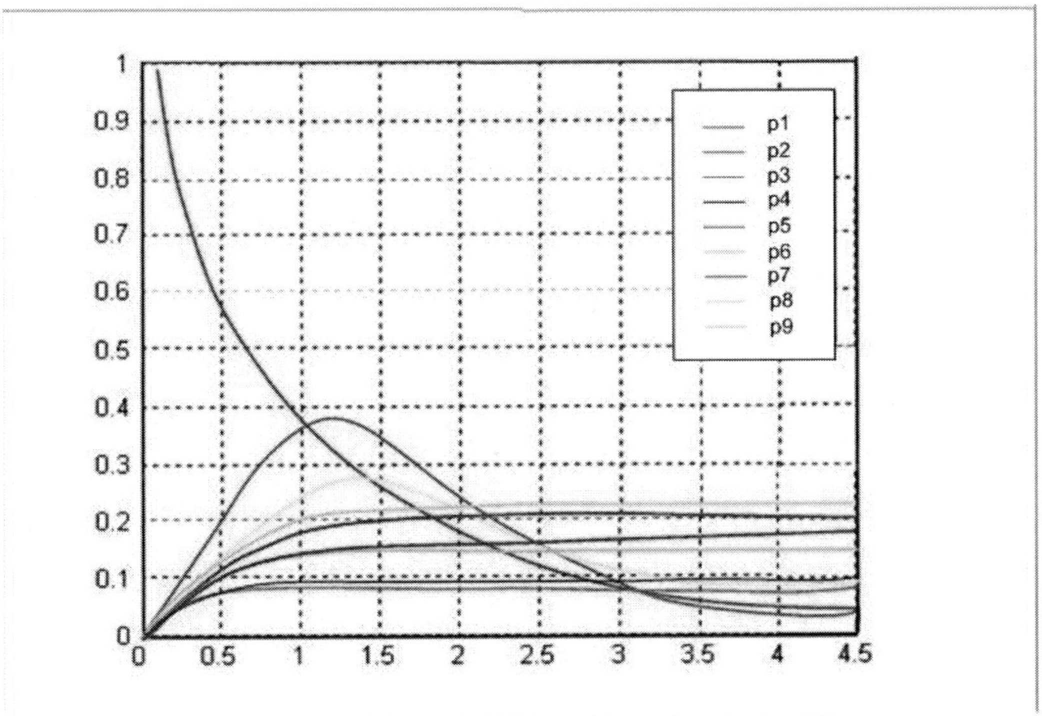

Fig. 3. Results of calculations of the probabilities of the train to be in different states.

Probability of the train with cars loaded with DG being between the intersections of all types (outside of settlements) consists of P1, P3, P5, P7 and P9 (0.556); probability of the train with the cars loaded with DG moving within residential places without stopping consists of P2 and P6 (0.262); probability of the train with cars loaded with DG being under technical operations at technical stations (change of locomotive) corresponds to P4 (0.093); probability for the train with cars loaded with DG to be under crossing is P8 (0.089); and probability of the train with cars loaded with DG to be under overtaking is 0.

Thus, calculation (7) will have the form

$$E(w_a) = E(w_1) + E(w_2) + E(w_3) = 0{,}556A_2 + 0{,}182\cdot(A_1 + A_2) + 0{,}262(A_2 + A_3) +$$
$$+E(w_2) + 0{,}355E(w_3)$$

The second (parallel) route of DG delivery from the point of departure to the point of destination is calculated similarly.

Results of calculations of the second route

$$E(w_a) = E(w_1) + E(w_2) + E(w_3) = 0{,}483A_2 + 0{,}231\cdot(A_1 + A_2) + 0{,}286(A_2 + A_3) +$$
$$+E(w_2) + 0{,}392E(w_3)$$

Thus, the first transportation route of DG is more expedient from the economic point of view.

3. Conclusions

Total average cost reduced to one emergency situation consists of average reduced cost related to damage to fixed assets (buildings, structures), cars, tracks, locomotives and third party's property; average cost incurred as a result of environmental harm; average reduced cost for monetary compensation for causing adverse consequences for a person (death, injury, disability). However, taking these components into account in full value will not allow correct predicting of consequences of emergencies during the transportation of DG in different routes. The given approaches give an opportunity to choose a transportation route for DG with the lowest possible cost. The promising directions of transportation of DG are shipment of such cars to ports and from ports by rail. Accordingly, using large ports (such as Odessa, Mariupol, etc.) as experimental sites is advisable for further study.

One of the conditions for using this approach is fully automated accounting system and the use of the decision support system (DSS) by the operational personnel.

This system will allow the station personnel to choose a route of the train during the operation depending on the following parameters: number of cars with DG of different classes, time of day, seasonal factor, irregularity in different days of the week, etc.

References

1. Y. Buts, E.. Kraynyuk, D. Kozodoy, V. Barbashin, Visnik Dnipropetrovskogo nacionalnogo universitetu zaliznichnogo transportu, **3 (75),** 27-35 (2018)
2. O. Ohar, O. Rozsokha, M. Kutsenko, Yu. Smachilo, Eastern-European Journal of Enterprise Technologies, **6/3 (90)**, 48-57 (2017)
3. A.M. Kotenko, D.S. Kozodoy, A.V. Svetlichnaya, P.S. Shilaev, Zbirnik naukovih prac UkrDAZT, **141**, 272-280 (2013)
4. P. Sukhov, V. Popov, V Filippov, MATEC Web of Conferences, TransSiberia, **239**, 1-7 (2018)
5. O. Lavrukhin, A. Kovalov, V. Schevcenko, A. Kyman, D. Kulova, Eastern-European Journal of Enterprise Technologies, **2, 3(98),** 25-31 (2019)
6. O. Rodkevych, Organizaciya perevezen i bezpeka transportu, 171-182 (2014)
7. N. Berezhnaya, Technical service of agriculture, forestry and transport systems, **7**, 64-76 (2017)
8. T. Bondaryeva, N. Grebenyuk, Ekonomika ta upravlinnya pidpriyemstvami mashinobudivnoyi galuzi, **3**, 91-102 (2014)
9. D. Dmytriv, O. Rogatynska, Y. Kapatsila, Matematichni metodi, modeli ta informacijni tehnologiyi v ekonomici, 123-130 (2016)
10. I. Carenkova, Zbirnik naukovih prac DNUZT im. akad. V. Lazaryana, **14**, 95-101 (2017)

Influence of tourist flows intensity on market of passenger air carriages

Yuliia Lola[1,*] and *Mariya* Aldoshyna[2]

[1]Simon Kuznets Kharkiv National University of Economics, Department of Tourism, Kharkiv, Ukraine.
[2]Simon Kuznets Kharkiv National University of Economics, Department of Tourism, Kharkiv, Ukraine

Abstract. This research deals with the impact of intensity of inbound and outbound tourism on the sphere of passenger air carriages, based on the dataset that covers the indicators across different economic regions over the period of 2015-2017. The results based on building a correlation-regression model have shown that there is a close connection between tourism development and passenger carriages intensity. Verification of the model by the criteria of Fisher and Student (F-test and t-test) have proved the hypothesis about its statistical significance only after excluding five countries, for which no direct connection is observed between the chosen indictors. Further research has shown that the level of the country economic development does not influence the passenger air carriages industry. Specific attention is paid to the state and main tendencies of passenger air carriages industry in Ukraine. Impact of tourism on this kind of activity is analysed. The basic measures to increase its efficiency are highlighted.

1 Introduction

Attractiveness of the country tourist industry depends greatly on the quality of transport services provided. The central place in tourism is taken by the sphere of passenger air carriages. The more branched and intensified passenger air carriages are, the bigger potential of tourism development in the country is (on condition of logistic approach to management and increasing economic efficiency of air companies activities).

The level of air carriages industry development is characterised by number of passengers carried, number of internal and international flights, number of city-pair routes. In 2018, airlines continued to increase the number of city-pair routes globally. Almost 22,000 city pairs are now regularly serviced by airlines. The real, inflation-adjusted cost of air transport was halved in the past 20 years and declined further in 2018 [1]. Though according to the statistical data of International Civil Aviation Organisation, the number of air carriages in the world has increased (in 2017 in comparison with 2010 by 51,4 %) and constitutes 3,979 billion passengers. Naturally, the intensity of tourist flows has increased quite dynamically as well. In comparison with 2010, in 2017 the outbound flow of tourists in the world increased by 40,3 % and constituted 1,341 billion tourists.

* Corresponding author: yuliia.lola@hneu.net

2 Theoretical development and hypotheses formulation

Tourism development affects the development of a range of transport infrastructure (especially air transport). Inbound and outbound tourism has bidirectional causality with air transportation [2]. Air transport and tourism are highly connected. The research shows that tourist-oriented airports may achieve higher efficiency levels than non-touristic ones [3].

Air transport development is a key driver for growth in national and international scale. The air transport has both a direct positive impact on key economic indicators and business sectors at a regional or national scale, and indirect impact in the form of side effects for tourism industry to the local or national economy [4].

Traffic flows decrease despite significant fare reductions. However, these data vary substantially based on the flight origin and ticket-type purchased [5]. The air passenger duty modifies the budget allocations of UK outbound tourists by increasing the relative share of transportation expenditure, while correspondingly decreasing the at-destination expenditures on items such as accommodation and food [6].

Air transport is highly connected with other modes of transport such as high speed trains and car travels for short time trips. Researchers compare development of travel time to show how the evolution of the air transport network has influenced reduction in travel time [7].

Regulatory systems of the air transport policy have an impact on passenger traffic and tourist flows. Restrictions on the air transport sector also have a substantial negative impact on bilateral tourist flows. Further liberalisation in this sector could help build up a stronger tourism industry [8].

The analysis of the mentioned resources has allowed hypothesising the following:

Hypothesis 1. There is mutual influence of industry development of passenger air carriages and intensity of inbound and outbound tourist flow of the country.

Hypothesis 2. Development of passenger air carriages is influenced by the development level of the country economy.

3 Research methodology

It is suggested to apply correlation methods of regression analysis to verify the hypotheses. In particular, a matrix of pair correlation coefficients has been constructed to detect the linear connection between factors. The Method of Ordinary Least Squares has been used to evaluate the parameters of linear equations of regression. The statistical criteria of Fisher and Student (F-test and t-test), Glejser test have been used to test hypotheses about statistical significance of regression models.

Dummy-variables have been used, that are variables with a discrete set of values that describe qualitative characteristics in a quantitative way, have been used to verify structural changes in number of air carriages for countries with different level of income. Dummy variables of the binary type «0-1» are usually used in econometric models.

The algorithm of the research involves the following steps:

1) choosing the period of the research;
2) forming the space of indicators;
3) choosing the form of correlation between factors;
4) parameterising the model;
5) verifying the model;
6) applying the model.

4 Results

The research period from 2015 to 2017 is chosen in the process of realisation of the first step of the proposed algorithm. The quantity of the world countries, which have out-coming data for each year by the chosen indices of Air transport, passengers carried (AT) and Number of arrivals (NA) equals 135. The data are taken from the site of the World Bank [9].

The second step presupposed outlining factors which characterise intensity of passenger air flow as well as intensity of inbound and outbound tourist flows: Air transport, passengers carried (AT), Number of arrivals (NA), Number of departures (ND). Hypothesis 1 presupposes a significant correlation between these indicators. The matrix of even correlation coefficients between the researched indicators for the given period looks as follows (Fig. 1).

	AT_15	AT_16	AT_17	NA_15	NA_16	NA_17	ND_2015	ND_2016	ND_2017
AT_2015	1,000	0,999	0,997	0,718	0,708	0,687	0,741	0,744	0,759
AT_2016	0,999	1,000	0,999	0,719	0,710	0,689	0,749	0,752	0,767
AT_2017	0,997	0,999	1,000	0,718	0,709	0,688	0,756	0,760	0,775
NA_2015	0,718	0,719	0,718	1,000	0,996	0,994	0,735	0,733	0,738
NA_2016	0,708	0,710	0,709	0,996	1,000	0,998	0,731	0,731	0,734
NA_2017	0,687	0,689	0,688	0,994	0,998	1,000	0,712	0,711	0,714
ND_2015	0,741	0,749	0,756	0,735	0,731	0,712	1,000	0,999	0,998
ND_2016	0,744	0,752	0,760	0,733	0,731	0,711	0,999	1,000	0,999
ND_2017	0,759	0,767	0,775	0,738	0,734	0,714	0,998	0,999	1,000

Fig. 1. Matrix of even correlation coefficients
Source: Own research

Let us pay attention to the highlighted diagonal matrix sectors given in Fig. 1. The values of the highlighted coefficients indicate a high level of auto-correlation connection between consequent values for each index that almost does not lessen in the course of time. It proves the essential inertness of the processes. Besides, the elements of the presented matrix indicate a close linear correlation connection between current values AT and NA and ND, as well as between lag variables. Influence of ND on Air transport, passengers carried (AT) is even bigger than influence of NA on Air transport, passengers carried (AT). However, absence of data by index ND for many countries has stipulated the choice of NA as an exogenous variable.

During realisation of the third step, the linear form of connection was chosen through high values of even correlation coefficients between the researched indices.

The following model has been built to test the first hypothesis:

$$AT_t = a_0 + a_1 NA_t, \quad (1)$$

where a_0 and a_1 – are unknown mdel parametres, estimated at the fourth step of the research with the help of the Method of Ordinary Least Squares, both for the whole sample frame in general (405 observations – data for 135 countries for three years), and each year t separately:

Verification of the built models was realised at the fifth step. All four models have turned out to be statistically significant on the whole upon Fisher criterion (F-test). However the values of determination coefficient R^2 for these models range within 0,39 – 0,43, that indicates not high enough quality of the model. Verification of the statistical significance of certain model parameters with the help of Student criterion (t-test) has revealed that the parameter a_0 is statistically insignificant for all four models. Thus the models of the following type have been built:

$$AT_t = a_1 NA_t \tag{2}$$

Models of the kind (2) have turned out to be more adequate and statistically significant. Let us consider in more detail the results of building the summarised model by 405 observations. The model is of the following type:

$$\widehat{AT} = 3{,}556 \cdot NA \tag{3}$$

The correlation coefficient $R = 0.682$; the determination coefficient $R^2 = 0.465$; $F(1;404) = 351.2$; $t_{a_1}(404) = 18.74$. Thus, the model is adequate and statistically significant both in general and by certain parameters. However, there is heteroscedasticity of surpluses that is proved by non-parametric and parametric tests *Goldfeld – Quandt.* The reason of dispersion of random errors is data non-homogeneity that can be noticed in the graph of factual and theoretical (predicted) values Air transport, passengers carried (Fig. 2).

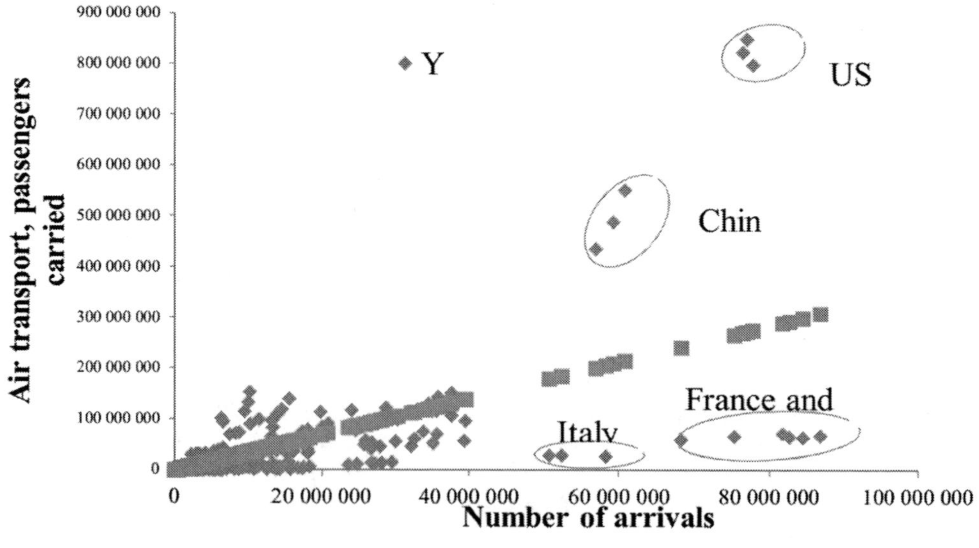

Fig. 2. Line Fit Plot
Source: Own research

As the diffusion diagram shows, the leading countries in the rating by indicators of tourists' incomes are outliers in the correlation-regression model. So, The USA and China are the largest countries in size of Air transport, passengers carried (AT) with a significant amount of inner carriages. The essential deviation upward in factual values of AT form the regression line indicates a significant amount of aviation passenger carriages, which are not connected with inbound and outbound tourist flows. It is resulted by the fact that the US passenger market has remained the largest in the world. The US domestic market, with almost 590 million passenger journeys in 2018, is still the world's largest single market. China is on the second place, with 515 million, although China's domestic market added the most passenger journeys in 2018 [1]. Such growth of international tourist aviation will make it possible for China to outrun the USA soon, and to become the largest market of air carriages by 2022 [2]. Chinese air companies propose low prices at the market of passenger air carriages. At the same time they increase quality of their services by investing and implementing international standards. It has become possible due to the policy of the state that subsidises the sphere of passenger air carriages and allows aviation companies to seize the world markets, often without getting profit from their activity. Moreover, China is facing a steady increase in the net of international airports. The nation well-being is increasing as well, that allows intensifying tourist activity of the most densely populated country in the world.

On the other hand, such countries as Spain, France and Italy are characterised by significant shift downwards in comparison with the regression line. It indicates that these countries are the most attractive for tourists, but an insignificant number of these tourists get to the countries not by air, but by traffic, railways and sea. Intensity of traffic flow into these countries promotes free transportation within Schengen zone to which these countries belong.

As the model with heteroscedasticity of surpluses could not be applied for analysis and forecast, then at the next step the enumerated countries were excluded from the model and a new model of the type (2) was constructed by the sample frame of 390 observations.

The diffusion diagram of the model after extraction of outliers is presented in Fig. 3.

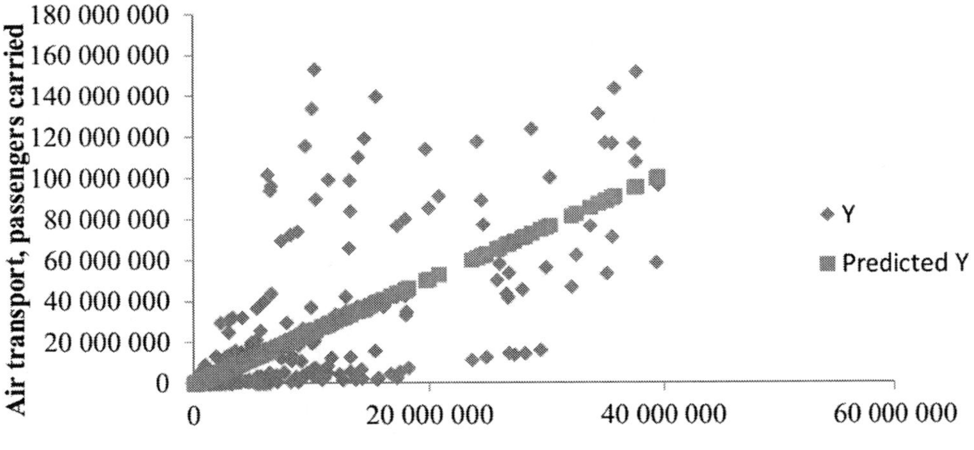

Fig. 3. Line Fit Plot after extraction of outliers.
Source: Own research

The model on data without anomalous observations looks as follows:

$$\widehat{AT} = 2{,}5473 \cdot NA\,. \qquad (4)$$

Characteristics of quality of the model (4) compared to the model (3) are better: the correlation coefficient $R = 0.779$; the determination coefficient $R^2 = 0.606$; $F(1; 389) = 598{,}8$; $t_{a_1}(389) = 24{,}47$. The decrease in parameter a1 to 1 indicates that for two inbound tourists the quantity of air carriages has reduced by 2 people. So, hypothesis 1 can be accepted.

To verify hypothesis 2 the model of the following type has been tested:

$$AT_t = a_1 NA_t + a_2 d_1 + a_3 d_2 + a_4 d_3, \qquad (5)$$

де d_1, d_2, d_3 – dummy-variables, that indicate the group to which a country belongs: with High income, Upper middle income and Lower middle income correspondingly. Adding dummy-variables to the model allows researching the extent to which the corresponding group of the country increases AT level in comparison with Low income countries.

Evaluation of the model parameters by the Method of Ordinary Least Squares has given the following result:

$$AT = 2{,}515NA_t + 1\,771\,330d_1 - 194\,7176d_2 + 1\,956\,443d_3 \qquad (6)$$

The quality of the proved model is acceptable: the coefficient of the multivariable correlation $R = 0.780$; the coefficient of determination $R^2 = 0.609$; $R^2_{adj} = 0.603$; $F(4;386) = 150{,}04$; $t_{a_1}(386) = 18{,}28$; $t_{a_2}(386) = 0{,}765$; $t_{a_3}(386) = -0{,}828$; $t_{a_4}(386) = 0{,}844$. The coefficient of regression near dummy-variables is statistically insignificant. That is why the effect of such a factor as a group of the country with a corresponding income level on the number of passenger air carriages is not significant. So, hypothesis 2 cannot be accepted as intensity of passenger air carriages does not depend on the level of the country economic development.

In 2018 the number of passengers carried constituted 12,529 mln., that is 98,8% more than in 2015. Besides, this index is practically not influenced by inbound tourist flow that after 2014 has become 6 times less [10]. Increase in intensity of outbound tourist flows influences the level of passenger air carriages, but it is not a critically important factor (fig. 4). Moreover, according to the data of Airports Council International only 4-5% of Ukrainians use services of air companies [11].

Fig. 4. Dynamics of passenger air carriages, thousand people.
Source: compiled by the authors based on [11]

The reasons of the market recovery of passenger air carriages are as follows:

– accumulation of postponed demand of the population resulted from the fact that in 2014-2016 many Ukrainians restrained from rest abroad;

– gaining the status of visa-free regime in 2017 between Ukraine and European Union;

– efficient activity of foreign and local air companies at the Ukrainian market of passenger air carriages;

– high performance of low-cost companies at the Ukrainian market of passenger air carriages: Wizz Air (Hungary), Ryanair (Ireland), Vueling Airlines (Spain), Ernest Airlines (Italy), which offer tickets to passengers at quite a low price; it is a motivating development factor, especially in conditions of low solvent demand;

– response to market demands and launching new regular routes by Ukrainian air carriers (17 such routes were launched in 2018), besides, foreign air companies have enlarged significantly the net of their carriages as well;

– stimulating transit carriages by the strategy of forming an international hub on the basis of Boryspil airport.

So, despite quite a dynamic development of air carriages in Ukraine in recent years, the air industry has not reached a desired position yet. It requires a clear vision and determined policy of the country concerning development of aviation and airports. Development of transport air infrastructure requires:

– orientation of the state policy at strategic integrated development of the sphere of air carriages and infrastructure of airports system in Ukraine;

– adjusting Ukrainian laws about air carriages to European standards (including Customs office rules, rules of carrying passengers and luggage);

– renewal of aircraft fleet of Ukrainian air companies, signing Agreement about common air space with EU;

– creating European common air space that can establish mutual security standards and liberalise market relations in the aviation sphere between EU and Ukraine;

– formation of integrated approach to reducing tariffs for air carriages;

– harmonisation of development in the system of regional airports, as by now 61 % [12] of common passenger flows has been concentrated in the international airport «Boryspil»;

– creating a favourable economic environment for development of this trend of the economic activity (tax reductions, credit rates);

– increase of collaboration between tourist companies and air carriers.

Realisation of the proposed events will allow increasing efficiency in the field of passenger air carriages.

5 Conclusions

Using empirical data from the tourism and travel industry as well as air carriages industry in 135 countries over the period of 2015-2017, this study reveals the effect of intensity of inbound and outbound tourism on market development of passenger air carriages. This statement was given as a hypothesis at the beginning of the research and has been verified with the help of the matrix of correlation paired coefficients. They allowed proving a close linear connection between the indicators and applying methods of the correlation-regression analysis. As in the built correlation model there was heteroscedasticity of outliers, it required data exclusion by such countries as the USA, China, Spain, France and Italy. These countries had significant shifts upwards and downwards compared to the regression line that is connected with especially intensive development of industry in passenger air carriages or tourism. The gained result does not contradict to the results of the previous research and proves that air transport and tourism are closely interconnected.

For further research the country groups have been considered by the level of their economic development (classified by the World Bank) with High income, Upper middle income and Lower middle income. The hypothesis about impact of economic development on air carriages has not been proved.

References

1. *IATA Annual Review* [online]. Available at: https://www.iata.org/publications/Pages/annual-review.aspx (2018).
2. S. A. Rehman, K. D. Qianli, W. SongBo, K. Zaman, Y. Zhang. JATM, **58**, 125-134 (2017).
3. X. L. Fernández, P. Coto-Millán, B. Díaz-Medina. The Spanish case. Utilities Policy, **55**, 52-58 (2018).
4. J. D. Dimitriou, C. J. Mourmouris, F. M. Sartzetaki. TRP. **25**, 5242–5254 (2017).
5. Sh. Corbet, J. F.O'Connell, M. Efthymiou, C. Guiomard, B. Lucey: The impact of terrorism on European tourism. ATR. **75**, 1-17 (2019).
6. H. Song, N. Seetaram, S. Ye. Journal of Destination Marketing & Management. **11**, 32-39 (2019).
7. W. Grimme, H. Pabst. TRP. **37**, 346–353 (2019).
8. Y. Zhang, C. Findlay: Air transport policy and its impacts on passenger traffic and tourist flows JAT. **34**, January 2014, 42-48 (2014).
9. *BBC News* [online]. Available at: https://www.bbc.com/russian/features-45206717 (2017).
10. *State Air Service of Ukraine* [online]. Available at: https://avia.gov.ua/pro-nas/statistika/periodychna-informatsiya/ (2019).
11. *Council International* [online]. Available at: https://espreso.tv/article/2018/02/26/kryla_v_modi_chomu_vzletily_aviaperevezennya_v_ukrayini_ta_chogo_chekaty_nadali .
12. *Airport «Boryspil»: Beforehand Year Results* [online]. Available at: https://kbp.aero/news/ .

The Organizational and Economic Policy of the state Administration of Development of Railway Transport of Ukraine

Oleksandra Mandych[1], Tetyana Suhorukova[2], Olena Butenko[3], Ivan Solomnikov[2] and Hanna Ostroverkh[2][*]

[1] Kharkiv Petro Vasylenko National Technical University of Agriculture Educational and Scientific Institute of Business and Management, Department of Economics and Marketing 61002, Kharkiv, Alchevskikh st. 44, Ukraine
[2] Ukrainian State University of Railway Transport, Department Economics and management of industrial and commercial business, 61050, Kharkiv, Feierbakh Square 7, Ukraine
[3] Kharkiv National University of Construction and Architecture, Department Management and Public administration, 61002, Kharkiv, Sumska st. 40, m. Ukraine

Abstract The article deals with the main directions of the organizational and economic policy of the state administration of the development of railway transport and provides suggestions for improving the management mechanism of the railway industry, to ensure the competitiveness of railway transport enterprises. Conceptual aspects of state administration of railway transport development are defined. Components of the organizational and economic policy of management development are provided. The necessity of reforming such components as an innovative renewal of production, overcoming structural deformations in industry, investment in the real sector is substantiated. the general mechanism of organizational and economic support of reform is provided.

1 General problem statement

Now in Ukraine, the governmental policy is aimed at minimizing government regulation where possible, focusing on the basic functions of the state, which includes the development and implementation of clear transport policy, monitoring implementation and evaluation of results. And it became a big problem for accurate functioning of railway transport.

The railway industry lost its competitiveness and is now inferior to road and air transport. State administration should make an effort to protect the interests of the country and society.

The main problems of the railway industry in Ukraine, that require state regulation of development, are as follows:

[1] Corresponding author: annakharkov8@gmail.com

– the presence of territorial and structural imbalances in the development of transport infrastructure;

– insufficient accessibility of transport services for the population, mobility of labor resources;

– insufficient quality of transport services;

– low level of export of transport services, including the use of transit potential;

– insufficient level of transport security;

– strengthening the negative impact of transport on the environment, etc

2 The analysis of recent researches and publications

Development of the theory and methodology of state regulation of transport systems became the subject of many studies of contemporary domestic and foreign scientists in different scientific areas, including: V. L. Dykan, M. V. Korin, V. A. Ovchinnikova, A.V. Tolstova, G. D. Eitutis.

Organization of state administration and regulation in various spheres of society, engaged in such scientists: Yu. S. Barash, V. M. Gumaca, O. A. Degtyar, O. V. Daugaliou, O. M. Ivanitsky, A. M. Novikova, M. L. Pogrebetskiy, A. I. Semenchenko, O. V. Shkidchenko, etc [1-7].

According to their scientific achievements in this area, should be noted that the research related to the functioning and reform of railway transport remains particularly relevant, due to its strategic importance in the development of the state.

3 Priorities of the organizational-economic policies state administration of development

3.1. Structural transformation and formation of the transport services market

Today, the main task for Ukraine is to form a competitive transportation market and change the management principle of Ukrzaliznytsia JSC as a Joint Stock Corporation.

In the European Union, many railway transport enterprises are fighting for the customer with automobile transport. The share of rail transport in the EU is about 30%. But in Ukraine, the share of freight by rail is about 60%.

In 2018, the Ukrainian business transported about 320 million tons of cargo, of which about 107 million tons for export.

From this transport, Ukrzaliznytsia has got UAH 16.6 billion in operating income, due to what has financed capital investment and to cover the losses from passenger transportation.

The World Bank estimates that the restructuring of railway industries takes 5 to 10 years. The Association agreement of Ukraine with the EU provides for 8 years.

Since the entry into force of the Agreement, 4 years have passed, but the main railway legislation has not yet been adopted.

That is why, the restructuring of the railway system should be aimed at the gradual formation and development of a competitive market for transport services. This will be a prerequisite for achieving an optimal ratio of price and quality of cargo and passenger transportation, which will lead to a reduction in transport costs and prices.

Having achieved this goal, Ukraine will be able to implement the requirements of the EU directives regarding the commercial separation of the transport service from the state infrastructure. It should be taken into account that in order to ensure free competition between operators, two important aspects for the railway industry need to be addressed –

different gauge between the Ukrainian and European network (1520 mm and 1435 mm, respectively) and high barriers to entry into the business of rail transport.

In Ukraine, the railway network primarily serves large volumes of demand in the mining and metallurgical industry both in the domestic and foreign markets.

In passenger transportation, railways had to compete with the growing share of air and road transport, but in recent years Ukrzaliznytsia JSC has not tried to improve services for passenger transportation, so the railway industry has lost a large share of passenger transportation.

To renew the environmental, economic and social benefits of rail transport and increase its transit potential, Railways will have to provide a quality level of service.

This can be achieved only through the creation of a new organizational and economic policy of state development management, which would include: the improvement of rolling stock, infrastructure, technology, and management.

3.2 Railway reform should include the following steps:

- making changes and additions necessary for structural changes in the current legislation of Ukraine: "On Railway Transport", "On Natural Monopolies", "On Objects of State Ownership that are not subject to Privatization", as well as the Charter of the Railways of Ukraine:
- development and adoption of the State target program of railway transport reform in Ukraine;
- separation of activities not directly related to road traffic, which will, in particular, reduce the financial burden on the cost of freight and transport element in the cost of production;
- implementation of the financial and economic model that will ensure clear and transparent accounting and distribution of financial flows by type of activity;
- phase-out of cross-subsidization of passenger transport with freight income;
- creation of favorable conditions for attraction of investments necessary for updating and modernization of technical and production base of Railways, if necessary, making appropriate changes and additions to the national legislation and industry regulations;
- improvement of the regulatory framework, which should establish uniform requirements and conditions for access to the infrastructure for any traffic operators with the simultaneous introduction of certification and licensing system to provide quality and safe services.
- improvement of the tariff policy on freight transport in accordance with the WTO requirements with the consolidation of the state regulatory functions on tariffs for socially important passenger transport and the introduction of effective compensation for unprofitable suburban, privileged and other modes of transport.

Today, the railway is in fierce competition with road and air transport. In order to achieve the goal of greater rail traffic, States and the business community must become more active.

First of all, it is necessary to create the correct organizational and economic policy of the state administration of development where the main goal will be the interaction of policy and the railway industry, which will further increase the attractiveness of railway transport in Ukraine. The choice of policy measures should be the subject of a financial reserve for further development [8].

The spheres of state administration in the railway industry should influence:

- infrastructure management;
- coordination between different types of transport infrastructure, including transport exchange points and logistics centers;

- implementation of agreed licensing and certification mechanisms for products and transport services;
- implementation of the agreed tariff and tax policy;
- management of Supervisory activities in the transport sector;
- cooperation with Central and local Executive authorities in the development of the transport sector;
- ensuring the harmonization of social standards of transport services for all population groups and regions of the country;
- implementation of harmonization of environmental standards and ensure their compliance [8].

The main principles of the organizational and economic policy of the state administration of the development of railway transport should be the following:

- development and technical improvement of transport infrastructure;
- providing a legal basis for transport activities and its supervision;
- structural reforms and institutional changes in transport;
- tariff policy for monopoly services;
- technical and environmental safety;
- ensuring state transport security;
- social and labor legislation, which should be clear and unique in the transport sector [8].

For the implementation of these processes of influence, organizational and economic policy, which will be based on the tools, makes it possible to make different decisions depending on specific tasks and situations.

It is important to emphasize that the list of specific instruments of influence is changing due to changes in external and internal factors, the emergence of new instruments, the improvement of the main elements of the regional management system (governance, structures, resources, technologies), which is an objective process of renewal.

4 Organizational and economic mechanism of state administration of railway transport development

The main objectives of the state policy mechanism should include improving the efficiency of rail transport, reduce costs and improve the quality of services.

To achieve these goals, it is necessary to act in two main directions: reforming the existing system of railway tariffs supporting competition in railway transport.

At the initial stage, it is proposed to separate the management of infrastructure and rail transport to create an incentive for new operators owning rolling stock to use the infrastructure of Ukrzaliznytsya JSC. And act as the leading operator of freight transport; while passenger transport is allocated to separate subsidiaries, which may eventually become the property of subjects of the state, local governments or may even be privatized.

In addition, the management mechanism should fully allow for competition between vertically integrated railway companies.

The main components of the mechanism of state administration of railway transport development:

• regardless of how the railway system was structured during the reform period, it is first necessary to establish an organizational and economic policy of state regulation – to ensure the development of competition (which is especially important given the high proportion of Railways compared to other modes of transport), and partly to ensure overall efficiency;

• regulation is needed in areas such as the structure of competition (especially non-discriminatory and competitive access to infrastructure) and freight tariffs in those segments where market mechanisms are in place (direct trade-offs against the nature of competition).

The relevant regulatory framework depends on the chosen structure of the railway industry and, in particular, on the relationship between the infrastructure management organization and the freight carriers.

Until now, the issue of access to infrastructure is regulated by the Cabinet of Ministers of Ukraine (wherever it is), and tariffs for freight transportation are regulated by the Ministry of the infrastructure of Ukraine. However, this regulatory framework may change with the new policy structure.

5 Conclusion

The ultimate goal of the organizational and economic state policy for the development of railway transport in Ukraine should be the formation of a privatized sector of rail transport, works on a commercial basis and is composed of many separate and competing enterprises that must use the state support infrastructure.

At the first phase e of the reform, which has already formally ended with the creation of Ukrzaliznytsya JSC, provided for the separation of functions of state regulation and economic activity through the creation of a quasi-independent operator company, which has in its composition separate interrelated activities.

In the second phase of the reform, these activities will be separated legally and institutionally, placed in an open competitive environment and, where possible, transferred to private sector.

In the third phase, incentives will be created to promote competition and, if possible, the privatization of all types of transportation.

References

1 V. L. Dykanj M. V. Korin Bulletin of the Economy of Transport and Industry, **61.** 9-19, (2018).
2 V.L. Dykanj, O.Gh. Kirdina Bulletin of the Economy of Transport and Industry, **28**, 13-20, (2009).
3 V.L. Dykan, M.V. Kondratyuk Bulletin of the Economy of Transport and Industry**, 26**, 13-18, (2009).
4 V.L. Dykan, G.E. Ostroverkh. Bulletin of the Economy of Transport and Industry, **62**, 11–19, (2018).
5 V.O. Ovchynnikova, *Stratehichne upravlinnya rozvytkom zaliznychnoho transportu Ukrayiny*, Kharkiv, 427 (2017)
6 A.V. Tolstova Bulletin of the Economy of Transport and Industry, **40**, 303-306, (2012).
7 National Transport Strategy of Ukraine 2030 URL: https://mtu.gov.ua/files/for_investors/230118/National%20Transport%20Strategy%20of%20Ukraine.pdf] .
8 V. Yanovska, Socio-economic effects of high-speed highway project "China-Ukraine-EU", Materials of the Round table of Association High-Speed Highways, 12 (2018).
9 V.N. Goncharov, E.V. Ivanova, Concept of strategic organization of innovative activity of enterprises, URL: www.maop.vorstu.ru/Gancharov.html. Accessed 07 May 2019.

10 Intelligent Information Systems in Management, https://rep.bntu.by/bitstream/handle/data/23986/%D0%A2..B8.pdf?sequence=11. (2019).
11 E.O. Naumenko On the model of management of the innovative process of the enterprise in modern conditions, URL: http://ej.kubagro.ru/2006/04/pdf/03.pdf. Accessed 03 May 2019 .
12 V.O. Zadoya, Problems of transport Economics, **9**, 71-76 (2015).

Prevention of Economic Losses with the help of the System of Control of Saving and Storing Bulk Cargoes in the Process of Train Movement

Vitalyi Sakhatsky[1], *Nina* Lyubymova[2], *Vladimir* Pusik[2], *Luydmila* Pusik[2], and *Ievhenii Chepusenko*[1]

[1]Kharkiv National Automobile and Highway University, Yaroslava Mudrogo str., 25, Kharkiv, Ukraine, 61002
[2] Kharkiv Petro Vasylenko National Technical University of Agriculture, Alchevskykh str., 44, Kharkiv, Ukraine, 61000

Abstract. In order to ensure the demand of the market for competitive rail transport services, the structure of automatic safety control for bulk cargo is proposed on the basis of the radar sensor and Wi-Fi technology, which allows to avoid the false alert signal caused by changes in the level of liquid due to various deviations of the railway track from the horizontal plane and to fix leakage of the liquid from damaged tank. The issues of increasing the reliability of the proposed control system are considered.

The economic effect of the railway transport system in many respects is determined by higher speed and volume of traffic, traffic safety during the delivery of goods to the consumer [1, 2, 3]. In [4] it is noted that with a modern traffic safety management system, a further increase in traffic volume by 20-30% can lead to an increase in traffic safety violations by 40-80%. The amount of losses from traffic accidents in Ukraine in 1996 amounted to about UAH 40,000 thousand. In 2006 it was about UAH 22,000 thousand, and in 2014 – UAH 7,525 thousand [2]. According to [5], the main causes of emergency situations include increased traffic flow density, failures in the operation of technical means, and untimely prevention of causes of defects. To reduce losses, economic methods of traffic safety management should be used which involve technical audit, a comprehensive assessment of railway transport, non-destructive monitoring and diagnostics [6]. Improving the safety and reliability of rail transport provides a reduction in the level of real transport costs in the price of products by 30% and an increase in the productivity of rail transport by 2-4 times [7].

A significant part of freight traffic falls on bulk cargo that is transported in tanks, in bunker open car and special containers. Bulk cargoes are divided into hazardous and non-hazardous. Dangerous goods are transported in special packages in covered cars and containers, liquefied gases and chemical goods are transported in specialized tanks. Any malfunction of the tank can lead to leakage of bulk cargoes in transit, which causes a threat to traffic safety and economic damage to both the railway and the owners of the cargo. Therefore, when damage to the tank is detected, it is detached from the train and transported to a safe place. The safety of the bulk cargo is usually checked during its loading and unloading [8].

In [9], it was stated that "one of the strategic directions for the development of rail transport is the creation of a technical diagnostics system with increased reliability of defect detection and predictive diagnostic systems based on fundamentally new ways to detect defects when a train is moving". In industry, many methods for controlling the level of fluid are used: buoyancy, visual, hydrostatic, float, electric, radio-wave, magnetostrictive, ultrasonic and others [10]. In railway transport, when monitoring the level of oil products in tanks, the measuring rod is widely used as a measuring tool. It is used, for example, to measure the level of fuel oil and the level of commercial water [11], as well as other bulk cargoes. However, this method of monitoring the status of bulk cargo is unacceptable when the train moves.

In [12] an ultrasonic level gauge is presented, which is used to measure the level of liquid substances in railway tanks. It is intended for solving problems of automation of loss control of bulk cargoes on railways, control of loading and unloading. When these sensors are used, information about the damage to railway systems or the leakage of liquid cargo from tanks can be obtained early and thereby train traffic safety is increased.

To monitor and record various parameters of liquids through the wall of tanks and closed tanks, Ultrasic Advanced Technologies LLC has developed ultrasonic level analyzing gauges for liquids: ANALIQ-M and ANALIQ-S. The devices can measure the level, mass, volume, density of various liquids (oil products, liquefied gases, oils, acids, alkalis, alcohols, poisonous liquids, etc.) that are in tanks, barrels, cylinders and other closed tanks. Level measurement range from 3cm to 50m with an error of ± 1mm. Video presentation of the operation of ANALIQ-M is presented in [13].

A promising method for measuring the level of a liquid is the radio wave method, the kind of which is the radar method [14, 15]. It is based on the measured time of passing the radio signal of a double distance from the emitter to the surface of the liquid.

For example, the Rosemount Radar Sensor Series 5600 allows measuring the level of such media as petroleum products, suspensions, solvents, aqueous solutions, cement, and fine particulate powder materials [16].

A principle of the method for determining the level of liquid of all radar gauges is that the surface of the reflection is taken over the surface of the liquid and the level of the liquid is identified by determining the time of propagation of the signal from the antenna to the surface of the liquid. This is quite true for dielectric environments and needs to be clarified for conductive media. Scientific works devoted to the question of reflecting radio waves from conducting media showed that for such media there is an effective reflection surface that does not coincide with the actual surface of the liquid [17]. Therefore, the result of measuring such a liquid with radar will have a methodological error, which has not been studied to date.

Therefore, the creation and improvement of reliability of radar methods and automatic safety control systems for a wide class of bulk cargoes is an urgent task.

The purpose of the work is to reduce economic losses by improving the safety of transportation of bulk cargoes using an automatic system for monitoring their safety.

We define the requirements that apply to the developed control system.

Currently, the railway transport uses a variety of security systems. They are based on the known technologies such as Ethernet, CAN, RS-485. Systems using radio signals, such as GPS/GLONASS, GSM-R, Wi-Fi and others, are being developed and implemented. The modern infrastructure of the train includes various systems: the CLUB-U system, which ensures the safety of the train on the hauls by monitoring the exceeding of the permissible speed and braking, recording the motion parameters, the diagnostic system, the wheel pair monitoring system, the locomotive telemetry system.

When monitoring the level of liquid goods when the train moves, information should be transmitted from the level sensors to the liquid level indication system in real time. In this case, if the tank or other container is damaged and the fluid leaks, the control and measurement system will timely generate an alarm signal. If the railway track has angles of deviation from the horizontal plane, the level of the liquid in the tank at the installation site of the level sensor will also change, which can cause a false alarm. To prevent such an effect, additional information is required that the change in the level of the liquid is due to a spatial change in the trajectory of the railway, and not a leak. Therefore, one of the controlled parameters should be the angle of inclination of the railway track relative to the horizontal plane.

Let us consider the structure and principle of operation of the automatic system for monitoring the safety of liquid cargo.

To determine the spatial position of various objects, use special inertial systems. Such systems may include an accelerometer, gyroscope and magnetometer. The accelerometer is used to measure linear accelerations, the gyroscope is for angular, the magnetometer measures the Earth's magnetic field and is used to determine the angle in azimuth. By combining these three sensors in one device, we get a complex system for determining the spatial position of the object.

Installing such a device directly on the locomotive will help determine the angles of inclination of the railroad bed relative to the horizontal plane during train movement and eliminate the false alarm signal.

One such device could be the BMX 055 sensor from Bosch Sensor tec. The BMX 055 sensor contains three high-precision and diverse three-axis sensors: a 12-bit low-noise accelerometer, a 16-bit gyroscope with programmable measurement ranges, and a low-noise wide-range geomagnetic sensor optimized to achieve a better performance/power ratio. Incorporating all of these devices, the BMX055 Absolute Orientation Sensor provides detection of acceleration, angular velocity, and geomagnetic measurements.

When choosing the means and method of transferring measurement information to the operator, the following was taken into account. To reduce the effects of electromagnetic interference in the transmission of information the most suitable is the range of the microwave.

Wi-Fi technology works at a frequency of 2400–2483.5 MHz, according to the IEEE 802.11 standard. One of the Wi-Fi modules that operates at 2.4 GHz is ESP-07S. It includes a microcontroller and a radio transmitter at the same time, which is a big advantage.

Wi-Fi ESP-07S module supports the standard IEEE802.11b/g/n protocol with WEP / WPA-PSK/WPA2-PSK security modes, supports network and packet protocols: IPv4, TCP / UDP / HTTP / FTP. It can function in three modes: client, access point, client + access point. Interface support: UART, GPIO, I2C, I2. SDIO, PWM, ADC, SPI. The transmitting device acts as a client and consists of a Wi-Fi module ESP-07S and a position sensor BMX055. The BMX055 sensor connects to the ESP-07S and communicates with it via the I2C protocol.

The ESP-07S module reads data from the BMX055 sensor registers for each of the axes of a 3-axis accelerometer, 3-axis gyroscope and a 3-axis magnetometer, processes the embedded microprocessor and transmits the processed information through a wire transmission line to the receiver.

The Wi-Fi module ESP-07S, which is configured as an access point, also acts as a receiver. This module is located in the locomotive driver's cab. The received information signal from the transmitter is displayed to the operator.

The block diagram of the measuring system for determining the parameters of the spatial position of the railway track is shown in Fig. 1.

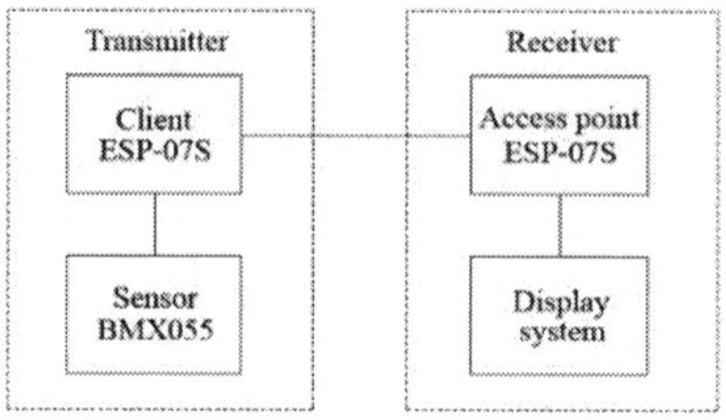

Fig. 1. Block diagram of the measuring system for determining the spatial position of the railway track.

The occurrence of leakage of fluid from the tank (tank) can be controlled using an additional data transmission system from a radar level gauge. Such a system, for example, can be implemented on the basis of devices developed by EMERSON, which is produced by the Industrial Group METRAN (Chelyabinsk, Russia) [18]. A Rosemount 5400 Series Radar Transmitter with a HART / 4-20mA output signal can be used as a sensor. The model of the radar of this series 5401 operates at a frequency of 6 GHz, and the model 5402 – at 26 GHz. A distinctive feature of the radar is the use of circular polarization of the wave, which allows increasing the measurement reliability by reducing by 50% the number of echo signals from the internal structures of the reservoir.

To transmit information over a wireless protocol, Wireless HART, you must connect a Rosemount 775 converter to the sensor, which is designed to transmit data from any wired sensor that has a HART / 4-20 mA output signal. The Rosemount 775 converter transmits data to the Rosemount 1420 wireless gateway, which allows the integration of a measuring device (gauge) into the Smart Wireless wireless network, as well as remote configuration of devices.

The 1420 Wireless Gateway connects Smart Wireless networks to any higher level. Data transmission is carried out via Modbus protocol via RS 485 interface or local Ethernet network, which are available on modern trains. Reception and transmission of data is carried out at the operating frequency of 2.4-2.4835 GHz via Wireless HART protocol. The Rosemount 1420 Gateway is the entry point for data transmission to the operator and can support up to 100 level gauges simultaneously.

During transportation of liquid cargo, the accuracy of the information transmitted from the liquid level sensors is of paramount importance.

Let us estimate the methodological error that occurs when measuring the level of conductive liquids. We choose three well-known fluids - hydrogen iodide (HI), for which σ = 10630 cm/m, ε_r= 2.9; nitric acid (HNO_3), for which σ = 1.5 cm/m, ε_r= 1.77 and sodium chloride (NaCl), for which σ = 3.4 cm/m, ε_r= 6.3 [19].

According to the method of equivalent mirror reflection for media with finite conductivity, the time taken to receive a reflected pulse increases (is delayed) for the time they pass the distance [17]:

$$d_E = \frac{-j\lambda}{\pi\sqrt{(\varepsilon_r - 1) - j60\sigma\lambda}},$$

which causes a methodological error when measuring the level of conductive fluid.

In [20] it was shown that the group velocity of propagation of a radio pulse in a medium at frequency can be represented through the derivative of the complex wave number $\gamma^I(\omega)$ as follows

$$V_{гp} = 1/\gamma^I(\omega),$$

$$\gamma' = \frac{1}{300}\left[\frac{1(2\omega\varepsilon_r - j36000\pi\sigma}{2\sqrt{\varepsilon_r\omega^2 - j36000\pi\omega}}\right]$$

In this case, ω is the maximum frequency of the spectrum of the radio pulse [21]. Then the delay time of the reflected liquid from the radiation is equal:

$$\Delta t = \frac{d_E}{V_{zp}},$$

When finding the distance to the surface of the liquid, the radar computational system does not take into account the full path of radio signal passage from its moment of radiation to the moment of its reception, but only half of this path, then the unregarded thickness of the liquid layer (methodical error of measurement), which is located above the equivalent surface of reflection, is determined by the following way:

$$h = Re\Delta t \cdot c / 2,$$

where c is the speed of light.

Since the level levels of the liquid level in most of them operate in the frequency range from 5 GHz to 25 GHz, the delay time was determined for the frequencies of this range with a pulse duration $\tau_i = 10$ μs.

The results of the calculation showed that for the selected fluids, the methodical error does not exceed 7.1 mm, which is higher than the accuracy of high-precision radar for measuring the level of dielectric liquids, which is equal to ± 1 mm. This error is systematic and should therefore be taken into account as an integral part of other errors in the level meter, which will provide greater reliability of measurement.

Findings

According to the results of the research the following conclusions can be drawn. A structure for designing a control and measuring system for the preservation of bulk cargoes is proposed. It consists of two subsystems: the subsystem for measuring the level of liquid in the tank and the subsystem for monitoring a false alarm signal that occurs when the railroad tracks deviate from the horizontal axis.

The choice of the element base of the system on the basis of devices that are produced in industry has been made.

This allows reducing the cost of production of the proposed control and measuring system by eliminating the need to develop special components and elements of the system.

The features of the interaction of radiation from radar sensors with conducting fluids are investigated. It is shown that the presence of fluid conductivity causes the appearance of a methodological error in measuring its level, which may exceed the known errors of radar sensors.

Is this error is taken into account the reliability of the control and measuring system as a whole increases.

References

1. Transportna strategiya Ukrayini na period do 2020 roku URL: http://zakon4.rada.gov.ua/laws/show/2174-2010-p.

2. A.I. Ogar, Yu.O. Pazoyskiy, A.V. Rozsoha, A.A. Sidrakov, Yu.V. Smagilo ZbIrnik naukovih prats UkrDUZT, **156**, 18-28 (2015).
3. Rasporyazhenie OAO RZhD # 987r ot 29.05.07 «Funktsionalnaya strategiya obespecheniya garantirovannoy bezopasnosti i nadezhnosti perevozochnogo protsessa OAO RZhD.-115c.
4. O.S. Burch. Vestnik nauki Sibiri. **1 (7)**, 139-146, (2013)
5. V.A. Olentsevich, V.E. Gozbenko. Sovremennyie tehnologii. Sistemnyiy analiz. Modelirovanie. **1 (37)**, 87-90, (2013)
6. V.I. Yakunina. Zheleznodorozhnyiy transport, **1**, 2-11, (2009)
7. A.N. Golovash Problemyi sovremennoy ekonomiki, **14**, 378-382, (2010)
8. V.P. Perepon Organizatsiya perevozok gruzov: Uchebnik dlya tehnikumov i kollezhdey zheleznodorozhnogo transporta, M.: Marshrut, 614, (2003)
9. Strategiya razvitiya zheleznodorozhnogo transporta v RF do 2030 goda Utverzhdena Rasporyazheniem Pravitelstva RF ot 17 iyunya 2008 g. # 877- [Elektronnyiy resurs]. – Rezhim dostupa: http://www.mintrans.ru/documents/detail.phpELEMENT_ID=13009
10. V. E. Nazarov *Analiz suchasnogo stanu metodiv i zasobIv vimIryuvalnogo kontrolyu rivnya ridinnih seredovisch :navch. posIb. severodonetsk* : NU Im. V. Dalya, 114, (2016)
11. Gost R8.787-2012 Gosudarstvennaya sistema obespecheniya edinstva izmereniy (GSI). Massa mazuta. Metodika izmereniy v zheleznodorozhnyih tsisternah. Obschie metrologicheskie trebovaniya.
12. I.P. Merkulova , P.A. Shirokolobova, V.A. Olentsevich. Molodaya nauka Sibiri: elektron. nauchn. zhurn. **2**, (2018), URL: http: // mnv.irgups.ru /toma/22-2018,
13. Video prezentatsiya rabotyi ANALIQ-M URL: http://www.yotube.com/watch.v.hLcbLzntyLk
14. V. K. Zhdankin *Priboryi dlya izmereniya urovnya: ucheb. posob.* M.: STA-PRESS, 19, (2002).
15. N. G. Farzane *Tehnologicheskie izmereniya i priboryi :navch. posIb.* M. : Vyissh. shkola, 456, (1989)
16. Kompaniya Emerson Process Management i Promyishlennoy gruppyi «Metran». Urovnemeryi. Moskva, **16**, 206, (2011), URL: :https://www.emerson.com/documents/automation/уровнемеры-каталог-ru61704.pdf
17. P. R. Banister. RadioScience, **17.5**,1095-1102, (1982)
18. Urovnemeryi- Vyipusk 2018 – Emerson URL:https://www.emerson.com/documents/automation/catalog-уровнемеры-каталог-metran-ru
19. B. P. Nikolskiy *Osnovnyie svoystva neorganicheskih i organicheskih soedineniy. Spravochnik himika.* M.:Himiya , 2, 1169. (1971)
20. E.S Birger., L. A. Vaynshteyn. Tehnicheskaya fizika. **11,** 2217- 2228, (1973)
21. G.I.Tyorina. Radiotehnika i elektronika. **9**, 611- 613, (1971)

Assessment of strategic development potential of the production units of enterprises of railway transport

Irina Tokmakova [1*], *Natalia* Panchenko [2], *Hanna* Obruch[1], *Anastasia* Kozlova [3], Hanna *Zaiets* [1]

[1] Ukrainian State University of Railway Transport, Department of Economics and Management of Industrial and Commercial Business, 61050 Feerbach Square 7, Kharkov, Ukraine
[2] Ukrainian State University of Railway Transport, Department of Higher Mathematics, 61050 Feerbach Square 7, Kharkov, Ukraine
[3] O.M. Beketov National University of Urban Economy in Kharkiv, Department of Tourism and Hotel Management, Kharkov, Ukraine

Abstract. The article reveals the theoretical and methodological basis of the resource approach to the development of the strategic potential of railway enterprises, which is recognized as progressive in matters of the formation of competitive advantages at the present stage of management. The component structure of the strategic potential of enterprises has been clarified and the reasonability of diagnosing the strategic potential of railway enterprises within the framework of a competitive approach in such areas as unique resources and basic competences, operational competencies, adaptive competencies, "proactive competencies", and the ability to create common values has been explained. A method of assessing the level of development of the strategic potential of the production units of the branches of Ukrzaliznytsya JSC has been developed.

1 Introduction

The processes of globalization, European integration and liberalization form new conditions for managing the Ukrainian railway industry, in which survival and further effective development, urgent improvement of the main parameters of its activities and positions before competitors should be ensured. However, the strategic thinking of the management of JSC "Ukrzaliznytsia" is still limited to the assets that the company owns today, so its innovation and investment activities are aimed at solving many of the current problems that have accumulated over the past three decades. At the same time, the features of post-industrial society form new approaches to ensuring competitiveness, based on knowledge and competences. The speed at which railway enterprises are able to create the necessary assets and develop them, demonstrating a high level of adaptability to a changing external environment is really important. This nature of development highlights the strategic potential of the production units of the branches of Ukrzaliznytsya JSC as the basis for the

* Corresponding author: it.tokm@ukr.net

formation of long-term competitiveness of railway transport.

2 Analysis of literature data and problem statement.

The methodological basis for assessing the strategic potential of business entities is formed in the works by J. Barney, B. Wernerfelt, G. Hamel and K. Prahalad, D. Tisa, G. Pisano and E. Shuen, R. Nelson, S. Winter [1- 5] and others. Researchers point out that the assessment of the strategic potential of enterprises is an extremely difficult problem, especially in terms of choosing the methodological basis for the analysis, since the theory of strategic potential is a new direction that has not yet fully formed. Its foundations are laid in the resource concept of strategic management, according to which the source of long-term competitive advantages is the resource base of the enterprise, as well as the process of effective management, creation and combinations of resources. According to the generally accepted logic of the resource approach, enterprises form competences, influence resources and transform them into a new state. The unique resources and organizational abilities (competencies and/or dynamic abilities) form the strategic potential, which is the source of sustainable competitive advantages realized in the form of economic rents that are inaccessible to rivals.

Although in the field of railway transport, the assessment and formation of the strategic potential is not given due attention, certain aspects of the problem are revealed in the works by V.L. Dikan [6], in particular, in the development of such researchers as M.V. Koren [7], V. Hoop [8], V. Ovchinnikov [9], I. Tokmakova [10] and others. However, the existing differences in approaches to assessing the strategic potential of enterprises and absence of a scientifically recognized solution in this area in rail transport require further study of this problem.

3 Substantiation of the approach to determining the components of the assessment of the strategic potential of the production units of railway enterprises

The management of the strategic potential of the production units of railway enterprises, and in particular the assessment, requires the determination of its component structure. The generalization of the views of scientists on this issue [11-23] indicates the predominant use of two main approaches to the formation of the strategic potential architecture, namely, the functional and the competitive ones.

The functional structuring of the traditional potential is based on the allocated functional subsystems of the enterprise and their components. With this approach, the component model of the aggregate strategic potential of enterprises is determined by the structure of the production system and local potentials of various functional orientations. The differences in this arrangement are due to different views on grouping of areas of activity, factors and resources. A significant disadvantage of this direction in research is the identification of the components of the economic and strategic potential.

The generalized functional structure of the strategic potential can be represented by three levels of formation of its local potentials: production and technological; organizational and managerial; financial and economic. In general, the functional approach is focused on determining the rational structure of the strategic potential of the enterprise and evaluates its level on the basis of progressive regulatory ratios set by the most advanced technologies of individual enterprise subsystems. Its disadvantages are due to focusing on the limiting possibilities of resources, the minimum necessary for the existence of an enterprise on the market and its ability to compete in the external environment.

Management of the strategic potential in this situation for enterprises will be reduced to supporting this minimum level and tracking the dynamics of the internal and external environment, which may affect the required minimum level.

At the same time, the management of strategic opportunities that enable the enterprise to become a leader determines the competitiveness of an enterprise. The source of competition is the combination of the necessary resources and competencies, which will ensure the implementation of the strategy and a reasonable choice of strategic alternatives in conditions of uncertainty, while maintaining the intra-balanced sustainable development of the enterprise. Thus, a competitive approach reveals the structure of strategic potential based on the identification of sources of ensuring competitive stability. The views on the structure of the strategic potential within the framework of this approach are systematized, supported by the scientific community, presented in the work of V.V. Vasilyeva [21], which defines such sources of competitiveness of strategic potential as resources and key competencies, where resources are tangible and intangible assets of an organization, and competences are a superstructure over resources, represent activities and processes by which an organization uses these resources. An example of the concretization of the above provisions is given by I.P. Otenko and A.S. Danevich [22] who define such components of the strategic potential of an industrial enterprise as: managerial competence, flexibility of the organizational structure, strategic position of the enterprise and its ability to ensure production and technological efficiency. In our opinion, A.D. Antonov [23] gives the most reasonable provisions containing a structured detailed description of key sources of enterprise competitiveness in the conditions of a post-industrial economy, which indicates the need to diagnose such elements of strategic potential: unique resources and basic competences; operational competences; adaptive competencies; "proactive competences"; ability to create shared values. The relevance of this list of components of competitiveness conditioned by the fact that it contains the factors that have a transformative effect on the fundamental characteristics of the behavior of enterprises in the post-industrial conditions of the development of society.

4 Development of a methodology for assessing the strategic potential for the development of production units of railway enterprises

The presented component structure of the strategic potential of the production units of railway transport enterprises and the advantages of using the expert method determines the approach and assessment methodology (Fig. 1).

The main assessment steps are as follows:

- formation of a system of particular indicators of strategic potential, taking into account the specifics of the production units of the branches of Ukrzaliznytsya JSC;
- determination of reference values of indicators by their rationing;
- standardization of indicators;
- determination of coefficients of significance of indicators; calculation of the integral indicator;
- interpretation of the results.

Taking into account the proposals on the components of the strategic potential of enterprises, the main indicating factor characterizing their condition at the level of production units of railway transport enterprises include:

1. Unique resources and core competencies:

1.1 The proportion of the active part of fixed assets, taking into account the age structure;
1.2 The proportion of qualified personnel;

1.3 Coefficient of validity of the active part of fixed assets;
1.4 Utilization of production capacity;

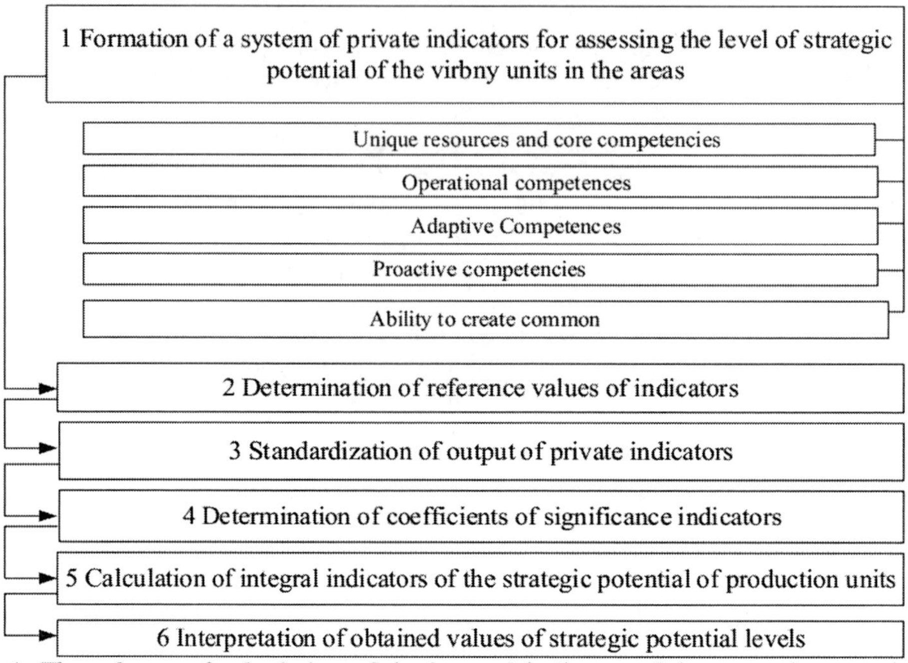

Fig.1. The scheme of calculation of the integral indicator of the strategic potential of the production units of enterprises of railway transport.

2. Operational competencies:
2.1 Indicators of resource efficiency;
2.2 Indicators of quality and reliability;
3. Adaptive competencies:
3.1 Share of investments in innovations in the total amount of expenses;
3.2 Proportion of universal fixed assets;
3.3 Level of progressiveness of production and management technologies;
4. “Proactive competencies”:
4.1 Proportion of workers who have improved their skills;
4.2 Proportion of proactive management personnel;
4.3 Coefficient of renewal of fixed assets;
4.4 Proportion of workers rationalizers;
5. Ability to create shared values:
5.1 Proportion of investments in the improvement of personnel and their living conditions;
5.2 Share of expenditures for charity,
5.3 Proportion of positive communication channels.

In order to carry out standardization of indicators of strategic potential, it is important to determine the ideal value, perhaps based on the rationing of indicators by applying expert assessments.

To assess the level of strategic potential, its characteristics such as critical, low, satisfactory, medium and high are offered. Although the assessment range is from 0 to 1, such intervals for assessing the level of development of strategic potential can be determined: critical level is from 0 to 0.2; low - from 0.21 to 0.4; satisfactory – from 0.41 to 0.6; medium - from 0.61 to 0.8; high - from 0.81 to 1.

Evaluation of the strategic potential of an enterprise with the help of the proposed methodology allows finding weak elements in its structure and making an appropriate strategic decision aimed at ensuring the competitive stability of railway transport.

6 Conclusion.

For the effective implementation of strategic decisions to ensure the sustainability of the functioning and development of the production units of railway transport in the long term, in unstable environmental conditions, it is necessary to identify areas and corresponding measures to increase the level of strategic potential. After all, the availability and quality of resources, competencies and abilities substantially determines and individualizes the configuration of possible vectors of the strategic development of enterprises.

The key stage of management of the strategic potential of the production units of the branches of Ukrzaliznytsia JSC determined the assessment of its condition, requires the identification of components and the development of methodological bases for their assessment. In the process of disclosing the main approaches to the structuring of strategic potential, the priority is given to the study of the state of sources of competitive advantage, taking into account the features of the external conditions of their functioning. The proposed system of indicators of strategic potential characterizes the ability of domestic rail transport to ensure its competitiveness in the long term. It has been substantiated that the obtained results of the assessment of the level of strategic potential allow us to identify promising reserves and develop effective measures to ensure the development of the production units of the branches of Ukrzaliznytsia JSC in accordance with the railway industry development strategy.

References

1. J. Barney. Journal of Management, **17 (1)**, 99-120 (2009)
2. B. Wernerfelt. Strategical Management, **5.2**, 171-180 (1984)
3. C. K. Prahalad, G.Hamel. Harvard Business Review, **68.3**, 79-91 (1990)
4. R. R.Nelson, I.S.J. Winter, *Evolutionary Theory of Economic Change.* Per. from English M., Delo, 536 (2002)
5. D.J. Teece, G.Pisano, A.Shuen. Strategic Management Journal, **18**, 509-533 (1997).
6. V. L. Dikan, M. V. Koren. Collection of scientific papers of the Ukrainian State University of Railway Transport, **154**, 98-103 (2015)
7. M. V. Root. *Ensuring innovative development of railway transport based on the formation of an industrial and logistic system.* Kharkiv, 21 (2013)
8. G. V. Hoop *Ensuring the development of the potential of competitiveness of car-building enterprises of Ukraine.* Kharkiv, 265 (2017)
9. V. A. Ovchinnikova. *Strategic management of the development of railway transport in Ukraine: monograph.* M.: USURT, 427 (2017)
10. I. V. Tokmakova *Ensuring the harmonious development of railway transport in Ukraine: monograph.* M.: USURT, 403 (2015)
11. V.I. Bitsa. Vestnik Khmelnytsky National University, **6.2,** 203 - 206 (2011)
12. T. N. Kibuk. *Strategic potential of an enterprise (on the example of enterprises in the metallurgical industry of Ukraine).* Kyiv, 20 (2010)
13. V.D. Markova. *Strategic management: a course of lectures.* M.: INFRA-M; Novosibirsk: Siberian Agreement, 288 (1999)
14. G. B. Kleiner, V. L. Tambovtsev, R. M. Kachalov. *Enterprises in an unstable economic environment: risks, strategy, safety.* Moscow: Economy, 425 (1997)
15. V.N. Gavva, *Estimation of Enterprise and Industry Potential: A Monograph.* Kharkov.

Aviation. Institute", 287 (2004)
16. O. A. Sushchenko. *Formation of the strategic potential of the enterprise.* Lugansk, 20 (2002)
17. S. B. Alekseev. European vector of economic development, **2 (15),** 17-25 (2013)
18. M.I. Kovach. Bulletin of NLTU of Ukraine, **23. (10)**, 207-212 (2013)
19. Yu. V. Samoilik. Technological audit and production reserves, **2 (2),** 32-35 (2011)
20. Ye.V. Lagunova. Problems of Management, **6,** 40-44 (2007)
21. V. V. Vasilyeva *Management of strategic potential of the organization.* Moscow, 26 (2010)
22. I. P.Otenko, A. S. Danevich. Business Inform, **9,** 302-307 (2014)
23. A. D. Antonov. Economy and Entrepreneurship, 3 **(26),** 139-148 (2012)

Stages of Formation of Railway Economic Safety Schemes

Tetyana Tymofyeyeva[1]1, *Natalia* Gritsenko[1], *Yujia* Myroshychenko[1], *Iryna* Motsna[1]

[1]Ukrainian State University of Railway Transport, Economy and Management of Industrial and Commerce Business Department, Feuerbach sq. 7, 61050 Kharkiv, Ukraine

Abstract. It is found that the railway economic safety system shall be formed taking into consideration current social changes in the society. There is the comprehensive approach to the formation of railway economic safety schemes. It is found that railway economic safety enforcement shall be the continuous cyclic process to be regularly improved. There is the scheme of execution sequence of railway economic safety management. It is found that while determining railway economic safety, in whole or in part, of each element, quantitative and qualitative evaluation methods shall be used and combined. There is the calculation of railway economic safety based on the determination of aggregate criterion. It is found that the selection of figures upon certain functional elements shall be made based on any threats.

1 Introduction

Under current conditions the matter of railway economic safety enforcement takes the first place. The preliminary conditions of the occurrence of some problems in railway industry became various transformations in national economy that caused transformations in railway business and railway transport, generally. Up to this point, sectors of national economy had stability and forecasting of phenomena of external and internal nature over an extended time frame. Sectors of national economy were protected by state against influence of all negative factors. For railway sector, this fact was the main condition of performance efficiency – supply of the sufficient volume of works as those types of transport to be used in sectors of national economy for transportation of own products were defined irrespective of business demands. Herewith, railway transport was in priority.

Transformations in economy of Ukraine cause both external and internal factors having influence with railway performance both in negative and positive terms.

Under such conditions there is the necessity of timely forecasting and response to any changes both in external and internal environment. The objective necessity of improving the existing system and developing any scheme for its enforcement is the basis of railway economic safety enforcement.

Study of the problem of railway economic safety has been made by such scientists and economists as T. Tymofyeyeva, N. Gritsenko, Y. Myroshychenko [1], T.H.Vasyltsev [2], S.V.Dovbnia [3], S.V.Kavun [4], O.A.Kirichenko [5], V.M.Haiets [6], M.O.Kyzym [6], T.S.Klebanova [6], O.I.Cherniak [6], O.I.Sudakova [7], S.M.Shkarlet [8] and et.al. In their

[1] Corresponding author: timofeeva_ta@ukr.net

work papers the authors describe the problem of corporate economic safety management both in general and in certain elements. The sense of economic safety, its elements, necessity of comprehensive research of this phenomenon and evaluation approaches are of greatest focus.

Without underreport of strong contribution of scientists named, it should be noted that this problem is understudied, thus, the focus is on the study of theoretical principles of the problem that refers to the detection, evaluation and forecasting of railway status development by means of the most essential figures and indicators of its functioning in terms of economic safety enforcement affected by various negative events of external and internal nature.

2 Description of basis

The important figure of railway economic safety is the economic safety level.

The railway economic safety level is the evaluation of resource usage status upon criteria of the railway economic safety level. In order to achieve the highest level, the railway service shall ensure maximum safety of basic functional elements of work.

Thus, the main goal of railway economic safety enforcement shall be the prevention of threats, provision of industrial activity protection and achievement of goals through solving this task: detection of measures and ways to prevent threats, to facilitate or eliminate any sequences of actions.

Any total figures of railway economic safety evaluation shall not be considered as comprehensive and complete. The definition of economic safety is constantly increasing in terms of deeping the knowledge of economic laws and social demands in a certain stage of social development, thus, the system of railway economic safety evaluation is to be improved on the basis of economic science improvement.

Every form of economic threats concerning railway industry has various modifications. Thus, the selection of optimal problem-solving method shall have forecasting nature. And, therefore, it should be concluded that railway economic safety enforcement is the constant cyclic process to be regularly improved. This process is given in the figure 1.

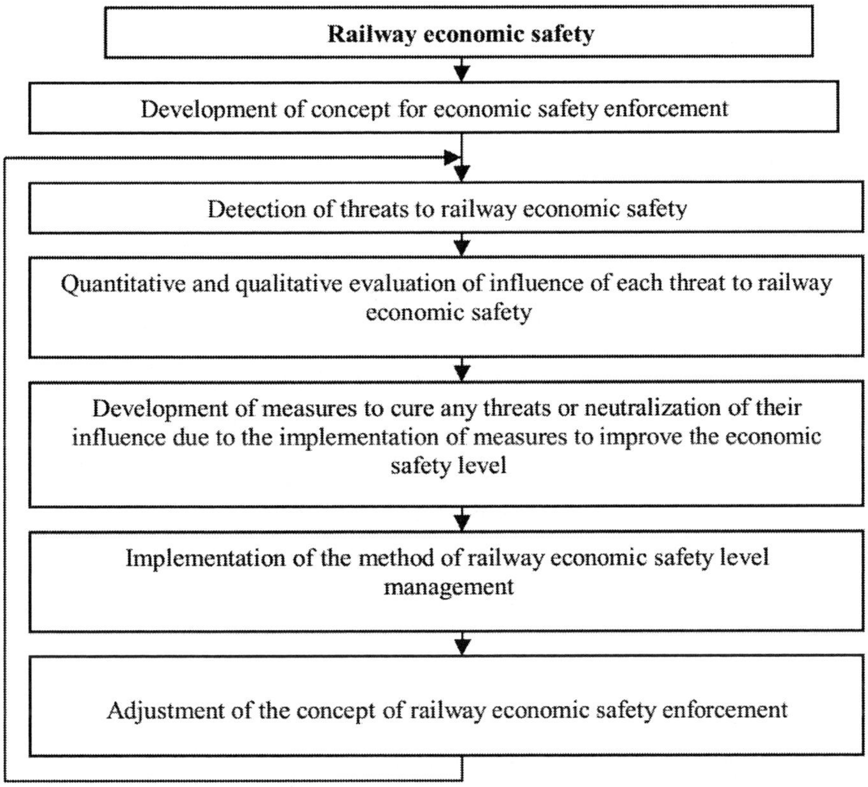

Fig. 1. Scheme of sequence of railway economic safety process management.

The first stage is the development of the concept for railway economic safety enforcement.

The second stage of enforcement is the detection of actual and potential threats to railway economic safety. As many existing threats took place [previously, they are unchanged and they are automatically referred to further periods. But some threats (those not being existed previously) may be detected by means of forecast.

Upon the detection of the list of threats, there is the third stage – consideration of factors that undermine the industrial stability and characteristics. Whereas threats may include several negative factors having influence with industrial operation, and the same factors may be repeated in some threats, it is required to determine all potential intercollerations between factors.

Based on the study actually made, the selection and development of optimal methods of economic safety enforcement is being carried out.

The optimal methods of railway economic safety enforcement are being a set of tools and the system of performance and monitoring that will enable to achieve the highest values of economic safety. It should be noted that the closer to completion the process of development and preparation is, the harder it is to make any significant changes.

Any economic process, object or event is being expressed in quantitative or qualitative form by means of figures. However, when it refers to the figures reflecting the level of railway economic safety, it should be noted that some of them are exposed to physical measurement and some are impossible to express in numerical value. Thus, in determination of railway economic safety, in general and separately upon each element, quantitative and qualitative evaluation methods shall be used and combined.

Quantitative figures shall mean such evaluation methods which enable to create numerical estimates, whereas qualitative figures shall not be reflected in numerical form.

Qualitative methods of analysis are based upon so called heuristic evaluations to be usually used by experts, expert groups and commissions, councils, staff based on experience, intuition and comparison. Qualitative analysis shall be carried out according to procedures the algorithm of which is not fixed forwards, i.e. the context of these procedures are not regulated and they as such do not have any calculative nature.

Quantitative analysis methods are reflected in calculation and computational procedures and determination of quality criteria in numerical form. Herewith, quantitative evaluations may be a result of the qualitative method if at the final stage of qualitative analysis experts determine numerical appraisal (for example, score, coefficient, and predicted value) of qualitative parameter.

Level of railway economic safety will be defined both by means of value judgement reflecting the opinion of certain heads, experts, and by means of objective appraisal based on existing data on railway service.

Level of railway economic safety depends on those figures that completely cover its service and its calculation will be based on the determination of an aggregate criterion (C_k) using a formula:

$$c_k = \sum_{s=1}^{n} K_j \alpha_j , \quad (1)$$

where k_j - figures of partial functional element of railway economic safety;

α_j - ratio of importance of partial functional elements of railway economic safety ($\sum\alpha_j$ =1);

n – number of selected functional elements of railway economic safety.

The ratio of importance of partial functional elements of railway economic safety shall be defined according to experts' conclusions.

Taking into consideration the fact that the aggregate criterion of railway economic safety depends on the figures of partial functional elements of railway economic safety level and ratio of importance of partial functional elements that include: financial (K_f); technical and technological (K_t); legal (K_p); personnel (K_k); environmental (K_e); informational (K_i) and security (K_o). It is defined using a formula (2).

$$C_k = K_f \cdot \alpha_f + K_t \cdot \alpha_t + K_p \cdot \alpha_p + K_k \cdot \alpha_k + K_e \cdot \alpha_e + K_i \cdot \alpha_i + K_o \cdot \alpha_o , \quad (2)$$

where K_f, K_t, K_p, K_k, K_e, K_i, K_o – values of partial functional elements upon certain elements of railway economic safety;

α_f , α_t , α_p , α_k , α_e , α_i , α_o – figures of the ratio of importance of partial functional elements of railway economic safety.

Ratio of importance of partial functional elements of railway economic safety shall be defined by the expert evaluation method through previous questioning of 12 experts under 10-score scale. The experts were as follows:

- heads of economic services with work experience of 15 years and above – 7 persons;
- specialists with work experience of 10 years and above – 2 persons;
- scientists with academic degree and academic title – 3 persons.

The aggregate criterion of railway economic safety shall be defined based on the received data of ratio of importance of each functional element by means of weighting coefficients using a formula 3:

$$C_k = K_f \cdot 0.217 + K_t\, 0.219 + K_p\, 0.123 + K_k\, 0.136 + K_e \cdot 0.126 + K_i\, 0.081 + K_o\, 0.098. \quad (3)$$

Prior to the determination of railway economic safety level, it is required to classify figures which will characterize each functional element of railway economic safety. The selection of these figures upon separate functional elements will be defined based on the threats.

The calculation of weighting figures shall be the basis for the determination of economic safety figures and potential upon each criterion and, generally, of economic safety and potential of railway transport.

3 Conclusion

Thus, under current conditions of operation, railway enterprises need to implement any structural transformations that would be consistent with up-to-date peculiarities of competition and ensure the proper level of economic safety and potential. According to the railway economic safety management scheme reviewed, those proposals concerning the improvement of integral part and enforcement of the high level of railway economic safety and, thus, of the state, will be developed.

References

1. T. Tymofyeyeva, N. Gritsenko, Y. Myroshychenko International Journal of Science "Internauka": Zbirnik naukovich articles. **14 (54)**, 46 -50, (2018).
2. T.Vasyltsev National Institution of Strategical Researches, 385, (2008)
3. S. Dovbnia. Finance of Ukraine, **4**, 91-97, (2013).
4. S. Kavun. Scientific review "Economics of growth". **3 (43)**, 97–101, (2013).
5. O. Kirichenko. Investments: practice and experience. **23**, 22-24, (2012).
6. V. Geets, M. Kyzym, T. Klebanova, O. Chernak. Monographs, 240, (2006).
7. O. Sudakovaia. Economy: problems of theory and practice: collection of scientific papers, **231(9)**, 165–171, (2012).
8. S.Shkarlet. Synopsis, 24, (2008).

Transit potential of Republic of Belarus: current state and development prospects

Vasylchenko Anastasiia [1,*]

[1]Polessky State University, Economical Department, Dneprovskoy flotilii, 23, 225710, Pinsk, Belarus

Abstract. The article is devoted to the development of transport infrastructure, the problems of logistics and transit potential of the Republic of Belarus. The review analysis of the logistics potential is given taking into account its development prospects. The monitoring of the dynamics of transit revenues was carried out, relevant conclusions are given, recommendations on its increase are presented. The article presents the dynamics of cargo transportation by type of transport indicating the structural elements of logistics centers and their localization in the Republic of Belarus. Efficiency indexes of the development of the country's logistic system according to the World Bank are analysed. Reasonable conclusions are given.

In the context of globalization and integration processes, there is an active development of foreign economic relations between countries, strengthening of trade cooperation, growth in foreign trade and transit traffic, improvement of the technical level of transport system and the quality of services provided. All these intensify and increase the role of transport system, which becomes the basis for the formation of domestic as well as international markets, thus providing the foundation for market economy processes in the world. The Republic of Belarus has an advantageous economic, geographical and geopolitical position, being located at the crossroads of the developing world business clusters – the countries of the European Union, the Russian Federation, the Asia-Pacific region (with its promising development of the new Silk Road). This will help to develop the logistics system, improve the transport infrastructure in the context of increasing the transit potential of the country in the international arena.

The following scientists were involved in the development of the transport infrastructure, logistics and transit potential of the Republic of Belarus: M. B. Gubsky, T. R. Kisel, R. B. Ivut [1], I. A. Elova, P. G. Nikitenko, I. I. Poleshchuk, A. A. Evsyuk, V. V. Yasinsky, A. V. Chernovalov, etc. However, many aspects have not been studied sufficiently due to ongoing changes in the global market economy, absence of the uniform approach to the definition of the main concepts, different basic approaches to the assessment of the transit potential.

In recent years, there have been significant changes in foreign markets due to geopolitical processes, socio-economic transformations and the reorientation of the major

[1] Corresponding author: vasichenko.2012@inbox.ru

market players from Western (the European Union) to Eastern countries (Asia-Pacific region). To analyze the transport system and its development prospects, together with the expansion of international trade relations of the Republic of Belarus, let's define the concept of "transit potential". In some sources it is defined as "transportation of goods or the movement of passengers from one place of the country or region to another through the third place on its way or from one point to another through intermediate points" [2]. Transshipment operations can be carried out in the transit territory, but, according to the author, they are not a decisive factor in defining this concept. At the same time, the underlying characteristic of transit is the location of the final point of delivery outside the transit territory. The ability of the local transport infrastructure to meet the needs of international transit operations for the movement of goods and passengers plays a huge role in transit.

The work on the development of the logistics potential of the Republic of Belarus was carried out under the Development Program of the Logistics System of Belarus for 2008-2015. In the course of its implementation, 20 logistics centers with a total area of more than 400 thousand square meters were built, over 4.6 million Belarusian rubles (BYN) of investments were attracted, more than 4.6 thousand jobs were created [3].

The volume of logistics services rendered in the Republic of Belarus in 2015 amounted to 1498 billion BYN, including 981.6 billion BYN provided by transport and logistics centers, wholesale and logistics as well as trade and logistics centers - 115.9 billion BYN. Revenues from logistics services for handling of transit cargo on the territory of the Republic of Belarus amounted to 462 billion BYN.

Experts predict that taking into account the current trends in the development of the logistics system, the volume of logistics services rendered in the Republic of Belarus will increase by a factor of 1.5 by 2020 compared to 2015 and will be about 2258 million BYN. At the same time, the total storage space of logistics centers of various functionalities will increase by a factor of 1.64 by 2020 and will amount to more than 656 thousand square meters [3]. Such forecast data are given in the Development Program of the Logistics System and Transit Potential for 2016-2020.

The income dynamics from the transit of the Republic of Belarus is very unstable and depends on a number of factors: the political situation, the dynamics of prices and currency fluctuations, the state of the transport and logistics system and the development of its potential, the economic situation of the shipping countries. At the same time, revenues from transit of the Republic of Belarus from 2006 to 2013 steadily increased, despite the Global economic crisis of 2009. However, since 2014 this trend has not persisted, the revenues have declined significantly. This happened for objective reasons, due to the introduction of economic sanctions by the United States and the European Union against the Russian Federation, as well as the trade embargo on a number of goods from the European Union countries. In 2016, the revenues fell to $ 1200 million, which is a record low. This happened due to unfavourable geo-economic and geopolitical factors prevailing in the global market. The Republic of Belarus depends on these factors owing to its geographical position and its location at the crossroads of world transportation routes. Nevertheless, in 2011-2015 transit revenues amounted to $ 13.9 million, which is almost 30% as high as compared to 2006-2010. It is still too early to judge by 2016-2020 indicators because of the need to end the reporting period.

In the item "Revenues from transit", the statistics of the Republic of Belarus includes the revenues from the transit of oil, natural gas and the revenues from transit in the field of transport. Meanwhile, the leading place in the structure of these revenues is occupied by the revenues from the transit of energy carriers and from the movement of transit cargo by rail. In 2015, the profits from the export of communication services amounted to $ 5.7 million; from the carriage of goods by rail — $ 378.3 million; from the transportation of passengers by rail — $ 22 million; from the maintenance of transit flights by air — $ 73.0 million; from fees on toll roads — $ 117.2 million; from the tolls of motor vehicles of foreign countries on public roads — $ 11.2 million, and $ 1.1 million from foreign heavy and/or large vehicles [4]. It is estimated that in 2016 the revenues from oil transit amounted to $ 150.5 million, from gas transit — $ 476.7 million, from transit in the transport sector (except for pipeline) — $ 879.2 million [5].

More than 13 thousand organizations of various forms of ownership and more than 17 thousand individual entrepreneurs work in the transport sphere of Belarus, employing about 275 thousand people – 6.2% of the total number of employed in the country's economy. On the territory of the Republic of Belarus there are about 100 thousand km of roads, 5.5 thousand km of public railways, of which one fifth is electrified, as well as 1.7 thousand km of inland waterways. In terms of freight turnover of road and rail transport per capita (6.88 thousand ton-km), Belarus is ahead of such European countries as Germany (5.22 thousand ton-km), Belgium (3.6 thousand ton-km), France (2.9 thousand tons-km), Italy (2.15 thousand tons-km) [6].

The state policy of the Republic of Belarus also promotes the development of logistics, as evidenced by the following regulatory legal acts: the State Development Program of the Transport Complex of the Republic of Belarus for 2016-2020, the Program of Social and Economic Development of the Republic of Belarus for 2016-2020, the Resolution of the Council of Ministers dated 12.05.2016 №372 "On Priority Activities (Sectors of the Economy) for Investments", etc. The latter outlines the priority sectors of the economy for investments, including the development of the logistics system, as well as rail and air transport. There are five standards for the development of logistics infrastructure in the Republic of Belarus: SSRB 2047-2010 "Logistics activities. Terms and definitions", SSRB 2046-2010 "Transport and logistics center. Requirements for technical equipment and freight forwarding services", SSRB 2133-2010 "Classification of warehouse infrastructure", SSRB 2306-2013 "Logistics services. General requirements and certification procedure", SSRB 2345 "Logistics activities. Requirements for professional competence of logistics service personnel and certification procedure".

According to statistics, the dynamics of goods carriage by types of transport in the Republic of Belarus is as follows (Figure 1).

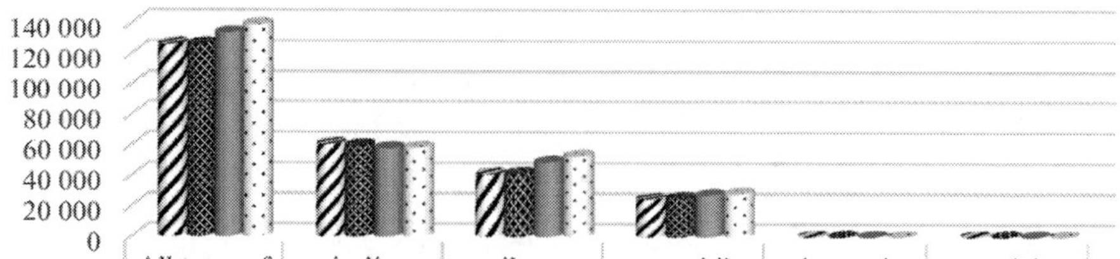

	All types of transport	pipeline	railway	avtomobile	internal water	aerial
2015	125 957	60 552	40 785	24 523	21	77
2016	125 820	59 345	41 107	25 239	21	108
2017	133 348	57 708	48 538	26 987	32	83
2018	138 838	58 071	52 574	28 082	37	75

2015 2016 2017 2018

Fig. 1. The transportation of goods by different types of transport (2015-2018), million tons [7].

The Development Program of the Transport Complex of the Republic of Belarus for 2011-2015 resulted in the building of 20 logistics centers with a total area of 400 thousand square meters. The implementation of these projects called for more than 4.6 trillion BYN, 4.6 thousand new jobs were created. The volume of logistics services rendered in 2015 amounted to 1498 billion BYN, of which: transport and logistics centers – 981.6 billion BYN, wholesale and logistics as well as trade and logistics centers – 115.9 billion BYN. The revenues from logistics services for handling transit cargo in Belarus amounted to 462 billion BYN [8].

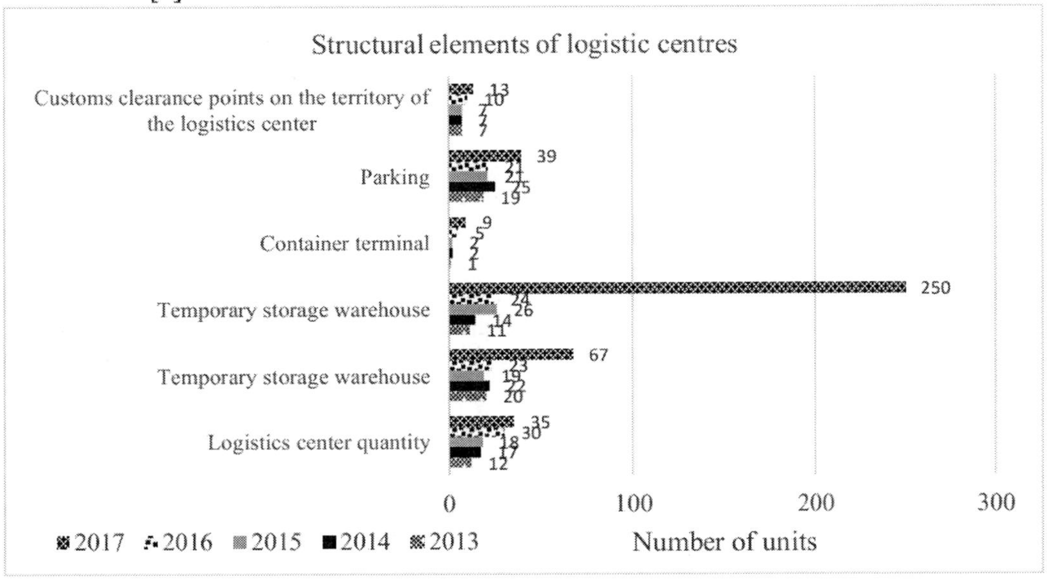

Fig. 2. Structural elements of logistics centers (2013-2017), units [9].

The dynamics of changes in the structural elements of logistics centers can be seen in Figure 2. On the whole, in the Republic of Belarus there is a tendency to increase their number (in 2013 there were12 units, in 2018 – 48 units), the growth rate is 400%. Altogether, this is a positive trend, indicating the increase in capacity in the field of logistics infrastructure. Furthermore, there is a positive trend in the context of the structural elements of logistics centers: the number of temporary storage warehouses is increasing (the growth rate of 2017 by 2013 is 335%); the number of public warehouses from 2016 to 2017 increased by a factor of 4.6; the number of container terminals increased by a factor of 9 compared to 2013; the number of parking lots for the same period increased by 20 units, and the number of customs control points doubled.

Experience shows that for the successful implementation of the transit potential of the Republic, not only the developed transport infrastructure is important, but also the optimal location of logistics centers near the main transport flow routes, which allows for timely, efficient and high quality logistics services. It can be seen [9] that the volume of logistics services provided by logistics centers is growing (it increased by 37.6 million BYN in 2017 compared to 2016).

The volume of logistics services for handling transit cargo in the Republic of Belarus for 2016-2017 increased by 32.3 percentage points, and the volume of freight forwarding services – by 51.9 percentage points respectively.

The analysis of the range of services provided by logistics centers (on the terms of voluntary certification) showed that these logistics centers provide no more than 15 of the 88 possible types of services established by the standard SSRB 2306-2013, which is certainly an obstacle to the development of foreign trade relations. Besides, a number of negative trends have been revealed in connection with the changes in world markets, namely: under the influence of external political and economic factors, Russian export cargo is reoriented from the ports of Lithuania, Latvia and Ukraine to its own ones. In general, changes in the foreign trade policy of the Russian Federation are due to the cooperation change from Western countries (EU countries) to the countries of the Asia-Pacific region and China. There is a trend for the Russian ruble devaluation. While 92% of transport by rail is export-import cargo of Russia and Kazakhstan, this factor is undoubtedly a negative trend for the expansion of the use of the transit potential of the Republic of Belarus.

Commodity flows passing through the territory of the Republic of Belarus are mainly formed at the expense of the EU and the Asia-Pacific countries. The change in global market processes leads to the redistribution of these flows and their reorientation from the Central European to the South European direction. Accordingly, the use of the transit potential of Belarus depends on the demand for certain types of products, including timber, mineral fertilizers, oil and petroleum products, as well as ferrous metals. The key point here is the low competitiveness of the existing logistics system along with unfavourable external conditions: unequal conditions for logistics centers within the Eurasian Economic Union, high competition with European Union companies, and complexity of technical regulations.

In 2007, the World Bank, in order to determine the efficiency of the country's logistics system, developed a methodology for calculating the LPI index (logistics performance index – LPI). This index is measured every two years, the results of the study were published in 2007, 2010, 2012, 2014, 2016, 2018. In 2010, the Republic of Belarus did not participate in the assessment. The LPI methodology includes a five-point scale of six evaluation criteria and calculating the average index value for each country.

The following evaluation criteria are used:
- efficiency of customs clearance;
- quality of logistics transport information technology infrastructure;
- simplicity and availability of international shipments;
- level of logistics quality and competence;
- tracking and tracing of international delivery;
- timeliness of deliveries.

The results of the research are shown in table 1.

Table 1. LPI index of the Republic of Belarus in 2007-2018 [10].

The evaluation criterion	**The values by year**									
	2007		**2012**		**2014**		**2016**		**2018**	
	score	**rank**	**score**	**rank**	**score**	**rank**	**score**	**rank**	**score**	**rank**
Customs	2,67	50	2,24	121	2,55	87	2,06	136	2,35	112
Infrastructure	2,63	54	2,78	65	2,55	86	2,19	135	2,44	92
International transport	2,13	127	2,58	107	2,74	91	2,62	92	2,31	134
Quality and competence	2,13	120	2,65	89	246	116	2,32	125	2,64	85
Cargo tracking	2,71	166	2,58	98	2,51	113	2,16	134	2,54	109
Timeliness of deliveries	3,00	78	2,87	114	3,05	93	3,04	96	3,18	78
LPI index	2,53	74	2,61	91	2,64	99	2,40	120	2,57	103

As can be seen from table 1, in 2018 the Republic of Belarus took 103rd place from 160 countries, in 2016 it was 120th. The statistics of the previous years show that in 2007 it occupied 74th place, in 2012 – 91st place, in 2014 – 99th place, which points to a large role of state support in the development of the logistics system of the Republic of Belarus. For comparison, in 2016 Russia ranked 99th, Kazakhstan – 77th, Ukraine – 80th, Lithuania – 29th, Latvia – 43rd, Poland – 33rd, Germany – 1st. In 2018 Germany took first place, Russia moved to the 75th place, Kazakhstan – 71st, Ukraine – 66th, Lithuania – 54th, Latvia – 70th, and Poland rose its position to the 28th [10].

The method of calculating this index is not perfect, which is acknowledged by its developers from the World Bank. It does not take into account the peculiarities of the geographical location, the access to the sea, the area of the country, etc. In addition, it is based on a survey of international logistics companies, while the opinion of consumers of logistics services is not taken into consideration. All in all, the market of logistics services of the Republic of Belarus is developing steadily, but not as fast as that of the neighbouring countries.

Therefore, the degree of geopolitical risks depends on establishing favourable relations with the Russian Federation, the European Union, the Asia-Pacific region and China. A number of factors will contribute to the use of the country's transit potential: improvement of the existing legislation system, training highly qualified specialists in logistics, modernization of the logistics infrastructure, optimization of logistics centers location for the convenience of transport links, further development of information and communication technologies, introduction of innovative developments in the field of logistics.

References

1. R. B., Ivut, A.F., Zubritskiy, A. S., *Development of transit potential of the Republic of Belarus in the conditions of formation of its transport and logistics system*, **1(32)**, (2015)
2. T. F. Efremova, New dictionary of russuan language, (2000)
3. Logistics system and transit potential development program for 2016-2020, Retrieved from http: //www.baifby.com/page/60
4. A.D., Molokovich, V.A., Luksha, Results of implementation of the State program of development of transit potential of the Republic of Belarus for 2011-2015, Retrieved from http://elib.bsu.by/bitstream/pdf
5. Belarus loses the income from the transit, Retrieved from http://belaruspartisan.by/economic/349400/
6. Transport complex of Belarus, Retrieved from http://www.baif.by/stati/transportnyi-kompleks-belarusi/
7. Belarus in numbers. 2019: Statistical Yearbook. (2019). *National statistical Committee of the Republic of Belarus.* Minsk (2019)
8. How to save Belarusian transit, Retrieved from http: ://neg.by/novosti/otkrytj/kak-spasti-belorusskij-tranzit/
9. Transport and communication in the Republic of Belarus. 2018: Statistical Yearbook. (2018). National statistical Committee of the Republic of Belarus. Minsk (2018)
10. URL: http://www.worldbank.org

High-Speed Railways is a Priority Direction for the Development of Railway Transport

Viktoriia Yanovska[1], *Olga* Levchenko[2], *Viktoriia* Tvoronovych[3], and *Anastasiia* Bozhok[4,1]

[1]State University of Infrastructure and Technologies, Head of the department "Economics, marketing and business administration", Kyrylivska str, 9. Kyiv, 02000, Ukraine
[2]State University of Infrastructure and Technologies, associate professor of the department "Economics, marketing and business administration", Kyrylivska str, 9. Kyiv, 02000, Ukraine.
[3]State University of Infrastructure and Technologies, associate professor of the department "Economics, marketing and business administration", Kyrylivska str, 9. Kyiv, 02000, Ukraine
[4]State University of Infrastructure and Technologies, associate professor of the department "Economics, marketing and business administration", Kyrylivska str, 9. Kyiv, 02000, Ukraine

Abstract. The article deals with the Chinese global strategy the Silk Road Economic Belt and the possibility of involving Ukraine in it. The authors pay special attention to the implementation of the China-Ukraine-EU High-Speed Railway Project within the framework of the Silk Road Strategy. The geopolitical position of Ukraine allows the authors to state that Ukraine may represent a special interest for the Celestial Empire as a logistical hub and a kind of "window" to Europe, as the central axis of the international transport corridors passes through Ukraine. The authors analyze the direct economic, as well as external economic and social effects of the construction of High-Speed Railways in Ukraine.

1 Introduction

China is the world leader in the production and export of most types of products. Its foreign policy is aimed at expanding the resource base and markets of the Chinese products [1]. With its flexibility and pragmatism, China established itself as a regional leader, consolidating the sphere of influence in the Southeastern and Central Asia. Due to the globalization of foreign trade and policies, PRC has expanded the range of its economic partners in Africa and Latin America. At the same time, China seeks to expand the geography of economic cooperation at the expense of EU countries. In this regard, Central and Eastern European (CEE) countries play an important role, which may be a perfect basis for a common EU market, as well as by convergence with the countries of the region China seeks to lobby for its interests at the EU level and to determine the agenda for cooperation between China and EU [2].

In this regard, in 2013, the Chinese President Xi Jinping announced the Silk Road Economic Belt Project as a new form of deepening cooperation between China, the states of Central Asia, Russia and Europe. The Silk Road Economic Belt Project provides a new

[1] Corresponding author: nastya.bozhok@gmail.com

model of regional cooperation based on an innovative idea, which connects the Silk Road and the Economic Belt. The Silk Road Economic Belt is considered not only as a transit transport project and a direction for accelerating the development of the western regions of China and Eurasia, but also as a strategy for the economic development of a number of countries, which includes infrastructure development, industry, trade and services [3].

The Great Silk Road provides several routes from the western regions of the People's Republic of China towards the key centers of economic activity in Europe and South Asia, covering from 8.5 to 11 thousand km. It is assumed that the North route will pass through the territory of Kazakhstan and the Trans-Siberian Railway. The sea routes will be operated by the Kazakh port of Aktau, as well as ports in the Caspian Sea (Makhachkala, Baku) providing access to the Caucasus region, Turkey and the Black Sea basin. The southern routes run along Kyrgyzstan, Uzbekistan, Turkmenistan, Iran and Pakistan, providing access to the Indian Ocean in the Arabian Gulf region [4].

2 Ukraine's Place in China's Global Strategies

Involving Ukraine in these projects will allow us to build the shortest link between the Eastern Asian and European countries, as well as significantly reduce the delivery and storage times. The Eurasian direction of economic interest for Ukraine and distinguish the economic corridors of the New Silk Road which are worth noticing include China-Pakistan corridor; Bangladesh-China-India-Myanmar Corridor; Indochina Peninsula Corridor; New Eurasian Land Bridge; China-Mongolia-Russia Corridor. For Ukraine, it is in particular China – Central Asia – West Asia corridor, through Turkey, the South Caucasus, and the countries of Central Asia [5].

The construction of the transcontinental railway is of strategic importance for the development of the transport industry of Ukraine, as it involves the construction of new international highways, the creation of infrastructure objects, including transport and logistics complexes, construction of high-speed railways, which allow reaching a speed mode of more than 200 km/h.

Ukraine can represent a special interest in the Celestial Empire as a logistical hub and a kind of "window" to Europe, as the central axis of international transport corridors passes through Ukraine, which include the main railway routes that can become part of the Ukrainian section of the Lisbon- Shanghai Railway:

1) Zernove (border with Russia) / Hornostaivka (border with Belarus) – Nizhyn – Kyiv – Zhmerynka – Kuchurhan (border with Moldova) / Odessa. The length of the route is 1159 km;

2) Chop (border with Slovakia and Hungary) / Mostynska 2 (border with Poland) – Lviv – Zhmerynka – Fastiv – Dnipro – Krasnoarmiysk – Debaltseve – Chervona Mohyla (border with Russia) / Yasynuvata – Kvashyno (border with the Russian Federation). The length of the route is 1792 km; and

3) Izov / Yagodyn (border with Poland – Kovel –Zdolbunov – Shepetovka – Kozyatin – Kyiv – Poltava – Kharkiv – Kupiansk – Topoli). The length of the route is 1167 km [6].

Thus, Ukraine can become a full member of the corridor with a potential of 10 million tons of cargo per year. At the same time, the Ukrainian-Chinese bilateral trade has great prospects, under the condition of effective foreign economic policy and rational use of the resource potential of bilateral cooperation. This is confirmed by the data of the State Statistics Service of Ukraine in the geographical structure of the foreign trade turnover between Ukraine and the countries that are part of the Caspian-Black Sea route of the New Silk Road: in 2017 from the PRC - $7,688 million, Poland - $6,039.4 million, Germany - $5,529.1 million [7]. At the same time, interstate trade between Ukraine and China is asymmetric: China is the third trading partner of Ukraine (after the EU and Russia), while

Ukraine is only the third partner in the post-Soviet space (after Russia and Kazakhstan). Due to the agrarian sector, Ukraine became one of the leaders among exporters of food products to the Celestial Empire [2].

Ukraine can be of particular interest to China as a large-scale logistics center, because its territory is the central axis of the international transport corridors. During 2008-2018, the volume of trade between Europe and China increased by almost 50% (Fig. 1), which confirms the necessity of construction of High-Speed Railways through Ukraine in order to accelerate trade between the EU and China.

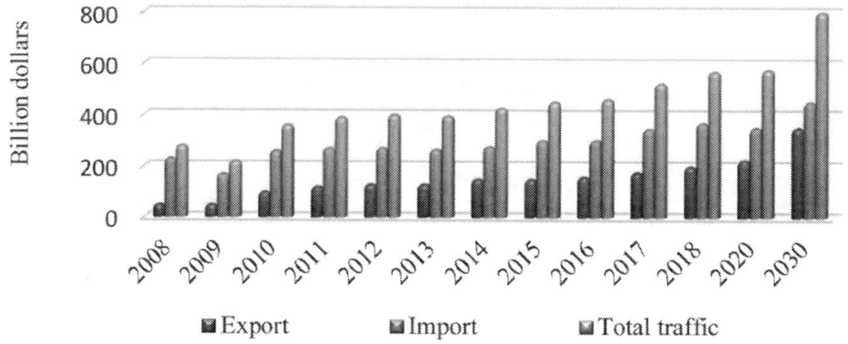

Fig. 1. EU-China Trade Turnover [8, 9].

3 Potential benefits from the implementation of high-speed railway projects in Ukraine

Ukraine has the opportunity to effectively use its geographic potential, which is capable of accelerating the trade between the EU and China. To this end, it initiated an international project "High-Speed Railways – 1435". It is proposed to build a High-Speed Railways through Ukraine from China to Europe. The railway will connect the Celestial Empire with 7 states - Kazakhstan, Uzbekistan, Turkmenistan, Azerbaijan, Georgia, Ukraine, and Poland. The total length of the railway corridor will be 5100 km (excluding sea distances), which is 200 kilometers shorter than the Eurasian Land Bridge (China – Kazakhstan – Russia – Belarus – Poland), which now passes containers from China to the European Union. These 200 km will save 4.5 billion dollars, at an average cost of 1 km of High-Speed Railway 21 million dollars. [2]. It is important to note that today the main flow of goods between the leading commodity markets of the world – China and the EU countries is carried by sea for 30-40 days, as well as by the Trans-Siberian Railway from China to Russia, Belarus, Poland in small but growing volumes of transportation. Therefore, the implementation of the High-Speed Railways project will significantly accelerate the delivery of goods.

Ukraine-China strategic partnership as a kind of bilateral relations is of a profound nature, which is not limited only to foreign trade relations and financial benefits, the influence on different socio-economic processes should also be taken into account:

1) Modernization of infrastructure (ports, motorways, railways) at the expense of the Chinese capital;

2) Additional component of the European integration of Ukraine;

3) Strengthening of the regional position of Ukraine in the CEE;

4) Reduction of pressure from Russia due to the inclusive and mutually beneficial nature of cooperation of all interested parties;

5) During 10 years of operation of the high-speed railway "1435", an increase in transit traffic of goods will be at least 18 million tons a year, in the context of a significant improvement in the quality performance indicators: two - three times the average speed growth, multiple reduction of downtime;
6) 450 thousand new jobs will be created during the construction phase, and at the exploitation stage - 70 thousand. And these are only direct effects. Thanks to the project, new jobs will be created in other sectors of the economy. In addition, the requirements for the quality of labor force, level of its qualification and professional education will be raised;
7) The high-tech production will be developing. As a result of the fact that the creation and operation of the high-speed railway "1435" is not a one-time measure and in the process of functioning of a new industry it cannot be limited to the purchase of several rolling stock, and technology transfer will take place. Obtaining access by the Ukrainian producers to world technologies will make it possible to create a market for own competitive products with high added value;
8) In the process of construction and operation of High-Speed Railway "1435" the large-scale orders of products, works and services will take place. Accordingly, the industry, the construction industry, the sphere of services, education, and science will be developing;
9) Expected growth of the industrial production during construction for 2 trillion hryvnias, for ten years of operation - 30 trillion hryvnias;
10) High-Speed Railways will play a significant role in the growth of the socio-economic level of the regions, especially the railway areas, namely, Kyiv, Odessa, Vinnytsia, Khmelnytskyi, Ternopil and Lviv regions, whose gross regional product, due to the construction of transport hubs, the creation of logistics terminals, the growth of economic activity of the population, will increase by 10% per year, and its growth will exceed UAH 800 billion for ten years of its exploitation; and
11) Extra tax revenues only from the freight traffic will be almost 80 billion UAH for ten years of operation of 1435 railways. In addition, additional tax revenues from high-speed passenger transportation, the growth of industrial production, the development of services, education, science, increase in the economic activity of the population in the regions [10].

Thus, the implementation of the High-Speed Railway 1435 Project, with the corresponding creation of a new industry and new productions, is oriented and objectively determines the socio-economic development of Ukraine and is very important in the modern conditions. The need for economic growth, integration into the global economic system requires accelerated development of the transport complex of the country. It is necessary to support the importance of Ukraine as a transit state, the Ukrainian manufacturer with new orders, the labor market and create conditions for the growth of the welfare of the population.

4 Silk Road Program and Financing

The development program of the "Silk Road Economic Belt" project consists of the following items:

1) Intergovernmental cooperation, strengthening coordination of the states of the region in the political sphere;

2) Infrastructure integration, intensification of construction of a unified road network;

3) Development and reduction of trade expenses by eliminating trade barriers, increasing the speed and quality of the economic operations;

4) Financial integration, that is, an increase in foreign exchange flows through the transition to payments in the national currencies; and

5) Strengthening the role of international diplomacy, scientific and cultural exchanges, tourism, health care, etc.

It should be noted that more than 30 countries have entered into agreements on joining the project "One Belt – One Way".

China is cooperating in the construction of individual industrial projects with more than 20 countries. It is supposed to provide financial support to China from the countries participating in this project. New institutions have been created: the Silk Road Fund, the Asian Infrastructure Investment Bank. The Silk Road Fund is made up of the Chinese funds (based on foreign exchange reserves, Chinese Investment Company, Chinese Import and Export Bank, State Development Bank of China) that will be used to build the Silk Road Economic Belt. Also, the PRC will provide support for training 20,000 professionals for five years to work on the implementation of the strategy.

The infrastructure is the primary and most cost-effective part of the project implementation. The Great Silk Road provides several routes from the western regions of the People's Republic of China towards the key centers of economic activity in Europe and South Asia, covering from 8.5 to 11 thousand km. It is assumed that the North route will pass through the territory of Kazakhstan and the Trans-Siberian Railway. The sea routes will be operated by the Kazakh port of Aktau, as well as ports in the Caspian Sea (Makhachkala, Baku) providing access to the Caucasus region, Turkey and the Black Sea basin.

The southern routes run along Kyrgyzstan, Uzbekistan, Turkmenistan, Iran and Pakistan, providing access to the Indian Ocean in the Arabian Gulf region [4].

Conclusion

It should be noted that in the external environment there may be certain circumstances that would impede the implementation of the Silk Road project, in particular the application of the political considerations in economic logic in the field of transport and transit of goods and energy; strict national requirements for the transportation of goods, which are the cause of slowing cargo flows; lack of necessary financial resources from the countries to implement large-scale projects; lack of security in some areas and their borders (Afghanistan, the Caucasus, Ukraine), which is the reason that some routes are not used.

However, the deepening of the Ukrainian-Chinese cooperation and the construction of the high-speed Ukraine-China-EU will create many opportunities, there are a significant acceleration of trade turnover, modernization of infrastructure (ports, roads, railways); and strengthening the regional position of Ukraine in the CEE among the largest ones. In addition, infrastructure-related projects are essentially development-oriented. Therefore, their implementation leads to the expansion of orders to enterprises in different industries, job creation, increase in household incomes, expansion of domestic consumption, payment of taxes, increase in budget revenues, as well as the appearance of many other effects.

References

1. Eurostat Statistics [online], Available at: https://ec.europa.eu/eurostat/statistics-explained/index.php/China-EU_-_international_trade_in_goods_statistics.
2. Ukraine's place in the China's global strategies. *International Centre for Policy Studies.* 5-22 (2015).
3. M. Shevchenko. From China to Ukraine: New Silk Road. *Unian* [online], Available at: http://economics.unian.net/finance/1552361-izkitaya-v-ukrainu-novyiy-shelkovyiy put.html.
4. V. Povoroznik, V. Perebyinis. International Centre for Policy Studies. **31**. (2015).

5. I. Shevyrev. Three years of Silk Road. *Political news* [online], Available at: http://izvestia.kiev.ua/blog/show/84135#disqus_thread.
6. V.L. Dykan. The economic zone of the Silk Road in the context of the strategic priorities of Ukraine's economic development. International transport corridors and corporate logistics. 6-7 (2015).
7. External economic activity. State statistics Service of Ukraine [online], Available at: http://www.ukrstat.gov.ua.the
8. World Trade Organization: world trade statistical review [online], Available at: https://data.wto.org/.
9. V. Yanovska, O. Pylypenko, V. Tvoronovych, A. Bozhok. . International Journal of Engineering & Technology, **7 (4.3),** 583-587 (2018).
10. Project concept high speed lines «China-Ukraine-EU». *Materials of the Raund table of Association High-Speed Highways*. 2018 p.

Technical and economic evaluation of justification for preventing negative effects of feedback in local elements of interaction of rail yards

Petro Yanovsky[1], *Sergiy* Lytvynenko[2]*, *Larysa* Lytvynenko[3], and *Volodymyr* Tkachenko[1]

[1] National Aviation University, Military Training Department, 03058 Kyiv, Ukraine
[2] National Aviation University, International Economy Department, 03058 Kyiv, Ukraine
[3] National Aviation University, Management of Foreign Economic Activity of Enterprises Department, 03058 Kyiv, Ukraine

Abstract. The operation of rail yards is characterized not only by direct links, but also by feedback in the local elements of interaction, resulting in a violation of interaction conditions between adjacent elements, which causes a negative effect on the state and functioning of the elements. The duration of the track occupation in the arrival yard depends on both its parameters: the marshalling hump and the operation of other elements of the yard located at the arrival yard in the direction of cargo traffic processing. Therefore, when choosing the rational parameters of the rail yards, attention should be paid to the existence of feedbacks in the local elements of interaction and conditions of their negative manifestation should be investigated in order to justify measures to eliminate them. In the article, the issue of improving the customer transport service quality by technical and technological modernization of railway enterprises was studied. Technical and economic evaluation on justification of measures for the elimination of negative effects of feedback in local elements of interaction of the rail yards is presented for practical use.

1 Introduction

The operation of rail yards is characterized by action of direct links and feedback in the local elements of interaction (LEI) resulting in a violation of interaction terms between adjacent elements arises, which has negative effect on the state and operation of the elements.

For example, it was determined that the duration of track occupation in the arrival yard depends on its parameters and parameters of the marshalling hump, but also on the functioning of other elements of the yard located at the arrival yard according to the direction of the railway car traffic processing. So, when selecting rational parameters of the rail yards, the feedback in the LEI should be taken into account and conditions should be studied for its negative manifestation aimed at justifying measures on its elimination.

Such research should begin with the departure yard (DY), where the duration of staying of trains is influenced by the following factors: capacity of adjacent sections, the intensity of locomotive conveyance to trains, the intensity of servicing trains by inspection station brigades and the number of tracks in the departure yard.

2 Literature review and defining the problem

Transport production is a solid basis of the market economy and ensures the efficient functioning and development of all economic sectors serviced by transport as a market relations subject in a state with qualitative activity. Economic development is provided by the rational interaction of railways and industrial enterprises in the context of establishing mutually beneficial economic and legal relations between transport and cargo owners. Thus, compromise management of the transport process is carried out on equal participation of all its participants, as a result of which the highest social effect of the production and economic activity of subjects of different ownership forms and size (large, medium or small business) is ensured in the state.

The peculiarities of transport production, as a multi-aspect issue, have been constantly studied in our country and abroad. Important results of such studies are published in the scientific works of F. Kochnev, M. Pravdin, M. Fedotov, M. Galatchenko, M. Stefanov, V. Negrey, I. Tikhomirov, M. Ilovaysky, J. Kulayev, N. Tereshina, A. Smekhova, S.

* Corresponding author: sllitvinenko1982@gmail.com

Razer, D. Lomotko, E. Nagorny, V. Shish, A. Kotenko, T. Butko, E. Alyoshinskyi, D. Kozachenko, V. Mironenko, V. Samsonkin and others.

Issues of reducing the cargo delivery duration by rail through decreasing the train procession time at border transfer stations were studied by Ye. Aloshynskyi, V. Naumov and O. Pestremenko-Skripka [1], resulting in offering risk management systems based on the selective survey operations principle to solve the problem.

In the paper [2] Ye. Aloshynskyi, H. Prymachenko and Yu. Shuldiner analyzed another aspect of improvement: basic directions of increasing the level of logistics principles implementation in Ukrainian conditions. Also, D. Lomotko, Ye. Aloshynskyi and G. Zambrybor [3] have come to the conclusion that methodological aspect of the logistics technologies formation in reforming processes on the railways is related to system-wide optimization providing synergy effect and considering the complex of all participants' interests in the transportation process.

An automated traffic capacity calculation method taking into account the operating reliability of the rail transportation system was developed by S. Panchenko, T. Butko, A. Prokhorchenko, L. Parkhomenko to increase the efficiency of transport logistics and ensuring more accurate evaluation of rational workload limits of a rail network [4].

T. Butko and O. Shander in [5] developed mathematical model formalizing the technological process of managing the cargo transportation fleet of operator companies which was solved with the use of mathematical apparatus of genetic algorithms.

Some aspects of the efficiency increase of the transportation process of the railway directions by improving the operational management of the locomotive fleet operation were considered by L. Yelnikova [6]. The study is focused on solving the task of developing an operational plan for railway locomotive fleet as a multi-criteria assignment problem.

In the transport process, in addition to manufacturing enterprises, workers of various railways structures are involved, whose activity is aimed at solving of one common task for the economy – the safe and high-quality transportation of passengers and cargoes. All components of a complex multi-sectoral railway system should work in a close interaction with each other. This over-complex task is constantly being studied by domestic scientists and practitioners. However, the introduction into the national economy of market relations reveal previously under-studied aspects of operation activity of railways in modern economic conditions in a new way, requiring the continuation of carrying out R&D in this direction.

3 Research aim and objectives

The purpose of the article is the development of methodical recommendations on the feasibility study of measures to prevent the negative effects of feedback in the local elements of interaction (LEI) of the rail yards. In general, the optimum mode of the rail yard is provided by the high level of operating technology and technical equipment of individual LEI, due to their precise interaction and coherence. Such a level of interoperability of all LEI of a specific yard can be achieved by the introduction of effective measures the economic feasibility of which is justified using given methodical guidelines.

4 Overview key findings

4.1 Detection and systematization of the feedback negative influence factors in local elements of rail yards' interaction

Under the specific operating conditions, it is necessary to identify and systematize all factors by their degree of action, revealing among them the most influential ones, in order to concentrate the main attention, as well as, if necessary, financial resources on them.

Then the state and features of LEI functioning should be analyzed (moving towards the processed railway car traffic): output sections (OS), departure yard (DY), dead-end tracks (DET), maneuver locomotives (ML), marshalling bay (MB), marshalling hump (MH), arrival yard (AY), inlet sections (IS). Such a back review of LEI will provide an opportunity to study feedbacks in detail and predict measures implementation of which eliminates their negative effect.

The total annual costs of the rail yard caused by the negative effect of feedback in LEI due to the increase in downtimes of trains in the station yards are determined by the formula:

$$E_c = 365 \cdot m\left[\Delta t_{dc}(N_{tr} + N_f) + \Delta t_f N_f + t_{ay} N_{ds}\right] C_{c-h}, \tag{1}$$

where m – average number of railway cars in the train; Δt_{dc} – average additional downtime of railway cars in the departure yard (DY) caused by the violation of the conditions in interaction of DY with adjacent sections; $N_{tr} + N_f$ – total number of trains departing from DY to all adjacent sections (transit and its formation); Δt_f – average additional downtime of railway cars in the marshalling bay (MB) caused by the violation of the conditions in interaction between stretches of formation (SF) and the departure yard (DY); N_f –number of formation trains; t_{ay} – average additional downtime of railway cars in the arrival yard (AY) caused by the violation of the conditions in interaction of marshalling hump (MH) and marshalling bay (MB); N_{dis} –number of trains arriving to disassembling; $C_{c\text{-}h}$ – present value of one downtime railway car-hour.

In the conditions of market transformations in Ukraine there is a significant shortage of financial resources. For the purpose of their rational use it is necessary to invest them in the most effective and important measures for accelerating the passage of railway cars through the rail yard. Therefore, in such difficult economic conditions it is important to identify rational implementation phasing of measures from a possible list. In addition, any acceleration in the turnover of railway cars gives positive benefits to cargo owners in terms of cargo delivery, and costs are borne by the railways. Unfortunately, the cooperation of railways and a number of clients is not yet based on the partnership principles with the implementation of compromise solutions. It's time to decide on a fair distribution of the non-transport effect among all the participants in the transport process at scientific and technical levels. But so far, railways are forced to solve their own problems independently without the help of authorities and business. Public administration bodies should change the consumer attitude to railway transport, the state of which is currently catastrophic [7]:

- share of main roads with overdue major repairs – 27%;
- 78% of road vehicles has expired service life;
- deterioration of traction sub-stations is 67% and contact network – 65%;
- deterioration of main electric locomotives is 92% and their operating fleet – – 66% of the inventory;
- deterioration of main diesel locomotive is 99% and their operating fleet – 42% of the inventory;
- deterioration of maneuver diesel locomotives is over 80% and their operating fleet – 62 % of the inventory;
- deterioration of cargo railway cars is more than 90% and their operating fleet – 64% of the inventory;
- deterioration of passenger railway cars exceeds 87% and their operating fleet – 66% of the inventory;
- deterioration of electric and diesel trains is over 90% and their operating fleet – 79 %.

These negative phenomena should be also complement by the permanent worsening of staffing of the transport process – the number of dismissed railway workers is increasing due to very low salaries and practical absence of the social protection system and provision of adequate working conditions. The average salary in Ukraine is the lowest in Europe (€190), which is 2.5 times less than in Russia and 24 times less than in Switzerland.

It is also worth mentioning the lowest (non-market) tariffs for cargo transportation in Europe and the unfair charges for the use of railway cars. Taking into account the given data, authorities should radically change the attitude towards the railway transport and to stop investing big business at its expense by introducing the transport process compromise management system (TPCMS) into the economy.

4.2 Determination of the optimal mode of rail yard operation

The optimal mode of rail yard operation is determined by one of the options of the operation organization and technical equipment of individual local elements of interaction, for which total present costs will be minimal [8–11]. If violations of interaction between the rail yard's adjacent local elements of interaction and the acute shortage of finances in determining the rational phasing of enhancing technical equipment and improvement of the elements' technology were suspected, it is desirable to comply with the following terms (in the implementation of measures for the development of LEI):

$$E_c \geq E_{dy} + E_f + E_{mb} + E_{ay}, \tag{2}$$

where E_{dy} – total annual costs of the departure yard; E_f – total annual costs of the formation subsystem; E_{mb} – total annual costs of the marshalling bay; E_{ay} – total annual costs of the arrival yard.

According to the results of the analysis [6], at rail yards average downtime of trains waiting for train locomotives in the departure yard exceeds 1.5 hours and is up to 3 or more hours, negatively affecting the operation of all previous elements of rail yards right the way to the arrival yard. Also, it is worth pointing out that part of the waiting time of a train locomotive in the total downtime of railway cars at rail yards ranges from 10% to 60% depending on the category of cars (with processing, without processing).

It should be noted that in justifying specific measures for the development of individual LEI under condition (2) it will not be possible to foresee simultaneous improvement of all elements of the rail yard. Therefore, the realistic option is the allocation of financial resources for improving equipment and technology only for the development of individual elements, starting with the train departure subsystems.

4.3 Methodology for determining costs of implementing measures for the development of individual LEI

Consequently, it is necessary to determine costs for the implementation of each measure separately. Additional number of train locomotives required for the removal of prepared trains from the rail yard delayed during the period due to violation of the condition in interaction between the departure yard and adjacent sections (the shortage of train locomotives) is determined by the formula:

$$\Delta M = \Delta N \cdot K_n, \tag{3}$$

where ΔN – the number of trains delayed in the departure yard per year during periods Δt_l^a due to absence of train locomotives; K_n – the coefficient of locomotive needs for a pair of trains:

$$K_n = \frac{1}{24}\left(\frac{2l}{v_g} + t_y' + t_y''\right), \tag{4}$$

where l – length of the design-section; v_g – sectional speed on the design-section; t_y', t_y'' – average time of finding locomotives at yards limiting the design-section.

Increased investments in a locomotive fleet are determined by the formula:

$$\Delta I_l = C_l \cdot \Delta M \cdot \alpha_l, \tag{5}$$

where C_l – cost of one train locomotive; α_l – coefficient taking into consideration the location of locomotives in repair and costs associated with the locomotive development.

Costs of purchasing additional locomotives and their maintenance are determined by the formula:

$$E_{dy}^l = \frac{\Delta I_l}{t_{pp}} + \Delta B_d^l \ , \tag{6}$$

where t_{pp} – normative payback period; ΔB_d^l – additional running costs for the maintenance of additional locomotives and related devices per year.

If needed to increase the inspection station staff, costs on salary are determined by the formula:

$$E_{dy}^{br} = 12 \cdot t_s \cdot n_s \cdot x \cdot S \cdot C_{mon}^{gr} \tag{7}$$

where t_s – duration of one shift (12 hours); n_s – number of shifts (4, 5); x – number of groups in the brigade; S – number of brigades; C_{mon}^{gr} – salary of one group of inspection station employees per month.

In case of insufficient number of tracks in the departure yard, annual costs of track-laying and maintenance of additional tracks are determined by the formula:

$$E_{dy}^t = \Delta Q_{dy}\left(\frac{I_t}{t_{rr}} + c_d^t\right), \tag{8}$$

where ΔQ_{dy} – number of additional tracks in the departure yard necessary to eliminate delays in the formation of trains from marshalling bay; I_t – capital costs for laying one track with all necessary devices; c_d^t – running costs of maintaining one track and necessary devices per year.

Moreover, it is necessary to take into account the costs associated with railway cars' downtimes in the departure yard under technological operations and in waiting for their execution as follows:

$$E_{dy}^d = 365 \cdot m \cdot t_{dy}(N_{tr} + N_f)C_{c\text{-}h} \ , \tag{9}$$

where t_{dy} – average duration of trains' downtimes in the departure yard.

Total annual present costs for the implementation of all measures to eliminate additional trains' downtimes in the departure yard due to violation of the conditions in interaction of adjacent elements are calculated by the formula:

$$E_{dy} = E_{dy}^l + E_{dy}^{br} + E_{dy}^t + E_{dy}^d . \tag{10}$$

When carrying out technical and economic calculations, it is necessary to take into account the restrictions on loading of all elements of the yard (not more than 100%).

At various rail yards the amounts of maneuver stretches and shunting locomotives are not the same that affects the time of railway cars being in the train formation subsystem. Annual costs for the construction and maintenance of maneuver stretch, and for the purchase and maintenance of shunting locomotives are determined by the formula:

$$E_{ef}^{el} = \frac{\Delta Q_{ef} \cdot I_{ef} + \Delta M_{man} \cdot C_{man}}{t_{rr}} + \Delta Q_{ef} \cdot b_{ef} + \Delta M_{man} \cdot b_{man}, \tag{11}$$

where ΔQ_{ef} – additional required number of maneuver stretches; I_{ef} – capital costs for the construction of one exhaust track; ΔM_{man} – additional required number of shunting locomotives; b_{ef} – costs of maintaining one maneuvering track; C_{man} – cost of one shunting locomotive; b_{man} – costs of maintaining one shunting locomotive.

Annual costs related to the railway cars' downtime in the train formation subsystem:

$$E_{ef}^{d} = 365 \cdot m \cdot N_{tf} \cdot t_{tf} \cdot C_{c\text{-}h} , \tag{12}$$

where t_{tf} – average duration of a train being in the train formation subsystem.

Then total annual costs of in the train formation subsystem amount to:

$$E_{ef} = E_{ef}^{el} + E_{ef}^{d} . \tag{13}$$

Rail yards have different number of tracks in the marshalling bay. In addition, the average duration of prepared trains being on marshalling tracks may vary.

In cases when there is a need to increase the number of tracks in the marshalling bay to provide a high level of interaction between the elements of rail yards, costs of their construction and maintenance will be:

$$E_{mb}^{t} = \Delta Q_{mb} (\frac{I_{mt}}{t_{rr}} + c_{d}^{t}) , \tag{14}$$

where ΔQ_{mb} – required number of additional tracks in the marshalling bay; I_{mt} – costs of construction of one marshalling track; c_d^t – running costs for the maintenance one marshalling track.

Costs of to railway cars' downtimes in the marshalling bay after their accumulation due to waiting formation and relocating of trains to the departure yard are determined by the formula:

$$E_{mb}^{d} = 365 \cdot m \cdot N_{f} \cdot t_{w}^{f.d} \cdot C_{c\text{-}h} \tag{15}$$

where $t_w^{f.d}$ – average downtime of one train in the marshalling bay waiting formation and relocating to the departure yard.

Total annual costs of the marshalling bay amount to:

$$E_{mb} = E_{mb}^{t} + E_{mb}^{d} \tag{16}$$

At rail yards marshalling humps have different technical equipment and differences in the operating technology reflected in the value of the important marshalling hump's operating indicator – marshalling hump interval t_{mh}. The duration of railway cars being in the arrival yard is affected by the number of tracks, the technology of technical maintenance of railway cars by the inspection station brigade, the intensity of dismantling of trains in the marshalling hump. Therefore, the list of options for the technology of the arrival yard's and its technical equipment's operation is formed depending on the above mentioned factors.

Costs of salaries of inspection station brigades E_{ay}^{br} is calculated by the formula (7), and costs of technical equipment of the marshalling hump E_{mh} – by the formula which is similar to the dependency (14).

To determine costs of building additional tracks in the arrival yard E_{ay}^{t} the formula (14) is applied, and costs of downtime of railway cars in the arrival yard E_{ay}^{d} – dependency (15). Then, total costs for the implementation of measures to improve the technical equipment and operating technology of the arrival yard and marshalling hump can be calculated as follows:

$$E_{s.\,mh} = E_{ay}^{br} + E_{mh} + E_{ay}^{t} + E_{ay}^{d} \tag{17}$$

4.4 Discussion of research findings

Thus, when selecting the optimal option of organizing of the rail yard's operation and reducing the influence of violations of the conditions in interaction between the adjacent local elements of interaction (LEI) on the operation of elements of the yard, it is necessary to take into account the variability of delivery of railway cars, as well as differences in technical equipment and technologies of rail yards. Moreover, it is also important to consider both direct and indirect mutual effects of parameters of yard elements.

In modern conditions of sharp reduction in loading of rail yards during a prolonged economic crisis will not require the need to perform technical and economic calculations to determine the cost of implementing all the activities listed in the research. Probably, it will be enough to calculate only one or maximum several measures to improve the feedback interaction of adjacent LEI.

5 Conclusions

The railway transport plays especially important role in market transformations of the state and has a significant impact on the proper functioning of industrial enterprises and agriculture. It is very important in providing foreign economic relations and defense of the state. At present the railway transport has the task of comprehensive development, speeding turnover of railway cars, increasing the speed of trains, compliance with the train schedule and ensuring the cargo and passenger transportation safety. These requirements are extremely urgent for the railways in the face of a large deficit in financial resources.

Serious technical condition of the infrastructure and the rolling stock of railways require a radical change in the attitude of public authorities towards solving these issues. To resolve them, it is necessary to provide the railways with the right to fairly earn money for their operation at the legislative level. First and foremost, it is urgent to introduce a market tariff system and payment for the use of railway cars. The railway transport needs to be transformed into a full partner in the economic system of the state through introduction of the transport process compromise management system and termination of investing large business at the expense of the railway transport. The implementation of such a statesmanship to the solution of the transport problem in Ukraine will enable railways to not only urgently carry out technical and technological modernization, but also to resolve an extremely important staffing issue by substantially raising salaries for employees not less than three times. Thereby, the financial state improvement will enable railways to reasonably make progressive decisions on their own development using the designed methodical recommendations, which will significantly improve the quality of transport servicing of national economy.

References

1. Ye. Aloshynskyi, V. Naumov, O. Pestremenko-Skripka, *Technical Transaction. Civil Engineering*, **6**: 43-54 (2018)
2. Ye. Aloshynskyi, H. Prymachenko, Yu. Shuldiner, *Perspectives of Science and Education,* **1**: 5-12 (2018)
3. D. Lomotko, Ye. Aloshynskyi, G. Zambrybor, *Transportation Research Procedia*, **14**: 2762-2766 (2016)
4. S. Panchenko, T. Butko, A. Prokhorchenko, L. Parkhomenko, *Scientific Bulletin of National Mining University*, **2**: 93-98 (2016)
5. T. Butko, O. Shander, *Eastern-European Journal of Enterprise Technologies,* **2/3**, 68: 55-58 (2014)
6. L. Yelnikova. *The efficiency increase of the transportation process of the railway directions by improving the operational management of the locomotive park operation*. Dnipropetrovsk: Dnipro National University of Railway Transport named after academician V. Lazaryan (2016) http://eadnurt.diit.edu.ua/bitstream/123456789/4713/1/aref_Yelnikova%20.PDF
7. Development Strategy of PJSC "Ukrzaliznytsya" for 2017-2021 [Online] from http://cfts.org.ua/files/Session_1_Kravtsov.pdf [Accessed on 27 May 2019]
8. P.E. Chestnov. *Railway Transport,* **9**: 64-65 (2013)
9. A.I. Verlin, N.I. Berezovyiy, R.V. Vernigora, A.M. Shepeta, *Transport systems and transportation technologies*, **3**: 15-19 (2012)
10. P.O. Yanovsky, G.G. Volkovska, *Transport systems and transportation technologies*, **4**: 103-108 (2012)
11. P.O. Yanovsky, *Railway transport of Ukraine*, **1**: 41-49 (2014)

Economic and legal aspects of transportation of cargo in Ukraine in the conditions of Euro integration

Nadiia Bocharova[1], *Victor* Popov[2] and *Evheniia* Tupytska[2]

[1]Kharkiv National Automobile and Highway University, Ukraine
[2]YaroslavMudryi National Law University, Ukraine

Abstract. The article is devoted to the study of legal and economic relations that arise in the field of cargo transportation, in particular, their transshipment and legal grounds for its implementation. The article analyzes the legal nature of cargo handling in both the technological process and civil law services. It is determined that transshipment is a separate technological operation in the course of transportation and has its own basis for implementation, which serves as an agreement and therefore confirms and ensures the existence of a legal relationship between individual actors in economic turnover. The authors list the criteria for the economic efficiency of cargo handling, including delivery of cargo to the area where there are no certain types of transport; economic expediency of certain types of transport; and speeding up the delivery process. The European experience in using the cargo handling operation, the basis of the activity of logistic cents and the purpose of their functioning are analyzed. The statistical indicators of economic efficiency in the EU member countries are determined. The authors draw conclusions on the impact of European integration processes on the Ukrainian economy, in particular on the development of the logistics sector.

In the era of reaching the peak of commodity circulation both inside the countries and in international directions, the issue of cargo transportation remains one of the most urgent for modern society. The production and supply of certain goods in a territory is impossible without the proper functioning of the transport system, the coordinated interaction of the relevant elements, as well as the harmonization of economic and legal regulation of transportation relations. Despite the geographical location of Ukraine and the important place in the world's transport flows, as well as the potential for transportation of goods, our country is only the 80th among the 160 countries in the World Bank's LPI ranking (Logistics Performance Index [1] (Logistics Efficiency Index)) One of the consequences of the rapid development of European integration processes in Ukraine is its gradual access to the European Union transport space (hereinafter - the EU). This, in turn, leads to the need to create an effective legal framework that will meet modern social and economic requirements, as well as ensure proper trade between Ukraine and the EU countries and transit freight across the territory of Ukraine.

[1] Corresponding author: bocharova.n.a.xnadu@gmail.com

Freight forwarding is an integral part of the direct shipment, irrespective of whether such transportation is carried out with the help of certain containers, tanks, etc. on the platforms of the relevant vehicles, or such cargo is loose or liquid and requires overloading or transfusion between separate capacities intended for transportation by separate types of transport. Cargo handling has a number of criteria for economic efficiency, both for carriers and for recipients of such services. In particular, they can be attributed to: 1) the delivery of cargo to the area where there is no type of transport which was the primary transportation; 2) even in the case of physical delivery of cargo to a certain area there is an economic expediency of transportation by a certain type of transport. This is the case, for example, of the air communication in which a cargo aircraft flies, for example, from New York to Kiev, while the goods must be delivered to Zhytomyr, in which case it is economically reasonable not to use air transport again, but simply to carry out the transshipment of goods to rail or road transport (if available); 3) Transshipment of cargoes can not only reduce the cost of transportation of cargo, but also accelerate the transport process. An important point, especially for perishable goods, is the possibility at any price to increase the speed of delivery of goods to the destination, changing the mode of transport at a certain stage of following the cargo can significantly accelerate the process of transportation, thereby reducing the likelihood of damage to goods.

With the cargo handling, a number of theoretical issues of the contractual sphere of economic relations are linked, as well as some aspects of the law enforcement practice. In particular, the issues of the subject structure of the relevant relations, distribution of the risks of loss or damage to the cargo, etc. are relevant. In this regard, it seems expedient to consider, within the scope of this study, the procedure for the transshipment of goods within the framework of relations with their carriage in direct and direct mixed traffic.

It is worth noting that the term "transshipment" is used in the laws and other special regulations on the transport of goods by different modes of transport. Part 3 of Art. 312 of the Commercial Code of Ukraine established the participation of the transshipment point in the transport of goods in direct mixed traffic. In this case, the relations of carriers during such transportation, as well as the conditions of operation of the transshipment points, are regulated by the agreement, the procedure of which is established by transport codes and statutes [2]. Thus, transshipment is a separate operation within the process of transportation, which should be considered as a legal relationship in which there are parties, an object, as well as content in the form of rights and obligations of the participants in the transportation process.

The answer to the question about the nature of the cargo handling operation and its place in the process of transportation is in the Service Provision Rules in seaports, where Clause 2.1. The definition of cargo operations is given. These include loading, unloading and overloading of cargo, loading and unloading of a vehicle. It is established that cargo operations are carried out on the basis of the relevant agreement [3]. In this regard, the question arises whether the transshipment process can include all the above operations? Clause 2.1 of the specified normative act stipulates that all types of operations are cargo and are carried out on the basis of the contract of transshipment, but along with this, in clause 3.3., it is assumed that the services related to the cargo operations that can be provided by the port operator, the terminal operator , include: load of goods; unloading of goods; clearing of holds of vessels, cars after unloading of cargoes; moving of cargo; technological accumulation of cargo, warehouse operations with cargo; fastening, special fastening of loads, etc. [3].

At present, the use of categories of transshipment, loading and unloading, in our opinion, is related to the designation of a specific part of the technological process carried out in the field of transportation. Any movement of goods by several modes of transport is associated with transshipment, but therefore the category of "load" is narrower and refers to the process of moving the goods to a particular type of transport, while the meaning of "transshipment" is wider and includes "shipment" to one vehicle and the actual load of another and is carried out for a specific purpose of delivering the goods.

As a cargo operation, transshipment, in this case, is an operation carried out at the stage of movement (transportation) of the cargo. Since cargo handling takes a clearly defined independent place in the transport process implemented on the basis of the separate civil law agreement, it can be considered as a service and as a technological operation provided (carried out) when transporting goods in direct combinations. At the same time, both legal and economic content of the relevant service are disclosed through the content of the operation. For example, the technological process of transshipment of petroleum products includes their acceptance at delivery points - at the railroad bridge or in the indicators of the system for measuring the quantity and quality of petroleum products at the delivery and receipt point (depending on the way of delivery of oil shipment to the seaport), oil transportation through oil pipelines to tank farms , formation of tanker batches, acceptance for storage, delivery to the pier, shipment to tankers and acceptance of oil stores from tankers. The above stages of the technological process, as O. Gavrilina points out, constitute a single comprehensive service for the handling of oil cargo through a specialized oil port with the purpose of their further removal outside the territory of the state. Taking into account the content of the said technological operation, the scientist states that the transshipment service involves the release of oil from railway tanks or the acceptance of the quantity of oil of the specified quality, delivery to the terminal tanks, accumulation and storage in the tanks of the terminal, delivery to the point of loading of oil cargoes and pouring them into a vessel (tanker) [4, p. 202, 206].

Thus, it can be concluded that the transshipment service provided in the course of the carriage of goods in direct mixed traffic is part of the legal relationship arising between the carriers on the basis of the carriage agreement and is absorbed by the legal relationship arising from the contract of carriage of cargo in direct mixed traffic. It should be noted that the content of the service in more detail can be disclosed through the means, methods and place of its provision. Transshipment services should be understood as material things, mechanisms and equipment, in particular, cranes, lifts, forklifts, overpasses, discharge systems, etc., through which the relevant service is provided.

Although the transshipment point is an integral part of the carriage of goods, these relations are not sufficiently regulated in domestic legislation because of obsolete approaches to property rights that arise in the field of transportation, since the Soviet legal system considered the infrastructure as an object of strategic importance. This also affected the fact that the probability of privatization of such objects was practically excluded. One can only recall what discussion arose about the possibility of privatization and its proper infrastructure after Ukraine gained independence. Although this type of private property is common in European countries. In practice it turned out that infrastructure objects began to be owned by separate legal entities using various mechanisms of privatization. As a result, due to the lack of a comprehensive program for the privatization of rail transport, they become new owners of newly created things. In addition, in uncomplicated cases, transshipment points are acquired by individuals, including by way of their privatization. Despite this, there is still no comprehensive approach to regulating ownership relations with transport infrastructure. Therefore, we can conclude that, against the background of fundamental changes in economic relations, since the 90's, the theoretical model of ownership has not undergone any changes [5, p. 32].

The domestic legislation provides for the possibility of staying transfer points in private ownership, as evidenced in particular by paragraph 9 of the Charter of Ukrainian Railways of [6]. The specified norm actually determines the possibility of staying of transfer points and access railroads in private ownership. In addition, in individual cases, the relevant objects are owned by individuals who are not registered in the Unified State Register of Legal Entities, Individual Entrepreneurs and Public Associations as Individual Entrepreneurs.

The current stage of development of the Ukrainian economy, especially in the context of European integration, gave a new impetus to the development of logistics in the middle of the country. One of the new directions of development of the domestic transport process is the creation of transport-logistic (logistics) centers. The practice of such centers is not new for Europe, since, for example, Interpol Bologna, which is currently one of the most important logistics platforms throughout Europe, has been active since 1971 [7, p. 203]. Starting in the 1990s, logistics centers are beginning to form in Europe as economic zones providing various transport and logistics services [8, p. 72].

The progressiveness and efficiency of the functioning of such centers is confirmed by the economic indicators of the European countries that have turned to the relevant practice. At present, the total volume of the European logistics services market is over EUR 600 billion. At the same time, logistics companies account for 30% of logistic functions in all branches of the economy. For example, in Holland the activity of transport and logistics centers brings 40% of the revenue of the transport complex, in France - 31%, in Germany - 25%. The average in the countries of Central and Eastern Europe - 30% [9, p. 32]. However, for the domestic legal system, this phenomenon is new.

In modern scientific literature there is a lack of a unified approach to the definition of a transport and logistics center. According to a study conducted by McMaster University at the request of the Ontario Ministry of Transport, there are more than 20 terms in the world that are intrinsically close and describe this phenomenon [10, p. 201]. Transport and logistics centers may have different names depending on local features of the place of operation. The literature determines the possibility of the existence of a transport and logistics center in the form of a transport node, which performs a wide range of tasks [11, p. 115]. In Europe, such centers are called freight centers (Gueterverkehrszentren), intermodal Hub (Russia), multimodal platform, logistics areas, transport centers, etc. [12, p. 77]. In Japan, the logistics center is called a logistics hub [13, p. 215]. However, their most common name in world trade practices is “freight villages” [8, p. 72].

In general, transport and logistics centers are regarded as certain infrastructures equipped with all means to provide a range of services related to all transport operations, and cargo villages - as a specific region of the country, in which different market operators are provided with a full range of services relevant to transport and logistics distribution of goods for compatriots, as well as for international transit [14, p. 33]. Also in modern literature the transport and logistics center is considered as the center, which coordinates the use of different types of transport, performs loading and unloading and cargo handling, provides short-term and long-term cargo storage, cargo processing, execution of necessary customs procedures, inventory and freight forwarding, provides a set of service and commercial and business services, including banking, information, consulting and analytical services [15, p. 279]. M. V. Kondratyuk considers the transport and logistics center as a territorial association of independent companies engaged in cargo transportation and related services (for storage, maintenance and repair), which includes at least one terminal suitable for intermodal transport [16, p. 31-33]. The European Association of Cargo Villages and Logistics Centers (Euro platform) defines a logistics center as a center located within a defined territory, within which, on a commercial basis, different operators carry out activities related to transport, logistics and distribution of goods for national and international transportation. In this case, operators can be the owners or tenants of buildings and other infrastructure (warehouses, distribution centers, storage areas, offices, automobile services, etc.), which was built there [17].

The legal category of logistics centers has started to be used in domestic legislation recently and is associated with events in the East of Ukraine. The order of movement of goods to the district or from the area of the anti-terrorist operation, approved by Resolution of the Cabinet of Ministers of Ukraine of March 1, 2017 No. 99, contains the definition of the humanitarian-logistic center as a site on the controlled territory in the area of anti-terrorist operation, where the retail and small-scale trade of food products is carried out, personal hygiene, clothing and footwear, household chemistry, as well as services, humanitarian and charitable assistance [18].

Although the transshipment point and the logistics center are not identical, they can perform similar functions from a functional point of view. The magnitude of the activities of transshipment points and logistics centers is different. Historically, the former are primarily engaged in the transshipment of goods, as well as, if necessary, their storage. At the same time, logistics centers provide a whole range of services, and when viewed as objects of transport infrastructure, they are considered as the unification of several such objects, for example, terminals with warehouse areas, platforms, offices, etc.

Thus, we can assert that the transshipment point and the logistics center are characterized by different approaches to the form of organization of transport processes. At the same time, in the current development of contractual relations, logistic centers exist as an economically and organizationally more progressive form of organization of cargo transportation, which provides the sender and carrier with a whole range of additional services. A logistics center can exist, in particular, in the form of a traditional transshipment point, but the transshipment point cannot be a logistics center due to the limited functionality defined by the provisions of regulatory acts [19, p. 129].

Summarizing the above, it should be noted that the gradual distribution of transport and logistics centers in our country creates a considerable basis for the development of the economy in the field of transport and logistics. Integration of Ukraine into the European space and the opening of new markets have created conditions for this development, especially as the commodity turnover constantly increases both within the country and in the foreign economy. The results of the introduction of new economic models can be noticed today, and consequently, the speed and quality of services provided in the transport sector are increasing. In turn, the provisions of the law regulating relations in the field of the operation of transshipment points require gradual changes. The legal basis in the form of legal acts regulating relations on the carriage of goods by various modes of transport, which was formed in the second half of the 20th century, is characterized by its efficiency and complexity. Evidence to this is that today the legal acts adopted during this period remain valid and they are effectively coping with the implementation of the regulatory function. However, the development of technological processes in the field of trade and the emergence of new transport infrastructure objects and new entrants in the relevant relations require the bringing of previously adopted regulations in line with the new requirements of the present.

References

1. *International LPI global ranking* URL: http://lpi.worldbank.org/international/global (access date: 03.05.2019).
2. *Hospodars'kyy kodeks Ukrayiny: Zakon Ukrayiny vid 16.01.2003 № 436-IV.* Vidom. Verkhov. Rady Ukrayiny, **18**, 144 (2003)
3. *Pravyla nadannya posluh u mors'kykh portakh Ukrayiny: zatv. nakazom M-va infrastrukt. Ukrayiny vid 05.06.2013 r. № 348.* Ofits. visn. Ukrayiny, **65**, 2352 (2013)

4. E.A. Gavrilina, *Sistema dogovornykh svyazey na rynke nefti i nefteproduktov*. M., 244 (2014)
5. M. Domashenko. Pravo Ukrayiny, 12, 32-39, (2010)
6. *Pro zatverdzhennya Statutu zaliznyts' Ukrayiny: postanova Kabinetu Ministriv Ukrayiny vid 06.04.1998 r. № 457.* Ofits. visn. Ukrayiny, **14**, 548 (1998)
7. *Postanova Vyshchoho hospodars'koho sudu Ukrayiny vid 8 lyst. 2016 r., sudova sprava № 914/595/16.* URL:http://reyestr.court.gov.ua/Review/62724626.
8. S. M. Bonyar, Y.R.Korniyko. Investytsiyi: praktyka ta dosvid, **7**, 71-72 (2012)
9. M. V. Kondratyuk, Visn. ekonomiky transportu i promyslovosti, **47**, 31-33 (2014)
10. *Postanova Vyshchoho hospodars'koho sudu Ukrayiny vid 8 lyst. 2016 r., sudova sprava № 914/595/16* URL: http://reyestr.court.gov.ua/Review/62724626
11. O. E. Sokolova. Naukoyemni tekhnolohiyi, **1 (21),** 114-118 (2014)
12. O. V. Nelipovych, S. A. Popel', K. D. Motyuk. Mytna bezpeka, **2**, 76-82 (2013)
13 K. S. Romanenko. Materials of III sciences conference (Odessa, 22-23 Nov. 2013). Phoenix, 214-217 (2013)
14. S. M. Bonyar, YA. R. Korniyko, Ekonomika ta derzhava, **3**, 32-35, (2012)
15. N. M. Dashchenko. Project management, system analysis and logistics. Technical series, **8**, 278–282 (2011)
16. N. A. Butakova, *Pravovoye regulirovaniye mul'timodal'noy perevozki gruzov: komparativnyy podkhod*, M., 494 (2016)
17. *Logistics Center Definition.* URL: http://europlatforms.eu/Logistic %20CenterDefinition.html (access date: 11.05.2019).
18. *Pro zatverdzhennya Poryadku peremishchennya tovariv do rayonu abo z rayonu provedennya antyterorystychnoyi operatsiyi: postanova Kabinetu Ministriv Ukrayiny vid 01.03. 2017 r. № 99.* Ofits. visn. Ukrayiny, **21**, 592 (2017)
19. Vozdushnyy kodeks Soyuza SSR : utverzhden Ukazom Prezidiuma Verkhovnogo Soveta SSSR ot 11.05.1983 g. № 9575-KH. URL: http://www.libussr.ru/doc_ussr/usr_11552.htm (дата обращения: 17.05.2019).

Evaluation of influence of the logistics chain infrastructure on the logistics costs formation

Andrii Galkin[1]*, *Yevhen* Kush[1] and *Nataliia* Davidich[2], *Yuriy* Davidich[3]

[1]O. M. Beketov National University of Urban Economy in Kharkiv, department of transport systems and logistics, (Kharkiv)
[2]O. M. Beketov National University of Urban Economy in Kharkiv, department of project management in municipal services and construction
[3]Ukrainian State University of Railway Transport, department of transport systems and logistics (Kharkiv)

Abstract. The article is devoted to revealing the influence of the infrastructure of logistics chain on expenses in it. Possible types of warehouses which can be used in the logistics chain were identified. The expenses on the functioning of the warehouse according to its parameters were analyzed. The variable costs estimation of the warehouse was modelled via the freight turnover. The fixed costs estimation of the warehouse was modelled via its area. Regression nonlinear models of variables and fixed expenses of own/hired warehouse were obtained.

1 Introduction

The problem of adjusting the process of continuous supply of goods to a retail network requires the solution of the problem of the infrastructure of the logistics chain [1]. The easiest solution for retailers is the option when the goods are delivered to the stores "directly", the control over the volume of stocks belongs to the management company on the outsourcing rights [2] or "fulfilment" [3]. According to the results of research [4], most suppliers are not able to provide such services, or the expenses of these services are too high. An analysis of the modern logistics market in Ukraine shows that not more than 15% of suppliers can fulfill their contractual obligations when delivering goods. With the increase of sustainable freight turnover, the question of having of own warehouse arises. Many experts and researchers agree with the idea of combined flows via a company-owned distribution center or warehouse, which can accumulate the required margins and reduce overal costs [5], that is, to introduce additional participants into the infrastructure of the logistics system.

On the other hand, as a result of falling purchasing power of end-consumers, retailers began to reduce the size of orders to decrease stock level. This fact has a negative effect both on the logistics chain infrastructure and in general on its key performance indicators, which leads to the need to find alternative ways to reduce their own costs to ensure product competitiveness and sales. The set of elements that perform logistic tasks and ensure the

* Corresponding author: galkin.tsl@gmail.com

implementation of logistics processes determine its infrastructure and affect the formation of logistics expenses.

2 Literature Survey and Purpose Statement

Analysis of studies has shown that in the field of logistics services, there are two options for using the warehouse in the distribution of material flows [5]:

1) construction or rental of premises and the creation of a warehouse at their base;

2) use of public warehouses or outsourcing.

According to researchers [7], warehouses by the form of use can be classified as follows (Fig. 1):

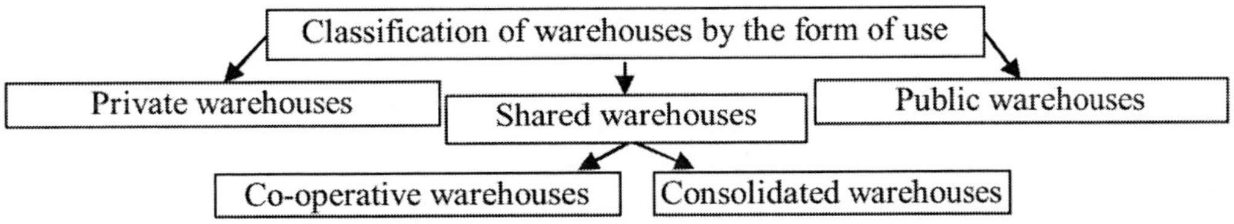

Fig. 1. Classification of warehouses by the form of use.

The authors in [2] propose to allocate two types of warehouses: co-operative warehouses and Public warehouses. The various types of warehouses by the form of use have a number of advantages and disadvantages [4]. The Company-owned warehouse has a fairly large number of benefits, but in terms of profitability, not every enterprise has the ability to hold it.

In the case of using an own warehouse, the costs of its operation consist of lease, wage costs, operating costs, utilities, depreciation, etc. [6]. The value of the processing of goods in the hired warehouse is influenced by the type of goods (their weight, volume, consistency) [8], the type of handler [9], the type of equipment for storing goods taking into consideration their properties [10], the technology of placement and stowage of goods during storage and processing [6]. Recently, "secure storage" of goods and equipment has become common. Secure storage of goods implies a basic list of operations, such as unloading and loading, receiving goods, picking and controlling orders, marking, repacking and other services.

The type of elements of the logistics infrastructure determines the costs necessary to deliver the material flow to the end-conusmer. The costs of the functioning of the warehouse scientists propose to determine on variable and fixed components [6]. Variables include manipulation costs (internal displacement costs) and costs that vary depending on the size of stocks and the level of material flow [6, 12]. Fixed costs include costs that do not depend on stocks level. These include leasehold depreciation, depreciation of equipment, salaries of managerial staff and specialists, utility bills and other [6, 12, 13].

Thus, the total costs of functioning of the warehouse can be determined by the following dependence[14]:

$$C_w^o = C_f S + C_{\text{var}} Q, \tag{1}$$

where C_f – fixed costs, money units/day; S – area of the warehouse, m^2; C_{var} – variable costs, money units/m^2; Q – freight turnover of the warehouse, t/day.

The expenses for the operation of the hired warehouse will be equal to:

$$C_w^r = T \cdot Q, \tag{2}$$

where T – fare for storage (processing) of one ton of material flow, c.u /t.

Provided that the manufacturer's costs for warehouse operations are known, the retail network has been formed and the maintenance costs for the material flow are fixed, the choice of the type of intermediate warehouse can determine the costs of the logistics system. The purpose of the article is to investigate the influence of the logistics chain infrastructure on the formation of logistics costs.

3 Methodology of conducting experimental researches

The infrastructure of the logistics system can be represented as a set of *M* producers of material flow, *W* of warehouses and *S* of retailers. The connection between the elements of the logistics system participants is carried out by means of transport on the site of the Manufacturer-Distribution and Distribution-Sales (Figure 2).

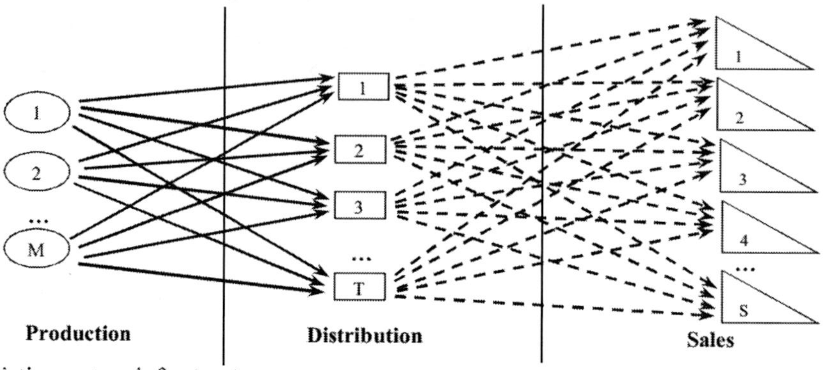

Fig. 2. Logistics system infrastructure:
⟶ - the movement of material flow from the sphere of production into the distribution sector;
- - -➔ - the movement of material flow from the sphere of distribution into the sphere of sales.

In this case, the total logistics costs for the promotion of the material flow in the logistics system will be equal to:

$$C_{LS} = \sum_{i=1}^{n} C_i, \tag{3}$$

where C_i expenses of the *i*-th participant of the logistics system, money units;
n – number of participants in logistics system, units.
In expanded form, the total logistics costs for the promotion of the material flow can be formalized as follows:

$$C_{LS} = \sum_{j=1}^{M} C_j^M + \sum_{k=1}^{T} C_k^{M-W} + \sum_{l=1}^{W} C_l^W + \sum_{m=1}^{D} C_m^{W-S} + \sum_{o=1}^{S} C_o^S, \tag{4}$$

where C_j^M – expenses for storage of material flow in the warehouse of *j*-th manufacturer, money units; M – number of producers, units; C_k^{M-W} – expenses for transportation of material flow from the sphere of production in the sphere of distribution by *k*-th transport participant, money units; T – number of transport participants in the section Production-Distribution, units; C_l^W – the cost of storing the material flow in the l warehouse, money units; W – number of warehouses, units; C_m^{W-S} – expenses for the transportation of material flow from the sphere of distribution into the sphere of marketing by *m*-th transport participant, money units; D – the number of transport participants in the Distribution-Sales section, units; C_o^S – the cost of storing the material flow of the *o*-th participant of the retail network, money units; S – number of participants in retail network, units.

The formation of the costs of the logistics chain for the material flow movment depends on the type of warehouse used in the system. In addition to the own and hired warehouse, the company may be offered to use a shared warehouse. The costs of using the shared warehouse include only the cost of renting the warehouse, depending on the area chosen and the location of the warehouse. They may depend on the size of the chosen area and location.

Research of the dependence of variable and fixed costs on its parameters was carried out using regression and correlation analysis [15]. Characteristics of model parameters were determined using methods of mathematical statistics, and the significance of model factors was determined on the basis of Student's t-test, information capacity of models – using F-ratio. The tightness of the relationship between dependent and independent variables was determined using the correlation coefficient and the approximation error. Further, a study was conducted of the dependence of the variable costs of own warehouse on the volume of storage of products using nonlinear regression analysis. A graphical representation of this dependence is shown in Fig. 3.

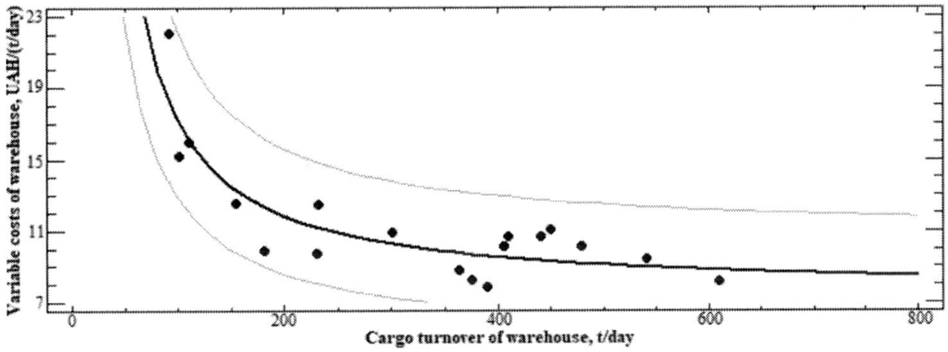

Fig. 3. Chart of variable expenses dependence in owned warehouse on its freight turnover.

Mathematically, this dependence can be described by a regressive nonlinear model:

$$C_{\text{var}} = (2,7343 + \frac{140,792}{Q})^2, \tag{5}$$

where Q – freight turnover of warehouse, t/day.

Statistical characteristics of the obtained model (correlation coefficient 0.77, F-ratio 53.31) testify to its adequacy. An increasing of the freight turnover in the warehouse leads to the use more handling machines and workers to meet the demand, and vice verse. Subsequently, a study was made of the fixed costs estimation of owned warehouse on the warehouse space (S) variation, fig 4.

Mathematically, this dependence can be described by a regressive nonlinear model:

$$C_{fix} = 20,5336 + \frac{10406}{S}. \tag{6}$$

Statistical parameters for the models obtained (correlation coefficient 0.92, F-ratio 190.99) testify to its adequacy.

With the increase in warehouse space the cost of its maintenance and wages and utilities also increases. However, at the same time, the specific cost of operatting 1 m^2 of warehouse space is reduced, which leads to a reduction in fixed costs.

Consequently, the cost of storing the material flow in the l-man will be equal to:

$$C_l^W = \sum_{l=1}^{W} Q_l \cdot (2,7343 + \frac{140,792}{Q_l})^2 + \sum_{l=1}^{W} S_l \cdot 20,5336 + \frac{10406}{S_l}. \tag{7}$$

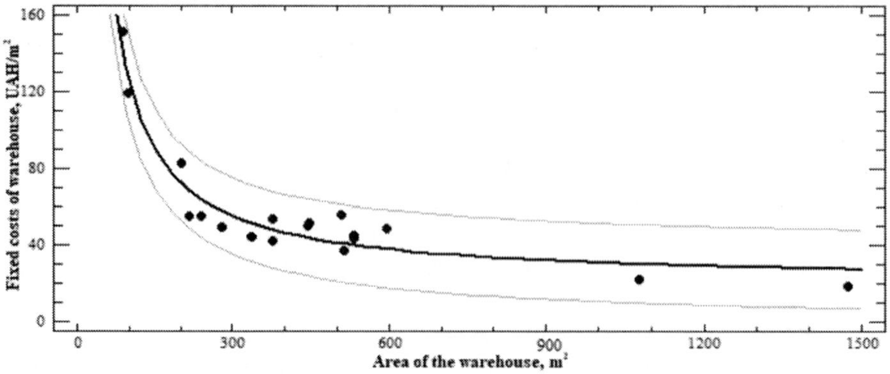

Fig. 4. Dependence of fixed expenses of own warehouse on its area.

The technological process of material flow processing at the hired warehouse is a set of consistently executed operations that are provided to the customer as services. These services are related to preparation for acceptance of products, receipt, transfer, unpacking, acceptance of products by quantity and quality, placing on storage, packing, selection, complectation, preparation for a delivery and delivery of products to the consumer. The effectiveness of managing processes in the warehouse essentially depends on the execution of separate operations [11]. The costs for technological performance of the warehouse can be found as the sum of the component costs of operations performed in it:

$$C_T^{W-R} = C_{st}^W + C_u^W + C_l^W + C_c^W + C_p^W + C_o^W \quad (8)$$

where C_{st}^W – the storage expenses of 1 ton of goods, UAH/(t/day); C_u^W – the unloading expenses 1 ton of product, UAH/(t/day); C_l^W – loading expenses 1 ton of product, UAH/(t/day); C_c^W – the complectation expenses of 1 ton of goods, UAH/(t/day); C_p^W – packing expenses 1 ton of goods, UAH/(t/day); C_o^W – other expenses assossiated with technology of freight processing, UAH/(t/day).

The main indicator for estimation the warehouse efficiency is the freight turnover, which depends on the parameters of the technological operation, namely: the processing time of one order, the working shift, the efficiency of the use of equipment and other parameters:

$$Q = f(M_1, M_2, ..., M_n), \quad (9)$$

where M_1, M_1, ..., M_n –parameters of technological operations in the warehouse.

Fig. 5 shows the dependence of the total costsc of servicing 1 t of material flow, depending on the area of the warehouse.

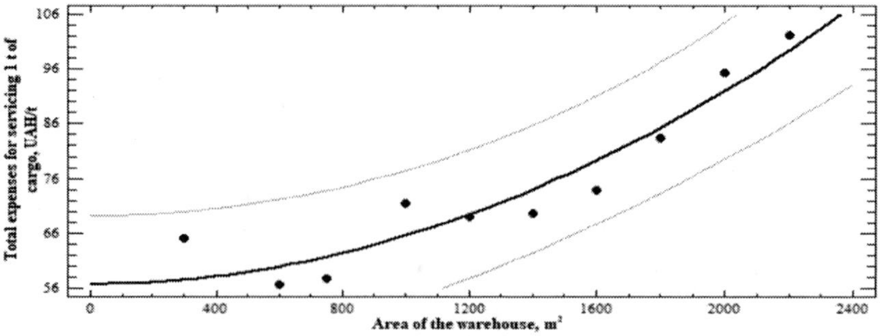

Fig. 5. The dependence of the total costs per 1 ton of MF depending on the area of the warehouse.

There is a feedback between the parameters of technological operations in the warehouse, which are dependent on freight turnover.

Within the presented research, an option was considered when the company selects to use of hired warehouse. In this case, it pays only the cost of services, which includes all the costs, as well as the use of the infrastructure of the complex.

Mathematically, this dependence can be described by a regressive nonlinear model:

$$C_T^{W-R} = 56,76 + 8,83 \cdot 10^{-6} \cdot S^2 . \tag{10}$$

Statistical parameters for the models obtained (correlation coefficient 0.91, F-ratio 79.53) testify to its adequacy. On the basis of the obtained dependence it can be concluded that with the increase of the warehouse area, the cost of services will increase.

4 Conclusions

The study of the influence of the logistics chain infrastructure on the formation of logistics costs in the article enabled to formulate the following conclusions:

- it was established that the infrastructure of the logistic chain (the types of warehouses), determines the logistics costs for the process of material flow progression;
- it was established that the total cost of storing depends on the volume of goods stored on it and the area of the warehouse. It is possible to formalize the total costs using a regressive multifactor model, in which the variables included the goods turnover in the warehouse and its area.

The results obtained in the article allow to determine the costs of functioning of the warehouse, depending on its type. This allows us to determine the total logistics costs, which consist of the costs of storage, transportation and distribution, which the system spends on promoting the methionic flow in it.

References

1. J. Fernie, *Handbook of Logistics and Supply-Chain Management,* (2008)
2. A. Galkin, C. Dolia, N. Davidich, TRP, **27**, 1187-1194 (2017)
3. G. Vaidyanathan, CACM, **48**, 89-94 (2005)
4. I. Litomin, I. Tolmachov, A. Galkin, TCP, **16**, 313-322, (2016)
5. R. D. Shapiro, J. L. *Heskett Logistics Strategy: cases and concepts*, (1985)
6. E. V. Krikavskiy, LogIstika dlya ekonomIstIv, (2014)
7. A. G. Kalchenko, LogIstika, (2003)
8. V. M. Kisliy, O. A. Bilovodska, O. M. Olefirenko, O. M. Smolyanik, Logistika: Teoriya ta praktika, (2010)
9. L. Douglas, M. Cooper, J. Pagh, IJLM, **9**, 1-19 (1998)
10. L. Filina-Dawidowicz, M. Postan, LF, **12** (2016)
11. A. Razzaque, C. Sheng, IJPDLM, **2**, 89-107, (1998)
12. M. Asadabadi, IJIEC, **6**, 305-314, (2015)
13. R. H. Wilson, HBR, **13**, 116-128, (1943)
14. Y. Kush, V. Skrypin, A. Galkin, K. Dolia, I. Tkachenko, N. Davidich, TRP, **30**, 216-225 (2018)
15. A. Frenkel, Mnogofaktornyie korrelyatsionnyie modeli proizvoditelnosti truda, (1966)

SHS Web of Conferences 67, 03003 (2019) https://doi.org/10.1051/shsconf/20196703003
NTI-UkrSURT 2019

Modeling and forecasting of cargo turnover at the railways of the Russian Federation according to statistics data

Petr Gerasimenko[1,1], *Valentin* Khodakovsky[1], *Sergey* Verteshev[2], *Sergey* Lyokhin[2] *and Alexander* Khvattcev[2]

[1]Saint Petersburg State Transport University,
[2]Pskov State University, Pskov, Russia

Abstract. The approach of modeling and forecasting of cargo turnover of transport is described. Based on statistical information on the transport activities of the Russian railway, a study of the dependence of the autocorrelation coefficients was conducted and a trend in the change in cargo turnover was established. The quality of the trend was evaluated and the point and interval forecasting of cargo turnover was carried out.

The role of railway transport in the Russian economy cannot be overestimated. A number of advantages make it the main mode of transport in the country. Losing to air transport in terms of speed of transportation, railway transport surpasses it in economic indicators in mass transportation. The advantage of road transport over rail transport with respect to transportation of bulk goods is mainly in the case of transport over short distances. River transport is inferior to rail transport because of the low speed of cargo delivery and the irregularity of traffic during the year.

Therefore, for Russia, the development of railway transport is of particular importance. It is the basis of the Russian transport system, and in some cases, not only the main, but also the only mode of transport that connects different regions and regions of our country, contributing to the development of the state economy. The work of railway transport can be judged by the development of cargo turnover, as an important indicator for the entire Russian national economy.

The study of the main indicators of cargo transportation in the Russian Federation for the period from 1920 to 2018 is carried out on the basis of statistical data. Table 1 presents statistical data on cargo turnover in billion tons of km [1], which means that the turnover only in 2018 began to exceed the level of 1990.

It should be noted that after the collapse of the Soviet Union, the turnover in Russia fell sharply, but later before the global economic crisis began to recover. With the onset of the crisis, there was a new drop in cargo turnover (Table 1).

Since cargo turnover determines the level of development of economic relations, improvement of trade and industrial activity, the degree of economic activity, due to the economic crisis, the drop in cargo turnover with the lag in one year is quite natural.

[1] Corresponding author: pv39@mail.ru

Table 1. Statistics

No.	1	2	3	4	5	6	7	8	9	10
Year	1928	1940	1950	1960	1990	2000	2005	2006	2007	2008
cargo turnove r	59	287	415	1047	2523	1373	1898	1951	2090	2116
No.	11	12	13	14	15	16	17	18	19	20
Year	2009	2010	2011	2013	2014	2015	2016	2017	2018	2018
cargo turnove r	1885	2011	2128	2196	2301	2306	2344	2493	2596.4	2506.17

Assessing further changes in cargo turnover in the Russian Federation, its point and interval estimates is interesting. As you know, forecast values can be obtained by constructing regression and trend dependencies. For correct forecasting of cargo transportation with the help of multivariate regressions, it is necessary to know well the need of the national economy in transportation and the existing cargo flows, as well as the main patterns of growth in transportation. The volume of cargo transport, measured by the number of tons of cargo transported and the number of ton-kilometers net, depends on many factors. Therefore, evaluation of the factors determining the functional dependence is quite a complex task and requires a large amount of information.

As a rule, not all factors are usually known and their values in most cases have a high error, especially in a crisis period. Typically, no factor regression model can include all the factors that affect cargo. In addition, the impossibility of their inclusion in the regression equation is explained by the limitations of the mathematical apparatus due to multicollinearity and heteroskedasticity, especially when the number of factors is close to the sample size. Therefore, more often carry out a forecast on the trend. For large systems such as Railways, the trend forecast for the previous years is quite effective.

It should be noted that the trend equation has an advantage, as it allows to take into account, although in an implicit form, the influence of all factors on changes in the levels of the projected indicator of cargo transportation. Moreover, the trend equation simulates a dynamic process, which is the logical basis for forecasting. With respect to multivariate regression, it should be noted that it determines variations in the level of the indicator in the static population, i.e. it does not explain the change in the time of the volume of cargo transportation, but its compliance with a certain combination of factors. Thus, the logical base of the forecast for multifactor regression is not adequate to the forecasting problem.

A significant disadvantage of trend forecasting is the inability to control the process of choosing the optimal mode of transportation, since it is not possible to determine different forecast options with different combinations of factor values, which usually allows you to perform a regression model.

The peculiarity of trend forecasting is related to the reliability of its parameters. If the parameters of the trend is not robust enough, it will be unreliable and prognosis.

According to the statistical information (Table 1), since the beginning of the crisis, namely since 2009, there have been significant changes in a number of economic factors that have a significant impact on the turnover of goods.

The analysis of Table 1 shows that there is a change in the structure of the transportation process and the turnover of goods will essentially change abruptly. It is obvious that a single trend will not provide qualitative and reliable forecasting. In this case, it is necessary to investigate the nature of the change in the trend of the dynamic series, namely whether it is continuous or piecewise continuous.

In the simulation of the dynamic series trend, the initial statistical data are divided into two sets: before the time of the jump and after that moment. Further, continuous regression functions are constructed (separately for each set).

On the basis of statistical data considered in the work as the values of the time series, the autocorrelation coefficients presented on the correlogram are determined (Fig. 1).

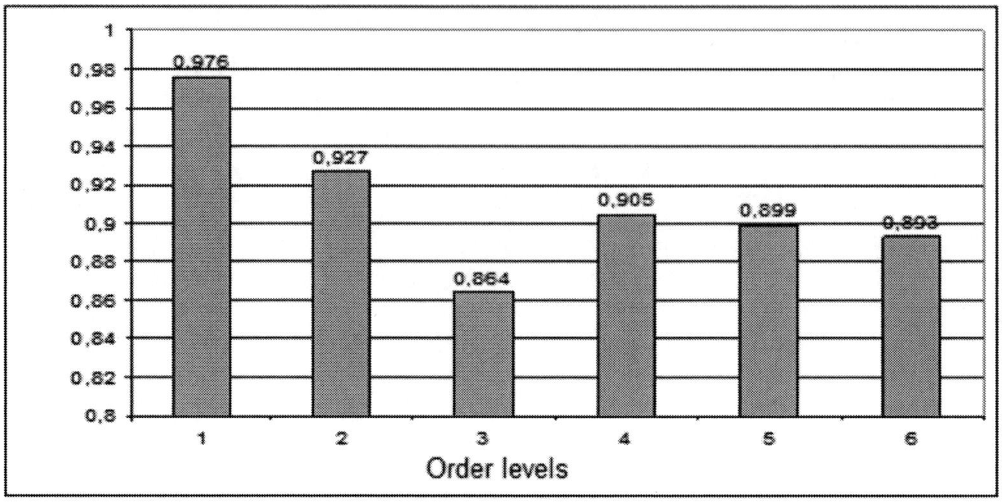

Fig.1. Correlation of autocorrelation coefficients.

From the correlogram follows the presence of an increasing trend and the absence of oscillatory changes in cargo turnover over time [2]. Based on this conclusion, a mathematical model of the dependence of cargo turnover on time (year).

Modeling in the work is performed using the least squares method [2], according to statistical data [1] (Fig. 2) and on the part of data (Fig. 3). The linear model is considered as mathematical models. The Regression tool of the Excel spreadsheet analysis Package add-in was used to model linear regression [3].

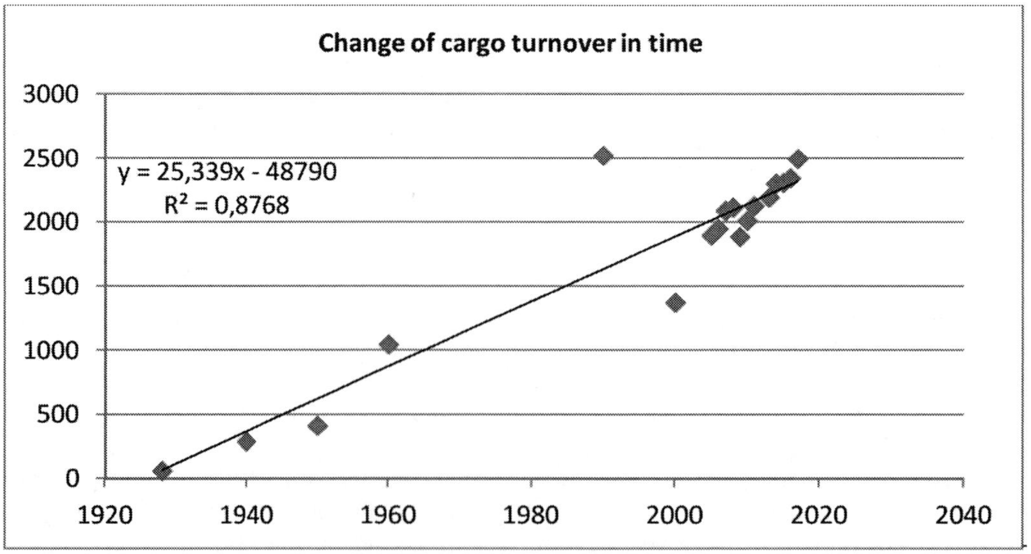

Fig. 2. “Trend line” based on data from 1920 to 2017.

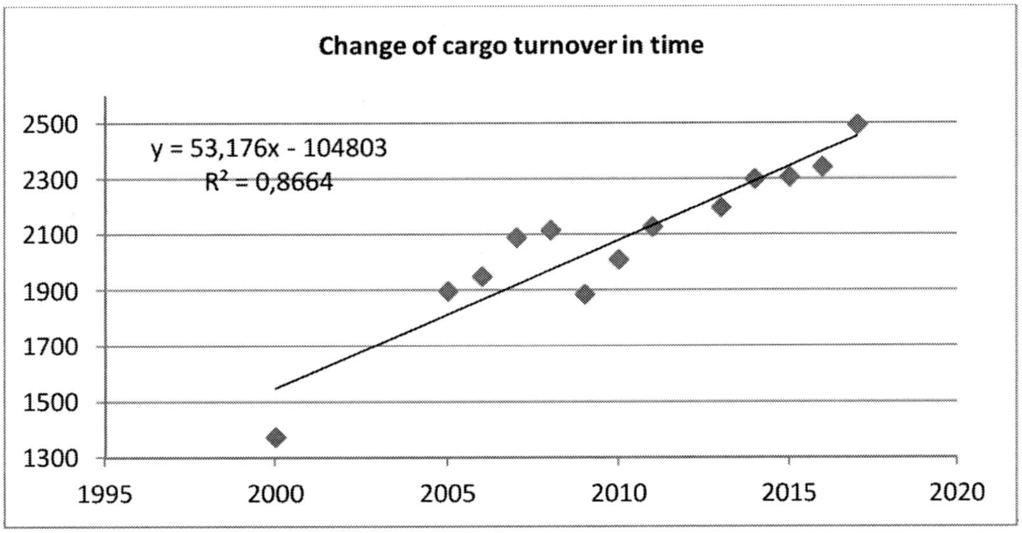

Fig. 3. "Trend line" based on data from 1998 to 2017

Trend line analysis (Fig. 2 and Fig. 3) and the values of their parameters allows us to conclude that the quality of the constructed regression functions is high. The coefficient of determination is 0.88 and 0.87. Further research was conducted on the dependence of rice. 3. The estimation of time series approximation errors showed that for the linear dependence the local maximum error is 12.8% and the integral average is 4.1%. Fig. 4 shows the graph and the error values in percent.

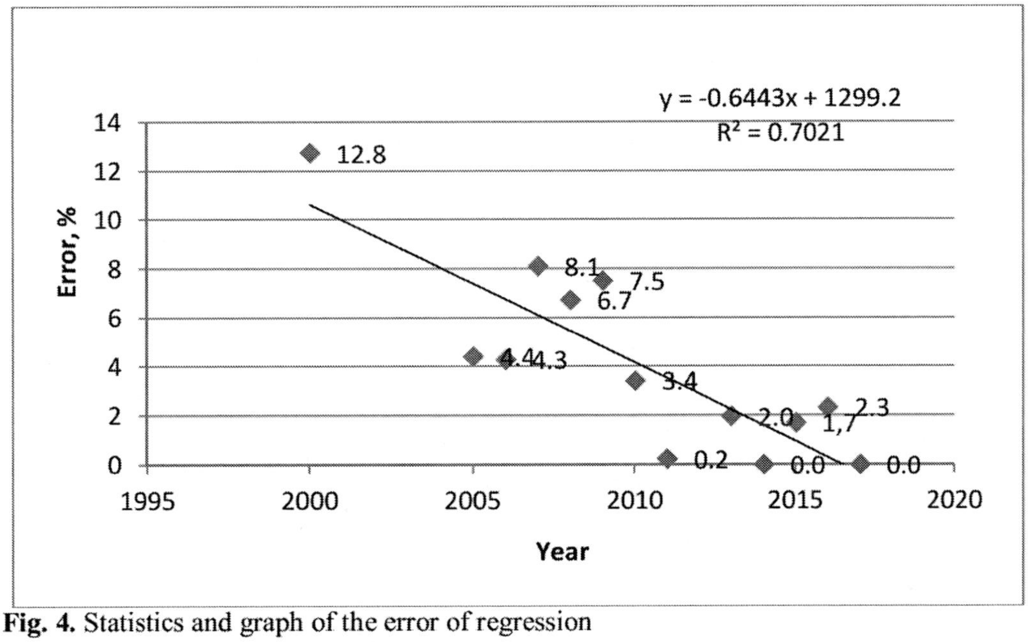

Fig. 4. Statistics and graph of the error of regression

To assess the adequacy of the regression equations obtained, the forecast for 2018 and its comparison with the real value were used. In Table 1 the penultimate value of cargo turnover at number 19 is real, and the last at number 20 is calculated according to the received model (Fig. 3). The error was 3.48%.

https://doi.org/10.1051/shsconf/20196703003

Assessment of the statistical reliability of the regression equation generally produced with the help of F-Fisher criterion. Evaluation of the models by Fisher's criterion allowed to reject the null hypothesis about the random nature of the equations and regression coefficients, and, consequently, for the estimated models, to accept an alternative hypothesis about the statistical significance of the regression equations [4]. On the basis of the linear model constructed and recommended for practical application, point and interval forecasting of cargo turnover was performed [5]. Fig. 5 and Fig. 6 show the graphs of the forecast.

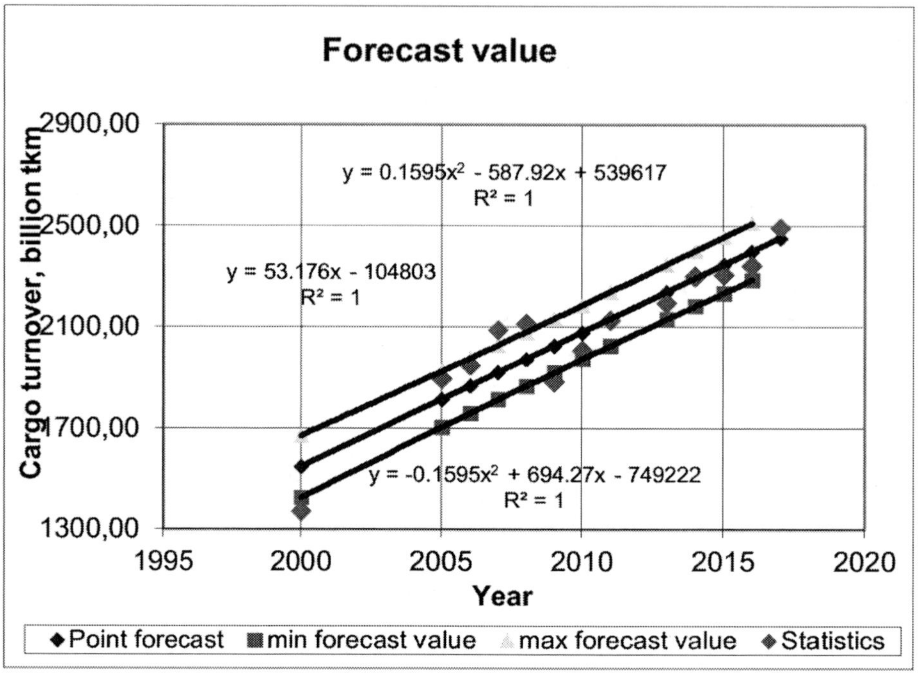

Fig.5. Graph of point and interval forecast values of cargo turnover

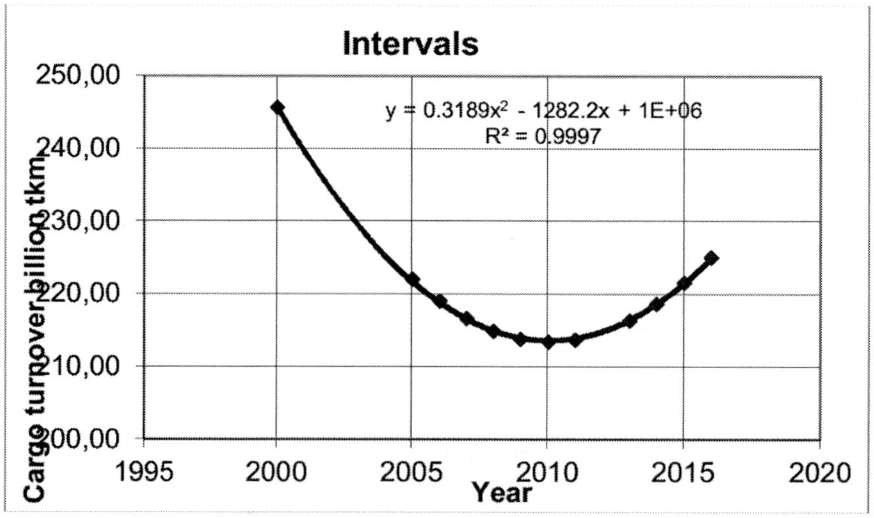

Fig. 6. The values of the confidence interval

Since the parameters characterizing the quality of the models and its adequacy have shown high performance, it is advisable to apply the described methodology of cargo turnover modeling in order to predict it in practice.

References

1. Russian statistical yearbook. 2018: Stat. SB.Rosstat, M., 694 (2018)
2. P. V.Gerasimenko, V. A. Khodakovsky. Sat. Tr. International scientific and methodological conference. St. Petersburg.: PGUPS, 84-88 (2014)
3. P. V. Gerasimenko, G. B. Staresina. In the book: Government and business. Modern problems of economy. Proceedings of the VII International scientific practical conference. North-Western Institute of Management Ranepa under the President of the Russian Federation, Faculty of Economics and Finance, 111 – 116 (2015)
4. P. V. Gerasimenko, G. B. Titov. Proceedings of St. Petersburg University of railway engineering, **3 (40),** 162-167 (2014)
5. P. V. Gerasimenko. Proceedings of the International Academy of Sciences of higher school, **56.2**, 64-70 (2011)

Optimization of agroecological efficiency of transport processes in agriculture

Serhii Kharchenko[1], *Oksana* Pankova[1*], *Mikhail* Shulyak[1], *Alexander* Anikeev[1], and *Kirill* Sirovitskiy[1]

[1] P. Vasilenko Kharkov National Technical University of Agriculture, 44, Alchevskykh street, Kharkiv, 61002, Ukraine

Abstract. Agricultural production is dispersed in huge areas and requires the transfer of a large number of technological and operational materials, intermediate and final products. Therefore, in the agricultural sector, transport occupies an important place: it begins and completes the processes of crops production and livestock products and carries out technological connections between individual states of work. With the development of agricultural production, the role of transport is steadily increasing: if at present, at the average 45...50 tons of different cargoes is the share of each hectare of ploughed field, then in the coming years, an increase of this volume is expected. For more than 35% of the volume of the main work types is the share of transport works which are carried out by tractors in the agrarian sector. Increasing the dynamic and energy indices of tractors during transport operations in the agrarian sector by ensuring their functional stability will lead to lower energy consumption, increase the ecological efficiency of transport processes and will ensure a steady increase in profits in the agro-industry. The concept of the formation of tractor functional stability during transport works, which is based on the reduction of amplitudes of accelerations (slowdowns) in a three-dimensional geometric space, which, unlike the known ones, provides reduction of additional energy costs, is substantiated in the article.

1 Introduction

In the process of economic development, production processes that are focused on providing different market needs, in particular, in crop production, are becoming increasingly relevant. Effective implementation of these processes is capable of providing a logistic approach that allows optimizing the entire chain of storage and marketing of agro-ecological products, ranging from material support to implementation [1-2]. One of the key tasks of the country's food security is to further develop the grain storage sector and ensure the efficient use of transport for its timely delivery to its target. All this can be achieved by introducing elements of optimization of agroecological logistics processes. The assessment of export opportunities of the Ukrainian agroindustrial complex has shown the necessity of expanding and optimizing the transport sector to meet the needs of production and

* Corresponding author: pankova_oksana@ukr.net

increasing the economic effect. The constant increase in agroecological requirements for the quality of products compels manufacturers to look for new pressing methods of storage and transportation [3-4]. The need for implementation of modern methods and optimization of existing ones was formulated by N. Prisiyazhniyk [5]. Modern European experience shows that ensuring the production of modern scientifically-based methods increases the level of profitability by 15-25% and it is one of the most effective vectors for the development of wholesale trade, which opens the prospects for the development of the agrarian sector and improves the quality results for the fulfillment of food targets [6-7]. The selection of rational criteria for the delivery of agricultural cargoes plays an important role in ensuring the marketing of products and rationalization of grain transportation technologies, which are not possible without the introduction of computer technologies and modern methods of calculating operating costs [8-10]. Also, an important role is played by agroecological quality control of products. One of the practical ways to increase the effectiveness of this control is to make it in a recycling scheme when a "qualitative object" is checked again [11].

2 Methodology

Based on the review of scientific and technical literature on the use of tractors during transport works in the agricultural sector, outstanding issues were identified (including: functional stability, energy conservation, and dynamics).

In order to increase the efficiency of tractor-trailer trains and the stability of their operation, various constructive solutions to this problem were proposed. For example, V. Shalyagin and V. Yasenevich proposed application of active drive wheels of the trailer to increase the pull and energy efficiency indicators of the tractor-trailer train. At the same time in the work of these scientists it was pointed to the need to justify the parameters of functional stability of tractors during transport works. For tractor-trailer trains Z. Zebelyshynsky recommended to take as the main functional parameter the motion roadholding, taking into account the effect of the distribution of propelling force between the axles of the tractor on resistance to skidding (jack-knifing).

The prospect of applying information about the acceleration of mobile machines in evaluating their functional parameters was substantiated. By analogy with related branches of science and technology, the functional stability of transport aggregates can be estimated according to their functional stability, which characterizes the deviation of the basic parameters of complex technical systems from the nominal values when the accelerated motion and normal motion.

When solving the energy saving problem it is recommended to evaluate the tractor efficiency during the transport work at effective work which is carried out by the transport unit per unit time of the cycle (delivery of cargo to place of consignment, return). In this case, the effective work is determined by the work of the tractor's driving forces, less the work spent on overcoming the forces of harmful resistance. Dedicated to the problem of energy saving of mobile agricultural units, in which it is concluded that there is a interconnection between the energy consumption of these aggregates from the stability of their high-speed motion. This problem for tractor-based transport units has not been resolved.

Different methods and devices are proposed for the estimation of functional stability, energy saving and dynamics of mobile machines in which attention was paid to the lack of control systems and executive system of the functional stability of the tractor operation.

The conception of the dynamic space of TA functioning is formulated, on the basis of which the functional parameters of the tractor during transport works are substantiated.

3 Empirical results

In general, the transport unit (Fig. 1) is represented in the form of a system at the entrance of which there are vector-functions of control $U(t)$ and excitation $F(t)$. Output variables, which characterize the functional parameters of the TA, are represented by a vector-function $Y(t)$. Dynamic properties of the TA are characterized by a transfer function $W_m = S_y/S_u$, where, S_y and S_u – the output and control parameters accordingly.

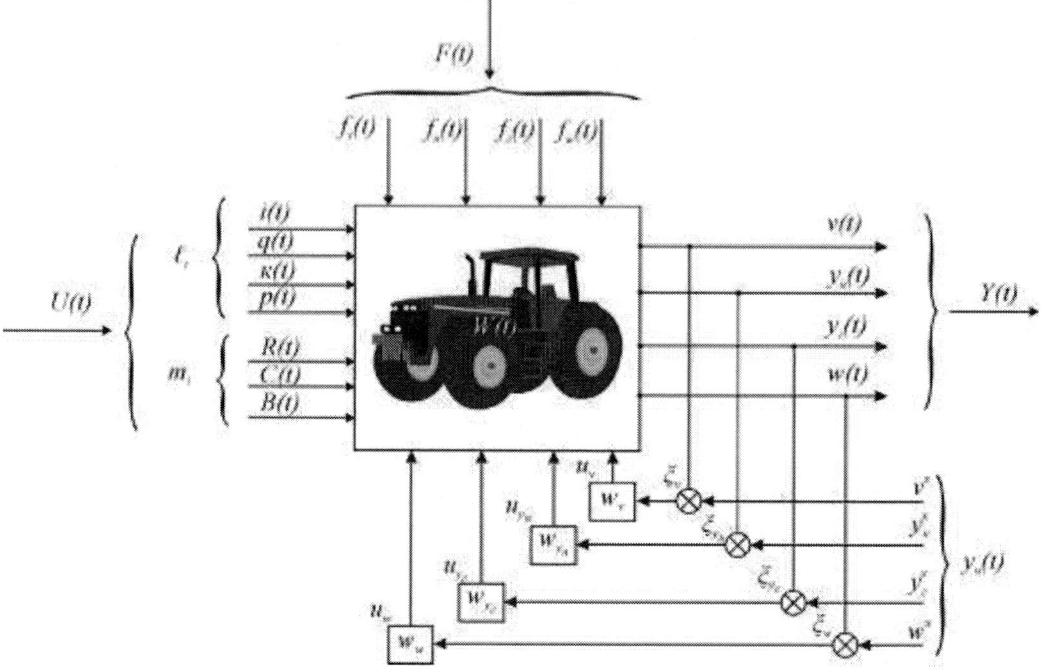

Fig. 1. Generalized functional diagram of the transport unit.

The TA is influenced by both controlled input vector-control functions li [$i(t)$ – gear ratio, $q(t)$ – fuel supply, $\kappa(t)$ – gear ratio of the steering, $p(t)$ – the brake control effort], and uncontrolled m_i [$R(t)$ – motion resistance, $C(t)$ – road relief, $B(t)$ – wind force, etc.]. Output variables $Y(t)$ are characterized by the rate of motion $v(t)$, motion-direction stability $y_н(t)$ and braking $y_г(t)$, by the stability of the dynamic properties $w(t)$ of the unit.

Usually, excitations $F(t)$ are determined by the instability of the output (functional) parameters $v(t)$, $y_н(t)$, $y_г(t)$ and $w(t)$. When comparing these parameters with their values v^x $y^x_н$ $y^x_г$ w^x and, with which TA functions stably, at the links of the comparison ξ_v, $\xi_н$, $\xi_г$ and ξ_w with the transfer functions w_v, $w_{yн}$, $w_{yг}$, and w_w the signals u_v, $u_{yн}$, $u_{yг}$, and u_w of the correction of the controlled input variables $u(t)$ are generated.

The functional parameter $v(t)$ is characterized mainly by the input control functions $i(t)$ and $q(t)$, and determines the TA running qualities. The parameters $y_н(t)$ and $y_г(t)$ are determined mainly by $v(t)$, excitations of the motion direction $f_н(t)$ and braking $f_г(t)$ and they are aimed at ensuring the motion safety of the TA. The dynamic parameter $w(t)$ characterizes the TA motion in the longitudinal, horizontal and vertical planes under the action of forces in these planes.

The dynamic space of the tractor's operation during transport works is based on the metric space of states, each element of which completely determines the state of the considered system in terms of functional parameters $v(t)$, $y_н(t)$, $y_г(t)$ and $w(t)$.

In the case of an unstable one or more of the tractor's functional parameters $v(t)$, $y_н(t)$, $y_г(t)$ and $w(t)$, which is characterized by a deviation from the nominal values y^x, $y_н^x$, $y_г^x$ and w^x, it is possible the loss of the functional stability tractor. In this case it will not perform the functions, which determined by the normative-technical documentation (NTD). In this case, the dynamic space of the tractor's operation is determined by the transfer functions W_v, $W_{y_н}$, $W_{y_г}$ and W_w, which characterized by the ratio $v(t)$, $y_н(t)$, $y_г(t)$, $w(t)$ from y^x, $y_н^x$, $y_г^x$, w^x.

When operating a tractor during the transport work, the task of ensuring its functioning in one of two areas is solved: OA – the operation area, in which the tractor operates according to the purpose (the necessary traction effort, stability of the driving direction and braking, etc. are provided); OO – area of optimal operation, for example, by the criterion of energy saving, in which the tractor operates with the allowable change of speed (change of movement acceleration within the permissible limits) (fig. 3). In this figure, by the points B_y and B_g marked accordingly to the tractor operation in the conditions of unsteady mode of operation without acceleration and permissible area according to energy conservation; Δ_v – reserve of optimal functioning; V_o, $V_ф$ – the functioning vectors is optimal, actual; ΔB – reserve of functioning in the area AF.

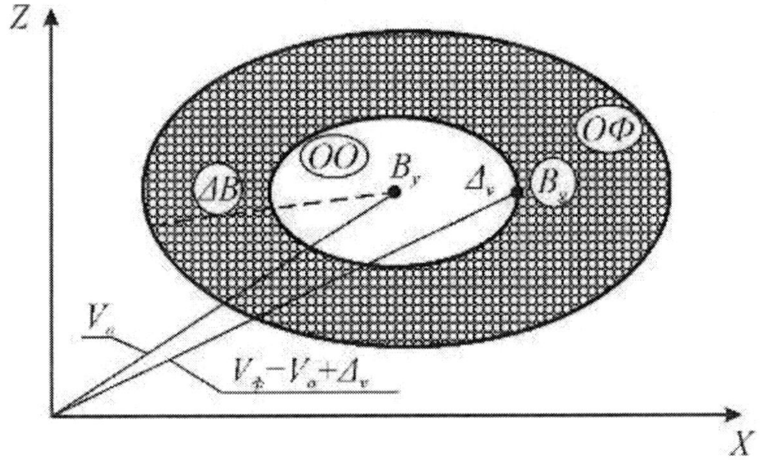

Fig. 2. Correlation of areas of optimal work OO and the tractor operation $OФ$.

In order to solve the scientific problem of ensuring the tractor functional stability during transport work, it has been provided the methodology of the choice of the nominal values of the functional parameters $v(t)$, $y_н(t)$, $y_г(t)$, $w(t)$ and also the tolerance zone on them, with which the tractor work in the zone OO is substantiated. Based on the fundamentals of theoretical mechanics, the energy analysis of the transport aggregate is carried out during the active and passive tractor operation in the dynamic space of operation. In this case, useful (active) work is determined by the work of the driving forces, less the work spent on overcoming the forces of parasitic drag, which characterizes the non-effective (passive) work of the tractor. The passive work of force is equal to the product of its module on the path and on the cosine of the angle α between the directions of force and movement.

Proposed indicators of tractors functional stability during the transport works, which include: rate of motion $v(t)$, stability of the motion direction $y_н(t)$ and braking $y_г(t)$, stability of dynamic properties $w(t)$. If one or more parameters $v(t)$, $y_н(t)$, $y_г(t)$, $w(t)$ are unstable, the functional stability of the tractor may be lost. In this case, the problem is solved on the basis of the justified methodology of providing the work of the transport aggregate in the zone of optimal work according to the criterion of energy conservation.

References

1. V. Melnik, A. Anikeev, K. Sirovitskiy, Eng. of Natural Management **2(8)**, 6-10 (2017)
2. S. Kornienko, V. Pashenko, V. Melnik, S. Kharchenko, N. Khramov, East.-Europ. JET **5**, 34-43 (2016)
3. S. Gritsenko, DonNTU. Series: economic **5**, 185-189 (2014)
4. V. Nelep, Economy of Ukraine **9**, 54-63 (2011)
5. N. Prysyazhnyuk, P. Sabluk, M. Kropivko, Economy of Ukraine **6**, 4-16 (2011)
6. E. Shubravskaya, N. Rindenko, Economy of Ukraine **8**, 77-85 (2012)
7. E. Shubravskaya, Economy of Ukraine **8**, 64-76 (2013)
8. V. Nefyodov, Yu. Tkachenko, East European Journal of Advanced Technologies **3(63)**, 13-15 (2013)
9. S. Kharchenko, O. Anikeev, M. Tsiganenko, O. Kalyuzhny, KhNTUSG (MAP) **156**, 174-179 (2015)
10. D. Avalos-Gonzalez, D. Hernandez-Balbuena, V. Tyrsa, O. Sergiyenko, F. Murrieta-Rico, V. Kartashov, M. Kolendovska, S. Sheiko, V. Melnyk, IECON 3159-3164 (2018)
11. V. Puzik, L. Puzik, N. Lyubymova, O. Pankova, V. Sakhatsky, IJET (UAE) **7**, 507-513 (2018)

DEVELOPMENT OF SPECIALIZED MODELS OF URBAN PASSENGER TRANSPORTATION

Evgeniy Kolbachev[1]*, *Natalya* Napkhonenko[2], *Maryna* Karayeva[3], *Dmytro* Maloshtan[4]

[1]ScD (Econ.) Platov South-Russian State Polytechnic University (Novocherkassk Polytechnic Institute) (Novocherkassk)
[2]PhD (Econ.) Platov South-Russian State Polytechnic University (Novocherkassk Polytechnic Institute) (Novocherkassk)
[3]PhD (Econ.) Platov South-Russian State Polytechnic University (Novocherkassk Polytechnic Institute) (Novocherkassk)
[4] PhD student, Kremenchuk Mykhailo Ostrohradskyi National University, Ukraine

Abstract. Peculiarities of transport services for the transportation of passengers, which must be taken into account when creating logistics models for organizing and managing passenger flows, are considered. Need of cost reduction of time of passengers for transport service is proved. The conclusion is drawn on need of the effective transport system creation, for the purpose of observance of intervals of the movement of buses on routes. The possibility of use of architectural approximations classes for the solution of transport logistics problems is analysed. The methods used for solving transport problems where the most widespread is modeling and algorithmization of the determined task formulation strategy with use of a classical algorithm of mathematical programming are considered. The prospects of a meta-heuristic method use is a genetic algorithm for the solution of tasks of the control of city passenger transport adapted to objective conditions are proved. The main features of genetic algorithms consisting in an opportunity by optimization to use criterion function and to consider the necessary number of restrictions are defined. The structure of a chromosome which represents the coded option of the movement of the bus in real time depending on a road situation is received.

Statement of problem. In strategic programs of the transport systems development of any city, special attention is paid to management of traffic flows, increase in capacity of a street road network and safety of the movement. At the same time the urban passenger transport (CPT) is always one of the main components of transport infrastructure and has significant effect on social stability [1]. It is known that a main type of GPT is bus transport which is capable to provide most necessary mobility of citizens and therefore improvement of the organization of its work has priority value.

Basic provisions on control of passenger transport and improvement of its functioning with use of logistic technologies were considered by many authors [3, 4, 5, 7].

* Corresponding author: kafedraturizma02@gmail.com

The theoretical bases of organizational economic activity of the passenger motor transportation enterprises (ATP) concerning ensuring the uninterrupted and safe movement of buses along all routes are investigated in works of domestic and foreign experts [1, 4, 6].

It is known that transport transportation services of passengers have a number of features, at the same time they are influenced by a set of various and often uncontrollable factors which need to be considered at creation of logistic models of the organization and management of passenger flows [8]. At the same time it is necessary to unite separate elements of transportation process in the uniform system capable to provide high-quality services to the population at the minimum expenses. A basis of logistic approach in relation to passenger transport is identification of the shortest communications between the main passenger created points, accounting of volumes of passenger traffics and requirements of comfortable journey when choosing type and calculating required number of vehicles [5].

Analysis of the main achievements and publications. The extensive analysis of publications [1, 4, 5, 8, 9] and the accumulated experience of the advanced passenger motor transportation enterprises suggests what remains so far insufficiently studied a problem of public passenger transport logistic service improvement taking into account specifics of technical and operational and economic indicators of ATP of all forms of ownership. It is necessary to emphasize that special attention at the solution of tasks of management of urban passenger traffic has to be paid to issues of cost reduction of time of passengers for transport service and consequently, observance of the schedule of buses on routes.

Allocation of unresolved parts of common problem. Despite the increasing number of studies on this subject, problems of control and management of urban passenger transport, especially in megalopolises, are insufficiently covered in scientific literature; besides, questions of formation of flows of passengers on stops and ways of impact of the passenger enterprises on losses from the services which are not provided in time are practically not considered in it [4, 9].

The purpose of this research is creation of passenger transport and the urban infrastructure work logistic model allowing to establish functional connection between the main passazhiroobrazuyushchy points. A necessary operating condition of model is the equipment of these points the corresponding constructions for accounting of volumes of passenger traffics and requirements of comfortable journey when calculating necessary quantity and choosing the rolling stock.

Statement of the main material of probe.

On the basis of Rostov-on-Don Trasservice trucking company-3 rolling stock functioning studying problems the main stages of creation of transportations management model in the conditions of unstable passenger traffics are allocated:

– selection of criteria for evaluating the performance of the trucking company;
– creation of logistic model of urban passenger traffic;
– the choice of mathematical methods for the solution of objectives;
– formation of basic data of an optimizing task;
– designing of control algorithms;
– calculation of model;
– realization of model and analysis of results;
– adjustment of the entering parameters and the choice of option of optimum performance of a system.

On the basis of the performed analysis the conclusion is drawn that on trucking company an effective functional system should be created which demands a certain material, technological, financial and information support [7]. At the same time, indicators of passenger traffic system efficiency of each participant have to be considered, while they often are contradictory. For example, reduction of waiting time of passengers is evidently connected with increase in number of the rolling stock on a route and consequently, with decrease in its load and an economic benefit from use [11]. But the aspiration to increase profit the transport organizations can lead to refusal of the population of services of carrier and emergence of the competing organizations.

The development of specialized models of urban passenger traffic even on individual routes is a complex process, and therefore they are used in practice, which allows us to answer the question of what will be the characteristics of the simulated system if the initial parameters acquire the planned values.

In the existing models such descriptions are carried out at the production and technological level in the integrated indicators [7, 11]. Need of development of more high-class models which not only describe process of transportation and separate stages of its organization is explained by it, but also reflect relationship of cause and effect between characteristics of process and output parameters.

The choice of a numerical method of the solution of management assumes realization optimization problem of the following stages:
– choice of the solution method of the differential equations, and/or functions integration;
– justification of a method of the solution of optimization the nonlinear algebraic equations and problem systems;
– drawing up simulation or mathematical model.

Application of the differential equations demands for all numerical methods of optimum control. In indirect approach numerical solutions of the differential equations are combined with methods of the nonlinear equations, and in direct – with methods of the solution of nonlinear problems of optimization.

The algorithms offered [3, 5, 6] for the solution of resources on transport dynamic distribution are reformulation major problems of the dynamic equations of programming in terms of "a condition of the after solution". They assume statement of approximations and algorithms of execution which correspond to architecture of approach to criterion function – minimization of costs for transportation.

Two classes of architecture of approximations were successfully used for the solution of various problems of transport logistics:

– linear architecture:

$$\bar{V}^x \ \ R_t, v_t \ \ = \Sigma_{a \in A} v_{ta};$$

– concave (nonlinear) architecture:

$$\bar{V}^x \ \ R_t, v_t \ \ = \Sigma_{a \in A} \ \ \Sigma_{i=1}^{|R_{ta}|} v_{tai} + R_{ta} - |R_{ta}| v_{taR_{ta}} \ \ .$$

Proceeding from it, linear approximations provide advantages in relation to algorithms of the corresponding parameters calculation. Linear approximations, as a rule, should be used at the initial stage of implementation of difficult strategy, but sometimes they can badly work. In such cases it is necessary to pass to more difficult architecture and approximation algorithms [6].

For a solution of transport and logistics problems the most widespread is modeling and algorithmization of strategy of formulation of the determined task with the subsequent use of a classical algorithm of mathematical programming. The determined tasks differ in big expenses of time for development of linear integer models which can be effectively solved. Besides, they completely ignore uncertainty and difficult dynamics which arise in actual practice.

The most widespread approach to a solution of the problem of modeling of distribution of resources in practice is setting of the determined task and application of the mathematical algorithms of programming developed for the selected parameter.

Dynamic problems of distribution of resources, as a rule, result in high dimension of a stochastic dynamic programming. There are approximation algorithms which were successfully applied in applications to transport and logistics [6]. They are constructed on several key ideas which solve dimension problems. Among these ideas: reformulation of classical problems of a dynamic programming around a variable recursion, in a status of acceptance of "after solution"; modeling on the basis of methods of stochastic approximation and continuous approximations of architecture of functional forms which facilitate software solutions for optimization of a solution of subtasks [6].

The analysis of methods of a solution of tasks of control of urban passenger transport showed that the most perspective are metaheuristic methods, to be exact genetic algorithms [5, 6]. The main feature which an opportunity is by optimization, first, to use target function, but not its estimates or approximations, and secondly, to consider the necessary number of restrictions and an unlimited number of parameters of a system.

Application of the genetic algorithm (GA) allows to receive a set of the solutions satisfying possible variations to target function and restrictions of model.
Process optimization of dynamically changing passenger traffics assumes use mastering of this method adapted to objective conditions. At the same time consideration of concepts of a chromosome, gene, population and operators of random changes assumes designing of an algorithm [6].

The coded way of solving the task represents a chromosome and consists of solution elements – genes. The set of such candidate solutions makes population. For designing of chromosomes there are different approaches, however the most effective is the following [3, 5, 6].

The first – is used when the solution has an appearance of set of the elements accepting values in continuous area. In that case to each element the gene presenting it in the binary form is put in compliance. For the purpose of reduction of length of a gene the solution can be coded.

It is reasonable to apply the second approach in case of a discrete solution for which the state space can be described in a binary look [6]. Then length of a chromosome corresponds to dimension of the state space of a system in a binary look.

It is revealed that for a solution of the considered task it is reasonable to use the second approach and consequently it is necessary to execute designing of genes for everyone, entered into model Boolean variable (x_{ij}, y_j, z_j, ...). These genes, according to the accepted approach, should have dimension of the corresponding variables. In this case use of such approach will lead to obtaining a large number of the discarded solutions received as a result of execution of operators of random changes as there are restrictions for the number of "units" in one gene. So, according to these restrictions the gene of □i, j may contain one unit in any j-y of position (0001000).

The chromosome A consists of (m+2) of a gene:

$$A = \alpha_1, \alpha_2, \ldots, \alpha_j, \ldots, \alpha_m, \beta, \gamma \ ,$$

where genes of α_i represent the vectors defining time intervals between stopping points, $\alpha_i = bin\ j^i$, here j^i – conditional number of time intervals of passing of a route, $x_{ij} = 1$. The *bin* function carries out conversion of an argument to binary form. Thus, the gene of α_i has an appearance $\alpha_i = \alpha_{i1}, \alpha_{i2}, \ldots, \alpha_{it}, \ldots, \alpha_{iq}$, $i = \overline{1,k}$, where q – the number of bits necessary for binary representation of number n time intervals: $q = \log_2 n$ rounding of a fractional part is made to the next bigger whole.

The gene of β defines $\beta = z_1, z_2, \ldots, z_j, \ldots, z_n$; gene $\alpha = \alpha^1, \alpha^2$ bears information on the number of runs; the gene bears information on a bus v-go start of motion in minutes for N-y run $v = \overline{1,z}$, where z – the number of the rolling stock on the line in time in a dress; in a gene $\delta^v = \delta_1^{v,N}, \delta_u^{v,N}, \ldots, \delta_h^{v,N}$ start time of the movement v-go of the bus in minutes for N-y run is coded $v = \overline{1,z}$.

Genes

$$\gamma_{1,2}^{v,N} = {}^l\gamma_{1,2,1}^{v,N}, {}^l\gamma_{1,2,2}^{v,N}, {}^l\gamma_{1,2,3}^{v,N}, \ldots,$$

$$\gamma_{j,i+1}^{v,N} = {}^l\gamma_{j,i+1,1}^{v,N}, {}^l\gamma_{j,i+1,2}^{v,N}, {}^l\gamma_{j,i+1,3}^{v,N}, \ldots,$$

$$\gamma_{k-1,k}^{v,N} = {}^l\gamma_{k-1,k,1}^{v,N}, {}^l\gamma_{k-1,k,2}^{v,N}, {}^l\gamma_{k-1,k,3}^{v,N}$$

act as vectors of separate stages passing time intervals on a v-m route as the run N-go bus where $i = \overline{1,k}$.

The gene $\mu^v = \mu_1^{v,N}, \mu_2^{v,N}, \ldots, \mu_i^{v,N}, \ldots, \mu_k^{v,N}$ sets the place of the bus v-go motion start, $v = \overline{1,z}$.

Comparison of chromosomes is carried out as follows: from the analyzed population of $P = A_1, \ldots, A_t, \ldots, A_z$ the best A_i chromosome with the smallest value of violation of restrictions $\Delta(A_i)$, is considered, and among chromosomes with equal violation of restrictions the chromosome with the greatest value of target function $F(A_i)$.is selected.

For determination of value of violation of restrictions of$\Delta(A_i)$ the following variables are entered:

$$d_1 = \left(\max\left(0, \left(z_j y_j \sum_{i=1}^{m} \sum_{k=1}^{m} x_{ij} \lambda_{ik} - P^{hub,j} \right) \right) \right)^2 ;$$

$$d_2 = \sum_{i=1}^{m} \left(\max\left(0, \left(\bar{y}_j z_j \sum_{k=1}^{m} x_{ij}\lambda_{ik} - P^{sw} \right) \right) \right)^2 ;$$

$$d_3 = \sum_{i=1}^{m} \left(\max\left(0, \bar{y}_j z_j \left(\sum_{i=1}^{m}\sum_{k=1}^{m} x_{ij}\lambda_{ik} - \sum_{i=1}^{m}\sum_{k=1}^{m} x_{ij}\lambda_{ik} \right) - P^{sw}_{маг} \right) \right)^2 ;$$

$$d_4 = \bar{\eta} \sum_{i=1}^{m} \left(0, z_j \max_j \left(y_j \sum_{i=1}^{m}\sum_{k=1}^{m} x_{ij}\lambda_{ik} + \bar{y}_j \left(\sum_{i=1}^{m}\sum_{k=1}^{m} x_{ij}\lambda_{ik} - \sum_{i=1}^{m}\sum_{k=1}^{m} x_{ij}\lambda_{ik} \right) \right) - P^{sw}_{маг} \right)^2 ;$$

$$d_5 = \sum_{i=1}^{m}\sum_{j=1}^{n} \left(\max\left(0, l_{ij} z_j - L^{\max} \right) \right)^2 ;$$

$$d_6 = \sum_{j=1}^{m} \left(\max\left(0, b_{\mu j} z_j - L^{\max} \right) \right)^2 ;$$

$$d_7 = \left(\max\left(0, \sum_{j=1}^{n} \left(\gamma_j - n \right) \right) \right)^2 ; \quad d_8 = \left(\max\left(0, \mu - n \right) \right)^2 ,$$

where all genes belong to Ai chromosome A_i.

The value of violation of restrictions is equal to the sum of the entered variables:

$$\Delta\ A_i = d_1 + d_2 + d_3 + d_4 + d_5 + d_6 + d_7 + d_8 .$$

Thus, the structure of a chromosome which represents the coded option of the movement of the bus in real time depending on a road situation can be received. The structure of the created chromosome allows to work without loss of accuracy with the reduced state space in comparison with initial model of optimization and also simplifies a type of some restrictions.

Implementation of this algorithm allows:

– define the optimum number of the rolling stock on the line for route operating time;
– set the necessary number of the rolling stock on the line in "rush hours";
– make the optimum schedule for each bus in time in a dress;
– assess a road situation at present of time;
– adjust the schedule of the rolling stock depending on a road situation;
– set motion speed on route stages;
– make effective decisions under force majeur circumstances;
– define the reasons of failure of the schedule.

Conclusion. The offered methods of passenger traffic management logistic models creation are approved at improvement of work of Rostov-on-Don JSC trucking company-3 Transservice rolling stock . The cost efficiency during introduction of a system was 10.83% that allows to draw a conclusion on expediency of practical application of the developed algorithm.

In development of the offered technique of creation of management logistic models of passenger traffic use of the rank analysis (cenological approach) which essence and the principles act as the new direction promising high practical results seems perspective.

References

[1] Stavnichy Yu.A. Transport systems of the cities / Yu.A. Stavnichy. – M.: Stroyizdat, 1990. – 224 pages.
[2] Karayeva M. R. of Managing Dynamically Changing Traffic Flows / Masculine Karayeva, N.V. Napkhonenko, V. Perevoznyuk//International Journal of Engineering & Technology. - 2018. - Vol. 7, Is. 4.3. - P. 222-227.
[3] Napkhonenko N.V. Agricultural Cargo Transportation Logistics System Development / N.V. Napkhonenko, D.M. Zagirnyak, M.R. Karayeva//International Journal of Engineering & Technology. - 2018. - Vol. 7, Is. 4.3. - P. 185-190.
[4] Karayeva M. R. Use of mathematical tools for the solution of tasks of the organization and management city passenger traffic / Masculine of Karayev//the Messenger of the Rostov state economic university (RINH). – 2015. – No. 1 (49). – Page 31-36.
[5] Efficiency of operation and functioning of the system of an indirect transport flow regulation and control / S.V. Zhankaziev [et al] // International Journal of Applied Engineering Research. – 2017. – Vol. 12, Is. 13. – P. 3645-3652.
[6] Karayeva M. R., Napkhonenko N.V. Optimization of the organization and management of the system of passenger traffic transport enterprise / Masculine Karayeva, N.V. Napkhonenko//Vestn. YuRGTU (NPI). It is gray. Sots. - экон. sciences. – 2015. – No. 2. – Page 29-31.
[7] Ureychik V.V., Ureychik V.M. About management on the basis of genetic search//Automatic equipment and telemechanics. – 2001. – No. 10.
[8] Remarks by Jesse Anderson, Board Member, Greater Cleveland Regional Transit Authority at APTA Transit Board Members Conference, Memphis, Tenn., July 2005.
[9] The Swedish Board of Housing, Building and Planning BFS 2004:15 ALM 1 Sweden, 2004.
[10] Gwynne Richards, Susan Grinsted. The Logistics and Supply Chain Toolkit: Over 90 Tools for Transport, Warehousing and Inventory Management. 2013. 313 P.
[11] Andreas Fink, Franz Rothlauf. Advances in Computational Intelligence in Transport, Logistics, and Supply Chain Management (Studies in Computational Intelligence Volume 144- Springer). – 2008. – C. 131.
[12] Jalel Euchi. Metaheuristics to solve some variants of vehicle routing problems: Metaheuristics algorithms for the optimization of some variants of logistics and transport problems Paperback. – 2012.
[13] Hans-Otto Günther, Kap Hwan Kim. Container Terminals and Automated Transport Systems: Logistics Control Issues and Quantitative Decision Sup-port. Springer 2010. – 359 p.
[14] Issa Baluch and Charles H.W. Edwards. Transport Logistics: The Wheel of Commerce. 2010.

Logistic model of operational planning works terminal

Evgeniy Kolbachev[1]*,, *Natalya* Napkhonenko[2], *Maryna* Karayeva[3], *Dmytro* Maloshtan[4]

[1]Platov South-Russian State Polytechnic University (Novocherkassk Polytechnic Institute)name of M.I. Platov, (Novocherkassk)
[2]Platov South-Russian State Polytechnic University (Novocherkassk Polytechnic Institute) name of M.I. Platov, (Novocherkassk)
[3] Platov South-Russian State Polytechnic University (Novocherkassk Polytechnic Institute) name of M.I. Platov, (Novocherkassk)
[4] Kremenchuk MykhailoOstrohradskyi National University, Ukraine

Abstract. The analysis of cargo terminals forwarding activity operational planning problems is provided and their main reasons are established; the existing approaches to improvement of operational planning system of transport infrastructure in theoretical researches of the Russian and foreign authors and also the accumulated experience of the advanced enterprises of the industry are considered; the main objectives which can be solved in technological processes of freights handling and forwarding on terminal technology with application of logistic and economic mathematical methods are formulated; specific features of optimization of the operating plan of forwarding activity of the terminal are considered; statement and creation of the terminal operation planning logistic economic-mathematical model is executed, the choice of model optimization criterion in the form of freight handling and long-distance forwarding costs taking into account indicators of the terminal operation planning quality is reasonable

Problem statement

The main objectives of operational planning include ensuring coordinated operation of all links of production, identification of additional reserves of the overworking, throughput and carrying opportunities of motor transport and technical means of the terminal, their correct use for successful performance of plan targets on freight handling and long-distance forwarding.

Analysis of the main achievements and publications

Solving problems of transport service of the branches of the national economy demands the analysis of domestic and foreign experience [1, 2].

In the system of the long-distance transportation of goods (LDTG), the value of

* Corresponding author: kafedraturizma02@gmail.com

terminals is defined by their role in implementation of technological process of transportations [3-6]. However, researchers note that process of cargo delivery is often accompanied by losses: deterioration and full or partial loss of goods; delayed transportation and warehousing, poorly organized loading and unloading; and excessive unproductive expenses [7-9].

The quality of transport logistics service which domestic carriers and terminal complexes can provide fail to meet the requirements of the European and international standards. In this regard, 70-75% of transit freight traffics are taken by foreign companies [10, 11]. Meanwhile foreign experience shows that warehouse business is being integrated into the large logistic companies and terminals. It allows rationalizing the process of the cargo movement, provides economy on systemic expenses and obtaining the maximum economic effect in the sphere of production, distribution and consumption of goods and services [12-13].

In transport literature issues of general performance of the enterprises are rather well studied [14, 15]. Optimization of traffic flows and stocks of freights generally is the result of logistic operations [16]. Development optimization of transport and logistics system development is necessary for the purpose of complete solution of problems of the concept transport problems [17]. Currently, a number of federal, regional and municipal programs and projects are being developed and implemented, the elements of which are macroscopic systems [18, 19].

Allocation of unresolved parts of common problem

The most characteristic reasons of deviations arising at implementation of the daily plan of transportation of goods are lack of vehicles for transportation, the insufficient number of the cargo handling gears (CHG) and personnel. Noted defects can be basically eliminated due to improvement of quality of operational planning of operation of the terminal.

The feature of the operating plan optimization of the terminal forwarding activity consists in choosing the best option of freight handling and forwarding organization which takes into account the technical condition of the terminal facilities, parameters of technological processes, freight traffics and volumes of forwarding, temporary characteristics of the carried out operations is considered.

The purpose of article is theoretical justification and development of effective methods of operation of terminals due to improvement of operational planning system.

Statement of the main material of probem

The formulated problem covers the entire process of transportations from arrival of freights on the terminal up to sending to consignees, unloadings and receipts of new consignments. The system of operational planning of operation of the terminal constructed on the basis of logistic approach is directed to the choice of the most effective forms of the organization of transportation process and performance of the complex of the transport-forwarding services (TFS) accompanying it, ensuring further drop of transport expenses of the national economy.

Cutting freight handling and long-distance forwarding costs is important reserve of increase in overall performance of the cargo terminal and therefore the minimum of costs of implementation of loading and unloading, warehouse works and performance of forwarding operations is accepted as optimality criterion of logistic model of the terminal operation planning problem. Besides, the criterion considers quality indicators of functioning of the terminal to which idle time of the rolling stock, PRM and personnel belong. The optimality of criterion function is reached on condition of full use of limited material and human resources [19, 20].

Thus, the purpose of problem of operational planning is the optimization of activity of the cargo terminal for rational use of the rolling stock, technological base and human resources to ensure the planned volume of freights handling and forwarding at minimum cost [21].

The general statement and creation of the terminal operation planning economic-mathematical model is as follows.

Terminals send shipment from the consignor's warehouses to warehouses of the consignee (direct option) and also from own warehouses. Look k freights (k = 1 … K) are transported by the small, average and large-capacity j-go cars of type (j =1 … J) q_j loading capacity on m route of transportations (m = 1 … M). Efficiency of run on route – β_m. The transported freight provides use of loading capacity of the rolling stock according to j_k coefficient. The operating plan of operation of the terminal represents the daily plan, namely N – the mass of freight which can be accepted from suppliers, is sent to consignees or left on temporary storage.

For cargo handling the terminal is equipped with type PRM s-go in number of B_s (s = 1…S), and also warehouses r (r = 1…R). P_r – corresponds to extreme lot of storage of freights on r-m warehouse. At this W_s – productivity of PRM s-go at loading unloading per day; $g_{1.rk}$, $g_{2.rk}$ – the mass of k-go of freight which is stored on r-ohm terminal warehouse for the beginning and the end of day according to (k = 1…K; r = 1…R).

The cars necessary for implementation of the daily plan of freight forwarding and handling, conditionally are divided into two streams: A_j – the type j-go cars (j = 1…J) perform work on delivery and export of freight. The entering and leaving car traffics, in turn, are divided as follows:

$x_{1.jkms}$ – the number of the j-go cars of type which have arrived with k-m freight on m to route of transportations and remained not unloaded by the end of day, unit (j = 1…J; k = 1…K; m = 1…M; s = 1…S.

$x_{2.jkms}$ – the number of the j-go cars of type which have arrived with k-m freight on m to route of transportations and unloaded by s the PRM type on warehouse of the terminal, unit (j = 1…J; k = 1…K; m = 1…M; s = 1…S);

$y_{1.jkms}$ – the number of the j-go cars of type intended for work on m to route of the transportations which are standing idle waiting for discharging of PRM s-go or receiving k-go of type of cargo, unit (j=1,…J: k=1,….K: m=1,….M: s=1,….S);

$y_{2.jkms}$ – the number of the j-go cars of type loaded by s-m the PRM type, in unit (j = 1…J; k = 1….K; m = 1….M; s = 1…S);

$y_{3.jkms}$ – the number of the j-go cars of type which are taking out k-y freight on m to route of transportations by direct option, unit (j = 1…J; k = 1…K; m = 1…M).

In time in dress of T_n the j-go car of type stands idle waiting for loading unloading of s-m the PRM t_{prj} type and transports on m route with V_{tj} speed, at the same time it carries out a_{jm} the goer.

For determination of total size of expenses the following designations are entered.

Expenses on:

C_s – loading unloading of 1 t of freight of s-m the PRM type, rub (s = 1 … S);

S_{skl} – warehouse processing of 1 t of freight in terminal warehouses within a day, rub;

S_{eksp} – performance of forwarding operations, rub;

S_{prj} – 1 h idle time of the j-go car of type, rub (j = 1 … J);

S_{prs} –1 h idle time of PRM s-go of type, rub (s = 1 … S).

Terminal costs of freight handling and forwarding consists of the following elements:

1. The sum of expenses on the terminal per day:

$$\sum_{j=1}^{J}\sum_{k=1}^{K}\sum_{n=1}^{N}\sum_{s=1}^{S} C_s q_j j_k a_{jm} x_{1.jkms} + \sum_{j=1}^{J}\sum_{k=1}^{K}\sum_{n=1}^{N}\sum_{s=1}^{S} C_s q_j j_k a_{jm} x_{2.jkms} +$$
$$\sum_{j=1}^{J}\sum_{k=1}^{K}\sum_{n=1}^{N}\sum_{s=1}^{S} C_s q_j j_k a_{jm} y_{1.jkms} + \sum_{j=1}^{J}\sum_{k=1}^{K}\sum_{n=1}^{N}\sum_{s=1}^{S} C_s q_j j_k a_{jm} y_{2.jkms}.$$

2. Sum of costs of warehouse transactions of the terminal per day

$$(\sum_{r=1}^{R}\sum_{k=1}^{K} g_{1.rk} + \sum_{j=1}^{J}\sum_{k=1}^{K}\sum_{n=1}^{N}\sum_{s=1}^{S} q_j j_k a_{jm} x_{2.jkms} - \sum_{j=1}^{J}\sum_{k=1}^{K}\sum_{n=1}^{N}\sum_{s=1}^{S} q_j j_k a_{jm} y_{2.jkms}) \cdot C_{skl}. \quad (2)$$

3. Cost of performance of forwarding operations.

$$\left\{ \sum_{j=1}^{J}\sum_{k=1}^{K}\sum_{n=1}^{N}\sum_{s=1}^{S} [q_j j_k a_{jm} (x_{1.jkms} + x_{2.jkms} y_{1.jkms} + y_{2.jkms})] + \sum_{j=1}^{J}\sum_{k=1}^{K}\sum_{n=1}^{N}\sum_{s=1}^{S} [q_j j_k a_{jm} y_{3.jkms} \right\} C_{eksp}.$$

4. The cost of losses from idle time of cars on the terminal per day

$$\sum_{j=1}^{J}\sum_{k=1}^{K}\sum_{n=1}^{N}\sum_{s=1}^{S} (x_{1.jkms} + y_{1.jkms}) \cdot C_{prj} t_{prj}.$$

5. The size of losses from idle time of PRM

$$\sum_{s=1}^{S} Z_s \cdot t_{pr_s} \cdot C_{pr_S},$$

where Z_s – quantity of idle PRM

The problem is solved at the following restrictions:

1. The number of the j-go cars of type with k-m freight which need to be unloaded on the terminal and the cars which have remained not unloaded from last day does not exceed total number of the j-go cars of type to which TEU on import have to be rendered per day:

$$\sum_{k=1}^{K}\sum_{n=1}^{N}\sum_{s=1}^{S} (x_{1.jkms} + x_{2.jkms}) \le A_j,$$

for $j = 1 \ldots J$; $k = 1 \ldots K$; $m = 1 \ldots M$; $s = 1 \ldots S$.

2. The number of the j-go cars of type which are taking out k-y freight on m to route from warehouses of consignors; cars which have to be loaded the s-m the PRM type in terminal warehouses according to the plan of the forthcoming day and cars which are standing idle waiting for discharging of PRM or receiving freight from last day does not exceed total number of cars which need to be loaded according to the plan.

$$\sum_{k=1}^{K}\sum_{m=1}^{M} (y_{3,jkms} + \sum_{s=1}^{S} (y_{1,jkms} + y_{2,jkms}) \le H j,$$

для $j = 1 \ldots J$; $k = 1, \ldots K$; $m = 1, \ldots M$; $s = 1 \ldots S$.

3. The quantity of PRM s-go of type used for loading unloading does not exceed their total on the terminal

$$\frac{1}{W_s} \cdot \sum_{j=1}^{J}\sum_{m=1}^{M} q_j j_k a_{jm} (x_{1.jkms} + x_{2.jkms} + y_{1.jkms} + y_{2.jkms}) + Z_s = B_s,$$

where $j = 1 \ldots J$; $m = 1 \ldots M$.

4. The rest of freights in terminal warehouses by the end of the planned day is equal to the sum of the passing rest of freights from last day and the arrived shipment minus the mass of the freights sent from terminal warehouses

$$\sum_{k=1}^{K} q_{2rk} = \sum_{k=1}^{K} q_{1rk} + \sum_{k=1}^{K} j_k \sum_{j=1}^{J} q_j \sum_{m=1}^{M} a_{jm} \sum_{s=1}^{J} (x_{1.jkms} + x_{2.jkms} - y_{2.jkms}),$$

for each $r = 1 \ldots R$.

$$\sum_{k=1}^{K} q_{2rk} = \sum_{k=1}^{K} q_{1rk} + \sum_{k=1}^{K} j_k \sum_{j=1}^{J} q_j \sum_{m=1}^{M} a_{jm} \sum_{s=r+1}^{J} (x_{1.jkms} + x_{2.jkms} - y_{2.jkms}),$$

for each $r = 1 \ldots R; j = 1 \ldots J; k = 1 \ldots K; m = 1 \ldots M; s = 1 \ldots S$.

The rest of freights in warehouses of the terminal is not negative and does not exceed the capacity of warehouses.

$$0 \le \sum_{k=1}^{K} g_{2.rk} \le P_r ,$$

for each $r = 1 \ldots R; k = 1 \ldots K$.

5. The volume of the freights (import plus export) handled by the terminal and warehousing do not exceed the planned building of the terminal for day

$$\sum_{j=1}^{j} q_j a_{jm} \sum_{k=1}^{K} j_k [y_{3.jkms} + \sum_{s=1}^{S} (x_{1.jkms} + x_{2.jkms} + y_{2.jkms})] = N ,$$

for each $j = 1 \ldots J; k = 1 \ldots K; m = 1 \ldots M; s = 1 \ldots. S$.

6. At the same time the requirement of nonnegativity of variables has to be fulfilled

$$x_{1.jkms} \ge 0;\ x_{2.jkms} \ge 0;\ y_{1.jkms} \ge 0; y_{2.jkms} \ge 0; y_{3.jkms} \ge 0 .$$

Vector argument in the formulated problem of x_j – the number of cars necessary for implementation of the program of transportations of N.

$$X_j = \sum_{j=1}^{J} \sum_{k=1}^{K} \sum_{m=1}^{M} [\sum_{s=1}^{S} (x_{1.jkms} + x_{2.jkms} + y_{1.jkms} + y_{2.jkms}) + y_{3.jkms}] + \sum_{s=1}^{S} Z_s ,$$

for each $j = 1 \ldots J$.

Criterion function of model has appearance:

$$F = \sum_{s=1}^{S} \{ \sum_{j=1}^{J} \sum_{k=1}^{K} \sum_{m=1}^{M} \{ q_j j_k a_{jm} [(C_{pr_s} + C_{eksp}) x_{1.jkms} + (C_{pr_s} + C_{skl} + C_{eksp}) x_{2.jkms} +$$

$$+ (C_{prs} + C_{eksp}) y_{1.jkms} + (Cpr_s - C_{skl} + C_{eksp}) y_{2.jkms}] + C_{pr\,j}(x_{2.jkms} + y_{1.jkms}) \} + C_{prs} Z_s \cdot t_{prs} +$$

$$+ \sum_{j=1}^{J} \sum_{k=1}^{K} \sum_{m=1}^{M} (q_j j_k a_{jm} y_{3.jkms}) C_{eksp} + \sum_{k=1}^{K} \sum_{r=1}^{R} C_{skl} \cdot g_{1.rk} \to \min .$$

Calculation of the task mathematical model is carried out on stages.

First of all, the car traffic which is subject to loading in the passing direction is allocated. Thus, drawing up the operating plan of operation of the terminal consists in determination of necessary quantity of PRM and cars for long-distance forwarding by direct option; in optimization of operation of the terminal on handling small batch and the container freights followed by the transfer through warehouse.

Conclusions

As a result of solving the problem of operational planning of operation of the terminal, the necessary number of cars for routes of transportations is determined, including the optimum number of cars which have to be sent on the corresponding loading and unloading points and are sent by the terminal by direct option, cargo balances in terminal warehouses are passed.

To optimal variant of functioning of the terminal there corresponds the minimum sum of freight handling and forwarding costs, including costs for forwarding cars in the direction of travel and also losses from different idle times of vehicles and PRM. The developed economic-mathematical model of process of operational management is applicable for any terminal systems. Results of calculation can be used for analysis and identification of reserves for cost saving for the purpose of increase in efficiency and quality of transport-forwarding services. The mathematical model of the task is carried out in stages.

References

1 L.B. Mirotin. *Integrated logistics of accumulative and distributive complexes (warehouses, transport hubs, terminals)*. M.: Examination, 448 (2003)
2 N.P.Zhuravlev, O.B. Malikov. *Transport and cargo systems*. M – the Route, (2006).
3 O.B. Malikov. *Warehouses and cargo terminals*. M.: – Business-Pressa publishing house. (2005).
4 A.M. Gadzhinsky. Modern warehouse. Organization, technology, steering and logistics. M.: – Shopping Mall Velbi, Publishing house the Avenue, – 2005.
5 I.M. Trunina, V.G. Zagoryansky, E.L. Zaroryanskaya, D.V. Maloshtan: International Journal of Engineering & Technology, 7., 4.3, 537-540 (2018)
6 N.V. Napkhonenko. Scientific thought of the Caucasus, 1, 82-94 (2004).
7 T.A.Prokofieva, O.M. Lopatkin. *Logistics of transport and distributive systems: Regional aspect,* M.: RKonsult, 400 (2003)
8 V.V. Dybskaya [etc.]. *Logistics: integration and optimization of logistic business processes for deliveries*. M.: Eksmo, 939 (2014)
9 A.V. Velmozhin, V.A.Beeps, L.B. Mirotin. *Theory of the organization and steering of automobile transportations: logistic aspect of formation of transportation processes*. Volgograd: RPK «Politekhnik», 179 (2001)
10 D. Shekhter. *Logistics. Management skill supply chains*. M.: Alpina, 452 (2013)
11 Yu. M. Nerush. *Supply and transport: effective interaction*. M.: Economy, (1999)
12 J.Evers, L.Loeve, D.Lindeijer. *New Logistic Control: Concepts, Architecture and Logistics Programming. 3-th Intermational Meeting for Researth in Logistics*. Trois-Rivieres, Canada, 15 (2000)
13 D. Bauersosk, D. Klos. *Logistics: the integrated chain of the deliveries*. M.: CJSC Olympe-business, 640 (2001)
14 E.A.Kravchenko, A.V.Woman's, E.N. Ushmayev. Basic researches, **12**, 61-65 (2007)
15 Element R. Basic, D.Cox, R. Brown. *Information and risk in the marketing*. M.: Finstatinform, 374 (1993)
16 A.A. Bakayev. *Mathematical methods in planning and economic calculations*. M.: Science. 223 (1987)
17 N.V. Napkhonenko, D.M. Zagirnyak. International Journal of Engineering & Technology. **7, 4.3**, 185-190 (2018)
18 I.Kabashkin. Logistics. **3** (2001)
19 N.V. Napkhonenko. Izv. higher education institutions. Sowing. - Kavk. region. Tekhn. Sciences, Special issue, 141-146 (2003)
20 H.Yu. Eldarkhanov. *Transport and logistics: monograph*. Tambov: Diploma, 214 (2008)
21 N.V. Napkhonenko. Scientific thought of the Caucasus, **1**, 82-94 (2004)

Railway freight transportation liberalization as a solution for improving client services quality

Yevhen Krykavskyy[1] , *Kateryna* Stasiuk[1,*]

[1]Lviv Polytechnic National University, 12 Bandera street, Lviv, 79013, Ukraine

Abstract. Scientific approaches for defining liberalization as a priority trend in rail transport development has been generalized. The main liberalization purpose and goals have been revealed and systematized. The international experience in rail freight traffic liberalization has been analyzed and conclusions about its efficiency have been made based on statistical data. The basic prerequisites for performing the reformation of rail freight transport in Ukraine by means of liberalization have been revealed. The existing form of transportation process organization at JSC "Ukraine Railways" has been described. A thorough analysis of statistical data characterizing the operational results of the main railway enterprise in Ukraine has confirmed existing problems in the industry that can be resolved through the liberalization process. A survey has been conducted for the main railway customers and a conclusion based on the obtained results has been made considering liberalization necessity for improving customer service quality. A plan of further actions for freight transport liberalization has been proposed.

1 Liberalization essence and goals

Liberalization is a priority direction for Ukrainian rail transport development in the near future. Currently, the railway industry is in reforming process in order to bring its activities in line with European Union requirements.

According to world experience, the market liberalization is primarily connected with its demonopolisation and deregulation. Liberalization process involves reducing the state's influence on the industry and creating favorable conditions for competition. The EU's actions are aimed at providing complete liberalization, where the state role includes exclusively legislative and regulatory aspects as well as the sphere of transport infrastructure management. Detailed rules and the whole process of liberalization are described in numerous directives and laws for European rail transport. The main purpose of the railway market liberalization is to create competition by stimulating new business entities entry into the market. Developing effective intra-industry competition involves rail transport preparation for competitive actions in the inter-industry market in future.

Liberalization involves separating the traditional organizational structure of rail transport into two zones: management and operational. The main tasks for the unit performing railway infrastructure management functions are infrastructure building and

* Corresponding author: kuzyk.katheryna@gmail.com

maintenance, responsibility for trains movement using available railroads, real estate included in the railway infrastructure management, railway transportation security, accompanying services provision for railroad travels. Operational area of railway transportations market is form by carriers in competitive conditions. Liberalization processes result is transparent competition, market pricing and improved reliability and quality for entire transportation chain.

1.1 Worldwide experience in railway freight transportation liberalization

Global liberalization of railway transportation began in 1990's and nowadays it is possible to sum up the positive effect of these processes.

Let us consider the results of rail freight traffic liberalization on the example of Germany. In 1994, Germany was one of the first to start liberalization process, and in 1995, the first private operator entered railways market. This result proves to be positive because in such a short time the country was able to introduce qualitative mechanisms for market competition creation. It is known that it was in 1990s when German state railway monopoly was in crisis, and that liberalization brought the expected profits (Deutsche Bahn AG's revenues grew by 174%). Railway transport market is increasing every year, and now there are about 450 private operators operating in the market. The main performance indicators for rail freight industry in Germany after liberalization are shown in fig. 1.

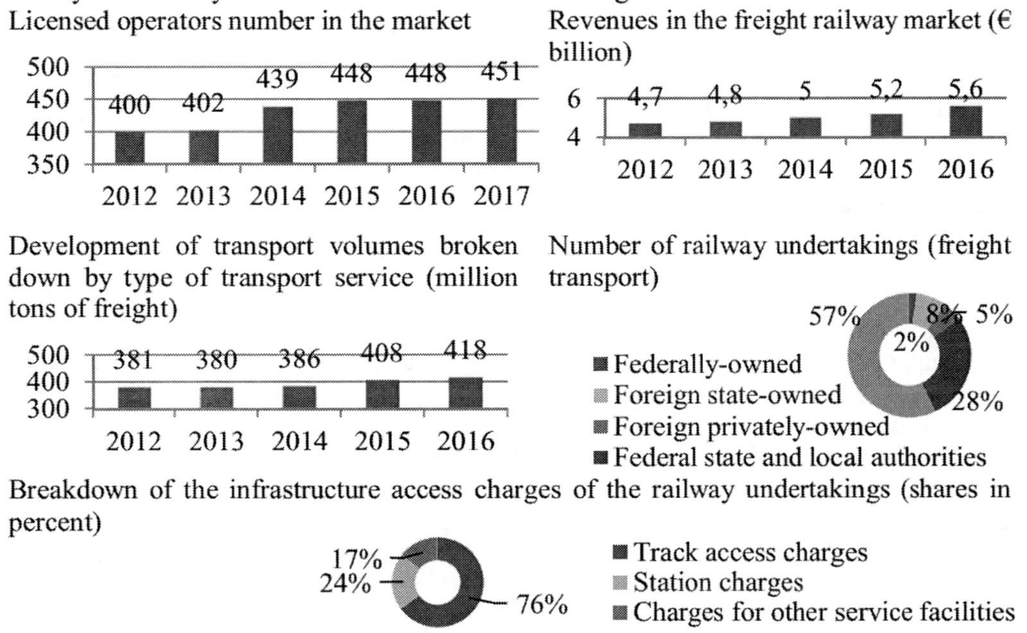

Fig.1. Main indicators for German rail freight industry after liberalization [systematized by authors on the basis of 1]

Data demonstrated in Fig. 1 shows positive results of liberalization in Germany, including: operators number on the market is 451; freight traffic revenue increases annually; cargo transportation volumes show positive dynamics; the market is represented by different private, state or foreign carriers; adjusted infrastructure usage mechanism and income derived from its usage and other services. A survey for German carriers about railway transportation market shows, that it is characterized by the following factors: customer orientation, non-discriminatory pricing system, access to infrastructure, international activities, industry development.

1.2 Rail freight liberalization importance for improving customer service quality

Liberalization process makes it possible to make rail transport more dynamic, increase investment and attract transit traffic. One more advantage is service quality improvement for shippers / cargo owners, because carriers will compete for the client. Today, in terms of economy and environmental friendliness, rail transport has advantages, but because of low supplies reliability and their regularity, railways are inferior to road freight transport. Customers often complain about rail transportation conditions. New rules will prevail on the market after liberalization providing possibility to change this situation. From the customer's point of view, these new rules mean lower shipping costs, service quality, client-oriented management, railway freight transportation scope expanding, existing problems elimination and railway development in general.

Today, rail transportation has to meet high requirements of customers in order to maintain a competitive position in transport services market. Customers are a valuable resource, and therefore working with them requires proper approach.

In this case, the object of research is freight rail transportation area of JSC "Ukrainian Railways". Certainly, the enterprise activities volumes in this area are very large, so only indicators important for liberalization are selected for this article.

2 The freight rail transportation area of JSC "Ukrainian railways" – results of the study

The project of sustainable development logistics strategy of Ukraine [2] identifies the main problems of the railway industry and proposes recommendations for addressing these problems. The main objective is rail freight transport liberalization. In other words, it is a process of establishing a public-private partnership through transferring some services that accompany the transportation process to suborder, contractual access to the railway infrastructure, encouragement for private operators to manage the transport process and repair the rolling stock. In addition, part of investment will come from the private sector, stressing the need to replace most current government decisions in favor of market development and competition, especially in the field of freight traffic.

Railways of Ukraine reforming is carried out in accordance with the State target program for rail transport reforming for 2010-2019 [3] and remains the main task for today and medium term. The main purpose for rail transport reforming is to develop competition in railway transport market and increase industry efficiency. In order to introduce a market model for rail transport functioning, it is necessary to create an effective mechanism of intra-industry competition based on liberalization policy, which involves access to potentially competitive sectors. Ukraine follows a European liberalization path, based on the principle of infrastructure management and transport process separation.

Reformation should relate to changes not only in the organizational but also in the financial model of the rail transport market, as this is a prerequisite for ensuring equal access to the railway infrastructure [4,5]. Open access to Ukraine's railways infrastructure should result in providing an opportunity to receive new and better-quality services from private market operators.

The main railway company operating causes fair complaints from customers, which undoubtedly indicates the existence of more complex, systemic barriers to its further development. A survey was conducted for the main freight transportation clients of the regional affiliate "Lviv Railways". 18 major customers attended this survey in 2017. The results are presented in Fig. 2.

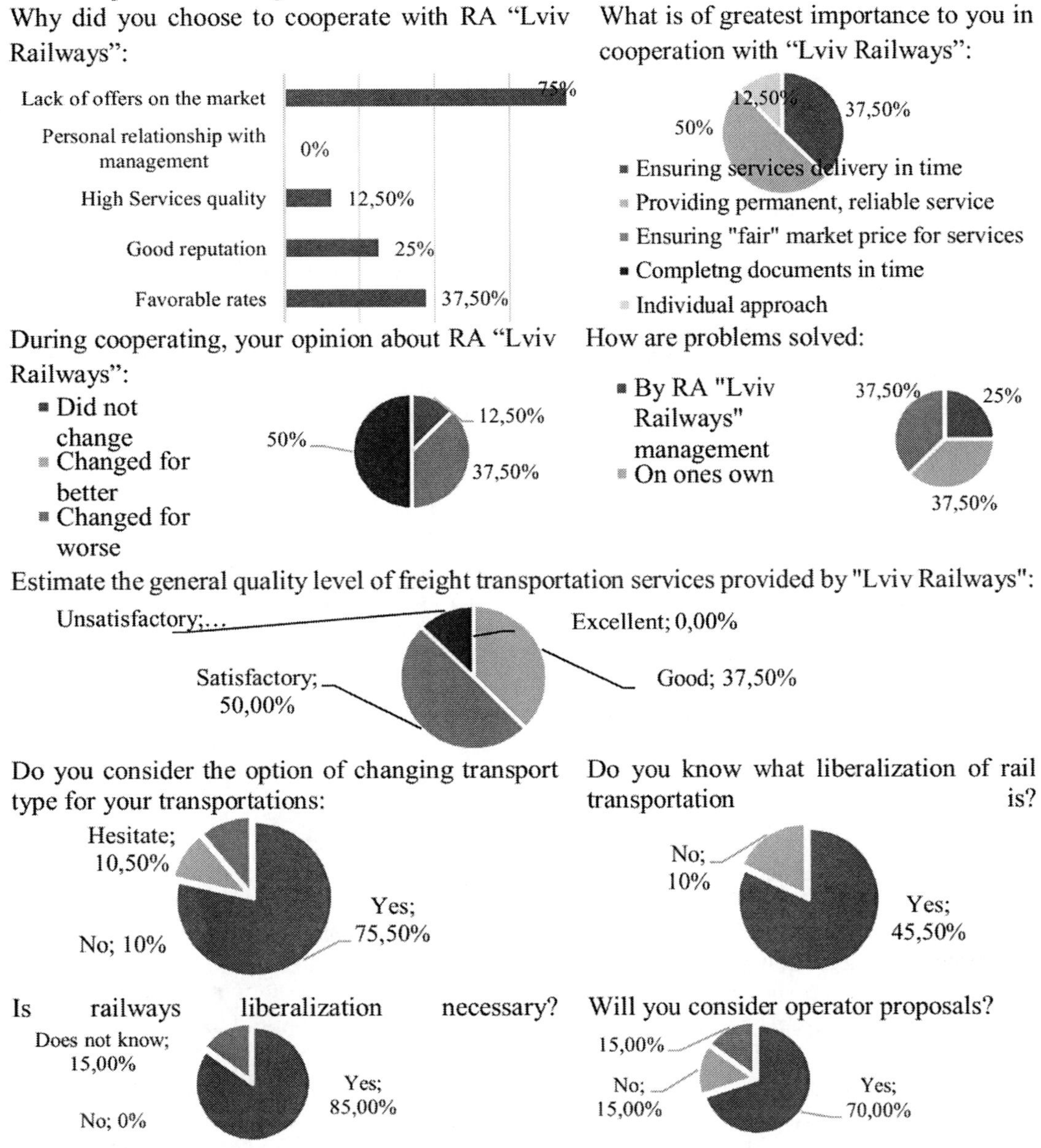

Fig.2. Poll results for "Lviv Railways" clients. [authors research]

The survey results revealed that most of the cooperation with RA "Lviv Railways" is chosen because of the lack of other offers on the market (75%). The most important reason for clients cooperation with RA is the provision of a permanent, reliable service (50%). The most often problems encountered by customers are: delayed delivery of cars, faulty equipment, insufficient equipment / employees to provide services in peak periods, employee incompetence. The following services are assesses as "excellent": employees' attitude and goods safety, a high appreciation was also expressed for information availability and documents processing process. During cooperation period, the opinion on RA "Lviv

Railways" is constantly changing for 50% of customers and changes to worse for 37.5%. It was revealed that in 37.5% of cases clients' problems are solved on their own. The general level of freight transportation services quality for RA "Lviv Railways" is estimated as satisfactory (50%). RA "Lviv Railways" clients are not satisfied with service quality, because 75.5% will consider alternative transport types. The study also aimed at clarifying the awareness level and customer attitude towards the liberalization process. Only 45% of clients understand exactly what liberalization is. Rail freight transportation liberalization necessity is estimated at 85%. 70% of respondents will consider private operator proposals. Clients consider these main advantages for liberalization: carrier choosing possibility, delivery reliability, state influence reducing, optimal tariffs, rolling stock condition improving and equipping. Polled clients' expectation for liberalization: market pricing, quality service, constant meeting demand for transportation, transportation process improvement, problems with rolling stock solution, reduction of bureaucracy in the course of transportation.

Therefore, the following conditions for liberalization can be formed based on the analysis of rail freight transportation area:

1. At present, JSC "Ukrainian Railways" is the only enterprise in Ukraine, having railway infrastructure legally fixed, the exclusive competence of which by laws include transportation process regulation in domestic and international communications, and owns both traction rolling stock (main locomotives) and wagons.
2. Transportation process management in domestic and international communications falls within the competence of JSC "Ukrainian Railways".
3. JSC "Ukrainian Railways" occupies a dominant position in the market of main cargo locomotives provision within Ukraine with a share of 100 %.
4. In recent years, JSC "Ukrainian Railways" has received appeals from business entities regarding the possibility to obtain permission to use their main cargo locomotives on public roads.
5. The methodology used by JSC "Ukrainian Railways" for calculating costs of their own cars using is imperfect.
6. Market participants - the operators of rolling stock and shippers, consider the main barrier for entering railway transportation market to be impossibility, according to the current legislation, to use own traction rolling stock (locomotives) for the carriage by railways of general use, what means no possibility for carrying out independent transportation of clients' goods.
7. JSC "Ukrainian Railways" is the only carrier to use general railways on the territory of Ukraine. Public transportation by railways with private locomotive draft is not allowed. All forwarding companies provide their services directly through JSC "Ukrainian Railways"; therefore, there can be no real competitors in the market of freight rail transportation for JSC "Ukrainian Railways".
8. At present, the state of JSC "Ukrainian Railways" rolling stock does not meet user requirements; therefore, demand for transportation is not fully satisfied.
9. Rail freight users are not satisfied with service quality of the transportation process and are ready to consider alternative transport types.

Liberalization is not a panacea for all railroad problems, but in addition to reducing the financial and time costs for railways, the improvement of the rail system will reduce the negative effects of motor vehicles excessive use, including traffic congestion, road degradation, emissions and accidents. In the long term, efforts may be directed towards improving intermodal services and developing comprehensive service offerings for shippers that could cover the entire supply chain within Ukraine and partly at the international level.

3 Conclusions

Current state of the industry has been described and analyzed, a research has been conducted resulting in forming preconditions and justifying necessity for reforming the industry by implementing liberalization process. A further action plan proposals:

- improve the new edition of the Law of Ukraine "On Railway Transport" in accordance with the requirements of the Railway Transport Directives of the EU (provisions should promote the introduction of competitive conditions for the market, identify the infrastructure operator, approve the mechanism for new entrants to enter the market, etc.);
- create an independent regulator in rail transportations field;
- adjust the reform terms and contents in accordance with the requirements of EU directives;
- carry out locomotive traction and fleet of freight cars market demonopolization (creation of proper legislative base and issuance of permits for private rolling stock usage);
- process business entities proposals regarding their participation in freight locomotives updating and repairing.

Further research will be associated with a more detailed liberalization consideration as one of the priority directions for rail transport development.

References

1. Market Analysis Railway 2017 /Section 702 —Rail Regulation Economic Policy, Market Watch, Statistics (2017)
2. Website «Ukrainian Railways». Retrieved form https://www.uz.gov.ua/en/ (05.06.2019)
3. State target program of rail transport reform for 2010-2019. Retrieved form https://zakon.rada.gov.ua (05.06.2019)
4. Łakomy-Zinowik, M. Public-private partnership as an alternative source of financing of public tasks. *Ekonomia i Prawo. Ecomomics and Law*, **16**(1): 75–86 (2017)
5. Turgut Ozkan, Gozde Yanginlar, Salih Kalayci Railway Transport Liberalization: A Case Study of Various Countries in the World *Journal of Management and Sustainability* **6**, 4 (2016)

Logistics and transport in industry 4.0: perspective for Ukraine

Igor Matyushenko[1] *, *Sergiy* Berenda[1], *Tatyana* Shtal[2], and *Larissa* Grigorova-Berenda[1]

[1]V.N. Karazin Kharkiv National University (Kharkiv, Ukraine)
[2]Simon Kuznets Kharkiv National University of Economics (Kharkiv, Ukraine)

Abstract. The elaboration of the fourth industrial revolution and the process of globalization of the world economy is *sine qua non* of the successful operation and transformation of logistics. The formation of Logistics 4.0 relies in particular on cutting-edge technologies and the digitalization of business processes. The complexity of the structure of world trade arouses interest to the complete range of logistics services provided within a single company. This logic of development requires the application of interdisciplinary approaches, combined with the efforts of engineers, researchers and those interested in the field of logistics.

1 Introduction

Comprehensive advanced manufacturing technologies of the Industry 4.0, including nine core technologies, such as cloud technologies, Internet of things, big data, modeling, system integration, cyber security, additive technologies, robotics or augmented reality, have a fundamental impact on how to manage and organize business and provide resources for it right here and now [1, 2]. These new technologies, which are enforced virtually ubiquitously and are kept with digital capabilities, increase the pace and scale of business change. At the same time, the development of the Industry 4.0 and the processes of digitalization of the world economy are impossible without the successful operation of logistics. The complexity of the structure of world trade arouses interest to the complete range of logistics services, which foregrounds *the problem* of formation of Logistics 4.0, which relies first and foremost on new technologies and the digitalization of business processes. This is confirmed in the works [3, 4].

This logic of development requires the application of interdisciplinary approaches, combined with the efforts of engineers, researchers and those interested in the field of logistics (in particular in the field of intelligent transport systems), which determines the relevance of the study.

2 Presentation of basic material of the research

* Corresponding author: imatyushenko@karazin.ua

The transport and logistics industry is characterized by a relatively small number of large international companies and numerous regional and small companies. Such structure of the industry is due to the fact that there are few barriers to entry, but significant investments are required in order to grow to a certain size. The limited number of companies that are of sufficient size to use integrated information supply chain systems has led to the formation of a system environment that has a few standards and even fewer large-scale specialized software and technologies.

Therefore, many large logistics companies have developed their own information systems, annually investing significant funds in supporting and upgrading these systems, thus forming an infrastructure of 3 PL, 4 PL and even 5 PL. Small players had to either use applications provided by local or niche software vendors or, as is often the case with small players in each industry, not use any applications or technologies outside the regular office of the program [5, p.12].

Leading Logistics Services Providers (LLPs), that are also called logistics providers (4PLs), have a vital role in the formation of a supply chain. They assume many of the same functions as Logistics Service Providers (3PLs), but have much wider responsibility, helping clients achieve their strategic goals.

Functioning as 4PL, the logistics provider becomes a true partner with the client, working on creating an economic supply chain.

4PL usually manages each moving part within the customer supply chain and serves as the single point for contact with all parties. The LLP's role usually involves managing and analyzing large volumes of data, controlling transport management, managing other 3PLs and transport service providers, overseeing warehousing operations, or managing any other part of the supply chain.

Transparency is key to 4PL, which can also function as 3PL. They should actively cooperate with clients to demonstrate the containment and counterbalances they have introduced, for instance, the conclusion of confidentiality agreements that prohibit the exchange of information and data from other 3PLs and transportation [6, 7]. 5 PL providers of logistics services are formed around a business model using e-Business.

Each single logistics company in the supply chain manages its own ICT systems that connect to external systems through bi-directional communications (EDI). This information and communication technology management structure has benefited the software industry as well as the domain members with the financial and technical resources needed to use this business model [8].

In Ukraine, large technology companies in the IT sector are torn off from the domestic market. In fact, each of the “big IT four” – Luxoft, Infopulse, SoftServe, Eleks – has worked in area 4.0. But only Eleks started offering these solutions in Ukraine (Kernel project), all others work exclusively on Western markets. That is, neither these companies nor other key players of the IT industry pay attention to the development of the customer market in Ukraine.

Among the 100,000 programmers in the IT industry, 90% work for export, of which 70% are outsourced (IDC-Ukraine data from 2017). And only 5% of the IT industry is involved in industrial applications [9].

Moreover, the latest data of 2019 indicate an unprecedented rise in this sector of the economy, in the city of Kharkiv alone there are officially employed about 30,000 IT specialists and another 10,000 are working on free freelancing serving export markets. Thus, it can be argued that there are IT opportunities for creating the appropriate infrastructure for 4PL and 5 PL operators in Ukraine. However, at the moment, only foreign 4PL operators are working in Ukraine. And the general engineering infrastructure and business environment shows that the creation of national 4 PL and 5PL operators is possible only in close cooperation with leading foreign companies.

However, the positive development trends of this market in Ukraine and in the world overlap. In Ukraine, there is a rapid growth in the development of 3 PL operators and their revenues, as a response to new opportunities offered by Internet commerce in the area of cost reduction, and as a result, a decrease in prices. Under conditions of low solvency in Ukraine, this factor is a key driver of the development of the entire market.

On the background of the appearance of 5 PL logistic operators, and the improvement of business models of 1PL-4PL providers, the crucial task for the further development of logistics services is to create the appropriate infrastructure. The European Commission has adopted regulations aimed at accelerating the implementation of intelligent transport systems on the European road system (C-ITS). These systems should improve traffic flow and safety.

The new technology will allow vehicles to "talk" to each other, with road infrastructure and other road users - for example, on dangerous situations, road works and time of traffic lights, making road transport safer, cleaner and more efficient, the so-called V2V approach (vehicle to vehicle). This solution gives car makers, road operators and others this long-awaited legal certainty required to start a large-scale deployment of C-ITS services in Europe, remaining open to new technologies and market development. The first C-Roads tests (intelligent roads with the C-ITS system) were conducted within the framework of the pilot program in 2018. Cars from France and Portugal easily “communicate” with the “intelligent” highway in Austria. At the moment, car makers such as Hyundai, Opel, Honda, Volkswagen, Renault, Volvo Trucks and Fiat have committed to equip their cars with C-ITS since 2019 [10, 11].

Therefore, it is quite logical to ask how the logic of Industry 4.0 and the Logistics 4.0 paradigm can conjugate, where its important part is its dependence on the past, the so-called “path depends” and how it is bound to affect future development.

There are five main elements of Logistics 4.0:

1. Transparency. One of the most important aspects of modern logistics is the increase in transparency resulting from the enhanced digitalization of the entire supply chain, which has the potential to make modern logistics much more efficient and comprehensive.

2. Smart Containers and Pallets. As for smart ports, there are many other "smart" versions of traditional logistics components that change the way the goods move from suppliers to customers.

3. Adaption to IoT (Internet of Things). When it comes to smart containers, pallets, vehicles, ports, etc., the question remains: how to maximize the ability of a person to obtain significant data from different adherent point of the value chain and turn these data into valuable, integrated ideas. In Industry 4.0, this role is filled by IoT devices that are installed in production facilities and warehouses.

4. Integration with Industry 4.0. One of the most important aspects of logistics 4.0 is its ability to integrate easily with Industry 4.0 systems, creating symbiotic, synergistic relationships between manufacturers and shippers. Despite the fact that the full implications of the relationship between logistics digitalization and intelligent manufacturing processes remain unresolved, it is clear that the benefits of such integration are much greater.

5. Collection of analytics. So far, many important things have been done to collect data in these new logistics frameworks, but the real value of this data is not limited to simplifying manual planning. Most likely, completing success in the area of transparency, visibility and data collection using sensors, NFC tags and RFID chips is the use of advanced analytical work. By providing a large amount of information to forecasting and planning algorithms, modern logistics providers can improve their forecast of demand and shipping, identifying potential loss zones or possible improvement of the market. This helps lay the foundations for a smoother version of the supply chain management, which is less prone to risks, failures and opacity [12].

3 Conclusions resulting from the study

Consequently, we can draw the following conclusions: in the leading economies of the world there is a synchronous and integrated development of Industry 4.0 and Logistics 4.0. Modern logistics services in the EU countries are commercially successful in exploiting the latest approaches, such as 5PL.

Ukrainian companies are only fragmentary ready to implement the principles of industry 4.0 in logistics, so the competitive struggle for this highly profitable market is losing by them. Meanwhile, this technological gap is only increasing. Therefore, the prospects for the emergence of 5PL operators and the corresponding market for such services in the next 3 years in Ukraine cannot be expected. The focus of further research is on the management of interdisciplinary approach, which enables the formation of competitive advantages of modern logistics services based on a combination of engineering, business and research experience.

Refernces

1. I. Matyushenko *Prospects for the development of convergent technologies in the countries of the world and Ukraine for solving global problems*: a monograph. Kharkiv: FB Liborkina LM, 448 p. (2017)
2. K. Shvab *The Fourth Industrial Revolution* / Trans. from English Moscow: Publishing house "E", 2017. 208 p.
3. K. Witkowski *"Internet of things, big data, industry 4.0–innovative solutions in logistics and supply chains management."* Procedia Engineering **182** (2017): 763-769.
4. E. Hofmann, M. Rüsch. *"Industry 4.0 and the current status as well as future prospects on logistics."* Computers in Industry **89** (2017): 23-34.
5. Information Systems for Interconnected Logistics. Research & Innovation Roadmap. https://www.etp-logistics.eu/wp-content/uploads/2015/08/W36mayo-kopie.pdf (2019). Accessed 6 Feb 2019
6. 4PL VS. 3PL. https://www.penskelogistics.com/solutions/supply-chain-management/lead-logistics-provider/4pl-vs-3pl-differences/ (2019). Accessed 10 Feb 2019
7. A hybrid data analytical methodology for 3PL transport provider evaluation using fuzzy multi-criteria decision making. https://doi.org/10.1080/00207543.2015.1022266 (2019). Accessed 22 Feb 2019
8. 5 Key Elements of Logistics 4.0. https://blog.flexis.com/5-key-elements-of-logistics-4.0 (2019). Accessed 6 Feb 2019
9. National strategy of Industry 4.0 Project for the Cabinet of Ministers of https://drive.google.com/file/d/1VU0LPVXUs6rUTjuct69Dptek8tqkNIsd/view (2019). Accessed 2 Mar 2019
10. Brussels: new vehicles must communicate with the road infrastructure. https://trans.info (2019). Accessed 12 Mar 2019
11. Information Systems for Interconnected Logistics. https://www.etp-logistics.eu/wp-content/uploads/2015/08/W36mayo-kopie.pdf (2019). Accessed 1 Feb 2019
12. S. Berenda, N. Zaporozhsky *Use of international fullfilment-warehouses in the formation of supply chains* // "Current state and prospects of economy, accounting, finance and management in Ukraine and the world": a collection of abstracts of the International Scientific and Practical Conference, (November 14, 2017 in Poltava, Ukraine) - Poltava: TSFEND, pp. 249-250. (2017)

Improving the efficiency of road transport companies by optimizing the costs of information and advertising activities

Yevgen Nagornyi[1*], *Olha* Shulika[2], *Oleksandr* Severyn[2], and *Oleksandra* Orda[3]

[1]Kharkiv Automobile and Highway University, Ukraine
[2]Kharkiv Automobile and Highway University, Ukraine
[3]Kharkiv Automobile and Highway University, Ukraine

Abstract. In this article, we propose to improve the efficiency of transport companies by managing information and advertising activities to promote material flow. In this research, within the framework of planning the supply of consumer goods in cities, we offer the following alternative supply schemes: the direct goods delivery and delivery through distribution centres. In order to choose a rational supply scheme as a criterion of efficiency it was proposed to use the total delivery cost including the information and advertising component. The research also provides the methodology for determining the maximum effect of advertising activities by road transport companies. In addition, an example of developed advertising packages with the minimum and maximum advertising costs is given. The measures proposed in the research will allow developing a program to stimulate demand for transport services and increase the competitiveness of road transport companies in the transport market.

Key words: delivery scheme, consumer goods, road transport, advertising activity

1 Introduction and literature review

The transition of the national economy of Ukraine towards the "green" model of development [1] requires from the city logistics the creation of efficient freight logistics systems. They have to take into account not only the costs of participants in the planning and organization of delivery of goods but also the consequences of vehicles' use in the context of the environment [2, 3]. An important role in this process belongs to an efficiently organized integrated logistics service using the latest innovative technologies "Industry 4.0". These technologies allow us to create a real-time planning optimization model focusing on energy efficiency of operations, the development of information and advertising management systems for companies [4]. Widespread implementation of these technologies also contributes to solving one of the key issues – the last mile problem [5, 6]. This problem of cargo delivery to the end user is the least efficient delivery stage in cities and is up to 28% of the total delivery cost.

* Corresponding author: shulika.khnahu@gmail.com

Each year, hundreds of millions of tons of different goods are delivered from suppliers to the country's distribution chains. According to researchers, today more than 60% of cargo delivery in large Ukrainian cities is carried out using the Internet for the placement delivery requests on transport portals [7]. One of the leading Ukrainian IT companies, EVO which develops online shopping space reports that in 2018, online purchases increased to a total cost of UAH 9.5 billion [8]. The Internet has changed the concept of purchasing. Logistics has such a convenient feature as an online order. As online stores have come, the customer has the opportunity to make purchases without leaving the house. At the same time, businesses that make online trading can reduce costs as salesroom lease is no more needed. Retail shops often need no warehouse. Cargo owners whose main activity was off-line sales were able to expand the market and find new customers [9].

The concepts of a virtual enterprise can vary in practice from the internet site to the full-scale production and logistics system in which the web representation serves as a link between buyers, sellers and manufacturers [7]. At the same time, an important moment in advancing the services of transport companies is the development of a unique proposal that will emphasize the competitive advantages of a transport company in the transport markets. The development of an advertising proposal consists of the following steps:

– outline of potential clients: a sample of companies that may need a cargo delivery service; sending them a commercial offer;

– local media analysis. For successful promotion, it is important to choose the appropriate supply chains for advertising messages through monitoring.

However, issues of information support, speed and high cost of goods delivery remain problematic. Thus, the optimization of total costs taking into account the costs of advancing transport services will increase the efficiency of transport companies in transport markets.

2 The mathematical model of choice a rational scheme for the delivery of consumer goods

In this research, the following alternative delivery schemes are proposed within planning the delivery of consumer goods in cities: direct delivery of goods; delivery through distribution centres.

In order to choose a rational delivery scheme, the total delivery costs (C_{1t}, UAH/t) under consideration of the advertising component (C_I, UAH/time period) are taken as the criterion of effectiveness in the research. The input parameters of the delivery process model are consignment volume (Q, t), the area of the service area (F, km^2) and the intensity of the advertising effort (I, days/period):

$$C_{1t}(Q,I,F) \to \min. \quad (1)$$

The given total costs are determined based on the total costs of the delivery of consumer goods (C_{del}, UAH/period):

$$C_{1t} = \frac{C_{del}}{Q}, \quad (2)$$

$$C_{del} = C_T + C_{LU} + C_S + C_I, \quad (3)$$

where C_T – transportation cost, UAH/ period; C_{LU} – loading and unloading cost, UAH/ period; C_S – storage cost, UAH/ period.

$$C_I = C_A \cdot Q - \delta_m , \tag{4}$$

where C_A – unit advertising cost, UAH/ t; δ_m – unit of margin of advertising per one advertising company, UAH/ period.

System of limitations and assumptions:

$$\begin{cases} \sum_{i=1}^{N} W_i \geq \sum_{j=1}^{J} Q_j; \\ 0{,}05 \leq Q, \text{t} \leq 25; \\ 50 \leq L, \text{km} \leq 1500; \\ 500 \leq I, \text{UAH/month} \leq 10000; \end{cases} \tag{5}$$

where W_i – carrying capacity of i-th vehicle of a road transport company, t/time period; N – number of vehicles of carrier, ones/ period; Q_j – the consignment volume for i-th request, t/ period; J –number of requests to deliver, unit/ period.

To optimize the use of information and advertising activities based on the Bayesian approach [10] we proposes to establish a zone of economic expediency of advertising costs and the volume of advertising measures. It determines the amount of advertising effort (the number of advertisements, the writing of text, the adjustment of advertising by geolocation, the adjustment of advertising for a list of contacts, etc.) and their intensity.

The following dependence of advertising costs is determined:

$$C_A = b \cdot I^2 , \tag{6}$$

where b – model parameter that characterize the intensity of cost changes, UAH/day.

The margin equation has the following form:

$$M_A = -c \cdot I^2 - d \cdot I , \tag{7}$$

where c, b – model parameters that characterize the intensity of margin changes , UAH/day.

The area of economic feasibility of advertising costs is limited to points A and B shown in Fig. 1.

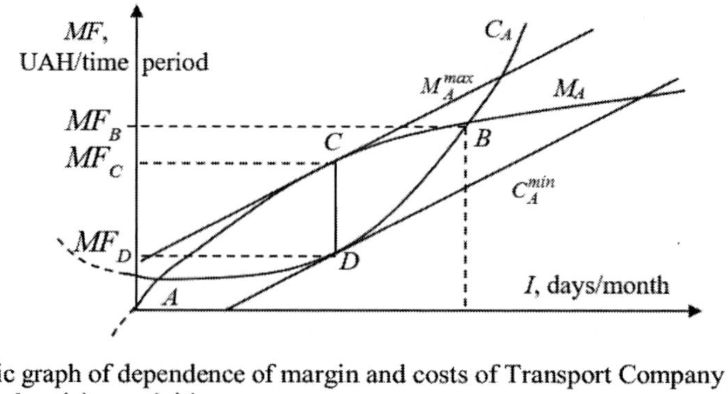

Fig. 1. Characteristic graph of dependence of margin and costs of Transport Company on the intensity of information and advertising activities.

In this case, point B characterizes a situation where further increase in the volume of advertising measures becomes inappropriate. The maximum revenue from advertising measures corresponds to CD segment, which represents the difference between MF_C and MF_D ordinates. The equation of tangent follows:

$$M_A^{max} = a \cdot I + e , \tag{8}$$

$$C_A^{min} = a \cdot I - f , \tag{9}$$

where a –parameter of straight lines, UAH/day; e, f – parameters of straight lines, $e>f$, UAH/period.

The approach to finding the ordinates of the points C and B consists in the fact that there are two and only two points on the cost and income curves, to which two parallel tangents can be made.

Abscise of B point is calculated on condition $\frac{dM_A}{dI} = \frac{dM_A^{max}}{dI}$, then $-2 \cdot c \cdot I_c - d = a$, so

$$I_c = \frac{a+d}{-2 \cdot c} . \tag{10}$$

Abscise of D point is calculated on condition $\frac{dC_A}{dI} = \frac{dC_A^{min}}{dI}$, then

$$2 \cdot b \cdot I_D = a , \; I_D = \frac{a}{2 \cdot b} . \tag{11}$$

Since the condition $I_D = I_C$,

$$\frac{a+d}{-2 \cdot c} = \frac{a}{2 \cdot b} , \tag{12}$$

$$d = -a \cdot (\frac{c}{b} + 1) . \tag{13}$$

Using the previous equations, one can calculate the ordinates of the C and D points:

$$MF_D = \frac{a^2}{2 \cdot b} - f , \tag{14}$$

$$MF_C = \frac{a - a \cdot (\frac{c}{b} + 1)}{-2 \cdot c} \cdot a + e = \frac{a^2}{2 \cdot b} + e . \tag{15}$$

Consequently, the length of CD segment corresponding to the maximum of the effect is

$$\Delta MF = e + f . \tag{16}$$

The coordinates of the point of intersection of the income and expenditure curves that characterize the margin of economic expediency of further increasing the volume of advertising activities (the amount of information on the market situation of the collected market) can be determined provided that the functions $C_A(I)$ and $M_A(I)$ at this point are equal. That is $C_A(I) = M_A(I)$, then $(b+c) \cdot I^2 + d \cdot I = 0$, so

$$I_B = -\frac{d}{b+c}. \tag{17}$$

The ordinate of *B* point follows:

$$MF_B = b \cdot \frac{d^2}{b+c}. \tag{18}$$

3 Case study

The area of economic expediency of advertising costs shown in Fig. 2 was determined according to the data of the Individual Entrepreneur E.V. Gayova (Kharkiv, Ukraine).

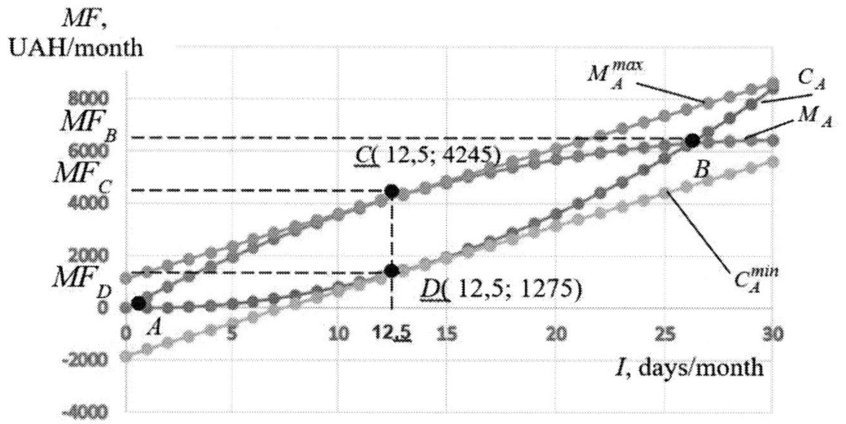

Fig. 1. Graph of dependence of margin and costs of "Gayova E.V." on the intensity of information and advertising activities.

The required amount of advertising measures that provide the optimal costs of the transport company for advertising should be 12.5 days/month, but no more than 26 days/month. At the same time, the maximum effect from the advertising impact on the consumer is equal 53,125 UAH.

The set amount of advertising costs affects the amount of advertising services. Two packages of advertising services are offered in the Table 1 for the specified enterprise proceeding from the defined zone of economic costs for advertising.

Table 1. Advertising packages with minimum and maximum advertising costs.

Advertising service	Value	
Package of advertising services, UAH / month	1,275	4,245
Number of ads, units	5	10
Text writing	+	+
Instagram advertising	+	+
Facebook advertising	-	+
Create a photo and process it	+	+
Number of taken photos, units.	5	10
Look-a-like audience setting	-	+
Facebook pixel settings	-	+

4 Conclusion

Thus, this approach allows, firstly, to determine the rational scheme of cargo delivery in cities, in which the total cost of delivery taking into account the costs of the transport company for advertising will be minimal. Secondly, it allows getting the maximum effect from the use of advertising activities of transport companies. It will stimulate demand for transport services and increase the competitiveness of road transport companies in the transport market.

References

1. Overview of the Ukrainian economy. *American Chamber of Commerce in Ukraine* [online]. Available at: http://chamber.ua/Content/Documents/1017431819ACC_CountryProfile2018_UKR.pdf (2018)
2. F. Russo, A. Comi, A classification of city logistics measures and connected impacts. *Procedia-Social and Behavioral Sciences*, **2**, 6355–6365 [online]. Available at: https://core.ac.uk/download/pdf/81950420.pdf (2010)
3. C. Navarro, Designing new models for energy efficiency in urban freight transport for smart cities and its application to the Spanish case. *Transportation Research Procedia*, **12**, 314-324 (2016)
4. O. Shulika, O. Glita, The role of information and advertising activity of cargo owners in searching for freight customers in large cities. *Collection of scientific works on the materials of the International scientific and practical Internet-conference "Modelling and information technologies in science, engineering and education"*, 246-249 [online]. Available at: http://dl.khadi.kharkov.ua/pluginfile.php/52741/mod_resource/content/4/IPM_CONF_SB_2018.pdf (2018)
5. L. Ranieri, S. Digiesi, B. Silvestri, M. Roccotelli, A Review of Last Mile Logistics Innovations in an Externalities Cost Reduction Vision. *Switzerland: Sustainability*, **10** (3), art. no. 782 (2018)
6. T. Bányai, Real-time decision making in first mile and last mile logistics: How smart scheduling affects energy efficiency of hyperconnected supply chain solutions. *Energies*, **11** (7), art. no. 1833 (2018)
7. A. Cherepakha, Formation of the virtual management system for delivery process of the consumer goods in large cities: aref... PhD, spec.: 05.22.01 – Transport System. KhNAHU [online]. Available at: https://dspace.khadi.kharkov.ua/dspace/bitstream/123456789/1860/1/%D0%A7%D0%B5%D1%80%D0%B5%D0%BF%D0%B0%D1%85%D0%B0_%D0%B0%D1%80%D0%B5%D1%84.pdf (2016)
8. Evo.company [online]. Available at: https://evo.company/category/media-about-us/evo/(2019)
9. A. Pakholkova, Analysis of new information technologies used in logistics. *Current issues of economics and management: Proceedings of the IV International. scientific conference,* 170-174 (2016)
10. O. Smekhov, *Marketing model of transport market.* - Moskow: Publishing groupe «Transport» (1998)

Accounting and analytical base for management of logistic flows in tourism business

Olena Prokopishyna[1*], *Natalia Kozubova*[2] and *Karina Satusheva*[1]

[1] Simon Kuznets Kharkiv National University of Economics, Department of accounting and business consulting, 61001 Kharkiv, Ukraine
[2] Simon Kuznets Kharkiv National University of Economics, Department of tourism, 61001 Kharkiv, Ukraine

Abstract. The paper grounds that the existing potential of the tourism industry of Ukraine can be realized through the introduction of efficient logistic approaches for the development of tourism and recreation and management of tourist enterprises. Based on modern knowledge of the theory and practice of tourism management, a logistic strategy in tourism business serves for the creation of tourism product which is highly competitive in the global market due to implementation efficient management decision upon each component of the logistics system of the company. The paper reveals the essence and major functions of accounting and analytical subsystem of logistic management of tourism companies that enables to shorten the duration of the operating cycle, improve the quality and expand the range of travel services, and increase the flexibility of the travel service. The paper concludes that efficient accounting and analytical subsystem of logistic management gives travel companies wide opportunities to gain competitive advantages on global travel market.

1 Introduction

Scientific guidelines on logistics are very useful for the development of tourism industry in Ukraine as well as of highly competitive world market of travel services. Application of logistic management principles can be extremely useful in the tourism and recreation industry, due to accumulation of experience in the creation, implementation, monitoring and control of chain supplies. In the case of the formation of a supply chain in the tourism industry, experienced and professional knowledge in the cargo and passengers transport system is vitally important. A highly competitive tourism market requires high quality service, which is a priority area in finding ways to improve tourism logistic.

2 Discussion

[*] Corresponding author: elena_prokopishin@ukr.net

2.1 The essence and significance of logistics for the management of a travel company

Modern theory treats logistics systems as managerial systems for material, information, finance and human flows. The human flows are not generally accepted in the field of logistics, but may be successfully qualified as the object of logistic science according to the theory of tourism resources management. The application of the term "logistics" in tourism management, is considered as an effective method for managing flow processes of tourists, as well as accompanying information and financial flows, with the possibility of optimizing the costs involved in the production of tourism product and customer service during travel.

Tourist logistics can be understood as a science of planning, controlling and managing operations that are carried out in the process of forming, bringing the tourist product to the consumer, as well as in the process of transferring, storing and processing information. In turn, the logistic management of tourism activities is a purposeful impact on the coordination and synchronization of the processes of formation of flows of tourists and their service [1].

Being in compliance with the general principles of management of human flows, tourist flows have their own peculiarities. Management of tourist human flows is carried by tourism business logistics, aimed at managing human and associated flows, in contrast to traditional logistics flows. In the tourist area, the priority is the human flow, which submits the analysis and is predicted, that enables planning and forecasting of necessary material, energy, financial and information resources.

Human flows are divided into flow of actors (providers) and flow of service objects (consumers). In the composition of the tourist flow 4 categories of tourists and the corresponding streams may be distinguished: domestic tourism (resident visitor), inbound tourism (non-resident visitor), outbound tourism (resident visitor), transit (non-resident visitor) [2]. The foreign tourism is essential for a national recreational complex. It contributed to the development of infrastructure, investments and improvement of the quality of travel services to satisfy high requirements of world class consumers.

The definition and study of the features of tourist flows is important for elimination of negative factors of tourism development as the decline in the quality of services, reduction of the number of regular customers; it also provides the opportunity to establish tourist flows, and develops mechanisms that can create cyclic and nonstop tourist flows, despite the qualitative characteristics of tourists.

All tourist movements form an uninterrupted human flow that combines subjects and objects, as well as concomitant flows. Distinctive feature of flows of consumers is that they simultaneously act as objects of management and service, carriers of information and material resources. As the management of human flows in tourism takes place during the provision of services provided throughout the period, and management should ensure that this procedure is carried out in compliance with authorities, legislation and moral principles of travel companies activities.

Management of human flows of tourism business requires the development and implementation of a comprehensive logistics strategy.

2.2 Logistic strategy for development of tourism business

The complex and multilevel logistics system of the producer and the organizer of tourism differs not only in terms of activities, but also in terms of the degree of integration of business, the volume of consolidation. Based on the analysis of the logistics management system in the subjects of the tourism industry, most enterprises should look for reserves and ways to improve logistics management, create a structured logistic model, taking into account modern scientific developments, international experience, managerial innovations that require certain enterprise managers to have certain knowledge and the ability to organize the process of providing people with knowledge, desires, opportunities, as well as creative behaviour in conditions of fierce competition.

According to the researchers, the logistics strategy is a long-term, qualitatively defined direction of logistics development concerning the forms and means of implementation in the enterprise of interfunctional and inter-organizational coordination and integration, formulated by the higher management of the enterprise in accordance with its corporate goals [3, 4] and aimed at optimizing the resources of the company when managing the main and concomitant flows [5, 6]. Strategic goals in logistics are achieved by establishing its connection with the basic principles of corporate governance.

In the case of developing a logistics strategy, the starting point may be the definition of a common corporate strategy, that is, a higher level strategy with respect to logistics. Conceptually, the principles and requirements of the corporate strategy are developed taking into account the public needs, needs of consumers, personnel, business owners and business partners. The higher-level strategies set goals and context, which takes into account all the logistics solutions [7]. The experience of countries with a developed market economy shows that logistics plays an important role in shaping a common corporate strategy. This is especially true for companies that practice advanced technologies based on integrated logistics networks.

The unceasing process of globalization, together with the liberalization and expansion of the international tourism market, has led to unexpected losses in market share for developed touristic countries. New directions are often counted on resources that are little known and, therefore, uniquely attractive. They are able to introduce new products and services competitive for the international market [8 – 10]. Due to the low level of development, new tourist countries were able to obtain a temporary competitive advantage in the form of low labour costs and favourable exchange rates of the national currency relative to convertible ones. Consequently, despite the benefits that traditional tourism countries derive from the high level of development of tourism infrastructure, specialization in tourism and competitive experience, their ability to compete at a low price due to high wages and exchange rates , which slows the economic development of tourism in these countries and prompts the invention and introduction of innovations.

Service Response Logistics is defined as a process of coordinating the logistics operations necessary to provide excursion and accompanying tourism services in the most efficient way in terms of costs and customer satisfaction. This approach is the main strategic element of management of many enterprises that provide tourist services. The critical elements of this approach are the acceptance of travel services and monitoring of the provision of tourist services. Flows of tourist services as well as material flows are distributed in a certain delivery environment, in which there are links in the logistics system, logistics channels, circuits, etc. This network needs to be built in such a way as to maximally meet customer requirements for service levels. For storage of goods and stocks at certain tourist enterprises designated warehouses. The travel company bears some expenses related to the storage of warehouses and warehousing operations. They are part of the logistics costs. Therefore, warehouse operations need to be considered not in isolation, but as an integrated component of the logistics chain. Only this approach is a guarantee of successful fulfilment of the main functions of the composition and achievement of a high level of profitability of sales of a tourist product. The purpose of creating warehouses in logistics systems is not the conservation of material resources, but the transformation of material flow parameters for their most effective use.

The marketing logistics of tourist products is an integral part of the overall logistics system, which ensures the most effective organization of distribution of products ready for use. The functions of marketing logistics are to distribute and promote the tourist product through the channels of commodity circulation, supply of the product oriented to the needs of consumers (according to terms of service, quality, quantity and price), optimization of costs associated with the operation of the logistics system and commercial marketing activities, the use of optimal information marketing systems.

Logistics should be engaged in the organization of the search and purchase of necessary goods of the appropriate quality and at the lowest prices. An analysis of possible logistics costs plays an important role in improving the logistics efficiency of logistics. The optimization of the logistics chain is influenced by both external and internal factors. External relations should include the relationship of entrepreneurs with creditors and suppliers. Internal factors are associated with a different understanding of the optimal activity of chain links, the need for economic compromises between different divisions of firms. Successful procurement involves the availability of a wide range of information on the state of the markets. For the market research, detailed information is regularly selected and evaluated to determine market capacity and create prerequisites for optimizing procurement.

The essence of information logistics in tourism is the organization of data flows that accompany the tourist flow, and is an essential link that connects the supply, production and sales. Information logistics operates with a set of economic information in the horizontal and vertical directions, in internal and external coordination and requires the appropriate infrastructure. Under the information infrastructure of the logistics system in tourism is understood a set of interrelated tools and methods that provide information preparation, namely: data collection and their initial processing, control of reliability, storage and transmission of information, its updating, correction and presentation.

The information flow in the tourism logistics is the main component, as it provides interconnection between the links of the logistics chain "travel agent - tour operator - material and technical base - recreational and tourist resources" through the control and accounting of tourism, material and financial flows. Rational and coherent implementation of all these functions necessitates the creation of a special accounting and analytical subsystem of management of logistics flows of tourism business, which should include functional and security subsystems. Consequently, the accounting and analytical subsystem of management of logistic flows of tourism business is a set of organized and interrelated processes regulating the activities of tourism enterprises. In the logistic information system of tourism, phenomena of vertical and horizontal integration are often observed. The essence of vertical integration is the availability of information links between the tour operator, travel agents and suppliers of tourist products. Horizontal integration represents the information links of tourist firms on the same level.

The functioning of the accounting and analytical subsystem of management of logistic flows of tourism is related to the implementation of certain functions, including planning, coordination, maintenance and control. Execution of these functions makes it possible to coordinate work between different logistic elements.

In the organizational structure of the accounting and analytical subsystem of management of the logistics flows in the tourism business, following subsystems may be distinguished: management of orders processes, implementation of scientific research and communication, the formation of logistics solutions, creation of initial forms and reports. These subsystems are interconnected and provide information and computer support for the whole logistics system at the micro and macro levels. The subsystem of process management of orders provides direct contact with the consumer in the process of formation and execution of the order; the subsystem of scientific research and communication performs analysis and predicts the influence of the external and internal environment of the organization; The subsystem of logistics decision making uses computer databases and analytical models to optimize decision making.

The functioning of the accounting and analytical subsystem of management of tourism logistics flows should be based on the following principles: reliability and completeness of information (for rational and reasonable decision-making by the logistics manager); the accuracy of the source information (is important in determining the amount of available and required resources); timeliness (information should be received on time in a timely manner, since it is necessary for all logistic functions); orientation (information in the logistics system should be aimed at solving certain problems); flexibility (information must be adapted for use in different forms and for different users of information).

Implementation of the accounting and analytical subsystem of management of logistics flows in the tourist enterprise enables: to increase the efficiency of its work, to achieve a high level of service for customers and to provide effective control; to provide a modern approach to the management of sales processes and organization of tourist services; monitor the effectiveness of tourist services; carry out an analysis of the profitability of the tourism business; to increase the efficiency of the pricing policy of the tourist enterprise, and so on.

The accounting and analytical subsystem of management of logistic tourism flows also provides an opportunity to quickly and qualitatively create a package of new services for organizing events of different nature in order to maximally meet the needs of the client. An important direction of its application is the optimization of the use of profit, which, in turn, gives the opportunity to gain competitive advantage and significant increase in revenues from the work in real time, constantly analyzing booking information, seasonal fluctuations, segments specifics, price levels and tariffs on market of tourist services.

Application of the accounting and analytical subsystem of management of logistic flows is possible both for separate tourist firms, and for the network of enterprises of this branch, which has a whole set of data.

The advantage of using the accounting and analytical subsystem of logistics flow management is the ability to conduct flexible marketing and pricing policies, as well as the maximum coverage and aggregation of all customer inquiries. So, using such a system, a travel company can effectively sell tourist products based on the analysis of trends and market demands, constantly monitor the conditions of their services sales, adjusting them in accordance with the market conditions in any segment, which, in turn, will be to promote the quality of services and the level of profitability of the activity.

The effectiveness of the logistics systems of the tourist enterprise in the most general form can be represented as a systemic economic task, aimed at solving the problem of achieving high quality of its work. Regarding logistics, systematic means that all participants in tourist flows are regarded as part of a single whole process, and the process itself is an integrated problem of all participants.

3. Conclusions

Implementation and use of the accounting and analytical subsystem of management of logistic flows in tourism business allows: to create a unified system of accounting and control of the formation and movement of a tourist product; reduce the intervals of formation, promotion and realization of the tourism product; to automate the processes of accounting and control of financial and information flows during tourist services; shorten the time and improve the quality of service of consumers of tourist product; reduce the amount of documentation and significantly reduce the number of errors in the accounting records.

The objects of the accounting and analytical subsystem management of the logistics flows of the tourism business are the processes of formation, promotion and realization of the tourism product, as well as related financial and information flows. These complex processes require the formation and analysis of information on: the estimated contingent of tourist product consumers (business, sports, recreational, educational and other tours); types of tourist products that are in greatest demand at present or in the near future (winter or summer season); conditions of accommodation (hotels, restaurants, holiday houses, camping sites); tourist services and their logistical support; components of marketing; operating tariffs; quality and reliability of the services rendered.

The main functions of the accounting and analytical subsystem management of the logistics flows of the tourism business are: operational analysis, control and management of the processes of formation, promotion and implementation of tourism product, as well as strategic planning and forecasting of tourism organizations; operative transfer (reception) of necessary volume and quality of information at the given time and in the right place; operational control of the movement (movement) of tourists and tourist product; Integration of separate functioning tourism information systems (local, regional, transcontinental networks, etc.) into a single logistic information system; calculation and organization of the financial reserve in case of force majeure situations on the basis of analysis and forecasting of the development of negative trends in the tourism sector.

Implementation of modern methods of logistic management in business practice can reduce inventory, reduce production and circulation costs, reduce the duration of functional logistics cycles, and improve customer service. The current state of the consolidation of the tourism business, globalization processes, restructuring, mergers and acquisitions, grouping into strategic alliances adds value to the corporate logistics strategy as a highly architectural managerial form, which in essence is innovative and requires significant intra-firm coordination. Thus, the management of human flows in the field of tourism is carried out through the management of material, financial and information flows. The main task of the tourism logistic is development and application of certain mechanisms to ensure the cyclicality of the specified flows and their efficiency in the competitive world market.

The paper was funded by a grant "Formation of the scientific and technological platform for the harmonization of the socio-economic interests of Ukraine in the world economic space". (No. 0116U004521) from the Ministry of Education and Science of Ukraine

References

1. R. Baggio, R. Sainaghi, Complex and chaotic tourism systems: towards a quantitative approach, IJCHM **6**, 840 (2011)
2. S. Cole A logistic tourism model: resort cycles, globalization, and chaos, ATR **4**, 689 (2009)
3. O. Prokopishyna, N. Kozubova Organizational support for internal tourism based on rational use of Ukraine tourist and recreational resources, NVIU **13**, 54 (2015)
4. J. Zhang, C. Jensen Comparative advantage: explaining tourism flows, ATR **1**, 223 (2007)
5. W. Zhao, J. Ritchie, C. Echtner Social capital and tourism entrepreneurship. ATR **38(4)**, 1570 (2011).
6. L. Oigenblick, A. Kirschenbaum, Tourism and immigration: Comparing alternative approaches. ATR 29(4), 1086 (2002).
7. T. Tyrrell, R. Johnston The economic impacts of tourism: a special issue. JT, **45(1)**, 3(2006).
8. D. Hughes, M. Shields Revisiting tourism regional economic impact: Accounting for secondary household employment. RRS, **37(2)**, 186 (2007)
9. S. Smith New developments in measuring tourism as an area of economic activity. TORLT **1**, 225 (2000).
10. U. Avci, M. Madanoglu, F. Okumus, Strategic orientation and performance of tourism firms: Evidence from a developing country. TM, **32(1)**, 147 (2011).

City Retail Network Influence on Transportation Expenses

Natalia Ptytsia[1,*]

[1]Kharkiv National Automobile and Highway University, Transport Systems and Logistics Department, Ukraine

Abstract. An approach to determining the influence of retail network parameters on the share of transportation costs taking into account the density of outlets is presented. The obtained results can be used to develop recommendations for the optimal area of operation of retail delivery systems, and to estimate cargo flow volume in cities.

1 Introduction

The current state of the Ukrainian economy requires retail companies to exhibit flexibility when faced with changes in market conditions. An integrated approach to the organization and planning of logistics and marketing processes, and the ability to quickly change the existing strategy to take advantage of new opportunities, is needed to achieve financial success. Transportation of small quantities of goods is the most costly for road transport, and precisely because of this the share of transportation costs considerably raises the final price of the goods. To improve competitiveness, it is necessary to focus on meeting the needs of end-users, because, when buying a product, they always pay for its delivery.

To reach a larger market share, retail networks open new trading points, which leads to a significant increase in transportation costs at the stage of the last mile. Transportation of cargoes at this stage is carried out by a rolling stock of small and medium cargo. The small size of orders, the daily delivery of products and the variety of outlet formats leads to the need for an objective assessment of the cost of product delivery, depending on the density of the trading network.

The active development of logistics systems for the supply of consumer goods in the retail chain has led to a lack of a systematic approach to problem solving. The paper [1] analyzes the components of the material flows in the logistics system. It has been determined that, depending on the structure of the supply chain, the type of goods and their prices, the share of transportation costs can reach 15% of the total shipping costs. This statement applies to monopoly companies that own distribution channels, and does not apply to relatively small retail chains, which are the overwhelming majority in the Ukrainian consumer market. Significant influence on the delivery cost of goods has such indicators as the density of the location of consignees, which is considered in [2, 3, 4, 5], and the size of the batch of cargo [2, 6, 7, 8]. However, proposals for the improvement of a

* Corresponding author: nataliya.ptitsa@gmail.com

company-specific situation are formulated in these works for only one case, which makes it impossible to determine changing trends and further forecasting.

2 Modelling of the share of transportation costs

The purpose of the study is to identify the patterns of change in the delivery parameters of the retail network based on the share of logistics costs, while taking into account the density of the logistics network. The object of research is the process of delivering consumer goods in the logistic system of the retail trade network. The subject of the study is the pattern of influence of the density of the logistics system of the retail trade network on the share of logistics costs for the delivery of goods.

The process of delivery to the logistics system of the retail network can be represented in the form of a cybernetic model "black chest" (Fig. 1).

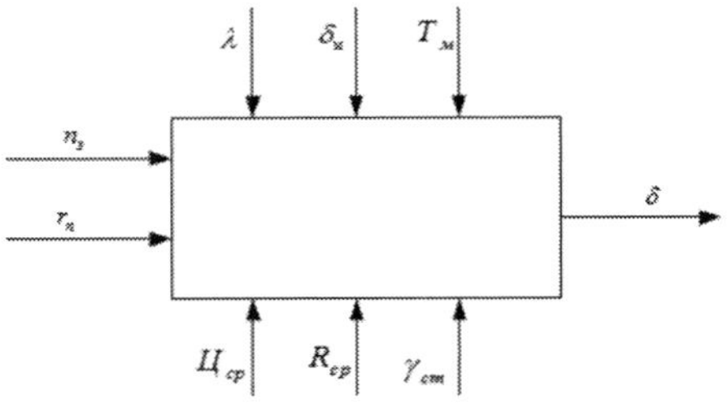

Fig. 1. Model of the process of delivery of consumer goods in the form of a "black chest"

The optimization factors are the selected number of points on the route ($n_з$), and the radius of half of the demand of the trading point (r_n).

The addition of new objects to the distribution system increases the operating costs and the costs of transporting the entire system. It has been proved that, when changing the number of objects in the system (outlets), part of the expenditures is increased, while a part of other expenditures is decreased, with an increase in income for the entire system. This allows researchers to set and solve the problem of finding the optimal correlation of these indicators.

The objective function in this case represents the share of logistics costs for the transportation of consumer goods to a new trading point (δ, %), and has the following appearance

$$\delta = [B_{mp} / (D_i + D_j)] \cdot 100 \rightarrow \min \tag{1}$$

where B_{mp} - is the cost of transportation to a new trading point of the network, UAH;
D_i - total revenue of the retail chain, UAH;
D_j - expected revenue of a new trading point, UAH;

The minimization of the objective function is achieved by optimizing the logistic parameters of the delivery process of the retail trade network. The influence of the parameters of the logistics system was considered under the following conditions: the number of points on the route is greater than one or equal to one; the order size is equal to the order amount for all trading points which are serviced, taking into account the actual number of points on the route; the sum of all arrival points on all routes equals the number of trading points in the service area, and the maximum carrying capacity of the car/vehicle is 10 tons.

The service radius actually represents the area of the city around the trading points, which "attracts" buyers, thus characterizing the density of the retail network. The structure of the trading network is complicated by the presence of its trading points in different formats, which leads to a significant change in the behavior of buyers, and this can lead to significant errors in determining the parameters of the cargo delivery process. The value that characterizes the ratio of buyers to the outlets of different formats is the radius of half of the demand, which represents the radius of customer service, at which the market share of the trading network is 50% of the market capacity. In an analytical form, the radius of half-demand can be described as follows [9]:

$$r_n = (Ц_{cp} \cdot R_{cp})/(Ц_{zp} + Ц_i) \tag{2}$$

where $Ц_{cp}$ - is the cost of the average check on the market, UAH;
R_{cp} - average radius of service for each trade point, km^2;
$Ц_{zp}$ - marginal purchase price, UAH;
$Ц_i$ - the cost of the average check of a specific trading point, UAH.

The average distance of the delivery of consumer goods (l_i, km) is a separate characteristic of each particular object in the network, because it depends on their mutual location and varies within the area of one settlement. Analytically, this can be represented as follows [10]

$$l_i = 2R_{cp}/3 \tag{3}$$

where R_{cp} – the average radius of the service area, km;

$$R_{cp} = \sqrt{\lambda^{-1}/\pi} \tag{4}$$

where λ - trade network density, unit/km^2;

The average mileage of cars between adjacent points of arrival on the route ($l_{(i-1)-i}$, km) directly depends on the density of the location points in the area of service, which can be represented as follows

$$l_{(i-1)-i} = 0,76\sqrt{\lambda^{-1}}\ . \tag{5}$$

As a result of the decomposition of the target function, the following mathematical model of the share of transportation costs is obtained:

$$\delta = \frac{\left(A_{зм} + B_{зм} \cdot q_н\right) \cdot \left[\frac{4 \cdot \sqrt{\frac{1}{\lambda}}}{3 \cdot \sqrt{\pi}} + 0,76 \cdot \sqrt{\frac{1}{\lambda}} \cdot (n_з - 1)\right] + \left(A_{пост} + B_{пост} \cdot q_н\right) \cdot \left[2 \cdot t_д + t_д(n_з - 1) + \frac{\frac{4 \cdot \sqrt{\frac{1}{\lambda}}}{3 \cdot \sqrt{\pi}} + 0,76 \cdot \sqrt{\frac{1}{\lambda}} \cdot (n_з - 1)}{V_T} + 2 \cdot q_н \cdot \gamma_{ст} \cdot \tau_{нр}\right]}{n_з \cdot \left(\frac{T_M \cdot r_П}{r_П + \frac{\sqrt{\frac{1}{\lambda}}}{\sqrt{\pi}}} + 1\right)} \tag{6}$$

where $A_{зм}$, $A_{пост}$, $B_{зм}$, $B_{пост}$ - are the coefficients of the regression model of the cost of road transport depending on the cargo capacity of UAH/km, UAH/h, UAH/tkm, UAH/tkm*h.
$q_н$ - nominal load-carrying capacity, t;
λ - trade network density, unit/km^2;

$n_з$ - number of points on the route, units;
$t_д$ - additional time for documents processing, h.;
V_m - technical speed of the car, km/h.;
γ_{cm} - static coefficient of use of cargo;
$\tau_{нр}$ - normalized time spent on the loading or unloading of cargo, h/t;
$T_м$ - the cost of commodity circulation of consumer goods in the region, UAH/t;
r_n - half-demand radius.

The cost of an average check on the market is a random value, as determined by the behavior of individual buyers. Its condition determines the way its distribution law is established, according to statistical data. This study is described in the paper [11]. As a result, it was found that the variation of values can be described by the Rayleigh distribution law, which was taken into account in experimental studies.

With a large number of factors in the model and the need to obtain an accurate result, a full-factor experiment is needed. This is an experiment that uses all possible combinations of factors at different levels.

On the basis of the obtained mathematical model, the following factors are selected, which are variables: trade network density (λ, unit/km^2), the share of turnover in the network ($\delta_м$, %), the cost of commodity circulation of consumer goods in the region ($T_м$, UAH/t), the cost of an average check on the market ($Ц_{cp}$, UAH), average radius of service in the region (R_{cp}, km), and the static coefficient of the use of cargo (γ_{cm}).

The constant factors of the system include all other factors: technical speed of the car (V_m,km/h), normalized time spent on the loading or unloading of cargo ($\tau_{нр}$, h/t), additional time for documents processing ($t_д$, h), the coefficients of regression models of the cost of road transport depending on the cargo capacity ($A_{зм}$, UAH/km, $A_{пост}$, UAH/h, $B_{зм}$, UAH/tkm, $B_{пост}$, UAH/ tkm*h).

An analysis of the influence of the number of points on the route and the half-demand for the share of logistics transportation costs in the retail chain indicates the nonlinear nature of the dependencies. This indicates that there are such values of the specified optimization parameters that will minimize the share of transportation logistics costs. The plan of a full factor experiment of type 2k allows the variation of factors on two levels (minimal and maximal), which are determined through the ranges of variation of the selected variables, and therefore it is necessary to establish these limits, relying on statistical research data (Table 1). The number of experiments was $2^6 = 64$.

Table 1. Levels of variation of factors

Variation level	λ, unit/km^2	$\delta_м$, %	$T_м$, UAH/t	$Ц_{cp}$, UAH.	R_{cp}, km	γ_{cm}
	x_1	x_2	x_3	x_4	x_5	x_6
Maximal	25,398	0,8	15326,2	470	16,434	1
Minimal	1,948	0,01	10382,7	100	3,569	0,4

The obtained after the transformation of the mathematical model of the share of logistics transportation cost expression needs to be solved by means of a numerical method. The method of generalized reduced gradient found solutions for/of optimization parameters when minimizing the target function. The "Search Solution" add-in in the Microsoft Excel application suite provides the ability to implement this method. The regression coefficients are obtained using the least squares method [12]. To approximate the dependence, and taking into account the nature of the change of the share of logistics costs on transportation from the selected factors, exponential and power functions are used as the most common non-linear models of regression. Coefficients of regression models were obtained for these functions.

The system of empirical equations is obtained through the transformation of the obtained coefficients for guidance in the natural form for the number of points on the route and the radius of half-demand

$$\begin{cases} n_з = 7{,}13 \cdot 0{,}005^{\lambda} \cdot 0{,}99^{T_м} \cdot 0{,}263^{\delta_м} \cdot 1{,}045^{R_{cp}} \cdot 1{,}79^{\gamma_{cm}}; \\ r_n = 0{,}129 \cdot 0{,}834^{\lambda} \cdot 0{,}078^{\delta_м} \cdot 1{,}01^{Ц_{cp}} \cdot 0{,}97^{R_{cp}} \cdot 1{,}45^{\gamma_{cm}} \end{cases} \quad (7)$$

and for the share of logistics transportation cost

$$\delta = 0{,}0034 \cdot 1{,}087^{\lambda} \cdot 0{,}999^{T_м} \cdot 0{,}013^{\delta_м} \cdot 1{,}253^{R_{cp}} \cdot 0{,}61^{\gamma_{cm}} \quad (8)$$

The information ability of regression models, defined by Fisher's criterion F-criterion [13] for the number of points on the route is 28.95, which provides sufficient accuracy for the practical needs of determining the share of logistics transportation cost. The results of the verification showed that all regression models obtained at a significance level α = 0,05 are statistically significant (Table 2).

Table 2. Assessment of the suitability of the models obtained

Indicator	Number of points on the route n, unit	Radius of half demand, r_n, km.	Share of logistics transportation cost δ, %
F-criterion calculated	28,95	2,6	32,4
F-criterion tabular	2,26	2,26	2,26
Error models, %	6,63	7,6	12,9

The obtained results show that the models can be used to determine the optimal parameters of the logistic network of retail trade in order to minimize the share of costs for the transportation of products.

During the experiment, the dependencies between the factors and parameters of the logistics system were determined (Figures 2, 3 and 4). At the chosen interval of research, the nature of the influence of the factors on the share of logistics cost for transportation is monotonous (Fig. 2). The function of the share of logistics transportations cost is the most sensitive to changes in the factors, the density of trading points and the average service radius at their maximum values. The market share ($\delta_м$) factor has a significant impact on the share of transportation costs at its minimum values.

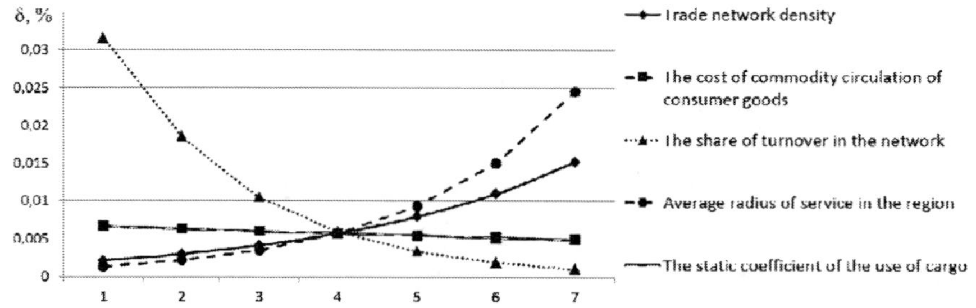

Fig. 2. Dependence of the share of transport costs from external factors

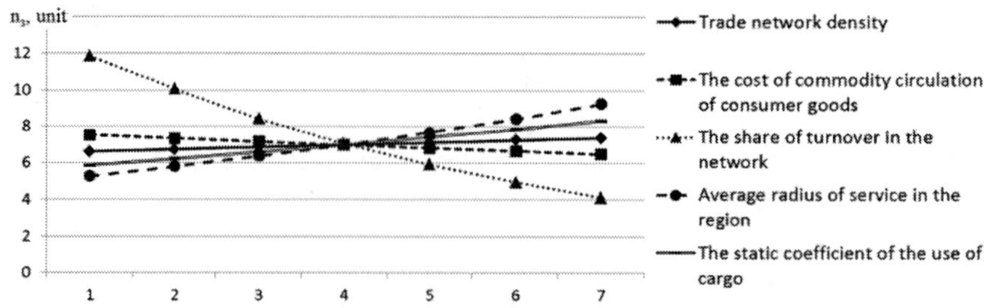

Fig. 3. Dependence of the half-demand radius on external factors

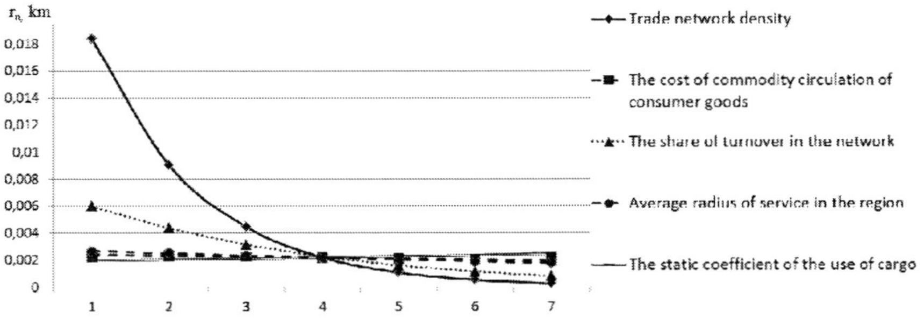

Fig. 4. Dependence of the number of points on the route on external factors

The analysis of the obtained dependencies shows that the indicator - the density of trading points - has a significant impact on the share of transport costs.

The obtained results testify that the models are suitable and provide sufficient accuracy of the parameters of the logistic system of the retail trade network for practical needs, namely the share of logistics transport costs, the number of points on the route and the radius of half-demand. These models will allow development of practical recommendations for the optimal area of the system operation and will make informational basis for prediction cargo flows on the city territory. As a performance measure, the minimum share of logistic costs for delivery was proposed. This will provide the opportunity to make a decision on the inclusion in the network of a new facility.

References

1. D. Bowersox, D. Closs, *Logistics: An Integrated Supply Chain* (2001)
2. F. Oliveira, Jr A. Oliveira, L. Bessa Rebelo, IJ PME, **5**, 480, (2014)
3. T. Crainic, B. Montreuil, Physical Internet Enabled Hyperconnected City Logistics, **12**, 383, (2016)
4. L. Ranieri, S. Digiesi, B. Silvestri, M. Roccotelli, A Review of Last Mile Logistics Innovations in an Externalities Cost Reduction Vision, **10**, 782, (2018)
5. D. M. Lambert, J. R. Stock, *Strategic Logistics Management,* 830, (2005)

A. V. Zyryanov, N. O. Zueva, V. I. Nabokov, *Household logistics and placement of commercial enterprises,* 242, (2010)

6. V. I. Sergeev *Logistic systems for monitoring supply chains*, (2003)
7. V. S. Lukinsky, V. V. Lukinsky, I. A. Plastunyak, *Transportation in logistics: Proc. manual*, 139, (2005)
8. L. B. Mirotin, *Transport logistics*, 512, (2003)
9. N. Nefedov, Bulletin of Kharkov State Automobile and Highway Technical University, **8**, 70 – 72, (1999)
10. Yu. I. Ryzhikov, *Inventory Management*, 343, (1969)
11. M. A. Nefodov, N. V. Ptitsa, Communal economy of cities, **146**, 20-24, (2018)
12. V. E. Gmurman, *Probability theory and mathematical statistics*, 479, (2003)
13. V. G. Galushko, *Probability-statistical methods on motor transport*, 232, (1976)

Management of the development of human capital of the airline using personnel logistics tools

Sergey Smerichevskyi[1]*, *Yuri* Kolbushkin[2], *Anatoly* Mikhnenko[3], and *Olga* Polous[4]

[1]National Aviation University, Department of Marketing, 1, Kosmonavta Komarova ave., Kyiv, 03058 Ukraine
[2]Department of economic planning and budget calculations NJSC "Naftogaz of Ukraine", 6, B. Khmeljnyckogho str., Kyiv, 01601 Ukraine
[3]National Aviation University, Department of management and administration, 1, Kosmonavta Komarova ave., Kyiv, 03058 Ukraine
[4]National Aviation University, Department of Air Transport Economics, 1, Kosmonavta Komarova ave., Kyiv, 03058 Ukraine

Abstract. The article deals with the main approaches to the definition of "personnel logistics". The study of approaches to the definition of "personnel logistics" can reveal the lack of a unified approach to its understanding. The expediency of using personnel logistics tools in managing the development of human capital of the enterprise is established. The human capital of the enterprise is in constant motion, varying in size and structure, it can be said that to create a system for managing the development of human assets of the enterprise appropriate to apply a logistic approach. The sectoral features of airline personnel management are presented. The system of management of development of human capital of the airline with the use of personnel logistics tools is offered. The process approach to the system allows constructing an adaptive personnel management system with feedback on the basis of the logistic concept, since these processes act as objects of management of this scientific direction. The technologies of management of development of human capital of the airline with the use of personnel logistics tools are systematized.

Introduction

The quick development of the transport sector in Ukraine requires the introduction of innovative technological, scientific and managerial achievements in the practice of domestic transport companies. For airlines in the context of operating in a changing global aviation space, entering the markets of other countries and the development of international business as one of the most important areas of activity are of particular importance. A key role in this process is played by the management of the development of human capital of the airline, which includes management of personal, monetary and occupational careers of the personnel.

* Corresponding author: os_f_smerichevsky@ukr.net

The combination of these types of management into a single effective system is impossible without the use of the latest personnel logistics tools.

The issue of the adaptation of personnel logistics tools to the needs of management of the development of human capital of airlines is not sufficiently investigated in modern conditions. Conducting research in this area will further develop the theories of human capital management and deepen understanding of the personnel logistics tools inherent in the airline itself.

The purpose of the article is to develop a management system for the development of human capital of the airline using personnel logistics tools.

Theory of the matter

The management of the human capital development of enterprises and individual workers was carried out at various stages within the concepts and schools of human resources management. Earlier concepts in this area of study are classical concepts (F. Taylor, L. Gilbert, M. Emerson, O. Gastev, P. Kerzhentsev, M. Vitko, F. Dunaevsky) [12], the concept of administrative management (A. Fayol, M. Follet, C. Barnard) [13], bureaucratic concept (M. Weber) [22], the concept of "human relations" (E. Mayo, F. Rotisserberger, P. Layckert) [20], the concept of human resources and the school of behavioral sciences (V. Scott, M. Folette, A. Maslow, D. McGregor, W. Ouchi) [19], the concept of a situational approach (J. Woodworth, V. Vroom, M. Stevenson, F. Fiedler, P. Gersy and C. Blanchard, R. Blake and D. Mouton), the concept of a systematic approach (P. Drucker, R. Lickert, T. Peters, R. Waterman, I. Ansoff) [17] and the concept of institutional approach (D. North) [9]. Modern concepts in this area include the theory of human capital (M. Becker, J. Minster, T. Schultz), the theory of strategic human resources management (S. Fombrun, M. Bir, P. Boksall, D. Gest, D. Story, D. Parcel, K. Legge, D. Ulrich, and others) and competency-oriented approach (J. Hotkinson, P. Sperrou, A. Farnem, M. Armstrong, etc.) [1].

It should be noted that the issue of personnel logistics at this stage remains not fully investigated. A number of scholars [4, 6, 7, 8, 11, 14, 21] define personnel logistics as one of the directions of logistics or as a term. The issue of staffing of airline companies is covered in a number of works [2, 3, 5], which consider its interconnection with other resource components of the companies.

Discussion of results

In domestic and foreign scientific literature, most logistics professionals consider logistics, above all, as the management of material, informational and financial resources. Only some of them include into the objects of logistic control flows of resources of a different kind: energy and human. But they see these flows through the prism of creating value during the production process, while, for example, human flows are directed not only to production processes. They function separately from the latter. So, in any organization, there are personnel flows, and during their movement they also create value, but the cost including a different kind, namely, capitalized human cost. In addition, under special conditions of functioning contributing to the development of human potential and increasing the motivation of personnel, there is a synergistic effect during the movement of human flow within the organization, which does not occur as a result of the movement of resource flows of another kind. In other words, the input staffing flow in the organization, as it were, increases with its potential under the influence of the enterprise management system as a whole and the system of personnel management in particular.

Today, the issue of the introduction of personnel logistics in the management system of human capital development of the enterprise is not paid enough attention. Thus, the study of approaches to the definition of "personnel logistics" can reveal the lack of a unified approach to its understanding.

Formation and development of human capital of an enterprise occurs during the organizational transfer of employees, which can be compared with the migration process at the macroeconomic level. It is precisely in the process of personnel movement at the enterprise that the exchange of information and abilities is intensified, the change of motivational units, which, based on the definition of human capital, is directly related to the process of its development.

Thus, the process of career management of employees today is transformed into a process of managing the development of human capital of the enterprise, since the human resources flow is a stream of human assets of the enterprise. Since the human capital of the enterprise is in constant motion, varying in size and structure, it can be said that to create a system for managing the development of human assets of the enterprise appropriate to apply a logistic approach.

Today, the aviation sector is extremely important for the stable development of our state in the conditions of intensification of the processes of integration and globalization. The volumes of passenger transportation by domestic airlines annually demonstrate steady growth, which entails the creation of new jobs, the use of the latest technologies of service provision and business, which, in turn, positively reflects on the Ukrainian economy. According to the results of 2018, the five largest passenger airlines carried 11620.6 thousand people, which is 20.2 percent more than in 2017, and accounts for almost 93 percent of the total passenger transportation volume of Ukrainian airlines (Fig. 1) [10].

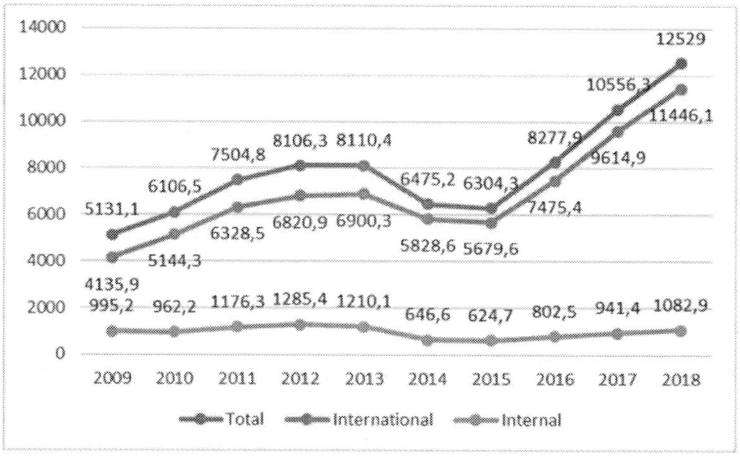

Fig. 1. Dynamics of volumes of transportation of passengers by air transport of Ukraine, ths. [10]

Such dynamic development requires airlines to quickly and qualitatively adaptation to changes in the environment and the needs of the domestic and foreign markets for providing aviation services, which is impossible without the development of human capital in accordance with the strategic objectives of the airlines and the aviation sector as a whole. These goals require the development of a new, innovative approach to managing the development of human capital of the airline using existing personnel tools and gradually transforming them in accordance with international standards for the provision of aviation services.

The main criteria for the formation of human capital and its further development, put forward by international aviation companies include [18]:

- age - depending on the requirements of the airline. The maximum age must be between 25 and 35 years old;
- level of education - university or college education;
- knowledge of English, which is a key prerequisite of hiring by an airline;
- health information - applicants must undergo a preliminary medical examination and provide a health certificate;
- work experience - although this is not a specific requirement, preference is given to candidates who have already worked in this industry.

Most airlines abroad, if needed, turn to specialized companies to hire and recruit personnel. Aeropeople and Aerotek are the leading companies whose expertise is the strength and staffing of airline personnel in the UK, the United States and Europe as a whole. Another area of their work is cooperation with leading organizations that help in recruiting employees at all levels, as well as offering candidates for positions according to the specifics of the work, the location of the company or the industry itself.

Aeropeople and Aerotek are members of REC, the Recruitment & Employment Confederation, known for its strict code of conduct, which ensures high quality work practices. Aeropeople and Aerotek are fully accredited to ISO 9002 (international quality system standard) [15, 16].

An important advantage of Aeropeople and Aerotek is the availability of an extensive network of more than 150 offices worldwide, and specialized units that provide aviation employment services, help airlines hire skilled workers, including administrative staff.

However, the main staffing and its further development are still engaged by the airlines themselves. We are invited to build an adaptive system for managing the human capital development of the airline based on personnel logistics. It involves the allocation of a single function of management human flows, the integration of management of human capital development into a single system of personnel management of the airline (Fig. 2).

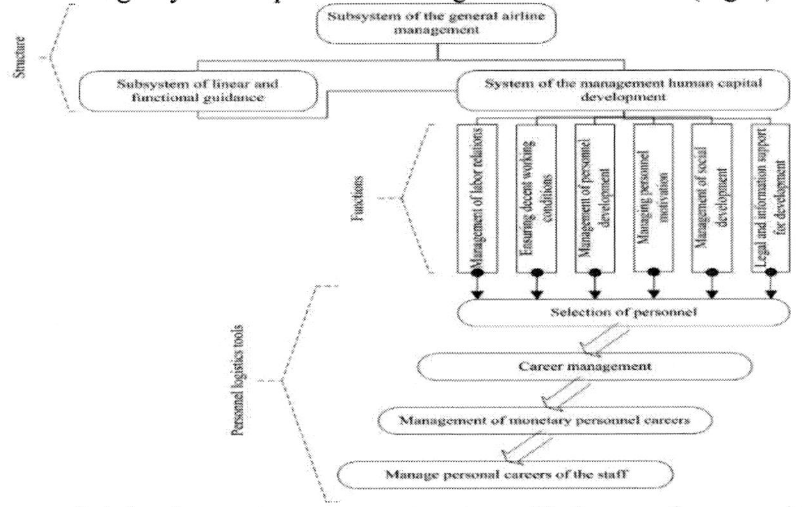

Fig. 2. Human capital development management system with the use of personnel logistics tools

Thus, HR processes can be defined as a separate set of actions aimed at transforming the human capital of the airline. It should be noted that they take place within any organization and, accordingly, any human resources management system, which is reflected in the ISO 9000 standard, where a number of human resources processes are presented as an organization's auxiliary business processes.

The functions of the management system for the development of human capital of the airline belong to the responsibilities that form the system of structural elements, namely, the personnel management units and managers of the units of general and linear control. In connection with this, the main functions of this system are: hiring, accounting, legal support, marketing of personnel, etc., presented on Fig. 2.

Personnel development, staffing and monetary transfer of personnel, and selection of personnel were considered as the main personnel processes. The presented system reflects the composition of the management subsystems for developing the human capital of the airline on the basis of a functional and process approach. From the scheme it is clear that a number of functions in the management of personnel are carried out by heads of various units and management of the entire company. In this case, the personnel management processes are mainly the responsibility of the personnel management systems, not of individual leaders. Each the departments and/or employee of the latter concentrate their efforts on a certain function of personnel management, which results in the loss of integrity of managerial influence on staff. At the same time, dynamism and constant change in the personnel assets of the airline allow building a system for managing human capital development on the principles of a process approach, when all functions of personnel management are integrated into unified processes of career management of personnel and selection of employees.

The process approach to the system allows constructing an adaptive personnel management system with feedback on the basis of the logistic concept, since these processes act as objects of management of this scientific direction. In general, the difference between the logistic approach to human resources management and human capital from the traditional one is to allocate a single management function in the beginning of isolated human flows and their management processes, the integration of individual units of the personnel management service and functional units of the airline into a single system of personnel management.

Thus, the use of personnel logistics tools enables to create an objective set of technologies for managing the development of human capital of the airline (Fig. 3), aimed at increasing and improving the utilization of human capital of the company, achieving its goals and staffing.

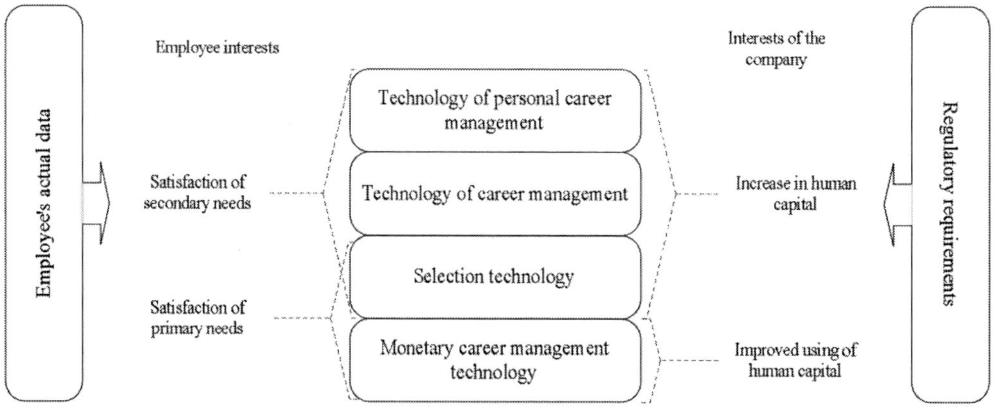

Fig. 3. Technologies for managing the human capital development of the airline using personnel logistics tools

This allows assessing the efficiency of spending on human resources both from the standpoint of the company and employees separately, and from the position of two entities taken together. Fig. 4 shows technologies in a hierarchical sequence, reflecting the orientation of each technology to meet the specific needs of staff. Thus, the technology of monetary career management is aimed at satisfaction of the basic needs of employees, and the technology of career and personal career management are aimed at meeting secondary needs. The location of selection technology in this hierarchical sequence is determined by the fact that during the selection, employees tend to take the position of career space which will allow them to improve their financial position, improve their status, and develop. In other words, this technology allows meeting both the primary and secondary needs.

Conclusion

Consequently, from the point of view of managing the development of human capital, the organizational movement of human assets of the company can be reduced to three main vectors: career, monetary and personal. The personal career vector is directly linked to the increase in the value of human assets due to changes in staff abilities. Moving along the vector of an official career means creating a favourable opportunity for the growth of human capital, as well as, in conjunction with the movement on the vector of a monetary career, increases the loyalty of employees by increasing their interest. Increasing motivation, in turn, leads to an increase in human capital. Consequently, during the career movement of human assets there is a change in the size of human capital, its development.

So, using the aforementioned technology complex, the airline will be able to increase the size of human capital (with the application of selection technologies, management of the public and personal careers) and improve its use (in the case of applying monetary management technology). Thus, in the case of the practical implementation of this set of technologies, it will be possible to meet the interests of both parties - the employee and his employer.

Subsequent research will be aimed at studying the management system of human capital development of a specific domestic airline with the development of solutions to existing problems with the use of personnel logistics tools.

References

1. Armstrong M. Strategicheskoe upravlenie chelovecheskimi resursami [Strategic human resource management]. Moscow: INFRA-M. (in Russian) (2002)
2. Vysotskaya M.P. Formirovanie kadrovoy politiki aviapredpriyatiy Ukrainy [Formation of personnel policy of Ukrainian aviation enterprises]. Kyiv: NAU. (in Russian) (2005)
3. Grabs-Uest L. Sotrudniki na vsyu zhizn': Uroki loyal'nosti ot Southwest Airlines [Employees for all life: Southwest Airlines loyalty lessons]. Moscow: Mann, Ivanov i Ferber (in Russian) (2008)
4. Kormin N.G. Bulletin of Chelyabinsk State University, **7(108),** 138-141 (2008)
5. Lashhenko O.Ju. Upravlinnja kadrovym zabezpechennjam ekonomichnoji bezpeky aviakompanij [Management of human resources provision of economic safety of airlines] (PhD Thesis), Kyiv: National aviation university (2010)
6. Leontjjeva I.O. Kadrova loghistyka v systemi upravlinnja vodnotransportnym pidpryjemstvom [Personnel logistics in the management of the water transport enterprise]. Scientific herald of Kherson State University. Economic sciences, **20, 1**, 111-114 (2016)
7. Meljnykova K.V. Scientific herald of Kherson State University. Economic sciences, **17, 3**, 58-60 (2016)
8. Mishyna S.V., Mishyn O.Ju. Economics and society, **12**, 492-499 (2017)
9. Nort D. Instituty, institutsional'nye izmeneniya i funktsionirovanie ekonomiki [Institutions, institutional changes and the functioning of the economy]. Moscow: Fond ekonomich. knigi «Nachala». (in Russian) (1997)

 https://doi.org/10.1051/shsconf/20196703012

10. Ministry of infrastructure of Ukraine. Pidsumky dijaljnosti aviacijnoji ghaluzi Ukrajiny za 2018 rik [The results of the aviation industry in Ukraine for 2018]. Available at: https://mtu.gov.ua/content/statistichni-dani-v-galuzi-aviatransportu.html?PrintVersion (accessed 05.05.2019) (2018)
11. Serbin V.D. Osnovy logistiki [Basics of logistics]. Taganrog: TRTU. (in Russian) (2004)
12. Teylor F. Nauchnaya organizatsiya truda [Scientific organization of labor]. Moscow: NKPS Transpechat'. (in Russian) (1925)
13. Fayol' A. Obshchee i promyshlennoe upravlenie [General and industrial management]. Moscow: Kontrolling. (in Russian) (1992)
14. Chemekov V.P. Management technology, **12**, 25-28 (2002)
15. Aeropeople. Official site. Available at: http://www.aeropeople.com/ (accessed 04.05.2019).
16. Aerotek. Official site. Available at: http://www.aerotek.com/ (accessed 04.05.2019).
17. Drucker P.F. Management: Tasks, Responsibilities, Practices. N.Y. Harper and Raw. 398 p. (1973)
18. International Civil Aviation Organization. Official site. Available at: https://www.icao.int/ (accessed 04.05.2019).
19. Maslow A.A. Theory of Human Motivation. Psychological Review, **50**, 370-396 (1943)
20. Mayo E. The Human Problems of an Industrial Civization. Boston: Graduate School Administration, Harvard University. 318 p. (1946)
21. Velychko O., Velychko L. Journal of international studies, **10, 3**, 206-219 (2017)
22. Weber M. The Theory of Social and Economic Organizations N.Y.: Free Press. 436 p. (1947)

The model of vehicle optimum quantity for grain crop harvesting under the conditions of farming in Poltava region

Volodymyr Zagoryanskii[1*1], *Olena* Zagoryanskya[1], *Tetyana* Haykova[1] and *Volodymyr* Khorolskii[1]

[1] Kremenchuk Mykhailo Ostrohradskyi National University, Transport Technologies Department, Pershotravneva str. 20, 39600 Kremenchuk, Ukraine

Abstract. The previously developed optimization model for determining the required number of vehicles as part of the harvest transportation complex during grain crop harvesting for the conditions of grain farming in Poltava region has been applied. A computational methodology was developed to determine the design parameters depending on the farm conditions included in the expression for the target function of the optimization model constructed, namely, average number of arrival rate per time and waiting time in queue (excludes service time) for each individual customer. Using the queueing theory methods, a ratio of the number of harvesting and transport equipment in the harvest transportation complex composition was found, which provides the minimum total costs (corresponding to the target function of the previously proposed model), which are due to the downtime of combine harvesters and the cost of vehicle maintaining. It has been established that with the increase in the vehicle number, the total costs of downtime of combine harvesters and vehicles decrease at first and then increase.

1 Introduction

In agricultural production, the most labor-intensive and complex in terms of organizational, material and labour costs is the process of crops harvesting, including grain crops [1, 2]. Transportation costs take a significant place in the total amount of agricultural production costs – they make up to 15 to 40% of the cost of agricultural products [3]. Road transport is the main mode of transportation in agriculture, which accounts for up to 80% of the total volume of transportation. Road transport for grain is in demand in the agricultural regions of Ukraine, including Poltava region, for delivery of harvested wheat, rye, barley, legumes from fields to elevators and to other places of storage.

It is known that combine harvesting methods prevail in agricultural production [4, 5]. There are two methods of combine harvesting, namely, direct and separate harvesting. With massive maintenance on road transport, there are many processes involved, the examples of which can be specified in the process of accepting applications for cargo transportation,

* Corresponding author: zagor_vlad@ukr.net

servicing the population with transport, vehixles running repairs, organizing technical assistance on the line, and much more [6-13].

In a mathematical model, a harvesting brigade can be represented as a system of mass service with anticipations, in which the servicing machine is a vehicle, and the served ones are combines.

Let us determine the number of vehicles in the harvest transportation complex based on the optimization model built using the queueing theory [14].

The model changes the vehicle number, which should be the criterion for minimizing the total costs of combines and vehicles downtime included in the harvesting complex for grain harvesting.

The calculated parameters depending on the conditions of the economy that are included in the expression for the target function of the optimization model constructed include average number of λ filled tankers (arrival rate) per hour and waiting time in queue W_q (excludes service time) for each individual customer, hours.

In the article, the composition of the source data to determine these calculation parameters is given. They are grouped together in three groups of parameters that characterize the conditions for grain harvesting in the given farming, determine the characteristics of the combine harvester, and the characteristics of the vehicle.

Substantiation of rational quantitative composition of machine complexes is the subject of numerous scientific works. With all their advantages, their general disadvantage is the complexity and calculations labour intensity [15]. It has been established [16] that the most rational number of machines in the complex is 8 to 16, and in the technological chain it is 3 to 5 units. The average value for the forest-steppe and more for the steppe zone is recommended [16]. With such a quantitative composition, the maximum productivity of machine tractor aggregates is reached at the least expenses for their servicing.

2 Model of determination of the vehicle number in the collecting-transportation complex

The following model is offered:

1. Select a model of a combine harvester from the number of available or based on comparative analysis of energy consumption of several models in the harvest.

Relative aggregate energy costs, MJ/ton, are expressed as the sum of the components:

$$\omega_C = \omega_K + \omega_{XX} + \omega_{Tn} + \omega_z + \omega_M \,, \tag{1}$$

where ω_K – relative energy consumption for moving self-propelled combines; ω_{XX} – relative energy consumption for idling; ω_{Tn} – relative energy consumption for the technological process; ω_z – relative energy costs of live labor of combiner; ω_M – relative energy consumption required for the production, repair and maintenance of machines.

Relative energy consumption for self-propelled combines, MJ/ton:

$$\omega_K = \frac{10^4 \cdot f_n \cdot g \cdot (m_a + m_o) \cdot (1 + \gamma_{xx})}{\eta_e \cdot \eta_T \cdot (1 - \eta_b) \cdot B \cdot A} \,, \tag{2}$$

where f_n – coefficient of rolling resistance in the movement of the machine ($f_n = 0.17$); g – acceleration of free fall, g = 9.81 m/sec^2; m_a – machine weight, kg; m_o –mass of grain transported in the harvester hopper during the operation, kg ($m_o \approx 0.5\varepsilon_0\gamma V_b$, where ε_0 is the coefficient taking into account the filling of the bunkers, $\varepsilon_0 \approx 0.9$ to 0.95, γ is the grain density, ton/m^3; V_b – the capacity of the bunker for grain, m^3); γ_{xx} – coefficient taking into account the proportion of idle passes ($\gamma_{xx} = 0.04$ to 0.25); η_e – effective efficiency factor of diesel engines ($\eta_e = 0.32$ to 0.36); η_T – coefficient of efficiency of mechanical transmission of energy from engine to propulsion ($\eta_T = 0.65$ to 0.85); η_b – coefficient of towing, for self-propelled combines, η_b = 0.04 to 0.06; B –width of the capture of the reaper, m; A – yield of grain, ton/hectare.

Relative energy consumption per idle of combines, MJ/ton:

$$\omega_{XX} = \frac{3,6N_{XX} \cdot g_{on}}{\eta_e \cdot W_g}, \tag{3}$$

where N_{XX} is the relative power required to drive the working bodies at idling, kW×(kg/sec)$^{-1}$ (for grain harvester combine N_{XX} = 2.7 (2.8) kW×(kg/sec)$^{-1}$); g_{on} – nominal capacity of the combine harvester, kg/sec; W_g – hourly productivity of the combine, ton/hour.

Relative energy consumption per technological process, MJ/ton:

$$\omega_{Tn} = \frac{3,6N_{Tn} \cdot g_{on}}{\eta_e \cdot Q_g}, \tag{4}$$

where N_{Tn} is the relative power required for the technological process, kW×(kg/sec)$^{-1}$ (for grain harvester N_{Tn} = 7.6 kW×(kg/sec)$^{-1}$); Q_g – milling of grain with a combine for an hour of work, ton/hour.

Relative energy costs of live labor combine harvesters, MJ/ton:

$$\omega_z = \frac{\alpha_z \cdot n_o}{W_g \cdot \tau}, \tag{5}$$

where α_z – energy consumption of live labor of combine harvesters, MJ/hour (according to FAO standards (UN Food and Agriculture Organization), equivalent α_z = 1.26 MJ/hour); n_o – number of hours of combine harvesters; τ – coefficient of use of working time change.

Relative energy consumption required for machine maintenance, MJ/ton:

$$\omega_M = \frac{\alpha_M \cdot M_M \cdot (a_a + a_p)}{T \cdot W_g \cdot \tau}, \tag{6}$$

where α_M is the energy equivalent which determines the energy consumption for the production of 1 kg of machine weight, MJ/kg (for tractors, self-propelled combines and cars, the energy coefficient α_M is taken to be 120 MJ/kg); M_M – constructive weight of the machine, kg; a_a, a_p – share of deductions from total energy consumption, respectively, for the production, repair and maintenance of machines (a_a = 0.111; a_p = 0.103); T – norm of annual loading of combine, hours.

Relative consumption of diesel fuel on aggregate energy consumption when harvesting 1 ton of grain by a combine, litre/ton:

$$q_T = \frac{\omega_C}{Q_T}, \tag{7}$$

where Q_T is the relative heat of combustion of diesel fuel, for calculations it is possible to take equal 35.9 MJ/liter (relative heat of combustion of diesel fuel – 42.7 MJ/kg [18]; according to the European standard EN-590 the average density of summer diesel fuel is 0.84 kg/litre).

Relative cost of diesel fuel for total energy consumption when harvesting 1 ton of grain by combine, UAH/ton:

$$C_T = q_T \cdot c_T, \tag{8}$$

where c_T – cost of a diesel fuel, UAH/litre.

2. Working speed of combine, km/hour [19]:

$$v_p = \frac{36 \cdot g_{on}}{B \cdot A \cdot (1 + \delta_c)}, \tag{9}$$

where g_{on} – nominal capacity of the threshing machine combine, kg/sec; B – width of the header capture, m (the short technical characteristic of the main models of combine harvesters is given in [9]); A – yield of grain crops, ton/hectare (shown in [16]); δ_c is the straw coefficient, the ratio of the mass of the nonzero mass of the crop (straw and half) to the mass of the grain (given in [19]).

3. Variable productivity of combine, hectare/shift [3]:

$$W_{ZM} = 0{,}1 \cdot B \cdot v_p \cdot T_{ZM} \cdot \tau , \tag{10}$$

where T_{ZM} – duration of shift (7 hours); τ – coefficient of time use shift (determined depending on the length of the race, see [16]).

4. Variable coefficient:

$$\alpha_{ZM} = \frac{T_{RD}}{T_{ZM}}, \tag{11}$$

where T_{RD} – duration of the working day (it is recommended to take T_{RD} = 10 hours on harvesting [21]).

5. In accordance with the given natural and production conditions of the economy set the required number of machine complexes and determine the number of n_K combine harvesters, units, by the formula:

$$n_K = \frac{F}{D \cdot W_{ZM} \cdot \alpha_{ZM} \cdot K_M \cdot K_G}, \tag{12}$$

where F is the harvest area of grains for the i-th complex, hectare; W_{ZM} – production rate per 7-hour shift, hectare [21]; α_{ZM} – coefficient of variability; K_M – coefficient taking into account meteorological conditions (in the autumn period take K_M = 0.7); K_G – coefficient of technical readiness (with a runtime of up to 15 days take K_G = 0.95, more than 15 days – K_G = 0.90).

There is also an dependence (an inverse dependence (12)) on determining the length of harvesting of cereals (the calendar harvest period), days:

$$D = \frac{F}{n_K \cdot W_{ZM} \cdot \alpha_{ZM} \cdot K_M \cdot K_G}, \tag{13}$$

where F – total harvest area of grain crop, hectare; n_K – total number of combine harvesters in the farm.

6. Choice of straw-harvesting machinery for variable productivity (see [16]).

7. Number of straw-cutting equipment:

$$n_c = \frac{F \cdot A \cdot \delta_c}{D \cdot W^c_{ZM} \cdot \alpha_{ZM} \cdot K_M \cdot K_G}, \tag{14}$$

where F – harvest area of grain crop, hectare; A – grain yield, ton/hectare; δ_c – coefficient of straw; W^c_{ZM} – variable productivity of a straw car, ton/shift.

8. Choosing a car to transport grain.

It is proposed to determine the estimated amount of grain bins that fit into the body of the used car:

$$n_b = \frac{q_a}{\gamma \cdot V_b}, \tag{15}$$

where q_a – nominal load-carrying capacity of the vehicle, ton; γ – grain density, ton/m^3; V_b – capacity of the hopper of the combine, m^3. The value n_b must be equal to or close to an integer. The capacity of the bunkers of the wide range of combine harvesters and the carrying capacity of the vehicles that correspond to their capacity is given in [16].

9. Time to fill the hopper of the combine with grain (loading), minutes [19]:

$$t_{ZK} = \frac{600 \cdot V_b \cdot \gamma}{B \cdot v_p \cdot A} \tag{16}$$

where v_p – operating speed of the combine, km/hour; B – working width of the combine harvester capture, m; A – grain yield, ton/hectare.

10. Duration of the vehicle's circulation, hours:

$$t_{ob} = t_P + t_{ZA} + t_G + t_D + t_V + t_X \ , \tag{17}$$

where t_P – time of the car in the field, hours; t_{ZA} – time of vehicle loading, hours; t_G – time of the car with a load, hours; t_D – time of documents processing, hours; t_V – time of unloading grain from the body of the car, hours; t_X – time of the car without cargo, hours.

11. Number of vehicles (cars) for maintenance of the machine complex:

$$n_a = \frac{n_K \cdot W_G \cdot A \cdot t_{ob}}{q_a \cdot \alpha_q \cdot K_G} \ , \tag{18}$$

where n_K – number of working combines, units; W_G – hourly productivity of combine, hectare/hour; A – yield of grain crops, ton/hectare; q_a – nominal load carrying capacity of a vehicle (vehicle), ton; α_q – coefficient of use of carrying capacity of a vehicle (vehicle); K_G – coefficient of technical readiness (with a duration of work up to 15 days take $K_G = 0.95$, more than 15 days - $K_G = 0.90$). It is recommended [12] to determine the number of vehicles (cars) separately for each type of transport, by its technical characteristics. To determine the calculated parameters λ and W_q, which are included in the expression of the target function of the developed optimization model, given in [14], the method shown in [16] can be used.

The value of λ corresponding to the number of harvester hoppers that will be filled in one hour depends on the loading time t_{ZK} and the discharge of the t_{VK} to the combine hopper:

$$\lambda = \frac{60 \cdot n_K}{t_{ZK} + t_{VK}} \ , \tag{19}$$

The duration of unloading of the hopper of the combine is determined depending on the density of the grain γ:

$$t_{VK} = \frac{V_b \cdot \gamma}{v_{rozv}} \tag{20}$$

For the system with expectation, the probability that the system is free, there are no applications for maintenance in the system, i. e. all service channels (cars) are idle, no combine is ready for unloading the grain:

$$P_0 = \left(\sum_{k=0}^{n} \frac{\rho^k}{k!} + \frac{\rho^{n+1}}{n!(n-\rho)} \right)^{-1} \ , \tag{21}$$

where n – the number of all channels of service in the system (the number of cars in the harvesting complex).

Determine the probability that the service occupied 1...n cars by the formula:

$$P_k = \frac{\rho^k}{k!} P_0 \tag{22}$$

Average queue length (number of combines waiting for bunker unloading):

$$L_q = \frac{\rho^{n+1} \cdot P_0}{n! \cdot n \cdot \left(1 - \frac{\rho}{n}\right)^2} \tag{23}$$

Average standby time of the combine, hours, which was queued at the start of service:

$$W_q = \frac{L_q}{\lambda} \tag{24}$$

As follows from the foregoing, calculations according to the proposed mathematical model indicate that, with the increase in the vehicle number, total costs from downtime of combines and vehicles are initially reduced and then increased [16]. That is, to determine the optimal value of the target function, it is expedient to erode the total costs received as per the number of vehicles, and with their increased number. The vehicle number for which the total costs will be minimal will be optimal for the considered harvesting complex.

References

1. D. Teodorovic, M. Janic. *Transportation Engineering: Theory, Practice, and Modeling*, Butterworth-Heinemann (2017).
2. N. Lütke Entrup, J. Oehmichen (Hrsg.). *Lehrbuch des Pflanzenbaus. Band 2: Kulturpflanzen*. Verlag Th. Mann Gelsenkirchen (2000).
3. A.A. Zangiev, A.V. Shpil'ko, and A.G. Levshin. *Ekspluatatsiya mashinno-traktornogo parka*. KolosS, 320 (2008).
4. I.P. Ksenevich (et al.), *Mashinostroyeniye. Entsiklopediya. Tom IV-16. Sel'skokhozyaystvennyye mashiny i oborudovaniye*, Mashinostroenie, 720 (2002).
5. D.A. Muzylev (et al.), Vostochnoyevropeyskiy zhurnal peredovykh tekhnologiy. **2/3 (80)**, 11–21 (2016).
6. V.A. Zyazev (et al.), *Perevozki sel'skokhozyaystvennykh gruzov avtomobil'nym transportom*, Transport, 288 (1979).
7. I.Trunina, V.Zahorianskyi, O.Zahorianska, D.Moloshtan, International Journal of Engeneering and Technology, **7, 4.3 (2018), Special Issue 3**, 537–544 (2018).
8. M.M. Moroz (et al.), International Journal of Engeneering and Technology, **7, 4.3 (2018), Special Issue 3**, 206–210 (2018).
9. Moroz Olena, Moroz Mykola, Actual Problems of Economics. **160, 10.** 239–246 (2014).
10. Moroz Mykola, Actual Problems of Economics. **166, 4**, 235–243 (2015).
11. M.M. Moroz (et al.) Actual Problems of Economics. **1 (175)**. 385–398 (2016).
12. M.M. Moroz, V.L. Khorolskyi, O.V. Moroz (et al.), International Journal of Engeneering and Technology. **7, 4.3 (2018). Special Issue 3**. 206-210 (2018).
13. V.I. Kravchuk (et al.), *Mashyny dlya zbyrannya zernovykh ta tekhnichnykh kul'tur*, UkrNDIPVT im. L. Pohoriloho, 524 (2009).
14. V.G. Zagoryanskii, T.V. Haykova, V.L. Khorol'skii, I.O. Kuzev, Visnyk Kremenchuts'koho natsional'noho universyteta imeni Mykhayla Ostrohrads'koho. **2/2019 (115)**. 146–151 (2019).
15. E.V. Zhalnin, A.N. Savchenko, *Tekhnologii uborki zernovykh kombaynovymi agregatami*. Rossel'khozizdat, 318 (1985).
16. M.E. Artemov, *Kursovoye i diplomnoye proyektirovaniye po ekspluatatsii mashinno-traktornogo parka*. Krasnoyarskij. gos. agrar. un-t, 326 (2008).
17. Ye.K. Markelova (et al.), *Prakticheskaya metodika opredeleniya energozatrat i energoyomkosti proizvodstva produktsii, a takzhe potrebnosti v energoresursakh*, FGNTS Rosinformagrotekh, 206 (2001).
18. L.S. Vasil'yeva, *Avtomobil'nyye ekspluatatsionnyye materialy*, Nauka-Press, 364 (2004).
19. A.V. Novikov, *Ekspluatatsiya sel'skokhozyaystvennoy tekhniki v primerakh i zadachakh*, Belorusskiy gosudarstvennyy agrarnyy tekhnicheskiy universitet, 344 (2010).

Inflation processes as determinants of development of the economic activity subjects: economic and legal aspects

Serhii Arefiev[1,1,], *Iryna* Miahkykh[2], *Samira* Piletska[3], *Iryna* Sopilko[3]

[1]Kyiv National University of Technology and Design, Kyiv, Ukraine
[2]Kyiv National University of Technology and Design, Kyiv, Ukraine
[3]National Aviation University, Kyiv, Ukraine.

Annotation: Today, inflation takes one of the leading places among the factors that give rise to social and economic destabilization. The purpose of the study is to determine the nature of inflation and its impact on the real sector of the economy, as well as factors that it generates in the process of distributing and redistributing financial resources and justifying measures to offset the negative effects of inflationary impact on the Ukrainian economy. The authors used the following methods of research: systematic approach, theoretical and empirical methods of scientific knowledge. The essence of inflation and its influence on the real sector of economy are investigated. The factors that generate inflation during the distribution and redistribution of financial resources are indicated: political and social factors contributing to anticipating inflationary expectations, which accumulates inflationary potential; the institutional support of economic development is to assume stabilization of inflation through the mechanism of managing the money supply; factors of inflation reflections, capable of generating further unmanaged negative behavior of economic agents, which can be disclosed through the concept of "inflation of the conflict"; inflationary expectations require the implementation of monetary policy of the state.

1 Introduction

An integral part of the economic development of business entities as social and production systems is the presence of inflation, which to some extent causes an increase in the value of investment resources, on the one hand, and, on the other hand, becomes a manifestation of unsustainability and violates the sustainability of the system itself. At certain (controlled) values, inflation can be viewed positively, as it stimulates economic growth, and at values beyond the permissible (defined) limits negatively affects the development of both economic and social processes, creating threats to the existence of the system. The process of forming financial support for the development of business entities is under constant influence of

[1] Corresponding author: lena-2009-19@ukr.net

inflationary processes, creating certain threats to their continuous functioning. The solution to the problem of managed inflation needs to be considered not only at the macro level, but also in inextricably linked with the activities of enterprises, small and medium-sized businesses, which first experience the effects of inflation and fall under their own initiative reduction, destabilizing the social sphere. Saving business units also requires appropriate regulatory influences on the part of government agencies. The analysis of causal relationships in this context also requires a scientific substantiation of the nature of inflation and factors that determine it, justifying its management at different levels of management of the national economy.

2 Literature references.

The economic literature generalizes theoretical approaches and empirical studies that explain the formation of inflationary processes, including the directions of the classical, monetarist, Keynesian, neo-Keynesian, theory of rational expectations, the new Keynesian and many other schools. In principle, the approaches of all of the above currents can be divided into two directions: the classical (the views of which representatives will further develop within the school of monetarism) and Keynesian.

Thus, the basis of the postulates of the classical school was the quantitative theory of money, which caused the causal link between the increase in money supply and the growth of the general price level [1]. The fundamental idea of Keynesianism was to consider stimulating effective demand, as the main direction of state regulation of the economy. The latter was achieved through increased spending on the state budget and liberal monetary policy. The Keynesians' views on explaining the factors of inflation were presented in a fairly detailed way within the framework of two competing theories: demand inflation and cost inflation, where the effects of exogenous shocks on the domestic price were first raised [2]. M. Malkin and I. Rosmaninsky, formulating the institutional theory of inflation, highlight difficulties and problems in explaining the causes and forecasting inflationary processes. In this case, the inflationary process is seen as the unity of open inflation and accumulated inflationary potential [3]. The accumulation of inflationary potential is seen as a "delayed" mechanism of inflation. D. Belousov considers this phenomenon as a manifestation of the "compensatory" function of inflation [4], defining it as the level of inflation, which, with the rise in cost of primary resources, provides commodity producers with the level of financial resources necessary for simple reproduction.

M. Yu. Malkin, considering this process in the framework of the institutional approach, notes that the accumulation of inflationary potential is a dynamic phenomenon, which is based on complex transformational changes of intra-system character, reflecting institutional and structural imbalances. At the same time, the transition of the inflationary potential to open inflation is usually associated with the accumulation of a critical mass of imbalances, which is accompanied by the crisis of the former regulatory system, its further transformation and adaptation [5]. This correlation is confirmed by the fact that inflation is not a phenomenon that is born solely as a result of economic factors and incorrect monetary policy of the state. Along with the economic factors of inflation, according to D. I. Ramazanov, it is necessary to investigate non-economic factors. Analyzing the non-economic group of factors and suggests distinguishing two major subgroups: political and social. The latter represent a set of institutions of political, legal, social and psychological nature, affecting economic processes, while these norms can have both a formal (legally stipulated) nature and an informal status [6].

A.V. Stepanov notes the need to reconcile social conflicts and economic interests in order to stabilize inflation. Among social factors, taking into account the psychological nature of inflation, inflation expectations are recognized as the most important psychological factor in inflation. At the same time, inflationary expectations are associated with the accumulation of inflationary potential, while the most sustainable expectations during periods of high inflation are dangerous by the development of inertial inflation processes and the formation of an inflationary spiral [7].

As a special psychological factor, it is necessary to allocate inflation expectations - assumptions about the level of future inflation, which are formed by economic entities. This results in a sharp increase in demand and slowing down the supply side, which are considered by the factors of demand and supply inflation, as a consequence of the corresponding expectations of consumers and borrowers or producers and creditors. This is also confirmed by A. Ulyukayev, pointing out that "inflation is a very psychologically motivated phenomenon, without disclosure of information, without dialogue with society, you cannot manage inflation expectations" [8]. According to K. Nagatani, the professor of the University of British Columbia (Canada), inflation is a natural result of the struggle between creditors and debtors. Permanent debtor in the modern world is the public sector and private business, and the creditor is all other entities. The institutional structure of a modern market economy in the developed countries, as it were specially designed to facilitate debtors use of the redistributive function of inflation. Most of the various agents are linked to each other through financial markets, and the bulk of financial assets is fixed in cash [9]. According to V. Popov, "the intensity of inflation is determined by the relationship of the real and financial sectors of the economy, as well as the parameters of the money supply established by the NBU" [9, p. 70]. That is, the level and rate of inflation determine the structural parameters. The authors of the world-famous "Economics" determine the causes of inflation on the basis of the equilibrium theory of "aggregate demand aggregate supply" [11, 170, 171].

At the same time, further research and detailed disclosure require methodological issues to determine the nature of inflation in changing the conditions of management and the impact of its factors on different levels of the national economy. This will allow for the formation of adequate solutions to minimize the threats that create unchecked influence of inflation on the external and internal environment of the respective levels, to develop methodological support for a thorough study of inflation processes in the economic and social spheres of the economy.

3 Results and discussion

The effect of inflation violates social reproduction, which prevents the formation of productive forces at the necessary level and the introduction of innovative technologies. Imbalances in these processes impede the creation and receipt of value added, which in turn reduces the national economy, reducing the consumption and export activities of business entities. In this regard, it seems that the study of the nature of inflation and its impact, as well as factors that generate it during the distribution and redistribution of financial resources is necessary: firstly, to assess the source and qualitative structure of the effects of inflation, namely, the part of conjunctural, partial, fundamental, as well as the allocation of leading factors of the inflation process by the levels of the economy, which require different in time, periodicity, degree of depth of measures for the formation of anti-inflationary policy; secondly, in order to calculate and justifiably choose the lower target of inflation, which has a stimulating effect and does not lead to unwanted losses in economic growth, since the latter is possible under conditions of a credit restraint that has a significant impact on the financial system and is negatively reflected on the budget process, monetary aggregates, activity of business entities. The confirmation of the redistributive function of inflation is the presence of inflationary surges that destabilize the economic interests of different levels and industries. In this context, institutional maintenance of economic development is considered to stabilize inflation through the mechanism of managing the money supply; thirdly, understanding the certain expectations, the inertia of the influence of the existing level of inflation, and its synergy is important to determine the fate of the factors of inflation reflection, capable of generating further unmanaged negative behavior of business entities, which can be disclosed through the concept of "inflation of the conflict" that occurs in the consequence of the struggle for the existing national benefits. This is reflected in the complex effects of inflationary processes, both on the state sector of the economy and on the private sector, while the creeping interest rate causes the emergence of a conflict between creditors and borrowers; fourthly, forecasting and estimating the possible effects of inflation, which causes imbalance in demand and supply, prices for raw materials and finished goods, payables and receivables of economic entities, on the one hand, as well as income and expenditure components of budgets at all levels of the national the economy, on the other, and the country's export-import balance; fifthly, in order to control the openness of inflation and to prevent it from conducting anti-inflationary policy, it is necessary to substantiate and form the provision of compliance, which will allow comparing the risks and costs of appropriate measures that arise in the process of economic activity of business entities. Since digitalization has changed the reproduction process in the branches of the economy, it is necessary to implement it while monitoring at all stages of creating value added.

Recently, in the conditions of a permanent financial crisis and unstable external environment, there is a qualitative change in the nature of the impact of inflation on economic processes, including the economic development. In connection with this, it seems expedient to consider inflation in terms of functional, process, synergistic approach, which should be used in accordance with the state of the economy of the country, the rate of its decline or development, general characteristics of the activity of economic entities, the system of state regulation in general, and prices, in particular. This creates some obstacles to ensure sustainability of the formation of necessary assets, creation of reserves and complicates the attraction of the population as the main investor of the country's economy. At the same time, it is impossible to completely exclude state interference in pricing processes, in particular, in order to prevent monopolies and create conditions for the development of competition. In no country with a developed market economy, there is no full freedom in terms of pricing, in most of these countries, as a general tactic, certain rules (principles, methods and norms) are used for pricing, which are formalized in the form of legislative acts, and are aimed at regulating the methodology and procedure of formation prices on commodity markets of the country. In addition, the state assumes the role of price setting for certain goods and services that are crucial for the development of the national economy and controls them. Thus, the sphere of regulated and controlled by the state pricing ranges from 10% to 40% [12, p. 206].

According to the Decree of the Cabinet of Ministers of Ukraine On the implementation of the pilot project on the temporary limitation of the implementation of the Cabinet of Ministers of Ukraine Decree of December 25, 1996, No. 1548 and the Resolution of the Cabinet of Ministers of Ukraine of October 17, 2007 No. 1222 dated September 22, 2016 No. 656, a pilot project has been launched whose task is to determine the degree of influence of the state on the real dynamics of food prices [13]. If the results of the project reveal that state regulation of prices only worsens this dynamics and raises the cost of products, the restriction will be abolished.

Implementation of state regulation of inflation requires the establishment of a mutual equilibrium with market laws and the necessary level of social security of the population [14]. Since inflation is one of the main indicators that characterizes macroeconomic instability. It characterizes adverse changes in prices, which indicate the occurrence of certain changes in commodity-money relations and the distribution of aggregate income. Therefore, it is expedient to use the Consumer Price Index (Lespereis index) to identify the tendency of retail prices to grow (Fig. 1), which reflect the effect of inflation both directly on food products and the indirect effect of changes in prices for raw materials, energy and other components of the cost of production.

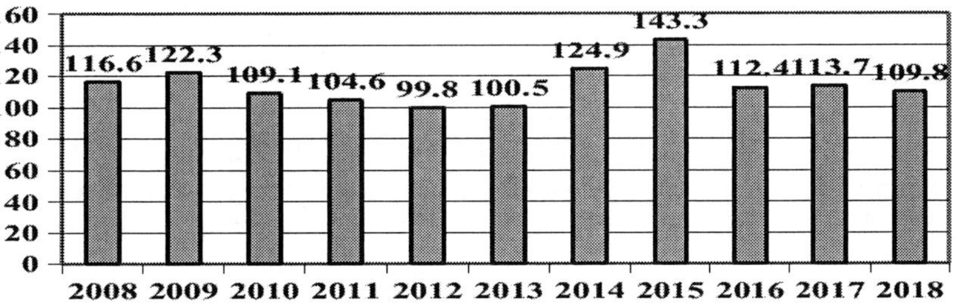

Fig. 1. Consumer price index in Ukraine in 2008 - 2018 [15].

An objective system analysis of the economic and legal positions of inflation processes requires the study of its impact on production and reproduction processes in the country's economy, which will help establish certain trends and will allow formulating conclusions and formulating a set of measures to prevent its negative impacts on the necessary development processes that for the most part, are carried out at enterprises through their financial support. As, according to V.I. Osciposhchev, "financial security is regulated on the basis of the appropriate system of functioning, which can be carried out in three forms: self-financing, lending, external financing [16, p. 59], one of the important components of the activity of enterprises is the formation of financial support from the relevant sources, the value of which is also determined by adjusting the interest rate on bank loans. Therefore, the ability to self-finance the current financial needs is the advantage of enterprises that operate with profits, which makes it possible to reduce the value of attracted capital in the context of the deployment of inflationary processes in the economy.

An illustration of the influence of the above-mentioned factors that determine inflation in the realities of the current state of the country's economy is the increase in the share of unprofitable industrial enterprises of Ukraine for the period 2010-2017 (Fig. 2).

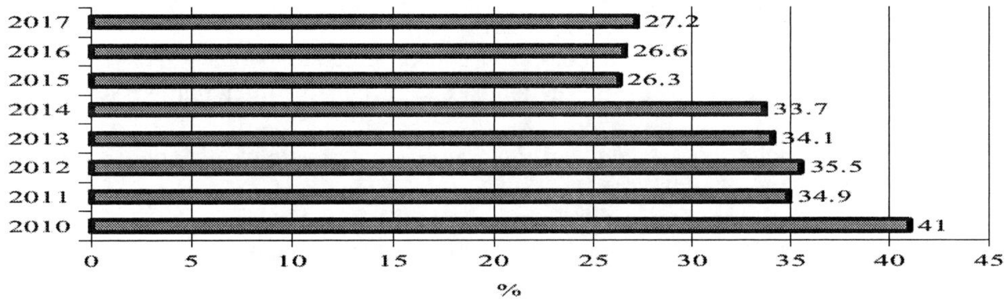

Fig. 2. The share of unprofitable business entities for the period 2010-2017 years, % [15].

The urgency of implementing the processes of development of the branches of the economy on an innovative basis requires the implementation of strategic and long-term directions of innovation duality. The legal, economic and organizational foundations for the formation and implementation of innovation activities in Ukraine are defined by the Law of Ukraine "On Priority Areas of Innovation Activity in Ukraine" dated September 8, 2011. Its main objective is the creation of a legal framework to provide an innovative model for economic development through the concentration of state resources on priority directions of scientific and technical renewal of production, increase of competitiveness of domestic products in the domestic and foreign markets [17]. The financial support of innovation development means "the totality of economic relations arising from the search, attraction and effective use of financial resources, as well as organizational and managerial principles, methods and forms of influence of these resources on the innovative development of the national economy." In the structure of financial security methods, special importance is attached to financial regulation, which occurs mainly through taxation [18, p. 48].

Thus, it can be noted that the level of profitability of economic entities during the survey period sharply decreases (from 4.0% in 2010 to -4.1% in 2014), which significantly limits them in the planning of investment projects and job creation. One of the main reasons for lowering financial indicators of enterprises is the reduction of production and sales (services). This is due to the fact that enterprises try to implement already manufactured products in warehouses without increasing the rate of production (provision of services), which, in turn, contributes to the increase of inflation and negatively affects the development of the economy of the country.

4 Conclusion.

The crisis caused by the influence of inflation has led to a reduction of the real sector of the economy due to the narrowing of the consumption market. This requires the development of mechanisms to prevent the rampant effects of inflation through the implementation of anti-inflationary regulation in Ukraine: to develop comprehensive state programs for economic development, based on the study of inflation expectations that will contribute to increasing the volumes and rates of GDP; to pursue a consistent antitrust policy to maintain competitive prices in the domestic market, taking into account existing inflation potential; to stimulate credit and investment activity of banks, ignoring the inflationary conflict of interests; limit the coverage of a cash deficit due to a bank loan; to improve the state regulation of the tax system by providing the formal status of investment-attractive branches of the economy through tax incentives; to create favorable conditions for the termination of import of inflation.

References

1. W. E. Bidwell, 1914. "Review: The Purchasing Power of Money by Irving Fisher". Bull. Amer. Math. Soc. **20 (7)**, 377-381 (1914).
2. Dzh. M. Keins Obshchaia teoryia zaniatosty, protsenta y deneh. Helyos ARV, 352 (2002)
3. M. Yu. Malkina & I. V. Rozmainskij. Zhurnal institucionalnyh issledovanij, **2**, 69-87 (2013)
4. D. R. Belousov *Mekhanizm inflyacii v sovremennoj ekonomike Rossii (finansovo-vosproizvodstvennyj aspekt)* (1998)
5. M. Yu. Malkyna. Fynansyi y kredyt. **24 (192)**. 2-7. (2005)
6. D. I. Ramazanov. Inflyatsiya: politicheskiy aspekt. Ivanovo: Ivan. gos. un-t, (2011)
7. A.V. Stepanov. Sotsyalno-ekonomycheskye yavlenyia y protsessu, **11**, URL: http://cyberleninka.ru/article/n/institutsionalnyy-aspekt-inflyatsionnyh-protsessov (2016)
8. A. Ulyukaev. Minekonomrazvitiya i TsB razoshlis v otsenke perspektiv inflyatsii. Gazeta «Vedomosti». 04.02.2012 g. URL: http://www.vedomosti.ru/finance/articles/2012/02/04/minekonomrazvitiya_i_cb_razoshlis_v_oc enke_perspektiv (2012)
9. K. Nagatani. *Political Macroeconomics*. Oxford: Claredon Press. (1989)
10. V. Popov. Ekonomika Ukrainy, **11**, 69-79. (2011)
11. K.R. Makkonell & Briu S.L. *Ekonomiks: printsipyi, problemyi, politika*. INFRA M, 974, (1999)
12. A.S. Bazdnikin. *Tsenyi i tsenoobrazovanie*. Yurayt-Izdat, 332 (2008)
13. Pro realizaciyu pilotnogo proektu shodo timchasovogo obmezhennya zastosuvannya postanovi Kabinetu Ministriv Ukrayini vid 25 grudnya 1996 r. № 1548 ta postanovi Kabinetu Ministriv Ukrayini vid 17 zhovtnya 2007 r. No. 1222: Postanova Kabinetu Ministriv Ukrayini vid 22.09.2016 № 656. Oficijnij visnik Ukrayini. **78. 18.**, 2602. (2016)

14. O.V. Arefieva, I.M. Miahkykh & S.T. Piletska. Ekonomichnyi visnyk Donbasu. Naukovyi zhurnal, **2 (52)**, 95 – 101 (2018)
15. Statystychnyi zbirnyk. Diialnist subiektiv hospodariuvannia za 2017 rik. Official website. http://www.ukrstat.gov.ua (2018)
16. O. P. Blyzniuk, L. I. Lachkova, V. I. Ospishchev. Finansy: Navch. posib, 415 (2006)
17. Zakon Ukrainy "Pro priorytetni napriamy innovatsiinoi diialnosti v Ukraini", 3715–VI. [in Ukrainian] (2011)
18. I. V. Osmirko. Biznes Inform, **7**, 47–49 (2012)

Research of the economic risk factors of the functioning of the Ukrainian transport system

Nataliia Bogdan[1*], *Sergii* Pogasiy[1], *Iryna* Pysareva[1] and *Iuliia* Krasnokutska[1]

[1] O.M. Beketov National University of Urban Economy in Kharkiv, 17, Marshal Bazhanov Street, Kharkiv, 61002, Ukraine.

Abstract. The purpose of this article is to research and to evaluate the risk factors of the transport system of Ukraine. The article examines the state of the transport system of Ukraine. The classification of specific risks of the transport system has been developed in two areas: the nature of occurrence and in relation to the transport system. The conceptual model of comprehensive research of the factors of economic risk has been proposed. It includes 4 modules. The approaches to the identification of the factors and indicators for assessing the economic risks of transport systems at the national and regional levels have been formed. In the result the comprehensive evaluation of the risk factors of regional transport system of Ukraine has been carried out. Identification of risks at the national and regional transport system is necessary for the effective strategy of the industry development through quantitative risk analysis which enables to give a real assessment of their own resources and capabilities in relation to the state (and needs) of the external environment. The rational choice of the strategies from a variety of options should be made on this basis.

1 Introduction

The strategy of the transport systems development at national and regional levels which are being established currently in most cases do not consider the economic risks emerging at all levels of management. The complexity, the vagueness and differentiation of the sources of risk manifested in the transport systems makes it difficult to assess the integral level of economic risk, which in turn leads to a decrease in the effectiveness of management decisions in the industry.

Recently the impact of economic risks on transport system at meso and micro levels has been reinforced. The volume and quality characteristics of transport, its infrastructure do not allow to solve the problem of the economy effectively. This requires a substantial restructuring of the transport sector which involves additional risks.

The problems of the development of transport system was considered in the researches of Ukrainian scientists: K. Andriushchenko [1], M. Bidniak [2], O. Boyko [3], P. Gorbachiov [4] A. Gorev [5], V. Dykan [6], O. Kravchenko [7], O. Pikulic [8], I. Rogozian [9] etc. The most

[1*] Corresponding author: bogdan.nataliya.67@gmail.com

famous foreign researchers of transport system are A. Balducci, L. Boelens, J. Hillier [10], P. Kellermann, A. Otto [11].

Despite the large number of research papers on transport issues, problems of identification, assessment and minimization of risks are not sufficiently considered.

The research papers on the risks of transport system functioning are not enough. Further research is required, in particular, the analysis of risk factors based on industry and socio-economic characteristics of the development of territories, their classification, quantification, degree of influence of risk factors on the competitive position of industrial complex and the formation of its development strategy and a number of others are not covered in modern scientific papers.

Insufficient development of theoretical and practical problems of the functioning of transport systems in risk conditions determined the choice of research topic, its aims and objectives, and logic of research. Thus, the aim of this paper is to study and to evaluate the risk factors of the transport system of Ukraine.

2 Methods

The system of indicators for assessing external and internal risk factors of a transport system has been established by the authors. These indicators were evaluated using the method of expert assessments. The integrated risk assessment of regional transport complexes of Ukraine is calculated.

3 Results and discussion

The transport sector includes all participants of transportation market. Risks in the transport sector involve the discrepancy between the actual parameters of the market and its expected values [12]. These parameters include the degree of concentration of capital, elasticity and other characteristics of supply and demand in the industry, inflation rates and stability of rates, the balance of the market, the deterioration of the material and technical base of the industry, aging rolling stock, the quality of the labor force and employment in the industry, the instability of tax policy, the liquidity and solvency of the motor transportation enterprises (organizations), reduction of investment attractiveness of the industry, imperfect legal support of the carriers.

The situation in the transport system depends on the general socio-economic situation in the country and the region.

At the present stage of Ukraine's economy development transport plays a crucial role in ensuring the economic growth and social development of the country. At the end of 2017, Ukrainian transport sector takes 6.4% in the sectoral structure of GDP. This rating has decreased significantly since 2010 – more than 37% which proves the efficiency reduction transport functioning

Analyzing the socio-economic development of Ukraine by the macroeconomic indicator of gross domestic product (GDP), it should be noted a steady tendency of growth and change of this indicator during 2010-2018: its value increased almost 3 times. However, a significant impact on the growth rate of economic development had a global financial crisis, which increased the level of economic risks. The domestic risks took place starting in 2014, when there was a decline in all sectors of the economy.

Economic risk in the transport sector is sectoral, national and regional in nature. It represents the probability of deviations of results of the activity of entities of transport system from the predicted values due to changes in the economic situation in the transport sector and in the socio-economic situation of the country.

The following risks for transport enterprises and the industry in the whole can be defined: macroeconomic, geopolitical, operational, social, technological, environmental.

The authors propose a classification of specific risks of the transport complex (Fig. 1).

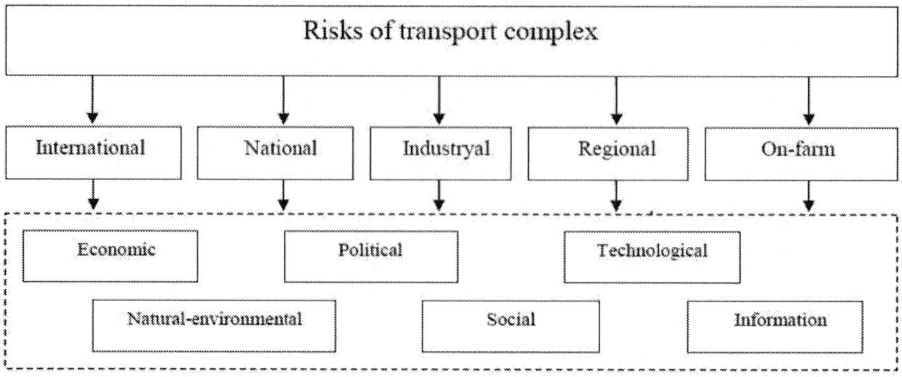

Fig. 1. Risk classification of transport system

Thus, it is possible to determine the economic risk of the transport system as the probability of deviations of the results of its entities from the predicted values due to changes in the economic situation in the transport sector and in the socio-economic situation of the country or region.

Analyzing the economic risk of the transport sector including the following types: general economic, financial, investment and innovation, socio-demographic, market, resource, logistics, production, management. Each of these sources of risks include general economic risk complex and requires analysis and evaluation [13].

Authors systematized the risk factors of the transport system and classified them according to the following criteria:

– the environment of occurrence (external, generated by the conditions of macro - and mezzo level of transport system of the region);

– according to the economic nature of occurrence.

Thus, the approaches to the identification of factors and indicators for assessing the economic risk of transport systems at the national and regional levels have been formed.

Considering the indicators for risk assessment of the transport system, the individual and integrated risk analyses of regional transport of Ukraine was undertaken [14].

Table 1. Weighting factors of the group risk factors regional transport.

Group	Group weights	Group	Group weights
External risk factors		Internal risk factors	
1	0.09	1	0.045
2	0.04	2	0.045
3	0.05	3	0.05
4	0.035	4	0.045
5	0.045	5	0.06
6	0.035	6	0.055
7	0.055	7	0.055
8	0.04	8	0.05
9	–	9	0.04

The data of the official website of the Ministry of infrastructure of Ukraine, materials of RA "Expert-Rating", "Transport' journal was applied for the evaluation [15, 16].

The experts from RA "Expert rating" (7 experts) and experts in the field of transport infrastructure, the authors of "Transport" journal (8 experts) were involved for expert assessment of risk factors of regional transport complexes, and for determining the weights of groups of risk factors (table. 1).

Based on statistical data, the scores for each risk factor for each region were given to experts, on this base the integrated assessment of regional transport were calculated. The results of the evaluation are presented in table. 2.

The evaluation of the risk factors of regional transport of Ukraine has been conducted in the result of the research.

The accounting system and the assessment of risk factors of the transport complex must be built into the management system, due to the high-risk nature of the transport industry [17].

Table 2. The results of evaluation of the risk factor of regional transport of Ukraine in 2017.

Region	2017	
	IR	Rank of the region
Vinnytsya region	0.435	13
Volyn region	0.282	19
Dnipropetrovsk region	0.988	5
Donetsk region	0.818	16
Zhytomyr region	0.349	11
Zakarpattya region	0.249	20
Zaporizhzhya region	0.971	5
Ivano-Frankivsk region	0.395	12
Kyiv region	0.852	3
Kirovograd region	0.382	10
Lugansk region	0.579	18
Lviv region	0.484	4
Mykolayiv region	0.701	7
Odessa region	0.767	6
Poltava region	0.907	7
Rivne region	0.322	17
Sumy region	0.399	8
Ternopil region	0.191	21
Kharkiv region	0.727	2
Kherson region	0.301	14
Khmelnytsky region	0.371	15
Cherkasy region	0.505	9
Chernivtsi region	0.188	22
Chernigiv region	0.335	16
Kyiv	0.991	1

The high level of risk in national and regional transport complexes requires the development of appropriate strategies of the activities for creating the competitive advantages through formal procedures, which are defined in stages and for application the techniques of analysis and modeling aimed at minimizing the risks.

The most important stage in the development of effective strategy of development of the industry is quantitative risk analysis, which enables to give a real assessment of their own resources and capabilities in relation to the state (and needs) the external environment. On this base it should be the rational choice of strategies from a variety of options.

4 Conclusions

The providing the territory with transport is an integral characteristic of its level of development, the most important condition of formation of competitive advantages and investment attractiveness. The risk in the transport sector is both sectoral and territorial in nature and represents the probability of deviations of the results of activity of subjects of a transport complex from the predicted values due to the changes in the economic situation in the transport sector and the socio-economic situation.

The classification of risk factors enables to identify the classification groups that allows to identify them precisely and develop anti-risk events correctly.

During the research, the system of 63 risk indicators (local factors) has been established. In the next stage, each of 63 indicators has been analyzed using the developed technique of the expert assessment of their impact on the functioning of the regional transport systems. The method of rank estimation has been applied for these purposes.

Identification of risks at the national and regional transport systems is necessary for the effective strategy of the industry development through quantitative risk analysis, which enables to give a real assessment of their own resources and capabilities in relation to the state (and needs) of the external environment. The rational choice of the strategies from a variety of options should be made on this basis.

References

1. K. A. Andriushchenko. Derzhavne upravlinnia: udoskonalennia ta rozvytok. http://www.dy.nayka.com.ua/?op=1&z=379 (2012)
2. M. N. Bidniak. *Production systems on a transport: theory and practice: monograph*. Laborator practical work, (2006)
3. O.V. Boiko, I.V. Didovets. *Potentsial rynku transportnykh posluh yak ob'iekt naukovykh doslidzhen.* http://www.nbuv.gov.ua/portal/Soc_Gum/Vcndtu/2010_41/13.htm.(2010)
4. P. F. Gorbachev, I. A.Dmitriev. Osnovy teorii transportnyh sistem. Izd-vo HNADU, Kharkiv, (2002)
5. A. Je. Gorev. Osnovy teorii transportnyh sistem. SPbGASU, SPb., (2010)
6. V.L.Dykan, M.V. Korin. Visnyk ekonomiky transportu i promyslovosti, **(33)**, 13-19 (2011)
7. O. P. Kravchenko, Ye. P. Medvediev. Avtomobil i elektronika. Suchasni tekhnolohii, **(4)**, 48-52 (2013)
8. O. B. Pikulik. Wide choice of Sciences. вісн. Well. National Un th. Lesya Ukrainka, **7**, 284-291 (2008)
9. Y. S. Rogozyan. Bulletin of the Eastern Ukrainian University named after V. Dahl, **9(163),** 169-173 (2011)
10. A.Balducci, L.Boelens, J.Hillier, T.Nyseth, C.Wilkinson, Transportation planning and policy in the pursuit of mega-events: Boston's 2024 Olympic bid Kassens-Noor. ETransport Policy, **74**, 239-245 (2019), doi: 10.1016/j.tranpol.2018.12.005
11. Otto, A., Kellermann P., Thieken A., Máñez Costa M. Carmona, M. Bubeck P. International Journal of Disaster Risk Reduction, **33,** 385-397 (2019), doi: 10.1016/j.ijdrr.2018.10.025
12. O.L.Vasyliev, O.M. Lukianova. Visnyk ekonomiky transportu i promyslovosti, **41**, 92-95 (2013)
13. A. I. Orlov *Prikladnaja statistika.* http://www.merriam-webster.com/dictionary/risk (2006).
14. O. Avci, O. Ozbulut. *Case Studies on Transport Policy*. **6, 4**, 663-673 (2018)

15. Official website of "Transport" journal http://transport-journal.com/about/
16. Official website of RA "Ekspert-Rating" http://www.expert-rating.com
17. N. M. Bogdan. Modeliuvannia regionalnoi ekonomiky. **2**(28), 36-51, (2016)

Optimization of planning as a means of improving efficiency of development management of resource potential of rail transport enterprises

Olena Chupyr[1], *Liudmyla* Kalinichenko[2], *Olena* Sychova[1], and *Yevgeniy* Burlaka[3,*]

[1] Department of Management and Public administration, Kharkiv National University of Civil Engineering and Architecture, 61002 Kharkiv, Ukraine
[2] Department of Economy, Kharkiv National University of Civil Engineering and Architecture, 61002 Kharkiv, Ukraine
[3] Department of Finances and Credit, Kharkiv National University of Civil Engineering and Architecture, 61002 Kharkiv, Ukraine

Abstract. This scientific paper presents a methodological approach to optimizing the planning process based on the analysis of existing scientific recommendations on this issue with the development of the resource potential of rail transport enterprises. The proposed method allows more accurately diagnosing of bottlenecks in the management of the resource potential of enterprises in the railway industry, which, in turn, allows us to choose more adequate measures for the development of resource potential.

1 Topicality

Railway industry of each country has a developed resource potential which is composed of various resources interacting between each other and are in permanent interaction. For this particular reason, resource potential development management of railway industry enterprises becomes one of the most important tasks which is set by the enterprises top management with respect to the industry strategic management. Planning process is the most important management element which allows determining lines of enterprise development, minimizing effect of accidental environmental factors and human factor, while ensuring minimizing of costs and irrational use of resources. The need to apply mathematical approaches to planning optimization is caused by the uncertainty with respect to choosing the best option for an enterprise development, since different options may be characterized by several different criteria that differ in their importance, each of them having their own advantages. The use of mathematical optimization methods will allow us to choose the best option for the enterprise development lines.

2 Analysis of the latest studies and publications

Resource potential development management issue was studied by the following domestic and foreign scientists as A.Ya. Bersutskiy [1], D.V. Volovyk [2], S.M. Isa [3], M. Kassab [4], O.V. Ulyanchenko [5], T. Khegazi [4], O.I. Shamanska [6].

Comparative characteristic of the above authors' models is provided in table 1.

*Corresponding author: burlaka89@ukr.net

Table 1. Comparative characteristic of the optimization models for resource potential management existing in the scientific literature.

Author	Model description
A.Ya. Bersutskiy [1]	Determination of a balanced set of resources ensuring the implementation of a generated operational production program while taking into account the priority vector values. The advantage of the model – it allows to plan the production program of an enterprise and forecast a possible resources shortage in the process of its implementation. The model's weak point is the lack of a procedure for the priority actions selection aimed at resource potential development.
D.V. Volovyk [2]	The method of multicriteria optimization in managing the resource potential development of agrarian enterprises. The advantage of the model is comprehensiveness of all resources types, considering external and internal factors, that affect resource potential management efficiency. Its weak point is the absence of an enterprise resource potential development assessment which could be used as output data for optimization.
S.M. Isa [3]	Holistic linear optimization model for resources leveling (single resource, continuous activity) in construction which guarantees optimal leveling. Its advantages is that the model target function minimizes absolute deviations between needs for resources and an even resources level, between consecutive needs for resources, or between requirements for resources and the desired uneven resources levels.
T. Khegazi, M. Kassab [4]	The method of combined simulation and genetic algorithms is employed for resources optimization in construction. The model determines the least cost-demanding and the most productive resources quantities which ensure the highest benefit/expenses ratio in specific construction operations.
O.V. Ulyanchenko [5]	Multi-criteria function optimization by the established limitations system. The advantage of this model is that it covers all types of resources and takes into account maximal number of internal context factors and environmental factors that produce their effect. Its weak point is the absence of mechanism for planning initial changes for development of the enterprise resource potential.
O.I. Shamanska [6]	The method of regressive analysis is employed with respect to resource potential of the industrial enterprises. Its criterion is profit maximizing depending on the resource potential composition. The model advantage is the possibility for an enterprise profit forecasting. The weak point is the absence of the procedure for selection of initial actions aimed at resource potential development.

Thus, the above scientists were engaged in solving resource potential optimization problems using different mathematical methods, such as regressive analysis, multi-criteria optimization, genetic algorithms and other. However, the issues of managerial procedure development remained unresolved; using this procedure a comprehensive approach to optimization of resource potential development planning would be carried out, as well as a more precise diagnosticating of the 'bottle necks' in the enterprise resource potential management.

3 Purpose of the article

The purpose of the article is to develop a comprehensive procedural approach to optimization of resource potential development planning to enhance a quality of resource potential development management of railway transport enterprises.

4 Presentation of the main material

When resolving the tasks of the market economy with the aim of determining efficient managerial decisions with respect to the object of management, a decision maker (hereinafter DM) is guided by their own system of preferences, based on which this person chooses the task solution, even if the presence of such system is not always comprehended. In any case, the DM preferences system and its related information are either specified or changed during the task study which implies finding additional data with regard to the tasks properties. Therefore, it is very important to have means in order to formalize DM advantages, or at least to approximate its advantages [7].

The theory of multi-criteria optimization serves as a basis for development of methods to support decisions taking in the case, when the solution choice is performed in accordance with several criteria, however, it does not substitute the methods for solution choice. This refers to methods of multi-criteria optimization as well, and to methods of selection from a few alternatives number. Generally, the multi-criteria task is viewed as a task of simultaneous optimization of several target functions on the set multiplicity of permissible plans [8]:

$$\left.\begin{array}{c} y_k = f_k(x) \rightarrow opt,\ k = \overline{1,p}, \\ x \in X, \end{array}\right\} \quad (1)$$

where p – the number of target functions, which are subject to optimization;

f_k– individual k-function from criterial set (k=$(\overline{1,p})$);

X– multiplicity of permissible plans, whose individual element is designated as x.

Planning is a process which includes determining of purposes or the organization's tasks, establishing a general strategy to achieve these aims and development of multilateral hierarchical structure of plans for unifying and coordination of the entire enterprise activity [9]. Need for optimization of planning resource potential development management is caused by the fact that during the planning process a choice of managerial decision is effected, which should be based on objective assessment of resource potential development condition, be the most effective under given conditions and to maximally comply with determined aims and tasks of the enterprise. Thus, a need for procedural approach emerges; and this approach should be a reliable tool in the enterprise top management hands and would allow resolving planning tasks on the enterprise resource potential development management in a comprehensive manner. In this context, the authors propose the following planning optimization procedure which is composed from several consecutive stages (Fig. 1).

At the first stage, diagnosticating of resource potential development level is carried out for railway transport enterprise using the author's procedure to assess resource potential development level [10]. Diagnosticating is carried out by determining of integral factor with application of hierarchies analysis procedure to determine specific weight of each resource potential component and of interpretation of integral factor values based on Harrington verbal-numerical scale.

At the second stage depending on the obtained value of resulting factor, a level of resource potential is studied using the universal Harrington verbal-numerical scale. This scale helps to establish correlation between physical (numerical) and psychophysical parameters (high/low, good/bad, big/small).

At the third stage, the direction of optimization subject to resource potential development is chosen: one-criterion optimization or multi-criteria optimization. For the mid, high and very high levels (0.37 ≤ K general level of development RP ≤ 1.00) one-criterion optimization direction is chosen, since for such obtained level no substantial measures are required for enhancing resource potential development level. For the low, and very low levels (0 ≤ K general level of development RP ≤ 0.37) – the direction of multi-criterial optimization, since these both levels require significant intrusions in the process of resource potential development management.

At the fourth stage, formation of possible development plans options occurs. It is expedient to carry out this process using answers to such key questions:

1. Where are we now? Condition and possibilities of the organization are assessed at the moment.
2. Where do we intend to move? Organization's intentions and circumstances of its environment that are capable of affecting its intentions implementations, are specified.

Stage 1. Diagnosticating of resource potential development level of the railway transport by calculating the comprehensive integral factor to assess resource potential development level of the railway transport: $K_{overall.lev.RP.dev.} = \sqrt[4]{K_m * \omega_m * K_t * \omega_t * K_f * \omega_f * K_i * \omega_i}$, where: K_m – group factor for material potential assessment; K_t – group factor for labor potential assessment; K_f – group factor for financial potential assessment ; K_i – group factor for informational potential assessment; ω_m – specific weight of the resource potential material component; ω_t – specific weight of the resource potential labor component; ω_f – specific weight of the resource potential financial component; ω_i – specific weight of the resource potential informational component.

Stage 2. Determining of resource potential development level based on generalizing factor with respect to the below scale:
0.63 ≤ K general level of development RP ≤ 1.00 – high – very high level; 0.37 ≤ K general level of development RP ≤ 0.63 – mid level; 0 ≤ K general level of development RP ≤ 0.37 – very low – low level

Stage 3. The choice of direction of optimization subject to resource potential development:

Resource potential development level:	**Optimization direction**
0.63 ≤ K general level of development RP ≤ 1.00 – high-very high	one-criterion optimization
0.37 ≤ K general level of development RP ≤ 0.63 – mid high	one-criterion optimization
0 ≤ K general level of development RP ≤ 0.37 – very low-low	multi-criteria optimization

Stage 4. Generation and analysis of possible options for determining potential possibilities of positive changes in railroad transport resource potential.

1) Answer to three key questions: 1. Where are we now? 2. Where do we intend to move? 3. How are we going to do this?
2) The choice of optimal strategy for resource potential development using formula:

$$j^* = \arg\max_j \left\{ \sum_{i=1}^{j} p_i(k_{ij}) \right\}, i = \overline{1,m}, j = \overline{1,n}$$

j* - optimal strategy; j – strategies available;
$p_i(k_i)$ – function that normalizes all partial assessment into a single interval (k_{ij}) and is determined as: $p_i(k_{ij}) = \left(\frac{k_{ij} - k_{i\Gamma}}{k_{iK} - k_{i\Gamma}}\right), i = \overline{1,m}, j = \overline{1,n}$,
where k_{ij}- value of correspondence to assessed j strategy of i purpose;
k_{iK}, $k_{i\Gamma}$ – the best and the worst value of partial criterion respectively, which it acquires in the field of permissible solutions

Stage 5. Alternative optimization options for the choice of the best option for development plan. Optimizations task generation:

One-criterion optimization

$f(x) \to max(min), x \in X$

f (x) – target function that describes criteria for solution choice
X – multiplicity of permissible plans, whose individual element is designated as X

Multicriteria optimization

$$\left. \begin{matrix} y_k = f_k(x) \to opt, k = \overline{1,p} \\ x \in X \end{matrix} \right\}$$

y_k – plan assessment by k-criterion
p – number of target functions that are subject to optimization;
f_k - individual criterial set k-function ($k=\overline{1,p}$))
X – multiplicity of permissible plans, whose individual element is designated as x

Stage 6. The search of effective X plan that complies with real permissible levels $\xi_k^{*0} \epsilon [y_k^0; y_k^*)$, $k = \overline{1,p}$ of all criterial factors. It is a solution for one-criterion task:

$$\left. \begin{matrix} u = \sum_{k=1}^{p} \frac{y_1^k - y_1^o}{y_1^* - y_1^0} * \frac{f_k(x)}{y_k^* - y_k^0} \to max \\ \frac{f_k(x) - \xi_k}{y_k^* - \xi_k} \geq t^*, k = \overline{1,p} \\ x \in X \end{matrix} \right\}$$

where, u is generalized value additive function; $f_k(x)$ is specific k-target function of criterial set ($k = \overline{1,p}$); ξ_k is permissible level of k-target function; y_k^* is optimal values of k-target function on the multiplicity of effective plans; y_k^0 is the worst value of k-target function on the multiplicity of effective plans

Stage 7. Assessment of financing actions for resource potential development

Do the actions aimed at resource potential development comply with financing possibilities?

$f_k(x) \leq A$, where A – a specified costs amount for resource potential development

$$x_k = \begin{cases} 1, & \text{if k – plan complies with financing capabilities} \\ 0, & \text{in the opposite case} \end{cases}$$

No (0) – return to Stage 4

Yes (1)

Stage 8. Confirmation of the final option of resource potential development

Stage 9. Implementation of the plan for resource potential development and carrying out control of its implementation

Fig. 1. Model of activity planning optimization for development of railroad transport resource potential in the industry enterprises

3. How are we going to do this? Possible ways, methods, means that are chosen to achieve the set goals, are determined.

Further, the choice of optimal strategy for the top priority changes is carried out using the group-based system of partial factors system (material, labor, financial, informational) for development of the enterprise resource potential using formula [10]:

$$j^* = \underset{j}{\operatorname{argmax}}\{\textstyle\sum_{i=1}^{j} p_i\,(k_{ij})\}, i = \overline{1,m}, j = \overline{1,n}, \quad (2)$$

where j* – optimal strategy; j – strategies available;

$p_i\,(k_i)$ – function that normalizes all partial assessment into a single interval (k_{ij}) and is determined as: $p_i(k_{ij}) = \left(\frac{k_{ij}-k_{i\Gamma}}{k_{iK}-k_{i\Gamma}}\right), i = \overline{1,m}, j = \overline{1,n},$

where k_{ij} – value of correspondence to j-strategy of-i purpose assessed

k_{iK}, $k_{i\Gamma}$ – the best and the worst value of partial criterion respectively, which it acquires in the field of permissible solutions.

At the fifth stage, alternative options for optimization are determined to ensure the choice of the best option for development plan and one-criterion and multi-criteria optimization is carried out subject to the choice made at the third stage of procedural approach.

One-criterion task is generated in the following way:

$$f(x) \to \max(\min), x \in X, \quad (3)$$

where $f(x)$ – target function which describes the solution choice criterion;

X – multiplicity of permissible plans, whose individual element is designated as x.

The multi-criteria task is viewed as a task of simultaneous optimization of several target functions on the set multiplicity of permissible plans:

$$\left.\begin{matrix} y_k = f_k(x) \to opt, k = \overline{1,p} \\ x \in X \end{matrix}\right\} \quad (4)$$

where p – the number of target functions, which are subject to optimization;

f_k – individual k-function of the criterial choice ($k = \overline{1,p}$);

X – multiplicity of permissible plans, whose individual element is designated as x [7].

Permissible plans are compared via correlation of their assessments, in which case, an arbitrary admissible plan assessment of multicriteria task is vectorial one:

$$\mathbf{y} = (y_1, \ldots, y_p). \quad (5)$$

A person who takes managerial decisions determines such permissible levels ξ_k, which he/she thinks to be satisfactory.

At the sixth stage, the search of such effective X plan that complies with real permissible levels of all criterial factors takes place. It is a solution to one-criterion task [7]:

$$\left.\begin{matrix} u = \sum_{k=1}^{p} \frac{y_1^k - y_1^o}{y_1^* - y_1^0} * \frac{f_k(x)}{y_k^* - y_k^0} \to max \\ \frac{f_k(x)-\xi_k}{y_k^*-\xi_k} \geq t^*, k = \overline{1,p} \\ x \in X \end{matrix}\right\} \quad (6)$$

A conclusion is made which is the result of the stage performance as to reality or unreality of initial permissible levels (value t^*) and real permissible levels $\xi_k^{*0} \in [y_k^0; y_k^*)$, $k = \overline{1,p}$, and an assessment $y^2 = f(x^2)$ with recommendation as to approval of x plan as a solution of the multicriteria task [7].

The reality of the determined criterial factors levels is effected using the solution to one-criterion task:

$$\left.\begin{array}{c} t \longrightarrow max, \\ \frac{f_k(x)-\xi_k}{y_k^*-\xi_k} \geq t, k = \overline{1,p} \\ x \in X. \end{array}\right\} \quad (7)$$

Case $t^* \geq 0$ testifies in favor of the permissible levles reality, while case $t^*<0$ – as to their unreality.

At the seventh stage an assessment of possibility for financing actions for resource potential development is carried out. If the target expenditures function is less or equal to A (A is costs amount specified allocated for resource potential development), the plan is chosen and approved at the eighth stage. In case the amount exceeds the planned one, return to stage 4 is performed and the choice of a new strategy of resource potential development is effected due to insufficient financing.

At the eighth stage the final approval of the plan chosen at a previous stage, is carried out.

At the ninth stage implementation of resource potential development plan is carried out and control of its implementation is effected.

5 Conclusions

Thus, the presented model allows implementing comprehensive approach to optimization of resource potential development planning of railway industry enterprises. These procedural recommendations are more precise compared with other existing procedural recommendations for resource potential optimization, since they provide for diagnosticating “bottlenecks” in the enterprise resource potential and choose more adequate actions from the enterprise resource potential development proceeding from its financial capabilities. Implementation of proposed procedural recommendations as to optimization of resource potential development planning would encourage raising efficiency of resource potential development management of railway transport enterprises.

References

1. A. Ya. Bersutskyi, *Models of decision-making on development management of resource potential of enterprise*, Donetsk, 16, 29-33 (2010)
2. D.V. Volovyk, Ukr. scien., Ternopil’, **19**, **1**, 11-16 (2014)
3. Said M. Easa, *Resource Leveling in Construction by Optimization.* Journ Constr. Engin Manag. **115, 2** (1989)
4. T. Hegazy, M. Kassab, Journ. Constr. Engin. Manag., **129**, **6**, (2003)
5. O.V. Ulyanchenko, *Management of resource potential in the agrarian sector* Kharkiv, 302-322 (2007)
6. O.I. Shamanska, Effect. econ., Dnipropetrovsk, **9** (2011)
7. V.R. Kihel, *Methods and models of decision making in a market economy*, Kyiv, Economic Institute of Management, (2003)
8. S.P. Robbins, M. Coulter, *Management.* Moscow, Williams Publishing House, (2007)
9. A.V. Lotov, I.I. Pospelova, *Multi-criteria decision making tasks.* MAKS Press, Moscow, (2008)
10. L.L. Kalinichenko, O.M. Chupyr, N.V. Yanchenko, V.M. Bredikhin, Ye. O. Burlaka, IJET, **7**, **4**, 5302-5312 (2018)

Theoretical aspects of determining the types of customer loyalty

Iryna Fedotova*, *Oksana* Kryvoruchko and *Volodymyr* Shynkarenko

Kharkiv National Automobile and Highway University, Department of Management and Administration, 61002 Ya. Mudrogo str., 25, Kharkiv, Ukraine

Abstract. An approach to identifying types of customer loyalty which is based on the customer loyalty matrix in terms of attitudinal and behavioral loyalty has been further developed. The proposed criteria are divided into three levels: high, medium and low. The improved customer loyalty matrix includes nine types of loyalty: true, active, latent, projected, neutral, basic, spurious, ultimate and no loyalty. Based on the loyalty matrix, a gradation of consumers according to the type of loyalty is made, which enables to develop a further strategy to increase the loyalty of an individual consumer.

1 Introduction

In modern business, the level of competition is quite high, so every company seeks to attract loyal customers to ensure their success. Information about the degree of customer loyalty is an indicator of how well or poorly an enterprise meets the needs of its customers. Therefore, it is strategically important to understand correctly what different consumers say. For the enterprise long-term relationships with the consumer together with an increased volume of transactions and their frequency are the most profitable; therefore, concentrating strategic efforts on building customer loyalty to the enterprise and its services (goods) is indisputably beneficial.

Before choosing a strategy of forming customer loyalty, it is necessary to determine the type of customer loyalty. The ambiguity of approaches to the formation of customer loyalty and contradictions in distinguishing its types require clarification of the conceptual framework of loyalty marketing.

2 Literature review

The issues of determining the type of customer loyalty were given considerable attention in the works of scientists [1-17] and others. In the marketing literature, three most common approaches to defining loyalty can be distinguished.

The first is based on the consideration of loyalty as a certain type of consumer behavior, expressed in long-term interaction with the company and making repeated purchases. This is the most popular point of view on loyalty, since the behavioral aspect of loyalty is closely related to the economic results of the enterprise. This type of loyalty is called behavioral (transactional) loyalty [1-4].

* Corresponding author: irina7vf@gmail.com

According to the second approach, loyalty is considered as the attitude or preference of consumers, which is formed as a result of the generalization of feelings, emotions, opinions about the service (or its supplier). This type of loyalty is called attitudinal (perceptive, perceived) loyalty [5-7].

The third approach suggests that loyalty is determined by a combination of behavioral and attitudinal characteristics. The combination of these aspects is called complex loyalty [8-13].

Most authors, when identifying loyalty types, take into account the behavioral and attitudinal characteristics of the consumer. J. Hofmeyr, B. Rice [8] introduce three types of loyalty based on the delimitation of the concepts of "loyalty" and "commitment": 1) loyalty without commitment; 2) commitment without loyalty; 3) loyalty + commitment. Here we consider separately the behavior and perception of the consumer, as well as the combination of these categories.

Within these two categories of loyalty, the author [9] highlighted four essential types of loyalty: contractual, transactional, functional

Categories of loyalty proposed by A. Dick and K. Basu [10] are defined by the relationship between attitudinal and behavioural aspects of loyalty. Antecedents to loyalty include cognitive, affective and conative factors. They described loyalty as the strength of the relationship between a customer's relative attitude and repeat patronage, four dimensions had been identified here: true loyalty, latent loyalty, spurious loyalty and no loyalty.

The author's classification [11] is based on three basic approaches to loyalty: loyalty as behavior, loyalty as attitude and loyalty determined by the circumstances of the purchase.

Several authors consider loyalty types in the form of successive stages. Customer loyalty development had been categorized into four sequential phases by R. Oliver [12]: cognitive loyalty, affective loyalty, conative loyalty and action loyalty. A. Bakanauskas and L. Pilelienė [13] identified four stages: neutral, potentially loyal, unconsciously loyal and sincere loyal.

Some authors distinguish consumer groups depending on their behavior and loyalty status. J. Rowley identified four types of loyalty: captive, convenience-seekers, contented and committed [14].

Philip Kotler [15] defines four statuses of loyalty:

1. Hard-core Loyals - who buy the brand all the time.
2. Split Loyals - loyal to two or three brands.
3. Shifting Loyals - moving from one brand to another.
4. Switchers - with no loyalty (possibly «deal prone», constantly looking for bargains or "vanity prone", looking for something different).

According to the neobehavioristic concept of customer loyalty, customer loyalty is expressed as: 1) behavioral loyalty, which reflects the actual brand-specific product / service and the number of purchases within a specified period of time; 2) approach loyalty, which reflects the preference for certain goods [16].

Comparing the considered approaches it is possible to make a conclusion that the selected types of loyalty correlate with each other to various extents. A review of existing classifications of consumer loyalty showed that the principle of identifying loyalty types is not entirely clear, there is no single approach in identifying classification indexes and criteria. Existing methods for determining the types of customer loyalty do not fully solve the problem of the possibility of their clear delimination. Each author divides loyal customers differently, there is no single method for identifying loyalty types, according to which the buyer can be classified.

Therefore, the classification of the types of customer loyalty should be systematized depending on the combination of behavioral and attitudinal aspects.

3 Purpose of the article

The purpose of the article is to improve the approach to identifying the types of customer loyalty by the criterion of the level of behavioral and attitudinal loyalty. The basis for solving this problem is ranging consumers according to the loyalty type by building a loyalty matrix based on behavioral and attitudinal aspects. Dividing consumers into loyalty types will help determine further strategic actions to build customer loyalty and improve the company's relationship with the consumer.

4 Results and discussion

Conducting visual definition of the type of customer loyalty on the basis of building a customer loyalty matrix is proposed. For the most complete and comprehensive assessment, using an approach that involves monitoring of both behavioral and attitudinal customer loyalties is most appropriate [17].

The approach proposed by the authors [11, 18] is of great interest. It highlights the types of customer loyalty based on building up a loyalty matrix, where the axes represent behavioral and attitudinal loyalty. The proposed criteria are divided into two levels – high and low, and at the intersection of these levels, four main types of loyalty are distinguished: true, latent, spurious, no loyalty. The disadvantage of this approach is the enlarged division of criteria into two levels, which entails the absence of a smooth transition of consumer loyalty from one type to another. This causes the difficulty of defining boundaries, since the consumer, having a behavioral or attitudinal loyalty value slightly below the average level, is considered disloyal, but with a slight excess of the average level of indicators his loyalty becomes absolute (true). In order to avoid these shortcomings in the classification of loyalty types, it is proposed to divide the indicators of attitudinal and behavioral loyalty into three levels – high, medium and low.

The matrix is built in a two-dimensional plane formed by indicators of attitudinal and behavioral loyalty which are applied to the coordinates of the final indicators characterizing the loyalty of each consumer. As a result of constructing the matrix (Fig. 1), five new types of loyalty are added to the four types of consumer loyalty described.

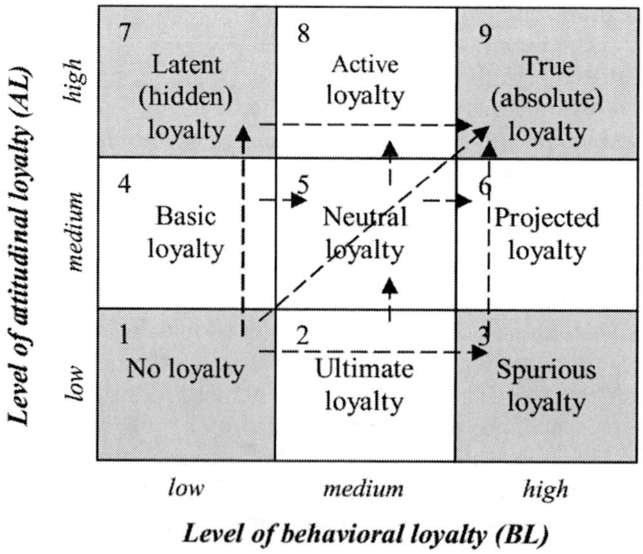

Fig. 1. Matrix of customer loyalty types.

The matrix space is divided into nine cells, the consumers who fall into one of these sectors can be grouped together, and, based on this, develop loyalty programs separately for each sector. The existing types of loyalty are flooded in gray, and the unpainted white cells are the proposed transitional types of customer loyalty. Let us consider them in more detail:

- *True (absolute) loyalty* is a situation in which a high level of attitudinal loyalty corresponds to a high level of consumer behavioral loyalty and is most favorable for the enterprise. This is the most stable part of the clientele, which is the least sensitive to the actions of competitors (price lowering, introduction of additional amenities, etc.). Customers with absolute loyalty are much easier to retain. Moreover, this may be sufficient for maintaining existing quality standards;
- *Active loyalty* is specific for the consumer when the consumer is emotionally attached to the enterprise, distinguishes it from a number of competitors and has an average level of consumption of services. The company needs to strengthen and maintain the position it has reached, and to increase behavioral loyalty through various activities;
- *Latent (hidden) loyalty* means that a high level of attitudinal loyalty is not supported by consumer behavior. That is, they differentiate this enterprise among the competitors, but purchase its products or services less frequently or not in such quantity as the absolutely loyal consumers. The reasons for this include, first of all, external factors, for example, insufficient income, high price, legal barriers (for example, there may be long-term contracts with other suppliers). In such a situation, the organization needs to strengthen its position through the development of behavioral loyalty. For this purpose, for example, price incentives can be used;
- *Projected loyalty* means that with a high level of consumption of services by the consumer, the attachment to the enterprise grows. It is quite favorable, since when it affects the level of attitudinal loyalty it becomes absolute loyalty;
- *Neutral loyalty* takes place at medium levels of attitudinal and behavioral loyalty, that is, the consumer is interested in interacting with the enterprise, which is supported by his behavior. This interest should be maintained and developed, the quality of services must be improved, and the client should be proposed some special offers;
- *Basic loyalty* is a type of loyalty in which the level of consumption of services does not change and remains low, but there is a tendency to improve consumer perception of the enterprise. In such a situation, an enterprise needs to find out the reasons for low behavioral loyalty and implement programs aimed at its improvement. Along with this, strengthening and developing the attitudinal customer loyalty is advisable;
- *Spurious loyalty* occurs when behavioral loyalty corresponds to a low level of attitudinal loyalty. This situation is threatening, since the consumer is not emotionally attached to the enterprise. Their purchases may be due to limited offers, seasonal or cumulative discounts, habits, lack of awareness of competitive offers, therefore, as soon as consumers find an enterprise that satisfies them more (for instance, a competitor offers more significant discounts), they will refuse from service. In order to retain consumers demonstrating this type of loyalty, the attitudinal loyalty needs to be bolstered;
- *Ultimate loyalty* is a situation in which the average level of behavioral loyalty corresponds to a low level of attitudinal loyalty. The emergence of this type of customer loyalty is not entirely desirable for the enterprise, since development in this direction leads to the appearance of false loyalty. The growth in consumption of enterprise services may mislead contact personnel about the true nature of the buyer's motives. In cooperation consumers may be interested in discounts or bonuses provided by the company, without experiencing emotional attachment the consumer remains indifferent, and can go to competitors as soon as they offer more favorable conditions of interaction. In this case, developing the attitudinal customer loyalty is necessary;

- *No loyalty* provides minimal retention. The enterprise should either refuse to keep this part of consumers (very beneficial for it, by the way), or take special measures to increase, above all, the attitudinal loyalty. This includes the competitors' customers that correspond to both true and false loyalty groups.

In Fig. 1 the arrows show the possible prospects for the development of customer loyalty. Based on the prospects of developing relations with each consumer, that are potential and desirable for the enterprise, the company develops a number of strategic actions (loyalty development programs) for the formation of consumer loyalty, given in Table 1.

Table 1. Determination of strategic actions of an enterprise according to consumer loyalty groups.

Loyalty type	Strategic actions to develop customer loyalty
1. No loyalty	Development of new ways and forms of stimulating consumer choice, individualization of the services offered, increased value of the service for the consumer
2. Ultimate loyalty	Marketing loyalty programs that create a permanent long-term incentive to buy and rely not only on financial but also on emotional motivation
3. Spurious loyalty	Building customer relationships based on emotions and trust, offering unique and highly valued privileges within the program
4. Basic loyalty	Implementing consumer incentive programs, replacing short-term stocks with a long-term partnership program with the consumer, increasing the value of the service for the consumer
5. Neutral loyalty	Introduction of programs to encourage customer activity of the client, the formation of the consumer's emotional attachment to the enterprise
6. Projected loyalty	Maintaining the achieved level of interaction, increasing the consumer's emotional motivation
7. Latent (hidden) loyalty	Programs that support the level of attitudinal loyalty, encouraging consumer to increase orders, expanding the range of purchased services
8. Active loyalty	Programs that encourage customers for loyalty to the company, material motivation of the consumer to increase orders
9. True (absolute) loyalty	Programs that encourage customers for loyalty to the company, maintaining the level of purchases, maintaining customer loyalty

The study will provide a basis for assessing the level of customer loyalty and identifying which type of loyalty each client belongs to, it will also enable to consider specific groups of consumers in terms of their contribution to the enterprise. Using the methodology for assessing the integral customer loyalty index of motor transportation services using the fuzzy set theory formalisms described in [19], the calculation procedures are carried out to determine the level of attitudinal and behavioral loyalty of each service consumer, and the proposed matrix will allow characterizing the type of loyalty of the client in question. In the interval [0, 1] the values of the membership function lie, which reflect fuzzy consumer assessments. Therefore, the loyalty matrix axes are in the interval from 0 to 1, and are divided into three loyalty levels in equal parts.

Grouping customers by type of loyalty enables to identify the most loyal customers who are most valuable to the enterprise, to develop measures to reduce the number of non-loyal customers and assess customer loyalty in more detail. According to the consumer's type of loyalty, the loyalty development programs are elaborated.

5 Conclusion

Thus, the approach to identifying the types of customer loyalty has been further developed by adding an average level of attitudinal and behavioral loyalty, resulting in five new transitional loyalty types. The customer loyalty matrix has been improved, which includes nine types of loyalty, based on it a gradation of customers according to the type of loyalty is made. With the proposed loyalty matrix, marketers can effectively segment their customers based on the type of loyalty that they show and, therefore, develop effective loyalty programs to elaborate and strengthen the existing loyalty. This is instrumental in developing a further strategy to increase customer loyalty, which improves the efficiency of operations and the competitiveness of the enterprise itself.

References

1. F.F. Reichheld, *The Loyalty Effect: The Hidden Force Behind Growth, Profits, and Lasting Value.* Harvard Business School Press (1996).
2. D.A. Aaker, *Managing Brand Equity.* New York: Free Press, (1991).
3. S.A. Taylor, G.H. Hunter, T.A. Longfellow. Journal of Consumer Satisfaction, Dissatisfaction and Complaining Behavior, **19**, 18-39 (2006).
4. J.-J. Lambin, *Strategic marketing: a European approach.* London; New York: McGraw-Hill, (1993).
5. R.F. Blanchard, R.L. Galloway. International Journal of Service Industry Management, **5, 4**, 5-23 (1994).
6. R. Hallowell. International Journal of Service Industry Management, **7, 4**, 27-42 (1996).
7. R.G. Javalgi, C.R. Moberg, Journal of Services Marketing, **11, 3**, 165-179 (1997).
8. J. Hofmeyr, B. Rice, *Commitment-Led Marketing.* John Wiley and Sons, (2000).
9. J. McKean, *Managing Customers Through Economic Cycles.* John Wiley & Sons, (2010).
10. A.S. Dick, K. Basu, Journal of the Academy of Marketing Science, **22**, 99-113 (1994).
11. M.O. Uncles, R. Dowling, K. Hammond, Journal of Consumer Marketing, **4**, 294-316 (2003).
12. R.L. Oliver. Journal of Marketing, **63**, 33–44 (1999).
13. A. Bakanauskas, L. Pilelienė, *Vartotojų lojalumas: teoriniai ir praktiniai aspektai.* Kaunas: Vytauto Didžiojo universitetas, (2009).
14. J. Rowley, Marketing Intelligence & Planning, **23 (6)**, 574-581 (2005).
15. P. Kotler, K.L. Keller, *Marketing management.* Global, Harlow, (2016).
16. K. Zikienė, V. Brazinskaitė, Organizacijų vadyba: sisteminiai tyrimai, **63(8)**, 113-124 (2012).
17. E.M. Kiseleva, M.L. Nekrasova, M.A. Mayorova, M.N. Rudenko, V.S. Kankhva. International Review of Management and Marketing, **6(S6)**, 95-103 (2016).
18. S. Tanford, S. Baloglu. Cornell Hospitality Quarterly, **54 (4)**, 333-346 (2013).
19. I. Fedotova, O. Kryvoruchko, V. Shynkarenko, N. Bocharova, L. Sotnychenko, S. Dimitrakieva. Eastern-European Journal of Enterprise Technologies. Control processes, **3/3(99)** (2019).

Modeling the effect of traffic jam on driver's level of fatigue

Nizami Gyulyev [1], *Oleksii* Lobashov [1], *Oleksii* Prasolenko [1], and *Igor* Bugayov [1]

[1] O. M. Beketov National University of Urban Economy in Kharkiv, Ukraine

Abstract. This study focuses on establishing the dynamics of changes in the level of fatigue of an average driver in a traffic jam, which affects road safety. Linear and nonlinear models of the effect of traffic jam on the level of fatigue of an average driver have been developed, which allow assessing their condition depending on the driver's age and the duration of the jam. Using a nonlinear regression model, the graphs of changes in the level of fatigue of an average driver of 25 and 65 years are plotted. The analysis of these graphs allowed us to establish the patterns of influence of road parameters on the functional state of the driver. The level of fatigue of the driver, which is determined by the change in his functional state, increases during his stay in the traffic jam. Therefore, the duration of the jam and the initial level of fatigue when entering the jam influence most the final level of fatigue of the driver. The driver's age affects the level of fatigue to a lesser extent, but still significantly.

1 Introduction

The increase in the level of motorization leads to saturation and oversupply of the urban transport system with traffic flows. This leads to traffic which excesses the capacity of the road network and the emergence of queues of vehicles and traffic jams, especially at the road intersections in the morning and evening peak periods. Basically this interaction is random. The traffic flow may be also considered as a complex system that needs to be constantly managed. The driver is the active element of this system, and his state determines the efficiency of the entire transport system of the city to a large extent [1-3].

The traffic flow consists of cars that have different dynamic characteristics and are driven by drivers with different psycho-physiological features [4]. Drivers manage vehicles at their discretion in accordance with the chosen goals and objectives. The presence of the human factor in the transport stream, as the main element of the system "human – technology – environment", and its manifestations, which are difficult to predict, determine the likelihood of the flow [5]. Traffic jams at intersections during peak periods significantly increase travel time and reduce the speed of traffic. Being in traffic jams adversely affects the psychophysiological state of the driver, causing deterioration of his functional state and a number of mental disorders [6–9]. Drivers of non-route vehicles try to choose such a route that would ensure minimum time and comfortable conditions of the trip [10, 11].

Negative emotions caused by excessive restriction of the driver's freedom of activity and the unsatisfactory state of the traffic organization lead to a violation of the driver's mode of work, his emotional stress and, ultimately, to the development of fatigue. Excessive level of fatigue leads to an increase in the driver's reaction time and an increase in the probability of a traffic accident [2, 3, 12–15]. The purpose of the study is to simulate the effect of traffic jam on the functional state of the driver. To achieve it, the following tasks were set: selection and justification of the method for assessing the level of fatigue and developing a model for changing the level of fatigue of the average driver.

2 Literature review

Road safety is determined by the reliability of the "driver – car – road – environment" system and above all the reliability of the most important element of this system – the driver. The authors of the works [1, 13] stressed the need to take into account the role of the human factor in the traffic organization. It was noted that the design and operation of the transport system must take into account the capabilities of a person in the performance of their duties. There are a number of factors that can lead to accidents, as well as measures to improve the road safety. One of the most significant factors that have a negative impact on the functional state of the driver is the road jam [16, 17]. Such qualities of a driver as reaction time, work experience, temperament were considered by the authors of works [2, 19] in their studies. The growth of emotional tension leads to a temporary breakdown of some mental functions of the driver, increasing the time of his reaction [17–19]. However, in these works the issues of the influence of the initial state of drivers on the change in their final functional state during their stay in traffic jams and the influence of the driver's age on the dynamics of changes in his state in a traffic jam are not fully considered.

3 Materials and methods of studying the impact of traffic jam on the driver's level of fatigue

3.1 The categories of drivers and equipment used in the experiments

The conducted experimental research involved drivers of all ages and categories. To develop a regression model of the average driver, the drivers of all types of the nervous system were examined.

Of all the psycho-physiological methods for solving described problem, the most suitable method for assessing the functional state is usage of electrocardiogram registering. In psychophysiology, an electrocardiogram serves as the main indicator of a person's emotional state during physical and mental stress [13]. The electrocardiogram is fixed by a "Cardiosens" portable electrocardiograph with autonomous power supply.

In the presented study, the general population consists of all the drivers of the city, who use transport means during the trip. The correspondence of the sampling structures and the general population was established according to the type of the nervous system. For the quantitative compilation of the sample and the determination of its conformity with the general population, the research results given in the works [20, 21] were used. The work [21] presents the results of studies on the type of the nervous system with percentage share of respondents: sanguine – 27.2%; choleric – 30.3%; phlegmatic – 27.2%; melancholic – 15.3%. The results close to these are also presented in [20], with the research sample: sanguine persons – 26–30%; choleric – 28-31%; phlegmatic – 25-27%; melancholic – 15-19%. These ratios were taken as the basis for the research.

3.2 Methodology of conducting experimental research

Experimental studies consist in fixing the electrocardiogram of drivers at the entrance to the traffic jam, in the jam and at the exit from it.

The sequence of cardiointervals of an electrocardiogram is coded information about the processes that take place not in the heart itself, but in different parts of the control system: the nerve plexuses; endocrine glands; nerve centers located deep in the brain. The structure of the heart rate can be judged by the state of the regulation mechanisms. This is an integral assessment of body fatigue. The methodology is based on the theory of directly proportional dependence of irregularity of cardiointervals and level of fatigue, which in literary sources is often called an indicator of the activity of regulatory systems of the body. It is calculated in conditional units (according to R. M. Baievskyi) using a special algorithm that takes into account statistical indicators, histogram indicators, and data from spectral analysis of cardiointervals. The level of fatigue allows us to differentiate various degrees of tension in the regulatory systems and evaluate the adaptive capacity of the organism [22]. It is calculated according to an algorithm that takes into account five criteria: the total effect of regulation; the automatism function; the vegetative homeostasis; regulation stability; activity of the subcortical nerve centers. Depending on the level of fatigue, it is determined what state a person is in: up to 3 points is the normal state; from 3 to 6 points is the state of stress; from 6 to 8 points is the state of overstrain [22].

4 Results of studies on the effect of a traffic jam on the level of driver's fatigue

The task of developing a regression model of the impact of a traffic jam on the functional state of the driver is creating a tool that will allow making correct, reasonable choice of the object of research and determine a set of factors that affect the behaviour of the object. Drivers of different temperaments act as an object within this research. Factors influencing the outcome of the study should be carefully selected. The following factors were taken into account when compiling the regression model: driver's age, driver's work experience, number of lanes on the road, comfort of the car, duration of stay in the traffic jam, the values of the driver's functional state before the stop.

The developed model has the following form:

$$L_{fex} = 0.037 \cdot D_a + 0.135 T_j + 0.61 \cdot L_{fen}$$

where L_{fex} is fatigue level of while exiting the traffic jam, c. u.; D_a is the driver's age in years; T_j is driving duration in the traffic jam, min; L_{fen} is level of fatigue while entering the traffic jam, c. u.

The results of calculations of the parameters of the model are given in Table 1.

Table 1. Characteristics of model of the change in the functional state of a driver in a traffic jam.

Factors	Marking, dimension	Measurement limits	Coefficient	Standard mistake	Student criterion
Driver's age	D_a, years	19-67	0.037	0.006	5.84
Duration of traffic jam	T_j, min	2-20	0.135	0.018	7.18
The level of fatigue at the entrance to the traffic jam	L_{fen}, conditional units	2.2-5.8	0.617	0.055	11.24

In order to improve the description of the impact of traffic jam on the functional state of the driver, a non-linear model was developed:

$$L_{fex} = 0.018 \cdot D_a + 1.278 \cdot T_j^{0,41} + 0.291 \cdot L_{fen},$$

where L_{fex} is fatigue level of while exiting the traffic jam, c. u.; D_a is the driver's age in years; T_j is driving duration in the traffic jam, min; L_{fen} is level of fatigue while entering the traffic jam, c. u.

The results of the calculation of the parameters of the model are given in Table 2.

Table 2. Characteristics of model of the change in the functional state of a driver in a traffic jam.

Factors	Marking, dimension	Measurement limits	Coefficient	Standard mistake	Student criterion
Driver's age	D_a, years	19–67	0,018	0,006	2,98
Duration of traffic jam raised to 0,41 degree	T_j, min	2–20	1,278	0,122	10,48
The level of fatigue at the entrance to the traffic jam	L_{fen}, conditional units	2,2–5,8	0,291	0,066	4,36

It is clear from Table 2 that in the developed mathematical model only three factors turned out to be significant. Their significance is indicated by the excess of the calculated value of the Student's t-test over the tabular.

The adequacy of the developed mathematical model was estimated using the average approximation error of 19.28%. This error is large and therefore the use of the developed model for calculations is possible only in the first approximation.

The non-linear model has approximately the same error as the linear one, so any of them can be used for calculations. This is explained by the fact that during their stay in traffic jams, the functional status of drivers with different temperaments changes differently.

When analyzing the experimental data, it was found that in some cases the values of the level of fatigue at the exit from the traffic jam turned out to be lower than the values of the level of fatigue at the entrance to the traffic jam. This indicates that for some drivers, mostly phlegmatic ones, the functional state in the traffic jam does not deteriorate. However, in the vast majority of drivers, the functional state deteriorates, which leads to an increase in the likelihood of an accident.

To assess the impact of traffic jam on the level of fatigue of the average driver, graphs were constructed with various combinations of the initial level of fatigue. Corresponding graphs were built for drivers aged 25 and 65 years (Fig. 1 and 2).

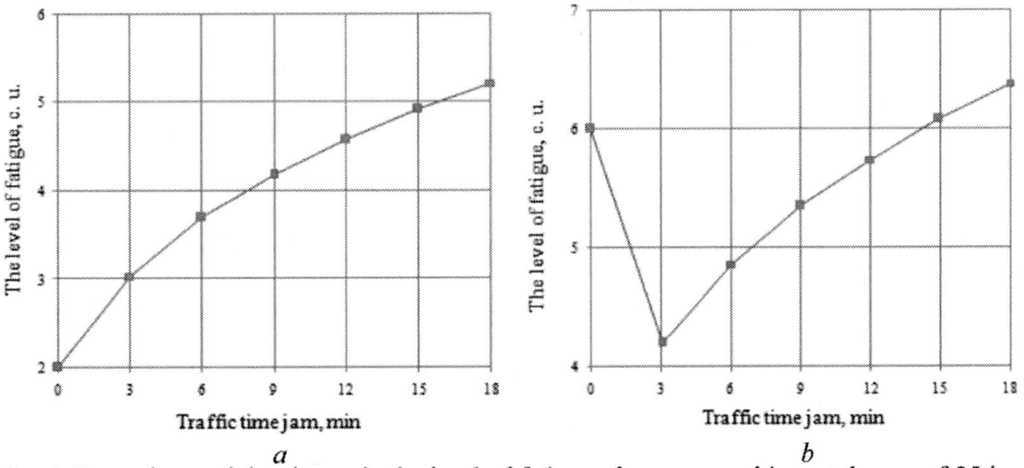

Fig. 1. Dependence of the change in the level of fatigue of an average driver at the age of 25 in a traffic jam at: L_{fen} = 2 c. u. (a), and L_{fen} = 6 c. u. (b).

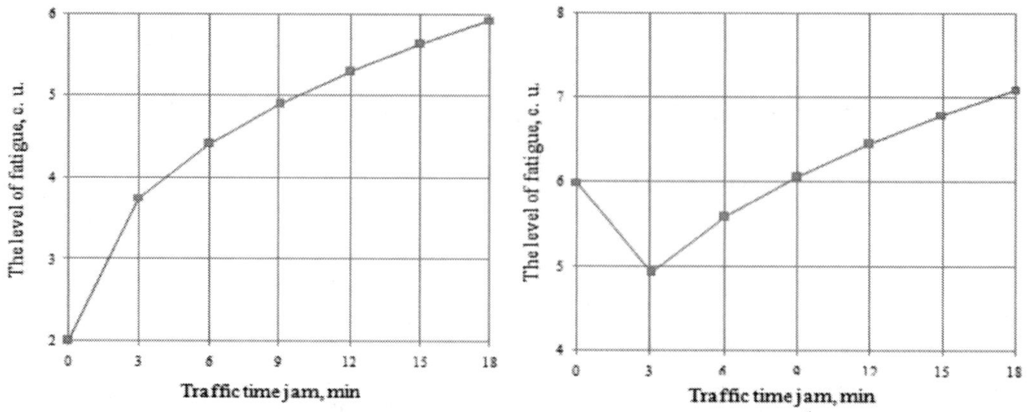

Fig. 2. Dependence of the change in the level of fatigue of the average driver of 65 years of age in a traffic jam with: $L_{fen} = 2$ c. u. (a) and $L_{fen} = 6$ c. u. (b).

It is clear from Fig.1 that the level of fatigue of the average driver at the end of the traffic jam increases significantly more than five c. u., which reflects the state of the expressed tension of regulatory systems.

If at the entrance to the traffic jam the initial level of fatigue is six conditional units, the functional state of the driver until the third minute of the jam first improves, and then gradually deteriorates (Fig. 1 (b)). In addition, the value of the level of fatigue of the average driver in Fig. 1 (b) exceeds six c. u., indicating a pronounced overstrain of regulatory mechanisms.

Dynamics of change of a functional state of the sixty-five-year-old driver on Fig. 2 is similar to the change in the functional state shown in Fig. 1, however, is more pronounced.

It follows from Fig. 2 (a) that the driver's fatigue level rises throughout the entire traffic jam, reaching six or more c. u. at the end of it. The driver's condition (Fig. 2 (b)) at the beginning of the jam slightly improves. The improvement in the jam is due to the fact that he entered the jam in a stressed state. Further, driver's condition deteriorates significantly.

In addition, the value of the level of fatigue of the average driver exceeds seven c. u., indicating a pronounced overstrain of regulatory mechanisms.

An analysis of Figures 1-2 shows that such a factor as the age of a driver has a different effect on his level of fatigue and, consequently, on the likelihood of a traffic accident.

Young drivers adapt faster to changes in road conditions than older drivers, including the impact of traffic jam. Their state and level of fatigue are more likely to recover from the adverse effects of jam due to their biological age.

5 Conclusion

The studies conducted and the regression models compiled testify to the objective and negative impact of the traffic jam on the functional state of the driver. The most significant factors affecting the condition of the driver in a traffic jam are identified. The developed models, adequately reflect the influence of traffic jam on the functional state of the driver, and can be applied for practical use in the first approximation. The value of the initial level of fatigue has a different effect on the functional state of the driver during his stay in a traffic jam. So, if a driver gets into a traffic jam with an initial state (L_{fen} = 5-6 conditional units), then in the first 3-5 minutes his functional state improves by 10-20%, and then it gets worse. These trends are characteristic for drivers of all ages. The most significant effect of the duration of the jam is observed among drivers of the older age. When analyzing the experimental data, it was found that in some cases the value of the level of fatigue at the exit from the traffic jam turned out to be less than the value of the level of fatigue at the entrance to the jam. This suggests that the functional state of some drivers, mostly phlegmatic ones, does not deteriorate in the traffic jam.

Therefore, to clarify the obtained patterns, further research should be carried out in order to simulate the effect of jam on drivers of different temperaments.

References

1. E. Gavrilov. SistemologIya na transportI / PId zag. red. Dmitrichenka M. F. – KiYiv : Znannya UkraYini,. – kn. 5 : ErgonomIka, 256 (2008)
2. N. Gyulyev, C. Dolia. American Journal of Social Science Research, **3(5),** 17–24 (2017)
3. N. Gyulyev, O. Lobashov, O. Prasolenko, D. Burko. International Journal of Engineering & Technology, **7(4.3)**, 308–314 (2018)
4. M. Zhuk, V. Kovalyshyn, Y. Royko, and K. Barvinska. Eastern-European Journal of Eenterprise Technologies, **2 (3),** 24-31 (2017)
5. P. Philip, P. Sagaspe, N. Moore, J. Taillard, A. Charles, C. Guilleminault, B. Bioulac,. Accident Analysis & Prevention, **37(3),** 473-478 (2005)
6. L. C. Davis, Physica A: Statistical Mechanics and its Applications, **391 (4),** 1679–1691 (2012)
7. Kalaižić M., Miloš K., Muić M. Simulation Model of Traffic Jam at Crossroads, Promet-Traffic & Transportation, **14(6),** 309–317 (2012)
8. D. A. Hennessy, D. L. Wiesenthal,. Ergonomics, **40 (3),** 348–361 (1997)
9. H. S. Song, Y. G. Sin, S. C. Gang. Journal of Korean Society of Transportation, **23(2),** 61–74 (2005)
10. Y. Li, J. Q. Wang, and J. Wu. Journal of Central South University, **24(6),** 1494-1502 (2017)
11. D. Ponkratov, I. Ivanov, O. Prasolenko, T. Grigorova, O. Lobashov, and V. Dolya,. International Journal of Automation, Control and Intelligent Systems, **1 (2),** 21-26 (2015)
12. O. Prasolenko, O. Lobashov, A. Galkin, International Journal of Automation, Control and Intelligent Systems, **1(3),** 77-84 (2015)
13. A. Maji, N. R. Velaga, & Y. Urie, International journal of injury control and safety promotion, **25(3),** 257-271 (2018)
14. E. Lagarde, J. F. Chastang, A. Gueguen, M. Coeuret-Pellicer, M. Chiron, S. Lafont, (2004). Emotional stress and traffic accidents: the impact of separation and divorce. Epidemiology, 15, 762–766.
15. J. W. Muttart. Evaluation of the influence of several variables upon driver perception response times. York, England. In Proceedings of the 5th International Conference of the Institute of Traffic Accident Investigators, 116–129 (2001)
16. N. Gyulyev, Vestnik Natsionalnogo tehnicheskogo universiteta «HPI», **10**, 140–144 (2011)
17. N. Gyulyev, Lyudskiy faktor I dorozhnI zatori : monograflya. HarkIv: HNUMG Im. O. M. Beketova, 252 (2016)
18. I. Afanasieva, A. Galkin. Assessing the information flows and established their effects on the results of driver's activity. Archives of Transport, 45(1), 7-23 (2018)
19. N. Gyulyev, V. Dolya. Vestnik Natsionalnogo tehnicheskogo universiteta «HPI». **26**, 47–50 (2012)
20. K. Stojmenova, J. Sodnik, Detection-Response Task - Uses and Limitations. Sensors, 18(2), 594 (2018)
21. A. G. Koveleva, *Psihologiya: ucheb. posobie*. Moskva, Prosveschenie, 451 (1966)
22. R. M. Bayevsky, O. I. Kirillov, S. Z. Kletskin, *Matematicheskiy analiz izmeneniy serdechnogo ritma pri stresse*. Moskva, Nauka», 177 (1984)

Growth Potential of the Market Force of Intermediary Structures

Olga Kibik[1]*, *Viacheslav* Kotlubay[2], *Iuliia* Khaiminova[3], and *Kateryna* Belous[4]

[1] National University "Odessa Law Academy", Department of National Economy », Ukraine
[2] National University "Odessa Law Academy", Ukraine
[3] National University "Odessa Law Academy", Ukraine
[4] Odessa National Maritime University, Ukraine

Abstract. The aim of the study is to determine the extent and prospects development of market potential of intermediary structures using an example of waterway transport enterprises. Economic liberalization, development of international economic activity, fragmentation and distantness of suppliers and buyers as preconditions for development of intermediary business structures are determined. The main factors hindering the development of intermediary activities are identified, for instance, imperfect regulatory framework, insufficient intermediary's professionalism, insecurity from intermediaries, low level of responsibility to counterparties, general distrust. Transport companies, in particular, the waterway transport enterprises, are identified as intermediaries that have a significant influence for development of individual economic agents and economic systems. Prospects for the development of sustainable mutually beneficial economic relations or the mutual benefit, the availability of effective information flow, a clear definition of roles, rights and obligations of the involved parties, mutual coordination of behaviour of intermediary structures and clients defined as the main conditions for development of effective interaction of intermediary structures of water transport and clients. In the matter of further studies, attention should be paid to improvement of strengthening mechanism for the responsibility of intermediary structures and clients.

1 Introduction

Present development conditions of economic activity are characterized by the growing role of intermediaries. Helping to identify and solve problems of economic agents' activity at the market, intermediaries are an important part of the infrastructure of modern markets. Half the turnover of all goods and services is the share of intermediaries in the United Kingdom, the United States, Japan, Germany, and Sweden [1]. In the European Union (EU), according to international statistics for 2009, trade intermediaries accounted for more than 45 % of total number of importers and about 39 % of all exporters. The share of trade intermediaries is about 21 % of the total value of the EU exports and 43 % of the total value of the EU imports. The institution of intermediation has a long history of functioning and numerous contemporary forms are applied in national and global economic systems. Progressive technologies allow intermediation not only in real life. An economic agent commissions intermediary services and expects to improve conditions and results of selling his product. It

*Corresponding author: kibik@ukr.net

is the basis of increasing competitiveness of goods and, accordingly, competitiveness of the economic agent. However, economic agent has both advantages and risks ordering intermediary services. The possibility of loss of market control, opportunistic behaviour of the intermediary, increasing the agent's dependence on the results of mediation activity, increasing the number of stages of the goods supply, that significantly increase expense of the economic agent, the impossibility of controlling the level of the final price of the goods should be noted as the main risks. Some Japanese distribution channels include more than ten intermediaries. Such number of intermediaries is leads to rise price by 500 % [2].

Transport costs are the integral component of the final cost of goods, besides, transport stipulates the qualitative characteristics of goods. That is, the level of competitiveness of the overwhelming majority of goods is largely determined by the level of transport companies. Transport companies such as waterway transport enterprises are intermediaries. Such intermediaries have important influence on individual economic agents and economic system development.

2 Analysis of recent researches and publications

Issues of types and features of intermediaries are studied by A. Zahorodnii, Z. Koval [3], N.N. Kazarian [4], V. Pavlova [5], V. Rieznikova [6]. Study of logistics as a kind of intermediary activity is devoted to works of L. Sosunova, N. Astafieva, O. Yudakova and N. Petrova [7]. Special features of return logistics and connection between points of sale and reverse activity of consumers were solved by F. Milichovskiy [8].

Such researchers as A. Kotlubay, V. Zhykhareva [9], L. Nikolaieva, V. Bereza [10], M. Dooms [11] and others devoted their works to identifying the role of Ukrainian waterway transport intermediary companies in the development of the Ukrainian economic system.

3 Unsolved aspects of the problem

However, the studies mentioned above reflect mainly the problem of the development of certain components of the waterway transport specific industries or sectors potential, e.g. shipping companies, ports, cruise companies, inland waterway transport and the like. Issues on development of the Ukrainian waterway transport intermediary companies are extremely relevant and should be investigated in connection with signing of the Free Trade Agreement between Ukraine and the European Union.

The aim of the study is to determine the extent and prospects development of market potential of intermediary structures on an example waterway transport enterprises in modern conditions.

4 Results

Ukraine has certain preconditions for the development of intermediary activities, in particular, transport companies which provide intermediary services. Liberalization in the economic sphere, the development of global economic activity, the fragmentation and remoteness of suppliers and buyers should be considered as main factors. The development of legitimate intermediary structures network, in particular in the transport sector, is restricted by such factors as imperfection of the regulatory framework, low level of intermediaries professionalism, the lack of guarantees from intermediaries, low level of responsibility to contractors, general distrust and others. Issues of standardization and licensing of certain types of intermediary activity is still relevant, due to the importance of reducing the risks levels in intermediary activities.

Let us consider the preconditions and prospects for the development of intermediary capacity through the example of maritime enterprises. Through transportation of cargo owned by residents of the country and passengers within the country, transport companies participate in the processes of internal integration. Transport, being a sector of the economic system of the country, while providing the international carriage of goods and passengers, is party to integration processes at the international level. All types of transport provide the international carriage of goods. However, the maritime and river transport traditionally takes an important place in such transportation, particularly through the export of transport products. The development of export services, particularly transport services, in modern conditions is a focal area of development of foreign economic activity for all economically developed countries.

The export of transport products can significantly increase foreign exchange earnings to the national budget through the creation and implementation of products with a considerable added value [12].

That is why the development of innovative capacity of the waterway transport is of top priority for Ukraine.

The main conditions for development of effective interaction of structures of waterway transport and clients include the availability of a general strategy of interaction; prospects for the development of sustainable mutually beneficial economic relations or the mutual benefit; strengthening of market positions; provision of balanced and sustainable economic growth; the availability of effective information flow between the intermediary structures and clients; a clear definition of roles, rights and obligations of the parties involved and mutual coordination of behaviour of intermediary structures and clients [13].

Market positions of waterway transport business structures can be strengthened due to increased participation in integration processes. The terms of the Deep and Comprehensive Free Trade Area within the Association Agreement between the European Union and the European Atomic Energy Community and their member states, as one party, and Ukraine as the other party, which entered into force on 1.01.2016, opened the coastal shipping market in the European Union for Ukraine. In conditions of development of the EU coastal shipping market, the said agreement provides Ukraine with the opportunity to enter the closed transport market, which is not yet fully formed. Accordingly, at present Ukraine is a country that has been granted the right for its merchant fleet to sail in the inland waterways of the European Union on a bilateral basis. This creates exceptional opportunities to enter and conquer the segment of the EU inland waterway transport market.

Recently, the annual transportation volume through the EU inland waterways is about 900 million tons of freight. More than two thirds of the above amount is transportation between ports of EU countries at an average distance of 200 km. Own inland fleet deadweight in EU countries is approximately 15 million tons. Price methods of competition and advantages of waterway transport are able to determine the situation, that about 30% of the cargo volume transported by motor transport on distance over 200 km, can also be refocused to transportation by inland waterway transport. The shortage of carrying capacity is estimated approximately at 4 million tons [14]. In a situation of such shortage, we should expect an increase in the participation of third countries carriers.

This situation creates real opportunities for integration of the Ukrainian waterway transport into the European internal water transport system. To take advantage of the chance, the process of the Ukrainian inland waterway transport reform should be speeded up by adoption of the Law "On Domestic Waterway Transport". Also conditions should be created to restore waterway E-40 that will link the Black and Baltic seas through the Dnieper-Vistula corridor provided for the terms of the European agreement on main inland waterways of international importance, signed on the 19th of January 1996 in Geneva.

To take the unique chance stipulated by the current business trend of the European inland waterway transport market, Ukraine urgently needs to establish the following conditions: formation of legal framework adapted to provisions of international and European laws, establishment of a system of administrative and economic instruments for the national maritime industry development; formation of adequate financial resources for the inland waterways system development, elimination of corruption schemes in the sphere of inland waterway navigation regulated by Ukraine, setting the system for promoting the shipbuilding industry development, ensuring of prerequisites for macroeconomic stabilization, which is the basis for growth in the cargo base volume of national transport companies, etc.

Conditions of navigation through the EU inland waterways have specific features and limitations that may in some cases create obstacles for Ukrainian shipping companies. A number of points to Chapter IV of the "Deep and Comprehensive Free Trade Area" of the Association Agreement between the European Union and the European Atomic Energy Community and their member states, of the one part, and Ukraine, of the other part completely exclude the access opportunities for shipping companies established in Ukraine, but not for those in the property of residents of Ukraine or other EU countries. However, the Ukrainian ship-owners whose vessels carry out transportation under flags of other countries will have access to the EU inland waterways.

Comply with the world experience, the most effective instruments of attracting investment and stimulating economic development are tax incentives, loans under favourable terms and subsidies from state budgets of different levels. These instruments are almost not use in Ukraine nowadays. This leads to the fact that the domestic ship owners have to register their vessels under so-called "flags of convenience".

Integration phenomenon in the field of maritime business can be considered also as a component of the overall process of transport and logistics system of Ukraine integration into the European and world economic spaces. Comply with the terms of the Deep and Comprehensive Free Trade Area within the Association Agreement between the European Union and the European Atomic Energy Community and their member states, of the one part, and Ukraine, of the other part, there are requirements on improving the regulatory framework for providing operations of maritime transport. Thus, the limited (3-6 years) period is being introduced for implementation of the EU Directives listed in Annex XXVII-5 Rules Applicable to International Maritime Transport. They concern primarily requirements of safety of navigation and ensuring the social protection of seafarers.

On the other hand, enterprises of maritime transport are active components of the integrated systems of coastal regions, clusters in particular. Creation of transport-logistic and tourist and recreational clusters are most promising directions of integration processes development for coastal regions. It has been proposed to consider the experience of cluster functioning in the European space while developing this form of integration. Successful port and industrial clusters are based on seaports of Rotterdam (Netherlands), London (United Kingdom), Antwerp (Belgium), Marseille-FOS (France), Singapore and others. Port and industrial cluster of Rotterdam combines the capacity of a seaport, petrochemical industry enterprises, shipyards and ship repairing enterprises, agricultural enterprises, insurance, financial and consulting companies, organizations engaged in trade of auto and agricultural industry products. The port industrial cluster of Marseille includes large refineries, petrochemical and metallurgical enterprises, and shipyards.

Based on results of leading scientists studies [10, 14, 15], there has been proposed a system of measures, the implementation of which is necessary for the development of potential of Ukraine in the framework of accelerating the integration processes at different levels: regulative framework improvement for the transport system functioning, the implementation of international legal instruments, particularly European Union directives, regulatory framework improvement for the integration activities, clusters operation in particular, creation of favourable legal, organizational and economic conditions aimed to attract investments, including through the establishment of free economic zones aimed to attract foreign investors to the development of transport system elements based on the experience of Asian countries, intensification of participation of Ukraine in the development of the general strategy for the development of international transport corridors network at the

European and Asian areas within the formation of new transit routes of continental importance, improvement of transit transport infrastructure, development of competitive advantages of domestic waterway transport enterprises, creation of competitive national transit transport operators by main types of TTS, development of logistic support system for transit goods, improvement of pricing system for transport services, formation and development of integrated structures based on transport enterprises, clusters in particular, consideration of environmental impacts when making management decisions in the field of integration transport activities, strengthening of socio-economic responsibility of members of the integrated waterway transport structures, etc.

5 Conclusions

Intermediary activity is an important element of the economic system of each country. Creating conditions for the growth of the market power of intermediary business structures is an important task for increasing the competitiveness of economic agents in different levels and fields of activity.

Intermediary companies play an important role in the development of foreign economic activity. Governmental support is appropriate for export-oriented intermediary business structures that contribute to the development of the country export potential. Support may be in the form of preferential taxation, preferential conditions for licenses process (in case of licenses introduction) etc. Transport companies that carry out intermediary services have a special place in the intermediary system.

References

1. N. Tishkina, Black Sea Economic Studies, **10**, 145-149 (2016)
2. O. Dima, Foreign trade: economics, finance, law, **4**, 72-82 (2014)
3. A. Zahorodnii, Z.O. Koval, Lviv Polytechnic, **484**, 72-79 (2003)
4. N. Kazarian, Herald of GGTU, **2(60)**, 40-42 (2012)
5. V. Pavlova, V. Hertsyk, Academic review, **2(39)**, 66-72 (2013)
6. V. Rieznikova, *Theoretical problems of regulating mediation in the economy of Ukraine,* Extended abstract of Doctor's thesis (2011)
7. L. Sosunova, N. Astafieva, O. Yudakova and N. Petrova, SHS Web Conf, **62,** 07001 (2019)
8. F. Milichovskiy, SHS Web Conf, **61**, 01016 (2019)
9. V. Zhyhareva, A. Kotlubay, O. Kibik, *Economics of Maritime Transport,* Burun Knyga (2012)
10. L. Nikolaieva, V. Bereza, *Problems of development of navigational on inland waterways of Ukraine and ways of their solution*, Economics, Management, Law: Realities and Perspectives, Paris, 203-207 (2016)
11. M. Dooms, *Crafting the integrative value proposition for large scale transport infrastructure hubs: a stakeholder management approach,* ASP / VUB Press (2010)
12. N. Garmash, Culture of the People of the Black Sea, **117**, 83 – 86 (2007)
13. O. Kibik, O. Podtserkovnyy, V. Kotlubay, & others, *Organizational and Legal Fundamentals of maritime complex of Ukraine*, Feniks (2012)
14. A. Kotlubay, *Transport Politics of Ukraine and Agreement with the EU,* TransportInform.com. http://transportinform.com/corridors/457-transportnaya-politika-ukrainy-soglashenie-eu.html (2017)
15. O. Kibik, I. Khaiminova, V. Kotlubay, I. Redina, E. Belous, International Journal of Engineering & Technology, **7 (4.3)**, 461-466 (2018)

Institute of corporate responsibility in the system of valuation for stakeholders

Sergey Kivalov[1], *and Olha* Kibik[2,*]

[1] National University "Odessa Law Academy", Ukraine
[2] National University "Odessa Law Academy", Department of National Economy, Ukraine

Abstract. The principles and terms for providing corporate responsibility of economic agents of different levels are determined. Formation features of modern system for providing corporate responsibility of economic agents are examined on the example of agents operating in the field of port activity. The purpose of the article is to study concepts of development of the mechanism for corporate responsibility in order to optimize the system of stakeholders' interaction in the process of creating values. The method of system analysis was applied in studying concepts and preconditions for the functioning of corporate responsibility management systems. The systemic structural method helped in studying basic measures to increase the corporate responsibility level of enterprises of port activity. The goal of creating a system of corporate responsibility should be associated with the achievement of the maximum possible effect of functioning of the economic agent. Needs and contributions of stakeholders should be interlinked in order to increase corporate responsibility. The main stakeholders in the field of port activity and criteria for satisfying their needs, contribution and opportunities in the field of corporate responsibility have been identified. The proposed system of criteria for the effectiveness will improve the mechanism of corporate responsibility management.

1 Introduction

The institute of corporate responsibility plays an important role in the system of value creation in modern economic activity. Active use of tools to increase the level of corporate responsibility allows optimizing the level of competitiveness of the economic entity due to the identification of inefficient activities and unrealized opportunities, which in turn leads to the development of innovation and investment potential.

The institute of corporate responsibility, an important element of which corporate social responsibility (CSR) is, has been established in economically developed countries.

At the end of the year 2011, the EU Strategy for Corporate Social Responsibility was adopted for the period 2011-2014. According to the strategy, corporate social responsibility

*Corresponding author: kibik@ukr.net

is the responsibility of the business for its impact on society. In the European Union, it is considered appropriate to have a law according to which large business entities in the form of annual accounts should include sections reflecting the results of implementing corporate social responsibility measures. In the absence of such initiatives, the business structure should provide an explanation. Currently, according to experts, less than 1% of companies report CSR [1].

Works of many scholars are devoted to the formation and development of corporate responsibility concepts. The most significant results are delivered in studies of A. Caroll [2], M. Porter, M. Kramer [3], M. Halme, J. Laurila [4], H. Aguinis and G. Ante [5] et al. The works of such Ukrainian scientists as A. Kotlubay [6], N. Primachov [7], V. Zhikhareva [6], L. Nikolaieva, V. Bereza [8] and others are devoted to the study of the process of the management system formation in maritime transport aimed at improving the potential for development of the maritime complex enterprises. Scientists from other countries, namely M. Acciaro [9], P. Alderton [10], Costas Th. Grammenos [11], P. Kieran [12], S. Kumar, J. Hoffman [13], M. Stopford [14] also paid an attention to the development of a corporate responsibility system in maritime transport.

However, issues of the corporate responsibility development as an element of the value system for economic agents of different levels and fields of activity, in particular in the field of maritime business, remain insufficiently studied.

Purpose of the article is to study principles for the development of the corporate responsibility mechanism in order to optimize the system of interaction of stakeholders in the process of creating values. The dialectical method allowed investigating the definition of a corporate responsibility. The method of system analysis was applied in studying concepts and preconditions for the functioning of corporate responsibility management systems. The systemic structural method helped in studying basic measures to increase the corporate responsibility level of enterprises of port activity.

2 Presentation of the core material

In recent years, the national economic system and its components are being integrated into the world socio-economic processes which make the application of tools increase the level of economic entities responsibility more active, taking into account the basics of institutional theory. The purpose of development of the corporate responsibility system is a positive impact on the economic entity environment to ensure its success and, accordingly, long period profitability. To achieve this goal, business structures must constantly analyze business processes and adapt them to modern requirements and needs of the environment, which often leads to significant costs. That is, corporate responsibility is a part of the value system of business structures, their corporate culture, and marketing policy [15].

The basic principles of corporate responsibility are defined as follows [16]:

- institutional principle of legitimacy. Comply with the current legislation, each stakeholder has certain rights and obligations, that is, has a certain degree of power. Under the condition of irresponsible attitude to their rights and obligations, economic entities may lose their power, in particular the market one;
- organizational principle of administrative liability, which provides for the organization of responsibility for those results that relate to areas of primary and secondary interaction of economic agents with society;
- individual principle of freedom of management choice, according to which economic agents are obliged to exercise the right of choice in order to ensure responsible results.

The state can be admitted as a responsible economic agent when favorable conditions are created for the comprehensive sustainable development of the socio-economic system as a whole and for its separate elements.

Micro level economic agents are considered responsible if the following conditions are met: compliance with requirements of international and national legislation, in particular in the field of payment of taxes, production and marketing of high-quality products, implementation of corporate programs for the development of labor potential, providing for professional development, improvement of working conditions and improvement of the system of staff incentive, implementation of charitable projects and participation in regional development projects, etc.

It is necessary to determine sources of corporate responsibility and determine their impact on the corporate responsibility level. It is expedient to maintain the point of view of [16], which allocates internal and external sources. In its turn, internal sources include main components of the corporate culture. The list of external sources includes universal factors, market forces, and a political component.

The issue of increasing the corporate responsibility level for business entities, which is an important component of logistics systems, is of particular relevance. Commercial seaports and other port enterprises occupy a special place among such entities. Most of port enterprises round the world, including commercial ports, play a significant role in the development of cities, regions and even states. Accordingly, such structures should take responsible decisions considering the public interest while ensuring transparency in the adoption of such decisions. Modern enterprises of port activity operate in conditions of rapid increase of competition intensity in the national and international markets of port operations and services. Port authorities are responsible for the effective state property management in seaports and for ensuring the stable and safe business operation. Port authorities are under considerable regulatory pressure [9]. Therefore, external sources of corporate responsibility are of special importance for port enterprises. Their list should include trends in the development of the global transport market and its segments, the situation of resource markets, international and national political and legal conditions, and codes of professional associations, etc.

Leading foreign maritime business structures actively use benefits attributable to the existence of corporate responsibility systems. The results of a study by foreign scientists on the practice of implementing a corporate responsibility policy suggest that 10 well-known and successful foreign port operators have effective corporate responsibility systems. Ports of Antwerp, Hamburg, Hong Kong, Jebel Ali, Rotterdam, New York, Los Angeles, Pusan, Shanghai and Singapore actively apply initiatives such as cooperation with foreign ports, the presence of a corporate responsibility mission, environmental problems solution, educational activities, employment problem solution, gender issues, energy saving, humanitarian aid and charitable organizations support, cooperation with local communities, availability of reports on sustainable development, transparency, etc. Implementation of such initiatives results in minimizing risks of functioning and maximizing the value of port enterprises [9].

Numerous enterprises of the maritime business of Ukraine, in particular in the field of port activity, currently form and implement a system of crisis response measures, among which the improvement of the structure and methods of corporate governance is mandatory. A modern port activity feature is the large number of its participants. The main stakeholders involved in the port activity are the owners and investors of port enterprises, cargo owners, ship-owners, various intermediaries, enterprises of other modes of transport, state bodies of various levels, participants of port communities, personnel of enterprises, suppliers of various types of resources, population of territories linked to the enterprises concerned, etc. Proceeding from the stakeholders diversity, the criteria for determining the level and peculiarities of their impact on the formation of the corporate responsibility level of port enterprises differ significantly.

Solution of the task aimed at improving the interaction between the numerous participants of port activities in the shortest possible time and at optimal cost is possible provided that the clustering processes are activated. The development of a mechanism for providing corporate responsibility allows optimizing the system of interaction of stakeholders. Moreover, the concept of stakeholders is considered an alternative of the corporate responsibility concept. Modern business structures, in particular port enterprises, are distinguished by the presence of numerous stakeholders. The main stakeholders involved in the port activity are as follows: state authorities of different levels, port authorities (port administrations), owners and investors of port enterprises, cargo owners, ship-owners, various intermediaries, enterprises of other modes of transport, participants of port communities, employees of enterprises, suppliers of various kinds of resources, etc. Based on a vast variety of stakeholders, the criteria for meeting their needs and requirements for ensuring corporate responsibility differ significantly.

The most important measures of corporate responsibility development projects for national port enterprises are considered as follows: initiatives to increase the level of resource utilization efficiency and improvement of interaction with stakeholders possessing their own economic interests. Therefore, it is necessary to consider the level of satisfaction of the stakeholders' needs, which ultimately determines the corporate responsibility level of the port enterprise. The criteria for satisfying needs of owners and investors may be the additional cost, income, profit, return on capital, other actual financial and economic indicators compared with the forecast. The criterion for customer satisfaction can be the ratio of quality of the performed port operations (services received) and the price paid. Public authorities of different levels should solve state issues, therefore criteria for them are the level of competitiveness of the port activity and the rate of impact on competitiveness of the state economy. Recently, port communities have been incorporating and developing rapidly, with the aim of increasing the competitiveness of individual actors and communities in general. Accordingly, the criteria are the level of competitiveness and profitability. Satisfaction of employees is related to the port enterprise activity in general. All employees of the enterprise of the port activity, first and foremost, are interested in the level of remuneration. Most employees strive for an interesting, creative work that allows creating and developing useful knowledge and skills. Therefore, the criteria may be tendencies of staff turnover, reasons for dismissal, level of dissatisfaction, etc. Suppliers are interested in increasing supply volumes, which is determined by the volume of port production. The criteria are the length of cooperation with the enterprise of the port activity and the volume of its purchases [17].

On the other hand, the state and other economic agents directly or indirectly involved in port activities should also adopt and implement responsible decisions in the field of cooperation with port enterprises. Especially in times of crisis, the state should take economically and politically responsible steps to support the development of the competitive potential of national port enterprises. First, the state should create a favorable regulatory environment for the functioning of port business structures in the national market and legal prerequisites for the protection of interests of domestic economic agents in the segments of the global transport market.

The contribution of stakeholders should not be ignored. The contribution means what the stakeholders can give to the enterprise of port activity in the field of providing corporate responsibility on a mutually beneficial basis. Owners and investors may contribute funds spent on the development of the port enterprise operations, in particular, business reputation, relationships with various elements of the external environment, including the clientele, state bodies, etc. The criteria for customers' input can be their loyalty to the port enterprise, the degree of readiness to confirm the positive business reputation of the enterprise, the level of customers' offers regarding the development of the enterprise, the level of promptitude in payment, etc.

The state contribution may be determined by the following criteria: the degree of intervention in the port enterprises activities, flexibility of the state policy in relation to the development of the port activity, the level of responsiveness to inquiries and the resolution of problem situations arising in the field of port activity, etc.

The contribution of port community participants is to support and implement a single development strategy aimed cat enhancing the entire community competitiveness. The criterion may be rough estimation of the contribution of the community in monetary terms.

The criteria for employees' contributions are the added value created by each employee, productivity, the number of proposals for improving port production, etc.

Offering attractive prices and fulfilling terms of delivery, suppliers of different types of resources create the basis for generating the desired level of responsibility of the port enterprise. The criterion for their contribution is the number of complaints regarding the quality of supply and provision of invoices.

Criteria for the effectiveness of implementing strategies aimed at raising corporate responsibility can be the number of customers, the volume and consistency of their cargo flows, trends in the volume of demand for port enterprise products, market share, attractiveness for investors, the level of satisfaction of employees of the enterprise, the contribution to improving the level of state economy competitiveness, etc.

The study of works of scientists [9] and the results of their own research activity [18] allow us to make the following conclusion. Implementation of projects to strengthen the responsibility of business structures of port activities will help to improve their competitiveness and optimize financial results through identifying inefficient areas of activity, business processes, structural units, services; identifying business opportunities that might otherwise have remained unknown; better management of stakeholders' behavior, at least the core ones; increase of innovative potential, which allows optimizing the technical and technological level of port production; solution of problems of the personnel potential development, etc.

On the other hand, other participants in the port activity will receive positive results.

Sustainable development of modern society is possible subject to the economic and political stability and increased responsibility of all economic actors, especially business structures. The external and internal environment of the modern business structure expects its responsible behavior in the field of social and economic activity. Therefore, the formation and continuous development of the corporate responsibility concept is the key to the formation of competitive advantages of modern economic agents.

3 Conclusion

There is a system of factors determining the development of the institute of corporate responsibility in various spheres of economic activity of society. So, the main factors of increasing the level of corporate responsibility for the sphere of port activity are strengthening of requirements for security, in particular environmental, strengthening the role of port business in the development of economic systems of different levels, development of port enterprises as logistics centers, deregulation of port activity and strengthening the intensity of competition, etc.

Every entity in the maritime business, in particular, port enterprises, must implement CR projects. Initiatives to increase corporate responsibility are prerequisites for the creation of long-term competitive advantages of the business structure of the port business in the current conditions of the world, regional and national markets development of port operations and services granted.

The purpose of formation of the corporate responsibility system should be associated with the achievement of the maximum possible effect on the part of the economic agent, in particular, the port enterprise. Needs and contributions of stakeholders should be interlinked in order to increase corporate responsibility, and there may be many in practice. The main stakeholders in the field of port activities can be defined as follows: owners and investors of enterprises, cargo owners, ship owners, various intermediaries, enterprises of other modes of transport, state authorities of different levels, members of port communities, employees of enterprises, suppliers of various types of resources and etc. Based on the diversity of stakeholders, the criteria for meeting their needs, contributions and opportunities in the area of corporate responsibility differ significantly. The proposed system of performance criteria to ensure corporate responsibility, considering interests of various participants in port activities, will improve the mechanism of corporate responsibility management based on the timely identification and consideration of factors for the external and internal environment of the enterprise of port activities.

References

1. *New Strategy of the European Commission on Corporate Social Responsibility (CSR) 2011-2014: faster, higher, stronger. Expert Commentary from the Center for CSR Development*, CSR Ukraine http://www.csr-ukraine.org (2013) Accessed 30 May 2019
2. A. Caroll, Business and Society, **38** (3), 268-295 (1999)
3. M. Porter, M. Kramer, Harvard Business Review, **84** (12), 78-92 (2006)
4. M. Halme, J. Laurila, Journal of Business Ethics, **84** (3), 325-339 (2009)
5. H. Aguinis, G. Ante, Journal of Management, **38** (4), 932–968 (2012)
6. V. Zhyhareva, A. Kotlubay, O. Kibik, *Economics of Maritime Transport,* Burun Knyga (2012) [in Ukrainian]
7. N. Prymachov, A. Dmytriyenko, The bulletin of transport and industry economics, **43**, 57-61 (2013) [in Ukrainian].
8. L. Nikolaieva, V. Bereza, Development of administration and economic management methods on transport: Scientific works, **28**, 5-13 (2013) [in Ukrainian].
9. M. Acciaro, International Journal of Logistics: Research and Applications, **18** (3), 291–311 (2015)
10. P. Alderton, *Sea Transport: Operation and Economics*, Tomas Reed (1998)
11. C. Th. Grammenos, *Handbook of Maritime Economics and Business*, LLP (2002)
12. P. Kieran, *World trends in Shipping and Port reforms,* CPCS, Transcom (2003)
13. S. Kumar, J. Hoffman, *Globalization: The Maritime Nexus,* MPG Books Ltd (2002)
14. M. Stopford, *Maritime economics*, Harper Collins Academic (1988)
15. S. Kivalov, O. Kibik, Baltic Journal of Economic Studies, **5** (2), 73-79 (2019)
16. L. Petrashko, *Comparative Evaluation of Corporate Social Responsibility BusinessPractices: EU and Ukraine (2015–2016): the Research*, KNEU (2013)
17. O.M. Kibik, *The theory and methodology of pricing in sea trading ports*, ONMU (2005)
18. O. Kibik, I. Khaiminova, V. Kotlubay, I. Redina, E. Belous, International Journal of Engineering & Technology, **7 (4.3),** 461-466 (2018)

Determinants of railway transport sustainability

Olha Kravchenko[1,*], and *Nadiia* Bohomolova[1]

[1] State University of Infrastructure and Technology, Kyiv, Ukraine

Abstract. The deepening of integration processes between the countries of Europe and Asia, the need for regular and mass passenger and freight traffic contributes to the increasing importance of transport systems. In this context, railway infrastructure, which ensures the uninterrupted operation of industrial complexes in many countries of the world and contributes to the development of international cooperation, acquires particular importance. This is also true for the railway transport of Ukraine, which is a connecting element between the transport systems of the EU, Russia and Asian countries. The paper substantiates that the sustainable functioning of railway transport is one of the necessary conditions for the development of national economies. The system of external and internal determinants of railway transport sustainability is highlighted. These determinants are analyzed on the example of railway transport in Ukraine. It is shown that external determinants form the basis of the functioning of railway companies, which also affects their internal determinants. Management of railway transport sustainability should be comprehensive, which will allow developing an effective set of preventive measures.

1 Introduction

One of the strategic directions for the development of the EU is to improve the relations between the countries of Europe and Asia. This is due to a significant turnover between them: in 2016, exports from the EU to Asia amounted to 618 billion euros (35%) and imports - 774 billion Euros (45%) [1]. Ensuring further interdependence and strengthening economic cooperation implies stable ties not only between countries, but also regions. This is impossible without a well-developed and efficient transport infrastructure that promotes economic growth, enhances the competitiveness of national economies, creates new jobs and, as a result, improves the quality of life of the population. In this context, systems involving international transport corridors are of a particular importance. Such systems include the railway transport of Ukraine, which can be considered as a connecting element between the transport systems of the EU, Russia and the Asian countries.

2 Problem statement

Railway transport forms the basis of the country's transport infrastructure and, thus, has a significant impact on the functioning of enterprises in other sectors. The study of the

* Corresponding author: kravch.olha@gmail.com

functioning of rail transport in the context of ensuring the security of both national economies and mesoregions devoted to the publication by many scientists. Thus, the majority of researchers (Dugin and others) are in favor of uniting the Eurasian space through a developed and efficient transport system against thalassocratia (maritime countries), i.e. against a coalition of the Western countries led by the United States [2]. It is noted that the formation of such a system in the Eurasian region will allow each country to improve their own welfare, reduce the threat of terrorism and drug traffic, as well as implement the geopolitical goals of all countries of the Eurasian space [2, 3].

Based on an analysis of the countries of South and Southeast Asia, Zingel shows that transport, including railway transport, helps to reduce the development risks. He also notes that countries recognized as "failing or even failed" usually have poor infrastructure [4]. Sokolov, Kozhevnikov and Mezhokh noted that the level of development of transport is one of the factors ensuring the economic security of the country, both in terms of the complete geographical coverage of the country's territory by means of communication, as well as in terms of the technical and technological level of development of certain types of transport, the harmoniousness of their interaction, compliance with the development of productive forces of the state [5]. Kharlamova also notes that the transport system provides basic living conditions for society and serves as an important tool for achieving social, economic and the foreign policy goals. In addition, historically, the development of transport largely predetermines the economic and spatial development of countries, helps to strengthen their integrity and international influence [6]. Polzin emphasizes that a safe transportation system is one of the decisive factors of national security, including in the fight against terrorism [7]. In addition, transport, including railway transport, is considered as an integral element of the national security strategies in many countries of Europe and Asia [8–10 and others]. This actualizes the problem of ensuring the sustainable functioning of both the transport infrastructure as a whole and its individual sectors. The solution of this problem should be based on the institutional features of a particular country, as well as the degree of their influence on the state and level of infrastructure development.

In the context of increasing interdependence between the countries of Europe and Asia, the transport system of Ukraine is of particular importance, which has significant transit potential due to its advantageous geographical location and developed infrastructure. The basic transport sector is rail transport, which is in the process of reforming and has a direct impact on the economic security of Ukraine, as well as of neighboring countries. Therefore, the purpose of this study is to identify and analyze the determinants of a sustainable functioning of the rail transport.

3 Finding

At the European Conference of EU Transport Ministers, the criteria for classifying the transport system as sustainable and, as a result, effective were identified. These criteria relate primarily to the production activities of individual transport sectors, namely the provision of [11]: (i) equal access to transport infrastructure and meeting the needs of individuals, companies and society as a whole in transportation; (ii) clear and effective work, as well as a rational tariff policy as necessary conditions for the functioning of companies, maintaining the competitiveness of the economy, as well as ensuring a balanced regional development; and (iii) minimize the harmful effects on the environment.

The implementation of these criteria also implies an increased attention to the issues of not only the economy of individual transport sectors as a reflection of the effectiveness of their operations, but also of financial support for their development. Therefore, it seems necessary to comprehensively study the factors (determinants) that affect their current functioning and development opportunities, taking into account the peculiarities of production, economic and financial activities.

3.1 External determinants of railway transport sustainability

Rail transportation has a number of advantages, which determined their primary development in the past and the "renaissance" in the present. Thus, railway transport is universal (the ability to transport almost all types of cargo and passenger categories), has a high carrying capacity (a double-track electrified railway can carry up to 150-200 train pairs per day and provide more than 100 million tons of cargo in each direction), the relatively low cost of transport (as opposed to air and sea), relatively free accommodation (the construction of railways in almost any territory), regularity (the ability to carry out rhythmically transportation at all times of the year and in fact not to depend on the natural conditions, unlike river and sea transport), a relatively high speed of movement, and reliability.

The external determinants of the functioning and development of the railway sector are formed both by the institutional and macroeconomic environment. The difficulty of accounting for these determinants is manifested in the fact that transport companies cannot directly influence them, but only adapt to the changes that are taking place. One of these determinants is the level of development of the national economy. This is due to the fact that transport products are transport services for the movement (transportation) of goods and passengers and is not a new product that can be separated from the production process and participate in the economic turnover in a tangible form as a commodity. Transport products act as a beneficial effect from the work of transport, which can be consumed only during the production process and does not exist as a consumable item different from this process. The consequence of this is a steady correlation between the dynamics of freight traffic and the growth rate of industrial output (Fig. 1), which can be explained by a fairly stable nomenclature of rail freight traffic.

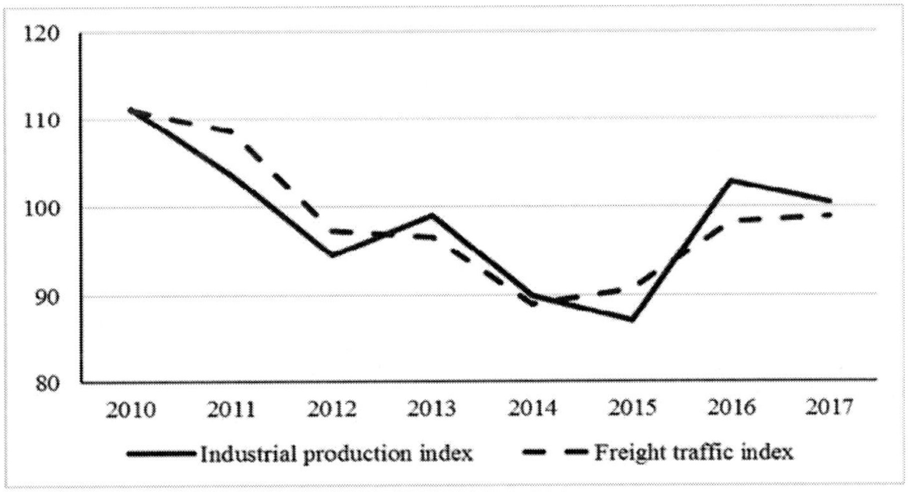

Fig. 1. Dynamics of industrial production and freight traffic (railway transport of Ukraine; correlation coefficient 0.912).
Source: built and calculated by the authors based on [12].

As for passenger traffic, it is also quite stable and its volume depends on the solvency of the population and employment opportunities. Thus, according to opinion polls, more than 45% of all passengers use rail transport almost daily, while more than 41% of commuter train passengers travel to work and study. In the long-distance (long-distance and international) traffic, 35% of passenger trips are related to visits from relatives and friends, 33% from business trips, and 21% from trips to places of recreation [13].

Another external determinant for the railway sector is the participation of the government in financing passenger transport and infrastructure maintenance. This is due to the fact that (iii) European railways are either owned by the state, or the state has a controlling stake; (ii) EU states actively regulate the activities of the national railway companies; (iii) states subsidize the functioning of the railway sector (subsidies for many railways account for more than 50% [14]); (iv) railway infrastructure is not attractive for private investors; (v) railway companies solve economic, social and military tasks, often without sufficient financial compensation of the state. The railway infrastructure of Ukraine is state-owned and the state shifts its social obligations for passenger traffic and the provision of preferences to individual companies and industries in general without adequate compensation. This not only has no contribute to the sustainability of rail transport, but also hinders its modernization.

The influence of the political and socio-economic situation in the country also significantly affects the sustainability of rail transport. An illustrative example is the example of Ukraine, which due to the hostilities in the Donbass lost 3.0% of the railway infrastructure [15], which served the most developed industrial region of Ukraine. This resulted in significant losses for the national economy due to a decrease in industrial production in 2014 by 11.1%, in 2015 - by 9.4%. This trend continues to this day.

The own losses of Ukrzaliznytsya JSC were also significant: the restoration of the destroyed railway infrastructure in 2014 was estimated at more than 900 million UAH. The situation is complicated by the fact that the loss of traffic on the Donetsk railway is multiplied for the entire Ukrzaliznytsya JSC, which led to a threefold decrease in performance across the entire railway network. The result was a loss of 15.4 billion UAH [15]. Since 2014, the situation continues to deteriorate due to the increased instability in the country, deterioration of relations with traditional partners, the lack of a rational government strategy in the development of the national infrastructure.

The external determinants form the basis of the functioning of transport enterprises, including the railway, which has an impact on their internal determinants.

3.2 Internal determinants of rail transport sustainability

The effect of external factors is also enhanced by internal ones, the negative impact of which at the moment is no less significant and at a certain moment can "override" positive changes in the macroeconomic environment. The management of Ukrzaliznytsya JSC recognizes 5 groups of risks that can have a significant impact on its activities, namely (1) economic risks (risk of a decrease in freight traffic); (2) significant investment needs (relatively high level of deterioration of infrastructure and rolling stock); (3) staff costs (a large proportion of staff costs in operating costs); (4) currency risk (part of expenses and liabilities denominated in the foreign currency); (5) refinancing risk and interest rate risk [15]. However, the results of the public audit and data of the consolidated reporting show that this list of threats does not correspond to the current situation in Ukrzaliznytsya, JSC.

At the moment, risk factors that have a key impact on the level of production, economic and financial stability of Ukraine's railway transport are: the lack of effective institutional transformations of the industry, the state of non-current assets, deterioration of financial condition, low investment activity.

Institutional changes in the railways of Ukraine should have taken place in the process of its reform. However, as repeatedly noted by researchers and practitioners, the changes that have occurred are inconsistent and irrational due to the lack of a state long-term transport strategy, excessive state intervention in the activities of Ukrzaliznytsya, ineffective top management, high levels of corruption etc. quality indicators of the industry. The result was a rapid deterioration in the quantitative and qualitative indicators of the industry.

Non-current assets of Ukrzaliznytsya JSC at the end of 2017 accounted for 94.3% of the total amount of its assets, including fixed assets - 69.2%, the depreciation of which according to the financial statements is on average 59.1% [15] with actual physical and moral wear and tear of more than 90% (electric fleet wear - 95.7%, diesel locomotives - 99.7%). In addition, 31% of the fleet of electric locomotives and 44% of the fleet of diesel locomotives are out of order [16]. Such a condition of fixed assets creates a real threat not only of emergency situations in the process of transporting goods and passengers, but also the actual withdrawal of certain sections of the network from service, which will lead to a violation of the integrity of the railway network.

In 2017, Ukrzaliznytsya JSC received a net financial result in the amount of UAH 114.5 million. However, this did not happen as a result of an increase in traffic volumes and the provision of services, but rather an increase in tariffs and a decrease in gross expenditures as a result of non-compliance with the capital investment plan. At the end of 2017, long-term liabilities and collateral amounted to 30.0 billion UAH, and short-term liabilities amounted to 23.7 billion UAH, that is, there is a problem not only of poor performance, but also of reduced solvency and financial stability, resulting in a technical default in 2016. This problem is aggravated by the fact that more than 90% of the capital investment of Ukrzaliznytsya JSC is financed with its own funds.

Under the conditions of a critical deterioration in the financial state, corruption, first of all, in the tender sphere becomes one of the factors reducing the level of sustainability of Ukrzaliznytsya JSC, since it leads to systematic leaching of funds from its economic turnover. The transition of tenders to the ProZorro system did not significantly change the situation.

In 2017, the management of Ukrzaliznytsya JSC presented an investment plan for the development of the railway sector for 2017-2021. This plan provides for investments in the amount of $5.8 billion, including the purchase of rolling stock - $ 3.2 billion, its modernization - $ 709 million, and infrastructure development - $ 1.9 billion (Table 1).

Table 1. Investment plan of Ukrzaliznytsya JSC.

	Current state, units	Elements of the investment plan		
		Purchase, units	Modernization, units	Cost, million dollars
Locomotives	3.9	262	798	1,667
Freight wagons, thousand	82.5 (working park – 62.9)	34.6	46.02	1,309
Passenger cars, thousand	4.6 (working park – 3.1)	419	596	541

Source: compiled by the authors based on [16].

Not only the future state of the railway network, but also its current sustainability will depend on the implementation of this investment plan. However, in 2018, Ukrzaliznytsya, JSC made capital investments in the amount of 16.9 billion UAH ($ 610 million), which made it possible to reconstruct 309 km of the track (at the same time, at the beginning of 2019, 3460 sections of routes with a total length of 1,857 km were closed due to extremely unsatisfactory condition), as well as overhaul of 130 locomotives and 3.5 thousand freight cars, 126 passenger cars were updated, and 15 locomotives were purchased. Such volumes of capital investments are insufficient, since the annual investment needs amount to about 30 billion UAH annually.

4 Conclusions

Ensuring production, economic and financial sustainability of rail transport should be comprehensive and involve the implementation of a set of measures aimed not only at improving the efficiency of its activities, but also the introduction of a rational reform program. Further study of this problem should be aimed at a model study of the degree of influence of both individual determinants and their interaction on the current and future sustainability of the railway sector. As a tool for such a study, simulation modeling can be used. Building and analyzing a set of interrelated simulation models will not only correctly assess the possible level of threats to the sustainable operation of rail transport and, as a result, the national economy as a whole, but also develop a set of preventive measures to level the negative impact of changes in external and internal determinants.

References

1. European commission. Connecting Europe and Asia: Building blocks for an EU Strategy. Retrieved from: https://eeas.europa.eu/headquarters/headquarters-homepage/50724/connecting-europe-and-asia-building-blocks-eu-strategy_en (2018).
2. A.G. Dugin, *Geopoliitika bases*. Moscow: "ARKTOGEYA-tsentr" (2000).
3. E. Zenkina, V. Kutovoi, Transport system and its impact on the sustainable development of the russian economy. Economy: Problems, Solutions and Perspectives, **2**, 95-100. DOI 10.26425/1816-4277-2019-2-95-100 (2019).
4. W-P Zingel, National Security and Economic Development: Securing Development – Developing Security. Retrieved from: http://www.bundesheer.at/pdf_pool/publikationen/eco_impacts_02_national_security_zingel.pdf (2009).
5. Yu.I. Sokolov, R.A. Kozhevnikov, Z.P. Mezhokh, Russian transport business, **6**, 10-13 (2016).
6. Yu.A. Kharlamova, Problems of National Strategy, **1(10)**, 113-128 (2012).
7. S.E. Polzin, Security Considerations in Transportation Planning. Retrieved from: https://www.planning.dot.gov/documents/SecurityPapers/SecurityConsiderations_Polzin.htm (2002).
8. National Security Strategy of the Republic of Poland Retrieved from: https://www.bbn.gov.pl/ftp/dok/NSS_RP.pdf (2014).
9. National Security Strategy of the Republic of Hungary. https://www.files.ethz.ch/isn/.../Hungary_English-2004.pdf (2004).
10. India's National Security Strategy. Retrieved from: https://manifesto.inc.in/pdf/national_security_strategy_gen_hooda.pdf (2019).
11. Sustainable Transportation and TDM. Planning That Balances Economic, Social and Ecological Objectives. Retrieved from: https://www.vtpi.org/tdm/tdm67.htm (2017).
12. State Statistics Service of Ukraine. Retrieved from: http://www.ukrstat.gov.ua/ (2010-2017).
13. N.P. Teroshina, V.G. Galaburda, M.F. Trikhunkov, *Economy of rail transport.* Moscow: UMTs ZhDT (2006).
14. J. Drew, European railway reform and performance. *The Voice of European Railways*, 25th March (2009).
15. State Administration of Railway Transport of Ukraine "Ukrzaliznytsia". Retrieved from: http://www.uz.gov.ua/ (2014).
16. Actual problems of Ukrzaliznytsia and possible ways of their solution. Public Audit. Retrieved from: http://publicaudit.com.ua/reports-on-audit/aktualni-problemi-ukrzaliznitsi-ta-mozhlivi-shlyahi-yih-virishennya/ (2017).

Formation of organizational support of the economic security management of the engineering enterprise: methodical and practical aspects

Andriy Shtangret[1*], *Taras* Shyra[1], *Khrystyna* Mandzinovska[1] and *Sofia* Bilous[2]

[1]Ukrainian Academy of Printing, Faculty of Media Communications and Entrepreneurship, Pid Holosko st., 19, Lviv, 79020, Ukraine
[2]Ivan Franko National University of Lviv, Faculty of Geography, Hryshevskolo st., 4, Lviv, 79005, Ukraine

Abstract. In the conditions of critically high uncertainty of changes in the operating environment for the most engineering enterprises, the relevance of ensuring financial security increases, not only for the effective implementation of economic activity, but also for the survival and preservation of business. The implementation of protective measures in the field of economic security provides for counteracting the negative impact of external and internal threats, and its effectiveness is determined by its ability to minimize potential losses. The development of scientific forecasts of changes in the impact of key threats is intended to form an analytical basis for adjusting tactics and strategies for managing economic security. A model for managing the economic security of a engineering enterprise has been developed that allows for the internal interaction of key elements to be carried out in advance, to concentrate the efforts of security subjects and to accumulate the necessary resources.

1 Introduction

In the report of the Munich Security Conference 2016, a forecast was made for the growth of challenges, threats and increased instability that the world community will face. According to the expert forecasts, the next few decades will be a period of elevated risks (including, in particular, the risk of international military confrontations), a period of fundamental transformation: the beginning of a more unstable era for the world community (Munich Security Report 2016). It can be argued that the problem of security has long historical roots, but its solutions are becoming increasingly relevant at every level, i.e., from international, national to regional, individual enterprises and individuals. So, A. Volfers consistently develops the thesis that security is a value, along with other values of

* Corresponding author: shtangret.am@ukr.net

the national state (The Insecurity Dilemma: National, Regime and State Securities in the Third World (1992). According to D. Kaufman, (1985) who revealed the concept of security in an objective and subjective aspect, in an objective sense, security reflects the absence of threats to the basic values of society, and in the subjective aspect - the absence of fear that these values may experience destruction. J. Lazowski and O. Lange (1977) have proved the relationship between security and risk.

An important role in the study of the evolution of economic security and its international and legal aspects was played by the works of domestic and foreign scientists, such as: R. Kupiecki (2014), L. Lima (2015), N. MacFarlane (2006), S. Huntington (1996).

A visual representation of all processes in a particular enterprise is reflected, above all, in the financial sector, and therefore it is important to find and implement effective solutions for managing economic security, which should provide flexibility and adaptation to changing conditions, and at the same time make it possible to achieve financial interests.

2 The periods of influence on the economic security of the engineering enterprise of external and internal threats

Ensuring economic security at the national level is not possible through the implementation of operational management, but primarily due to the prediction of changes in the level of security, including due to the increased influence of a certain external or internal threat. That is, from our point of view, the main prerequisites for ensuring the economic security of engineering enterprise is the ability not only to recognize, but also to predict the change in the influence of external and internal threats, makes improvements to the system of enterprise protection as a socio-economic system. At the same time, the basic position for forecasting should be information on the current level and changes in the economic security of a certain business entity. So, practice shows (Beshelev, Gurvich, 1973) that forecasts, not at all related to past experience and the existing situation, usually do not come true. The numerous varieties of scientific predictions can be divided into three main groups:

forecasts characterizing trends and prospects for the development of a specific phenomenon for a specific time in the future. Such forecasts will be called realistic (in the future - and a variant of the situation).

forecasts characterizing the most probable state for some specific time in the future. Such a forecast will be considered pessimistic (option II).

optimistic forecasts, that is, forecasts characterizing the desired state of the phenomenon in the future (option III).

In order to improve the management of economic security of the engineering enterprise, we propose to forecast the possible change in the impact of key external and internal threats through the formation of three scenarios (realistic, pessimistic and optimistic). To further specify the methodological basis for improving the management of the economic security of the engineering enterprise, forecasts of changes in external and internal threats will be considered separately.

It should be noted that the prevailing forecasts are based on and take into account changes in the nature of each of the identified threats which occurred in two base periods: 2010-2012 and 2013-2017. (Shtangret, Mandzinovska, Kotlyarevskyi. Melnykov, 2016).

Based on the characteristics of the impact of external threats on the economic security of an engineering enterprise during the years 2018-2025, the matrix of dependence of external threats from each other is built. Below the dependencies of threats for a realistic development of the situation are given. Next, based on the dependence matrix, a directed graph is constructed, at the vertices of which there is an element (q_1,... q_7), whose arcs connect the vertices for which the connection is defined (Table 1). In order not to increase the volume of the article we omit the graph. Next, a accessibility matrix is built on the basis of the graph (Table 2).

Table 1: Matrix of dependence of external threats to economic security of engineering enterprise during 2019–2025 pp. (I scenario)

q_{ij}	q_1	q_2	q_3	q_4	q_5	q_6	q_7
q_1	0	1	0	0	0	0	1
q_2	0	0	0	0	0	0	1
q_3	1	1	0	0	0	0	0
q_4	1	1	1	0	1	0	0
q_5	0	1	1	0	0	0	0
q_6	1	0	1	1	1	0	0
q_7	0	0	0	0	0	0	0

Table 2: The matrix of the reach of external threats to the economic security of a engineering enterprise during 2019–2025 pp. (I scenario)

q_{ij}	q_1	q_2	q_3	q_4	q_5	q_6	q_7
q_1	1	1	0	0	0	0	1
q_2	0	1	0	0	0	0	1
q_3	1	1	1	0	0	0	1
q_4	1	1	1	1	1	0	1
q_5	1	1	1	0	1	0	1
q_6	1	1	1	1	1	1	1
q_7	0	0	0	0	0	0	1

On the basis of the matrix of reach of external threats to the economic security of a engineering enterprise (Table 2), we build an iterative table and calculate the hierarchy level of the threats identified. Calculations for options II and III are performed as described above and are further omitted.

On the basis of the calculations, we have built a Table 4, reflecting the change in the priority impact of a certain external threat on the economic security of a engineering enterprise. It is important that the proposed forecasting options (realistic, pessimistic and optimistic) make setting clear benchmarks in relation to which the situation may develop, allowing you to further specify both the amount of resources needed and the tasks for each security subject.

The results obtained (Table 3) show that the level of development of the national legal system and compliance with the current legislation q_2 is crucial for the economic security of a engineering enterprise in recent years (2010-2018) and for the future (I scenario). Logical (due to its significant deterioration) is also a growing influence on the development of the situation of the state of the monetary financial system of the state in 2014-2018 compared to 2010-2013.

Compared with the above threats, it seems quite logical to stay first (2010-2013) in the middle of a rating, and then (2014-2018) at its end threats related to illegal activities, because the general unsatisfactory economic situation in the country leads to a gradual collapse of production, and this in turn, makes it possible to suggest hypothetically the existence of such a situation, when there is nothing to seize (q_4) and spy (q_5).

According to the results of the analysis of the impact of internal threats in selected periods (2010-2013, 2014-2018 and 2019-2025). Three possible options were developed (by analogy with external threats) to change the impact of threats, the source of which is the internal environment of the engineering enterprise.

We omit all intermediate calculations, as a result of restrictions on the volume of a scientific article.

Table 3: Results of ranking the importance of influencing the economic security of a engineering enterprise of external threats in different periods

<table>
<tr><th colspan="5">Periods</th></tr>
<tr><th rowspan="2">2010–2013</th><th rowspan="2">2014–2018</th><th colspan="3">forecast for 2019–2025</th></tr>
<tr><th>scenario I</th><th>scenario II</th><th>scenario III</th></tr>
<tr><td colspan="3">level of development of the legal system state q_2</td><td>q_3</td><td>q_7</td></tr>
<tr><td>q_3</td><td>state of the monetary and financial system q_1</td><td colspan="2">q_7</td><td>q_2</td></tr>
<tr><td>prerequisites raider capture q_4</td><td>q_3</td><td>q_1</td><td>q_5</td><td>q_3</td></tr>
<tr><td>activity of criminal structures and competitors q_5</td><td>q_7</td><td>illegal activities of government representatives q_3</td><td>q_2</td><td>q_5</td></tr>
<tr><td>volume of domestic market and increased competition q_7</td><td colspan="2">q_5</td><td>q_1</td><td>q_4</td></tr>
<tr><td>q_1</td><td colspan="2">q_4</td><td>q_6</td><td>q_1</td></tr>
<tr><td colspan="3">interest rates on loans from commercial banks q_6</td><td>q_4</td><td>q_6</td></tr>
</table>

The data obtained for the three scenarios in the years 2019-2025 just as for the situation of recent years (2010-2018), has it own internal logic:

the value of the level and quality of expenditures g_6 (I scenario) compared to other internal threats, provided that the current problems in the economy are preserved, primarily due to the need to optimize them;

accordingly, for the deepening of economic difficulties (scenario II), the level of dependence on external sources of financing will increase g_7;

in the optimistic scenario (scenario III), the amount of proceeds from the sale of g_5, availability of working capital g_2 and profitability of the main type of activity g_4, that is, indicators determining the economic efficiency of the engineering enterprise in terms of a favorable economic environment;

the amount of revenue from the sale of g_5 (for scenarios I and II) is less important than for the desired development of events (III scenario) because, given the accumulation of problems, the amount of revenue is no longer a decisive threat to the economic security of the enterprise.

In the list of threats (for scenarios I and II) the quality and size of accounts receivable and payable g_1, availability of working capital g_2, profitability of the main activity g_4 and liquidity of assets of the enterprise g_3 remain unchanged. Such analysis results can be explained by the fact that when the situation on the market of engineering products becomes catastrophic it is difficult to talk about asset liquidity (g_3), profitability (g_4), working capital (g_2) and debt (g_1), because when enterprise is on the verge of survival, these indicators in spite of their importance, with favorable developments, they no longer play such a role in its negative development.

3 Model of economic security management of the engineering enterprise

Our analysis is based on the application of a systematic approach, because each threat that affects the course of the analyzed process is evaluated in conjunction with all the others. The system approach allows comparing possible solutions to the problem. Each problem is known to be characterized by the need to change the state, and the solution establishes how the transition will be made from the existing state to what is assumed. Predictions allow you to prepare in advance for future events, take into account their positive and negative consequences, to a certain extent regulate the course of events. Based on this, we consider it appropriate, on the basis of the prevailing predictions about the possible change in the influence of external and internal threats, to offer a scientific and methodical approach to the formation of a tactics and strategy to ensure the economic security of the enterprise (Fig. 1).

The tactical and strategic tasks that we formed in the process of implementing the economic security management of the engineering enterprise are based both on a hierarchical streamlining of external and internal threats in 2010-2018. They also based on the results of forecasting changes in the impact of threats with a realistic, optimistic and pessimistic option. Therefore, this can be considered a fairly reasonable basis for improving the management of the economic security of each engineering enterprise.

The economic security management strategy is aimed at solving a number of tasks, the implementation of which should ensure the formation of conditions for the safe development of an enterprise while meeting financial interests to the maximum, based on financial possibilities and by coordinating with the key positions the strategy of the enterprise's economic development.

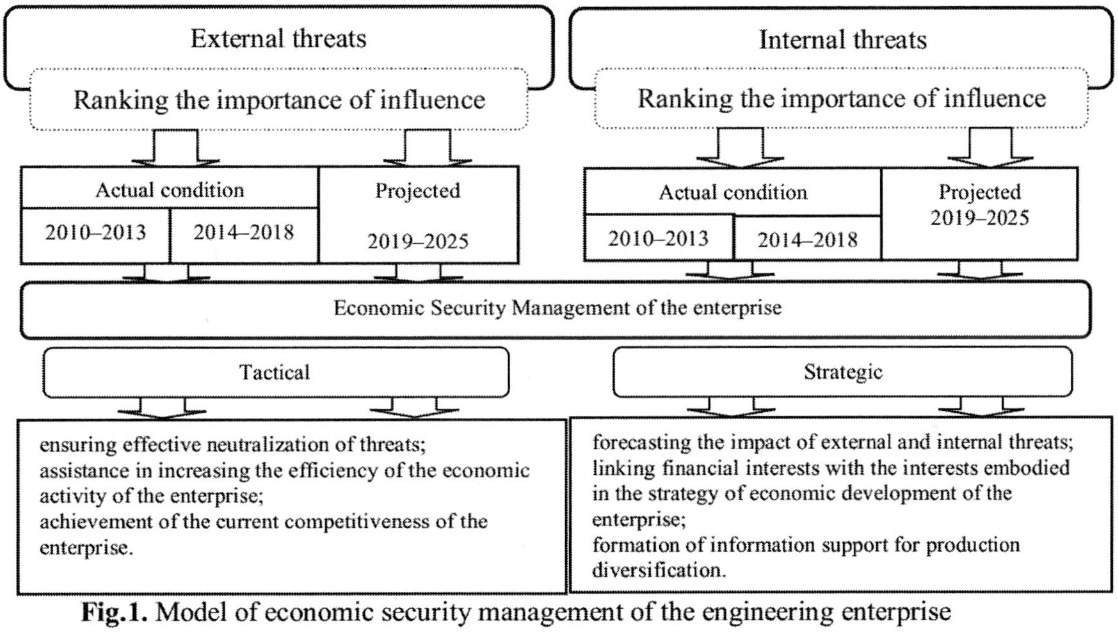

Fig.1. Model of economic security management of the engineering enterprise

4 Conclusions

Summing up, it should be emphasized that the increase in the efficiency of economic security management of domestic engineering enterprises to a large extent depends on the ability to predict changes in the level of economic security under the influence of key external and internal threats. Forecasting allows the concentration of available financial, material, labor, information and other resources to implement the most optimal tactical and strategic decisions, and provides improved management of economic security.

References

1. Munich Security Report. Boundless Crises, Reckless Spoilers, Helpless Guardians URL : https://espas.secure.europarl.europa.eu/orbis/document/munich-security-report-2016-boundless-crises-reckless-spoilers-helpless-guardians (2016)
2. The Insecurity Dilemma: National, Regime and State Securities in the Third World edited by Brian. L. Job. London, Boulder : Lynne Rienner (1992)
3. Kaufman D. US National Security: A Framework for Analysis. Lexington, MA. Lexington Books (1985)
4. Łazowski J., Lange O. Wstęp do nauki o ubeypieczeniach. Optymalne decyzje / Dzieła. Tom 6. Wyd. 2. Warszawa: PWE, 204 (1977)
5. Kupiecki R. Strategia Bezpieczeństwa Narodowego RP 2014 jako instrument polityki państwa. Uwarunkowania zewnętrzne i aspekty procesowe URL: https://www.bbn.gov.pl/ftp/dok/03/KUPIECKI_33-2015.pdf.
6. Lima, L. WorldingBrazil: Intellectuals, Identity and Security. NY, 179. (2015).
7. MacFarlane, N. Human security and the UN: a critical history. United Nations intellectual history project (illustrated ed.). Indiana University Press. (2006)
8. Huntington S. The Clash of Civilizations and the Remaking of World Order. N. Y. : Simon & Shuster, 368
9. Beshelev, S. D., Gurvich, F.G. Expert assessments. M.: Science, 160 (1973)
10. Shtangret, A., Mandzinovska Kh., Kotlyarevskyi. Ya., Melnykov, O. Financial security of a engineering enterprise: Methodical bases of formation and provision. Lviv: UAP, 240 (2016)

Modeling the effects of key threats on the economic security of industrial enterprises

Svitlana Shynkar[1] *, *Bohdan* Brynzei[1], *Natalia* Rozumovych[1] and *Marian* Kurliak[2]

[1]King Danylo University, Faculty of Management and Business, Konovaltsa st., 19, Ivan-Frankivsk, 76000, Ukraine
[2]Lviv State University of Internal Affairs, Faculty No.3, Horodotska st., 29, Lviv 79007, Ukraine

Abstract. The current business environment of industrial enterprises is characterized not only by the complexity of transformational processes in the national economy, political and social tensions, increased competition from foreign producers and the introduction of protectionist measures from importing countries, etc., but above all the unpredictability of changes in the environment. This aspect should be considered as the cornerstone in the development of a new approach to the formation of a strategy to ensure the economic security of industrial enterprises. The purpose of the study is to examine the theoretical and methodological aspects of the formation and implementation the strategy to ensure the economic security of industrial enterprises in order to create the basis for their safe development.

Introduction

In fact, one can speak of the presence of three key approaches, which differ in their interpretation of the essence of strategy and strategic management. According to the first approach, supported by J. Higgins, (1983) strategic management is defined as a decision-making process that combines internal organizational capabilities with threats and favorable opportunities provided by the external environment to interpreting strategic management and forming a modern definition of the management process in order to fulfill the organization's mission through management the organization's interaction with its environment.

According to the second (target) approach, strategic management is a set of tasks, the achievement of which helps the company to maintain its position in the market in the long term. Representatives of this direction are J. Pearce and R. Robinson (1985).

In accordance with the third (effective) approach, strategic management is a process that helps in the implementation of long-term plans of the enterprise. D. Schendel and K. Hatter (1972) characterize it as a process of identifying and establishing the connection of the enterprise with its environment, consisting in the realization of selected goals and in

* Corresponding author: drvel@ukr.net

attempts to achieve the desired state of relations with the environment through the allocation of resources, allowing the company and its divisions to operate effectively and efficiently.

The problem of developing and implementing a strategy of ensuring the economic security at the level of individual enterprises has been little studied. This is found in the scientific works of R. Inglehart and R. Abramson (1994), H. Poirson (1998), which actualize the importance of conducting scientific research contributing to the development of the most appropriate strategies for ensuring the economic security of the enterprises.

The rationale for choosing a particular strategy option

During the building possible future scenarios and making management decisions, the decision maker usually encounters complex interdependencies of events that need to be analyzed. To solve such problems, they usually use the method of analyzing hierarchies that is also used in solving economic problems (Saati, 1993; Sylkin, 2018; Sgurovsky 2007; Chernyak, 2016; Shtangret, 2016).

Using the method of analyzing hierarchies, we will justify the choice of a strategy for ensuring the economic security of the industrial enterprise. Therefore, we formulate the essence of the problem as follows: the need to choose a strategy to ensure the economic security of the industrial enterprise in accordance with the hierarchy of scenario choices (Fig. 1), the top of which is the goal "Choosing the economic security strategy of enterprise". The first level of this hierarchy is formed by four levels of economic security (criteria): "High level of economic security», "Average level of economic security", "Low level of economic security" and "Critical level of economic security".

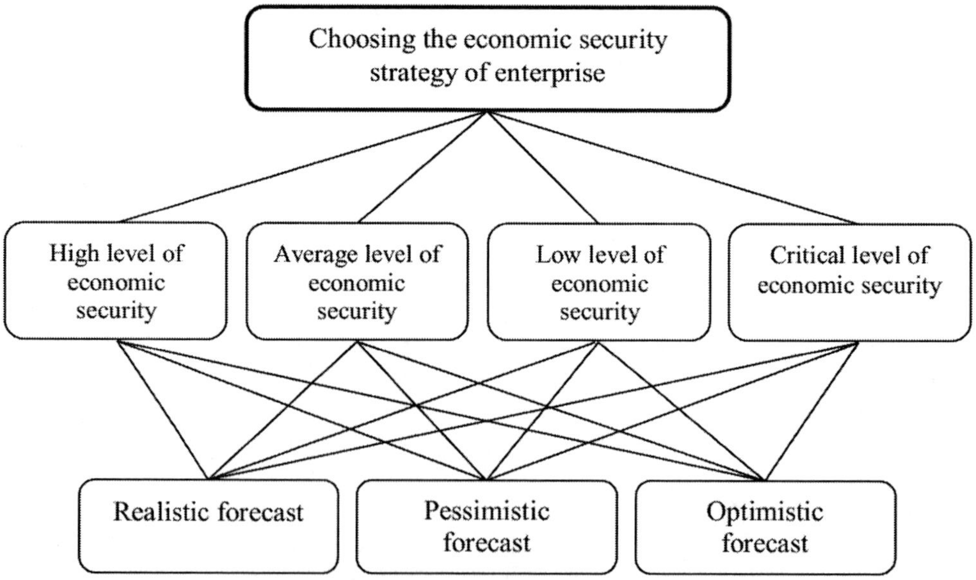

Fig. 1. The hierarchy of the problem of choosing a strategy to ensure the economic security of industrial enterprises

At the last level there are three alternative scenarios: "Realistic forecast of the situation development", "Pessimistic forecast of the development of the situation" and "Optimistic forecast of the development of the situation".

Next, the expert needs to make $\frac{n \times (n-1)}{2}$ comparisons, where n is the number of criteria on the same level. In our case, four levels of economic security, respectively, it is necessary to conduct six comparisons of different pairs of security levels.

Since there are three alternatives and four criteria in the hierarchy, it is necessary to carry out $n \times \frac{m \times (m-1)}{2}$, where m is the number of alternatives, that is 12 comparisons of pairs of alternative scenarios.

Approbation of the theoretical and methodological approach was carried out to select the strategy for ensuring the economic security of Natural Resources CJSC, which was based on analytical materials on the actual level of economic security of this enterprise in 2010-2018. And the forecast changes in the impact of key threats to the oil and gas industry.

Building a matrix of pairwise comparisons

During the building a matrix of pairwise comparisons of levels of economic security, they were compared in pairs with respect to a common goal (the choice of a strategy to ensure the economic security of the enterprise). The results of the paired comparisons are decorated in the form of a table 1.

During the comparing scenarios for the development of the situation, not one, but four tables are formed, because these options are compared for each of the four levels of economic security (Tables 4).

For each of the matrix of pairwise comparisons of forecasts of changes in the impact of threats on the economic security of industrial enterprises at different levels of economic security of the enterprise, the results of the ranking of forecasts are given (Table 5) and the levels of inconsistency of the matrix of pairwise comparisons are determined. The calculations were performed using a computer program (Gileta, 2012).

Table 1: Matrix of pairwise comparisons of the levels of economic security of industrial enterprises

Level of security	High	Average	Low	Critical
High	1	3	5	7
Average	1/3	1	3	5
Low	1/5	1/3	1	3
Critical	1/7	1/5	1/3	1

Table 2: The sums of the columns of the matrix of pairwise comparisons of the levels of economic security of industrial enterprises

w_i	w_1	w_2	w_3	w_4
S_j	0.563	0.263	0.117	0.055

Table 3: Determining the level of inconsistency of the matrix of pairwise comparisons is presented in the form of Table 1

Indicator	The largest proper number, k_{max}	Coefficient of consistency, CI	Inconsistency level, CR
Calculation results	4.117	0.039	0.043

Table 4: Matrix of pairwise comparisons of forecasts of changes in the impact of threats on the economic security of industrial enterprises for a high level of economic security

High level	Realistic	Pessimistic	Optimistic
Realistic	1	5	1/3
Pessimistic	1/5	1	1/6
Optimistic	3	6	1

Table 5: The sums of the columns of the matrix of paired comparisons of forecasts of changes in the impact of threats on the economic security of industrial enterprises for a high level of economic security

u_{1j}	u_{11}	u_{12}	u_{13}
S_j	0.287	0.077	0.634

Table 6: Determining the level of inconsistency of the matrix of pairwise comparisons is presented in the form of a Table 4

Indicator	The largest proper number, k_{max}	Coefficient of consistency, CI	Inconsistency level, CR
Calculation results	3.094	0.047	0.081

By analogy, calculations were made for the average, low and critical levels.

The calculations were performed correctly, since the value of the largest eigenvalue k_{max}, the coefficient of consistency, CI and the level of inconsistency, CR for each of the matrices (Table 4,) are within the normal range (Table 3, 6).

The choice of the strategy of ensuring the economic security of the enterprise (synthesis of priorities) U_i is carried out according to the formula:

$$U_j = \sum_{i=1}^{4} w_i u_{ij}; \; j = 1,2,3, \qquad (1)$$

where w_i is the priority of the i-th level of economic security of the enterprise ($i = 1,\dots, 4$);

u_{ij} is the relative weight of each of the forecasts ($j = 1, 2, 3$) of changes in the impact of threats on the economic security of the enterprise according to the i-th level of economic security of the enterprise ($i = 1,\dots, 4$).

In accordance with (1), we have the following options for calculating the values of priorities for all alternative scenarios:

$$Un = w_1 \times u_{11} + w_2 \times u_{21} + w_3 \times u_{31} + w_4 \times u_{41}; \qquad (2)$$

Substituting the corresponding values from Tables 2-6 in formulas (2), we obtain the following values of priorities for all scenarios that were considered:

$$U_1 = 0.563 \times 0.287 + 0.263 \times 0.236 + 0.117 \times 0.279 + 0.055 \times 0.319 =$$

$$= 0.161 + 0.062 + 0.033 + 0.017 = 0.273;$$

$$U_2 = 0.082;$$

$$U_3 = 0.640.$$

The obtained results allow us to state that the third strategy of ensuring the economic security of industrial enterprises is the most acceptable one. That is, according to the third scenario of the development of the situation for CJSC «Natural Resources», it is necessary to carry out measures for adapting the activity of the enterprise to the effects of threats, providing for the implementation of the adaptation strategy.

On the basis of the above positions, we have formed a model that includes, in addition to the main stages of formation, the procedure for implementing the strategy to ensure the economic security of industrial enterprises (Fig. 2).

Briefly describing the process of implementing the strategy of ensuring the economic security of industrial enterprises, it is expedient to emphasize that with a quick change of the functioning environment, the role of systematic control and adjustment increases both by clarifying certain parameters of the chosen strategy and implementing tactical protective measures.

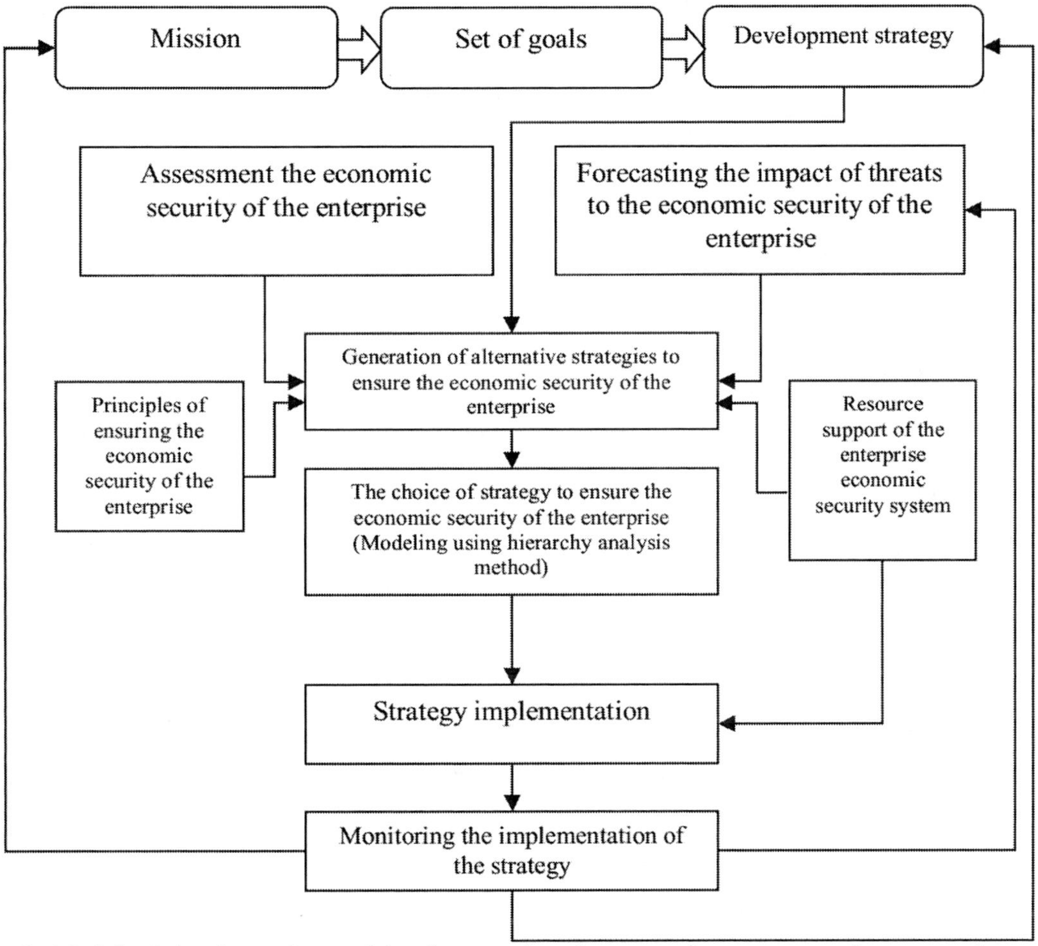

Fig. 2. Model of the formation and implementation the strategies to ensure the economic security of industrial enterprises

We consider the opinion of individual representatives of the business community to be erroneous regarding the expedient use of exclusive tactical control in the process of ensuring the economic security of industrial enterprises. This is explained by the fact that the essence of key threats to industrial enterprises does not allow organized the effective protection in short periods of time and with the available resource support. The developed strategy makes it possible to adapt gradually economic security system of the enterprise to new tasks and changed business conditions, therefore, to ensure that it can perform all the tasks to create safe conditions for the development of industrial enterprises.

 https://doi.org/10.1051/shsconf/20196704010

Conclusions

Summarizing, it can be argued that the discussed methodological and practical aspects determine the procedure for developing and implementing the strategy to ensure the economic security of industrial enterprises are governed by its subordination to the mission and set of objectives, the relationship with the development strategy. Strategic orientation in ensuring the economic security of industrial enterprises contributes to the timely improvement of the economic security system of the enterprise in accordance with changes in the development strategy and level of influence of the operating environment. Constant monitoring and adjustment of the implementation the strategy ensures the stability of the enterprise, mutual consistency and the possibility of achieving economic interests.

References

1. Higgins J. 1983. *Organizational Policy and Strategic Management.* Text and Cases. 2nd ed. Chicago: The Dryden Press, 237. (1983)
2. Pearse J., Robinson R. *Strategic Management. 2nd ed. Homtwood III* : Richard D. Irwin, 268 (1985)
3. Schendel D., Hatter K. Business Policy or Strategic Management. A Broader View for an Emerging Discipline. Academy of Management Proceeding. (1972.)
4. Inglehart R., Abramson P. American Political Science Review. **88**. 336–354. (1994)
5. Poirson H. Economic security, private investment, and growth in developing countries. Washington: International Monetary Fund, African Dept., 31 (1998).
6. Sylkin O., Shtangret A., Ogirko O., Melnikov A., Business and Economic Horizons, **14, 4,** 926-940 (2018.)
7. Zgurovsky M. Z., Pankratova N. D. *Technological Foresight.* Kiev: Polytechnic, 156, (2007)
8. Chernyak G. M. Effective economy. **3**. URL: http://www.economy.nayka.com.ua/?op=1&z=4863 (2016).
9. Khalina O.V., Shantreget A.M. Theoretical and Methodical Principles of Formation of Organizational Support for the Management of Economic Security of the Machine-Building Enterprise. Lviv: UAP, 252. (2016)
10. Gileta, V., Senkivsky, V., Melnikov, A. Simulation modeling in the system analysis by the method of binary comparisons: Computer program: registration certificate author. rights to a work 41832 Ukraine / № 42107; declare 11/17/2011; publ., Bull. **27**, 88 (2012).

Competitiveness Management in Motor Transport Enterprises

Volodymyr Shynkarenko, *Oksana* Kryvoruchko and *Iryna* Fedotova[1],

Kharkiv National Automobile and Highway University, Department of Management and Administration, 61002 Ya. Mudrogo str. 25, Kharkiv, Ukraine

Abstract. The theoretical rationale for the model of developing a strategy for competitiveness managing based on a combination of the positioning and provisions of the resource approach theory is given. The management strategy is being developed at the upper (strategic) and medium (tactical) levels. At the upper level, there is a positioning of the enterprise environmental requirements potential. At the level of individual strategic business units (SBU) – analysis by the method of M. Porter, the possibilities are determined on the basis of provisions of the resource approach. Formation of a common competitiveness management strategy is a compromise of a common strategy project with SBU enterprise strategies. The composition of the competitiveness management system of enterprise is validated: the main subsystems, their functions and interrelations between them are defined. Particular attention is paid to the management subsystem, which consists of three subsystems: structural and functional, self-development and informational and behavioral. The main functions of the decision, transforming and information controlling blocks at all management levels are defined. A model of structure of the enterprise competitiveness management system is proposed. Under strategic management, the development of functional strategies is proposed to be transferred to the upper and middle management levels. Using the matrix management structure of the functional units of the enterprise is suggested.

1 Introduction

Competitiveness management of the enterprise is the current interest problem of the economic objects development. Dissertations of many Ukrainian and foreign researches are devoted to the study of this problem. On many issues of this problem, there is no consensus among scientists. These contradictions begin with the definition of the concept "competitiveness management of the enterprise", with its separation into an independent type of management, with the definition of its content. Changes in economic science over the past five years, in terms of the theory and methodology of developing an enterprise strategy, have put on the agenda the issue of changing of an enterprise and its competitiveness management methods.

The development of the strategic management theory led to the need of clarifying the functions distribution between levels of government. The noted circumstances require bringing the methodology of competitiveness management to the accordance with the today's circumstances. Therefore, the problem of improving the enterprise competitiveness management is relevant and is the purpose of this study.

[1] Corresponding author: irina7vf@gmail.com

The object of the study is motor transport enterprises, while the subject of the study is theoretical and methodological guidelines for improving the competitiveness management of the motor transport enterprises. It is assumed that the implementation of the results of the study will improve the scientific level of managing the enterprises competitiveness, the level of occupying a decent position in the market.

2 Review of the literature and theoretical foundations

The central place in the economic systems management, including enterprises, competitiveness, etc., is taken by strategic management, it is proposed to use two approaches: market and resource in its implementation.

Nowadays, a market approach has become more widely used, the main ideologists of which were the Harvard Business School scientists and M. Porter, each of them presented their own versions of the approach to the enterprise strategy development. It is generally accepted that 1960 is the beginning of the company's research strategy, when three books of the management theory founders were published: "Strategy and Structure" by A. Chandler [1], "Business Policy" by K.R. Andrews [2], the fundamental work by I. Ansoff "Corporate Strategy" [3].

These fundamental results enable to form the current methodology for the overall enterprise strategy development. These results are outlined in many textbooks and teaching aids about strategic management. In accordance with them, the enterprise's strategy is enlarged in the following sequence: analysis of the external environment – establishing the presence of opportunities and threats for the enterprise in the environment – identifying its strengths and weaknesses; SWOT-analysis – the establishment of alternatives for the enterprises strategic development, the selection of the best ones.

The second variant of the market approach to the development of a competitive strategy for an enterprise was proposed by M. Porter – in the 1980s he promoted the idea of a "value chain". M. Porter found that the position of an enterprise – the results of operations and their sustainability– is determined by five competitive forces operating in the industry: competitors operating in the industry, potential competitors, suppliers, buyers, and substitute goods [4]. Profitability in the industry as a whole is determined by the action of the five forces listed above, since they affect prices, costs and the level of required investment. The efficiency of enterprises in the industry depends on the ability of the enterprise to resist the market power of suppliers and consumers, threats from potential competitors and substitutes, competition between competitors in the market.

Thus, with a market approach in the first case, the strategy of an enterprise is determined by adapting its potential to environmental factors, in the second case by adapting the potential of the enterprise to the conditions of the industry and competition. A common feature of the market approach is the direction of the strategy developing process from top to bottom – from the upper to the lower management level.

Since the mid-1990s, practical and theoretical conditions have arisen for the further transformation of the strategic management theory, changes in the content of an enterprise strategy and the approach to its formation, called the resource approach.

The theoretical basis for the development and use of the resource approach, the transformation of the strategic management theory, changes in the content of the strategy and the approach to its definition, were mainly the studies by Chamberlain and Rubenson (1930), Ed. Penrose (1959), J. Barney (1991), C. Wernerfelt (2006), G. Hamel and I. Prokhalada (1994).

Further development of the resource approach was K. Wernerfelt's organization theory [5], in which the author substantiated the feasibility of analyzing the company in terms of resources rather than products, and called it possibly a new paradigm in the theory of strategic management. Therefore, the starting point of the resource approach application is the establishment and evaluation of those resources characteristics that will allow the enterprise to increase its competitiveness and obtain a upper level of profitability.

G. Hamel and C. Prahalad made the most significant contribution to the development of the strategic management theory and the resource approach development, who proposed the theory of corporation key concepts, the methodology for developing an enterprise strategy based on its root competencies [6].

Taking into account the resource approach prerequisites: organizations within the same industry differ significantly from each other according to the possession and control of the resources available to them; resources are not mobile and therefore the resource heterogeneity of different enterprises within the same industry can persist for a long time; methods of the market approach and the characteristics of the study's object – the used resources homogeneity and the working conditions of enterprises are proposed to combine a market and resource approach to develop a strategy.

At the upper level, there is a potential positioning of the external environment enterprise requirements. At the level of individual strategic business units (SBU) – analysis by the method of M. Porter, the possibilities are determined on the basis of the resource approach provisions. Formation of a common competitiveness management strategy is a compromise of a common strategy project with SBU enterprise strategies.

3 Results and discussion

Enterprise competitiveness management is a specific function that is implemented by a certain composition of the management system. Its main components are: subsystem, elements, components, structure, and goals.

Currently, there is no unambiguous definition of the "control system" concept. The following definition of a management system is proposed: it is a set of subsystems, elements and interconnections between them, that ensure the functioning of the management process in the direction of an object (system, enterprise) achieving its goals.

Managed subsystem (control object) – a subsystem that is experiencing the control actions and executes management commands. By object of management, we mean an organized part of the environment – processes, phenomena, systems that give rise to a problem situation and solve it under the influence of management influence.

A management subsystem or a management entity is a person or a group of individuals who set themselves management tasks, detail them in the form of tasks, bring to the management object and control their implementation. It consists of three subsystems: structural and functional, self-developed, and informational and behavioral.

Structural functional subsystem is designed to implement the methods, principles and functions of management. It is represented as a control loop that combines all control functions in a single cycle.

Structural functional management subsystem (SFS) is a set of management bodies, departments and executors performing the functions assigned to them and solving the set tasks, as well as a set of methods by which management influence is implemented. Previously, it was decided to consider SFS as the actual management system. The structural-functional subsystem acts as the backbone of the control system.

At all levels of management, the SFS consists of three blocks of basic management functions: decisive – D, conversion – R, and information and control – I. On the extended cycle, decision-making functions are performed in the decisive unit. In the conversion unit, the decisions made in the decision unit are implemented. In the information and control unit information is collected, the managing object control and regulation are done.

The final step in developing an enterprise management system is kinematics management. Two more subsystems should be distinguished in the structure of the control subsystem: information-behavioral and self-development.

A managed system is a group of particular management objects that stand out from a common management object for various reasons. So, on the basis of the structural subdivisions of they are distinguished on such particular objects as – motor convoy 1, motor convoy 2, etc.; by types of resources used – fixed assets, labor resources, etc.; according to the results of activity – the volume of work, the competitiveness of the enterprise, the competitiveness of services, etc.

In the managed subsystem, the management process is carried out – the sequential execution of management functions, allowing to coordinate a set of actions that allow the management object to achieve its goals. Based on the control cycle, separate control loops are formed – a closed set of control system elements. In the control loop, each element of the control system influences the next one, receiving influence from the previous one. Each control loop is aimed at solving problems of a specific control object – a structural unit of resource use or the achievement of certain results. The difference between the abstract cycle and the real control loop is in following: the management cycle reflects the logical sequence and interrelations of the main stages of control of any object. Control loops are always specific, as they are formed for individual objects. They are characterized by various goals, forms and methods of performing traditional management functions at all levels and stages of the management process. Therefore, each such circuit can be considered as a control system of the considered control object.

Thus, the management loop, which is formed using the management cycle as a model, focused on solving the problems of managing the enterprise and its services competitiveness, can be considered as a competitiveness management system. Depending on the object of management, there are systems for managing the competitiveness of an enterprise and the competitiveness of services.

The enterprise competitiveness management system is the outline of an enterprise management system aimed at establishing and achieving an planned level of competitiveness of the enterprise through the cyclical sequential execution of management functions (analysis, forecasting, planning, organization, motivation, regulation and control) by the enterprise's activities as a management object [7].

The system of competitiveness management of transport services is the outline of an enterprise management system aimed at achieving a business unit (type of transportation) of its services competitiveness established level through the cyclical sequential execution of the functions of managing the supply process and providing services.

The design and description of such systems begins with the establishment of management levels. Currently, there is no consensus among scientists regarding the number of enterprise management and their content levels, including competitiveness management. E.g., M.Kh. Meskon proposes to distinguish three levels of the management structure: top management, middle management and lower management [8].

Representatives of the second approach are scientists, authors of textbooks and technical aids on strategic management: F.R. David, D. Thompson, R. Lynch, I. Ansoff, Z.E. Shershencheva, B.M. Mizyuk, S.A. Panov, V.N. Parakhina. They offer a four-level model of the organizational structure of management: the upper level – the general strategy of the enterprise, the middle level – the business strategy, the level of functional strategies management, and the lower level.

Considering that the first approach reflects an objectively existing hierarchy, the following levels of management in a diversified enterprise are assumed: upper, middle and primary. So we should provide the implementation of the strategic management functions at the upper and middle level. At the middle level, business strategies should be developed. In the motor transport enterprise, the development of business strategies is necessary for each bus route, for each customer served. For the structure of a diversified enterprise, a model of the management system structure is proposed in Figure 1.

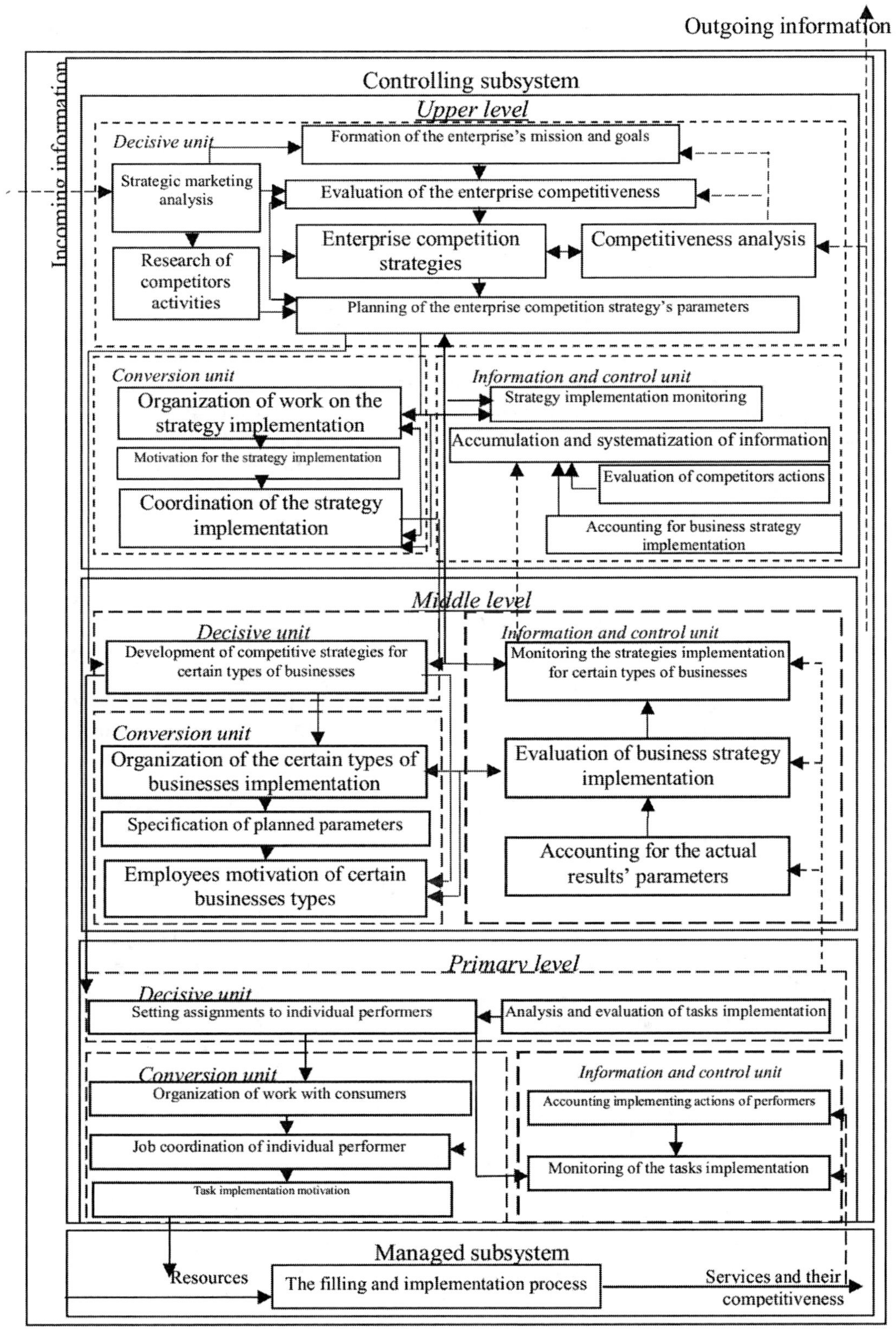

Fig. 1. The model of proposed structure of the enterprise competitiveness management system

Experts in the field of strategic management in their works provide for the presence of functional strategies in the enterprise strategies composition. Their goal is to prepare solutions in the areas of finance, marketing, personnel, investment, etc. In our opinion, it is not expedient to single out a separate hierarchical level of management decision-making on the previously considered issues, because of the following reasons: solutions, prepared by functional service workers, are not always take into the account all the diversity, especially the external conditions of the control object operation; final decisions on them are made at each level by the first person; the amount of functional services management work at the first and second level is different.

The solution of this problem is seen in the use of a three-level organizational management structure and the creation of functional services at each management level, which are vertically subordinate to the upper level, and horizontally to the line manager.

Heads of functional units on the basis of proposals of all SBU within the limits of available resources form the functional strategy of the enterprise. The last one, coordinated with the head of the enterprise, is submitted for discussions with line managers. After discussion and final approval, the functional strategy, together with other documents, is approved as the enterprise development strategy is adopted for implementation and implementation [10].

4 Conclusion

The following results of the study were obtained: a model of strategic competitiveness management of the enterprise was further developed, which, unlike the existing one, is based on a combination of a market and resource approach in the enterprise strategy development; concepts such as "competitiveness management of the enterprise and its services", "management system" were clarified; the methodology for elaborating a competitiveness management system was further developed, the novelty of which is the generalization of existing developments in this area; the enterprise competitiveness management structure in terms of the redistribution of the activities of functional units management work between the upper and middle management levels using a matrix structure was improved.

References

1. A.D. Chandler, *Strategy and Structure.* Cambridge, MA: MIT Press, (1962).
2. K.R. Andrews, *The Concept of Corporate Strategy.* Homewood, Il: Irwin, (1980).
3. H.I. Ansoff, *Corporate Strategy: An Analytical Approach to Business Policy for Growth and Expansion.* McGraw Hill Book Co.: N.Y., (1965).
4. M. Porter, *Competitive Strategy.* New York: The Free Press, (1980).
5. B. Wernerfelt, Strategic Management Journal, **5(2)**, 171-180 (1984).
6. G. Hamel, C. Prahalad, *Competing for the Future.* Boston, Massachusetts: Harvard Business School Press, (1994).
7. V. Shynkarenko, N. Popova, Economycs of transport complex, **30**, 5-14 (2017).
8. M. Meskon, M. Albert, F. Hedouri, *Fundamentals of management.* Moscow, Delo Publ., (1999).
9. I. Fedotova, V. Shynkarenko, O. Kryvoruchko, International Journal of Engineering & Technology, **7 (4.3)**, 445-450 (2018).

Enterprise competitiveness in the digital economy

Olga Sunigovets[1,*]

[1]National Technical University "Kharkiv Polytechnic Institute", NTU "KhPI" 2, Kyrpychova str., 61002, Kharkiv, Ukraine

Abstract. Digital economy is the direction of innovation, competitiveness and economic development of corporate structures. The rapid development of digital technology covers all areas of socio-economic activity. On the basis of digital technologies, new digital business models are being created, existing operational models are being transformed, and fundamentally new types of business are being formed. Digital capital becomes a factor in business competitiveness. The main factor of economic development is associated with the digital economy. Digital economy contributes to the formation of new business models.

According to the World Economic Forum, in 2015 the share of the digital economy in the global economy exceeded 20% and is growing rapidly. The digital economy acts as the accelerator of the socio-economic life of society in the modern world, as capable of rapidly increasing the GDP of the country. Successful experience proves the effect of integrated digital economy development at 20% of GDP over five years, while the ROI of digital conversion investment reaches 500%.

Real gross domestic product of Ukraine (GDP) in 2018 grew by 3.3%, after increasing it in 2017 by 2.5 % and in 2016 by 2.4%. Further acceleration of economic and social development is associated with digital transformations [1].

The development of the digital economy is a priority for Ukraine. Ukraine has adopted a digitalization program. Digital transformation of existing and creation of new conditions for the formation of new industries is central to solving the problem of ensuring sustainable development of the economy, increasing its growth rates with potential growth in labor productivity.

The share of the digital economy is increasing, giving significant business benefits. Large-scale digitalization of all sectors of the economy provides for maximizing investment in the development of digital infrastructure, innovation and modern technology.

Modern economy is innovative, it is economy of knowledge, competence and network interaction. The main factor of economic development is associated with the digital economy. The emergence of new technologies is indicative of a new stage in the development of an innovative economy. Digital technologies are the most acknowledged among the innovative directions of transformation.

* Corresponding author: oms.ukr.kpi@gmail.com

They are a powerful tool for transforming the activities of companies and entire industries, encouraging market players to fundamentally rebuild their usual business processes.

According to Global Human Capital Trends 2017, 90% of surveyed executives from all over the world (from 10,000 people in 140 countries) believe that their company is confronted with destructive changes driven by digital technologies, 70% say their organization has no adaptation skills. Scale and pace of digital transformation are key indicators of a country's development.

Information and communication technologies are used to ensure optimal interaction between business partners and create an integrated value chain.

Digital business develops new business models combining the physical and digital realms. Digital business transformation involves optimizing business processes through information and communication technologies, as well as developing new ideas and business models.

Digital maturity goes beyond technology - a matter of synchronizing talent, culture and organizational structure with the digital environment.

The development of digital technologies is based on information transformations and processes of globalization. Digital transformation, digital culture are important in the modern activities of entrepreneurial structures.

Digital transformation realizes the company's transition to digital business through changing the organization's culture, introducing new information technologies, expanding the boundaries of the organization, and also allows it to create its own ecosystem. The EU has introduced an updated Digital Competence framework (DigComp 2.0), consisting of 21 competencies. According to estimates, companies that are able to restructure their business models for growth in the context of the disappearance of borders between industries will win.

The identified processes involve the introduction of fundamentally new business models and technologies, including digital platforms, digital ecosystems, in-depth analytics of larger data sets.

The basis for a more effective policy of growth and welfare in the digital era, the idea of the state of digital transformation provides a measure of digital transformation. Thus, the digital conversion can be made visible.

In the digital economy, the efforts of independent businesses are consolidating, and the business models of each partner are transformed.

International cooperation in the business of micromultinational companies by organizing an optimal supply chain from independent producers and managing it on the basis of operatively obtained information provided a significant scale of activity for small and mobile companies, which can be implemented by various enterprises of various industries in modern conditions in the form of outsourcing model. Modern business representatives in this field are already working in the B2C-services market, and new businesses of this type are now actively emerging in the B2B segment.

In digital models, the client is actively involved. The product model transforms to the service model; information is used as a commercial entity. In terms of new business models, the service model is undoubtedly central to the digital economy. Such a model as Product as a Service is widely used in the transport industry, heavy industry, is used by the most advanced manufacturers of railway rolling stock and commercial vehicles. In accordance with it, the world's leading companies do not sell, but actually rent out their products to the operating companies for an indefinite term.

The competition of manufacturers occurs at the level of aspirations to provide the user with the final result.

In the situation of using the Product as a Service model, in an effort to sell the result, of course, various forms of interaction with the customer become much stronger. The key is the knowledge of how the customer uses the service or product. Companies are moving from a product development policy to a policy of improving its operation, which leads to a transformation of the business model.

The Product as A Service scheme leads to maximizing product utilization.

There is a change in the business ecosystem and the corresponding changes in business models of enterprises. In a digital economy, the business model extends to multiple partners from other industries.

Technology and innovation can activate the economy and even change some of its sectors. Innovative early stage companies (startups) are temporary organizations whose main goal is to find a replicable business model that can be scaled. The accelerator business model involves investments in leading projects and industry teams that show results in technology and business. Business accelerators are recognized as one of the most effective ways to select and grow innovative projects in the early stages, acceleration programs can significantly accelerate the growth of start-ups and help in finding a scalable business model.

The platform economy is also one of the most important areas of digital technology development, the scope of application of platform business models is beyond information technology. Business models are becoming more flexible and adaptive, stimulating the search for new sources of competitive advantages. In the era of digital business transformation, "platform" business models have become widespread in a number of ways due to the use of the latest scientific achievements. The main goal of the platforms is the spread of technology, leading to cost reduction. The conditions of their development determine the high share of technology in the cost of goods, a high level of competition, the network effect of the bilateral market, a high degree of infrastructure development. The ability to quickly launch new products on the market, high adaptability to customer needs meets modern requirements for the high cost of a time resource. Increasing the possibility of creating value through the involvement of various players in networks with a high degree of development is complemented by the formation of mutually beneficial advantages for users who create added value for each other. The development of networks is provided by the possibility of interaction in the following areas: work, movement, separation, cooperation.

The platform solution ensures the rapid growth of companies (e.g., startups grown in the company), mastering ways to create added value from outside, scaling business models and making profit in the long tail of the distribution schedule, facilitating the process of creating a unique value proposition by combining approaches from different areas. The advantage updated business models is the speed of change, decision making, the interaction of a huge number of participants. The creation of innovative platforms that provide a technological environment for independent developers should provide the opportunity to use specific solutions without the need for independent development or acquisition of patents. An innovative process was the active development of multilateral platforms, organizations with infrastructure that allows combining and coordinating actions of participants, where the process of creating value is directly dependent on the result of the interaction.

Platform firms exhibit high values of competitive advantages. Ecosystem business models, providing the use of external resources (ideas, information, customers, labor, etc.) with minimal costs, provide a new dimension of business openness, and in addition, a new dimension of competition and corporate strategy. Business leaders seek to incorporate digital platforms into their traditional business models. On the basis of computing resources, modern cloud technologies, effective digital platforms can be created. Digital platforms for outsourcing are a new way to assess the effectiveness and analyze the stages of business processes.

Digital transformations is among the key strategic priorities of many enterprises. Now many companies are set to drastic changes related to the development of digital technologies, aimed at transforming the businesses in order to adapt to new digital realties. Digital transformation are seen as an opportunity to increase enterprise productivity, reduce the time it takes for a product to enter the market based on the introduction of new business models based on the latest technologies. Improving business results acts as an incentive for digital transformation. For these purposes, new business models are also being developed and new technologies are being introduced. The success of the transformation process is associated with the ability to innovate, increase the income of the enterprise and reduce costs, the ability to form a system of interconnection of employees, business processes and technologies. According to research Boston Consulting Group (BCG), digital transformation can reduce costs by 30% and increase revenue by 20%.

Investment transformation and development using different formats for the consideration of proposals and different rationales should facilitate the introduction of fundamental changes. Business transformation is built on a culture of innovation. Transformation processes affect almost all aspects of activity, are implemented in flexible structures, based on trial and error, and should contribute to customer-oriented behavior. Making change everywhere involves strategic partnership and coherence. Modern companies producing high-tech products, along with the use of customer relationship management systems, are actively implementing stakeholder relationship management systems (SRM).

Digital transformation is mostly incompatible with existing processes and structures. Many businesses are already striving to enter the digital world, but not all have a digital transformation strategy.

Digital economy contributes to the formation of new business models. Digital economy determines the development of innovation processes, investment, knowledge sharing and the implementation of innovative ideas. Digital innovation creates competitive advantage. Most large, medium and small enterprises are lagging behind in digital innovations and associate the existing difficulties with their implementation with the rapid changes in digital technologies.

Stimulating private investment in digital innovation in all industrial sectors is the main task at various levels. The development of digital innovation centers will provide access to the latest technology to accelerate innovation. Cooperation with the participation of stakeholders throughout the value chain, the pooling of resources can ensure the development of digital technologies and platforms of lagging sectors, high performance cloud infrastructure for science and innovation. High digital readiness is determined by the formation of digital competencies, access to centers of competence and digital clusters are important for industrial modernization. Business access to the latest digital technologies, synergies between digital and other key technologies provides a network of digital innovation centers [2].

Digitalization changes the way business is organized and run, providing business with resources, production and transaction costs, the manifestation of a network effect and a scale effect.

According to a survey ESG and Dell EMC "IT Transformation Maturity", 71% of enterprises admitted that they lose competitiveness without digital transformation, 95% agreed that they lose out in businesses to competitors who have passed the IT transformation stage. Successfully transformed companies are 7 times more likely to call information technology a competitive advantage and source of benefits.

Depending on the stage of the IT transformation cycle, enterprises are subdivided into using outdated information technologies, developing, progressing and transformed, i.e. those that conduct timely digital transformation and such enterprises are only 5%. Transformed organizations have noted their progress in introducing innovations, bringing products to the market, in automating “manual” processes and tasks, managing IT as a center for increasing profits, not costs.

Business leaders have noted a serious lag in the pace of transformation around the world. Digital transformation is a meaningful, strategically controlled process of change in business, a set of transformations through the introduction of an innovative culture in a company, the adaptation of business models, widespread use of data, customer centricity and value management. The main areas of digital transformation are: customer –centricity, partnership and collaboration, use of data, search and implementation of innovations, value proposition, HR-strategy and digital culture.

The vision and the specific long-term plan for achieving the goals are determined by the strategy. Digital strategy is in the system of business strategy, marketing and communication strategy.

Stages of strategy development include setting digital goals, analyzing, defining the target audience and its values, formulating insights, studying the key consumer needs, building a communication model defining relationships and consistency, determining the KPI setting indicators that measure goals, and as a result the whole creative process.

In the digital economy, there is a change in competition. Intensified competition generates tendencies of in-depth relation with the buyer, a sensitive reaction to the change in his preferences. Competition from costs cutting moves into creativity. Competitors can become partners by integrating on the basis of digital platforms and sharing [3].

The emergence of competitive advantages in a beginner becomes possible due to a start-up or access to global digital platforms for research, development, distribution sales. Such companies overtake the leaders in speed, cost and quality of services. Competitive superiority is provided by digital technologies due to the ability to cross the boundaries of regions, the possibilities at the interdisciplinary level to use client bases, infrastructures and technologies. Enterprise size can also be a competitive advantage. Changes are also influenced by market demand. Digital technologies create transparency by defining new models of consumer behavior based on access to mobile networks and data.

The spread of complex innovations based on a combination of different technologies in new ways determines new forms and methods of business that form the opportunities for creating value in new market segments or developing new centers for creating value in former industries. Competition makes enterprises be constantly innovative, constantly restructured and changed. Integration of opportunities helps create new value. Collaborative innovation is a form of business collaboration in the digital economy. Digital technologies radically change operating models, making them mobile. The new operating model involving sharing is the platform. Digital platform brings together producers and consumers.

References

1. *World Bank. World Development Indicators: 2018 Edition: [sayt]. URL: http://wdi.worldbank.org* (2018)
2. *Word Economic Report. System Initiatives. Digital Transformation of Industries. February. URL: https://www.weforum.org/ reports/ditital-transformation-of-industries* (2016)

3. *The Global Competitiveness Report, 2018 : https://www.weforum.org/reports/ the-global-competitveness-report-2018* (2018)

Implementation of innovative marketing technologies for higher efficiency of the marketing communication complex

Olena Syvolovska[1*], *Tetiana* Neskuba[1], *Olena* Aleksandrova[2] and *Olena* Mkrtychyan[1]

[1] Ukrainian State University of Railway Transport, Department of Marketing, Feierbakh Square 7, 61050, Kharkiv, Ukraine
[2] Ukrainian State University of Railway Transport, Department of Economic Theory and Law, Feierbakh Square 7, 61050, Kharkiv, Ukraine

Abstract. The authors analyzed the basic trends in development of marketing communications in Ukraine and defined efficient innovating marketing methods, one of which is neuromarketing. The article generalizes experience in implementing neuromarketing techniques in Ukraine and all over the world, systematizes methods, technologies and tools of neuromarketing and defines the areas of usage for business activity in Ukraine. Besides, the article analyzes faults, advantages and prospects of neuromarketing development, classifies target audiences in terms of their response to neuromarketing methods and techniques. It also touches a need to implement on-line neuromarketing tools.

1 Problem Statement

Due to oversaturation of the infosphere, the customer is becoming more demanding and ignores any marketing techniques drawing their attention.

The product choice is based on the subconscious customer motivation. Therefore everything presented in the advertising, namely "rational arguments, attractive colors and high quality" do not increase the sales volume due to low efficiency. And in order to solve these problems innovating techniques are being integrated into the marketing communication complex thus influencing an emotional state of the customer. A simple tune and text in the Coca-Cola advertisement are an example of such technologies. Among highly effective innovative marketing techniques are neuromarketing, word-of-mouth marketing, native advertising, viral marketing and others.

A lot of experts in the field studied the scientific basis of the use of neuromarketing as an innovative and highly efficient trend in marketing activity. Among them are D. Ariely and G.S. Berns [11], G. Zaltman [6], M. Lindstrom [7], A. Traindl, [9], D. Ogilvy [14], V.C.R. Fortunato, J.V.E. Giraldi, J.H.C. de Oliverira [10], O. Zorina, K. Zybina, I. Syvolovskiy [2], M. Oklander, Yu. Gubareva [5] and A. Kislov [3]. A great number of such publications testify to the fact that development of a set of neuromarketing tools for Ukraine is demanded and needed.

* Corresponding author: alenasvl3@gmail.com

In spite of their high efficiency, neuromarketing techniques are not used for business activity in Ukraine. It implies that the problem is urgent and requires investigation.

The objective of the article is systematization and generalization of global experience in using neuromarketing techniques, classification of the target audiences in terms of the impact and development of on-line neuromarketing tools.

2 Basic material of the research

The analysis of trends in marketing communication development in Ukraine demonstrated that the advertising market increased by 25% in 2018. And an increase of about 27% has been forecasted for 2019. The statistical analysis of Ukraine's adverting market is given in Table 1 [1]

Table 1 Growth rates of Ukraine's advertising market [1]

Marketing communication tool	Totals of 2017 mln hrn	Totals of 2018 mln hrn	Change rate of 2018 to 2017	Forecast for 2019 mln hrn	Change rate of 2019 to 2018
TV advertising, totals	**7 329**	**9 269**	**26%**	**11 526**	**24%**
Direct advertising	*6 355*	*8 071*	*27%*	*10 089*	*25%*
Sponsorship	*974*	*1 198*	*23%*	*1 438*	*20%*
Press advertising, totals	**1 355**	**1 612**	**18.9%**	**1 843**	**14.4%**
Radio advertising, totals	**480**	**578**	**20%**	**715**	**24%**
National radio	*348*	*418*	*20%*	*518*	*24%*
Regional radio	*47*	*54*	*15%*	*65*	*20%*
Sponsorship	*85*	*106*	*25%*	*133*	*25%*
Media advertising, totals	**2 692**	**3 493**	**30%**	**4 307**	**23%**
Outdoor advertising	*2 263*	*2 923*	*29%*	*3 601*	*23%*
Transportation advertising	*327*	*444*	*36%*	*553*	*25%*
Indoor advertising	*102*	*127*	*24%*	*152*	*20%*
In-theatre advertising	**40**	**48**	**20%**	**58**	**20%**
Internet advertising	**2 145**	**2 520**	**17%**	**3 772**	**50%**
Advertising market, totals	**14 041**	**17 520**	**25%**	**22 221**	**27%**

Research [2] demonstrates that the advertising market suffers from a shift in emphasis from conventional market technologies to innovative tools, namely: neuromarketing, contextual advertising, targeted advertising, sms and SEO promotion.

All these methods are directly aimed at a targeted audience and rather popular, but the efficiency of neuromarketing accounts for 100 %.

The neuromarketing studies the subconscious customer response to certain stimuli. Neuromarketing appeared when it was impossible to use models of classic economic theory in the modern business reality. It refers to the expected utility theory of von Neumann-Morgenstern described by formula (1); but it does not work today.

According to the theory, the customer has to choose an alternative of the maximum mathematical expectations. However, under risks and uncertainty, while taking decisions about the product choice, the customer tries to achieve the absolute reliability from the product without maximizing the expected utility. It is called the Allais Paradox (after the French economist Maurice Allais [3]):

$$EU = P(W) \times U(W) \qquad (1)$$

EU is the expected utility of the customer choice;
P(W) is the probability of the customer choice;
U(W) is the utility of the customer choice.

Thus, market specialists could not understand what the actual motivation for a customer in terms of taking decisions was; as far as all attempts to open "the black box of awareness" with standard methods were useless, the specialists decided to soak into the customer's mind. That was the way how neuromarketing appeared.

The neuromarketing concept was developed in Harvard University in 1990. The technology is based on the model according to which the basic part of thinking activity (over 90 %), including emotions, takes place in the subconscious part located below the controlled realization level. The understanding of the fact what parts of the brain respond to certain decisions made it possible to manipulate the customer [4].

This method is called the Zaltman metaphor elicitation technique (ZMET). The founder, professor Zaltman, believed that studying gestures and other nonverbal communication forms could present considerable potential to obtain additional information. The brain processes the information subconsciously and at a high speed, and, therefore, decisions are taken subconsciously [5].

The ZMET model [6] implies the analysis of the customer behavior according to the brand images generated in the customer consciousness. It includes a set of some theme pictures for obtaining a positive emotional response, thus activating hidden images, for example, metaphors stimulating a customer choice. This model is used by Coca-Cola, GM, Nestle, P & G.

A ZMET model provides research into the customer's subconsciousness; the next stage was implementation of techniques to research the brain.

In 2002 in Atlanta within the project "The Brighthouse Institute for Thought Sciences" the MRI scanning and EEG monitoring techniques were implemented for the first time [5]. The test patients were exposed to visual images, smells, voices, while a magnetic resonance tomography and an electroencephalograph recorded active reactions of the brain areas.

Nowadays neuromarketing evaluates reactions of organ of senses (vision, scent, touch, taste, hearing) to various stimuli (sound, light, look, image, photo, smell, touch sensibility). How it works? Martin Lindstrom, a famous neuromarkting specialist, proves that a pleasant sound accompanying a purchase increases the amount of goods purchased by 65%; a pleasant taste – by 23%; a pleasant smell – by 40%; pleasant tactile sensing – by 26%; aesthetic appearance – by 46% [7]. Such aromatic neuromarketing (odor of flowers, mostly lilies) is used in the fashion shops HugoBoss. And genuine leather flavor is in compulsory use at leather shops.

The analysis of world experience of using the neuromarketing technologies made it possible to distinguish basic ones and characterize them (Table 2).

The global business makes ample use of neuromarketing techniques. In 2016 over 180 enterprises united in the Neuromarketing Science & Business Association. According to market research [8] today's neuromarketing is estimated by about $ 1 billion, and up to 2025 it is going to increase up to $ 2.2 billion.

Table 2. Description of the basic neuromarketing techniques.

Neuromarketing tools and techniques	Description
Facial Expressions Scanner	Obtaining data on the responder's emotions by analyzing the dynamic facial response to external stimuli
Eye-tracking	Tracking the movement of the customer's eyes and determining the areas of interest
Electroencephalography (EEG) and topography	Studying the processes in the cerebral cortex by recording the bioelectric activity.
Biometrical sensors (cardiac rates, breathing rates, galvanic skin responses)	Measuring changes in the physiological state, i.e. cognitive, vision, auditory and somatic sensory reactions.
Magnetic resonance imaging (MRI)	Monitoring brain processes

Leading international corporations widely use neuromarketing techniques, which is confirmed by business practices. Microsoft Corporation implements the EEG monitoring techniques to understand the reasons of excitement, satisfaction, disappointment when users interact with computers. The Frito-Lay company uses the MRI scanning techniques for determining verbal structures or symbols undesirable when addressing female customers. The ŠKODA company uses eye-tracking to study the efficiency of the video advertising CarBakers with making a Skoda Fabia cake [9].

On the other hand, neuromarketing techniques are often criticized due to lack of academic approach and its low practical value at very high costs [10]. The generalized analysis of advantages and disadvantages of neuromarketing tools is presented in Table. 3.

The analysis of global practices in usage of neuromarketing tools made it possible to define their prospective areas for Ukrainian businesses.

1. Marketing research into recording the responder's neurologic processes. Study into the customer's subconscious reactions to certain stimuli. High vividness in contrast to the conventional marketing research. The studies may lead to producing an ideal product.

Table 3. Advantages and disadvantages of neuromarketing tools.

Advantages of neuromarketing tools	Disadvantages of neuromarketing tools
High quality and objectivity of the data obtained in neuromarketing research	High costs and a pressing need for qualified specialists in psychiatry, neurophysiology and other fields [11].
High impact efficiency of neuromarketing stimuli on the customer's choice, brand loyalty and positive advertising perception.	Ethical incorrectness and breach of confidence [11]. For example, a scandal with the use of personal data of Facebook subscribers in Donald Tramp presidential campaign.
Image data obtained through neuromarketing research.	Physiological discomfort for responders, possibility to manipulate the customer consciousness.
High speed in obtaining data and processing efficiency	Ambiguous readings of the results of research due to various customer reactions to

	stimuli
A clear marketing concept for a new article based on the data obtained	Complex calculation of economic efficiency for neuromarketing techniques.

2. Branding. Creation of the basic brand truth with sensor, tactile, audio and aroma indicators which together form highly effective Net Promoter Scores (brand loyalty).

3. Marketing communication complex. Development and rollout of highly effective video and audio reels, outdoor, printed and ambient advertising, P.O.S materials, promotional production and others.

4. Internet-marketing with neurotechnologies. Creation of an online content with neuromarketing tools. For example, neuromarketing helps create webpage heat maps based on data about eye movements, basic objects of interest and other reactions through Web surfing.

5. Entertainment. It is used while producing movies, series and show programs for effective rating forecasting. For example, a spin-off of a super popular serial can be created with neuromarketing tools.

Implementation of neuromarketing techniques should obviously consider different reactions of target groups to neuromarketing technologies [12]. The problem can be solved with consideration of the results obtained in the research [4], where target audiences are classified according to impacts from neuromarketing techniques. 1. "Customers of safety". Their priorities are family, customs and stability. They respond to the product price rather than a brand; "aggressive" sellers cause them discomfort. They negatively react to bright colors, shocking behavior, loud music and emotional pictures. They should be excluded from the focus group in researching new products for young people, trend clothes, innovative technologies.

2. "Alfa people". Their choice is based on status, prestige and success. The brand is important for them, and only a very high price can be disappointing. They negatively react to the mass-market products. They should be included for the research of upscale goods.

3. "Social people". They adore bright shopping and entertainment complexes and huge supermarkets. They positively react to new products or improvements in the mass market. It is these people who must be included in such research as the focus group.

4. "Pioneers". They know much about innovations; they know what brand new product is going to enter the market and know it better than the sellers. They are ready to spend all their money for a latest model and drive the seller distracted with their questions. They form an ideal target audience for researching innovating technologies.

The advertising market statistics in Ukraine [1] proves a need to implement innovating marketing technologies via Internet, as it is Internet demonstrates the highest rates of development of trade and advertising. Therefore, there is a need to implement on-line neuromarketing methods.

Among on-line basic neuromarketing methods are:

1 The social proof effect. 70% of buyers study opinions about goods, and 63% of buyers make their purchase only at the sites providing reviews, ratings, comments and photos from the customers.

2 The free-of-charge basis and awareness of debt. Most people feel discomfort when they owe anybody money. It happens when a person gets something free-of charge. For online sales it refers to rendering a free-of-charge service. For example, Nova Poshta provides free delivery for regular customers.

3 The door-in-the-face technique. It refers to offering unsolicited goods at unreasonably high prices. 90% of offers are rejected, and only after that less costly products at reasonable prices are offered.

4 The anxiety effect. The seller has to find the stimulus attractive for the client and, after that, "to activate" it. The information on limited amount of the product causes anxiety. For example: "Only two leopard Zara skirts are left".

5 The suggestion or placebo method. On the basis of the placebo effect the seller proves that this product will achieve a needed result. It works only when the customer is not an expert in the field; those proposing placebo should be thought leaders and the product should not be of importance for the customer ("not costly… why not try… even if it turns out to be useless…").

6 The technique: Pain – More pain – Hope – Solution [14]. Prospective clients must realize the understanding of their problems and eagerness to solve them. But nowadays such slogans are considered to be poor tactics: "You haven't lost your excess pounds? And only three days left before your journey?" That's a way to get a bad name. As the famous copywriter David Ogilvy said: "The customer is not a moron. She's your wife".

The research conducted made it possible to conclude the following. Despite a negative social opinion about neuromarketing techniques, today it is the only marketing activity which answers the question regarding the advertising/product the customer is expecting. And for the manufacturer it is the possibility to create "an ideal product and service" with the maximum efficiency level. But for the customer it is "the ideal satisfaction of wants". As far as neuromarketing is a new marketing technology, it is not actually formalized in legislation throughout the EU. Neuromarketing can have ethical and unethical techniques, as a majority of marketing methods. The advertising activity is regulated by the Consolidated ICC Code of Advertising and Marketing Communication Practice. Similarly, neuromarketing requires a legislative concept limiting unethical methods. Neuromarketing is a branch of medicine, and its motto must be of medical nature: "Primum non nocere" – "First, do no harm!"

References

1 About advertising and commercial market of Ukraine and the forecast for 2019. Available at: http://vrk.org.ua/ad-market (2018)

2 O.I. Zorina, O.V. Sivolovska, K.V. Zybina, I.M. Sivolovski, *Vyperedzhajuchyj innovacijnyj rozvytok: teorija, metodyka, praktyka.* Sumy: Trytorija. 404- 423, (2018)

3 A. Kislov. *Forbes* Available at: https://www.forbes.ru/tehnologii/352423-signal-mozgu-kak-neyromarketing-mozhet-izmenit-nashu-zhizn (2017)

4 Neuromarketing. *Wikipedia.* Available at: https://en.wikipedia.org/wiki/Neuromarketing

5 M. Oklander, Ju. Ghubarjeva.. Economist. Available at: http://ua-ekonomist.com/archive/2014/11/Oklander.Pdf (2014)

6 Gerald Zaltman, *How Customers Think: Essential Insights into the Mind of the Markets* Boston: Harvard Business School Press, (2003)

7 Martin Lindstrom, *Buyology: Truth and Lies About Why We Buy Paperback.* Currency, 272, (2010)

8 Global Neuromarketing Solutions Market: Snapshot. Available at: https://www.transparencymarketresearch.com

9 A. Traindl, *Neuromarketing. Visualization of emotions.* Alpina Business Book, (2007)

10 V. C. R. Fortunato, J. M. E. Giraldi, J. H. C. de Oliveira, Journal of Management Research, **6**, 201-220 (2014)

11 D. Ariely, G. S. Berns, Nature Reviews Neuroscience, **11**, 284-298 (2010)

12 C. Morin, Society, **48**, 131—135 (2011)

13 David Ogilvy, *Ogilvy on Advertising Paperback*. Prion, (2007)

Priority-partnership approach as a mechanism for encouraging investment activities at the railway transport

Oleg Vasyliev[1], *Volodymyr* Volokhov[1], *Inna* Volokhova[1 1,], *Olena* Lukianova[1], and *Hanna* Zhovtiak[2]

[1]Ukrainian State University of Railway Transport, Kharkiv, Ukraine
[2]O. M. Beketov National University of Urban Economy in Kharkiv, Kharkiv, Ukraine

Abstract. Modern economic conditions and the integration of the national economy into the global economy require structural changes and a search for a new model of relationship between the public and private sectors. The relevance of this article is determined by the fact that the existing scientific concepts have not formed a general understanding of the theoretical aspects for the implementation of public-private partnership at the railway transport, have not established development priorities and tasks of the state in the implementation of the mechanism of public-private partnership at the railway transport. Realization of the projects capable of maintaining implementation of the development priorities of the railway industry requires significant investments of private business and financial, organizational and legal support of the state. The authors substantiate the necessity of using public-private partnership at the railway transport of Ukraine for solving the most urgent current problem – renewal of the material and technical resources and bringing them in line with the requirements of the leading European countries. The proposed priority-partnership approach and the obtained results open up new opportunities for further research into the solution of the problem regarding the implementation of public-private partnership at the railway transport.

1 Introduction

One of the most pressing problems in the development of Ukrainian railways is bringing their material and technical resources in line with the technical, economic, social and environmental requirements of economically developed countries. However, the limited own resources of the railways and the low investment capacity of the state constrain the implementation of such large-scale projects. According to foreign experience, one of the most effective tools for the development of the transport infrastructure is the mechanism of public-private partnership (PPP).

The authors share V.P. Gudkova's point of view that the railway transport of Ukraine is experiencing hard times when the deterioration of the financial state of railways, the lack

[1] Corresponding author: innavolokhova76@gmail.com

of budget funding for the implementation of infrastructure projects and the full-value compensation of reduced-fare passenger transportation by the state lead to the impossibility of its self-financing and maintaining innovation and technological branch transformations [2].

2 Problem Statement

The key problem that endangers the implementation of PPP projects is the low investment opportunities of the state and significant constraints of the financial resources of the railway industry. Taking into account the current state of the transport infrastructure objects (at the beginning of 2015, wear of rolling stock was 99% [2]), emergency conditions of the railway tracks, the steady and successful development of the railway transport can hardly be expected. All this requires intensification of investment activity and attraction of private capital in the implementation of socially significant infrastructure projects. The works by the leading scientists (V.Varnavsky, V. Tsvetkov, A. Medkov) are devoted to the conceptual foundations of public-private partnership [3-6]. Y. Sich, Yu. Vdovenko, I. Zapatrina dealt with the implementation of PPP projects in the transport complex [7-9]. The problem of cooperation between the state and business at the railway transport remains insufficiently studied and requires additional research, especially, in the conditions of limited budget receipts for the development of the railway industry and the existence of crisis phenomena in the economy.

3 Research Questions

The main issues of this research are as follows:

- further study of the theoretical aspects of the implementation of public-private partnership at the railway transport;
- identification of development priorities and state tasks in designing a mechanism for public-private partnership at the railway transport.

4 Purpose of the Study

The purpose of the study is to substantiate the need for the application of public-private partnership at the Ukrainian railway transport with the aim of increasing the investment attractiveness of the railway industry under further reform.

5 Research Methods

In studying the theoretical and methodological foundations of public-private partnership, the authors used the dialectic method of cognition for the formation of prerequisites for the development of public-private partnership and the possibility of its introduction at the railway transport; principles of system analysis ¬ to identify the priority conditions for the operation of economic entities in the railway transportation system; methods of statistical processing and data analysis - to estimate the costs of economic entities; fundamental provisions of the economic theory ¬ to clarify the meaning of the notion "public-private partnership", the substantiation for the notion "mechanism of public-private partnership"; the laws of social development - to substantiate the prerequisites for the implementation of public-private partnership projects; systematic approach – to analyze the processes of interaction between the state and private business in a market-transforming economy and to substantiate the priority-partnership approach for the implementation of the mechanism of public-private partnership at the railway transport; structural-functional analysis ¬ to research the structural-functional relations in the system of railway transportation.

6 Findings

The problem of government financing for construction and maintenance of socially significant infrastructure projects in modern Ukraine is quite acute. Budgetary funds are usually not sufficient for such infrastructure objects, but the private business has a required level of funds. On the one hand, enterprises of infrastructure industries, in the first place their monopoly segments, cannot be privatized due to the strategic, economic and socio-political significance of infrastructure objects, but, on the other hand, there are not enough funds in the state budget to provide such enterprises with reproduction processes. Just for the solution of the issue regarding direct private investment in social infrastructure objects, the state's efforts in public-private partnership projects are concentrated.

The problems facing modern society cannot be solved without a partnership between the authorities and business. The authorities can no longer cope with complex problems that arise in the field of economy, social maintenance, ecological environment, urban development, etc., neither financially nor methodologically or organizationally. Business, implementing strategic infrastructure projects, is not capable of overcoming extremely high risks existing in the relevant fields without the corresponding liabilities of the authorities regarding the consistency of legal regulation.

The situation is complicated by the fact that the state and society are generally distrustful and very cautious about the private sector.

The matter in hand is mainly due to the negative experience regarding the management of state and communal property by private operators. But there is also some fault of the state that has created unfavourable conditions for private business, which, along with political instability and corruption, repel private partners. Under such conditions, it is impossible to rely on the development of public-private partnership in Ukraine without stimulating these processes by the state. Budgetary support and administration of state guarantees are one of the most widely used opportunities for such stimulation worldwide [9].

Public-Private Partnership (PPP) is an alternative to the privatization of vitally important, strategically significant state-owned objects. According to a contemporary view, PPP is an institutional and organizational alliance between the state and business that pursues the goal of the implementation of national and international, large-scale and local, but always socially significant projects in a wide range of fields: from the development of strategically important industries, research and development activities to providing public services [10].

As stated in the National Transport Strategy of Ukraine for the period up to 2030 [11], most of the investment funds for improving the infrastructure are expected from external sources, primarily from international financial institutions, private investors and public-private partnership projects.

Within this framework it becomes necessary to define more exactly the priorities of the development of the railway system and the state's tasks in the field of transport development. The solution of the problems regarding the improvement of efficiency, complex renewal and modernization of the railway transport of Ukraine may be carried out by the system methods, combining the complex of normative legal support of its activities with the creation of a favourable investment climate, by consolidation of budgetary and non-budgetary sources of investment with the aim of completing a number of transport structural reforms.

According to the authors, the implementation of the mechanism of public-private partnership makes most sense to be applied taking into account the priorities that belong to the state or private sector.

The notion "public-private partnership" or "private-public partnership" should be used depending on who will have a key role in management, responsibility and risk-taking. Thus, according to the Rules of registration and maintenance of private freight cars, the infrastructure of railway transport is considered to be a technological complex that consists of public railway tracks, railway stations, power supply devices, communication networks, signals and interlocking systems, information systems and train control systems, and buildings, structures, devices and equipment maintaining the operation of this complex [12]. In this instance, the state should reserve a key role in preserving the state form of property for infrastructure objects and in controlling the safety performance with the aim to maintain the unified technological process, the integrity of the property complex and the strategic importance of the railway industry. Attraction of private investments in a monopoly sector is possible, provided that it will not hinder efficient operation of the railway complex. For the development of infrastructure projects, private investments will work only in that case when private companies begin to receive their dividends. Under the restriction of state financial resources, the development of railway infrastructure, according to the authors, can be ensured on the basis of public-private partnership.

The potentially competitive sector at the railway transport is represented by the freight and passenger transportation, repair of rolling stock, etc.

Railway freight transportation is profitable; therefore, in this field the free competition among operator companies for freight transportation should develop. Besides, for the transportation of strategic goods, for the retention of the key role in maintaining the defence potential of the state, it is necessary to create the National Freight Transportation Operator. Along with the National Operator there will be private operators in the freight market that have to compete with each other.

Passenger transportation is socially significant because it performs the transportation of low-income social groups. At the initial stage, the mechanism of private-public partnership in the passenger transportation sector can be the most efficient because the private sector will have a key role and the state will support the attraction of private capital by means of the development of appropriate mechanisms for supporting private investors, assumption of risks and liability, and exercise of influence at the legislative level. Further development of private-public partnership has to be aimed at the transition to the free competition in the field of passenger transportation.

This approach is essentially a priority-partnership one. The substantiation of the priority partnership approach is presented in Fig. 1. The subject of cooperation between the public and private structures will become the modernization of the railway infrastructure and the development of internal industry competition, and the entities of cooperation will be railway enterprises, private investors and the state. The parties of the public-private partnership should respond in proper time to the increase in demand for public services and meet public needs for quality services. The parties also attract additional financial resources for technical development (modernization) of the infrastructure, rendering of services and may allocate funds of the state and local budgets for the implementation of social and economic programmes. Due to the introduction of public-private partnership, the expansion of the competitive sector takes place and, consequently, there is a choice of an optimal private partner.

Therefore, the mechanism of partnership will stimulate the efficiency increase of railway companies, the quality improvement of transportation due to the competition in the market of transport services, and, based on it, the competitiveness increase of the railway industry.

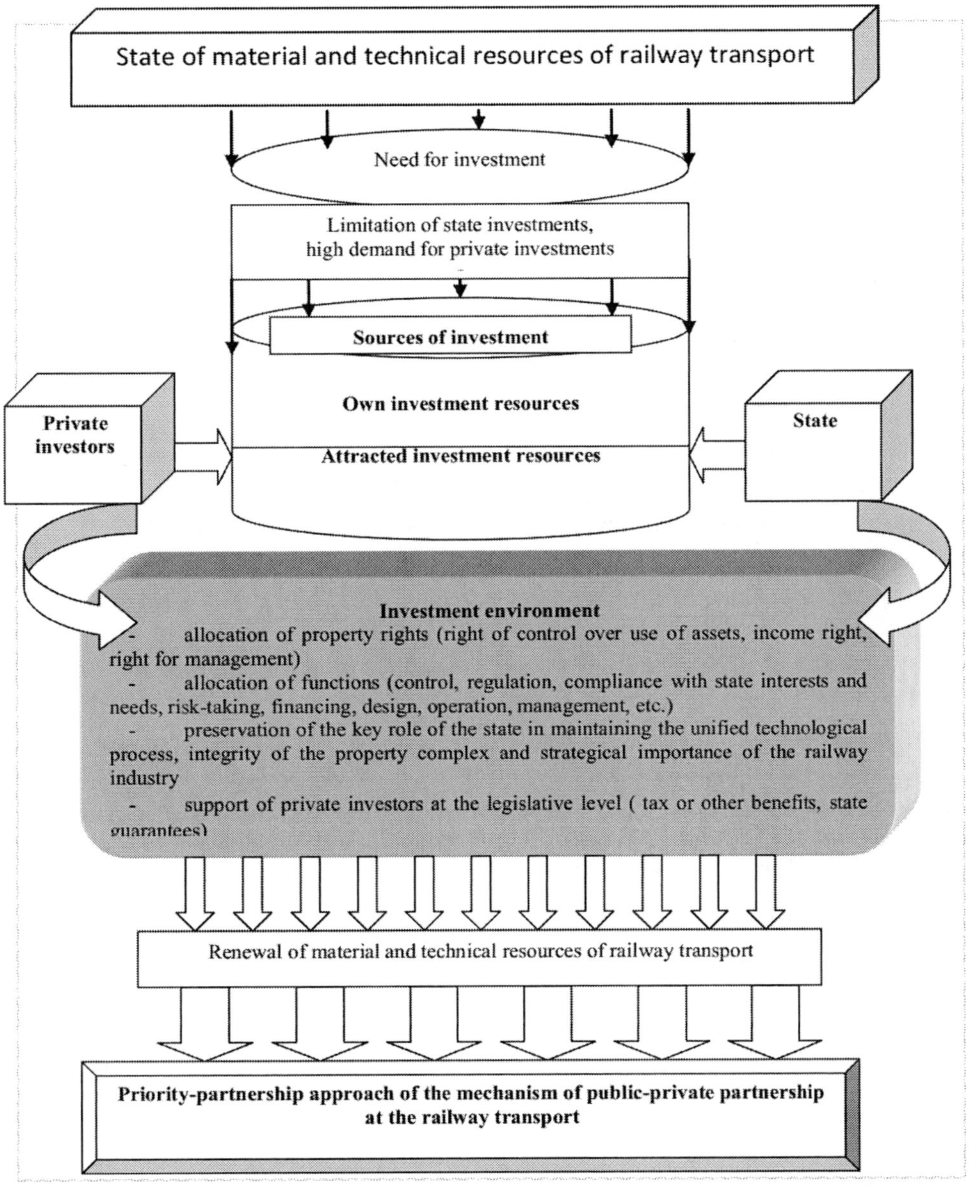

Fig. 1. Substantiation of priority-partnership approach

7. Conclusion

The proposed priority-partnership approach has to be a step forward towards further structural changes in the railway branch. The attraction of private investments will create opportunities for additional revenues for renewal and modernization of the material and technical resources of the railway transport and will promote the competitiveness of transport services.

References

1. V. Kruhlov, Publichne administruvannia: teoriia ta praktyka [Public administration: Theory and practice], Dnipro, **2** (2013)

2. V.P. Hudkova, O.V. Yarmolitska, Ekonomichnyi forum [Economic forum], **1**, 129-137 (2016)
3. V. Varnavsky, Otechestvennye zapiski [Domestic proceedings], **6(21)** (2004)
4. V.G. Varnavsky, *Kontsessionny mechanizm partnerstva gosudarstva i chastnogo sektora* [*Concession mechanism of partnership between state and private sector*] (MONO: IMEMO RAN, Moscow, 2003)
5. V.G. Varnavsky, *Partnyorstvo gosudarstva i chastnogo sektora: teoria i praktika* [*Public-private partnership: theory and practice*] (Nauka, Moscow, 2005)
6. V.A. Tsvetkov, A.A. Medkov, Zhurnal ekonomicheskoi teorii [Russian Journal of Economic Theory], **1**, 170-182 (2008)
7. E.M. Sych, Y.S. Vdovenko, Zbirnyk naukovykh prats KUETT [Collection of Scientific Papers of KUETT], **9**, 198-202 (2007)
8. Y.S. Vdovenko, *Pryvatno-derzhavne partnerstvo v avtodorozhnii sferi* [*Private-public partnership in the road sector*] (ChDTU, Chernihiv, 2009)
9. I.V. Zapatrina, Ekonomika i prohnozuvannia [Economy and forecasting], **4**, 62-86 (2010)
10. *Public-Private Partnerships in Infrastructure Days: Edited by Periklis Saragiotis. PPP Program, World Bank Institute, Day 2*, Washington (2008)
11. *Pro skhvalennia Natsionalnoi transportnoi stratehii Ukrainy na period do 2030 roku* [*On the Approval of the National Transport Strategy of Ukraine for the period up to 2030*], Kyiv (2018)
12. *Pravyla reiestratsii ta ekspluatatsii vlasnykh vantazhnykh vahoniv, zatverdzheni Ministerstvom transportu ta zviazku Ukrainy 28.09.2004 r. No.856* [*Rules for registration and operation of private freight wagons*], Ukrzaliznytsya, Kyiv (2004)

Methodical approaches to the distribution of actual costs of railway infrastructure in conditions of fuzzy indeterminacy

Olena Zorina[1*], *Iurii* Kalabukhin[2], *Nataliia* Kameneva[3]

[1] Ukrainian State University of Railway Transport, Department of Marketing, Feierbakh Square 7, 61050, Kharkiv, Ukraine
[2] Ukrainian State University of Railway Transport, Department of Economics and Management of Industrial and Commercial Business, Feierbakh Square 7, 61050, Kharkiv, Ukraine
[3] Ukrainian State University of Railway Transport, Department of Economic Theory and Law, Feierbakh Square 7, 61050, Kharkiv, Ukraine

Abstract. The authors propose methodical approaches to the distribution of actual maintenance costs for the station infrastructure of the track facility, as well as the signalling and communication facility of each station, as well as the distribution of cost measurements according to the type of work and the implementation costs in the organization of freight transportation.

1 Problem Statement

Currently, the development of market relations in the railway transport and intensification of competition raise the question of using the inefficient infrastructure, including such vivid examples thereof as low density areas and loss-making stations. Their operation requires a constant increase in the maintenance cost of outdated fixed assets, increased cost of materials, etc. This largely contributes to the increase of the primes cost of railway transportation, which, in turn, significantly reduces the level of socio-economic development and competitiveness of railways.

Today, when the reformation of the railway transport is in progress, and the competition in the transport services market increases, the railway industry should be adjusted to operating in the market economy, when in the transport services market the competition between different modes of transport intensifies, the volume of governmental orders for transportation is reduced, and the budget provides almost no funds for the development of the railway transport. The above facts make issues of ensuring sustainable socio-economic development of the Ukrainian railway transport under conditions of indeterminacy of its provision particularly relevant. Those cannot be solved without identifying ways to improve the efficiency of railway sections with a small volume of works, which accounts for about 60% of the total length of the railways.

* Corresponding author: market.kart.edu.ua@gmail.com

The operation of low density railway sections and loss-making stations of the regional branches of Ukrzaliznytsia JSC, providing services for population and enterprises located in the area around these sections, causes quite significant losses to Ukrzaliznytsia and is accompanied by considerable organizational and managerial problems (such as maintenance and repair, staffing, traffic safety, etc.). Therefore, the essence of the problem is reducing the dependence of efficiency of the operation of railway transport on the operation of low density sections and loss-making stations.

At present, there is only an outdated definition of a low density section, while there is not at all a legal definition of a low density loss-making station is absent. There are also no methodologies for cost accounting and calculating for individual sections and stations. Therefore, the relevance of the topic of the study is doubtless. Many researchers focus on the problems of the formation and development of a transport potential; the following Ukrainian and Russian economists considered economic problems of providing effective operation of low density railway sections: I.M. Aksenov A.I. Butenko, B.V. Burkinskiy, O.O. Vasiliev, O.M. Gnennniy, OI Zorina, G.D. Eztutis, V.I. Kopytko, O.M. Kotlubay, M.I. Mischenko, V.I. Pasichnyk, P.A. Smagliy, I.S. Starchenko, Ye.M. Sich, V.V. Zhikhareva, M.Ye. Yushkov et al.

However, in view of the evident scientific interest to the issues of improvement of performance of the railway transport and a wide range of studies conducted in this line, the problem of ensuring economically effective operation of low density railway sections has not been considered in detail and needs to be addressed.

The purpose of the article is improving the methodological approaches to the definition of operating costs for the maintenance of freight stations, including loss-making ones, at the present stage of operation of the Ukrainian railway transport.

2 Basic material of the research

Operating costs of railways are determined using data provided by the structural unit of the regional branch of Ukrzaliznytsia JSC, to which a particular station belongs. In this case, the financial statement for the full calendar year preceding the year when the event is considered is used.

The presented data are combines according to the aggregated cost indicators that describe the estimated capacity volume of the freight station and the adjacent track areas it maintains. The composition of the analyzed aggregated performance indicators is presented in Table 1.

Table 1 Formation of cost measurements according to types of work and implementation cost in the organization of freight transportation

Name of aggregated measurements	Cost item name	Item No. in the list
1	2	3
Consignment	Receipt of cargo for delivery and delivery of cargo	3001
	Maintenance of scales and weighing instruments	3002
	Maintenance of buildings, structures, equipment and inventory of the cargo and commercial management facility	4010

continuation of the table1

1	2	3
	Current repair of fixed assets of the cargo and commercial management facility	4011
	Overhaul of the fixed assets of the cargo and commercial management facility	4012
	Receipt and departure of trains at freight and marshalling stations	4032
	Maintenance of buildings, structures, equipment and inventory of the transportation facility	4040
	Current repair of the fixed assets of the transportation facility	4041
	Overhaul of the fixed assets of the transportation facility	4042
Station rails	Current maintenance of station tracks and permanent devices	4102
	Current maintenance of access tracks	4103
	Replacement of individual sleepers in station tracks	4105
	Replacement of individual sleepers in access tracks	4106
	Replenishment and replacement of ballast in station tracks	4108
	Replenishment and replacement of ballast in access tracks	4109
	Replacement of individual rails, replacement and replenishment of anticreeper braces and other elements of the upper structure of the station tracks	4111
	Replacement of individual rails, replacement and replenishment of anticreeper braces and other elements of the upper structure of the access tracks	4112
	Protection of tracks, crossings and artificial structures	4113
	Maintenance of artificial constructions	4114
	Maintenance of shelters	4115
	Overhaul of station tracks using old suitable materials of the upper track structure	4125
	Overhaul of access tracks using old suitable materials of the upper track structure	4126
	Overhaul of the ballast bed of the station tracks	4128
	Overhaul of the ballast bed of the access tracks	4129
	Overhaul of artificial structures of the station tracks	4131
	Overhaul of artificial structures of the access tracks	4132
	Medium repair of the station tracks	4134
	Medium repair of the access tracks	4135
	Complex improving repair of the station tracks	4137
	Complex improving repair of the access tracks	4138
	Overhaul of machines, equipment mechanisms	4153
	Depreciation of the upper structure of the station tracks	4164
	Depreciation of the ballast bed and artificial structures of the station tracks	4166
	Depreciation of machines and mechanisms used in all types of track repair	4168
	Welding, grinding and other processing of rails	4180
	Current repair of fixed assets of rail-welding trains	4181

continuation of the table1

1	2	3
	Overhaul of fixed assets of rail-welding trains	4182
	Depreciation of fixed assets and other permanent tangible assets of rail-welding trains	4183
	Current maintenance and routine maintenance of automatic block signal devices	4301
Shunting diesel hour	Shunting at freight and marshalling stations	4036
	Shunting operation of electric locomotives in freight transportation	5022
	Depreciation of dedicated shunting electric locomotives operating in freight transportation	5025
	Maintenance of shunting electric locomotives operating in freight transportation according to program TO-2, TO-3, TO-4, TO-5, respectively	5027, 5029, 5031, 5033
	Equipment of shunting electric locomotives	5041
	Shunting operation of diesel locomotives in freight transportation	5242
	Depreciation of dedicated shunting diesel locomotives operating in freight transportation	5245
	Maintenance of shunting electric locomotives operating in freight transportation according to program TO-2, TO-3, TO-4, TO-5, respectively	5247, 5249, 5251, 5253
	Equipment for shunting locomotives	5281
	Current types of repair of shunting electric locomotives operating in freight transportation according to program PR-1, PR-2, PR-3, PR-3P, respectively	6241, 6243, 6245, 6247
	Current types of repair of shunting diesel locomotives operating in freight transportation according to program PR-1, PR-2, PR-3, PR-3P, respectively	6441, 6443, 6445, 6447
	Overhaul of shunting electric locomotives operating in freight transportation	6249
	Overhaul of shunting diesel locomotives operating in freight transportation	6449

Actual maintenance costs of the station infrastructure of the track facility for each station are determined depending on the total length of the station infrastructure and the actual specific maintenance costs per 1 kilometre of the station infrastructure track within an hour by the formula

$$B_j^{t.f} = \frac{b_{t.f} \cdot L_j^{st.\inf r} \cdot 365 \cdot 24}{1000}, \tag{1}$$

where $b_{t.f}$ is actual specific costs for the maintenance of 1 kilometre of track of the station infrastructure per one hour, UAH;

$L_j^{st.\inf r}$ is total length of the station infrastructure of the j-th station, m.

Actual maintenance costs of the station infrastructure of the signalling and communication facility for each station are determined depending on the total length of the station infrastructure and the actual specific maintenance costs per 1 kilometre of the station infrastructure within an hour by the formula

$$B_j^{s.c.f.} = \frac{b_{s.c.f} \cdot L_j^{st.\inf r} \cdot 365 \cdot 24}{1000}, \tag{2}$$

where $b_{s.c.f}$ is actual specific costs for the maintenance of 1 kilometre of the station infrastructure per hour, UAH.

The actual operation costs of shunting locomotives for each station are determined depending on the actual workload of the shunting locomotives by the type of traction (loc-hrs.) The actual specific costs per one loc-hour of operation of a shunting locomotive by the type of traction can be calculated using the formulas:

- diesel locomotive traction

$$B_j^{sh.d} = b_{sh.d.} \cdot \sum Mt_j^{sh.d}, \tag{3}$$

- electric locomotive traction

$$B_j^{sh.el} = b_{sh.el.} \cdot \sum Mt_j^{sh.el}, \tag{4}$$

where $b_{sh.d}, b_{sh.el.}$ are actual specific costs per locomotive-hour of work, respectively, of a shunting diesel locomotive and a shunting electric locomotive, UAH;

$\sum Mt_j^{sh.d}, \sum Mt_j^{sh.el}$ are actual volume of shunting workload of shunting locomotives and shunting electric locomotives, respectively, for the j-th station, loc. hour.

The results of calculations are proposed to be included in a table containing the provisions summarized in Table 2.

Table 2. Results of calculations of actual cost for a railway station of the regional branch, thous. UAH

No.	Name of the station	Total amount of expenditures	Including									
			labour expense	social deductions	Material expenditures			depreciation	Other costs	operation costs of shunting locomotives	Maintenance cost for the station infrastructure	
					materials	fuel	electric power				track facilities	signalling and communication facility
1	2	3	4	5	6	7	8	9	10	11	12	13

References

1 Law of Ukraine "On Railway Transport" - available at: zakon.rada.gov.ua.
2 Draft Law of Ukraine "On Rail Transport"] - available at: http://w1.c1rada.gov.ua/pls/zweb2/webproc4_1?pf3511=62929.

3 Law of Ukraine "On Features of Creating a Public Joint Stock Company of Common Railway Transport".
4 Order "On Approval of the Ukrainian Railway Operating Rules" No. 411 of December 20, 1996 - available at: https: //zakon.rada.gov.ua/laws/show/z0050-97
5 Order "On Approval of the Operating Rules of Intersectoral Industrial Railway Transport of Industrial Enterprises" No. 1014 of October 1, 2009. - available at: https://zakon.rada.gov.ua/laws/show/z1166-09
6 Cost list for Main Types of Economic Activity of the Ukrainian Railway Transport (LLC Poligrafservis, 2011)
7 O.I. Zorina Effective Economy, **5**, 2012, Access Mode: URL: http: //www.economy.nayka.com. ua / index.php? nomer_data = 3 & year_ data = 2012 /
8 O.I. Zorina Internauka. **13(35),** 2017,43-45

Structure of transformation of the road motion parameters in the control system

Liudmyla Abramova[1], *Yevhen* Nahornyi[2], and *Henadii* Ptytsia[1*]

[1]Kharkiv National Automobile and Highway University, Department of Organization and Road Safety, Kharkiv, Ukraine
[2]Kharkiv National Automobile and Highway University, Department of Transport Technologies, Kharkiv, Ukraine

Abstract. Over the past decades, the world has witnessed an increase in the number of vehicles. According to the accident analysis and to the existing transport problems, the development of automated traffic control systems using adaptive management methods is the most optimal way to improve road traffic quality. From the technological point of view, the system must function according to the requirements for the traffic flow level and to the assessment of traffic efficiency, so a clear comparison of systems is impossible under the conditions of various principles for identifying managerial impacts and the designation of management system. The authors' analysis of the most commonly used traffic management technologies proves that experts choose the system architecture that affects its functions and the ability to implement one or another method of management. In accordance with the ITS approach and management tasks at each hierarchy level, we believe that the traffic management system of the intelligent transport system should be based on the principle of multi-level architecture using ring Ethernet technologies, which corresponds more to the diamond-shaped system structure; it allows to distribute methods of traffic management for purposes and to separate the informational and technological elements of the system.

Transport is one of the basic branches of the Ukrainian economy. But the current state of the transport system does not fully meet the requirements for the effective implementation of the European integration course of Ukraine and the integration of the national transport network into the trans-European transport system.

According to the National Transport Strategy of Ukraine for the period up to 2030, it was identified that the creation of competitive and efficient transport system was chosen among the strategy implementation priority directions, according to the world industry trends, namely, the development of intelligent transport systems. In order to implement actions in this direction, common problems were identified: high mortality and injury rate caused by road traffic accidents; imperfection of transport security management systems; low level of introduction and practical application of the latest technologies and technical devices for road traffic organization.

Therefore the development of the intelligent traffic management system foundations is considered appropriate and topical. Over the past decades the world has witnessed a rapid

*Corresponding author: gennadij.ptitsa@gmail.com

increase of the vehicles quantity and traffic, which leads to an increase of accidents number and their negative consequences. The data determining the urban transport network quality level were obtained with the help of entropy levels as the measure of its parameters inconsistency [1]: St. Petersburg – 3.25; Kyiv – 2.2; Kharkiv – 2.2; Moscow – 1.75; Kaliningrad – 1.6; Riga – 1.55; Poznan - 1.3; Vilnius - 1.3; Kaunas - 1.1; Minsk - 1; Warsaw - 0.9; Bratislava - 0.8; Prague – 0.6; Edmonton – 0.5; Helsinki – 0.45, Toronto - 0.4, Berlin - 0.4; Stockholm - 0.2; Oslo – 0.1.

According to the analysis of accidents in the large cities above, the development of automated traffic control systems (ATCS) with the use of adaptive management methods is the most promising and optimal way of the traffic quality improving, as evidenced by the experience of developed countries. At the XXIV Intelligent Transport Systems World Congress held on October 21- November 02, 2017 (Montreal, Canada) it was determined that the ATCS is intended to provide an efficient traffic management in cities with the help of traffic signaling for road traffic safety.

The accumulated world experience in the creation and implementation of ATCS shows that the development of activities in the traffic management systems design allows the following results: reduction of transport delays – 15-40 %; the street-road network capacity increasing – 10-15 %; reduction of the number of accidents – 15-20 %; and reduction of environmental pollution – 20-25 %.

From the technological point of view, ATCS should function according to the requirements of the traffic flow level and the estimation of traffic efficiency, so a clear comparison of the ATCS is not possible under the conditions of various principles for identifying management impacts and designating a management system. The main characteristics of systems are [2]: designation of the ATCS and its zone of operation; architecture (decentralized or centralized); place, method and devices of motor vehicles (MV) monitoring on the city's highway network (CHN); method of control parameters optimizing in the control system; configuration (modularity) of subsystems; selection and definition of road traffic performance criteria (local or systemic); way of regulating traffic.

The authors' analysis [3] of the most commonly used traffic management technologies and well-known traffic control systems such as UTOPIA, SCOOT, SCATS, BALANCE, MOTION, ITACA, ROSE proves that experts choose the architecture of a system that affects its functions and the ability to implement one or another method of management as the main hallmark of management principles. It has been discovered that the hierarchical architecture has the advantage, since it affects the flexibility and mobility of the management system to the coverage area, the configuration of the hardware and subsystems, and also has the ability to respond to the redistribution of traffic efficiency criteria. All this points to the formation of the principles of intelligent traffic management systems, so the adaptive real-time ATCS may not use transport models as a management algorithm and implement digital monitoring methods by means of telecommunications and positioning of online e-mapping (GPS, GPRS).

Well-known studies of intelligent transport systems both in our country and abroad [5] allow us to keep improving the road quality and safety by creating a cyber-physics system for computer online interaction of the driver with cloud services, managing MV on the basis of transport infrastructure digital monitoring and traffic with the use of smart sensors, telecommunication and navigation facilities [6]. However, this innovative approach to the creation of the latest technologies of intelligent traffic control systems, which are aimed at increasing the capacity of the CHN and the time of the MV movement, does not exclude the application of traffic management methods and measures implemented by the ATCS.

At the same time, it should be understood that the transition to each next generation of control systems is associated with significant costs, both for the development and for operation of systems. And in our country at present, in most cases, the existing traffic management systems belong to systems of the first or second generation, where traffic is managed without taking into account the change of transport situation on the CHN. Also speaking about the design features of ATCS, it is necessary to attribute conformity of the latest developments to the current standards and political directions of the state.

The design of a traffic management system based on the determination of dynamic object management (traffic) levels requires the formation of the theoretical foundations for the development of information technologies for the intellectual management of distributed control functions by levels based on the development of the structure of the control system parameters transformation.

In the intellectual management system, knowledge about unknown characteristics of the management object and environment are formed in the process of adjustment, and the information obtained in this case is used in the process of automated decision making, which improves the quality of management. At the same time, adjustment can occur in isolation from the management process or directly in the process of automatic control in real time. Integrated intelligent control system is a modification and development of the traffic management system, which implements the technology of situational management of traffic flows in the street-road network of cities.

It should be noted that intellectual and information technologies are to be distinguished. Information technology functions as intellectual support. Intellectual decision-making technologies deserve consideration – they provide an opportunity, along with a solution (or in the course of obtaining a solution) to seek new knowledge and accumulate intellectual resources. Information technology only creates information resources. Thus, knowledge, formalized in explicit form, can be part of the knowledge base in traffic control systems and can be used to make decisions about the choice of driving influences on road traffic.

In most of the existing traffic control algorithms in our country, control influences are aimed at eliminating obstacles to traffic, that is, there is a post-effect effect, because there is no biased management for their prevention. Under the current state of the road, the number of the urban street network sections is constantly increasing, which can be considered as a bottleneck of the transport network and there is a mismatch of traffic parameters with the parameters of the transport network. Consequently, there is a spatial relationship between abnormal and dangerous situations throughout the transport network as a whole, which implies the existence of a mutual relationship between controllers of the network.

One of the most important aspects of the traffic management problem is not only the determination of the totality of management influences and the sequence of their implementation by the management system, which should be distributed with a rational degree of decentralization and function in real time, but the most effective centralized architecture of the system. The indicated differences influence the differentiation in the architecture of the system, the composition of the subsystems for the management levels and the choice of technical support elements, taking into account the fact that intelligent control systems belong to the type of automatic systems.

According to the ITS approach [4], as well as to the management tasks at each level of the hierarchy, we believe that the traffic control system (TCS) of the intelligent transport system should be based on the principle of multilevel architecture using ring Ethernet technologies that more closely corresponds to the diamond-shaped form of TCS structures. It allows to distribute not only methods of traffic control (TC) for purposes, but also to separate informational (i, n^{inf}) and technological (i, n^{T}) elements of the system (fig. 1).

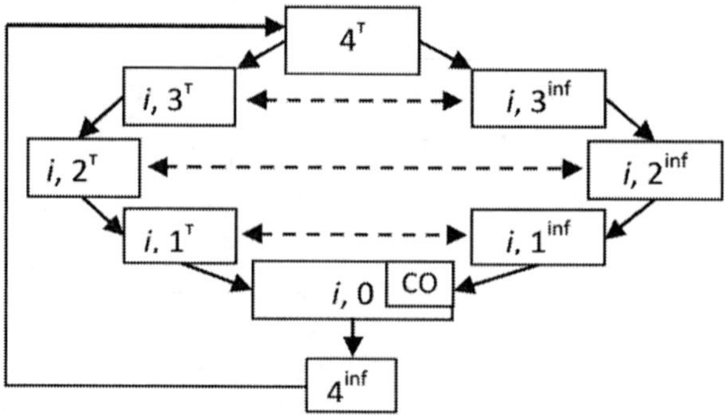

Fig. 1 The structure of the hierarchical traffic control system with two centers.

The term "diamond-shaped structure" of the control system is used to describe the systems under the control of two control centers, namely, information (4^{inf}) and technical (4^T), which are interconnected, we suggest this approach to use in traffic management systems [6]. Each link in a structure with number $i = 1, m_j$ located on the j-th hierarchical level ($j = 0, n$) can be displayed as:

$$S_{ij} : U_{ij} \times Y_{ij} \times H_{ij} \rightarrow X_{ij} \quad (1)$$

where $U_{ij} = \{u_{ij}\}$ – the set of managing influences of a higher level;
$Y_{ij} = \{y_{ij}\}$ – the set of feedback signals from the lower levels;
$H_{ij} = \{h_{ij}\}$ – the set of perturbing influences of this link;
$X_{ij} = \{x_{ij}\}$ – the set of output signals of the link i to the lower level.

For the uppermost part, the expression has the form:

$$S_{ij} : Y_{m,4} \times H_{m,4} \rightarrow X_4 \quad (2)$$

And for the control object (CO), a system of mappings is formed as:

$$S_{m,0} : X_{i,1} \times H_{i,0} \rightarrow X_0 \quad (3)$$

But there is a plurality of disturbing influences in expressions (2) and (3), and in TCS we have two types of disturbing influences – external (climatic and weather conditions, the quality of the roadway coverage, the coefficient of adhesion of the wheel with the road, etc.) and internal (change of the traffic flow mode, change of service levels of the transport network). Internal disturbances are of greater importance for the decision-making level, and external disturbances are more important to the center of system management. Therefore, the presence of two vertices in the proposed structure has the basis for the distribution of management functions – managing and decision making, which respectively apply the technological and informational components of the management system. Then each link in the proposed structure can be summed up by two interconnected parts: the first participates in the formation of control influences, taking into account external perturbations (H^E_{ij}), and the other influences the assessment of the current state of traffic or performance indicators (H^I_{ij}). Therefore, the mapping in a diamond-shaped structure for each link will consist of two parts:
- for technical devices

$$S^T_{ij} : U^T_{ij} \times Y^T_{ij} \times H^E_{ij} \rightarrow X_{ij} \quad (4)$$

- for the information components of the system

$$S^{inf}_{ij} : U^{inf}_{ij} \times H^I_{ij} \rightarrow X_{ij} \quad (5)$$

Thus, the efficiency of transition to a diamond-shaped structure during the transformation of TCS parameters has advantages in management and allows dividing the control and information channels of the system at all levels of the hierarchy; allocating local dynamic control contours in the system; considering the typical problems of static calculations of control effects taking into account deterministic and fuzzy parameters of traffic; and creating a "convenient" mathematical description of the individual links and the system as a whole.

We propose an approach to the structural formation of management levels. At the first (lower) level the solution of the tasks is provided, based on the principles of automatic regulation (AR), that is, the software changes the parameters of the object in accordance with the settings that are defined and set at the highest tactical level (fig. 2). It includes: control object (CO), technical regulation devices (⊗), local road controller (RC_L) and control algorithm (CA) for traffic lights (T_C).

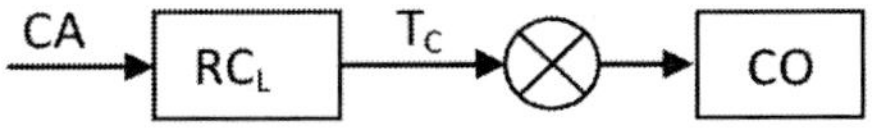

Fig. 2 The structural scheme of automatic regulation.

The principle of an AR is "rough" (robust) to change its input parameters. Robust management is a set of methods of the management theory, whose purpose is to determine the controlling influence in the presence of not complete and not detailed description of the management object. A special place among the concepts of the robust control theory is the uncertainty of the input parameters; it reflects the inaccuracy of control parameters. The work involves a small and timely change in the output of the control system open loop (T_C) when the parameters of the control object, such as intensity (*N*), composition (*C*) and speed (*V*) of the traffic flow, are changed. This is a main disadvantage of this kind of control.

This disadvantage can be avoided with the help of adaptive management. Thus, the management processes for objects whose parameters are unknown from the beginning of the system's operation are formed, and the property of adaptation is achieved by forming a mathematical model of the control object to determine the impact on it. Therefore, the peculiarity of adaptive control systems (ACS) is the possibility of obtaining input information for the formation of control impacts and the availability of a priori information regarding the uncertainty in the system. This is a fundamental distinction between adaptive control and robustness at the second tactical level of traffic control (fig. 3); they include transport detectors (TD) and road system controllers (RC_S) forming control influences $u\ (t)$.

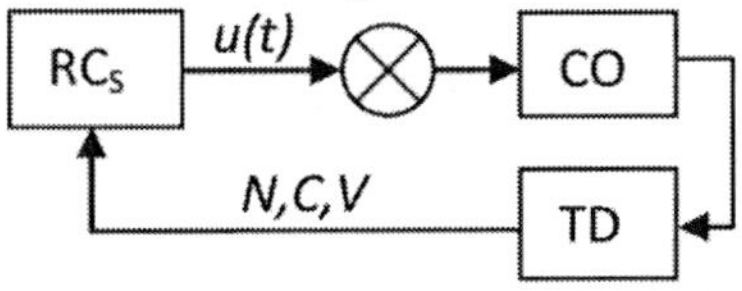

Fig. 3 The structural scheme of adaptive control systems.

The ACS level is designed to manage a limited number of objects subordinated to the corresponding local control units. Criteria for the purpose of management, acting on this level, may differ from the general criterion of the whole system functioning.

At the third strategic level, coordination and optimization of management are being implemented to coordinate the work of local lower level control devices to achieve the overall goal of the operation of the entire management system as a whole. Determining the optimal driving influence on traffic in urban conditions is a complex task with a large dimensionality of parameters. The mutual influence of the control object parameters, the technical means of regulation and the street-road network (parameters of lower levels of management) forms the conditions and limitations to the mathematical apparatus in the development of the management algorithm and decision-making process, in relation to the choice of control influence, carried out at the highest level of management.

The level of operational management and decision-making includes, in addition to the TCS technical support center, a governing body (a team of experts or a decision maker), equipped with modern computer technology to computer possible solutions and choose the most effective option for resource allocation management between individual subsystems and decision making in various emergency situations. The ideal variant to ensure the effective operation of the TCS is to use "fast models" to lower levels of the hierarchy and control object using the elements of the information control system (the formation of bases of control models, databases) for the operational level of management. This approach stipulates the requirements for the design of a hybrid traffic management system based on the received structural and functional description of the methods distribution and algorithms for traffic management.

Thus, the proposed approach to the distribution of management functions on the basis of the developed diamond-like structure gives the system the ability to clearly distinguish and link the task of two types of analysis and synthesis of dynamic control circuits: which can be structured in the course of "cross sections" of the vertical structure and the tasks of static calculation, and that can be formalized when performing horizontal "cuts" of the structure at each of the levels of the hierarchy of the information or control part of the system. In the conditions of incomplete, fuzzy and inaccurate knowledge of the characteristics of the street-road network, parameters of the traffic flow and environmental characteristics, the use of adaptive control methods is a rather promising direction for the construction of the TCS in the real-time mode. These methods are based on the search for expert knowledge of management and the use of fuzzy simulation, where the approximate model of the object is configured using simulation models that describe the possible states of the system, or the use of self-tuning control using hybrid systems.

References

1. D.V. Kapskiy, D.V. Navoy. Development of the automated road traffic control systems in Minsk as part of the intellectual city transport system. Science & Technique. **16 (1)**, (2017) 38 (in Russian) Available at: https://doi.org/10.21122/2227-1031-2017-16-1-38-48
2. Traffic Control Systems Handbook, U.S. Department of Transportation Federal Highway Administration, available at: https://ops.fhwa.dot.gov/publications/fhwahop06006/index.htm
3. L.S. Abramova, Y.V. Nahornyi, CONCEPTUAL APPROACH TO THE TRAFFIC CONTROL SYSTEMS DESIGN. Avtomobil' i elektronika. Suchasni tehnolohiyi [Car and electronics. Modern tehnology]. **12**, (2017) 94 (in Ukrainian) Available at: http://old-www.khadi.kharkov.ua/fileadmin/P_SIS/AE17_2/index.html
4. V. Hahanov, W. Gharibi, B. A. A. Abbas, S. Chumachenko, O. Guz, E. Litvinova. Cloud traffic monitoring and control, Proceedings of the IEEE 7th International Conference on Intelligent Data Acquisition and Advanced Computing Systems (IDAACS), Berlin, Germany, 244 (2013) Available at: https://ieeexplore.ieee.org/ abstract/document/6662681
5. A. Ziarmand, D. Kuchenenko, T. Soklakova Transport monitoring and control systems, Proceedings of the IEEE East-West Design & Test Symposium (EWDTS), Yerevan, Armenia, 474 (2016) Available at: https://ieeexplore.ieee.org/document/7807662
6. L.V. Derbunovich, L.S. Abramova. Hierarchical structures of traffic management systems, Visny`k NTU KPI, Avtomatyka ta pryladobuduvannya [Bulletin of the National Technical University KhPI, Automation and equipment]. **31**, (2008) 40 (in Russian) Available at: https://archive.kpi.kharkov.ua/View/28226/

Usage of information technologies for increasing quality of transport operations

Mikhail Shulyak[1], *Oksana* Pankova[1*], *Ivan* Kolesnik[1], *Kirill* Sirovitskiy[1], and *Evgen* Gaek[1]

[1] P. Vasilenko Kharkov National Technical University of Agriculture, 44, Alchevskykh street, Kharkiv, 61002, Ukraine

Abstract. Among the whole cycle of interconnected technological operations for crop and livestock production, transport operations play a significant part: transportation of seed grain, forage, fertilizers, product transportation to storage and selling places. Transport operations account for 15-25% of the production cost. According to the analysis of recent research, the development and introduction of new methods for ensuring the quality and safety of production. For a tractor on transport works the task of increasing the functional precision at which it is solved its deviation from the roadway configuration (traffic corridor) during the movement is assessed. In this case the problem for short-term one-time and repeated, long-term discrete and continuous impact of the driver on the tractor steering control is solved. For these operating modes of a tractor during transport works, the methodology of functional stability providing of a hydro-level steering control is substantiated. Based on the reasoned analytical model of the tractor spin during transport works, an express method of diagnosing the technical state of the tractor steering has been elaborated. It is based on the comparison of the angular accelerations of the tractor and the steering wheel when the motion direction is changed or adjusted.

Introduction

With the increase in agricultural production and entering new markets by Ukraine, the issue of quality and environmental friendliness is becoming more and more relevant [1]. According to the analysis of recent research, the development and introduction of new methods for ensuring the quality and safety of production by means of the study of mathematically grounded models, practical measures and modern information technologies proves that without the symbiosis of science and practice it is impossible to obtain competitive products of crop sector and livestock [2-6].

Among the whole cycle of interconnected technological operations for crop and livestock production, transport operations play a significant part: transportation of seed grain, forage, fertilizers, product transportation to storage and selling places. Transport operations account for 15-25% of the production cost [7]. Tractors are mostly used for transport operations in crop and livestock production. The development of theoretical

* Corresponding author: pankova_oksana@ukr.net

foundations and methodological principles for efficiency improving of the technical state control of the tractor steering in transport works plays one of the most important roles in ensuring the timeliness and reliability of transport operations carrying out.

In these research works it is noted that the problem of ensuring the dynamics of mobile machinery (including tractors on transport works), in the corridor of motion, depends essentially on the technical state of the steering, the variable mass of the transport unit, the periodicity of the impact on the steering wheel, etc. Vehicles used in agricultural production for the freighting differ significantly from those used in other industries. This connects with a number of specific features: the variety of transported load, seasonality of works, urgency, overall dimensions, etc. This problem is not solved in the direction of increasing the functional precision of tractor movement at transport work.

Methodology

The task of increasing the functional precision is solved for the tractor during transport works. With it, its deviation from the desired path within the traffic corridor is estimated.

In this case the problem for short-term one-time and repeated, long-term discrete and continuous impact of the driver on the tractor steering control is solved. For these operating regimes of a tractor during transport works the methodology of functional stability providing of a hydro-level steering control is substantiated. The first type of impact is characteristic when a tractor tripping-over an obstacle, pit, etc.; the second one - with a short-term one-time impact on the steering system with maintaining its operability if the movement of the transport unit is provided in the traffic corridor, in which its diametrical deviation (output parameter) $\Delta z(t)$ from nominal value Z_x is in the permissible range:

$$\alpha \le \Delta z(t) \le \beta, \tag{1}$$

where α, β are limits of the operability area: $\alpha < 0, \beta > 0$.

In the general case, the steering control error in the case of short-term one-time impact is a random quantity. As a result, the efficiency of this system is also a random quantity. The complete probability of realization of the assigned task by the steering control in the point of time t is found as mathematical expectation of the effectiveness function:

$$F(t) = \int_{-\infty}^{\infty} \Phi(\Delta z / t) f(\Delta z / t) d\Delta z \,, \tag{2}$$

where F $(\Delta z/t)$ is probability of realization a functional task at a time t by steering; f $(\Delta z/t)$ is the probability density of the steering control error at the moment t; Δz is functional steering control error.

In the case of different configurations of the roadway, we can accept that the boundaries (1) of a system functionality scope in the general case are random variables with some common probability density ω_t (α, β) [8-10]. Then, for all possible values of the functionality scope limits, the realization probability of target goal of the transport unit steerability (mathematical expectation of the solution of the assigned task) by steering control is written in the form:

$$F(t) = \int_{0}^{\infty} \int_{-\infty}^{0} \int f(\Delta z / t) \omega_t(\alpha, \beta) dz d\alpha d\beta \,, \tag{3}$$

where ω_t (α, β) – the probability density of the system in the range α and β.

https://doi.org/10.1051/shsconf/20196705002

For the case of a short-term one-time effect on steering system, the carrying out probability by a system of an assigned task (the effectiveness function) is determined by the dependence:

$$F(\Delta z/t)=\begin{cases}F_{\alpha}(\Delta z/t)=\int\limits_{-\infty}^{\Delta z}\omega_t(\alpha)d\alpha \text{ at } \Delta z\le 0;\\ F_{\beta}(\Delta z/t)=\int\limits_{\Delta z}^{\infty}\omega_t(\beta)d\beta \text{ at } \Delta z>0,\end{cases} \quad (4)$$

where $\omega_t(\alpha)$, $\omega_t(\beta)$ – accordingly, the probability density of the lower and upper limits of the operability area.

By the way of differentiation (4) by the experimental uncertainty Δz, we obtain:

$$\omega_t(\alpha)=\left.\frac{dF(\Delta z/t)}{d\Delta z}\right|_{\Delta z=\alpha};\ \omega_t(\beta)=-\left.\frac{dF(\Delta z/t)}{d\Delta z}\right|_{\Delta z=\beta}. \quad (5)$$

Analyzing dependences (4) and (5) it can be concluded that, for practical calculations, the accuracy of a transport unit movement with a short-term one-time impact on steering can be estimated with a uniform distribution of the traffic corridor limits.

The more compound interdependence between the efficiency of the steering control F(Δz) and functionality area exists in the case of short-term multiple impact. To increase the probability of solving this problem, short-term impact can be applied consistently n times, at some point of time t_1, t_2, …, t_n. The probability of carrying out task in each i-moment is equal to $F(t_i + \tau_i) \approx F(t_i)$, where τ_i – the action time of i-operation that was initiated at the moment t_i. To carrying out task it is enough that the impact on the steering solve it at least once at any n moment.

If the short-term impact on the tractor's steering does not provide accuracy in the traffic corridor, then it is necessary to switch over a mode of continuous impact on the steering.

This operation mode of the transport unit is characterized when its movement on long turns. In this case, with a long impact τ on the tractor's steering from the driver's side, the system changes its state with a probability close to one, as well as in systems with prolonged discrete effects. In this case, the accuracy of the unit movement by time τ will not be equal to the motion accuracy without influence, i.e. $F(t_0 + \tau) \neq F(t_0)$, where – t_0 the beginning of continuous control. Failure to carry out of a task that is solved at least one of the time point (t_0, $t_0 + \tau$) leads to a movement violation of the transport unit in the traffic corridor.

The problem of functional precision of tractor steering control is solved by the way of estimating the deviations (errors) of functional parameters from their calculated (nominal) values that arise under the influence of various destabilizing factors.

The level of functional precision of the steering control during the transport unit operation does not remain constant due to changes in its technical condition, disintegration, etc. In this case, during a transport unit operating, it is necessary, as a rule, to ensure minimization of the initial operation error and to maintain this parameter during prolonged operation of the transport aggregate.

On domestic and foreign tractors installed mainly hydropower steering, in which there is no mechanical connection between the steering and steerable wheels. They are connected hydraulically. The statistics of hydraulic actuator faults of tractors steering shows that according to the failures number, the failure of the main elements of the hydraulic actuator (hydraulic pumps, hydraulic distributors, hydraulic cylinders, etc.) prevail. Hydraulic circuit of the hydro-power steering of the tractor ensures its operation in a working order and emergency operation (volume pump exceptionally). Faults of the of the tractor hydrostatic steering elements, which lead to an increase in the leakage of the working fluid, is the main reason for the violation of the functional precision of the steering control. In this case, the functional parameters of the tractors steering control, in accordance with DSTU ISO 10998: 2013, include steerability and motion stability. When controlling steerability, the tractor's ability to react to the driver's action, aimed at preserving or changing the motion direction is estimated, and during the stability control – the preservation of a given motion direction is estimated. To take error of function of the steering control as the ability to perform the specified functions with a certain proximity degree to the ideal model, the functional error of the steering control at the current x and nominal $x_н$ values of the functional parameters is evaluated according to:

$$\Delta x = x - x_н. \quad (6)$$

The probability of controlling the error of function and operability of the steering control can be estimated according to:

$$Д = 1 - (P_1 + P_2), \quad (7)$$

where P_1 and P_2 – probability of errors of the first type (failure pass) and the second type (false refusal).

Empirical results

The probabilities of P_1 and P_2 errors depend on the partition law of values of controlled functional parameters and measurement errors, the time of the measuring process and the characteristics of the tolerance zone on the value of the measured parameter.

Steering, as a unit under test, will be operable, that is, suitable for further operation, when the measurement result satisfies the conditions:

$$c \le y \le d, \quad (8)$$

where c, d – limits of the tolerance zone of the controlled parameter y, $2\delta = d - c$; $y = x_к + \Delta x_к$; $x_к$, $\Delta x_к$ – the actual value of the monitored parameter and the error of its measurement.

Table 1. Probability of control of hydrostatic steering according to functional parameters.

Functional parameter, Д	**Probability of error**	
	P_1	P_2
Steerability:		
Operating fluid pressure	0.030	0.015
Steerable wheel slip	0.006	0.004
Steerable wheel lock at an extreme positions	0.012	0.006
Leak resistance of the hydrovalve of the dosing pump	0.016	0.008
Motion stability:		
Stability of uncontrolled movement	0.014	0.007
Rate of impacts on the steering wheel	0.003	0.002

The sum of P_1 and P_2 indicates that in order to guarantee a given probability of control (85%) of the tractor hydrostatic steering, the error of the technical state control of its aggregates and units according to the functional parameters should not become apparent with a probability of more than P = 0,005 (impact rates on the steering wheel) and P = 0,045 (operating fluid pressure).

The proposed methodology for controlling the error of function and operability of a hydrostatic tractor steering allows to substantiate the technology of finding faults and failures with the given probability.

Based on the reasoned analytical model of the tractor spin during transport works, an express method of diagnosing the technical state of the tractor's steering has been worked out. It is based on the comparison of the angular accelerations of the tractor and the steering wheel when the motion direction is changed or corrected.

Tractor steerability characterizes its ability to react to operator actions which are aimed at storing or changing the motion direction. The controlling impact causes the transient process from one state of the tractor balance to another, which is accompanied by a change in the traverse speed. Acceleration that arise in this case, characterize not only the tractor's steerability, but also the change in its technical condition, which is provided by the parameters instability.

It has been offered the error of function of tractor according to steerability – transfer function of the steerability:

$$W_k = 1 - \frac{1}{K_k} - \frac{v}{m} \cdot \frac{dm/dt}{\dot{v}_k}, \tag{9}$$

where $K_k = -v_k/v_c$ – steerability coefficient; v_k, v_c – the steering acceleration and acceleration of the resistance forces of the tractor withdrawal from equilibrium accordingly; v – traverse speed of the tractor; m – mass of the transport unit at a given point in time; dm/dt –rate of change of the transport aggregate mass of the variable mass.

It is proved that the tractor has perfect steerability in the conditions when $W_k = 1$. However, this motion mode cannot be obtained, since it is impossible to get $K_k = \infty$ and $v/m = 0$ when the tractor is moving. The approximation W_k to a unit is possible due to providing the functional stability of the tractor steering.

An express method of assessing the technical state of a hydrostatic tractor steering has been worked out. The essence of it is to determine the turning time of the steering wheels or half-frame of the tractor on-site and to evaluate volumetric efficiency of the control system according to nomograph.

In order to determine the dependence of the volumetric efficiency of the steering control on the time of turnover from the extreme to the extreme position, it is necessary to conduct preliminarily an experimental study of determining the time dependence of the rotation on the operating fluid pressure in the steering system and obtain an approximative dependence.

Instrumentation of the express method is a measuring complex of VDVMM 4-001 was developed by the cooperation of the Kharkiv National Automobile and Highway University and the P. Vasilenko Kharkiv National Technical University of Agriculture to assess the dynamics of mobile machinery, which has been modernized by the authors for its usage in the study of tractor steerability. The measuring complex allows to carry out technical diagnostics of the hydrostatic tractor steering by the express method: the calculation of the rotation angle of the steering wheel, the angular acceleration of the tractor half-frames rotation, the transferor function of steering, the volumetric efficiency of the hydrostatic steering, the comparison of the obtained values with normative.

The express method involves the accelerated carrying out of technical diagnostics operations of the hydrostatic steering control. At the same time, as the structural parameter of the diagnostics the volumetric efficiency is taken, which is the determining parameter of the technical condition. And as the diagnostic parameter the angular acceleration of the tractor's turning during the transport works is taken.

The dependence of the volumetric efficiency of the hydrostatic tractor steering on the turnover time has been proved (fig. 2).

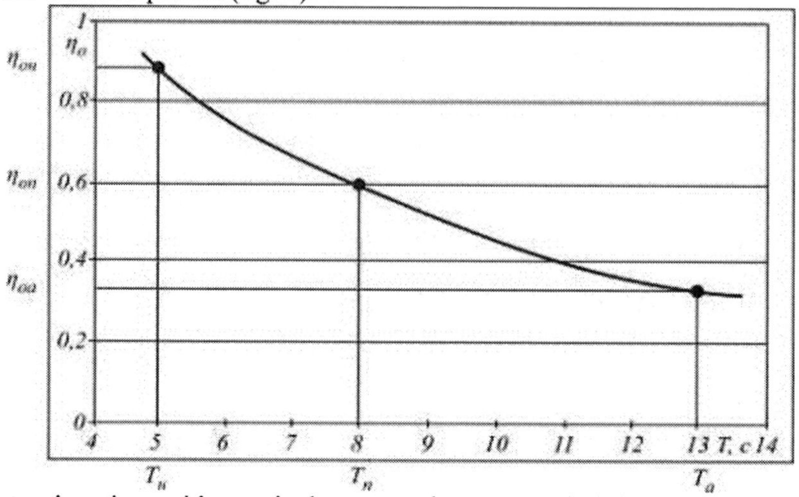

$T_{н}$, $T_{п}$, T_{a} turning time with nominal $\eta_{он}$, maximum permissible $\eta_{од}$ and emergency η_{oa} volumetric efficiency accordingly

Fig. 2. Nomogram for volumetric efficiency determination (η_o) of hydrostatic tractors steering of HTZ-170 against turnover time (T).

Thus, the essence of the express method for the technical state assessment of the hydrostatic power steering consist in identification, with the help of the measuring complex of the VDVMM 4-001, the turnover time of the steering wheels or half-frame of the tractor and the estimation according to the nomogram of the volumetric efficiency. It should be said that for different classes of tractors, the nomogram will be different, so the plotting of such nomograms should be carried out at the stage of research tests, with the entry of the results in the technical documentation.

References

1. V. Puzik, L. Puzik, N. Lyubymova, O. Pankova, V. Sakhatsky, IJET (UAE) **7**, 507-513 (2018)
2. S. Kharchenko, O. Anikeev, M. Tsiganenko, O. Kalyuzhny, KhNTUSG (MAP) **156**, 174-179 (2015)
3. S. Kornienko, V. Pashenko, V. Melnik, S. Kharchenko, N. Khramov, East.-Europ. JET **5**, 34-43 (2016)
4. I. Palamarchuk, O. Tsurkan, V. Palamarchuk, S. Kharchenko, East.-Europ. JET **2**, 79-85, (2016)
5. L. Tishchenko, S. Kharchenko, F. Kharchenko, V. Bredykhin, O. Tsurkan, East.-Europ. JET **2**, 63-69, (2016)
6. O. Nanka, O. Chalaya, S. Nagornij, O. Chalyi, East.-Europ. JET **5**, 31-39, (2018)
7. E. Shubravskaya, N. Rindenko, Econ. of Ukr. **8**, 77-85 (2012)
8. I. Kolesnik, UkrSRIPCT **21**, 169-174 (2017)
9. M. Shulyak, SoCNC **42**, 64-67 (2017)
10. M. Shulyak, EoNM **6**, 99-104 (2016)

Digital Transformation of the Ukrainian Economy: Digitization and Transformation of Business Models

Viktoriia Yanovska[1], *Olha* Levchenko[1], *Viktoriia* Tvoronovych[1], and *Anastasiia* Bozhok[1]

[1]State University of Infrastructure and Technologies, Department "Economics, marketing and business administration", Kyrylivska str. 9, Kyiv, 04071, Ukraine

Abstract: The studies are devoted to the transformational processes associated with digitalization, which takes place in the Ukrainian economy. Economic development is impossible without widespread introduction and full use of the digital technologies. The rate of digital innovation is growing every year. The transformation of the landscape of interpersonal connections and economic relations leads to the digitalization of the operational processes and the change in business models, which generally results in large-scale digital transformation of companies. Meanwhile, it should be noted that the digitization and transformation processes of business models, taking place mainly at the level of private and state enterprises, are expanding to the level of government and have a significant impact on the competitiveness of the economy. Today, there is a significant number of projects of digital transformation in Ukraine. The general idea is to create a unified digital communication system. It is nice to note that not only private companies but also state structures have joined this system, which testifies to their interest in self-development and the desire to keep up with the modern technologies, focusing on clients.

1 Introduction

The modern economic development is impossible without widespread introduction and full use of the digital technologies. The rate of digital innovation is growing every year. The number of end-users is increasing. Hootsuite's Global Digital 2019 research suggests that, on average, more than a million new Internet users appear on the Internet every day. In 2019, the number of: (1) mobile users has been increased by 100 million (or 2%) to 5.11 billion; (2) active Internet users – by 366 million (or 9%) to 4.39 billion; (3) users of social networks – by 288 million (or 9%) to 3.48 billion; persons using social networks on mobile

[1] Corresponding author: nastya.bozhok@gmail.com

devices - by 297 million (or 10%) to 3.26 billion [1]. In October 2018, the number of active users of Internet resources in Ukraine was 63% of the population. According to Factum Group Ukraine, 70% of users constantly use smartphones to access the Internet, 36% - laptop, 3% - working computers [2]. The transformation of the landscape of interpersonal connections and economic relations leads to the digitalization of the operational processes and the change in business models, which generally results in large-scale digital transformation of companies.

2 The Main Directions of Digital Transformation

At present, many scientific and applied researches of foreign specialists are devoted to digital transformation, but the Ukrainian scientists have begun to deepen in the study of the mentioned problems relatively recently. Globally recognized digital expert David Rogers (2016) argues that digital transformation is not about updating your technology, but about upgrading your strategic thinking [3]. But transformation of business models in most cases leaves behind the research area, since the main interest is aimed at the introduction of the technological innovations. According to the data of 2019, in Ukraine, technological landscape 4.0 included 62 companies, which are distributed in 16 segments +1 on system integration, which is the most numerous (21 companies). Among the technological segments that include IoT Plaform, Blockchain, IoT devices, Drones, Robotics, Machine Vision, Mobile Technologies, etc., the Big data/AI/ML segment (15 companies) is leading, the second largest is IoT devices segment (12 companies), and the third largest is AR/VR (11 companies) [4].

Meanwhile, it should be noted that the digitization and transformation processes of business models, taking place mainly at the level of private and state enterprises, are expanding to the level of government and have a significant impact on the competitiveness of the economy. On the one hand, digital transformation, by creating the conditions for gathering the necessary information about the business environment and operating activities, making informed decisions and their rapid implementation, is inevitable and economically feasible (contributes to lower costs, improved customer service, business scalability, development and flexibility of the companies) [5]. As a result, the prospect of the current opportunities for the future of business, the state and consumers is changing. On the other hand, it requires from the same companies the creation of flexible organizational conditions and the introduction of digital business models necessary for rapid adaptation to the changing realities of the dynamic ecosystems.

According to current studies (IMD, World Digital Competitiveness (WDC), 2018) [6], the level of digital competitiveness, which evaluates the ability of countries to implement and explore digital technologies, depends on three main factors: knowledge (with sub-factors: talent, training and education, scientific concentration), technologies (normative legal base, capital, technological basis), readiness for the future (adaptive relations, business flexibility, IT integration). According to the world digital competitiveness ranking, Ukraine occupies 58th place (index 51,292) out of 63 countries, though in 2017 it occupied the 60th place. At the same time, the factor of knowledge - 39th place, and the factors of technology and readiness for the future – 61st place, indicating a sufficiently high potential.

According to the Global Competitiveness Index [7], compared to 2017, Ukraine improved by six places, but the rating methodology has changed. Some positions dropped over the year, and Ukraine ranked 83rd out of 140. Ukraine scored the most points for such components as Skills (45th place), Market Size (47th place), Infrastructure (57th place) and Innovative Capacity (58th place). And by the indicator "level of electrification" even managed to take the first place. The components that pull Ukraine down are Macroeconomic Stability (131st place), Financial System (117th place) and Institutions (110th place).

Because of such low indicators of Ukraine's rating and in order to stimulate the economy and attract investments, transform the domestic industries into competitive and efficient ones, solve the digital gap issue, create new opportunities for human capital, develop innovative, creative and digital industries and business; develop export of digital products and services (IT outsourcing), the Cabinet of Ministers of Ukraine approved the Concept for the Development of the Digital Economy and Society of Ukraine for 2018-2020 [8].

The adoption of the Concept was preceded by a presentation in late 2016 of the Digital Agenda 2020 – a document aimed at endorsing priority areas, initiatives, digitalization projects by 2020 [9], developed under the Europe 2020 Strategy with the objective to create additional opportunities of implementation of Ukraine's development together with the EU. It is this document that initiated the principles of Ukraine's development in digital space on the state level and became the basis for the development of the digital economy. In accordance with it, the basic principles of digitalization of Ukraine are as follows:

- Any member of the society has the right to access digital technology, the Internet of knowledge;
- Digital technologies should become an instrument for achieving certain goals of development of various sectors of the national economy;
- Possibility of additional attraction of investments;
- Creation of primarily Ukrainian content to meet national needs, which will promote economic development not only, but also social and cultural one;
- Integration of Ukraine into the European and global communication system;
- Need to develop standards for digitalization, which will allow to increase competition and reduce expenses and cost of production;
- Compliance with the international standards in the financial area;
- Avoiding the orientation of the digital economy to private business only;
- Need to provide cyber security measures at the state level;
- Protection of the rights of Internet consumers; and
- The government should occupy one of the key places in the preparation, development and promotion of national digital strategies.

Today there are numerous digital transformation projects in Ukraine. The general idea is to create a unified digital communication system. It is nice to note that not only private companies but also state structures have joined this system, which testifies to their interest in self-development and the desire to keep up with the modern technologies, focusing on clients. In particular, the Digital Transformation Road Map in Agri-Food Project [10] has been created, which helps food and processing industry enterprises to be engaged in global processes of digital transformation.

Taking into account the above, it should be noted that the development of IT technologies has a constant tendency to increase and by 2025 will double twice, as shown in Fig. 1

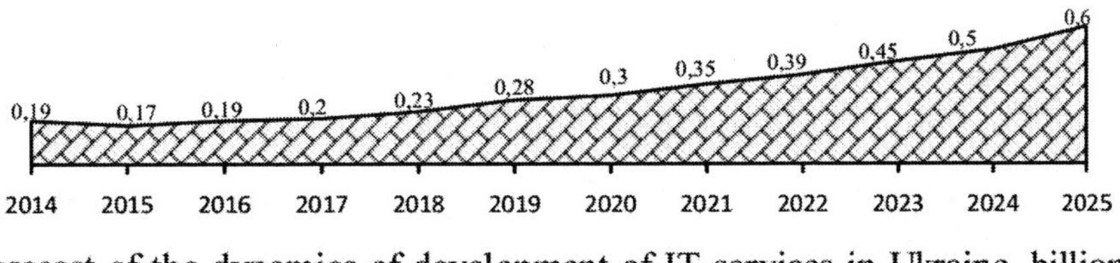

Fig. 1. Forecast of the dynamics of development of IT-services in Ukraine, billion dollars (the forecast is compiled according to [11]).

3 Opportunities and threats

Given the objective nature of the processes of digitization and transformation of business models, the ambiguity of future results should be taken into account. On the one hand, digital transformation processes are preceded by the emergence of additional opportunities, which, according to the results of the SWOT analysis, include the possibility of saving money and obtaining additional profits, increasing the competitiveness of products and companies, simplifying work with the database, building a positive image, increasing customer demand to the brand. On the other hand, there are threats to data security and cyber attacks. The strong points of these processes are the constant development in social networks, primarily due to the advertising of products or the company as a whole, as well as simplification of sales procedures (meeting the needs of consumers at anytime and anywhere). The weaknesses are imperfect data privacy systems, software imperfections (program crashes), temporary lack of proper computer and software. The weak sides of Ukraine as a whole include inadequate protection of intellectual property rights, the level of development of banks and financial services, rather high investment risks, the need to strengthen cybersecurity and the presence of piracy, the strong sides - the relative ease of starting a business, the rather high speed of Internet connections, e-democracy, the use of large data and the relatively high quality of training. In Ukraine, the ratio of teachers and students, many women researchers and graduates of scientific specialties are important.

In the process of digital transformation, digitalization and business model changes, the company has to go through several stages, namely, in general, to determine what to strive for and clarify the development strategy (future interaction with customers, methods of competition, tools for working with large data, directions and pace of innovation, it is possible even to change the basic values), then digitize marketing and automate the operational processes.

It should be understood that digital transformations would require personnel changes. According to the research [12], in the coming years, business structures around the world will face the situation of so-called personnel famine. A survey by more than 600 senior executives conducted by the Boston Consulting Group (BCG, 2017) showed that many companies nowadays lack the skills of qualified professionals to address key technological challenges. Trying to form a pool of “digital talents”, hire, educate and retain highly skilled specialists in the business areas, develop new digital skills from the existing personnel. During the study, interviews were conducted with a number of recruiters and experts in the digital sphere and about a million lists of requirements to the candidates was analyzed. Based on identified needs, BCG developed a staffing strategy that suggested: (1) identify areas where experts – so-called digital talents – would be most important, and (2) identify ways of finding, engaging and learning.

It is determined that the experts critical for the digital transformation of any business in any industry are: a digital enterprise strategist, who must perform a guiding function at all stages of the functioning of the digital-model; specialist in marketing automation, which, with the help of artificial intelligence technology, provides interaction with consumers on-line; user interface designer – focuses on creating the interface, user experience of software products and development of the digital potential of the organization; data-scientist, working as an analyst group and detecting hidden relationships and important patterns in the data; SMM manager who must know the basics of marketing, be able to analyze potential consumers, know the basics of sales psychology, evaluate the activity in the social media, use different methods in promoting services through planning, advertising, collaboration with bloggers, etc.

The leaders of change should understand that digital transformation is the construction of a new strategy in which all processes (clients, competition, data, innovations, core values) are digitized and/or transformed in order to improve labor productivity, increase company value, create new jobs, improvement of personal efficiency of employees. All these processes lead to an increase in the competitiveness of the economy and, as a result, an increase in the standard of living in Ukraine.

4 Conclusion

Summarizing the above, it should be emphasized that Ukraine, having a significant human IT potential, works below its digital capabilities. The measures stipulated by the adopted concepts lay the foundations for the digital transformation, aimed at digitizing business processes. However, accelerated digitalisation, in parallel with the transformation of business models, could add many millions of revenues to the economic growth, additional international investment, and increased international competitiveness. It is necessary to understand that for today the development of digital technologies is inevitable and it depends on us only how effective it will be and in what future we will exist.

References

1 *Global Digital 2019 reports* (Hootsuite, 221 p., 2019) https://wearesocial.com/blog/2019/01/digital-2019-global-internet-use-accelerates
2 Internet Asotsiatsiia Ukrainy [online], Available at: https://inau.ua/news/oglasheny-dannye-yssledovanyy-ynternet-audytoryy-ukrayny-za-3-kvartal-2018g
3 David L., *Rogers The Digital Transformation Playbook: Rethink Your Business for the Digital Age,* Columbia Business School Publishing, Columbia University Press, 296, (2016)
4 *Ukrainian Landscape Industry 4.0 Technologies* (2019) [online], Available at: https://industry4-0-ukraine.com.ua/2019/05/27/ukrainian-landscape-industry4-0-druga-vers%D1%96ya/
5 D. Ivanov, A. Dolgui, B.Sokolov. International Journal of Production Research, 57, (2019)
6 *IMD World Digital Competitiveness Ranking* (IMD World Competitiveness Center, 179, (2018).
7 *The World Economic Forum.* The Global Competitiveness Report (2018) [online], Internet Asotsiatsiia Ukrainy: http://www3.weforum.org/docs/GCR2018/05FullReport/TheGlobalCompetitivenessReport2018.pdf
8 Rozporiadzhennia Kabinetu Ministriv Ukrainy «Pro skhvalennia Kontseptsii rozvytku tsyfrovoi ekonomiky ta suspilstva Ukrainy na 2018–2020 roky ta zatverdzhennia planu zakhodiv shchodo yii realizatsii» vid 17 sichnia 2018 r. № 67-r. (2018) [online], Available at: https://zakon.rada.gov.ua/laws/show/67-2018-%D1%80
9 *Tsyfrova adzhenda Ukrainy – 2020 («Tsyfrovyi poriadok dennyi» – 2020) Kontseptualni zasady* (HiTECH Office, 90 p., 2016) [online], Available at: https://ucci.org.ua/uploads/files/58e78ee3c3922.pdf
10 *Digital transformation road map in agri-food* (2019) [online], Available at: https://agri-food.appau.org.ua/
11 Businessviews [online], Available at: http://publications.chamber.ua/2018/IT/the-infographics-report-it-industry-of-ukraine-2017.pdf
12 *Digital BCG. A Digital-People Strategy is Critical in Today's Changing Workplace* (2017) [online], Available at: https://www.bcg.com/d/press/19july2017-how-to-gain-digital-talent-skills-165720.

Risks of Traders on the World Stock Market on the Example of Ukraine

George Abuselidze[1], *Olga* Mohylevska[2], *Nina* Merezhko[3], *Nadiia* Reznik[4], *Anna* Slobodianyk[4]

[1]Batumi Shota Rustaveli State University, Georgia
[2] Kyiv International University, Ukraine
[3]Kyiv National University of Trade and Economics, Ukraine
[4] National University of Life and Environmental Science, Ukraine

Abstract. The article reveals the essence and features of the development of the stock market in Ukraine. It was established that the vigorous activity of countries in the world financial markets means that they also face a risk of global financial turmoil (the so-called "domino effect"). It is determined that the impact of global financial instability on the country depends on the openness of its economy that will lead to significant external "shocks". The possibility of providing effective influence on domestic stock market activity with taking into account the changing world situation, development of perfect trading strategies for each participant is substantiated. The conducted analysis of the world market conditions of stock markets in recent years has made it possible to assess the real risks for new participants in the stock market and become the basis for the development of an appropriate effective trading strategy. The practical significance of the results is that they allow for a measurable approach to assessing the existing risk when choosing one or another trading strategy to move to the world stock market.

Introduction

Necessary condition for success in an area of activity which is characterized by increased risks is the creation and improvement of risk management systems to identify, assess, locate and control risks. The decision-making mechanism should not only identify the risk, but also allow you to assess which of the risks is acceptable and define whether the expected returns will justify the appropriate risk. Justified or acceptable risk is a necessary component of the strategy and tactics of effective management.

An effective strategy of risk-taking behavior enables financial market participants to successfully carry out its activities and achieve its goals with the least losses. A strategy that can be called a risk management strategy consists of:

- awareness of risk, identification of causes of its occurrence and risk areas;
- analysis and risk assessment;
- selection of risk management methods;

[4] Corresponding author: nadya-reznik@ukr.net

- use of selected methods;
- control.

The first and most important component of the risk management strategy is the practical data that will be based on the construction of the entire trading strategy of the trader. Therefore, the development of qualitative analysis of the market with the dynamics in recent years is a prerequisite for the creation of a true trading strategy for entering the stock market.

The world stock trading market is still a rather new, but very attractive and promising arena for Ukraine, but the country's own experience is insufficient to react in a timely manner to changes in world market conditions, which determines the relevance of the topic of the article.

Theoretical basis

Market conditions are a system of indicators and conditions that characterize the current state of demand, supply, price and level of competition in the market (commodity, stock, etc.) in general or in its individual segments; for the market situation there are four stages (business cycle):

- rise of the market conditions;
- conjuncture boom;
- weakening of the conjuncture;
- decline in market conditions.

Table 1. Dynamics of trade volume on the world stock market, billion transactions. (authors' calculations on the base of dates [2])

Indicator	2014	2015	2016	2017	2018	Average over the last 3 years	Change 2017 to 2016, +/-%	Change from 2017 to the average indicator, +/-%
Total volume of deals	2498	21.19	21.55	21.83	24.78	22.72	2.94	0.13
Chain growth rate, +/-, %	x	-15.18	1.71	1.30	13.48	x	x	x
Basic growth rate, +/-, %	x	-15.18	-13.73	-12.60	-0.82	x	x	x

Analysis of the activities of the world's stock exchanges (Table 1) shows that they have not been trading for the long term, which, as a result of the evolutionary development of stock trading, has been ousted on the over-the-counter market. Instead, modern world exchanges serve as an indicator of world pricing, including the main types of agricultural products (cereals and oilseeds). Today, virtually all world exchanges are integrated into stock exchanges with electronic trading platforms.

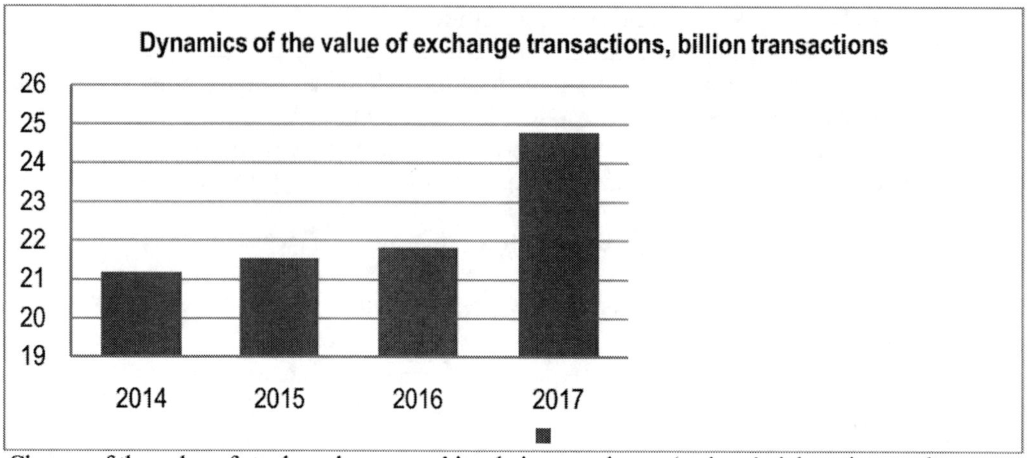

Fig. Change of the value of stock exchanges and its chain growth rate (authors' elaboration on the base of dates [2])

Dynamics of growth volumes continues from 2013 (Table 2). Since 2006, trade in futures and options has risen faster than any other sector of global derivatives markets. Currently, the global commodity derivatives market is characterized by a boom.

Table 2. Structure of world trade in derivatives, billion transactions (authors' calculations of changes on the base of dates [2]).

Stock sector	2013	2014	2015	2016	2017	Average indicator	Change 2017 to 2016, +/-%		Change from 2017 to the average indicator, +/-%	
Commodity derivatives	2.59	3.05	3.61	3.79	4.65	4.02	22,59	22.59	0.63	15.78
Part,%	10.36	14.41	16.73	17.37	18.76	x	x	x	x	x
Financial derivatives	22.16	17.88	17.60	17.69	19.31	18.20	9.16	9.16	1.11	6.10
Part,%	88.72	84.40	81.66	81.01	77.93	x	x	x	x	x
Other types	0.23	0.25	0.35	0.35	0.82	0.51	131.56	131.56	0.31	61.71
Part,%	0.92	1.19	1.61	1.62	3.31	x	x	x	x	x
Total quantity	24.98	21.19	21.55	21.83	24.78	22.72	13.48	13.48	2.06	9.05

The financial derivatives market significantly exceeds the commodity derivatives market, but in real time markets are analyzed considerably better with volumes of trade. Analysis of the structure of world trade in commodity and financial derivatives (Table 2) showed that the growth rate of derivatives of the commodity market significantly exceeds the growth rate of financial derivatives. These indicators characterize the commodity derivatives market as a rapidly developing market. The structure of commodity derivatives trading by types of underlying assets is quite stable and harmonious (Table 3).

At the global level, the total number of futures and options traded on exchanges around the world increased to 24.78 billion contracts in 2016. This figure is not a record, but compared to 2015, it shows an increase of 13.5%, giving the industry its highest growth rate since 2011.

Table 3. Commodity derivatives trade structure by types of basic assets (authors' calculations of changes on the base of dates [2]).

Commodity derivatives	**2012**	**2013**	**2014**	**2015**	**2016**
Commodity derivatives, billion deals	2.59	3.05	3.61	3.79	4.65
Part in the sector by the basic asset, %					
Agricultural products, %	38.51	41.07	33.56	36.60	35.27
Metals, %	30.01	28.59	29.96	32.79	34.46
Energy resources, %	31.48	30.33	36.48	30.61	30.27

One of the main reasons was a surge in trading activity in Asia. Recent annual surveys have shown that North America grew faster than the rest of the world in 2014 and 2015, but the model changed in 2016. On the Asian-Pacific exchanges, total trade volumes rose by 33.7% to 9.7 billion contracts in 2015, the highest level for this region since 2011.

Table 4. Dynamics of trade volume on the world stock market by geographic regions, billion deals (authors' calculations on the base of dates [2]).

Geo-graphic regions	**2012**	**2013**	**2014**	**2015**	**2016**	**Aver-age indi-cator**	**2016 to 2015, +/-%**		**2016 to aver-age, +/-%**	
North America	8.185	7.227	7,83	8,195	8,216	8,080	0.021	0.26	0.136	1.68
Part,%	32.76	34.11	36,33	33,08	37,63	x	X	x	x	x
South America	1.603	1.731	1,683	1,451	1,517	1,550	0.066	4.55	-0.033	-2.15
Part,%	6.42	8.17	7.81	5.86	6.95	x	X	x	x	x
Europe	5.017	4.389	4.359	4.77	4.41	4.513	-0.360	-7.55	-0.103	-2.28
Part,%	20.08	20.71	20.23	19.25	20.20	x	X	x	x	x
Asia-Pacific	9.825	7.526	7.302	9,701	7,257	8,087	-2.444	-25.19	-0.830	-10.26
Part,%	39.33	35.52	33.88	39,16	33,24	x	X	x	x	x
Other stock ex-changes	0.351	0.317	0.377	0.658	0.433	0.489	-0.225	-34.19	-0.056	-11.51
Part,%	1.41	1.50	1.75	2.66	1.98	x	X	x	x	x
Total vo-lume of world stock trad-ing	24.981	21.19	21.551	24.775	21.833	22.72	-2.942	-11.87	-0.887	-3.90

However, the increase in activity in the Asia-Pacific region was due to the presence of large exchanges. Out of the 28 exchanges in the region covered by our study, 18 had double digit growth in 2016 and only six were down. In addition, growth has been extended to a large number of countries, including China, India, Korea, Singapore and Taiwan. Even Japan, which has suffered from a drop in trade for several years, has seen growth. All exchanges of the studied regions had an increase in the growth rate of trade activity in 2016.

Results

The research of the last year is based on data collected from 78 derivatives traded exchanges. The volume includes futures and options traded on the stock exchange and presented for clearing, as well as contracts traded on the stock exchange. The geographical location is determined by the jurisdiction in which the exchange is regulated and not by the location of the parent company. Contracts based on stock indices, exchange rates, and commodities are the three categories with the highest rates of growth in 2016. The total number of futures and stock index options is increased by 13.7% to 8.34 billion contracts, almost reaching a maximum of 8.47 billion in 2012. Some of this growth came from markets where the size of contracts is relatively small, such as the National Stock Exchange of India. But there was also an increase in wholesale markets dominated by institutional investors, such as Eurostoxx 50 futures and options traded on Eurex. Foreign-exchange contracts jumped by 31.2% to 2.78 billion in compare with 2013, when the volume in this category fell by 15%. Here, much of the growth came from relatively small contracts traded on exchanges in places like India, Russia, Turkey and Argentina, and in most cases, contracts were tracked at the local exchange rate against the US dollar. But there was also growth in wholesale markets, such as futures and options on Euro FX, quoted on the Chicago Commodity Exchange. Futures trading in Euro FX increased by 25.2% to 65.4 million contracts in 2016, while Euro FX options increased by 62.1% to 12.4 million contracts.

The state of the Ukrainian stock market agrarian market in recent years shows its low level of development. At the same time, China's interest in Ukraine's capabilities increases every year.

The share of Chinese investment in the total volume of foreign investment in Ukraine suggests the need for risk assessment based on existing methods and world experience (Figure 2).

Fig. 2. The share of China in general quantity of direct investments in economy of Ukraine (authors' elaboration on the base of dates [4]).

A prerequisite for success in an area of activity characterized by increased risks is the creation and improvement of risk management systems to identify, assess, locate and control risks. The decision-making mechanism must not only identify the risk, but also allow it to assess which risks are permissible, and to determine if the expected returns justify the corresponding risk. Justified or acceptable risk is a necessary component of the strategy and tactics of effective management.

Conclusion

The presence of the stock market in Ukraine contributes to the integration of the national financial market into the world, which has a positive effect on the economy, but at the same time, it increases its dependence on the uncertainty of the world financial market. Greater integration of countries on global financial markets means that they are at greater risk of transmitting the effects of shocks from abroad (the so-called "domino effect").

The impact of global financial instability on the country depends on the openness of the economy, and accordingly – on the probability that financial instability will cause significant external shocks for the economy. Developed countries have created mechanisms to counter these negative impacts, but most developing countries, without internal safeguards, are becoming their victims. The research proves that risks have a significant impact on market participants. Functioning of the stock market is most often in the conditions of increased instability of the market situation. Consequently, it causes high risks and low income. For a competitive operation of Ukrainian traders on the world stock market, a thorough analysis of the world market situation in recent years has been conducted, which will help develop a successful trading strategy. With the help of statistical data, the analysis of the most active participants of the world stock market was conducted and the dynamics of trade volumes according to geographic regions is shown. This information can be used by scientists to develop practical recommendations for shaping trading strategies and risk assessments for entering the global stock market.

References

1. Müller, Janis and Posch, N. Peter, Wrong-Way-Risk in Tails http://dx.doi.org/10.2139/ssrn.3124863 (2017)
2. Official website of Futures Trading Association, https://fia.org/
3. Grant L. Kenneth. *Trading Risk: Enhanced Profitability Through Risk Control*. New Jersey, NJ: John Wiley & Sons, (2014)
4. Official website of State statistical service of Ukraine. External activity, http//ukrstat.gov.ua.
5. M. Rusnakova, Agricultural Economics, **61**, 149-157 (2016)
6. M.O. Solodkyy, V.O. Yavorska *Birzhova torhivlya tovarnymy deryvatyvamy: svitovyy dosvid ta perspektyvy rozvytku v Ukrayini*. Kyiv: TSP Comprint, (2015)
7. M.O. Solodkyy, V.O. Yavorska, *Birzhovyy tovarnyy rynok*. (Kyiv: TSP Comprint (2015)
8. Ramzan Imran, Role of Financial Derivatives in Risk Management. https://ssrn.com/abstract=3264962 (2018)
9. Kim, Myunghyun, How the Financial Market Can Dampen the Effects of Commodity Price Shocks. http://dx.doi.org/10.2139/ssrn.3256554 (2018)
10. Thakral, Mehak and Chander, Ramesh, Does Derivative Trading Facilitate Price Discovery and Risk Management? https://ssrn.com/abstract=3220294 (2017)

Comprehensive staff evaluation

Khalida Ahaverdiieva [1,*]

[1]Simon Kuznets Kharkiv National University of Economics, Department of Economics and social Sciences, avenue Nauki, 9-a, 61166, Kharkiv, Ukraine

Abstract. Evaluation is indispensable for the staff management system of any company, as according to evaluation data any reasonable managerial decisions can be made in the field of staff rotation, training, establishment of salary amount, etc. The evaluation results constitute an important factor of performance motivation and positive change in approaches towards professional experience. Due to staff evaluation the efficient company can be established that ensures better conditions for development and growth of the employees. Performance evaluation of various officials varies depending on tasks, importance, values, difficulty of scoring. Thus, staff evaluation may be considered as a part of the staff management system and independent evaluation system to be applied within any company in a particular form. The article proposes a method of personnel assessment using a comprehensive assessment. It proposes the scale of evaluation, which is used to implement actions for measuring of the bonuses and personnel developing of the enterprise. A number of criteria for evaluating staff have been set. It distributes the share of each criterion in total amount, certain levels and their scores are also determined.

1 Introduction

The efficient evaluation system enables to develop, clarify and confirm selection criteria and, thus, to improve recruitment methods to be used. The results of periodic performance evaluation confirm (or do not confirm) the correctness of professional successfulness forecast made on the basis of selection methods and procedures used.

Interest in methods of the employee competence evaluation developed at the beginning of the previous century. The scientific approach to workplace management caused the occurrence of algorithms that assisted the companies to evaluate employees of various qualifications. Due to the analysis of efficiency, their professionalism has been defined and performance values have been calculated. Currently, professional skills of any employee may be insufficient to solve the current corporate purposes.

2 Problem Statement

As of today, one of the main criteria of efficient performance is staff evaluation. Staff evaluation ensures the receipt of comprehensive accurate information on the

* Corresponding author: agaverdieva@ukr.net

employee. It enables to receive all required information on the certain employees and determine how one or another employee complies with the office requirements, as well as to improve the efficiency of staff management and development, and to improve psychological climate within the collective. Staff evaluation enables chief executives or staff management services to conduct more efficient human resources management, to improve and involve new employees instead of those who fail to perform their duties and, thus, to perform efficiently.

3 Research Issues

Staff evaluation priority refers to properly selected tools, staff evaluation methods, employee evaluation methodology, evaluation of personal and behavior features, demands in training, performance efficiency.

Staff evaluation is treated as part of management. Herewith, a method is required to study the qualitative composition of corporate staff, its advantages and disadvantages, and the basis for improving personal labor skills of the employee and further training [1]. As the evaluation results determine the employee position and their potential rotation, they are the important motivating factors for the labor improvement and attitude to the work.

4 Purpose of the Study

The purpose of the study is to conduct a comprehensive assessment of personnel and establish the criteria by which the assessment of personnel of the enterprise will be carried out.

Depending on purposes, according to scientists I.P. Migus and S.A. Chernenko, staff evaluation solves the following tasks: ensuring feedback between the employees and management; searching for measures to promote required specialists within the company; developing an efficient labor motivating system; obtaining psychological reference of required specialists; career planning and promotion; determining specialist competence level; determining the measures of satisfaction and interest of the employees in the company [2]. Thus, staff evaluation is the stable process of determining the compliance of qualitative staff features (skills, knowledge, motives) with the requirement of office or workplace [3, 4].

Staff evaluation provides for the comparison of certain human features – professional and qualification level, business proficiency, performance with the certain parameters, requirements and standards. Thus, it can be stated that staff evaluation means the conclusion on the basis of comparison of “perfect” employee with that real human working for certain company and holding certain office. Staff evaluation refers to the determination to which extent each employee achieves any expected results and complies with the requirements derivative from his production tasks. Therefore, staff evaluation is the procedure to be conducted in order to determine the measures of compliance of personal employee features, quantitative and qualitative performance with standards, objectives to be achieved [5].

5 Research Methods

One of the basic difficulties which the company faces while evaluating the employees is the evaluation of managers. Often, it is due to interdependency of business and professional features of manager with the evaluation of final performance. Moreover, such evaluation takes a lot of time. Thus, it is reasonable to demonstrate the interdependency using the comprehensive evaluation. Parts of comprehensive staff evaluation are given in Figure 1.

It is known that the selection of evaluation system is subject to various factors: office, work experience, activity of subdivision and company in general, corporate culture existing within the company, as well as staff motivation.

Therefore, such problems exist: waste of time, imperfection of evaluation system and probability of personalism.

To avoid such situations, staff evaluation shall be conducted on an anonymous basis by independent experts so that the level of personalism is on the minimum level.

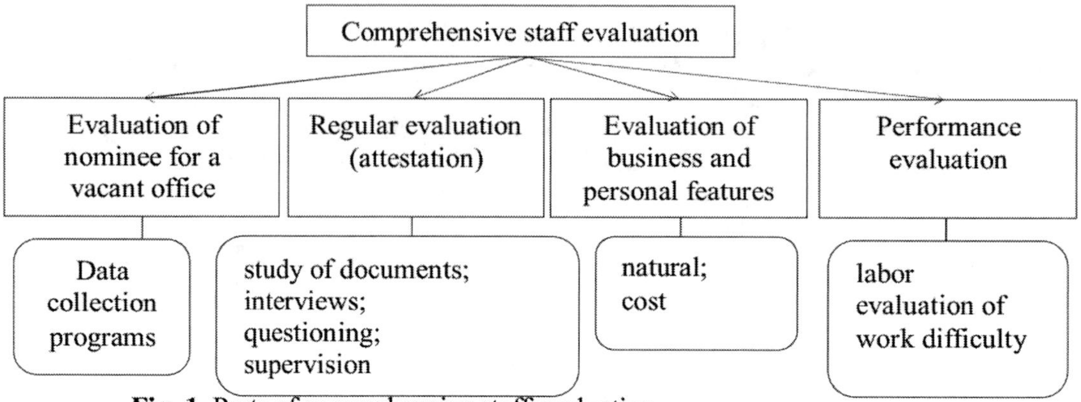

Fig. 1. Parts of comprehensive staff evaluation.

Quantitative figures may include as follows: standards established; performance of the best employees; plan of work performance. With regard to qualitative figures, they include as follows: knowledge, skills, abilities; employee's responsibility; employee's loyalty; ability to work in a team; leadership skills; ability to make decisions independently; independency in work execution.

With regard to heads and managers, there are qualitative figures to be evaluated as they indirectly influence over the results to be achieved; they should not be established by the number as they characterized the employee by criteria consistent with "perfect" beliefs how official duties and functions should be done, basic parts of such office. Thus, while evaluating the employees, it is reasonable to consider quantitative figures [6].

As of today, there are three requirements to the current manager: professional, business and personal skills. Relying upon basic requirements to managers, as well as reference and practical experience of leading national companies, there is the staff evaluation system for the company. Therefore, staff evaluation shall be conducted by business and professional skills.

Selection of evaluation criteria is based on reference in terms of specific features of corporate business. With regard to work experience, the employees may be divided into groups with the relevant level: work experience less than 3 years is evaluated as low level; 3 – 5 years – average level; 5 – 10 – above the average; above 10 years – high level.

Educational level of the employees shall be evaluated as follows: high level – the employees with academic degree; above the average – the employees with higher education; average level – the employees with advanced education; low level refers to those employees with secondary education. Upon all other criteria, evaluation shall be conducted by the direct manager. Herewith, each feature of professional and business skills and performance shall have the own level to be scored as follows: low – 0.5; average – 1; above the average– 1.5; high – 2 scores.

Score from 0.5 to 2 shall be awarded to the employee for each feature in terms of specific value. Herewith, the ratio of professional level of the employee shall be established using a formula:

$$R_{pr} = \sum L \times SV_c, \quad (1)$$

where L – relevant level;

SV_c – specific value of relevant criterion.

The employee shall comply with the requirements to be put against him upon official duties, context and work nature, and corporate requirements and corporate culture. The evaluation shall include not only potential abilities of the employees, professional level, their competence, but also the implementation of such possibilities in the process of the fulfillment of obligations assumed and achievement of corporate performance standards [7].

Through evaluation, criteria of staff evaluation shall be clearly distinguished. Herewith, for the employees the execution of work volume and products quality shall be of the great importance. Performance, but not personality shall be evaluated.

In table 1 there are the criteria of evaluation and specific value of corporate performance.

Table 1. Performance evaluation.

Evaluation criteria	Specific value	
	Employees	Managers
Volume of work executed	0.25	0.2
Quality of works executed	0.25	0.3
Knowledge and compliance with safety procedure	0.15	0.1
Compliance with technological process	0.15	0.1
Timely work execution	0.2	0.3
Total	1	1

Thus, evaluation of performance for the stated period shall be determined as the amount of evaluation of feature levels in terms of specific values. It shall mean the difficulty of work.

The comprehensive evaluation of the employees shall be calculated using the formula:

$$CE = Sh \times S_s + Sh \times P, \quad (2)$$

where, Sh – share of relevant skills per total evaluation, %;

Ss –scores of professional and business skills of the employee;

P –scores for performance.

The evaluation results are the basis for further decision of the certification commission and they are also focused on the evaluation of labor potential, career development, feedback and etc.

Moreover, the improved staff evaluation system, efficient evaluative technologies developed subject to corporate features, methodologies, unique skills and evaluation methods – all these in total may play a role of intellectual capital of the company. The staff evaluation system as the intellectual capital of the company occurs in such forms as: human capital – abilities, knowledge, skills, motivation of specialists in the field of staff evaluation. Therefore, we should say that the evaluation of labor activity of the employee as a part of total performance of the collective is quite difficult task. Those evaluation methods to be used as of today are quite subjective. The human evaluates the environment, including the behavior of other people, not through personal assessment, but through interpretation. Thus, the companies may use any evaluation method of individual performance. But the chosen method shall be simple and clear for the employees. Therefore, it should be noted that while evaluating performance the proper determination of quantitative and qualitative values reflecting final purposes of subdivision and company in general shall have practical importance.

The selection of performance evaluation criteria depends on affiliation with one or another category of the employees: heads, specialists, employees, workers.

The head performance is stated through progress of the company or branch or division: increase in profits, market share, increase in clients and etc., and through social and economic working conditions of the employees (for example, increase in salary, increase in discounts, etc.). The criteria of performance evaluation shall be subject to certain requirements. They shall: have direct and final influence with performance of the company; cause the achievement of corporate purposes; influence over at least 80% of performance of the employees. The criteria of performance evaluation shall be at least 5-6.

The selection of criteria shall be determined through the establishment of evaluation criteria, i.e. which tasks shall be solved through the used evaluation results. Thus, for example, if man purposes are the increase in performance and determination of salary amount, the evaluation shall be conducted upon the criteria of productivity. If the purpose is the promotion, there shall be criteria to determine the potential productivity at new workplace and etc [8]. The scheme of staff evaluation is given in Figure 2.

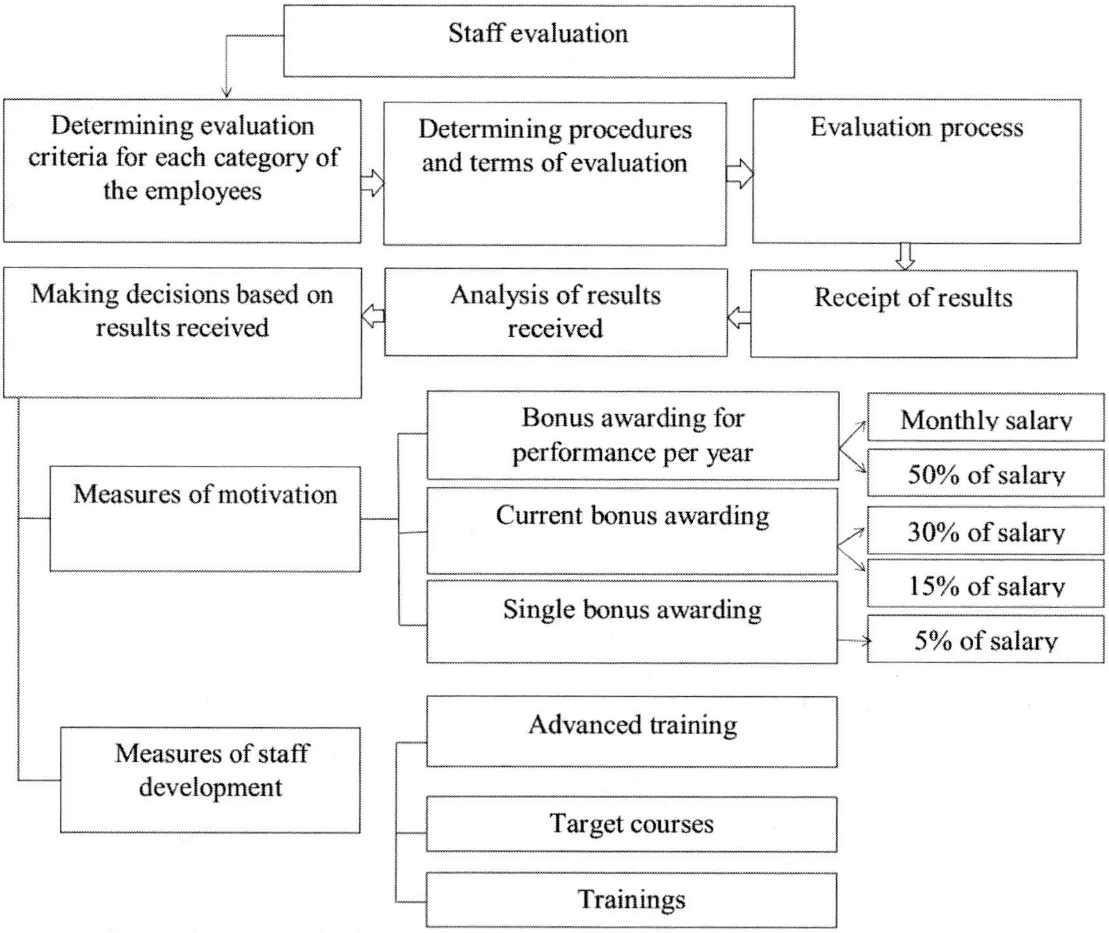

Fig. 2. Scheme of staff evaluation within the company.

The timing of evaluation depends on various factors: specific features of work, term of labor agreement, period of working cycle, significant change in productivity, work completion per each certain task and etc.

6 Findings

Therefore, it is recommended to conduct the final evaluation of the heads and managers twice per year. Scores received in a current evaluation shall be summarized for certain periods and converted into average values. As a result, the decision on bonus awarding and staff training will be made.

The evaluation shall be conducted at the following scale:

1.7 – 2 – the employee holds the office. Such employee shall be included into reserve for the promotion and at the end of the year he will be paid with 100% monthly salary.

1.4 – 1.6 – the employee holds the office and at the end of the year he will be paid with 50% of one monthly salary. He may be in reserve for the promotion.

1 – 1.3 – the employee holds the office. He may be included into reserve provided he passes training or further training or education. Bonus award shall not be paid.

0.9 and less. The employee holds the office provided he passes further training. Then the employee shall have the repeated evaluation. In case of negative result, the employee shall be reduced. Bonus award shall not be paid.

The selection of evaluation method for the company studied is the unique task to be solved only by the head. Thus, the implementation of the staff evaluation system demonstrated in this paper work enables to justify all long-term and short-term corporate purposes.

7 Conclusion

Thus, it should be concluded that staff evaluation conduct within the company gives the possibility of: evaluating potential for the promotion and reduction of the risk of non-competent employee promotion; arrangement for feedback with the employees on quality of their performance; determining the necessity of staff training; determining costs for training; developing plans and programs for staff development; determining potential increase in the employee's qualification; considering potential changes in salary amount; making amendments to the staff motivation system.

References:

1. N.N. Kozak, *Staff planning*. Top-manager's library, 60 (2017)
2. I.P. Migus, S.A. Chernenko, Business inform, **10,** 357-362 (2013).
3. L.V. Balabanova, *Staff management: textbook*, 512 (2006)
4. A.A. Vuchkovych-Stadnyk, Staff evaluation: clear algorithm of actions and quality practical solutions, 192 (2010)
5. O.A. Hrishnova, *Human resources management: problems, theory and practice*, **7,** 42–50 (2005)
6. Staff evaluation: [Electronic resource]. – Available at: https://hrliga.com/index.php?module=profession&op=view&id=1831
7. H.V. Nazarova. *Growth and use of managerial staff – strategical approach: monographs,* 216 (2012)
8. A.Ia. Kybanov, *Corporate staff management: actual technologies of employment, adaptation and attestation: textbook*, 360 (2016)
9. O.A. Havrysh, *Staff management technologies: monographs*, 528 (2017)
10. O. Krushelnitska, *Staff management: Textbook,* 304 (2005)
11. V.Ia. Chevhanova, Efficient economy, **4** (2014)

Economic security potential management mechanism of enterprises in the system of strategic decisions

Olena Arefieva[1], *Yuri Kopcha*[2], *Volodymyr Arefiev*[3], *Iryna Miahkykh*[4]

[1]National Aviation University, Department of Air Transport Economics, 1, Kosmonavta Komarova ave., Kyiv, 03058 Ukraine
[2]National Aviation University, Department of Air Transport Economics, 1, Kosmonavta Komarova ave., Kyiv, 03058 Ukraine
[3]Prima Leader Group, 3, Bohdan Khmelnytsky ave., Kyiv, 01030 Ukraine
[4]Kyiv National University of Technology and Design, Kyiv, Ukraine

Abstract. The article is devoted to the development of an economic security potential management mechanism of an enterprise in a dynamic external environment. Theoretical approaches to defining its essence are investigated, their systematization is carried out and the concept of "economic security potential management mechanism of an enterprise" is suggested, based on a two-pronged approach and, based on it, it is possible to make adaptive management decisions. The characteristic features and the main purpose of the economic security potential management mechanism of an enterprise are determined, its subject and object are substantiated. The basic approaches to forming the economic security potential management mechanism of enterprises are summarized, namely: systemic, process, resource, strategic, target, synergistic, functional and complex. It is established that the economic security potential management mechanism of enterprises performs both universal functions, namely: organizational, control, motivation, planning, and specific, including protective, regulatory, preventive, information, innovative, compensatory and social. It is suggested to consider a strategic approach as a management approach, as managing economic security potential is a complex task with many variables, random factors and various constraints. Errors in managerial decisions lead to bankruptcy.

Theory of the matter.

Management of activity of enterprises and their individual types is carried out with the use of various mechanisms, based on certain resources at the appropriate stages of development and state of the environment. E.g., such authors as N. Nizhnik, G. Lelikov, S. Mosov, I.V. Zhukova works [10, 6] consider the organizational and economic mechanism as a set of states and processes that make up a certain economic phenomenon that causes certain movements of the system. However, it should be noted that this definition does not reveal the main components (elements) of the mechanism and their relationship. According to the considerations of works by L. Tranchenko, Yu.M. Osipova, A.A. Malitsky, T.V. Polozova, Yu. V. Ovsyuchenko [15, 8,

12], organizational and economic mechanism is a set of organizational forms and economic levers, the interaction of which will allow to ensure the successful performance of the enterprise. In the definitions presented, the authors pay attention to the motivational methods that are included in the system of organization and management, their specific features are noted, but not specified. It should be noted that the "set of forms, methods and tools" does not always affect the change of others, not always determined by the content of the blocks and the corresponding elements that are part of the control mechanism. The more substantiated positions are followed by S.T. Piletska, O.M. Shkilnyuk, V.O. Fedorovich, A. Sokolova, A. M. Bukreeva, A.G. Goncharukin in [11, 16, 14, 3, 5], which consider organizational and economic mechanism as a system of interrelated, interdependent and clearly ordered organizational and administrative actions, implemented through economic methods, controls and incentives. It should be noted that the authors consider in more detail the concept by specifying methods, controls, incentives and management methods by defining their interaction.

Discussion of results

Fluctuating economic environment, conditions of fierce competition, intensified by the global financial and political crises, led to a decline in economic activity. Many machine-building enterprises operating in unfavorable economic conditions are experiencing a deep decline in production or find themselves in a critical situation. In general, economic security determines the ability of an enterprise to develop, increase the competitiveness and competitiveness of its commodity products and maintain its position in competitive markets. At the same time, the economic security of the enterprise reflects its ability to protect itself against any threats that are achieved through qualitative diagnostics and use of existing and potential resources of the enterprise. Diagnosing the economic security of an enterprise, it is a comprehensive multilevel study of the external and internal environment, and the company, which gives the opportunity to develop strategic directions, on the basis of which security is provided a system development strategy, a strategic plan for its provision. [1, p. 15]. High moral depreciation and wear and tear on fixed assets, lack of financial resources, and breaking of cooperative ties put most machine-building enterprises in a difficult financial position. As a result, there is the problem of managing the potential of economic security, which is may lead to conditions for reducing the level of threats to the activity of enterprises on the basis of strategic decisions, which is one of the most important tasks of science and practice. An important element of the enterprise security system is the mechanism of its management, which ensures the achievement of security goals and tasks. In the system of economic security of the enterprise under the mechanism should be understood in a certain way the order of the sequence of states and processes that ensure the economic security of the enterprise [7, p. 123-124].

The analysis of approaches to defining the concept of "organizational and economic mechanism" and the essence of the potential of economic security suggested a mechanism to manage the potential of economic security of enterprises as a dynamic element of the management system, which is, firstly, the scientific concept of managing the potential of economic security of enterprises in the direction, which is determined by their strategy, with a view to achieving appropriate competitiveness in the internal and external markets, secondly, it is a set of practical organizational and economic instruments of influence of the management entity on the managed object for the further development through qualitative assessment of the components of economic security potential, as well as the use of objective strategic recommendations by the relevant managers (Fig. 1).

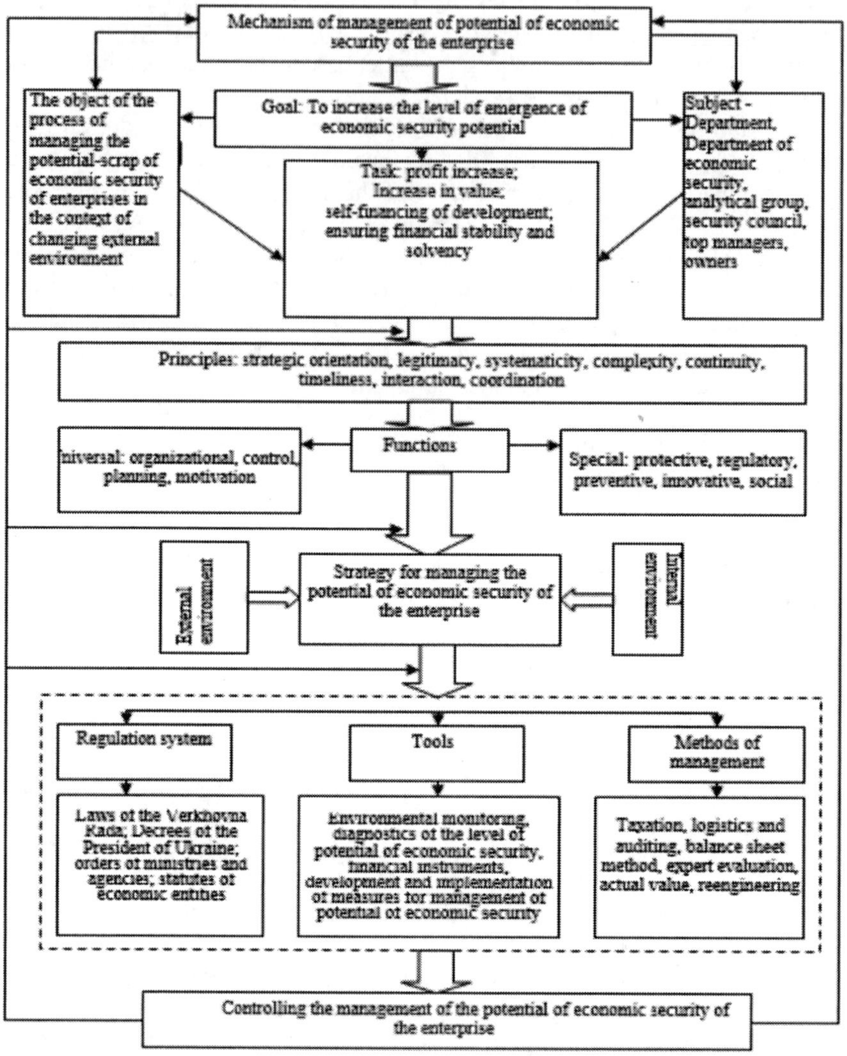

Fig. 1 Economic security potential management mechanism of the enterprise

The economic security potential management mechanism of the enterprise has the following features: dynamism and constant improvement of elements of the mechanism based on the analysis of the effectiveness of the adopted strategy; ability to change according to the external and internal environment of the enterprise functioning; effective use of economic security potential to reduce the level of threats to the enterprise; integrity and hierarchy of the structure, which allows to consider economic security as a whole and at the same time to analyze its structural subsystems and their relationship.

It should be noted that the main purpose of the economic security potential management mechanism of the enterprise is increasing the level of emergence of the potential when external environment is changing, which is achieved by maximizing the use of available resources during formulating and implementing the appropriate strategy to ensure sufficient working capital strategic measures.

The object of the mechanism is the process of managing the potential of economic security of enterprises in a changing external environment. The subject of the mechanism is a set of principles, methods, tasks, and practical aspects of managing the potential of economic security of enterprises. However, it is advisable for the entities of the mechanism to select the structural elements responsible for managing economic security potential: department or office of economic security, analytical group, security council, crisis expert group, specialists, top managers, owners, etc. (depending on the organizational form of the enterprise and its economic security policy).

The economic security potential management mechanism of enterprises should be based on certain principles: providing strategic orientation, which involves the use of strategic information in the analysis of the environment, identifying key problems that significantly affect the activity of the enterprise, the analysis of strategic alternatives to development, identifying the possibility of changing existing and emerging trends and etc .; orientation on the use of the potential of economic security, which helps to counteract negative changes in the external and internal environment of the enterprise; systematicity and complexity involves the use of systematic analysis and comprehensive consideration of factors in decision making. All elements of a managed system are considered in the aggregate, integrity and unity; legality involves the development of an economic security management system in accordance with national law; continuity, that is, continuous management of the potential of economic security; the need to justify the time horizon for solving strategic management problems regarding the utilization of the enterprise's potential of economic security; timeliness, namely, the development of preventive measures that will avoid or minimize the negative effects of potential threats; economic feasibility involves comparing the costs of managing economic security potential with potential losses; interaction and coordination of all business units.

The identification of the economic security potential management mechanism of enterprises should be implemented through universal and specific functions. The universal functions include organizational, control, motivation, planning [4]. The organizational function is to provide the process of managing the potential of economic security with the necessary labor, intellectual resources, means of labor, coordinating the activities of interested units and individuals. The control function is to identify trends of external and internal factors that negatively affect the process of managing the potential of economic security, and to develop suggestions for their elimination or elimination of action. The motivation function encompasses the development and use of incentives for effective interaction between the subjects of joint activity, to ensure the performance of work by members of the enterprise in accordance with the duties and plan. Planning as a function of the mechanism is to determine the guidelines for the future state of the enterprise; contributes to solving problems that arise in the process of strategy formation; makes necessary control of the execution, accounting and analysis. The economic security potential management mechanism of enterprises performs a number of important special functions, among which there protective, regulatory, preventive, information, innovative, compensatory and social. The performance of the protective function implies the prevention of internal and external threats to the enterprise and assumes that it has sufficient resource potential. In order to implement the protective function, the following conditions must be met: availability and effective utilization of the personnel and economic potential of the enterprise; optimization of available resources; development of human potential, management and human resources of the enterprise; protection of the enterprise from negative external influence, neutralization of threats and risks. Achieving a uniform motion of the enterprise (i.e. perfect interpenetration) becomes possible if the same pace and density of managerial influences are achieved. In such circumstances, the main purpose of managing dynamic changes in structural transformations is to increase the velocity per unit time in the planning period, since this will allow accelerating the main competitive processes [2, C 129-130]. The regulatory function of the economic security potential management mechanism involves regulating the state of the environment of the enterprise. The most important elements of such regulation are most often indicators that are influenced by the conditions of the external (market characteristics, competition, price level, etc.) and the internal environment of the enterprise. The preventive function of the economic security potential management mechanism of the enterprise is aimed at forecasting the situation and preventing the occurrence of critical situations, internal and external threats and risks. The implementation of this function is based on the development of many measures of economic, technological and organizational nature. The most important of these are measures that minimize the negative impact on the potential of economic security of the enterprise and the formation of an information security system.

The development of forecasts and preventive measures is based on the fulfillment of the information function, since the qualitative, adequate, reliable and timely information is needed for the qualitative prediction of risks, threats and the definition of effective measures. The information function is to centralize, systematize and process the incoming information, to form recommendations based on it to ensure the emergence of the potential of economic security of the enterprise. The innovative function of the economic security potential management mechanism is based on the development and further implementation of innovative solutions and measures to overcome emerging and already existing threats. The purpose of the compensation function is threefold: to eliminate the threat - to eliminate the consequences - to compensate for the loss. The social function of the economic security potential management mechanism is trying to achieve the highest level and quality of life of the owner and employees of the enterprise, through mutual partnership. The effectiveness of the financial security system of the enterprise depends on the mechanisms of its provision. A financial security mechanism is a set of tools, such as methods, resources, regulatory framework and managerial incentives, which, while interacting with each other, are intended to achieve the goals set to address the issue of ensuring the financial security of the enterprise. Formation of the economic security potential management mechanism of enterprises occurs through the implementation of certain actions on its objects, including profit, sources and volumes of financial resources, capital structure, structure of cash flows, structure of assets, investments, financial risks, system of financial innovations, etc. [17].

Elements of the economic security potential management mechanism are methods, tools, rules and regulations (regulatory system), information and analytical support system. The methods of the economic security potential management mechanism are the set of techniques and operations by which the enterprise manages the potential of economic security through the use of the following methods: technical and economic calculations, tax optimization, logistics and audit methods, balance sheet method, asset depreciation, economic and statistical method, expert assessments, present value, re-engineering, etc. In order to organize the process of managing the economic security potential, it is advisable to use the following tools: environmental monitoring, diagnostics of the level of economic security potential of the enterprise, financial instruments (payment, credit, deposit, etc.), development and implementation of measures to manage the potential of economic security. The system of regulation includes: Laws of the VerkhovnaRada; Decrees of the President of Ukraine; orders of ministries and departments; statutes of economic entities. Information support - economic, commercial, financial and other information - can be used as business reports, Internet; media and others.

Thus, the organizational and economic mechanism of economic security potential management of the enterprise involves the achievement of effective parameters of functioning, preservation of financial, production and personnel potential, formation of conditions for reliable and competitive functioning. The suggested structure of the organizational and economic mechanism for managing the potential of economic security of the enterprise reflects the economic and organizational components where a special place belongs to controlling.

Conclusion

The main purpose of the economic security potential management mechanism is to ensure the effective functioning, productive operation of the operating system and the economic use of resources, to ensure a certain level of labor potential of the personnel and the quality of economic processes of the enterprise, as well as constant stimulation for increasing the existing potential and its stable development. The general concept of operation of the suggested economic security potential management mechanism of the enterprise involves the creation of a system that provides recognition of threats, forecasting their consequences, determining methods of localization and submission of information to the general management system, where operational decisions to eliminate threats, or adjust the strategic goals of the enterprise. The implementation of the suggested mechanism requires the creation of an economic security monitoring system that allows providing adequate information support of the activity of the mechanism taking into account not only the external, but also the internal environment.

References

1. Arefieva Olena V., Miahkykh Iryna M., Shkoda Mariana S. *Bulletin of the Kyiv National University of Technologies and Design. Series: Economic sciences*, **3**, 8-17 (2019)
2. Arefiev S. O. *Economic Bulletin of the National Mining University*, **1**, 124-132 (2016)
3. Bukreev A. M. *Organizatsionno-ekonomicheskiy mekhanizm antikrizisnogo upravleniya: teoriya i praktika.*Voronezh (in Russian) (2000)
4. Herchikova I. N. *Menedzhment: uchebnik.* M.: UNITY. (in Russian) (1997)
5. Honcharuk A. H. *Formuvannia mekhanizmu upravlinnia efektyvnistiu pidpryiemstva (na prykladi kharchovoi promyslovosti).* (PhD Thesis), Odessa: Inst. market and eco-ecologist. research. NAS of Ukraine (2010)
6. Zhukova I.V. Power and management in the east of Russia, **4**, 43-49 (2010)
7. Ladyko L. N. Scientific Bulletin of Poltava University of Economics and Trade. Series: Economic Sciences, **3**, 123-126 (2014)
8. Malytskyi A. A. Proceedings of the *XIV International Scientific Internet Conference ADVANCED TECHNOLOGIES OF SCIENCE AND EDUCATION* (Ukraine, Kyiv, December 20-28, 2018). Available at: http://intkonf.org/malitskiy- aa-organizatsiyno-ekonomichniy-mehanizm-upravlinnya-pidpriemstvom-sutnist-ta-struktura
9. Mykhailenko V. M., Arefiev S. O. *Formuvannia orhanizatsiinoho potentsialu rozvytku ekonomichnoi bezpeky pidpryiemstv: monohrafiia.* Lviv: Ukr. Acad. Printing. (in Ukrainian) (2010)
10. Nyzhnyk N., Lelikov H., Mosov S. Bulletin of the Civil Service of Ukraine, **3**, 80-87 (2001)
11. Piletska S. T. *Antykryzove upravlinnia ekonomichnoiu stiikistiu promyslovoho aviapidpryiemstva.* K.: Center for Educational Literature. (in Ukrainian) (2014)
12. Polozova T. V., Ovsiuchenko Iu. V. Vesntnik VSU, **1**, 63-65 (2005)
13. Prokhorova V.V., Yatsenko B. I. Business Inform, **11**, 386-391 (2016)
14. Sokolova A. O. KNETE Bulletin, **2**, 88-92 (2006)
15. Tranchenko L. V., Balastryk L. O. Problemy udoskonalennia orhanizatsiino-ekonomichnoho mekhanizmu upravlinnia pidpryiemstvom. K.: Kiev University, vol. 7, pp. 188-195 (2005)
16. Shkilniuk O. M. Orhanizatsiino-ekonomichnyi mekhanizm rehuliuvannia innovatsiinoi diialnosti (PhD Thesis), K .: Economic Research Institute (2008)
17. Shtanhret A.M., Kotliarevskyi Ya.V., Karaim M.M. *Ekonomichna bezpeka pidpryiemstva v umovakh antykryzovoho upravlinnia: kontseptualne vyznachennia ta mekhanizm zabezpechennia: monohrafiia.* Lviv: Ukr. Acad. Printing. (in Ukrainian) (2012)

Analytical tools for ensuring economic security in the conditions of global competitiveness: theoretical aspect

Nina Avanesova[1], *Olena* Kirdina[2], *Irina* Volovelska[3], *Valentina* Maslova[3], and *Yuliya* Utkina[31]

[1] Kharkov National University of Civil Engineering and Architecture, Department of Management and Public Administration, 61002 Sumska st. 40, Kharkiv, Ukraine
[2] Ukrainian State University of Railway Transport, Department of Accounting and Audit, 61050 Feierbakh sq. 7, Kharkiv, Ukraine
[3] Ukrainian State University of Railway Transport, Department of Economics and Management of Industrial and Commercial Business, 61050 Feuerbakh sq. 7, Kharkiv, Ukraine

Abstract. Global competitiveness, based on the international division of labor, the use of factors of production and the country's place in the world market ensures the ability of the state to participate in the competition and the full use of its own competitive advantages. Globalization processes have negative consequences for functioning of national economies, which actualizes the problem of ensuring the economic security of countries. The economic security of the state is the criterion of the effectiveness of the state economic policy and characterizes the overall level of the country's economy; the presence and level of threats to the state, society or person; the effectiveness of state regulation of the economy; the full implementation of the functions of the state. At the same time, its analytical tools are based on the most important parameters of the state of the economic system, its stability and mobility: GDP growth, the level and quality of life of the majority of the population, inflation, unemployment, the structure of the economy, the property stratification of the population, criminalization and shadowing of the economy, the state of the technical and technological base of the economy, spending on research, import dependence, economic openness, internal and external debt.

1 Introduction

The growing trends of globalization and integration of the world economy make it necessary to ensure competitiveness in the global dimension. Now, according to the world economic forum, which is held annually in Davos (Switzerland), in the world ranking on the global competitiveness Index 2017-2018, Ukraine occupies the 81st position among 134 countries participating in the evaluation. Switzerland traditionally tops the rating, and up to 10 most competitive economies included the USA, Singapore, the Netherlands, Germany, Hong Kong, Sweden, the UK, Japan and Finland [1].

Global competitiveness means the ability to perceive new knowledge and trends of global development, adapt to them and adapt them to themselves in such a way as to be included in global processes without destroying the internal integrity and preserving their national and cultural identity, as well as to compete and win a worthy place in the system of international territorial division of labor [2]. It is provided through the formation of competitive advantages in the areas of social structure and the structure of the economy, regional accessibility and culture, innovation, environment, skills of the workforce. Within

[1] Corresponding author: yuliyautkina@ukr.net

the limits of actual competitiveness, GDP growth is observed by increasing the efficiency of labor and employment of the population under the influence of attracting foreign investment, the development of small and medium-sized enterprises, research and technological development, infrastructure and human capital, intellectual and social capital. At the level of global competitiveness of the national economy, GDP growth correlates with the quality of life of the population, providing a stable competitive position and the well-being of the population in the long term.

It is the global competitiveness based on the international division of labor, the use of factors of production and the country's place in the world market that ensure the ability of the state to participate in the competition and the full use of its own competitive advantages. At the same time, globalization processes have negative consequences for the functioning of national economies, which actualizes the problem of ensuring the economic security of countries.

2 Main part

Global competitiveness is the potential or realized ability of a subject of international competition to operate in the global economic space, which is based on the use of global competitive advantages and aimed at achieving the goals of its international business.

With the increasing globalization of competition, the importance of the development and implementation of global competitiveness strategies in the practice of world market players is also increasing. A global approach to the formation of international competitiveness strategies is now defined as an effective way that provides economies of scale, assimilation and response to international market demand, and an effective combination of resources – capital, labor, raw materials and technology – worldwide.

Most aspects of global competitiveness strategies are based on the formation of a certain type of competitive advantages, which are determined by the characteristics of the industry, sources and factors of competition in it. At the same time, global competitiveness strategies are based on a fundamentally different paradigm of competitive advantage formation, which does not recognize borders between countries. This "panoramic approach" to the management of the processes of formation and realization of global advantages in a much more complex geo-economic space requires the use of traditional principles and methods of management, and a new version of it – the so-called goal of management, or meta-management.

For industries and industries of primary sectors of the economy, the international economic activity of which is based on the exploitation of mainly resource-factor advantages, non-economic factors of global competitiveness are created, first of all, by the purposeful policy of the governments of the countries on the basis of:

- control of the world's raw materials and diversification of sources of reliable supply;
- control of global transport communications;
- influence on political and economic systems of the countries-suppliers of raw materials on the world markets;
- introduction of environmental, social and other norms and standards of safe business in basic industries.

The advantages that are created by maintaining the stability of global commodity markets and enhancing the security of global business strengthen the competitive position of transnational structures.

Global competitiveness is not a problem only for the primary sectors of the economy. Active formation of global advantages is manifested, first of all, in industries and industries that exploit resource and technological and innovative advantages.

Governmental assistance in the formation of global advantages of these industries is provided primarily through:

- formation of innovative infrastructure of the economy;
- facilitating the flow of highly skilled labour from other countries of the world;
- creation of a global business information system;
- ensuring the secrecy of technological innovations and their uncontrolled spread;
- protection of copyright and intellectual property through international organizations.

Innovation and technological advantages are the main means of entering the strategic flow of modern international economic exchange. Generation of a qualitatively new state of technology – "metatechnology" – creates a monopolistically advantageous global competitive position of the developed countries of the West. Among the non-Western countries, only ten of them – China, Japan, India, Brazil, Turkey, Poland and some others – are considered today to be integrated into the global economy. Among them there is not a single country of the post-Soviet space, including Ukraine, because a country with a dominant development of primary sectors of the economy cannot enter the system of the global economy [4-5].

In the context of non-proliferation of destructive reforms of the industrial sector of the economy, the dominance of imports, the uncertainty of the political situation in the country, as well as under the influence of military aggression in Eastern Ukraine and the annexation of Crimea, the state in recent years has lost more than 20% of its economic potential [3], which had significant negative consequences:

- destruction of the productive capacity of the country and the breach of cooperative ties between enterprises of individual regions;
- narrowing of export potential;
- reducing the investment attractiveness of the economy and curtailing investment activity;
- reduction of the internal resource base of the fuel and energy complex;
- reducing the competitiveness of the national transport system and the loss of transit capacity;
- growth of social tension and protest moods in society [6]

In numerous researches of scientists it is defined that economic security characterizes a state of national economy at which economic stability to internal and external threats remains and needs in system "person-society-state" are satisfied. Given the ability of economic security to ensure and maintain a stable competitive position of the national economy in the world economic system, to self-development and self-improvement, we note that economic security is not only an indicator of the internal state of the national economy, but also quite fully characterizes its external relations and interaction with the external environment.

Integration processes of recent times put before the economy of Ukraine a number of issues to ensure a sufficient level of foreign economic security of the state, because it is foreign economic security is the "guarantor" of full participation in the international division of labour, the system of world economic relations and the system of protection and countering global challenges and threats. During the integration of the state into the system of world economic relations, it faces the problem of contradiction between the need, on the one hand, to integrate into the world economy, and on the other – to ensure the harmonious development of the national economy, the protection of national economic interests, the domestic market and domestic producers. Given this, it should be recognized extremely ambiguous impact of globalization on the state of socio-economic security along the hierarchy – from the level of the state to the level of the citizen [7-8].

At the same time, it should be noted the indisputable role of foreign economic security in ensuring the global competitiveness of the country, which is part of the world economic space. One of the fundamental processes of market transformation of the national industrial complex of Ukraine is its integration into the world economic relations, as a result of which it gradually becomes part of the world industrial system. The formation of such a system is the result of the process of globalization of the economic space, characterized by the growing interdependence of national economies, increasing integrity and unity of the world economic system on the basis of increased openness of national markets and deepening of the international division and integration of labour [9].

In the context of sustainable post-industrial development of society and state institutions, when the state is able to confront the emerging internal and external threats to stability in all areas of economic security in the global competitive environment is a really achievable process. At the same time, in our opinion, it is expedient to define and establish security standards, the violation of which leads to destabilization in the economy or social sphere, and their thresholds. At the same time, the deviation of the actual levels of indicators from the threshold values to determine the system of priorities of economic security of the state in the global dimension.

According to [10] the main indicators and threshold values of indicators of macroeconomic security of Ukraine are as follows:

- the level of "shadowing" the economy – no more than 30% of GDP;
- the ratio of GDP to the average value in the EU – at least 75%;
- ratio of GDP per capita to the average value in the EU – at least 50%;
- the ratio of GDP per person to the global average - not less than 100%;
- changes in inventories should be in the range of -1.5 to +1.5% of GDP;
- the ratio of the balance of payments of Ukraine to GDP should be in the range from -1 to +1%;
- share of available income of non-financial corporations in gross available income-not less than 14-15%;
- the share of the sector of public administration in the available income – no more than 20%.

3 Conclusion

Globalization processes taking place in modern society are reflected in the change of functional components at various levels of economic security. The above-mentioned components change from ensuring the primary needs of the citizen through the financial independence of enterprises to improving the efficiency of the object of study at the industry level. With the further consolidation of the scope of application of the concept of economic security to the regional and national level, the functional components used to characterize the primary needs, to which the financial, technological, innovative and energy parameters are added, come to the first place. At the same time, only international competitiveness, a set of international conditions for coexistence and inter-state relations are taken into account at the global level. Globalization trends in the economy directly affect economic security and are aimed primarily at meeting the needs and economic interests of the individual or household. Thus, an attempt at the state, international and global level to bring the dependence of economic security on the functional components of the higher order, facilitate the task only for the highest state functionaries, narrowing and simplifying the strategic goals that they should achieve.

Economic security of the state is the criterion of the effectiveness of the state economic policy and characterizes the overall level of the country's economy; the presence and level of threats to the state, society or person; the effectiveness of state regulation of the economy; the full implementation of the functions of the state. At the same time, its analytical tools are based on the most important parameters of the state of the economic system, its stability and mobility: GDP growth, the level and quality of life of the majority of the population, inflation, unemployment, the structure of the economy, the property stratification of the population, criminalization and shadowing of the economy, the state of the technical and technological base of the economy, spending on research, import dependence, economic openness, internal and external debt.

References

1. Pozycija Ukrainy v rejtynghu krajin svitu za Indeksom ghlobaljnoji konkurentospromozhnosti 2017-2018. URL: http://edclub.com.ua/analityka/pozyciya-ukrayiny-v-reytyngu-krayin-svitu-za-indeksom-globalnoyi-konkurentospromozhnosti-2
2. N.Ya. Kalyuzhnova. *Konkurentosposobnost rossiyskikh regionov v usloviyakh globalizatsii*, TEIS, 526 (2003)
3. O.V. Sobkevich. Собкевич О.В. Ocinka vtrat ta mekhanizm vidbudovy realjnogho sektoru ekonomiky skhodu Ukrajiny. [URL: http://www.niss.gov.ua/articles/1832/
4. T.A. Kocko T.A. *Ekonomichna bezpeka terytorialjno-vyrobnychykh kompleksiv: energhetyka, ekologhija, informacijni tekhnologhiji*, MP Lesja, 256 (2015)
5. O.M. Lyashenko. Economy. Management. Business. **17(II)**, 179-187 (2007)
6. Metodyka rozrakhunku rivnja ekonomichnoji bezpeky Ukrajiny. URL: http://www.kmu.gov.ua/control/ru
7. G.V. Kozachenko, V.P. Ponomapev, O.M. Lyashenko. *Ekonomichna bezpeka pidpryyemstva: sutnist ta mekhanizm zabezpechennya*, Libra, 280 (2003)
8. N. Avanesova, I. Volovelska, V. Maslova, T. Sukhorukova, Yu. Utkina. Inetrnational Journal of Engineering & Technology. **7(4.3)**. 393-397 (2018)
9. V.L. Dykan. The bulletin of transport and industry economics. **34**, 141-147 (2011)
10. A.V. Tolstova, K.V. Khomenko. The bulletin of transport and industry economics. **63**, 187-195 (2018)

Labor market: problematic aspects of informal employment in Ukraine

Alla Bielova [1]*, *Nataliia* Zhuravska [2], *Gheorghij* Olijnyk [3], *Svetlana* Koval[4], and *Alona* Kochedykova [5]

[1] Kyiv Institute of Innovation Education, National University of Construction and Architecture, Department of Economics and Management of the GSP, Vozduhoflotsky Avenue 31, Kyiv, 03680, Ukraine
[2]National University of Construction and Architecture, Vozduhoflotsky Avenue 31, Kyiv, 03680, Ukraine
[3]Open International University of Human Development "Ukraine", Ukraine
[4] State Higher Educational Establishment "Pereyaslav-Khmelnyckyi State Pedagogic University named Grygoriy Skovoroda", Sukhomlinsky str. 30, Pereyaslav-Khmelnitskiy, Kiev region, 08401, Ukraine
[5] Boryspil Institute of Municipal Management at the Interregional Academy of Personnel Management, Kiev Region, Ukraine

Abstract. The subject of the article is the study of specific problems of informal employment of people in the labor market of Ukraine by analyzing the statistical estimation of dynamics and structure of some indicators in 2015 and 2017. The study used general scientific and special statistical methods: analysis of dynamics and synthesis, analogy and comparison, expert assessments, etc. On the basis of the study of the numerical characteristics of the main indicators of the labor market, an analysis of certain aspects of the problems of the informal sector in the field of employment and unemployment in Ukraine was conducted, as well as the main prospects for their possible future solution in the modern economic conditions. It is determined that coordination of efforts of the authorities at all levels in the implementation of the employment policy and reduction of its shadowing scale is impossible without a scientifically based comprehensive national program for promoting employment aimed at ensuring a balanced supply and demand for labor in the labor market.

Introduction.

Problem statement. At present, the development of market relations in Ukraine comes along with significant changes in the employment situation, leading to new features of the functioning of the labor market. One of the biggest obstacles to creation of a new model of socio-economic development of the country currently is the high level of economic shadowing, which makes it impossible to accelerate economic reforms, increase social standards of living of the population and integration into the European community. One of the most important issues in this area is labor market transparency. The urgency of the problem of legalizing employment and wages today is due to its scale and extremely negative and devastating consequences.

* Corresponding author: nzhur@ ua.fm

Analysis of recent research and publications. The issue of the functioning of the labor market and the principles for its regulation have been highlighted in numerous papers by scholars, researchers, economists, sociologists, and specialists in public administration: Z.P. Baranik, D.P. Goddess, O.V. Bokova, VS Vasilchenko, M.I.The lower one, T.M. Kirian, G.I. Kupalova, E.M. Libanova, V.V. Onikienka, L.G. Rozhdestvenskaya, V.A. Savchenko, A.A.Shenchuk, G.G. Trofimova, O. M. Umansky, V.M. Chamoty, L.S. Shevchenko, L.M. Yakushenko, L.D. Yatsenko and others [2, 3].

However, at present, Ukraine lacks fundamental scientific works which would have had a comprehensive approach to studying the problems of shadow (unreported) employment in Ukraine, i.e., the use of hired labor without an employment contract and shadow wages that have been rapidly spreading in recent years, covering a large number of jobs at enterprises and organizations of the formal and informal sector of the economy.

The purpose of the article is to review and statistical analysis of processes of informal employment of the population of Ukraine in terms of its comparison with actual official employment in the labor market and further developments in this direction.

The main part.

It should be noted that under the conditions of market economy in Ukraine, informal employment is spreading rapidly, covering all informal workplaces both in the informal and formal sectors of the economy. Usually informal employment is inherent in the population of certain socio-economic groups.

Thus, in 2015 the number of informally employed population of Ukraine amounted to 4.3 million people, or 26.2% of the total employed population. Informal labor relations predominated in the self-employment sector, where the share of the population working in informal work places was 73.2%, and among those employed by such persons was about 17.3% [1].

Among informally employed persons, more than one third (38.6%) worked in agriculture, forestry and fisheries. Other common types of economic activity of the population of this category are wholesale and retail trade (22.1%) and construction (16.1%) [1].

In 2017, the number of informally employed population in the country decreased by 642 thousand people compared to 2015 and amounted to 3.7 million people, or 22.7% of the total number of employed population.

Thus, informal labor relations predominated in the self-employed sector, where the share of population with informal jobs was 74.3%, while the number of employed persons was almost 13.1% (Table 1) [1].

Currently, people with a low level of education and those who are engaged in physical or unskilled labor are informally employed. In particular, the proportion of informally employed among the workers of the simplest professions in 2017 was significantly higher than among the officially employed (51.5% vs. 9.6%). Agriculture, forestry and fisheries are the main activities of the informally employed people (42.7%). Other common types of economic activity of the population of this category are wholesale and retail trade (19.5%) and construction (15.3%) [1].

During 2015-2017, changes in the status of employment among the formal and informal population took place in the structure of the employed population, the increase in the share of employees was (84.3% versus 84.1%).

Table 1. Employment of Ukrainian population by employment status and place of residence in 2015, 2017

	Employed population		Therefrom informally occupied the population	
	total,	including worked	total,	including worked

	thousand people			thousand people	for hire		not for hire	
		for hire	not for hire		thousand people	%	thousand people	%
2015								
Total	**16443.2**	**13823.4**	**2619.8**	**4303.3**	**2385.2**	**17.3**	**1918.1**	**73.2**
Urban area	11309.0	10311.3	997.7	2118.3	1697.8	16.5	420.5	42.1
Countryside	5134.2	3512.1	1622.1	2185.0	687.4	19.6	1497.6	92.3
2017								
Total	**16120.9**	**13585.1**	**2535.8**	**3661.3**	**1777.4**	**13.1**	**1883.9**	**74.3**
Urban area	11108.5	10148.5	960.0	1680.8	1248.0	12.3	432.8	45.1
Countryside	5012.4	3436.6	1575.8	1980.5	529.4	15.4	1451.1	92.1

At the same time, reverse processes occurred among informally employed population (Figure 1).

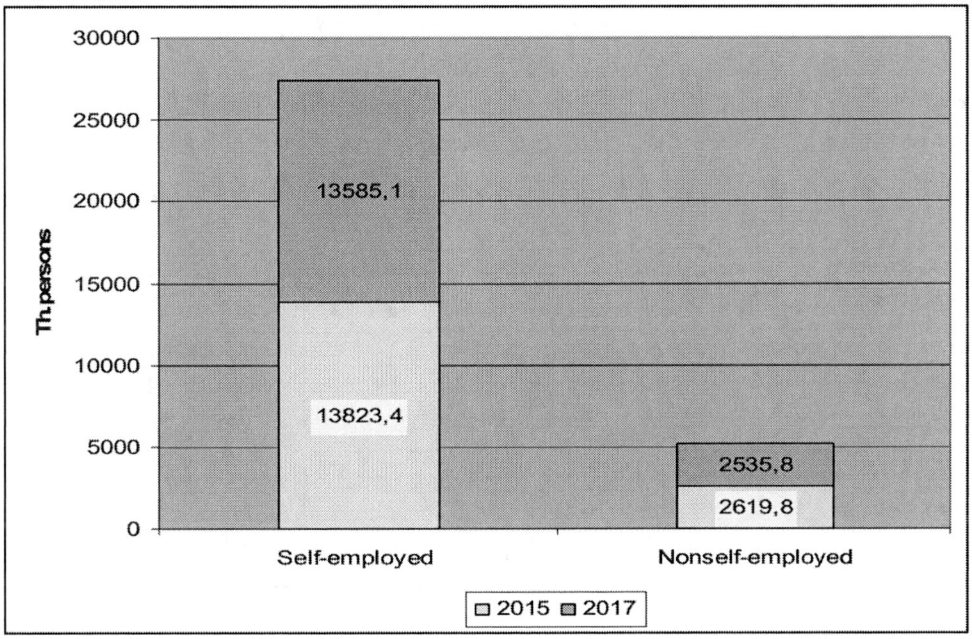

Fig. 1. Formal employment of Ukraine's population by employment status (whether employed or not) in 2015, 2017

In 2017, among the informally employed population, two thirds (66.6%) comprised persons who had vocational and full secondary education, and those with higher education (including full, basic and incomplete higher education) dominated the officially employed population (61.5%).

More than half of informally employed people (50.8%) were rural residents. Men were also more likely to work informally (58.8%), compared with women. The analysis of informal employment by age groups shows that its highest level is characteristic for persons aged 15-24 (36.1%) and at the age of 60-70 years (37.2%).

Table 2. Informal employment of Ukraine by age group, gender and location in 2015, 2017

	The number of informally employed population		Including by age group (years), in% of the total number of the employed population of the corresponding age group							Able-bodied age, thousand persons
	thousand people	in% of the total number employed	15–24	25–29	30–34	35–39	40–49	50–59	60–70	
	2015									
The whole population	**4303.3**	**26.2**	**36.1**	**26.8**	**25.0**	**25.6**	**24.0**	**24.0**	**37.2**	**4046.3**
women	1772.2	22.5	30.5	22.0	20.5	21.6	20.0	22.1	39.3	1626.5
men	2531.1	29.5	40.4	30.5	28.6	29.2	28.1	25.9	35.0	2419.8
urban area	2118.3	18.7	26.6	20.3	19.9	19.8	17.5	15.8	14.9	2052.5
countryside	2185.0	42.6	49.4	42.0	40.6	39.7	37.6	41.5	74.1	1993.8
	2017									
The whole population	**3661.3**	**22.7**	**34.7**	**23.8**	**20.6**	**20.2**	**20.9**	**21.4**	**33.1**	**3436.3**
women	1506.2	19.4	28.0	19.9	15.2	17.7	17.3	19.9	33.8	1387.7
men	2155.1	25.8	40.1	26.5	25.0	22.6	24.6	23.0	32.5	2048.6
urban area	1680.8	15.1	23.6	18.0	15.3	14.6	14.1	13.1	11.7	1631.3
countryside	1980.5	39.5	50.0	35.7	35.9	35.1	35.7	39.5	68.5	1805.0

In most cases, the youth nature of informal employment is manifested in the need for additional earnings during education. In 2015, more than two-thirds (67.5%) of the informally employed population consisted of those who had vocational and general secondary education, and those with a higher education, including full, basic and incomplete higher education, were prevalent among the officially employed population - 61.8%.

In 2017, the most widespread informal employment was among young people aged 15-24 years, as well as persons aged 60-70. These categories of the population are more vulnerable and socially unprotected in modern economic conditions.

The spread of informal employment among the retirement age is primarily due to the low level of pensions, which prompts them to seek informal sources of additional livelihoods. Among women, the share of non-employed people exceeded the corresponding indicator among men (76.1% vs. 73.0%), and in rural areas it was twice as high as in urban settlements (92.1% and 45.1%) (Table 2) [1].

In the total number of the employed population of Ukraine in 2017 in the twelve regions of the country, the proportion of informally employed persons did not exceed the average in Ukraine (22.7%). The largest value of this indicator was observed in Chernivtsi region (50.9%), and the smallest in the Kiev region (9.4%), (Figure 2) [1].

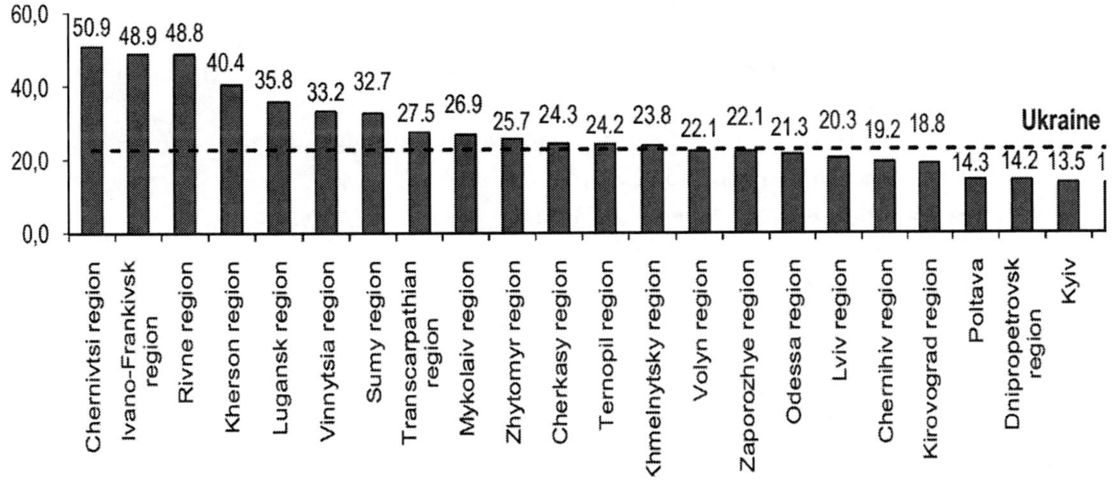

Fig. 2. Informal employment of the population of Ukraine by regions in 2017 [1] (in % to the number of employed population in the region concerned)

Conclusions

Consequently, the current state of the labor market of Ukraine is characterized by unfavorable trends in the level of employment and ineffective use of labor force. At this time, shadow occupation has become extremely widespread, with negative consequences for both the state and employees.

Coordination of efforts of the authorities at all levels in implementing the employment policy and reducing the scope of its shadowing is impossible without a scientifically based comprehensive national program to promote employment, aimed at ensuring a balanced supply and demand for labor in the labor market. Therefore, one of the main tasks of the Ukrainian state policy should be the concentration of efforts of all branches of power on the complex implementation of drastic economic, administrative and organizational measures to legalize the employment of the population in the Ukrainian labor market.

References

1. Official site of the State Statistics Service of Ukraine. URL: http://www.ukrstat.gov.ua/.
2. A. I. Belova, A. E. Kočedikova. Economics and Society. **7**, 201-205 (2016) URL: http://economyansociety.in.ua.

3. B. M. Danylyshyn, V.V. Mykytenko - In 2 t. - Vol. 2. - K.: RVPS of Ukraine, National Academy of Sciences of Ukraine, Vyd. CJSC "Nichlava", 210 (2008).
4. Senchuk A. The Strategy of Sustainable Development "Ukraine-2020" - the Way to a Better Future. External affairs. **8,** (2014). URL: http://uaforeignaffairs.com/ua/zs/anglomovna-versija/view/article/ strategy-stalogo-rozvitku-ukrajina-2020-shljakh-do-kr /.
5. Yatsenko L. D., Yakushenko L. M. *Analytical report of the Department of Social Policy*, Kyiv, 2016. URL: http://ndei.me.gov.ua/index.php?option=com_content&view=article&id =203&Itemid=89/.
6. Nesterova T.Yu., Sviridova S.V. *System Concept of Management of Industrial Enterprise in the Development of Information Technologies. Innovative Bulletin Region, 2011.* - No. 2.6. Bruce Stokes. Deep Cuts: What the Age of Austerity Means for Global Public Goods. - Brussels Forum, March, 2011. - Access: http: blogs.cgdev.org.
7. B. M. Andrushkiw, N.B. Kyrych, O.B. Pohaydak. Journal of economics of Ukraine: Scientific journal. **1 (21)**., 3-5 (2012).
8. *Modernization of state support of business entities in Ukraine: analytical note.* - C.: Izd NISS under the President of Ukraine,128 (2013).
9. C. Fellenshtein. *Exploring E-commerce, Global E-biness and E-societies*. Upper Saddle River: Prentice-Hall, 2000. – 269 p.
10. Global Internet and E-Commerce Trends 2017. Stats.com GmbH & Co. KG, 251, (2018).
11. Smolyanyuk P. Economic Analysis. **5**. 332, 336. 2010. Ternopil. TNEU

Strategic management accounting as an information platform for measuring innovation of the enterprise

Mykola Bondar[1], *Natalia* Iershova[21], and *Tetyana* Chaika[2]

[1]SHEI "Kyiv National Economic University named after Vadym Hetman", 03057, Kyiv, Ukraine
[2]NTU "KhPI", Institute of Economics, International Management and Marketing, 61002, Kharkiv, Ukraine

Abstract. The study provides an integrated analysis system of the innovative activity of the enterprise, generalized mechanism of information processing in the accounting and analytical system of the enterprise. The role of information of strategic management accounting and analysis in the accounting and analytical system of the enterprise is substantiated. The bimodal methodological approach to measuring innovation of the enterprise on the basis of an integral performance exposure, calculated on a set of partial indicators, using additive method was introduced.

Introduction.

Long-term development of the enterprises in the dynamic market conditions directly depends on the state of innovative activity. The measuring innovation is a prerequisite for the production of innovations and is of interest to potential investors. Most of the specialists involved in the innovation field associate the innovation activity with the growth of the profitability level of the enterprise and the achievement of competitive advantages in the future, reflected in the Sustainable Development Strategy "Ukraine-2020". Measurement of innovative activity is the initial stage of planning of innovative activity and evaluation of innovative development of the enterprise as a whole. Therefore, its methodological and information provision is of particular importance.

Methods. The research methodology was based on the theory of innovation management, management accounting and strategic analysis. The main methods that formed the conceptual basis of the study are: comparison, analysis and synthesis, as well as expert, empirical generalization and statistical processing. Telephone interviewing, analysis of publicly disclosed information, data from the Internet and insider information were used. The methodical approach to measuring innovation of the enterprise is developed using the method of expert measurements, as well as polling.

Results and discussion

[1] Corresponding author: iershova.ny@gmail.com

1 Analysis of innovative activity of the enterprise - special aspects of implementation

At present, there is no single methodology for analyzing the innovative activity of the enterprise due to the complexity of the development of the methodologies of the study, the availability of different approaches to defining the essence of the concept of "innovative activity of the enterprise", the problematic allocation of factors that affect the innovative activity of the enterprises [1-4]. The peculiarity of the analysis of the innovative activity of the enterprise is the coverage of all its stages, the orientation towards identifying the achievement of the main objectives of the enterprise, the close relationship with the system of strategic management and the adoption of long-term managerial decisions, the integration of quantitative and qualitative methods of analysis, methods of mathematical, logical analysis, methods of expert measurements (Fig. 1).

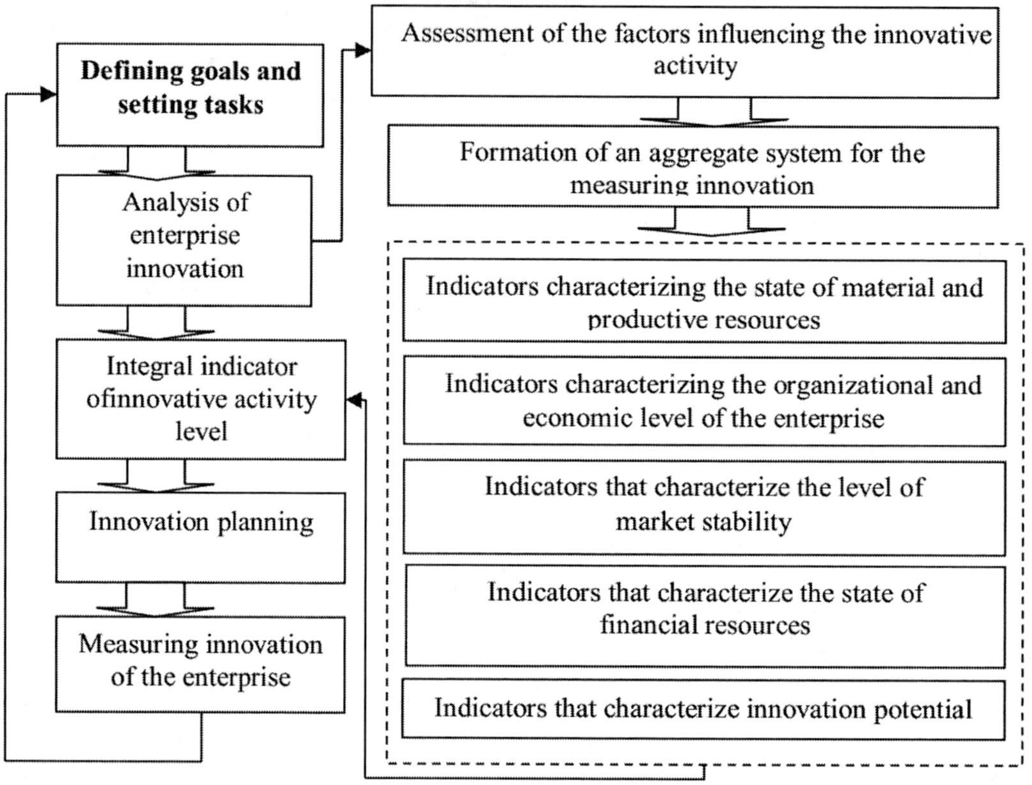

Fig. 1. Integrated analysis system of innovative activity of the enterprise

The complex model of the analysis of innovative activity of the enterprise characterizes innovation processes and investment decisions associated with them, therefore its results are useful in making managerial decisions concerning long-term sustainable development of the enterprise and achievement of the best results of production activity using available resources [5, 6].

2. Strategic management accounting as an information platform for the analysis of innovative activity of the enterprise

The main focus of strategic management accounting is the formation of financial and non-financial information when making strategic management decisions, including for a comprehensive analysis of innovative activity of the enterprises [7-9]. The attention has focused primarily on its external orientation, and information formed in the contour of strategic management accounting is different in nature, so for its processing used both accounting and special, including mathematical tools. Strategic management accounting is a component of the accounting and analytical system implemented in the enterprise (Fig. 2).

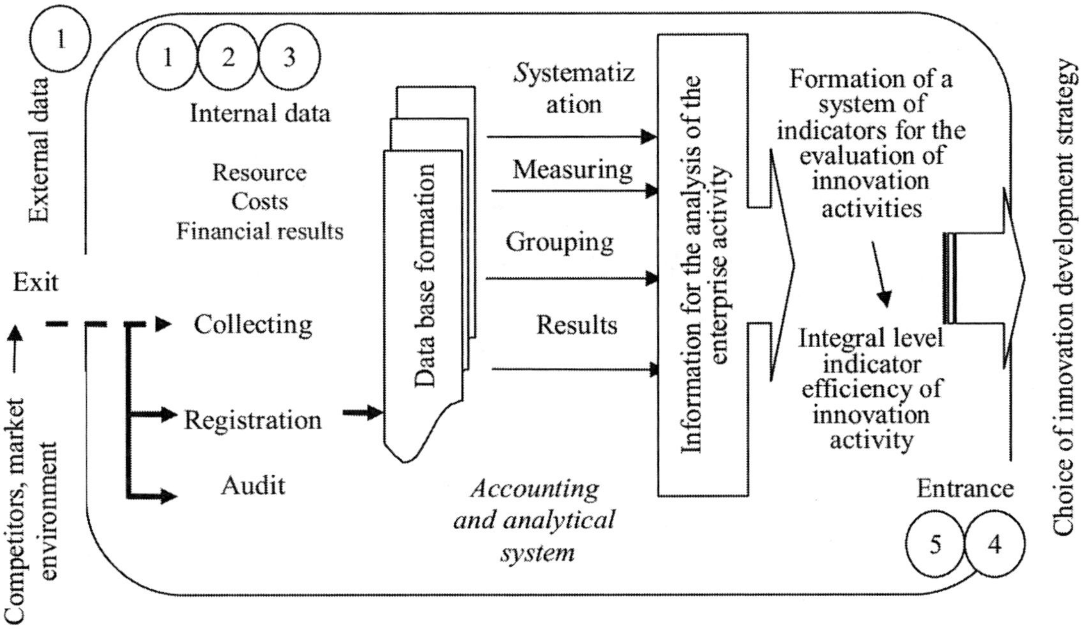

1 - strategic management accounting, 2 - accounting and management accounting, 3 - financial accounting, 4 - financial reporting, 5 - strategic management reporting

Fig. 2. The mechanism of information processing in the accounting and analytical system of the enterprise

In such a system, various content, quality, sources of formation of information are used and formed, which requires its timely collection, verification, analytical processing and preparation of analytical work outputs [7, 8].

3. Measuring innovation of the enterprise

This study proposes a bimodal measurement of innovation of the enterprise. The first level is the formation of aggregated indicators, *Ai*. Each aggregate indicator is represented as a group of partial indicators that characterize its state, *Gi*. The second is the calculation of the integral index, *I*. The aggregate indicators are: the level of material and technical provision (A_1), market stability (A_2), organizational and economic level of the enterprise (A_3), the state of financial resources (A_4), innovative potential (A_5).

The partial indicators used for the measurement should be simple, understandable and consistent with the goals of information users. Therefore, at their formation, we suggest taking into account the practical aspect. At the same time, information congestion should be avoided, because too many components complicate the evaluation procedure. The choice of partial indicators, the formation of aggregate indicators is based on the method of expert measurements [10]. The score scale is used to measure partial indicators: 0 points - the indicator is well below the standard level (or higher if the excess of the norm is negative) or the dynamics of change during the reporting year is negative; 0.3 points – the indicator is close to the limits of normative values, or the dynamics of its change during the year is unchanged; 0.6 points – an indicator within the normative value, or the dynamics of its change during the year is insignificant; 1 point – an indicator within the normative value, or a significant positive dynamics of its change during the study period. The rates of growth of indicators are divided by the relative 5%. Normative values of partial indicators are established on the basis of official statistical data, financial analysis techniques, expert experience and financial and managerial reporting data of the enterprise. Balance score, G_i

$$A1 = \sum_{i=1}^{5} G_i \times k_i \quad (1)$$

$$I = \sum_{i=1}^{5} A_i \times K_i, [0;1] \quad (2)$$

Since the partial indicators as aggregated have different significance in the formation of indicators of a higher level, in the study, the authors calculated their importance. In calculating weight coefficients a method of questioning is used, according to which the impact of each indicator on the levels of innovation of the enterprise is estimated on the scale as follows: 0 – no impact; 1 – very weak impact; 2 – weak impact; 3 – medium impact; 4 – strong impact; 5 – very strong impact. Expert interviews are conducted in two stages: first, experts express their opinion on aggregate indicators, and then measure their components. Expert judgments are recorded

As a result, a matrix is formed, on the basis of which averaged expert opinion and weight coefficient are calculated for each indicator. Averaging the results obtained during the expert evaluation is carried out using methods of mean arithmetic or geometric mean. The modified Harrington verbal numeric rating scale was used in the study to interpret the results (Table 2).

Table 1. Description of the levels of innovative activity of the enterprise

Activity level	Integral indicator	Description
low	0–0.24	The organization of production processes is inefficient. There is a lack of fixed assets and a shortage of own sources of financing. Investments are almost used to finance current activity
minimum allowable	0.25–0.49	Due to lack of own funds, it is difficult to implement innovative projects, which can not be financed without borrowing capital. Innovation projects are characterized by low profitability and a large number of them remains unrealized
normal	0.5–0.74	Innovative projects of the enterprise are quite profitable, with an acceptable payback period. The enterprise uses both its own and borrowed funds to finance innovation. The dynamics of economic indicators is characterized by low, but stable growth. Innovation activity can be characterized as successful, but not sufficiently sustainable
high	0.75–1	The introduction of innovation projects is accompanied by a tangible increase in economic performance and increased competitiveness. The enterprise actively uses its own funds and has opportunities for building up its production potential

The approbation of the introduced methodology for measuring innovation of the enterprise was carried out within the implementation of the scientific theme "Development of theoretical and methodological foundations for the formation of a management accounting system of the enterprise" (No. 0117U004814) by the Department of Economic Analysis and Accounting of NTU "KhPI". The sample includes four enterprises of Kharkiv, which are engaged in innovative activity, since the specifics of their activity requires continuous improvement of not only the production process, but also the system of management and quality of accounting and analytical provision. The information basis was the data of strategic management accounting, financial and internal management reporting. The expert group is composed of top managers, representatives of the analytical department of the enterprises selected for research and the lecturers of the Department with the total number of 11 people.

Based on the developed questionnaire, experts have developed a balanced system of partial and aggregate indicators. Calculations are carried out by the baseline method, intermediate calculations and calculation of the integral indicator of the levels of innovation of the enterprise by the introduced method. The results are presented in Fig. 3.

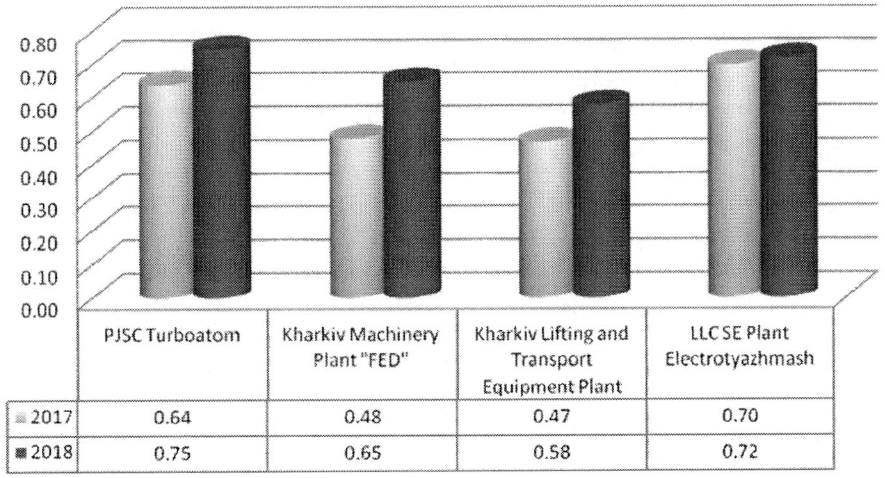

Fig. 3. The integral indicators of the levels of innovation of the enterprises

According to the results, none of the analyzed enterprises reached high levels of innovation. The normal levels of innovation are at PJSC Turboatom, SE Plant Electrotyazhmash. Each of the industrial enterprises that participated in the research suggested recommendations for the choice of further direction of innovation development. The developed method can also be used to forecast the innovative activity development of the analyzed enterprises.

Conclusions

The formed integrated analysis system of innovative activity of the enterprise clearly reflects the relationship of stages during its conduct with an orientation to achieve the main objectives of the enterprise, the integration of quantitative and qualitative methods of analysis. The presented mechanism of information processing in the accounting and analytical system of the enterprise reveals the role of information of strategic management accounting for the analysis of innovative activity of the enterprise. The introduced methodological approach to measuring innovation of the enterprise is based on a bimodal approach, which makes it possible to measure not only the levels of innovation of the enterprise as a whole, but also its individual components – the state of material and production resources, organizational and economic level of production, market stability, state of security financial resources, innovation potential. An approbation of the measuring innovation of the enterprise has been implemented on the data of industrial enterprises of Kharkiv.

References

1. O. Gadzevich, I. Matviychuk, Economic Journal Eastern Eur. National Univ. of L. Ukrainka, *Methodology of analysis and evaluation of enterprise innovation*, **3**, 100-104 (2017).
2. O. I. Tarasenko, O.M. Korolko, K.S. Belyavskaya Current problems of the economy, *Estimation of Innovation Activity of the Enterprise in the System of Strategic Management,* **9 (99)**, 133-141 (2009).
3. A. Kvitka, A. Kramarenko, Problems and Perspectives in Management, *From big business to small and medium-sized enterprises: factors and prospect,*. **2**, 42-48 (2018).
4. G.V. Brik, Global and national problems of the economy, *Accounting and analytical support of innovative activities of agricultural enterprises*, **14,** 905-909 (2016).
5. O. V. Portna Actual Problems of Economics, *Assessment methodologies for aggregate financial potential of a country,* **3(165)**, 106–112. (2015).
6. O. V. Portna, Actual Problems of Economics, *The essence and the synergy effects of the architectonics of cumulative financial potential of the country*, **10(172)**, 404–409 (2015).
7. M. Bondar, N. Iershova, Baltic Journal of Economic Studies, *Strategic management object as an object of scientific research,* **1**, 47-54 (2015).
8. N. U. Ershova, Actual Problems of Economics, *The quality of accounting information: methodological approach to assessment* **8(158)**, 368-374, (2014).
9. I. Melnichuk, Global and national problems of the economy, *Problems and Prospects of Improvement of Accounting and Analytical Support of Innovation Activities*, **7**, 845-848 (2015).
10. N. Liba Economic analysis, *Accounting and analytical support of innovation activity,* **11,** 247-251 (2012).

Social and economic efficiency of development and use of alternative power energy

Yuriy Borovik[1*], *Yuriy* Yelagin[1], *Natalia* Gritsenko[1], *Nikolay* Kondratyuk[1] and *Vladimir* Nakonechniy[2]

[1] Ph.D. in Economics, Ukrainian State University of Railway Transport, Ukraine

[2] Doctor of Science in State Administration, Department of Economics and International Relations of Kharkov Regional State Administration

Annotation. Research has been performed on the state of development and trends for alternative energy sources. Advantages and disadvantages of renewable energy have been considered. Main types of the effects due to the implementation of renewable energy sources have been investigated. Main aspects of the actual effects upon society as a whole have been determined, in order to perform the objective assessment of the social efficiency of implementation of renewable energy sources. Methods to determine the social and economic efficiency of using the renewable energy sources have been suggested.

1 Problem statement

The depletion of traditional energy sources, growing negative impact of power engineering upon environment, as well as the increased environmental requirements, significant fluctuations in energy prices, trends for increasing energy and economic security, political character of energy supply and another related factors caused the revision of the current state of the sector of power engineering and search for the opportunities to update and "reset" it.

In the course of its activities, Ukrainian alternative power engineering has been based upon a series of Acts and Resolutions of the Cabinet of Ministers of Ukraine [1-5].

The availability of energy resources has been traditionally considered as one of the main indicators of the energy security of a certain country. The development of renewable power engineering all over the world has been caused by two kinds of global circumstances: limited character of the world energy reserves and insufficient availability of organic fuel reserves, along with the dependence of majority of the countries on fuel import operations; as well as significant negative impact of traditional (fuel) energy supply upon environment and nature [6].

*Corresponding author: ytbx@ukr.net

According to the data presented by the State Agency on Energy Efficiency of Ukraine, during the first six months of 2018, they installed the energy generating facilities for about 270 MW in electrical capacity, based upon the renewable energy sources (RES) according to the "green" tariff in Ukraine. This amount exceeds the double amount in the first half-year period in 2017 (127 MW), and it overcomes the capacity of the installed facilities for the whole year 2017 (257 MW). However, as per the total power generation amount in Ukraine, the share of renewable energy sources still remains insignificant (1,644 MW) and it slightly exceeds the amount of 1.2% of the total capacity of power generation. Nowadays, nuclear and thermal power engineering facilities are mainly operated in the country [7].

It goes without saying that renewable energy sources can never replace the solid fossil fuels completely, though their use will provide a huge positive economic and energy effect, as well as guarantee its social impact. Therefore, further research work related to the state of development in the sectors of the alternative energy sources in Ukraine, the issue to overcome the energy dependence of our country and the study of the social and economic efficiency by means of creating a "pool" of alternative energy sources are the core subjects of development of the national economy.

2 Review of the Recent Studies and Published Works.

Considering the problems of development of alternative power energy, in particular, we refer to the aspects of feasibility study of the projects using renewable energy sources, assessment of their competitiveness based on traditional energy systems and the prospects for their implementation have been presented in the works by such researchers as A. Galchinsky, G. Geletukha, T. Zheleznaya, G. Zabarnyi, S. Kudrya, A. Prokop, G. Sytnik, A. Sukhodolya and others. Such foreign scientists as V. Weiss, G. Erdmann, M. Ragwitz, P. Samuelson, G. Scheer, I. Schumpeter, V. Streicher and many others investigated the issues of energy saving, development and state support of alternative power energy. The researcher B. Savenko suggested the environmental and economic assessment of the efficiency of using the alternative energy sources [8].

However, yet notwithstanding the significant scientific contribution to research and development of the above-mentioned problem, the issues of social and economic efficiency and the balanced use of alternative energy sources, in particular, in the context of political instability and energy dependence of the country, still remain the core subjects and enhance the actual and urgent character of the further scientific research in this field.

3 Problem Statement.

The purpose of the study is to provide certain more profound methodological approaches to the social and economic efficiency of using alternative energy sources.

4 Presentation of Principal Research Materials.

The requirement to implement a high priority policy related to the energy saving issues has been mainly associated with a shortage of own fuel and energy resources in the country, dependence on the gas and oil exporting countries, growing extracting costs, as well as global environmental issues.

Speaking about the importance of formulation of the energy development strategy in order to provide the sustainable development of society, we should underline that energy production affects such sectors as health and life of people, provision of sustainable development of settlements, protection of atmospheric and biospheric environment, rational use of natural and energy resources.

Renewable electricity production has the following advantages:

- inexhaustibility of resources;
- absence of extra carbon dioxide emissions;
- absence of harmful emissions;
- keeping the thermal balance of the Planet;
- availability of resources (sun, wind) to use;
- option of using lands not suggested for economic purposes (installation of solar and wind facilities);
- absence of any need for water (solar and wind power stations).

Meanwhile, using the renewable energy sources also has some disadvantages:

- low density of energy;
- unstable and "probabilistic" nature of energy supply (sun and wind resources);
- need for accumulation and reservation of electric power (sun and wind resources);
- absence of developed industry and lack of wind energy infrastructure in Ukraine;
- flooding of fertile lands and local climate alterations (large hydroelectric power plants).

The global advantage of renewable energy over the traditional fuel energy is that using fuel and application of the related fuel costs are not required. This concept can help both reduce the fuel dependence of the electric power industry and decrease the negative impact of fuel energy production upon the atmosphere of Earth and environment where the humans live.

The main types of effects from due to the introduction of renewable energy sources (RES) are considered to be economic and environmental factors. The economic effect is determined in monetary value as the difference between the benefits received and the costs incurred, taking into account the related discounting of such values. The resulting difference ratio is called the profit or the economic effect of the project performance.

Speaking about the efficiency, according to the key concept of any social and economic activity, the focus is traditionally made on economic efficiency; its concentrated value is associated with the "costs vs. effects" ratio. Since nowadays any public activity has been performed using the continuously increasing amount of the limited resources, the approach should be similar to that economic activity.

However, any economic activity is social by its nature, thus, the so-called "social efficiency" requires research, accounting and assessment.

The problem of assessing the impact of power energy upon the environment and society is directly connected with the measurement of external costs. The external costs mean the incidental costs related to the manufacturing of the goods and services that are not borne by producers and consumers, but incurred by the population and society as a whole, without any eventual reimbursement.

It should be noted that while performing the comparative analysis within the framework of feasibility study of financing the development of any particular type of electrical power production for society, we should consider not only the investments into the infrastructure of power engineering sector, but the estimated costs and effects for the whole society as well. The effects of using the renewable energy received by the society mean the social effects that help us solve the actual social and economic task – i.e. improving the living standards and quality of life of the population.

Nevertheless, while realization of the "green" economy principles and implementing the alternative energy sources for the purposes of energy saving, we believe that only economic and environmental indicators should be emphasized. The resulting social effect due to the use of renewable energy sources could become a significant incentive, in order to promote the concept of "green" economy on the regional level. When preparing the investment project authorization for renewable energy production, special attention should be paid to the social efficiency resulting from their implementation. In case of using the renewable energy sources (RES), the social effect would reflect the option to solve the urgent social and economic task – i.e. improving the living standards and quality of life of the population.

The assessment of social efficiency, unlike the eventual economic and environmental efficiency, is rather a less studied issue, because they haven't developed a consistent approach to its definition until nowadays. Making the assessment of social effect in monetary values and natural items incur considerable hardships for economists. Anyway, the estimated social efficiency should be taken into consideration, while implementing the measures related to the "green" energy.

The social effect of the implementation of investment projects has been divided into its direct and indirect types. The direct effect is related to the construction, reconstruction, modernization and further operation of the designed facilities. The indirect effect can be estimated in the form of tax revenues, additional investments for the project realization. The investments in a particular project can stimulate the additional increase in investments within the various sectors of domestic economy. Thus, the social effect of implementation of the renewable energy sources (RES) would be interpreted as a value to be obtained on the basis of identification and economic evaluation of qualitative characteristics that affect the social changes.

In order to perform the objective assessment of the social efficiency of implementation of renewable energy sources, first of all it is necessary to identify the main sector where the effects could be obtained for society as a whole.

The social effects of renewable energy mean those which the society receives while using the renewable energy in comparison with the traditional fuel energy.

We identified the main effects of renewable energy upon society in the following areas:

- environmental effects mean the actual effects of reducing the costs related to elimination of influence of power industry upon atmosphere and biosphere.
- effects of protecting the human life and health mean the actual effects of reducing the costs of protecting life and health from the impact of activities of the energy enterprises, which deal with the renewable energy, in comparison to the fuel energy, respectively.
- the infrastructure effects mean the actual effects of reducing the costs of society related to maintenance and operation of the infrastructure applied for the extraction of fuel raw materials, its preservation, delivery to electric power plants and disposal of wastes of the fuel power production.

The assessment of efficiency to protect human life and health, unlike the eventual economic and environmental efficiency, does not have a consistent approach and definition methodology.

The social effect of protection of human life and health means reduction of the incidence of population, improving their work and rest conditions, improving people's health; thus, it incurs a number of social effects: savings in social insurance assents and treatment costs for patients. In general, the social effect of protection the human health can be determined through such economic indicators:

- effect of reduction in payments from the Social Insurance Fund (under a sick leave certificate), as related to the incidence of population due to environmental pollution by the power energy enterprises;

- effect of reduction in payments for treatment of the population due to the same above-mentioned reasons;
- effect of increasing working capacity and income level due to the normalization of social and environmental situation.

Thus, the social efficiency of protection of human life and health means that very part of social efficiency, which reflects the cost effectiveness related to improving the human living conditions.

The society receives a long-term economic effect through investing into the development of power energy infrastructure which involves low external costs.

5 Conclusions.

While implementing the energy sector development strategy and the related feasibility study of investments into the projects for the further implementation of alternative energy sources, not only financial and environmental performance factors should be emphasized.

Ukraine has a sufficient renewable energy potential, it could completely satisfy the eventual demand for energy resources and services, even in case of maintaining a high proportion of energy-intensive industries (such as metallurgical engineering, chemical industry, etc.) in the country. The domestic production of structural elements required for the technological equipment for alternative power engineering in Ukraine should be arranged; it will help solve not only the issues related to the energy and economic sectors, but it would also provide a program for the development of socially efficient energy infrastructure within the country.

References

1. Law of Ukraine "On Power Engineering" dd. October, 16 1997 № 575/97-VR, as amended. Available at: http://kodeksy.com.ua/pro_elektroenergetiku.htm
2. Law of Ukraine "On Energy Saving" dd. October, 16 2012 № 5463-VI, as amended.– Available at: http://zakon2.rada.gov.ua.
3. Law of Ukraine "On Alternative Energy Sources" dd. February, 20 2003 № 555-IV, as amended. Available at: http: //kodeksy.com.ua/pro_al_ternativni_dzherela_energiyi.htm
4. Resolution of the Cabinet of Ministers of Ukraine dd. October 05, 2004 No. 1307 "On Procedure of Issue a Certificate for Fuel being Classified as the Alternative Source", as amended. Available at: https://zakon.rada.gov.ua/laws/show/1307-2004-%D0%BF
5. Resolution of the Cabinet of Ministers of Ukraine dd. August, 18 2017 № 605-r "On Approval of Energy Strategy of Ukraine until year 2035 - Safety, Energy Efficiency, Competitiveness" Available at: https://zakon.rada.gov.ua/laws/show/605-2017-%D1%80#n6
6. Yu.T. Borovik, Yu. V. Yelagin. Bulletin of Economy of Transport and Industry. **61**, 103-110 (2018).
7. Scientific and Manufacturing Enterprise "VD MAIS", website. Available at: http://vdmais.ua/rynochnye-trendy-alternativnaya-energetika-ukrainy-v-2018-godu
8. B.V. Savenko. Environmental Safety. **1**, 136-142 (2017).
9. "Transitional Stage for Ukraine to Come to Renewable Energy before year 2050" / O. Dyachuk, M. Chepelev, R. Podolets, G. Tripolska, et al.; under general editorship of Yu. Ogarenko and O. Aliyeva. Representative Office of G. Boll in Ukraine. – Kyiv: Publishing House "ART BOOK" LLC, 88 (2017).

10. Energy Efficiency and its contribution to energy security and the 2030 Framework for climate and energy policy. Available at: https://ec.europa.eu/energy/sites/ener /files/documents/ 2014_eec_communication_adopted_0.pdf.

Neo-industrialization of the economy in diffusion conditions socio-economic environments

Tetyana Calinescu[1], *Ganna* Likhonosova[2,*], and *Olena* Zelenko[3]

[1] Professor, Doctor in Economics Science Finance Department, 61070 National Aerospace University "Kharkiv Aviation Institute", Kharkiv, Ukraine
[2] Associate Professor, Doctor in Economics Science Finance Department, 61070 National Aerospace University "Kharkiv Aviation Institute", Kharkiv, Ukraine
[3] Associate Professor, Doctor in Economics Science, International Economic and Tourism Department, 93406 Volodymyr Dahl East Ukrainian National University, Severodonetsk, Ukraine

Abstract. Socio-economic achievements of Ukraine are currently characterized not only by positive but also by negative trends, such as the creation of political confrontations, provoking international conflicts and the emergence of military aggression. Therefore, there is a need to indicate the directions of development and competitiveness of the social and economic potential of Ukraine in the conditions of the established European integration tendencies. The authors discovered the patterns of emergence of the main risks of socio-economic development of the country, which threaten the free access to a large raw material base, which creates the possibility of reducing the stiffness of pressure on Ukrainian producers on the need to increase labour productivity and technological modernization. The authors conclude that integration aspirations in the world social and economic system in Ukraine should be complemented by facilitating cooperation within the country itself. The lack of national socio-economic development program, which basis on the formation of the platform for socio-economic potential, combined with the effect of synergy effects can lead to unpredictable destructive consequences. Therefore, further research of these processes will be connecting to the exposures of multi-disciplinary connections, to found same trends and to do its interpretation.

1 Introduction

While most of the world's leading countries have achieved the sixth technological development, Ukraine still cannot cross the fifth. The resistance to innovative development significantly impedes the development of the country's competitiveness on the world market. In spite of the improvement of the situation in comparison with 2017 (83 in comparison with 89), in 2018 the country could not cross the barrier and enter the top ten global ranking among 140 countries (Fig. 1). The highest negative impact on the place in

* Corresponding author: a.likhonosova@gmail.com

the rating is made by the level of macroeconomic stability (131 place), the state of the financial system (117th place).

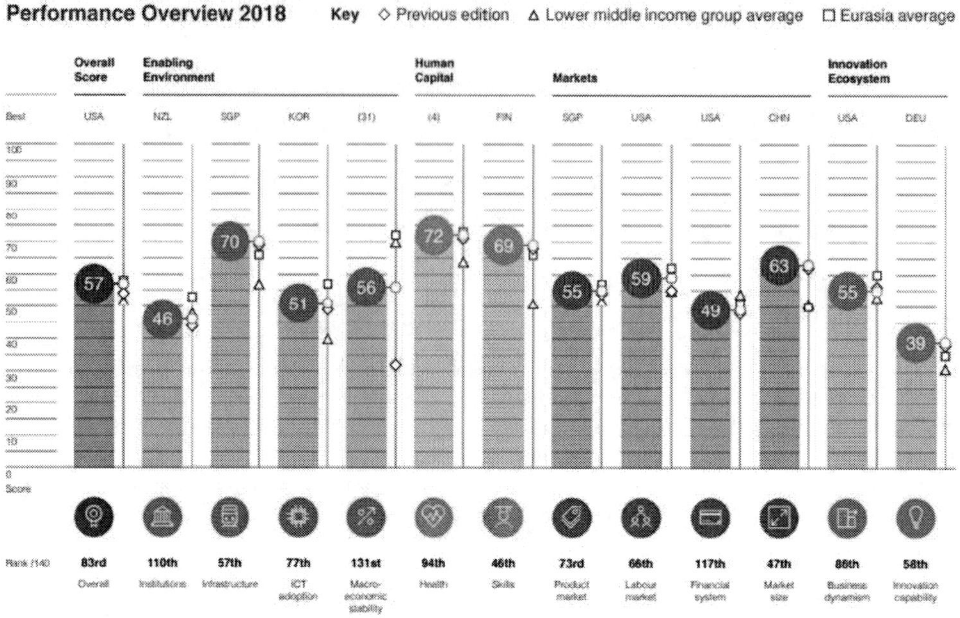

Fig. 1. Ukraine's place in the global competitiveness rating
Source: [1]

Indicators of the world ranking of competitiveness are also confirmed by negative tendencies of socio-economic development, therefore the question of finding ways to further effective development is more than ever relevant for Ukraine [1]. Neo-industrialization can be seen as a process with a certain purpose, methods, tools, technology and development strategies, sources of funding, and socio-economic implications that require a separate thorough study, in which a specified collective of authors attempts to develop regulators of reflection of national interaction in conditions of diffusion of socially -economic environments.

However, such socio-economic achievements are accompanied by negative consequences, such as the creation of political confrontations, provoking international conflicts and the emergence of military aggression. And these phenomena are accompanied by modern integration processes in Ukraine. Therefore, there is a need to indicate the directions of development of the socio-economic potential of Ukraine in the conditions of the established European integration tendencies and corresponding diffuses.

2 Methodological analysis of recent research

Problems of neo-industrialization became widely publicized both in developed countries of the world and in countries that only started this way. The most famous in this area is the publication of Peter March "The New Industrial Revolution: Consumers, Globalization, the End of Mass Production" [2], which was unveiled in 2012.

Among domestic scientists, supporters of neo-industrialization are D. Bodrova [3], V. Reshetylo [4] and others. Both domestic and foreign researchers emphasize the prospect of neo-industrialization and advancement of the services sector, which corresponds to the tertiary sector of the economy. According to many experts, special attention should be paid to the field of information services, which is sometimes separated into the quaternary sector of the economy.

In addition, the world has accumulated considerable experience and diversity of forms of international economic cooperation focused on integration tendencies. Questions of the relationship between economic integration and the dynamics of potential expansion are discussed in the context of the integration of financial markets and the growth of industrial specialization [5]. D. Brou and M. Ruth have built a model that shows that in cases where economic integration is accompanied by cooperation at the political level, it intensifies innovation activity, as firms in such conditions face less rigorous competition [6]. On this basis, the potential of socio-economic development, which will be based only on better access to resources, raw materials, commodity markets and financial capital, is negligible.

3 Features of neo-industrialization process in Ukraine

Neo-industrialization processes have begun to gain momentum within the last decade, and after the Fourth Industrial Revolution, they are actively implementing the system of economic relations in the level of developed countries and internationally.

Ukraine still cannot boast of the consistency of the above processes, according to the rating of the global index of innovations it occupies only 43rd place. Institutions and infrastructure remain the weakest link in innovation development (89th and 107th in the country ranking, respectively) [7]. The level of knowledge intensity of the main macroeconomic indicator – GDP – has decreased by 1.7 times over the past 8 years, with participation of the state. Changes in the structure of the domestic industry in recent years indicate an increase in disproportions and an increase in the share of metallurgy and energy with a significant reduction in the share of processing industries.

3.1 Integration trends as the basis of neo-industrialization of the economy

Intensification of international cooperation can provide new opportunities for accelerating social and economic development at certain levels (Fig. 3).

Among the main positive trends in the integration processes in the field of socio-economic potential formation [8], the country can be distinguished: ensuring the competitiveness of national industries as a result of their use of innovative behaviour and the development of new technologies and the accumulation of innovative knowledge and empirical skills by economically active subjects. The general market provides opportunities not only for acquisition of a greater spectrum of new production technologies and equipment, but also the training of a large number of organizational innovations. With the intensification of interaction its participants learn faster [9]. At the level of enterprises and the international exchange of knowledge-intensive resources and goods, the integration of markets has not yet become an incentive for breakthrough development. The actual reflection of this is the figures in Table 1. In 2017 the share of the research workers (researchers, technicians and auxiliary staff) in the total number of employment was 0.58%, including researchers – 0.37% (Official website of the State Statistics Service of Ukraine). According to Eurostat, in 2015 the highest share of this share was in Finland (3.21% and 2.35%), Austria (3.10% and 1.92%) and Sweden (2.97% and 2.33%); the lowest is in Romania (0.53% and 0.33%), Cyprus (0.83% and 0.61%), Poland (1.0% and 075%) and Bulgaria (1.0% and 0.65%) [10].

3.2 Educational expectations from the behaviour of the population in the context of neo-industrialization

One of the main directions of stimulation of new industrialization is the formation of an environment conducive to business development in the country. In a context of a shortage of investment resources in the real sector of the national economy, limited opportunities for public investment lending and the availability of socio-economic and political risks [11].

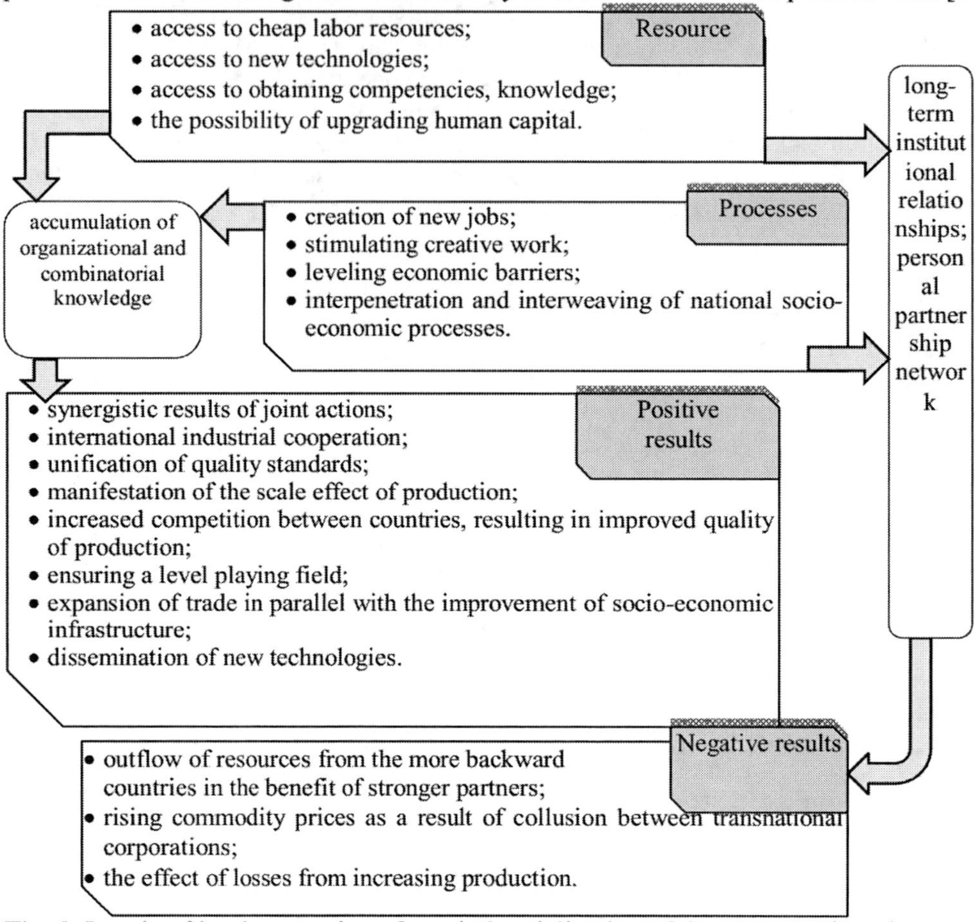

Fig. 2. Levels of implementation of neo-industrialization of the economy based on integration trends
Source: Developed by authors

At present, investments in industry, agriculture, transport and logistics, the development of hotel and restaurant business and public administration are the largest percentage of investments in the Ukrainian economy [12].

And if the percentage of investments in all positions gradually increases during 2007-2018 (the total investment in the economy of Ukraine increased by 35.7%), then the investment in public administration is significantly reduced. In addition, during the period of 2007-2018, the share of investment in mining and processing industries, electricity generation, gas and the production of associated equipment for servicing it substantially grows in the structure of investments into industry [13].

At the same time, investments in the production of rubber, plastic and non-metallic products decreased. In total, investing in industry increased over the years by 38.8%.

Further neo-industrial development of Ukraine is determined by the ability to find answers to a number of structural, technological and managerial challenges that have emerged as a result of global trends and transformations in the national economy during times of crisis. And this implies qualitative changes in the political and socio-economic spheres, which ultimately affects the results of general socio-economic development.

Table 1. The number of employees involved in the implementation of research and development by category (persons)

Year	Number of employees – total	Including				
		Persons that have a degree		researchers	technologist	auxiliary staff
		doctor of sciences	doctor of philosophy (Ph.D)			
2010[1]	182,484	11,974	46,685	133,744	20,113	28,627
2011[1]	175,330	11,677	46,321	130,403	17,260	27,667
2012[1]	164,340	11,172	42,050	122,106	15,509	26,725
2013[1]	155,386	11,155	41,196	115,806	14,209	25,371
2014[1,3]	136,123	9,983	37,082	101,440	12,299	22,384
2015[1,3]	122,504	9,571	32,849	90,249	11,178	21,077
2016[2,3]	97,912	7,091	20,208	63,694	10,000	24,218
2017[2,3]	94,274	6,942	19,219	59,392	9,144	25,738

[1] Data for 2010-2015 include permanent and temporary workers (contractors and persons working under civil-law contracts, including scientific and pedagogical workers).
[2] Data for 2016 are given without taking into account scientific and pedagogical workers who did not conduct research and development
[3] The data are given without taking into account the temporarily occupied territory of the Autonomous Republic of Crimea, Sevastopol and parts of the zone of the anti-terrorist operation.
Source: Official site of the State Statistics Service of Ukraine

Undeniably, the structure of the socio-economic environment can interfere or promote the spread of neo-industrialization processes, which determines the obtaining of different effects and synergistic efficiency. The defining of this is the established norms, standards and rules of conduct of members of the social system, which then point to individual individuals whose behaviour or perception of the results of neo-industrialization is expected from them. Proceeding from this, can propose the following system of measures and directions for the formation of the synergistic effect of the country's socio-economic potential in the integration processes.

1) Restructuring of the institutional framework for socio-economic cooperation: development of an integrated strategy of social and economic cooperation at different levels. 2) Development of intellectual property rights protection system: introduction of educational programs for the study of patent experts with the study of international aspects of intellectual property rights management; development of special schemes for the protection of intellectual property rights and the provision of consulting services in area. 3) Development of domestic demand for innovation: to introduce mechanisms of support of industrial enterprises of all forms of ownership and of all sizes in the development of innovative technologies and equipment. 4) The intensification of trade in high-tech goods and technologies: to develop the competence of Ukrainian economic subjects in the marketing of knowledge-intensive goods and management of socio-economic systems; to stimulate alliances of enterprises to jointly bear the costs of patenting science-intensive products. 5) Targeting of attracted foreign investments: development of adequate schemes for resolving socio-economic conflicts; development of sectorial investment strategies, in which strategic socio-economic objectives should dominate financial goals. 6) Activation and stimulation of the mobility of young scientists: development of special programs for bilateral projects of young scientists' cooperation with their counterparts from other countries where Ukrainian scientists could conduct research in cooperation with foreign colleagues. 7) Implementation of schemes for the harmonization of education systems: institution based on the classical universities of centres for the preparation of the required

specialties in cooperation with leading foreign universities within the framework of accredited programs; development of direct partnership links between universities of Ukraine. 8) Implementation of social dialogue at the regional and national levels: audit and updating of the existing regulatory and legislative framework for the formation of social dialogue with a view to involving a wide range of participants.

4 Conclusions

In the socio-economic environment, we should speak about the consequences of neo-industrialization and changes affecting the level of diffusion, which are not always measurable. Therefore, it can be argued that some effects can be measured by known methods and methods, but not all are subject to traditional measurement, and even do not yield positive results, and therefore can deepen the gaps, inequalities, imbalances of the entire socio-economic environment. Thus, the level of diffusion (access to information) of the socio-economic environment can contribute to the development of further priorities of neo-industrialization or to deepen the problem aspects, and then special measures should be taken to align the situation and its compliance with generally accepted standards or norms of the world's society.

References

1. K. Schwab. The Global Competitiveness Report. Geneva: World Economic Forum (2018).
2. P. Marsh. Levelling out: emerging markets and the New Industrial Revolution. Financial Times, **27** (2012).
3. D. Bodrova. Institutional factors of the neo-industrialization of the national economy. Intelligence XXI, **2** (2017).
4. V. Reshetilo. Neo-industrialization and its influence on the development of national and international economy. Scientific Bulletin of Uzhgorod National University. Series: International Economic Relations and the World Economy, **18** (2018).
5. A. Bommier, L. Bretschger, Le Grand. Existence of equilibria in exhaustible resource markets with economies of scale and inventories, Economic Theory, **63** (2017).
6. D. Brou, M. Ruta. Economic integration, political integration or both? Journal of the European Economic Association, **9(6)** (2011).
7. S. Dutta, B. Lunvin, S. Wunsch-Vincent. Global innovation index 2018: Energizing the World with Innovation. Cornell INSEAD WIPO (2018).
8. *Стан розвитку науки і техніки, результати наукової і науково-технічної діяльності за 2017 рік*: аналітична довідка. Київ: Український інститут науково-технічної експертизи та інформації (2018).
9. T. Calinescu, G. Likhonosova, O. Zelenko. Estimation of conditions the realization of democratic mechanisms the transformation of society: tearing away and social dialog, Montenegrin Journal of Economics, **1(14)** (2018).
10. A. Pobol. Economic potential of innovative development of countries in the integration processes of the CIS, EurAsEC and the CES, Eurasian Economic Integration, **1 (14)** (2012).
11. Official site of the Statistical Service of the European Union (Eurostat). Retrieved June 8, 2019, available at: https://ec.europa.eu/eurostat/
12. T. Calinescu, G. Likhonosova, O. Zelenko. Mechanism of Transformation Socio-Economic Potential of Ukraine: Tearing Away and Social Dialogue. *Proceeding of the 32nd International Business Information Management Association Conference (IBIMA), 15 – 16 November 2018, Seville Spain. Vision 2020: Sustainable Economic Development and Application of Innovation Management from Regional expansion to Global Growth.* Seville Spain: IBIMA (2018).
13. L. Shavaliuk. The mirror of development. What the structure in the economy of Ukraine signals about. The Ukrainian week, **3(133)** (2019).

The labour market in the context of the growth of globalization processes: evidence from Ukraine and countries of the European Union

*Tatiana V.*Chatchenko[1*], *Yurii O.*Miriasov[1], *Irina A.*Davidova[1], and *Vistoriia V.*Mykytas[1]

[1] V. N. Karazin Kharkiv National University, School of Economics, Svobody Sq. 4, 61022 Kharkiv, Ukraine

Abstract. The labour market is a special market of means of production, which requires special attention, since the subject of exchange is the ability of people to work, skills and knowledge of a person. The welfare of the population, its purchasing power, level of income and, in the end, the standard of living depends on the efficiency of the exchange at the market. The issue of unemployment is one of the key issues in the state regulation of the labor market, which causes a decrease in household incomes and the general population, a decrease in the welfare of the nation and in general the gross domestic product. Scientists distinguish several types of unemployment on various grounds: reasons, structure and period, etc. But one of the topical problems of the present, with a tendency towards rising retirement age, observed in both developed and developing countries, is youth unemployment. That entails a set of problems: the inadequacy of education with the current requirements of the labour market, the lack of vacancies, and the increase of retirement age employees delays the term of retirement, thus not freeing jobs, etc.

The problem is that more active, more skilled young people immigrate, and less competitive people stay in their native country, that also has a certain negative impact on the state of economic development of the country. At the same time, people who work abroad create financial flow into the native countries in the form of support to their relatives, which is a positive moment. As can be seen, the decision to reduce unemployment and especially the youth cannot be easy. State regulation of the labour market is a complex process and should be researched and it is relevant.

It is proposed to consider the structure of the labour market presented in Figure 1.

* Corresponding author: o.y.diachek@gmail.com

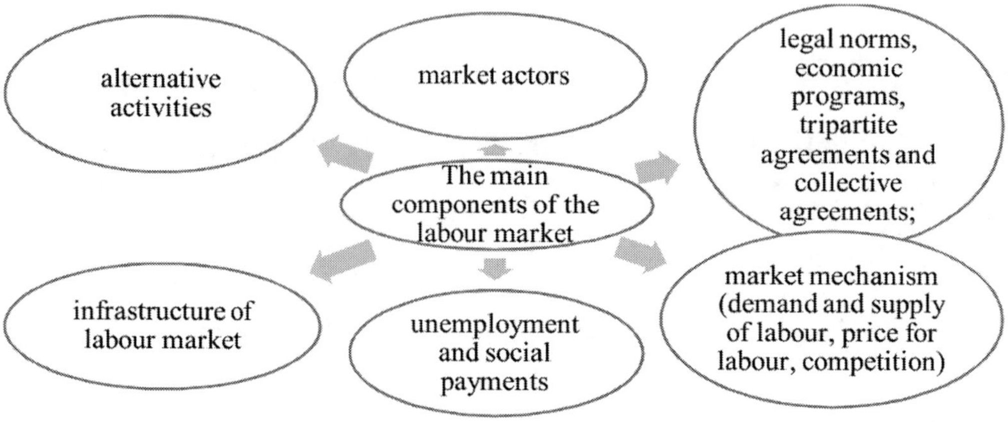

Fig. 1. The structure of labour market (compiled[1]).

As it is shown in Fig. 1, we have analysed the labour market by the components. The structure of the employed population is presented in Figure 2, which shows that changes in the structure of GDP and, accordingly, in the structure of employment are slow, but we still see a decline in industry, construction, financial and insurance activities, real estate operations.

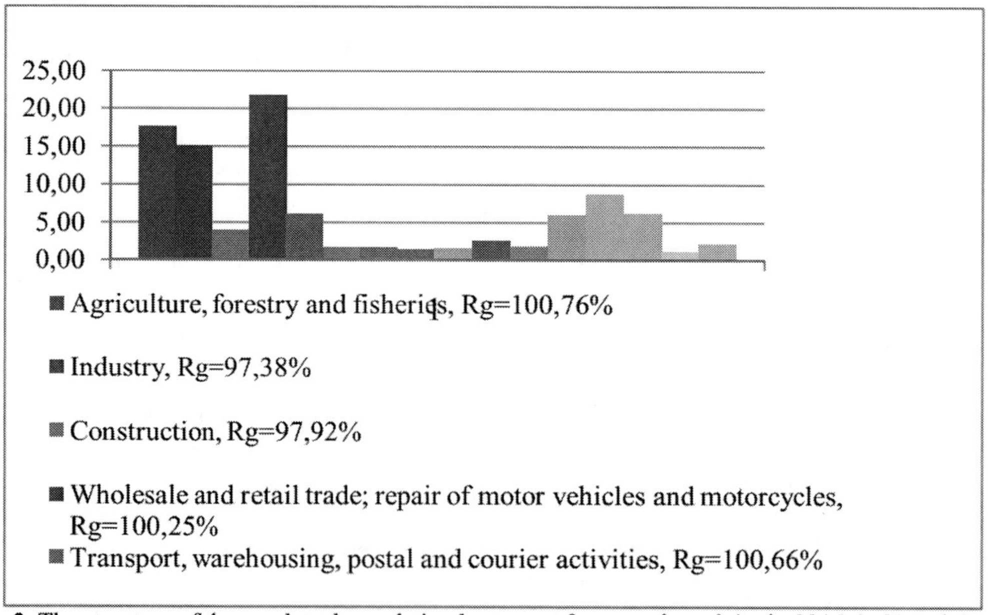

Fig. 2. The structure of the employed population by types of economic activity in 2016, in % and the average annual growth rate of the proportion of the type of activity for 2012-2016 (compiled by the authors [2]).

At the same time, employment has increased in the following sectors of the economy: information and telecommunications, public administration and defence, compulsory insurance, there is also slight growth in the arts, sports, entertainment and recreation. The growth of employment in defence is fully justified as a result of the anti-terrorist operation and operations of the combined forces in the east of the country. Relevant trends in the field of information and telecommunications also observed long ago, because of the orientation of the Ukrainian economy to the development of IT industry, the share of employment in agriculture is unchanged while the industry continues to develop and grow, which may indicate its intensive development. Let us consider in more detail unemployment in terms of the length of the job search (Fig. 3) and the causes of unemployment (Fig. 4).

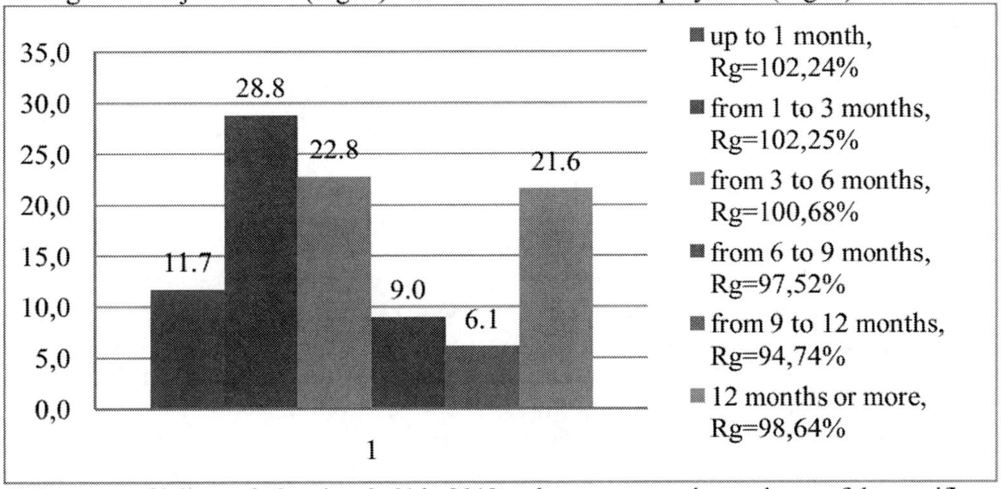

Fig. 3. Structure of job search duration, in % in 2018 and average annual growth rate of the specific gravity of the period 2010-2018 (compiled by the authors [2]).

From the analysis of the results presented in Figure 3, we can conclude that most people are looking for a job 1 to 3 months and 3 to 6 months. Also, there is a rather large proportion of those who are looking for a job for more than 12 months, but their relative share is decreasing (average annual growth rate is 98.64%). But this shows that for such people it is quite difficult to "find themselves" in the structure of the national economy. That may be due to the above mentioned change in the structure of employment and the need for training or retraining of the unemployed.

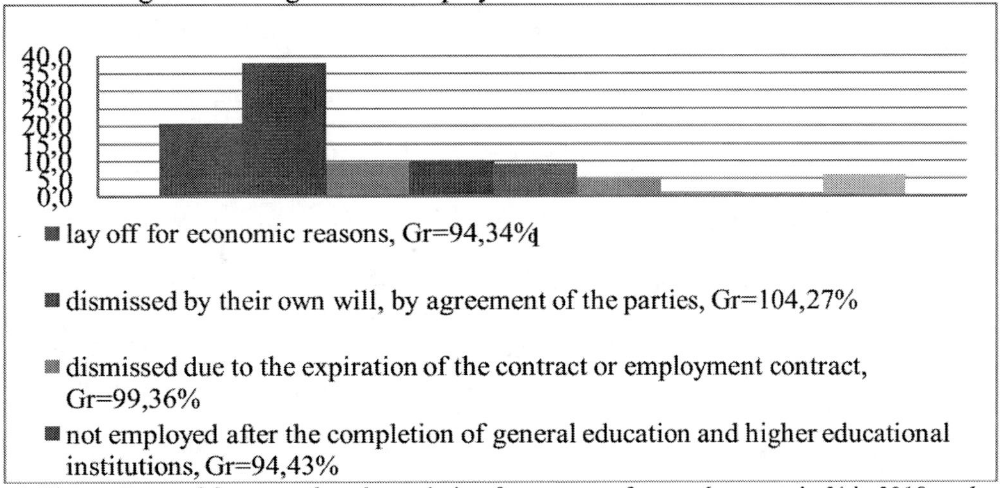

Fig. 4. The structure of the unemployed population for reasons of unemployment, in % in 2018, and the average annual growth rate of the proportion of the period 2010-2018 (compiled by the authors [2]).

The analysis results presented in Fig. 4 indicate changes in the structure of employment and GDP, a significant part, almost 21%, is released for economic reasons, 38% are released on their own, but this form of release is most acceptable in Ukraine for the release and reduction of staff. The high proportion of unemployed after high school and university level is high, that is, the high number of graduates who can not find work due to their unsuitability to the labour market or the lack of relevance of their knowledge and skills.

Today, unemployed youth is a difficult category with a high risk of social degradation, which threatens the development of marginal qualities, deformation of consciousness and public opinion.

The degree of realization of a person's own potential is directly dependent on the possibilities of functioning of the labour market: increasing the efficiency of the labour market will contribute both to the realization of the person of his potential and the growth on this basis of the economy of the country. However, the global financial and economic crisis and its implications mean that the world economy faces a prospect of rising long-term unemployment, an increase in poverty and inequality, and pre-crisis employment will resume only a few years after the improvement of the economic situation [3, p. 10].

Crisis processes that occur in the context of employment relationships require a thorough psychological reflection and a conceptual review of the most pressing problem areas. First of all, it concerns unemployment as a key phenomenon of the employment crisis, since in the conditions of a permanent socio-economic decline of the economy it accumulates in itself as potential opportunities for the development of the adaptive potential of the individual, as well as the most dangerous risks for its integral and full functioning.

The analysis of the essence of globalization suggests the following trends: reducing the boundaries of national states, increasing the role of supranational authorities and large transnational enterprises, etc. Negative features of globalization can be eliminated by well-organized regulatory instruments. Analysis of the experience of overcoming crisis phenomena in Europe shows, firstly, that labour resources play one of the key roles in the modern world, and secondly, certain measures aimed at activating different segments of the population in different kinds of activities give positive results to the development of the economic system [4,5,6].

The European labour market has also undergone some changes with the financial crisis of 2008, to overcome it faced certain challenges [7,8]:

1. Reforming labour market in four directions:

1.1. Attracting retired workers to work, as is the case in Scandinavian countries and the Netherlands;

1.2. Reducing structural unemployment through reforms in Denmark and the UK;

1.3. Reducing unemployment among young people by copying the successful experience of the Netherlands;

1.4. Balancing partial and full employment for women is one way to increase the average level of employment.

2. Disclosure of the full potential of service sector growth:

2.1. Deregulation of the scope of services, which is still sufficiently regulated

2.2. Increasing productivity through better regulation of commodity, resource and labor markets, support for greater efficiency and professionalism in such sectors of the economy as wholesale and retail trade, construction and transport.

2.3. Accelerating development through the introduction of new standards, educational services and infrastructure

2.4. Ensuring development throughout the European Union.

3. Eliminating growth and innovation policies - Applying opportunities for growth and innovation in high-tech and manufacturing sectors such as exports to emerging markets, green technologies, long-term technological innovations (nanotechnology, biotechnology, etc.) by:

3.1. View the priority of funding and fund their development in support of fundamental and applied research and innovation.

3.2. Development of large-scale clusters.

3.3. Improving communication between science, education and business.

3.4. Promoting entrepreneurial spirit.

These paths were highlighted in 2010 and apparently affect the labour market in the first place. These measures have yielded some fruits - the unemployment rate as a whole decreased in the EU [9].

The general reference point for the development of employment services in developed countries is the principle of activation, which involves the introduction of incentive mechanisms and incentives for employment and providing better job opportunities for jobseekers. Its key elements are strengthening the ability of the unemployed to go to work, encouraging more active inclusion in job search, and motivating effective employment instead of targeting unemployment benefits.

As a result the experience of EU countries is useful and can be applied to other countries and regions [10].

Conclusions

From the analysis of the employment structure of Ukrainian population by sectors of the economy, we came to the conclusion that the role of information technologies and telecommunications, public administration and defence is growing, which is substantiated by the trends of the country's development in general. Knowledge and skills in a particular population are irrelevant or obsolete, as evidenced by the period of job search and the causes of unemployment. Therefore, it becomes urgent to seek employment opportunities for the pristine population, especially in the context of globalization.

To overcome the negative features of the labour market in Ukraine that were discovered during the study, we are invited to use the following measures in the areas:

Direction 1: Professional guidance. The aim is to direct the professional interests of the youth by the formula: informing - learning - testing - attracting, forming conscious motives of choosing a certain type of activity that will meet the material and spiritual needs of young people.

Direction 2: Youth Entrepreneurship. The direction should promote the development of youth entrepreneurship, self-employment and the effective promotion of young people in the business environment.

Objective 3: Volunteerism. Objective: to conduct activities aimed at attracting young people to volunteer activities.

Direction 4: Promoting employment. Work placements and involvement in socially significant activities as a way of securing secondary employment are foreseen.

References

1. Mercy Corps. Why youth fight? Making sense of youth political violence in sub-Saharan Africa. Portland, OR: Mercy Corps. https://www.mercycorps.org/sites/default/files/Why%20Youth%20Fight.pdf (2014)
2. Official site of the State Employment Service of Ukraine[Electronic resource]. - Mode of access: http: // www.dsz.gov.ua.
3. O.V. Volkova. *Labour market.* Center for Educational Literature, 10 (2017)
4. A. Acarmos. *Labour market that works connecting talent with opportunity in the digital Mckinsey & Company (2015)*
5. C. Andrews, A. Kryeziu. Public Works and the Jobs Agenda: Pathways for Social Cohesion? *Background paper for the World Development Report* (2013)
6. P. Arni, R. Lalive, J.C. Van Ours. How effective are unemployment benefit sanctions? *Looking beyond unemployment exit.* J Appl Econ. 28(7). 1153–1178 (2016)
7. D.N.F. Bell, D.G. Blanchflower. Young People and the Great Recession. *Oxford Review of Economic Policy*. 27.2. 241–67(2011)

8. World Development Indicators [Electronic resource]. – Available at: http://databank.worldbank.org/Data/reports.aspx?source=world-development-indicators (2019)
9. World Economic Forum Addressing the 100 Million Youth Challenge: Perspectives on Youth Employment in the Arab World in 2014, Geneva. *World Economic Forum* [Electronic resource]. – Available at: https://www.weforum.org/reports/addressing-100-million-youth-challenge (2014)
10. C. Cramer. Jobs, unemployment and violence. *GSDRC Professional Development Reading Pack No. 10.* Birmingham: University of Birmingham. http://www.gsdrc.org/docs/open/readingpacks/jobsunemployment.pdf (2015)

Formation of economic information resources as a basis for raising the level of development of enterprises

Viktoriia Chobitok[1*], *Tetiana* Obydiennova[2], *Alina* Fandieieva[3] and *Anna* Cherniaieva[4]

[1] Ukrainian Engineering Pedagogics Academy, Department of Economics and Organization of the Activity of Business Entities, 61003, 16 Universitetskaya st., Kharkiv, Ukraine
[2] Ukrainian Engineering Pedagogics Academy, Department of Management, 61003, 16 Universitetskaya st, Kharkiv Ukraine
[3] Kharkiv National Automobile and Highway University, Department of Foreign Languages, 61002, 25 Yaroslava Mudrogo st., Kharkiv, Ukraine
[4] Kharkiv Institute of Finance of Kyiv National University of Trade and Economics, Department of Economic Theory and International Economy, 61003, 5 Pletnyovsky Lane, Kharkiv, Ukraine

Abstract. The paper discovers the issue of formation of economic information resources, which is a component of increasing the level of development of enterprises. The authors give a description of such concepts as "information", "information resources", "information support", and "economic information resources". The classification of the concepts of "information" and "information resources" is given separately. It is proved that the main goal of economic information resources is providing the information on ensuring the adoption of effective management decisions at any stage of the company's life cycle to the company's managers. The authors suggested the scientific and practical approach for the formation of economic information resources of the enterprise to increase the level of their development. In the work it is proved that availability of reliable and assured economic information resources is the basis of effective sustainable development of enterprises.

1 Introduction

Modern business conditions require the quick adaptation of domestic enterprises to the permanent effects of internal and external environment. For effective formation of management decisions and in order to enhance the competitiveness of enterprises, it is necessary to use the information resources that are the basis for a comprehensive analysis of the impact of internal and external environment. The receiving of the necessary, timely, assured and up-to-date information enables to structure it and to use it optimally.

In the scientific field, the topic of use of information resources in ensuring the effective operation of the enterprises and in raising the level of their development is relevant in modern economic environment.

*Corresponding author: vika_chobitok@ukr.net

The purpose of the study is the formation of economic information resources as the basis for increasing the level of development of the enterprises considering theoretical and applied aspects.

2 Presenting main material

The active development of the information age is accompanied by a permanent change in the spheres of society's life, and the progressive globalization of space, and so on. Important factor that leads to the transformation of the current state of activity of domestic enterprises and to the formation of axiological paradigm of information culture, is the phenomenon of information.

In the usual sense, the term "information" derives from the Latin word "*informatio*", which has several meanings: explanation, statement of facts, events, interpretation; representation, concept; studying, education.

According to The Law of Ukraine "On Information" the essence of the notion of "information" is that it is any information and/or data that may be stored on tangible media or may be displayed electronically. By the content information is divided into the following types: information about an individual; information of reference and encyclopedic character; information on the state of the environment (environmental information); information about the product (work, service); scientific and technical information; tax information; legal information; statistical information; sociological information; other types of information [8].

In the practical activity of modern enterprises, the use of timely, reliable and relevant information plays an important role in assessing the impact of internal and external factors on their development and helps to respond quickly to changes in the competitive environment. Realization of the analysis of the environment and formation of estimated figures of the development of the enterprises are the basis for the development and implementation of the effective management decisions at each stage of development of their economic relations.

Sources of information by the origin, necessary for making the decisions during the management of an enterprise, can be divided into two groups: internal and external.

Internal sources of information are represented by information and analytical departments and groups of employees of enterprises (their activity is connected with various information communications and periodic reporting). Internal information is the data formed inside the enterprise, namely: the state of organizational and technological opportunities, production capacity, level of qualification of personnel, availability of material and financial resources, as well as sources of their origin, etc.).

External sources include information that is publicly disclosed or comes from other business entities. This information is provided by mass media, advertising and information agencies and resources, suppliers, customers, buyers, competitors, and so on. External information includes information on innovations in the technological process of production, the level of competition in the domestic and foreign markets, prices for products of the enterprises – competitors, investment opportunities, markets, raw materials, etc.

External information, which is regulated and mandatory, should be distinguished separately. Information about the mechanisms of regulating the activities of enterprises through the economic and administrative levers of the state is established by the relevant regulatory acts. This information concerns the mechanisms of state regulation of pricing, state investment, taxation, protection of the rights of producers and consumers, etc.

Information resources are transmitted by oral, written or other method (exchange of information between people, man and machine, two automatic machines, signal exchange in the animal and plant world, transfer of signs from cell to cell, and so on). Information resources exist in time and space, and can be transmitted between generations.

Information can be presented in various forms, both electronically and in non-electronic form; however, during the development of communication systems and the Internet, the preference is increasingly given to electronic information, while creating an information base of resources.

Main directions of use the Internet by national enterprises and organizations of Ukraine in 2017, according to the data of the Office for National Statistics of Ukraine [8], are summarized in Table 1.

Table 1. The directions of use the Internet by the enterprises and organizations of Ukraine in 2017

The direction of use		Number of enterprises
Number of enterprises that had the access to the Internet network		**39,582**
From this amount of enterprises the Internet network was used for:		
1.	sending or receiving emails	**38,929**
2.	making phone calls via Internet / VoIP connection or video conferencing	**12,048**
3.	obtaining information about goods and services	**34,663**
4.	using instant messaging and BBS	**18,704**
5.	receiving information from the state authorities	**31,571**
6.	conducting various operations with the state authorities (with the exception of obtaining information)	**20,158**
7.	conducting banking operations	**38,227**
8.	access to another financial services	**15,535**
Number of enterprises that had a website which was in operation in the Internet		**16,240**

According to the results of the Table no. 1. the diagram of the directions of use the Internet network by the enterprises and organizations of Ukraine has been built (Fig. 1).

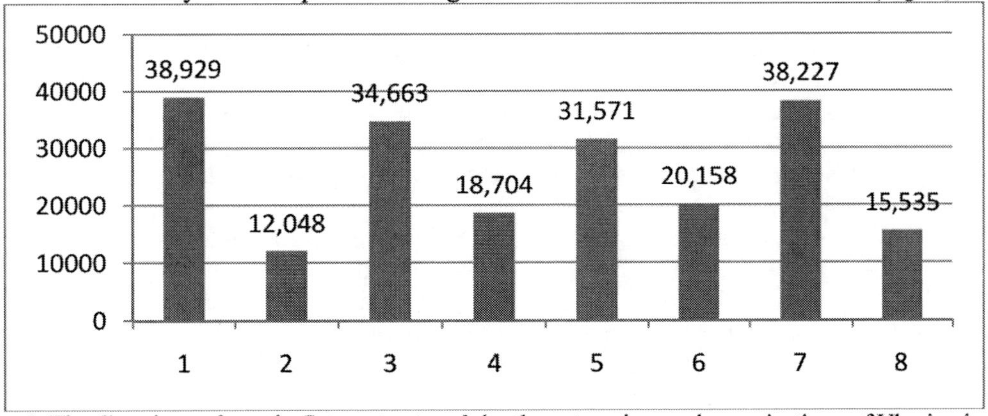

Fig. 1. The directions of use the Internet network by the enterprises and organizations of Ukraine in 2017

Information support of management activity at enterprises is an integrated communication system whose main tasks include the collection, storage, processing and transmission of information in order to analyze the obtained results for the preparation, justification and making the management decisions. The main functions of information support of management activity at enterprises are: computational, communication, information, optimization, regulatory, analytical one, and so on.

Therefore, the main goal of information resources is to provide managers of different levels of management with information in order to ensure the production of effective management decisions.

The management of the enterprise must make every effort to ensure that the received information is impersonal and reliable before final approval of the management decision. Information about the activities of the enterprises can be analyzed in several groups, namely: by its content, by the form of presentation, by the requirements set for it, and so on. Such grouping of information is necessary for the evaluation of objects in the process of activity, which will enable to structure the obtained data and to create the necessary conditions for their effective use.

Effective management decisions regarding the further development of enterprises will be the result of the analysis of the obtained information. Consequently, the information support of the process of enterprise management is of great importance, since any information that is necessary to make the management decisions must be an assured, complete, reliable, timely and relevant one.

Information support should be in the form of information resources, that is in a certain way, a structured and organized set of sources of information that can be applied for continuous use by a business entity in its economic activities. In general, information resources are all types of information produced in a society, and that are stored in data banks, archives, libraries, on Internet sites, and so on. Information resources are of value for different areas of social activity, and can be used for solving certain problems, for educational purposes, for entertainments, for other forms of consumption, and so on. Information resources to a wide extent are a collection of data, organized for the effective obtaining of reliable information.

The essence of the system of information resources is in the ability to be processed in accordance with the needs of the person, the task of a specialist, and so on. The bulk of information resources is collected, transmitted and processed by means of signs (signals that can transmit information in the presence of an agreement on their notional holding between sources and receivers of information).

Thus, not only technological, but also the logical aspect of data processing manifests itself in the field of social communication [7].

The system of information resources is a set of logical data processing that is used in social and historical practice by applying different forms, methods and means, and can be represented in various forms (sound waves, radio waves, pulses, light signals and so on) [1,3,5].

Information resources can be grouped according to different characteristics, the main of which are: content, time, source, quality, relevance to needs, method of fixation, language, cost, and so on.

National information resources are created both by the state and the entities of all forms of ownership irrespective of their content, form, time and place of creation, forms of ownership. They are intended for use by individuals, society, and the state. The Concept of Formation of a National Electronic Information Resource System approved by the Cabinet of Ministers of Ukraine in 2003, adopted this classification of information resources by the criterion of ownership: government, municipal and private information resources [6].

Information resources are characterized by various forms of information representation, organizational and technological solutions, the quality and quantity of which is influenced by the development of information technology.

The system of information resources is applied in all spheres of public practice (Fig. 1).

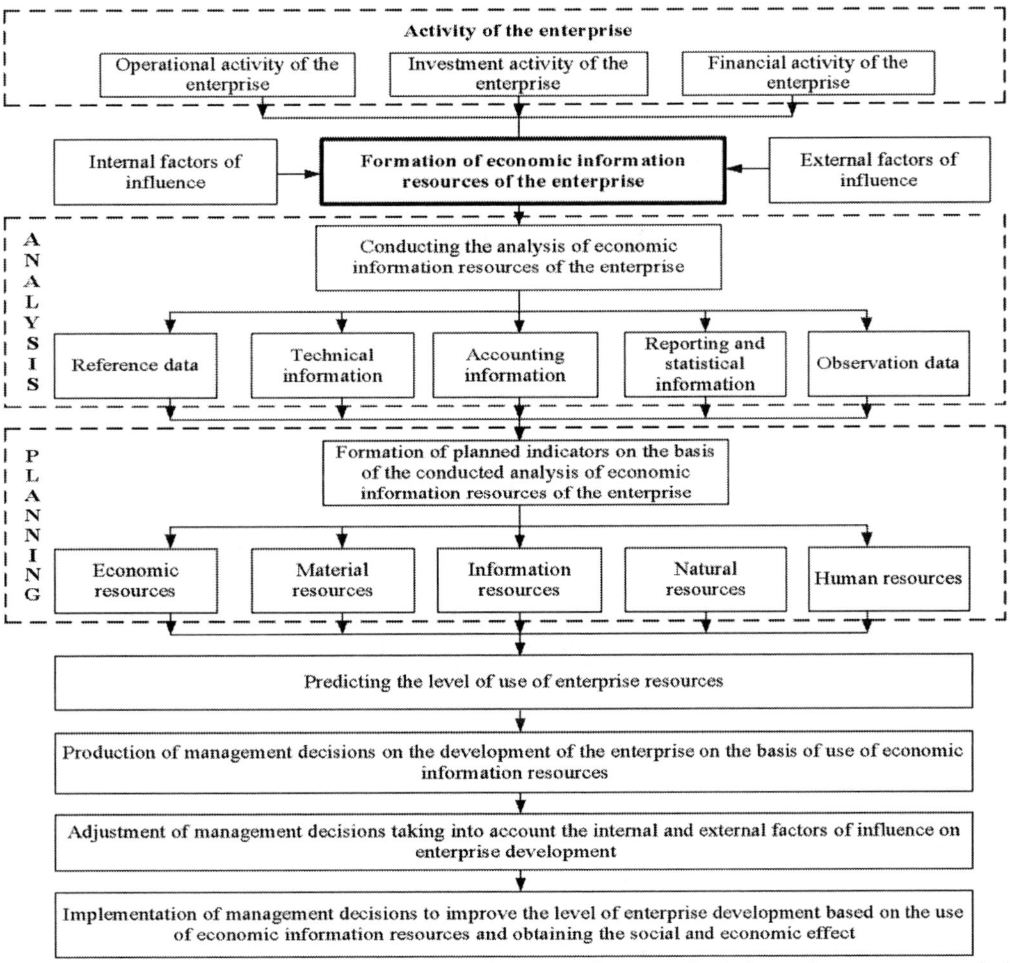

Fig. 2. Scientific and practical approach to the use of economic information resources of the enterprise for increasing the level of their development

One of such significant directions is the economy. Economic information resources are formed as a result of organizational and economic activity. It reflects the economic objects and economic processes that occur at these economic entities. The area of application of the system of information resources goes beyond the scope of the economy through the inter-branch information exchange. At the same time, information that is not economical, for example, scientific, technical, legal, etc., can be generated and used at economic entities.

According to their characteristics economic information resources are the peculiar potential of economic activity of enterprises. On their basis a complex of tasks and functions of economic entities is realized. At the same time, due to its cross-sectoral nature, economic information is widely used in other spheres of socially useful activity. Economic information resources are an integral part of the information potential of the economy of the entire society. Information resource of the economy is a system of information, data, and knowledge which are generated in the process of social and historical practice of people and which are used in economic activity. It should be noted that economic information occupies a significant place in the information resource of society due to not only its volume, but also due to its meaningfulness.

Assimilation of the system of information resources is characterized by the fact that at a certain stage of solving any problem a set of new data which it compares with the system of own ideas, concepts, attitudes and assessments, manifests in the human consciousness. Comparison, analysis, synthesis, classification, induction, formalization, modeling, and so on can be distinguished among the methods of processing the system of information resources. As a result of this analysis, new improved economic information emerges. Knowledge about objects or processes is renewed. In the further activities this updated information can be comprehended, structured and recorded (registered) in the form of a text by means of a certain sign system (the information can be documented).

3 Conclusions from the conducted research

Thus, in current economic environment, the system of information resources plays an important role in the activities of any enterprise and meets such requirements as relevance (the property of the data which characterizes the current situation); efficiency (the property of the data, which is that the time of its collection and processing corresponds to the dynamics of the change of a situation), identity (the property of the data to correspond to the state of the object), and security (avoidance of a negative influence of internal and external factors). Society is interested in the effective use of economic information resources through the development of international relations and the sustainable development of modern enterprises.

It is discovered that the availability of assured and reliable information in relation to all social and economic components is the basis of forecasting and rational strategic planning.

Enterprises that effectively organize and accumulate in the general system the information and economic flows, through which they predict and plan a complex aggregate of economic and social development on the basis of achievements of scientific and technological progress, will be able to increase their competitive advantages by changing the level of economic independence, expansion of markets and strengthening of positions on the domestic and foreign markets.

References

1. M.P. Denysenko. Information support of an effective management of the enterprise. *Economy and the state,* **7**, 19-25, (2010)

2. V. B. Ostaletskyi. Theoretical approaches to the classification of information systems of management of an enterprise. *Economic Bulletin of the National Technical University of Ukraine "Kyiv Polytechnic Institute",* **12**, 294-299 (2015)

3. P.R. Putsenteilo. Information support of analytical activity in the management of an enterprise. *Scientific and Information Bulletin of the Ivano-Frankivsk University of Law named after King Danylo Halytsky*, **11,** 224-232 (2015)

4. S. Ya. Babinska. The concept of information support of the innovation project and stages of its realization. *Business Inform*, **1,** 61-66 (2017)

5. P. I. Melnyk. Information support of the company's finances. *Problems and prospects of economy and management*, **1,** 132-143 (2015)

6. M. O. Chuprina, M. M. Hohot. The use of information systems in human resources management. *Actual problems of economy and management*. **11**, 236-238 (2017)

7. The Law of Ukraine "On Information". *Information from the Verkhovna Rada*, 48. – Art. 650-651 (1992)

8. *The use of information and communication technologies at enterprises in 2017.* [Electronic resource]. - Mode of access http://www.ukrstat.gov.ua

9. O.O. Grechishkina Information provision of management of social and economic development of the enterprise. *Scientific and technical collection. Utilities of cities,* **89**, 83-89, (2009)

10. V.D. Parkhomenko. Scientific and organizational problems of information resources management. *Scientific and technical information*, 3, .31-36, (2007).

Methodical aspects of the neutralization of risks of innovative managing the development of hotel and restaurant enterprises

Oksana Davydova[1*], and *Valentina* Protsenko[2]

[1] Kharkov State University of Food Technology and Trade, Hotel and Restaurant Business' Department, Klochkovskaya str. 333, 61051 Kharkov
[2] Licensing Department of the Ministry of Education and Science of Ukraine, Department of Control Over the Implementation of License Conditions, 10 Peremohy avenue, Kyiv, Ukraine

Abstract. The article proves the necessity to intensify the activity of hotel and restaurant busness which will allow to avoid risks or to minimize their impact in order to increase the level of competitiveness in the domestic and foreign markets. It is proved that HACCP system can provide requirements of the modern world in the manufacture of safe products and provision of safe services. The main advantages provided by the hotel and restaurant enterprise are the development and implementation of HACCP system in Ukraine. A structural logical scheme of evaluating one of elementary risks of HACCP system implementation to the activities of the hotel and restaurant business is constructed in accordance with the proposed approach. The method of neutralizing risks of innovative management of the development of hotel and restaurant business is worked out.

1 Introduction

In modern conditions, increasing the probability of emerging risks under the influence of internal and external factors in the activities of hotel and restaurant business stipulates the necessity to intensify activities that will allow avoid risks or minimize their impact that will enable to increase competitiveness in the domestic and foreign markets. The issues of reducing the influence of risks during the formation of the system of innovative management of the development of hotel and restaurant business gain particular relevance.

It is necessary to manage risks effectively applying measures, which will make it possible to predict the occurrence of risk situations in order to avoid or minimize them.

The scientists pay much attention to the use of methods for reducing negative impact of risks on the activities of hotel and restaurant business [1-6].

However, the degree of uncertainty of the process of substantiating innovative projects of managing the development of hotel and restaurant business requires development of measures for its reduction that presupposes adequate assessment both of possible losses and the probability of their occurrence, as well as the influence of individual factors on the overall risk of innovation development. Let us consider a practical example of analyzing the implications of the introduction of HACCP system into the activities of hotel and restaurant business and possible risk factors as one of the directions of innovative management of the development of hotel and restaurant business.

* Corresponding author: davydova_oks@ukr.net

2 Main body

2.1 Conceptual aspects the implementation of the HACCP system in the hotel and restaurant business

Requirements of the modern world in the manufacture of safe products and provision of safe services are able to provide HACCP (Hazard Analysis and Critical Control Points) system. It is a scientifically based, systematic approach to the identification, analysis and control of hazardous factors in the manufacture, sale and consumption of products in order to ensure its safety. HACCP system involves identification of risk factor and critical control points in the manufacturing process and provision of services.

It is worth noting that HACCP concept was originally created to ensure food products quality, but application of HACCP system is not limited at that.

The comprehensive analysis of functioning of HACCP system allowed to develop organizational measures for the implementation of HACCP system in the activities of hotel and restaurant business as one of the directions of innovation management in their development [7].

The strategic goal of HACCP system in the hotel and restaurant industry is to study each step, each stage of the manufacturing process and provision of services, identification of risks, hazards and introduction of effective methods of control and monitoring procedures.

Implementation of HACCP system in hotel and restaurant establishments involves process control of the parameters of implementing technological process of providing services, the evaluation of information and materials used for the provision of services, and also implements a strategy to guarantee their security and is based on three principles:

1. Legislative requirements to ensure safety of hotel and restaurant services.
2. A profound scientific approach to decision making.
3. Control over compliance with the requirements of the legislation.

This is a proven system that provides confidence that there is a successful product safety management. It permits to focus on security and product quality as the highest priority, as well as to systematically prevent misunderstandings, rather than expect them or to control after their occurrence. This methodology has proven itself as an effective tool of preventing the possibilities of inconsistencies in the technological process, as well as to identify and solve problems before inappropriate finished products become a source of poisoning or deterioration in the consumers' health.

Application of HACCP system can help during the inspection of authorities and promotion of international trade by enhancing the security of services. HACCP system is validated and audited. HACCP concept is recognized by the Codex Alimentarius Commission and the European Union, and as a consequence, HACCP principles are recommended for practical application. These principles can be applied to quality programs and policies, development of new products, services and technologies, control of raw materials and ingredients, production processes, maintenance, supply and use, inspection and training, etc. Codex Alimentarius leaders guide twelve steps to the implementation of seven HACCP principles. The implementation of these steps will lead to the development of a specific HACCP plan, which is a documented evidence containing details of all critical safety and service points [8-10].

2.2 Advantages of the HACCP system implementing in the hotel and restaurant industry

For Ukraine, as well as for other countries, the most acceptable means of creating security guarantees for services is the introduction of modern concept of HACCP system, which will be applied at all levels of the chain of providing services to the consumer from booking to departure. Based on modern scientifically substantiated requirements, this system provides a preventive approach and an opportunity to manage security of services effectively.

It is worth noting that the conceptual priorities of the security system development in Ukraine are at the stage of harmonization with international priorities concerning the main strategic course of our country for joining the EU, implementation of the state policy aimed at introducing innovative structural adjustment and economic growth.

There is no doubt that mechanical copying of European requirements on the Ukrainian basis is impossible because of significant inconsistencies that have arisen historically.

Results of the research allowed to reveal the main advantages offered by the hotel and restaurant industry to the development and implementation of HACCP system in Ukraine:

- control of safety of services throughout the whole life cycle;
- increased consumer confidence in the safety of products and services;
- effective management of the hazards that may arise during the provision of services;
- transition from corrective actions to preventive measures to ensure the quality of services;
- creation of conditions for the participation in European and international trade;
- existence of documented evidence of process control;
- prevention of negative situations in the field of security of services, which can negatively affect the image and activities of the enterprise;
- ability to changes caused by the use of innovative technologies, advanced equipment, procedures for sanitary and industrial practices, training of personnel, etc.

Like any other system, HACCP is a dynamic, evolving system, and should be such for maintaining efficiency. A periodic review of the HACCP plan is required to identify the points in which the plan is relevant, after any changes at the enterprise. Implementation of HACCP system will, of course, provide the hotel and restaurant business safe, high-quality and competitive services, self-assurance, guarantees, brand protection, as well as marking of services by the reputable HACCP mark possessing consumer confidence.

2.3 Analysis of risk impact from the introduction of the HACCP system in the activities of hotel and restaurant business

With the beginning of the development and implementation of HACCP system into operation of hotel and restaurant business, provision of the proper level of quality and safety of products should be ensured. In this case, it is necessary to consider the perspective. It is necessary to create permanent food control service, which will check not only the finished products for sale, but also necessarily all raw materials and semi-finished products that are being prepared for cooking. The experience should be borrowed from the world leaders in the food system. It is necessary to use efficient modern equipment, modern reagents, and a highly qualified staff. This will increase the quality of food products.

Consequently, in connection with this, there are problems with additional innovative and investment development of the hotel and restaurant business. The structural logical scheme of evaluating one of the elementary risks from the introduction of HACCP system into hotel and restaurant business was designed in accordance with the proposed approach (Fig. 1).

Creation of the structural logical scheme of risk analysis from the introduction of HACCP system into activity of hotel and restaurant business confirmed the necessity of the formation of a system, which will enable creation of the additional development, additional jobs, increase of demand and additional profit.

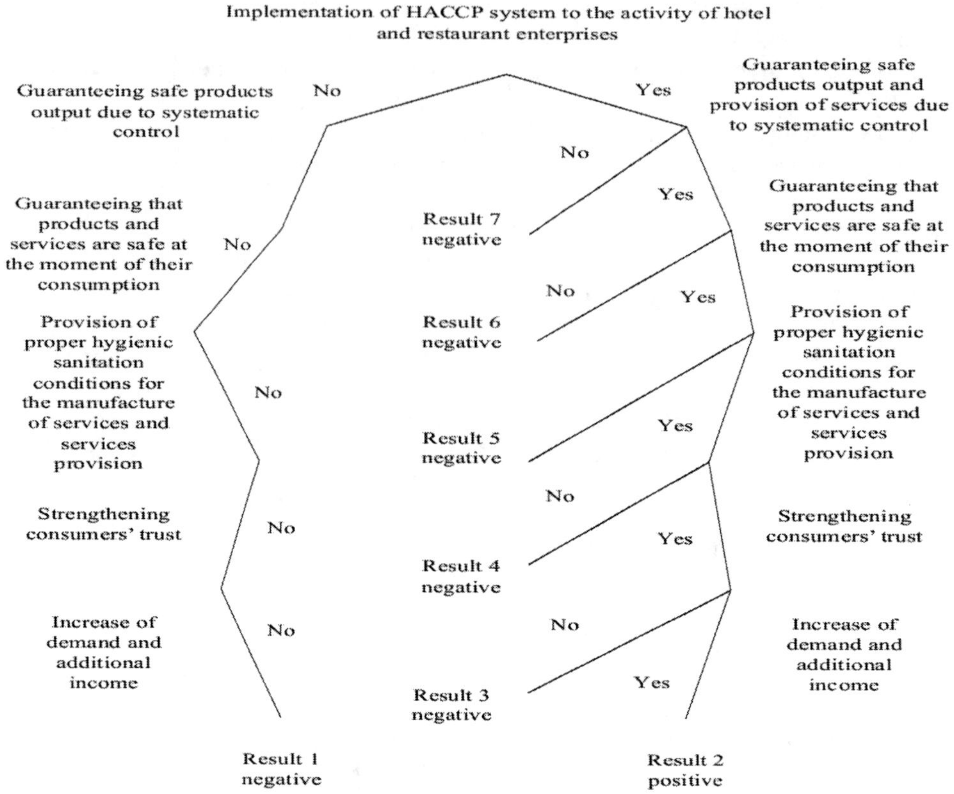

Fig. 1. Structural-logical scheme of the analysis of risk impact from the introduction of the HACCP system in the activities of hotel and restaurant business.

2.4 Neutralizing the risks of innovative management of the development of hotel and restaurant business

Managing risks of innovative management of the development of hotel and restaurant business at the stage of their neutralization involves implementation of a system-generating set of creative preventive managerial decisions in order to reduce the negative consequences, the possibility of which is determined by the occurrence of risks.

Neutralization of risks of innovative management of the development of hotel and restaurant business should be carried out in compliance with a set of certain criteria and principles, among which the following should be distinguished:

– informational authenticity, which presupposes timely and complete provision of information necessary for adequate neutralization of the risk of innovative management of the development of hotel and restaurant business;

– informational authenticity, which presupposes timely and complete provision of information necessary for adequate neutralization of the risk of innovative management of the development of hotel and restaurant business;

– adequacy as a form of the reasoned use of the method of neutralizing the risk of innovative management of the development of hotel and restaurant business in accordance with its initial characteristics;

– efficiency of the risk neutralization at hotel and restaurant business by preventing negative consequences, the probability of occurrence of which is determined in space and time;

– other principles, determined by the specific model of risk management of innovative managing the development of hotel and restaurant business.

The method of neutralizing the risks of innovative management of the development of hotel and restaurant business (Fig. 2) was developed. Provision of the neutralization of risks of innovative management of developing hotel and restaurant business based on the development of strategies of avoiding risks presupposes adoption of innovative managerial solutions through the comparative analysis of financial and mathematical models of alternative scenarios in order to determine the risk-free scenario followed by the adoption of its core.

3 Conclusion

It should be noted that the formation of a risk-free operation scenario in terms of permanent market environment is problematic, which means relative zero risk. In some cases, it is impossible to neutralize risks of innovative management of the development of hotel and restaurant business through internal stabilization mechanisms. In this case, in order to avoid losses from the onset of risks to management of hotel and restaurant business, it is advisable to turn to insurance services. Consequently, efficient and rational use of risk assessment methods can ensure stability of the development of hotel and restaurant business, increase the validity of the adoption of innovative managerial decisions in risky situations, improve financial position as a result of all types of their activities.

References

1. O. Ju. Davydova, *Kljuchovi faktory upravlinnja ryzykamy na pidpryjemstvakh restorannogho ghospodarstva,* Naukovyj visnyk KDU, **25**, 128–132 (2017)
2. K. M. Meljnyk, *Adaptyvne upravlinnja pidpryjemstvom v umovakh nevyznachenosti,* Naukovyj visnyk KDU, **3**, 81–85 (2013)
3. S. O. Mykhajlenko, S. O. Arefjev, *Formuvannja orghanizacijnogho potencialu rozvytku ekonomichnoji bezpeky pidpryjemstv*: monoghrafija. Lviv, 224 (2010)
4. V. Prokhorova, O. Davydova, *Organizational basis for the introduction of safety systems in the establishments of hotel and restaurant economy*, European Journal of Scientific Research, **1 (13)**, 749–757 (2016)
5. Ju. Senejko, *Suchasni pidkhody do traktuvannja kateghoriji "ryzyk",* Reghionaljna ekonomika, **1**, 206–211 (2006)
6. Dimitrios P. Cafetzopoulos, Evangelos L. Psomas, Panagiotis D. Cafetzopoulos, *Measuring the effectiveness of the HACCP Food Safety Management System*, Food control, **33**, 505-513 (2013), https://doi.org/10.1016/j.foodcont.2013.03.044
7. V. Prokhorova, O. Davydova, V. Protsenko, *Methodological Aspects of Implementation of the Safety System in the Activity of Hotel and Restaurant Industry Enterprises,* International Journal of Engineering & Technology, **7 (4.3)**, 502–506 (2018). Website: www.sciencepubco.com/index.php/IJET
8. FAO/WHO (1983) *The Role of Food Safety in Health and Development*, Report of a Joint FAO/WHO Expert Committee on Food Safety, Technical Report Series 705 WHO
9. FAO (1998) *Food Quality and Safety Systems: A Training Manual on Food Hygiene and the Hazard Analysis and Critical Control Point System, Food and Agriculture Organization of the United National Rome*
10. CAC (1997) *Hazard Analysis and Critical Control Point System and Guidelines for its Application*, Alinorm 97/13 A, Codex Alimentarius Commission, Rome

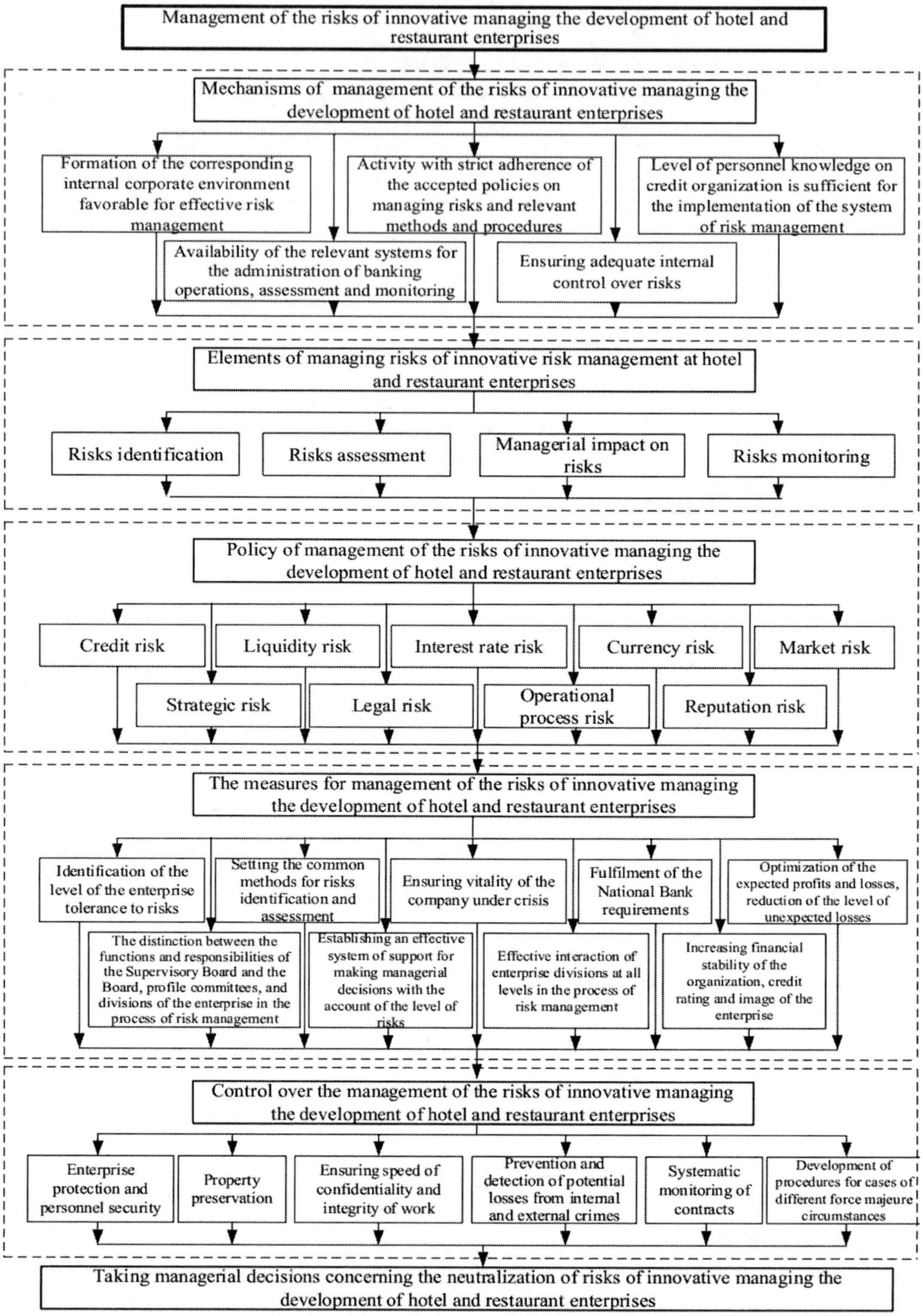

Fig. 2. The method of neutralizing risks of innovative management of the development of hotel and restaurant business.

Economic activity of enterprises: methodical aspects of assessment

Oksana Davydova[1*], *Nataliya* Chebanova[2], *Nataliia* Kashchena[3], *Hanna* Chmil[4], and *Valentyna* Protsenko[5]

[1]Kharkiv State University of Food Technology and Trade, Department of Hotel and Restaurant Business, Klochkivska St. 333, 61051 Kharkiv, Ukraine
[2]Ukrainian State University of Railway Transport, Department of Accounting and Audit, Feuerbach Square 7, 61050 Kharkiv, Ukraine
[3]Kharkiv State University of Food Technology and Trade, Department of Finance, Analysis and Insurance, Klochkivska St. 333, 61051 Kharkiv, Ukraine
[4]Kharkiv State University of Food Technology and Trade, Department of Marketing and Commercial Activities, Klochkivska St. 333, 61051 Kharkiv, Ukraine
[5]Licensing Department of the Ministry of Education and Science of Ukraine, Department of Control Over the Implementation of License Conditions, 10 Peremohy avenue, Kyiv, Ukraine

Abstract. Sustainable development of entrepreneurial structures requires the reformation of existing tools and the search for new effective tools of their economic activity managing. The aim is substantiation of the methodological provisions for assessing the economic activity of enterprises and developing applied recommendations for its improvement. The methodology of the enterprises' economic activity analysis is developed and the priority vectors of its intensification are identified in the research paper. The results of the research paper show that the determinant of national competitiveness and sustainable development is business decisions and technologies which determine the quantitative and qualitative renewal of individual enterprises potential and the combination of their business strategies with global development priorities. The proposed methodology of enterprises' economic activity assessment and applied recommendations for its strengthening are scientific basis for the development and implementation of schemes for rapid economic growth.

1 Introduction

Formation of grounds for increasing level of economic growth is extremely important task in the context of the guidelines implementation under achieving by Ukraine of the sustainable development global goals which are set [1] and approved at the UNO Summit in 2015 [2]. Business decisions and technologies which can provide quantitative and qualitative individual enterprises potential renewal, combining their business strategies with the global development priorities play leading role in achieving the goals of sustainable development. In this research paper we will consider the main methodological and organizational aspects of assessment of economic activity of enterprises as tool for determining the trends of competitiveness increasing and providing the sustainable development of the entrepreneurial structures of the national economy as a whole.

* Corresponding author: davydova_oks@ukr.net

The purpose of the research paper is substantiating the methodological provisions for assessing the economic activity of enterprises and developing applied recommendations for its increasing in the modern business environment.

2 Methodology

General scientific and special research methods, in particular: analysis and synthesis, systematization and generalization, system-activity approach and graphic methods are used for purpose achieving. The research is based on the assumption that important prerequisites for the transition of Ukrainian economy to inclusive development is improving the quality of the business environment, establishing mutually beneficial cooperation between the state and private business in all spheres and at all levels, and revitalizing the economic activity of entrepreneurial structures. For these purposes the expediency and methodology of analyzing the economic activity of enterprises as determinants of national competitiveness providing are substantiated and the priority vectors of its increasing are identified.

3 Literature review

The scientific literature review shows the active interest of domestic and foreign researchers in providing the economic activity of enterprises, the scientific substantiation of methodological approaches to measuring its level, determination of dependence on institutional and non-institutional uncertainty, influence on social and economic development and elaboration of mechanisms for stimulating economic activity of entities. Theoretical basis of economic activity of enterprises detailed in the works of domestic scientists I.Ya. Kulyniak [3], A.M. Turylo [4] and N.L. Meshkova [5]. Applied aspects of measuring economic expectations and economic activity in globalizing environment are presented in studies of L. Kilian [6], O. Claveria, E. Monte and S. Torra [7], S.V. Smirnov and N.V. Kondrashov [8]. The dependence of economic activity on the uncertainty of fiscal and economic policy of the state is determined by J. Fernández-Villaverde, P. Guerrón-Quintana, K. Kuester and J. Rubio-Ramírez [9], W. Kim [10] and P. Sorić and I. Lolić [11]. Channels of factors which stimulate economic activity in developing countries are presented by J. Caballero, A. Fernández and J. Park [12], and finally, the economic activity forecasting in the context of income receiving is considered by P. Maio and D. Philip [13]. At the same time, the issue of providing formation and sustainable development of competitive environment of the Ukrainian economy by stimulating the economic activity of business structures requires in-depth study.

4 Presentation of the main material

Solving the problems which are associated with providing the desired level of economic activity of enterprises requires strengthening of information management support through the formation of effective mechanism for its assessment. The authors propose the methodology for analyzing the economic activity of enterprise, which, unlike existing ones, determines the sequence of analytical procedures implementation and is based on defined methodological principles and methodical tools (system of indices and methods of analysis) for assessing the results and finding opportunities for increasing and stimulating economic activity of enterprises (Fig. 1).

The method of analysis of economic activity should be clearly subordinated to its methodological principles, namely: scientific, systemic, complexity, research in dynamics, purpose priority, concreteness and practical utility for providing the relevance of analytical information [14].

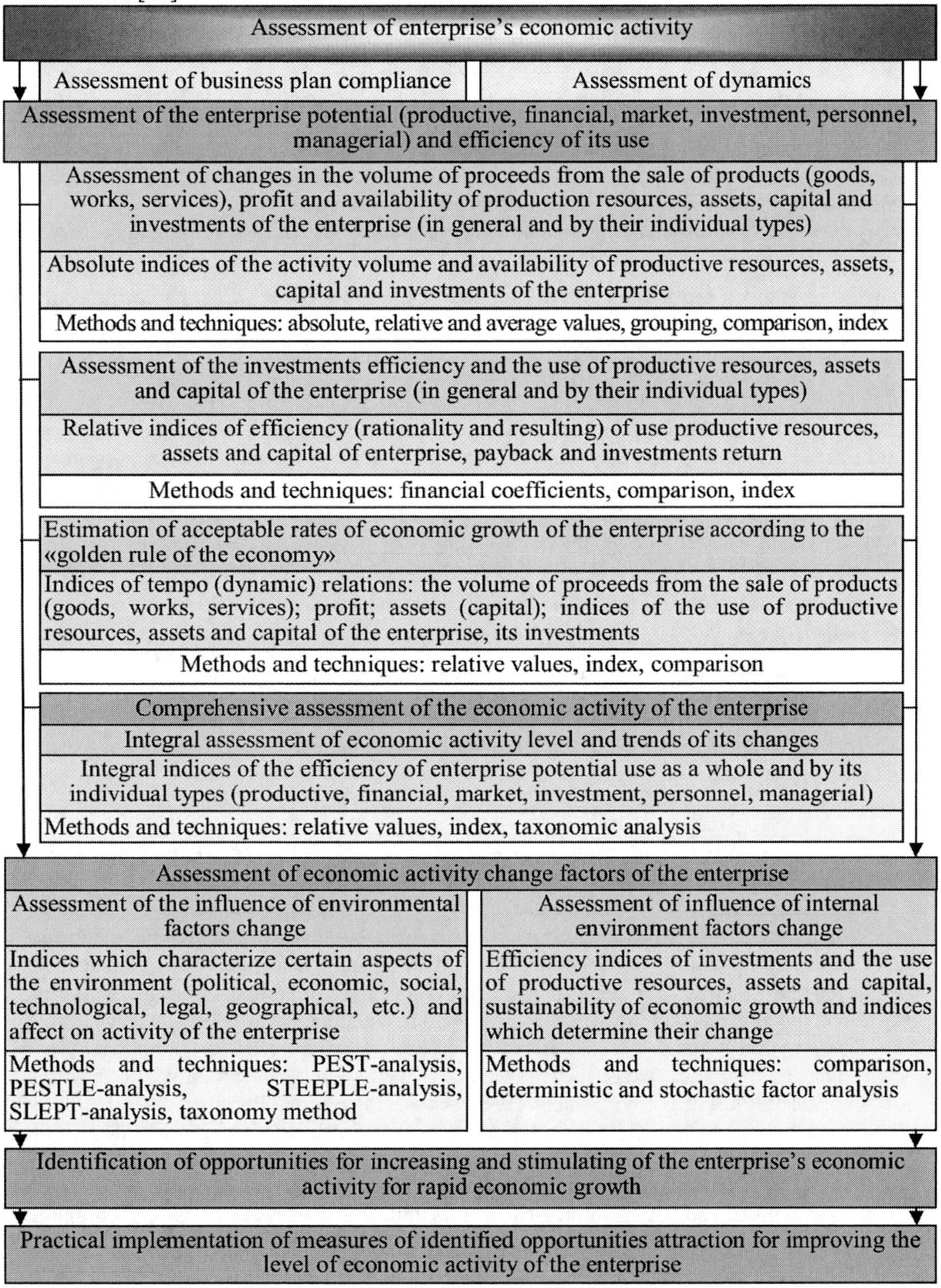

Fig. 1. Reference model for assessing the economic activity of the enterprise.
Source: Developed by authors.

Compliance with these principles allows both complex and independent realization in space and time of the main functions of analysis (assessment, diagnostic, search) which are interrelated and subordinated to the main objective of the analysis and the solving the analytical research trends which are defined for its achievement.

We determine the aim of enterprise's economic activity analysis as objective assessment of its level, identifying the causes of change and potential enhancements for rapid economic growth providing, value and cost of the enterprise.

According to the developed model, the initial stage of the analysis is the enterprise's potential assessment and the efficiency of its use. It is aimed at:

– preliminary assessment of economic activity on the base of determining the acceptable rates of economic growth in accordance with the "golden rule of the business economics", which is advanced profit growth rate (R_g^P) over the growth rate of net revenues from the sale of products (goods, works, services) (R_g^{NRP}) and the value of the enterprise's property (R_g^{VE}): $R_g^P > R_g^{NRP} > R_g^{VE} > 100$;

– detailed assessment of the economic activity of the enterprise on the base of absolute and relative changes of indices which determine it, namely, the indices of activity results, availability of investments, production resources, assets and capital of the enterprise (in general and by their individual types) and their turnover (in number of turnovers and days of turnover), profitability (in %) and payback;

– comprehensive assessment of the economic activity of the enterprise on the base of the proposed integrated index of economic activity (I_{EA}),which takes into account the dynamics of changes in individual components of the enterprise's potential (financial (I_{FP}), productive (I_{RP}), personnel (I_{PP}), managerial (I_{AP}), market (I_{MP}), investment (I_{IP})), which are formed on the base of indices of turnover and profitability of the enterprise's productive resources, its assets and capital, investments efficiency and business value:

$$I_{EA} = \sqrt[6]{I_{FP} \times I_{RP} \times I_{PP} \times I_{AP} \times I_{MP} \times I_{IP}}\,. \qquad (1)$$

In the process of further analysis, the quantitative assessment of the enterprise's economic activity change under the influence of external and internal environment factors is carried out. The trend and depth of the research of factor interconnections are determined by the aim of analysis and the needs of those who make managerial decisions.

The research proves that significant decrease of enterprises' economic activity is consequence of the negative influence of external factors under uncertainty and dynamism of the modern business environment. We propose to use the PEST-analysis method or taxonomy method for quantitative assessment of their influence strength and determining favorable conditions for enterprises' economic activity improvement.

The PEST analysis method allows comprehensive assessment of the influence of economic, political, social, technological factors and, if it is necessary, legal, geographical, social and demographic, ethnic factors etc. and involves calculating the integral indicator in the following sequence:

1) goals determining and tasks' statement for assessment of the influence of enterprise activity external environment factors;

2) formation of the assessment components which affect on the favorability of promotion of economic activity and rapid economic growth;

3) determination of assessment criteria for the i-th component of the environment factors of the enterprise's activity;

4) criteria assessment (P_{ij}) according to the determined indices;

5) calculation of the weight (λ_{ij}) indicator according to j-th assessment criterion in such way, that $\sum_{j=1}^{k} \lambda_{ij} = 1$;

6) calculation of the integral index for the i-th component of the influence assessment level of the environment factors of the enterprise's activity:

$$I_i = \sum_{j=1}^{k} \lambda_{ij} \times P_{ij} . \quad (2)$$

7) calculation of the weight coefficients (β_i) of i-th integral index of the total integral index component in such way that $\sum_{i=1}^{n} \beta_i = 1$;

8) calculation of the total integral index of the influence strength of external environment factors and its qualitative assessment:

$$\bar{I} = \sum_{i=1}^{n} \beta_i \times I_i . \quad (3)$$

The degree of experts' coordination is assessed with use of Kendall concordation coefficient. Harrington's five-point scale is used for interpretation the influence strength of external environment factors on the enterprise's economic activity level.

Algorithms for calculating of coefficients of productive resources rates of return, investment returns, turnover and profitability of assets and capital (in general and by their individual types) and sustainability of economic growth are usually used as factor models of dependence under the study of the influence of the internal environment factors on economic activity level of enterprise. They can be expanded depending on the depth of research and management needs.

We propose to carry out factor analysis of the economic growth sustainability coefficient (C^{gs}) for identifying the possibilities of rapid and stable economic growth on the base of the following factor model:

$$C^{gs} = C^{ec} \times NRS \times D^{rpp} , \quad (4) \text{ or } \frac{NR^{reinv}}{EC} = \frac{NRP}{EC} \times \frac{NP}{NRP} \times \frac{NR^{reinv}}{NP} \times 100 , (5)$$

where C^{ec} – turnover coefficient of equity capital, turns;

NRS – net profitability of sales, %;

D^{rpp} – share of reinvested profits in the net profit of the enterprise, coef.;

NR^{reinv} – sum of reinvested net profit, UAH;
EC – the average size of the enterprise's equity capital, thousand UAH;
NRP – net income from sales of goods, UAH;
NP – net profit of the enterprise, thousand UAH.

We propose realization of practical aspects of the procedural side of the analytical process with use of analytical and statistical methods, namely: sampling observations, groups, averages, series of dynamics, comparative and integral analysis, methods of elimination (complete chain substitutions, absolute and relative differences, integral, longitudinal participation, etc.), expert assessments and stochastic factor analysis (correlation analysis, regression analysis, etc.).

The enterprise's economic activity analysis is completed by the search and practical implementation of measures of identified opportunities attraction for improving and stimulating its economic activity for sustainable economic growth providing. The base of the search for internal opportunities is factor analysis results, and the measures for their practical implementation are conditioned by the general management policy which corresponds to the chosen strategy of enterprise development, and form priority vectors for strengthening its economic activity. External opportunities for increasing of enterprises' economic activity level which is aimed at updating of the domestic economy competitiveness and improving living standard of the population, are determined by the state policy of social and economic development. We defined improvement of state policy in the financial and innovation and investment spheres, creation of special territories with preferential terms, land market formation, abolition of income tax and introduction of the withdrawn capital tax, creation of favorable conditions of doing business, support of small and medium entrepreneurship, development of measures complex for cooperation between authorities and business for achieving priorities of smart, sustainable and inclusive growth as the main vectors of state provision of the desired level of enterprises' economic activity.

5 Conclusions

During this research work we obtain scientifically substantiated results, which prove that creation of favorable business environment for increasing of enterprises' economic activity is priority task of the present at all levels of state administration. It is proved that the achievement of the desired level of economic activity of enterprises determines the necessity of developing an effective mechanism for its assessment.

The authors propose the methodology for analyzing the economic activity of the enterprise on the base of certain methodological principles and methodical tools for assessment the results of enterprises' economic activity, which specifies the analytical procedures implementation sequence and allows identifying internal possibilities of its activization. The developed scientific and practical recommendations for assessing the economic activity of enterprises will increase of this process organization efficiency; improve its methodological support and provide scientific substantiation of options for management decisions for sustainable economic growth providing, value and cost of the enterprise.

References

1. Sustainable Development Goals: Ukraine (National baseline report) (2017). http://www.un.org.ua/images/SDGs_NationalReportUA_Web_1.pdf.
2. Sustainable Development Goals 2016–2030. http://www.un.org.ua/tsili-rozvytku-tysiacholittia/tsili-staloho-rozvytku.
3. I. Ya. Kulyniak. *Teoretychni aspekty traktuvannia poniattia "ekonomichna aktyvnist pidpryiemstva"*. Visnyk KNUTD, **4**, 69–79 (2017).
4. A. M.Turylo. *Teoretyko-metodychni pidkhody do vyznachennia poniattia «ekonomichna aktyvnist pidpryiemstva»*. Finansy Ukrainy, **10**, 79–84 (2017).
5. N. L. Mishkova. *Ekonomichna sutnist dilovoi aktyvnosti promyslovoho pidpryiemstva*. Ekonomichnyi Visnyk Donbasu, **1 (31)**, 36–39 (2013).
6. L. Kilian. *Measuring global real economic activity: Do recent critiques hold up to scrutiny?* Economics Letters, **178**, 106–110 (2019). Doi: https://doi.org/10.1016/j.econlet.2019.03.001
7. O. Claveria, E. Monte & S. Torra. *A new approach for the quantification of qualitative measures of economic expectations.* Quality & Quantity, **51(6)**, 2685–2706 (2017).
8. S. V. Smirnov & N. V. Kondrashov. *Indices of regional economic activity for Russia.* In Business Cycles in BRICS, Springer, Cham. 363–375 (2019).
9. J. Fernández-Villaverde, P. Guerrón-Quintana, K. Kuester & J. Rubio-Ramírez. *Fiscal volatility shocks and economic activity.* American Economic Review, **105(11)**, 3352–3384 (2015). Doi: 10.1257/aer.20121236
10. W. Kim. *Government spending policy uncertainty and economic activity: US time series evidence.* Journal of Macroeconomics, **61**, 103124 (2019). Doi: https://doi.org/10.1016/j.jmacro.2019.103124
11. P. Sorić & I. Lolić. *Economic uncertainty and its impact on the Croatian economy.* Public Sector Economics, **41(4)**, 443–477 (2017).
12. J. Caballero, A. Fernández & J. Park. *On corporate borrowing, credit spreads and economic activity in emerging economies: An empirical investigation.* Journal of International Economics, **118**, 160–178 (2019). Doi: https://doi.org/10.1016/j.jinteco.2018.11.010
13. P. Maio & D. Philip. *Economic activity and momentum profits: Further evidence.* Journal of Banking & Finance, **88**, 466–482 (2018). Doi: https://doi.org/10.1016/j.jbankfin.2018.01.013
14. V. V. Sopko, N. B. Kashena & H. L. Chmil *Theoretical aspects of performance assessment of company capital management.* Financial and credit activity: problems of theory and practice, **2(25)**, 350–356 (2018). Doi: https://doi.org/10.18371/fcaptp.v2i25.136829

Government guidance on the promotion environmental enterprise security

Olena Dykan[1], *Volodymyr* Nakonechny[2], *Maryna* Mashchenko[31], *Olena* Klimenko[3], and Ga*nna* Obruch[4]

[1] Ukrainian State University of Railway Transport, Department of Management and Administration, Feuerbakh sq. 7, 61050 Kharkiv, Ukraine
[2] Kharkiv Regional State Administration, Department of Economics and International Relations, Sumska str. 64, 61002 Kharkiv, Ukraine
[3] Simon Kuznets Kharkiv National University of Economics, Department of Economic Theory, Statistics and Forecasting, Nauky av. 9A, 61166 Kharkiv, Ukraine
[4] Ukrainian State University of Railway Transport, Department of Economics and Management of Industrial and Commercial Business, Feuerbakh sq. 7, 61050 Kharkiv, Ukraine

Abstract. Environmental security of the enterprise characterizes the state of protection of the enterprise and the environment, which allows you to maintain resistance to internal and external threats, providing the ability to solve problems of balanced development of the enterprise and improve the environment, contributes to the implementation of national concepts of socio-ecological and economic development and mechanisms to ensure it. The main tasks of the state to provide environmental security must be: substantiation of choice and decision-making in relation to the limited resources of nature and the environment as a specific public good; evaluation and comparison of costs and benefits in the environmental sphere, the rationale for the criteria and indicators of efficiency of nature protection actions; the analysis of market power for the environmental sector; consideration of external environmental effects (externalities); the development of methods for the estimation of an economic damage from pollution of natural environment; development of models of socio-economic development that meet the requirements of environmental security.

1 Introduction

In modern management models, more and more attention is paid to the safety of enterprises. Problems of environmental security are the object of study both at the global and at the personal level. This problem is also actualized for enterprises, especially for natural resource users. The principles of sustainable development, green economy, environmentally friendly production became the impetus for theoretical research. However, the enterprise is a complex socio-ecological and economic system, which determines the complexity and diversity of development goals. This requires their coordination, search for compromises, which necessitated the combination of economic and environmental components in the justification of the concept of safety of industrial enterprises. The practice of their management requires appropriate tools for planning and implementation of measures aimed, on the one hand, at reducing the negative anthropogenic impact on the environment, conservation, reproduction and rational use of the natural resource potential of ecosystems, on the other, ensuring effective, profitable activities, which allows for

[1] Corresponding author: mmashchenko@ukr.net

enhanced reproduction, competitiveness, implementation of the goals [1]. Therefore, environmental security becomes one of the basic criteria for assessing the effectiveness of enterprise management, and its provision is considered as one of the key processes of the enterprise.

2 Main part

To achieve the environmental security of the state, it is necessary to ensure the environmental security of the enterprise. The definition of "investment security of the enterprise" is broader than "economic security of the enterprise", as it includes three components: economic, social, environmental. The transformation of the socio-economic structure of Ukraine has led to changes in the mechanisms of enterprises and their management. To date, the provision of socio-ecological and economic security is an important issue. Entities in the course of their activities face various risks, the impact of which can not only lead to negative consequences, but also to bankruptcy. In turn, the knowledge of the nature of environmental security, its structure, quantitative and qualitative indicators to assess the level of environmental security will allow managers, economists and other professionals to ensure the safety of the enterprise. The need to ensure sustainable environmental safety of the enterprise is due to the stable functioning and success of each entity. The analysis of enterprises shows that financial, industrial, technological, investment and information sectors face acute problems related to activities. Their effectiveness also affects macroeconomic instability and structural imbalance [2]. Therefore, the enterprise, as the main link of the national economy, should demonstrate a significant interest in economic security in order to stabilize the financial situation of the country, the macroeconomic situation. The loss of production in the country, which changes the economic functions of the state, has led to a revision of the definition of "economic security" and emphasizes the need to address socio-environmental and economic threats of enterprises.

Environmental security of the enterprise characterizes the state of protection of the enterprise and the environment, which allows you to maintain resistance to internal and external threats, providing the ability to solve problems of balanced development of the enterprise and improve the environment, contributes to the implementation of national concepts of socio-ecological and economic development and mechanisms to ensure it. At the same time, socio-economic development has not only positive consequences: it is associated with the emergence of acute environmental problems and problems of the organization and implementation of socio-ecological and economic cooperation, the solution of which is in close relationship with changes in trends of natural character of economic development, without exception, in all sectors of the national economy.

It is important to identify the greatest threats and risks to the safe life of humanity in a timely manner. A comprehensive analysis and assessment of all processes related to the national security of Ukraine is needed. Adverse environmental conditions and the implementation of environmental safety precautions require appropriate legal, institutional and other measures. It is known that the complex impact of social, economic, environmental and other factors worsens human health and adversely affects the demographic situation in the country as a whole. At the present stage, it is necessary to prevent a global environmental disaster, the causes of which are the increase in the material needs of people, the expansion of economic activity and, thus, the increase in the impact of human activity on the environment. As a result, global environmental pollution, climate change, increasing the number of man-made disasters. In the context of globalization, the environmental and economic development of the social life of mankind requires recognition of the problems of conservation and rational use of natural resources, and at the same time requires constant harmony of the dynamics of their existence. The implementation of the mechanism of sustainable development brings together human (population), production (economy) and nature [3, 4]. Therefore, in the present economy along with the social, technological and economic deals with environmental objectives, which must satisfy the conditions of preserving the ecological balance.

Environmental security is considered as a condition of the most effective use of corporate resources to prevent threats and ensure stable business activity in the present and future. In the resource-functional approach there are 7 functional components, as the main spheres of environmental security: intellectual; financial; technical and technological; political and legal environment; information; power.

The most significant impact on the state of the enterprise is carried out by the state. The state directly and indirectly regulates the activities of economic entities. Indirect regulation creates a system of incentives that creates stimulating conditions for the activities of enterprises (effective tax system, price policy, financial and credit policy). Direct state regulation includes the development of laws, regulations, measures that directly affect the activities of the market economy. The main directions of state regulation of the enterprise:

– implementation of Antimonopoly policy;
– providing a legal framework for enterprises;
– market regulation;
– investment policy;
– effective tax system;
– formation and development of business infrastructure;
– regulation of innovation processes;
– financial support in the form of state guarantees for foreign loans [5, 6].

That is why the state has one of the main roles in ensuring environmental security. In the form of indirect influence, the state formulates a system of incentives to create motivational conditions for rational socio-ecological and economic activity (effective tax, financial, credit, price, social and environmental policies) [7]. Direct state regulation provides the legislative and legal basis for the practical implementation of all aspects of the socio-ecological and economic activities of enterprises. At the same time, the state guidelines for ensuring the environmental security of enterprises are aimed at implementing the concept and mechanisms of sustainable development.

The main tasks of the state to provide environmental security must be [8, 9]: substantiation of choice and decision-making in relation to the limited resources of nature and the environment as a specific public good; evaluation and comparison of costs and benefits in the environmental sphere, the rationale for the criteria and indicators of efficiency of nature protection actions; the analysis of market power for the environmental sector; consideration of external environmental effects (externalities); the development of methods for the estimation of an economic damage from pollution of natural environment; development of models of socio-economic development that meet the requirements of environmental security.

The main tools of the complex mechanism of formation and implementation of state policy in the field of safety of socio-ecological and economic systems are [10]: environmental legislation; strategies, programs, plans for social development and environmental protection; licensing system in the environmental sphere; environmental assessment and assessment of the impact of environmental impact assessment on the state of the environment; environmental audit, environmental management systems, environmental labeling; social and environmental insurance; technical regulation, standardization and accounting in the field of environmental protection, nature management and ensuring environmental security; education and scientific support for the formation and implementation of state socio-environmentally oriented policies; social protection and control in the sphere of environmental protection; cross-sectoral partnership and intersectoral cooperation in social development and environmental protection; the "hot" line on social and environmental issues in public authorities at all levels; conferences, lectures, research, surveys, advertising, media and Internet coverage of social and environmental problems and ways to solve them; taxes, benefits, subsidies, credit rates, customs tariffs, depreciation rates to promote environmentally sound production and the introduction of environmentally sound technologies; international cooperation in the field of environmental protection and social security.

3 Conclusion

The concept of “economic security” within the framework of sustainable development is a narrower concept that explains security. The concept of security has a number of features inherent in the whole system of nature as a form of interaction between nature and society. Under the embodiment of environmental security at enterprises is understood not just the process of their gradual adaptation to external conditions, but a conscious purposeful change in the foundations of the functioning of one, a group or all enterprises, carried out within a relatively short period of time. The embodiment of environmental security at the enterprise is considered as a process of complex radical transformation of the organization, financial and economic, production and technical, information, management structures, property relations of economic entities in order to ensure the viability and development of the enterprise.

The priority directions of environmental security should be: resource-saving focus of technologies used in production in all industries, reducing the resource intensity of production; the creation of a unified system of environmental monitoring; reducing the level of environmental pollution throughout the country to environmental and health standards; deepening cooperation with the world community in solving environmental problems; the development of environmental education, culture and training of the population; prevention and elimination of environmental disasters, natural disasters, consequences of emergencies and accidents; development of scientific and technical potential in the field of ecology; accelerated development of industries that can increase export potential; improvement of the legal mechanism for regulating the interaction of public authorities at all levels; improvement of investment attractiveness; improvement of the business climate; ensuring energy independence; effective personnel policy in the environmental and economic sphere; formation of a sustainable system of national values and interests in the environmental and economic sphere.

References

1. M.A. Mashchenko, A.V. Lytvynenko. The bulletin of transport and industry economics. **61**, 47-53 (2018)
2. M.A. Mashchenko. Upravlinnya rozvytkom. **2**, 64-68 (2018)
3. V.L. Dykan. The bulletin of transport and industry economics. **34**, 141-147 (2011)
4. G.V. Kozachenko, V.P. Ponomapev, O.M. Lyashenko. *Ekonomichna bezpeka pidpryyemstva: sutnist ta mekhanizm zabezpechennya*, Libra, 280 (2003)
5. O.Yu. Chubukova, O.V. Olshanska. Efektyvna ekonomika. 2, URL: http://www.economy.nayka.com.ua/?op=1&z=3765 (2015)
6. A.O. Cherchyk. Visnyk Nacionaljnogho tekhnichnogho universytetu “Kharkivsjkyj politekhnichnyj instytut” (ekonomichni nauky). **28(1200)**, 51-54 (2016)
7. N. Avanesova, I. Volovelska, V. Maslova, T. Sukhorukova, Yu. Utkina. Inetrnational Journal of Engineering & Technology. **7(4.3)**. 393-397 (2018)
8. O.S. Vlasyuk. *Ekonomichna bezpeka Ukrajiny v umovakh rynkovykh transformacij ta antykryzovogho reghuljuvannja*, Academy of Financial Management,

474 (2011)
9. A.V. Tolstova, K.V. Khomenko. The bulletin of transport and industry economics. **63**, 187-195 (2018)
10. M.A. Mashchenko. Efektyvna ekonomika. 11, URL: http://www.economy.nayka.com.ua/pdf/11-2018/65.pdf (2018)

The concept of management of economic security of the railway transport in the context of overcoming the manifestations of a systemic crisis

Volodumur Dykan[1], *Iryna* Tokmakova[1], *Volodymyr* Pavlichenko[2], *Oleksandr* Cherednychenko[3] and *Myroslava* Korin[1*]

[1] Ukrainian State University of Railway Transport, Department of Economics and Management of Industrial and Commercial Business, Feuerbach sq. 7, 61050 Kharkiv, Ukraine
[2]Kharkov Institute of Finance Kyiv National University of Trade and Economics , Pletnevsky lane 5, 610500 Kharkiv, Ukraine
[3] Yaroslav Mudryi National Law University, Department of Juridical Personnel Training Institute for the Security Service of Ukraine, Pushkinskaya st. 77, 61024 Kharkiv, Ukraine

Abstract. The article analyzes the trends in the functioning of railway transport in Ukraine, which allowed to identify the manifestations of the systemic crisis in the industry. On this basis, the risks and threats to the economic security of rail transport are highlighted. The concept of management of economic safety of railway transport is formed. The object and subject of management of economic safety of railway transport, principles and purpose are defined. The tools of management of economic safety of railway transport in the following directions are revealed: formation of institutional and organizational basis of development of railway transport; providing innovative transformations to improve technological capabilities and raise the level of security; ensure social development and competencies of staff to increase loyalty; create communications to maintain effective communication of railway transport and development of partnerships.

1 Introduction

Today, rail transport is going through perhaps the most difficult period in the history of its own operation. Long-term processes of reforming the industry in conditions of aggravation of macroeconomic instability in the country led to the immersion of Ukrzaliznytsia JSC in the state of economic crisis and deterioration of their economic security. For the most part, the trends in the functioning of rail transport, reflecting the accumulation of problems over many years, are extremely threatening to maintain the sustainable development of the industry. The decline in transportation volumes, depreciation of fixed assets, staff outflow and loss of investment, the decline in profitability

* Corresponding author: miraslava87@ukr.net

and solvency, the growth of dependence on foreign currency borrowings became the basis for the loss of stability of the industry to threats, opportunities for development and maintaining the image among key stakeholders.

2 Analysis of research and publications on the management of economic security of railway transport

The problem of economic security management of enterprises, including rail transport, is carefully studied by specialists and scientists of foreign and Ukrainian scientific schools. The theoretical basis of economic security management and its tools are thoroughly reflected in the works of such scientists as L. Abalkin, O. Arefieva, E. Vaughan, L. Donets, P. Drucker, Yu. Zhalilo, O. Lyashenko, C. Peterson, M. E. F. Seligman, N. Reverchuk [1-10]. Scientific search for tools to support the economic security of rail transport devoted their research V. Dukan, G. Eitutis, I. Volovelsky, V. Ovchynnikova, I. Nazarenko, I. Tokmakova and others [11-17]. In general, assessing the significant contribution of the developments and proposals of these scientists to the development of the theory and methodology of economic security of railway transport in the context of aggravation of manifestations of the systemic crisis in the railway industry, there is a need to develop the concept of economic security management.

3 Study of manifestations of the systemic crisis in the railway transport of Ukraine

Trends in the functioning of rail transport in recent years indicate the continued instability and unpredictability of the economic situation in the industry. Despite some progress in reforms, the restoration of positive dynamics of results of operations in General, rail transport has long stretched production capacity. Destabilizing impact on the economic situation of JSC «Ukrzaliznytsia» have a variety of risks and threats that exist in the environment of its functioning.

Rail transport is made economically vulnerable by industry-specific internal threats and risks to its operation (Fig. 1). The most threatening impact was the fall in the volume of work of Ukrzaliznytsia JSC, which created the risk of loss of income and investment and provoked the deterioration of solvency and credit ratings of railway transport. Turning to the data of statistics, it is worth noting that although the level of profitability of railways from freight traffic is significantly increasing in dynamics (growth in five years amounted to UAH 28.2 million), but the rate of increase due to the rise in tariffs for transportation. Thus, only for the period 2014-2017, the average annual tariff for transportation of 1 ton of goods in the industry increased by 72.27 UAH /t or 94.7 % [18]. In fact, the support of financial and economic stability in railway transport is not due to the growth of the volume of work of the industry, but to the increase in the cost of services for consumers.

The increase in the profitability of freight transportation, provided by the increase in tariffs for services of ukrainian railways, contributed to the revival of investment activity of railway transport in the field of updating the technological base. However, the problem is that the increase in the size of capital investments of JSC «Ukrzaliznytsia» in the renewal of fixed assets was also due to the increase in the size of foreign currency borrowings of railway transport, which in the conditions of increasing negative dynamics of volume indicators of the industry created a threat of loss of solvency and the situation of technical default on rail transport. Due to the accumulation of foreign currency borrowings (in 2017, the foreign currency component of the loan portfolio amounted to 98 %, while in 2014 – 84%, in 2015 – 83 %) [19-20] and the financial insolvency of railways to pay for their loan obligations, there was a deterioration in the investment and credit ratings of rail transport and provoked a threat of loss of confidence on the part of key investors.

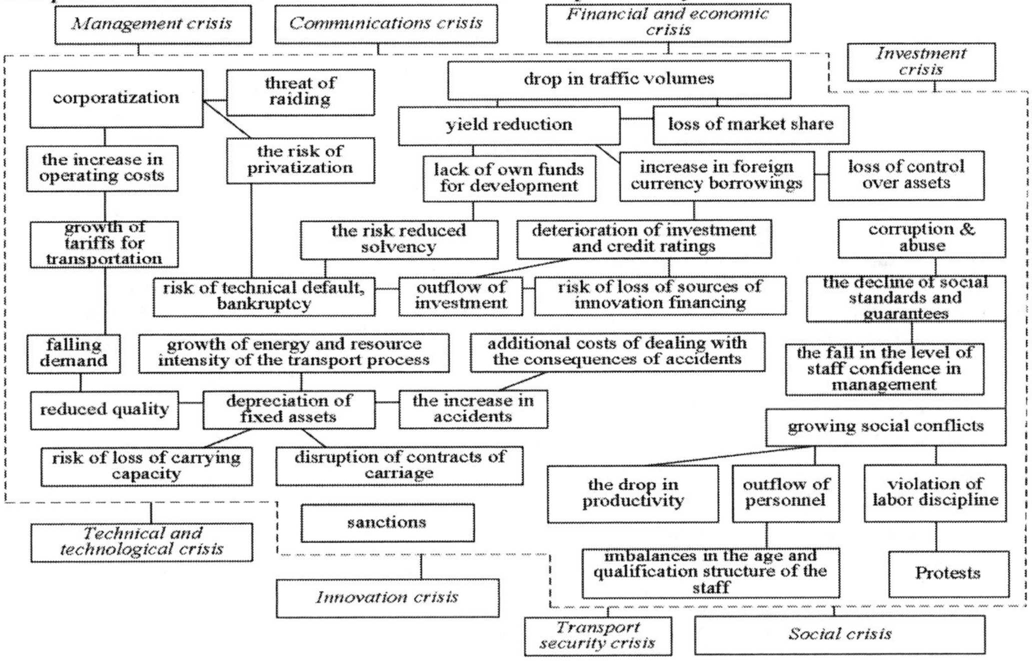

Fig. 1. Risks and threats to the economic security of railway transport

Increases and the threat of losing a share of the transport market by reducing demand for services of JSC «Ukrzaliznytsya» in the sphere of cargo and passenger traffic. The reorientation of freight and passenger traffic to road transport is provoked by the low quality of railway transport services, which in the conditions of raising tariffs does not meet either the requirements of consumers or service standards. On the one hand, the low quality of services of JSC «Ukrzaliznytsia» is the result of a deep technical and technological crisis that has taken place in the industry for decades. As evidenced by the official data of JSC «Ukrzaliznytsia» to date, the technical and technological capabilities of the company are on a critical verge, and their condition and quantity is not able to meet the existing demand in the country for transportation. First and foremost there is a huge shortage of locomotives (in use is 1857 units, representing 51.4 per cent), trucks (inventory Park 64 thousand units) and passenger cars (inventory of Park – 4220 units, the average age is 31.6 g), and railcar rolling stock in particular, electricity (the wear rate of 87.5 %) and diesel trains (the wear rate of 97,2 %) [19-21]. Vray is critical and the situation of railway infrastructure, most of which have long been operated with overdue repairs. Even despite the revival of the investment policy of JSC «Ukrzaliznytsia» in the direction of renewal and acquisition of rolling stock, in General, the industry has a situation of further accumulation of technically worn-out fixed assets. Plans in the field of capital investments of JSC «Ukrzaliznytsia» for the coming years are expected to purchase 110 freight and 90 passenger electric locomotives, 70 freight locomotives and more than 40 double-feed electric locomotives [21], and a significant (about 650 units) renewal of the wagon fleet is announced.

This state of Affairs with the main funds of railway transport provokes the growth of unreliability and distrust of JSC «Ukrzaliznytsia» from the leading partners in particular, manufacturers of industrial and agricultural products, forcing the latter to form their own car fleet to prevent violations in the supply. This led to a situation in which the inventory fleet of ukrainian railways is less than 1 thousand units of private freight cars (65 thousand units) [21].

Along with this, the situation with the lack of rolling stock provokes the industry to implement economically harmful measures both for railway transport and the economy as a whole. So, implementing the model of corporatization, JSC «Ukrzaliznytsia» is considering the possibility of the admission of private locomotives, leaving behind the industry, the function of the operator of the railway infrastructure. This, in turn, creates the risk of insolvency and bankruptcy of railway transport, because the appearance of private operators in the transport market means for the industry, further tightening of reforms and even more growth of operating costs of JSC «Ukrzaliznytsia», the concentration of low-income areas of activity.

The processes of social transformations in the industry also pose a threat to the effective functioning of ukrainian railways. On the one hand, already today, through the deterioration of the level of social security of workers in the industry, there is an increase in the manifestations of the social crisis. Social tension in the labor collectives of JSC «Ukrzaliznytsia», which was the result of inaction and systematic violation of the legislation and the requirements of workers, led to the most large-scale for the entire period of existence of the ukrainian railways outflow of personnel. Over the period 2014-2017, the average number of employees in core activities decreased from 293.7 thousand to 219.1 thousand persons, that is, by 74.6 thousand or 25.4 % [22]. The reduction in the number of employees provoked a drop in the level of labor productivity in the industry and led to an increase in qualification and age disparities in the structure of personnel. The reduction of social standards and guarantees for the majority of employees against the background of a constant increase in the salaries of management staff has provoked an increase in dissatisfaction and conflict in the staff of railways, causing the risk of loss of qualified human resources and the aggravation of the crisis of trust of railway transport management.

So, we can say with confidence that despite the positive dynamics in the field of reforms and implementation of the corporate management model, in General, the situation in railway transport is of a crisis nature and leads not only to the loss of competitive positions of ukrainian railways in the transport market, but also predetermines a sharp deterioration in the level of their economic security, making it impossible to implement strategic initiatives to support the economic development of railway transport. In the context of the reform and variability of the external environment of railway transport, the strategic directions of increasing the level of economic security should be focused on neutralizing the impact of existing threats and increasing the adaptability of the industry to changes in the transport market.

4 Determination of directions and development of the concept of management of economic safety of railway transport in the conditions of system crisis

The level of economic security of railway transport as well as any enterprise today is closely connected with such trends of economic development as intellectualization of work and socialization of personnel, Informatization of business processes and systems, their innovative transformation and integration. The ability to respond effectively to the challenges and threats to its economic security depends on the level of adaptation of the enterprise to such global development trends. In accordance with this, and taking into account the manifestations of the systemic crisis in railway transport, the management of economic security of the industry should provide for its adaptation to changing external conditions and be accompanied by the implementation of effective measures aimed at overcoming negative trends in intra-industry development. Based on the theoretical basis of the studied problems of management of economic security of railway transport will be considered in terms of a combination of management tools of institutional and organizational changes, innovation,investment and personnel management, as well as the introduction of tools of integrated marketing communications to maintain effective relations and development of partnership with key stakeholders and leveling on this basis of threats of systemic crisis (Fig. 2).

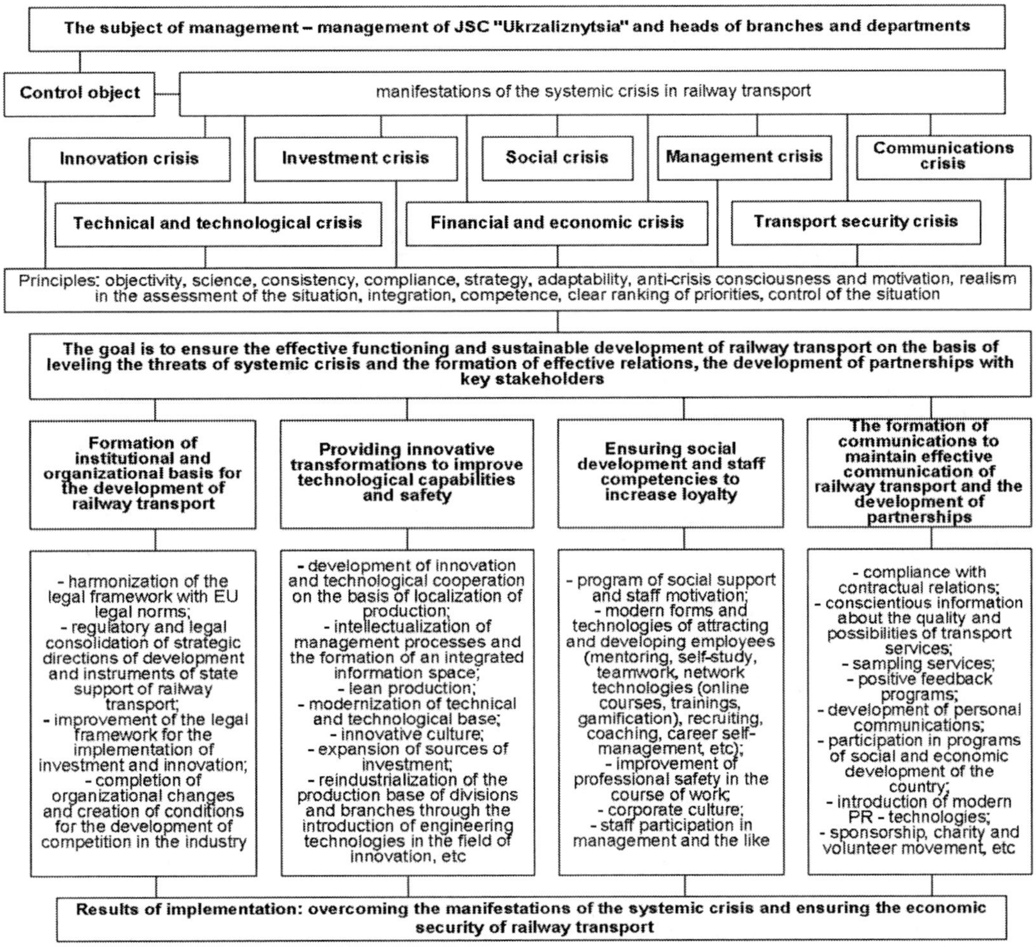

Fig. 2. The concept of management of economic safety of railway transport in the conditions of system crisis

Accordingly, it should be based on tools in the following areas: the formation of institutional and organizational basis for the development of rail transport; providing innovative transformations to improve technological capabilities and improve safety; ensuring social development and staff competencies to increase loyalty; the formation of communications to maintain effective rail transport links and the development of partnership. Their implementation will allow us not only to overcome the symptoms of a systemic crisis in railway transport, but also significantly increase the level of its economic security, creating the conditions to support the financial, economic, social, environmental efficiency of Ukrzaliznytsya JSC, the strengthening of the business position and image of the industry in the transport market.

5 Conclusion

The modern stage of development of railway transport is characterized by the aggravation of crisis processes in the industry. Technological degradation as a result of low level of innovative activity, loss of investments and qualified personnel in the conditions of deterioration of volume indicators of work of branch led to deterioration of level of economic safety of railway transport and loss of competitive positions in the market of transportations. The concept of management of economic safety of railway transport establishes the subject and object, principles and objectives, and also reveals the tools of management of economic safety of railway transport in three directions, the implementation of which will create a basis for leveling the existing risks and threats in the industry, ensuring its sustainable development.

References

1. L.I. Abalkin, Voprosy ekonomiki, **12**, 4-13 (1994)
2. O.V. Arefyeva, T.B. Kuzenko, *Planuvannya ekonomichnoyi bezpeky pidpryyemstv*, Kyiv, 172 (2004)
3. E.J. Vaughan, *Fundamentals of Risk and Insurance*, New York, 723 (1986)
4. L.I. Donets, N.V. Vashchenko, *Ekonomichna bezpeka pidpryyemstva,* Kyyiv, 240 (2008)
5. P.F. Drucker, *Management: Tasks, Responsibilities, Practicies*, New York, 840 (1974)
6. YA.A. Zhalilo, Stratehichna panorama, **3**, 97-104 (2004)
7. O.M. Lyashenko, *Kontseptualizatsiya upravlinnya ekonomichnoyu bezpekoyu pidpryyemstva*, Luhansk, 400 (2011)
8. C. Peterson, M.E.F. Seligman, Journal Psychological Review, **91**, 347-374 (1984)
9. N. Y. Reverchuk, *Upravlinnya ekonomichnoyu bezpekoyu pidpryyemnytskykh struktur*, Lviv, 195 (2004).
10. V.L. Dykan, I.V. Volovelskaya, Visnyk ONU imeni I.I. Mechnykova, **1**, 82-85 (2016)
12. H.D. Eytutis, D.V. Popova, Visnyk ekonomiky transportu i promyslovosti, **62**, 29-36 (2018)
15. V.O. Ovchynnikova, *Stratehichne upravlinnya rozvytkom zaliznychnoho transportu Ukrayiny*, Kharkiv, 427 (2017)
16. I.L. Nazarenko, S.V. Chernenko, Visnyk ekonomiky transportu i promyslovosti, **62**, 89-96 (2018)
17. I.V. Tokmakova, K.V. Dudka, Visnyk ekonomiky transportu i promyslovosti, **48**, 194-197 (2014)
18. *Zmina taryfiv na perevezennya vantazhiv zaliznychnym transportom u mezhakh Ukrayiny na 2017 rik.* https://publications.chamber.ua/2017/.../Session_2_Tarifs.pdf ahhh!. Accessed 28 Apr 2019
19. *Intehrovanyy zvit 2017: dlya investoriv*: ofitsiynyy sayt AT «Ukrzaliznytsya». https://uz.gov.ua/about/investors/issuer/. Accessed 08 May 2019
20. *Intehrovanyy zvit 2018: dlya investoriv*: ofitsiynyy sayt AT «Ukrzaliznytsya». https://uz.gov.ua/about/investors/issuer/. Accessed 08 May 2019
21. *Forum Rail Expo 2018 buv maydanchykom dlya obhovorennya onovlennya lokomotyvnoho parku.* https://info.uz.ua/articles/forum-rail-expo-2018-stav-maydanchikom-dlya-obgovorennya-onovlennya-vitchiznyanogo-lokomotivnogo-parku ahhh!. Accessed 20 Apr 2019
22. *Dovidnyk osnovnykh pokaznykiv roboty rehionalnykh filiy PAT «Ukrayinska zaliznytsya» (2002-2017 roky):* Publichne aktsionerne tovarystvo «Ukrayinska zaliznytsya». Upravlinnya statystyky, Kyyiv, 36 (2018)

Implementation of Smart City Concept in Ukraine

Valeriia Dykan[1], *Mariia* Ieromyna[2], *Uliana* Storozhylova[3],and *Liliia* Bilous[1*]

[1] V.N. Karazin Kharkiv National University, Management and Administration Department, Myronosytska str. 1, 61022 Kharkiv, Ukraine
[2] Ukrainian State University of Railway Transport, Department of Public and Corporate Finance Management , Feierbakh Sq., 7, 61050 Kharkiv, Ukraine
[3] Ukrainian State University of Railway Transport, Management and Administration Department, Feierbakh Sq., 7, 61050 Kharkiv, Ukraine

Abstract. The article deals with development of Ukrainian cities within the Smart City concept based on the experience of foreign countries. The experience of such countries as Sweden and Denmark has been analyzed. The main attention is paid to those elements of the concept that can be implemented in Ukraine at present. The analysis identified the following conditions for the development of the Smart City concept: the development of an effective e-government, transparency, inclusion in international economic processes, attracting new players and new sources of funding, service integration, strategic planning of the new business model, leadership and championship. These elements are relevant for implementation in Ukraine. Two regions of Ukraine, namely, the Black Sea and Slobozhansky regions, were selected to analyze the pace of implementing the Smart City concept. The geographic location, climatic conditions and infrastructure potential of these regions create good prospects for their attractiveness in the global world. The research has identified elements of the Smart City concept that has already been introduced and provided recommendations for possible next steps in Smart City development in Ukraine.

At the time of the actions of oppressive thinking and digital globalization, the instant development of the Internet and technological progress, the key precondition for Ukraine's attractiveness is the global transformation of the entire infrastructure of each of its regions. The introduction of the new generation of intelligent networks, innovations and investments foresees the further development of the cities of all regions of Ukraine on the principle of Smart City and focus on ensuring their comfort and security, restructuring of existing enterprises and organizations of the country in highly effective. These challenges can be solved by the concept of a comprehensive strategic approach to implementing smart concepts and the interaction of power, science and business.

* Corresponding author: l.bilous@karazin.ua

The works of such scholars as A.Yu. Berko, P. Voloshin, V.V. Dibskaya, O.I. Zaitsev, V.I. Sergeyev, O. Senkevich, D. Parks and others were devoted to the theoretical and practical questions of the implementation of Smart City technologies and the restructuring of existing Ukrainian companies in highly effective. However, despite the large number of projects, the problem of sustainable and dynamic development of regions in the direction of the introduction and development of modern technologies, including Smart City, remains open.

In an age when the world's population lives and operates in the digital world, more and more countries are rebuilding their digital development path, providing enormous benefits to the cooperation of private and state-owned companies that introduce Smart innovation into their business. The share of the traditional economy is decreasing, and the digital one is increasing. The globalization and the incredibly accelerated pace of modernization of technology constantly require countries and companies in the world to be flexible, to constantly seek out innovative solutions and semantic approaches. Expanding the capabilities of IT technology and technical progress, cross-border and interregional cooperation enable companies to be global and local at the same time, turn the regions of countries into perspective ecosystems with the further development of cooperation and management systems.

Smart City - a complex of measures aimed at modernizing the city management system. For each city it is individual, because each city has its own problems and features. However, it is possible to identify the main general directions of modernization of the city management system, without which any city can not function effectively: security, utilities and energy efficiency of the city buildings.The concept of Smart City can be offered as simple solutions, such as: urban transport management, automated parking, street lighting control; and more complex, the electronic health system, the automatic detection of pipeline failures in public utilities or electronic security system through surveillance cameras.

The implementation of Smart City is supported by international technology companies, as evidenced by IBM's Global City Business Vision [1] and the European Union over the years [2]. With regard to sustainability and climate change, such initiatives promise more efficient use of information and communication technology resources in energy infrastructure [3]. One of Smart City's methods of implementation is urban intelligence networks that are experiments with Smart City technologies that have shown potential for restructuring the relationship between energy companies, energy consumers and other participants by changing the configuration of the dynamics of demand and supply of energy [4]. Ukraine is taking the first steps in implementing such systems, drawing on the experience of countries already benefiting from the Smart City, and studying their experience to avoid mistakes. Let's consider the experience of the successful implementation of the Smart City concept in the case of countries such as Sweden and Denmark.

Sweden is one of the leading energy technology countries, and energy conservation is a priority for it. Thus, from 2011 to 2016, two projects were implemented in Malmo, which indicated that it is important not only to use advanced technologies but also to ensure a reasonable functioning of the administrative apparatus, which, in order to lobby for its own interests, can reduce the effectiveness of the changes, or even at all level them up. These projects have helped to institutionalize the technologies of urban intelligence networks [5]. Such experience is useful for Ukraine to implement its own projects.

Another example of the implementation of intelligent networks in the city's infrastructure system is the experience of the city of Aarhus, Denmark, where the e-government system, smart transport, eHealth, and the widespread data transmission system have been introduced under the Smar tcity concept. To implement the latest, the whole city was equipped with a free Wi-Fi network. The Smart City concept can be implemented in a variety of ways, for example, the Open Data Aarhus (ODAA) system, an online platform that promotes democracy, transparency and economic growth through open data, has been created. The platform provides everyone with access to the data they can use to create services and initiatives [6]. ODAA is based on the participation of stakeholders: citizens, students, researchers, and companies, for example, through a working group with members from the municipality of Aarhus, the Central Denmark region, the University of Aarhus, as well as private companies such as The Alexandra Institute [7], IBM [8] and Creuna [9]. This makes the system not only an example of easy access to the necessary information, but also an example of the interaction of power, science and business.

In accordance with modern trends and directions of development of European society, the regions of Ukraine have favorable conditions for the implementation of the elements of the Smart City concept. These regions are industrially developed regions, have a wide transport connection and are export-oriented. So, in our opinion, one of the most progressive regions of Ukraine is the Black Sea and Slobozhansky. The attractiveness of export activity in the Black Sea region was due to the development of a developed industrial sector in the region and a stretch sea border that allows receiving and shipping offshore transport from all countries. In the Slobozhansky region, the geographical advantage is the border.

To increase the level of attractiveness and ensure the sustainable growth of the regions, we consider it necessary: to provide an assessment of the potential use of the region's priorities for implementing the Smart City concept in the regions; ensuring of the development of cities on the principle Smart City; restructuring of existing companies in highly effective in the field of computerization and initiation of IT technologies; to carry out a major recovery of highways that are crucial for national and regional economic growth; to provide decentralization of routes; implement traffic safety programs; to update the infrastructure of railway transport, ports; to increase investment attractiveness of regions by attracting strategic investments in priority sectors of the region and other activities. The development of cities based on the principle of Smart City provides an opportunity to ensure sustainable economic development in the regions, and a high quality of life for the population. For the most part, in the development of the world's cities by the Smart City conception there are five main components that need to work together to achieve the common goal, namely, Smart Energy, Smart Water, Smart Buildings, Smart Government and Smart Transportation.

Smart Energy is a solution in the areas of energy supply and energy conservation, demand management, energy efficiency and renewable energy integration. Smart Water is a water resource management: water systems modernization, consumption monitoring, environmental safety systems and flood management. Smart Buildings are buildings in which all engineering and information systems are integrated into a single management system (BMS). Smart Government is the use of information technology to provide public services to a wide range of people and optimize the work of different departments. Smart Transportation is intelligent transport and logistics systems, monitoring and traffic management, payment of tolls, emergency response, intelligent parking and integrated traffic management, building intelligent logistics networks [10].

Creating Smart City in five sectors is crucial for sustained global growth and the attractiveness of the regions. An example of implementing the concept of Smart City, namely its elements of Smart Transportation, is the construction of a bicycle raft improvement facility on the Health Road in the city of Odessa, which is located in the territory of the Black Sea region. The trestle should connect all areas of the Odessa coast - from Cape Langeron to Great Fontana. Thus, a single cycling route, the length of which will be more than 15 km, with division of pedestrian and athlete movements and restoration of the pedestrian track, is organized. Complex improvement of route includes arranging descents, lighting, equipment for sanitary toilets, and renovating sports and children's playgrounds. The trestle is only for pedestrians and cyclists.

IAccording to the concept of cities development of Slobozhansky and Black Sea regions of Ukraine on the principle Smart City in cities and towns it is necessary to install intelligent sensors that will facilitate management of city through the system of intellectual video analytics. The priority strategy of socio-economic development and attractiveness of all cities of the Slobozhansky and Black Sea regions requires the renewal of urban electric transport: rolling stock of trams, trolleybuses and buses, as well as transport infrastructure.

Electric transport is always environmentally friendly, fast and economical. When designing new streets, roads and solutions, it is necessary to take into account the cyclists' movement and to install electronic displays with a timetable for urban transport. Infrastructure development is one of the key factors of attractiveness of Slobozhansky and Black Sea regions of Ukraine for investments. To implement such changes, a complete upgrade of the existing companies' system of work and a focus on the digitization and robotization of technological processes is required. Modernization of high-performance companies in the country should be carried out with a speeding up to the speed of reality.

Highly-performing companies should ensure the attractiveness and competitiveness of their products and services. Today, more flexible companies create strong ecosystems and partnerships. Modernization of companies in the Slobozhansky and Black Sea regions of Ukraine should include models of rapid adaptation of the entire company to meet the new trends in the world through information systems, mobility of company employees and cooperation between companies. For example, in Mykolaiv, a new, most powerful wind power plant in Ukraine (VEU 001) was commissioned at 3.2 MW Tuzlovskaya VES, "Vitroviy Park "Prychornomorsky" LLC.

The wind turbine was made at the Fuhrlaender Wind Technology factory in Kramatorsk. The Kramatorsk plant was used in the manufacturing cycle of wind power plants manufacturing: Kramatorsk heavy machine tool plant (KHMTP), Energomashspetsstal (EMSS), Novokramatorsk machine-building plant (NKMBP), Mykolayv Factory ESTA Ltd. [11]. The cost of the construction of traditional and alternative power generating capacities of the NPP unit in times exceeds the cost of construction of green energy facilities. In world practice, energy production with minimal impact on nature is only increasing. Taking into account the specifics of the Slobozhansky and Black Sea regions, modernization of existing companies in highly efficient and access to foreign markets can be achieved through: multifunctional development of regional management; "Reasonable" specialization; regional and cross-border cooperation; improvement of the investment climate; participation in international programs; a continuous dialogue between local authorities and companies and enterprises in the region; facilitating the establishment of IT transparent; improvement of the ecological potential of the region. Taking into account the fact that the export share to European markets from the Slobozhansky and Black Sea regions is increasing, high performing companies should pay more attention to penetration into new markets of manufactured products, optimization of logistic business processes, expansion of global sales and marketing opportunities, improvement of the quality of customer service, hedging of economic recession. An example of such modernization in companies is the enterprises of the Kherson region, which trade with more than one hundred countries of the world.

The attractiveness of the Slobozhansky and Black Sea regions mainly depends on revealing the main approaches to the implementation of the concept of Smart City and the restructuring of existing companies in highly effective, taking into account the orientation of development to innovation and exclusivity. Slobozhansky and Black Sea regions development projects under the applied Smart City concept, unites the following: the inhabitants of the city are in the center of attention and take part in the life and management of the city; expansion and improvement of city services; a friendly and attractive system for business creation; association through means of networks of citizens, authorities and business; universal and free access to culture, education and health.

The features of the Smart City concept are quality of life of citizens, mobility, socialization, virtualization, infrastructure modernization, personalization, smart technologies and urbanization. In Ukraine, the concept of Smart City began introduction in 2017 and is gradually gaining momentum, platforms are being launched, forums and round tables are being held and new technologies are being introduced. So, in place of Kharkiv in cooperation with Vodafone Ukraine, the Smart City UA project was launched [12]. By the end of 2020, the city plans to use technologies that will affect all areas of public life. The following projects will be implemented: Smart Transport, Smart Energy, E-medicine, Safe City, Ecology, Garbage Disposal, E-Government, and Tourism. 20 projects is planned to implement in total, some of which are in the process of implementation. So, bicycle tracks and bike rental with online payment, a single system of payments in public transport and "smart" street lighting were appeared in the city. By 2020, “smart” parking lots, pedestrian crossings and traffic lights, “smart” stops on which traffic can be tracked through online applications, alternative energy sources are planned to be realized.

The analysis of the concept of Smart City allowed to develop the table that identifies the various subsectors of the Smart City concept and allows solving the development of the Slobozhansky and Black Sea regions of Ukraine by implementing projects for the development of modern urban infrastructure, using information and communication technologies of infrastructure and services, cooperation between key stakeholders (citizens, universities, government, industry), integration of different areas (environment, mobility, governance, community, industry, services), investment in social capital.

Here are the main concepts of Smart City The concept of "smart" government, the provision of public services through the Internet to a wide range of people and optimizing the work of different departments, business development. Public safety, modern innovative development of comfort for the population and optimization of city life. Environmental protection, air and water quality control. Rational and ecological waste management.

Concepts of highly effective companies work. The concept of work of highly effective companies. Companies are advanced with the idea of introducing advanced world innovations and exclusivity, adaptability to the future. The organizational structure is based on projects and schemes. Targeting for new markets for products and new customers.

There are three types of subsections of the Smart City Concept for the development of the Slobozhansky and Black Sea regions of Ukraine. They must comply to such Essential principles of creation as: integration of subsystem, development of services and infrastructure of the city, investing, cooperation and should implement the following: high-performance management, high living standards, mobility,careful attitude to the environment, optimize resources usage ,solving security issues, economic growth support.

To do this, it is necessary to use the goals of the SWOT analysis of the Slobozhansky and Black Sea regions of Ukraine, where the strengths are considered as advantages in the implementation of the city's intellectual project. Weaknesses are obstacles to implementation, while opportunities are factors that can be acquired during the implementation of Smart City. The main opportunities are the availability of accessible and mature technologies suitable for solving environmental problems, support for the tourism sector, the use of intelligent systems.

Factors such as urbanization and industrialization not only draw the infrastructure of cities but also create growing problems for the effective management of the quality of life of citizens and the resilience of resources. The development of an intellectual and sustainable city is a way forward, a strategic process that requires novelty in approach, planning, operations, networking and management of urban enterprises.The SWOT analysis of the Smart City concept was used as a strategic tool for identifying and evaluating the city's intellectual transformation. In the future, work needs to be expanded in order to determine the degree of dependence and independence of the various factors of the transformation of an intellectual city based on a SWOT matrix created to assist decision makers based on a different combination of four elements. This can all help not only Ukraine in implementing such projects, but also worldwide to find more correct and rational ways to resolve the development of the Smart City concept. Taking the experience of foreign partners and international companies, Slobozhansky and Black Sea regions of Ukraine confidently introduce new technologies into their own cities to improve the quality of life of citizens.

Conclusions from the conducted research.

The development of an effective e-government, transparency, inclusion in international economic processes, attracting new players and new sources of funding, service integration, strategic planning of the new business model, leadership and championship are essential for the development of the concept of Smart City. The geographical location, climatic conditions and infrastructure potential create good prospects for the attractiveness of the Slobozhansky and Black Sea regions of Ukraine in the world. The acceleration of the modernization of the Slobozhansky and Black Sea regions of the country offers many new innovative solutions for the development of cities on the principle of Smart City and the creation of high performance companies and requires: identifying the priorities of the region's development, understanding the choice of strategic modernization, developing a set of measures, implementing selected measures. That is, it is necessary to develop the attractiveness of the infrastructure, to cover all the priority directions of international cooperation, to promote maximum convergence of the whole country with the European Union and the world.

References

1. IBM Global Business Services. *A vision of smarter cityes. IBM Corporation*. URL: https://u.to/DoS8FQ (2009)
2. Government of Sweden *Collaborative groups appointed to develop the government's innovation policy* URL: https://u.to/h4S8FQ (2016).
3. C. Harrison, I.A. Donnelly, *A theory of smart cityes*. Proceedings of the 55th Meeting of the ISSS – 2011, Hull, UK, 55(1), Retrieved from https://u.to/c4W8FQ (2011).
4. Mc. Lean, A. Bulkeley H., Crang M. *Negotiating the urban smart grid: Socio-technical experiment ationin the city of Austin*. Urban Studies, 53(15), 3246–3263. doi: 10.1177/0042098015612984 (2016).
5. D. Parks *Energy efficiency left behind?* Policy assemblages in Sweden's most climate-smart city, European Planning Studies, 27:2, 318-335, URL: https://www.tandfonline.com/doi/abs/10.1080/ 09654313 .2018. 1455807(2019)
6. *Smart Aarhus* URL: https://www.smartaarhus.eu/
7. The *Alexandra Institute* URL: https://alexandra.dk/uk
8. *International Business Machines Corporation* URL: https://www.ibm.com
9. *Creuna* URL: https://www.creuna.com/en/
10. M. Zyabkin *Technologies of "smart" cities and forecasts of their development*. URL: https://vc.ru/26713-smart-city.
11. *Talent Management. Highly efficient production organizations*. Chapter three of the report DELOITTE 2018. URL: https://u.to/7YW8FQ.
12. *Kharkiv in Ukraine by the project "Smart City UA"* URL: https://u.to/eoa8FQ

Economic growth as a key factor in the dynamics of level and quality of life of Ukrainian population

Alexander Frolov[1]*, *Marina* Kosich[1], *Elena* Aleksandrova[1], *Olga* Mozgovaya[1] and *Olga* Komarenko[2]

[1] Ukrainian State University of Railway Transport, Department of Economic Theory and Law, 7, Feuerbach Square, Kharkov, 61050, Ukraine
[2] V. N. Karazin Kharkov State University, PhD student, 4, Svobody St., Kharkov, 61022, Ukraine

Abstract. The objective of this article is conducting the analysis of economic growth, being a key factor in the dynamics of level and quality of life of the population, as well as to develop the related suggestions, in order to accelerate and stabilize such processes in Ukraine. This research presents the analysis of the real GDP's changing rate in recent years, along with the assessment of another evidence of the economic growth - i.e. the income of the population. The related cross-country comparative analysis of the level of population's well-being has been performed. The reasons for Ukraine's significant lagging behind its neighboring regions and other European countries in terms of main economic indicators have been determined. We've defined the "poverty trap", the authors believe that it is the current state of Ukrainian Economy. The related suggestions aimed to accelerate and stabilize the economic growth in Ukraine have been developed.

Introduction.

Nowadays, the actual improving of the population's level and quality of life is supposed to be the main goal and final result of the reproductive activity of Society. On the one hand, it is the main criterion for development of social production. Taking into consideration the achieved level and quality of life, we can assess the "quality" of the national economy, the satisfaction degree of various requirements of Society. Providing the conditions to guarantee the decent life and free personal development is a required pre-condition for the proper functioning of the Ukrainian Economy, along with the increase of its competitive ability and maintaining the social and political stability in Society. On the other hand, the current state of the population's level and quality of life depends upon the degree of development of social production and, because it is the final result at the time, if we evaluate it from this point of view.

In order to improve the level and quality of life, it is extremely important to define the factors of their dynamics. The authors used the following approach for the problem under study: it was supposed to identify and analyze the main factor in dynamics of the level and quality of life of Ukrainian population, as well as to compare it with the factors of such

* Corresponding author: afrolov27@ukr.net

activities in another countries and to reveal the trends which affect the alteration of the factor itself.

The objective of this research is the analysis of such a fundamental factor in the dynamics of the population's level and quality of life as the economic growth and development of suggestions aimed at its acceleration and stabilization in Ukraine.

Results and their Discussion.

The achieved rates of economic growth characterize the state and actual rate of social and economic development of the national economy. Speaking about the worldwide tendencies, the economic growth is determined by the dynamics of changes in real GDP in dollar terms, it shows the actual increase in the amount of consumed goods and services. This factor serves as a baseline for worldwide statistics and ratings, including the living standards.

The annual data on the real GDP's dynamics are regularly published. In Ukraine this factor decreased by 11,6 % [1] in 2017, as compared to the period of the year 2013. In 2018, the actual GDP growth rates have already reached 3,3% [2]. However, this growth factor turned to be absolutely insufficient for Ukrainian Economy. Under such conditions of development rates, the country will be able to reach the level of the year 2013 in 15 years only, not earlier yet.

The experts believe that the countries which belong to the group of the "emerging markets" (including Ukraine) would need to provide the economic growth rates from 5% to 10% [3, 4], in order to succeed in solving their social and economic problems. However, according to the forecast presented by the World Bank, in 2019 the actual GDP growth in Ukraine is expected to achieve 2,9 % only, while in 2020 it could reach the level of 3,4 % and in 2021 it would achieve 3.8 % [5], respectively. It's evident that even if Ukraine could have managed to achieve such growth rates, those results would not be sufficient to guarantee the sustainable development of its Economy and improve the level and quality of life of its population, either.

Nevertheless, the country's economic growth is characterized not only by the volume and dynamics of the actual GDP, but also by the size of the income value of its citizens.

In 2017 the worldwide ratings of the countries demonstrated that the results achieved by Ukraine turned out to be as follow: 1) in terms of GNI per head of population in the group of medium-income countries Ukraine was classified to the 158th position (based on 216 countries, respectively), its index amounted to \$ 2.310. In order to compare the results, it's worth mentioning that: in Georgia GNI value per head of population amounted to \$ 3.810; in Belarus it amounted to \$ 5.600; in Bulgaria it amounted to \$ 7.470; in Russia it amounted to \$ 9.720; in Hungary it amounted to \$ 5.170; in Poland it amounted to \$12.680) [6]; 2) in terms of GDP value per head of population based on purchasing power parity (PPP) Ukraine was classified to the 112th position (based on 185 countries, respectively); 3) in terms of GDP value per head of population in constant prices as of year 2010 (US dollars) Ukraine was classified to the 124th position (based on 187 countries, respectively) [7]; 4) by nominal value per head of population (USD, \$) Ukraine was classified to the 132nd position (based on 190 countries, respectively) [8].

In our opinion, such significant Ukraine's lagging behind its neighboring regions and other European countries, related to the key indicators that characterize the state of its Economy, which are also supposed to be the general indicators of the population's level and quality of life, first of all happen due to the orientation of its economy to be supplied with raw materials and, consequently, its considerable dependence on the world conditions of energy markets, metal products, grain crops.

The export - orientated character of the economy has been considerably justified by the export-GDP ratio in Ukrainian economy, which amounted to 47,9 % in 2017 [9]. Meanwhile, nowadays the 2/3 ratio of commodity export value are the raw materials or products with low added value. Analysis of the dynamics of the structure of Ukrainian commodity export values demonstrated that the share of export values of machinery, equipment and transport (goods with the highest added value) in 2017 was 3,5 times less than the share of export values of agricultural products and foodstuffs, while in 2013 it was lower by 1,6 times [10] only.

In the meanwhile, the country's export structure makes significant impact upon the dynamics of indicators of its population's level and quality of life. In other words, the more products with high added value are concentrated within the country's export structure, the greater is its GDP value per head of population.

The agricultural products make nearly one third of our export turnover. The export turnover of the agricultural goods makes nearly 40% of the overall foreign exchange earnings of Ukraine. It means that the maintenance of macroeconomic stability in the country is significantly affected by the level of world prices for agricultural - industrial goods. Meanwhile, the actual ratio of the agricultural - industrial complex (AIC) within Ukraine's GDP continues to grow almost during all recent years, and nowadays it has been approaching the level of 20% already [10].

However, the higher is the level of economic development of the country, the smaller is the share of agricultural production within its GDP. For example, in terms of agricultural output, France takes the third position among the developed countries of the world and the first position in Europe, but the share of its agricultural production amounts to 2,4 % of its GDP [11] only. On the contrary, the poorest countries in the world have the largest share of agricultural production. Consequently, there is a clearly evident relation between the ratio in the country's agricultural GDP (being the industry with a low level of added value) and the well-being of its population - the higher this ratio is, the lower the level and quality of life have been demonstrated.

The evident consequence of the problems of the economy oriented to the raw materials is the fact that Ukraine has been turning into one of the poorest countries in the world. In particular, according to the annual data on the world well-being presented by the Global Wealth Report, Ukraine was classified to the 123rd position, based on 140 countries, in terms of personal wealth of its population ($ 1.563 as of mid-year 2018) [12].

Nowadays the main characteristics of Ukrainian economy are the existence of self-generating mechanisms for its inner degradation and conservation of low level and quality of life of its population [13]. This factor means that the population has been involved into the "poverty trap".

The well-known Norwegian economist E. Reinert believes that in order to provide the country with an opportunity to escape from the "poverty trap" and get involved into the spiral movement of the "self-accelerating" growth of well-being, the manufacturing industry should become the dominating branch of its Economy, where the concept of "increasing returns to scale" should be applied in its sectors. This concept means that any increase in production under the circumstances of unmodified technology causes the eventual decrease in average operating costs and growth of increasing returns from additional production factors, i.e. this concept helps provide the economy of scale in production. In case if some developed manufacturing industry makes part of certain Economy (Reinert calls it the "real goldmine" [14]), it becomes the main condition for increasing the wealth of any country.

Due to the actual results of development of those branches which provide the "increasing returns to scale", the Economy undergoes such positive changes as the decrease in unemployment, income growth of the population, technological progress in some other branches of industry, along with the impact of synergy, stabilization of currency exchange rate, increase in business profitability and increase in budget receipts. We believe that economic development nowadays has been actually determined by the growth in those branches of industry which provide the "increasing returns to scale". This very concept "contains a huge potential that can increase the welfare of the whole mankind" [14].

In case if the economy has been composed of the branches of industry with predominantly decreasing "returns" of production factors (in particular, the agricultural sector is meant), "...

then even under the actual GDP growth, the country is getting involved into a deeper "poverty trap" and is sure to stay blocked in its industrial backwardness even more. Under such circumstances, the GDP growth becomes unsustainable and can be easily transformed into eventual economic collapse, along with the depreciation of the national currency, escalation of inflation, increasing stratification of society depending of the people's income, because the bulk of the population becomes even poorer, and, consequently, we evidence those crises and revolutions that happen with greater frequency" [13].

Ukraine needs a new concept to provide it with rapid and sustainable economic growth, with obligatory revival of manufacturing branches of industry and new industrialization. This condition for successful implementation of this strategy will help it escape from the "poverty trap".

The essential first step in this direction means the further refusal from the dominating liberal paradigm, because it stands for refusal of the required and widespread governmental intervention into Economy. This system of economic views which forms the basis for the recommendations provided by the International Monetary Fund (IMF) and the World Bank, would not meet the actual modern realities, so, it has been rejected by the industrialized countries long ago.

The experience of the developed countries of Europe and the USA demonstrates that they have made a dramatic "strike" in their industrial development despite (and not due to) those liberal theories.

For decades, the developed countries used to implement the policy of banning or restricting the export operations dealing with raw materials, rejection to deal with the goods of "decreasing returns to scale", as well as they used to establish, protect and develop the domestic manufacturing industry. It is precisely because of this economic policy that they have managed to make the industrial "breakthrough" and become wealthy and prosperous countries.

It goes without saying that, first of all, the economic policy of Ukraine should reject the liberal theoretical recommendations and become oriented to the proposals based on the Keynesian and institutional theories of government regulation of its economy. A new concept of sustainable economic growth in Ukraine implies the pro-active economic government policy.

Nowadays, the countries that have achieved a high level of competitiveness and sustainable economic growth are those ones which provided the basis for their innovative economic development, yet they are not those countries which are the current leaders in the fields of extraction and sale of natural resources. The innovations have become the main source of economic growth. According to the opinion by E. Reinert, those "innovations, and not the savings and the capital itself, promote the well-being" [14]. The "trigger" for the innovations in contemporary economy is the manufacturing industry, with its core factor – i.e. the engineering, being the value-added activity.

Firstly, the significance of engineering has been specified by the fact that, being the most innovative branch of industry, it plays a key role in determining the level, dynamics and nature of development of productive powers of Society. Any achievements in technical and technological progress in Economy, consequently, improve the population's level and quality of life, and they would be impossible without developing the engineering sector.

In terms of economic development in the fields of innovations and technological achievements, Ukraine significantly lags behind both the industrialized and actively developing countries. The extremely low social productivity of labor is the logical consequence of the current situation. According to this factor, our country at this time lags behind not only the developed European countries (in 4 - 6 times), but its development is lower than the dynamics of Bulgaria, which is one of the poorest countries in the European Union (in more than 2 times) [15]. It should be noted if such a "gap" is not going to be "closed" in the coming 15 years, then, obviously, Ukraine would have never get any opportunity to catch up with not only industrialized European countries, but with actively developing countries, either.

The engineering branch serves as a basis for technical re-equipment of the entire Economy. It comprises the high-tech and knowledge-based industry to its maximum extent and, besides, it is able to create the advanced instruments of labor and technologies to be applied not only in all the other branches of industry, but also in the leading sectors of the whole national economic complex. It is precisely the way how the level of development of engineering makes its basic influence upon the growth of social productivity of labor.

Under the current circumstances, the engineering branch plays its "locomotive" role in the development of Economies of the leading countries of the world. The structure of industrial production of those countries contains the engineering/machine-building ratio equal up to 35 - 50% (in Germany it amounts to 53,6 %, in Japan it amounts to 51,5 %, in England it amounts to 39,6 %, in Italy it amounts to 36,4 %, in China it amounts to 35,2 %) [16]. (In order to compare: in Ukraine the ratio of engineering/machine-building ratio complex within the structure of industrial production is less than 14 %) [17, pp. 98-99].

The governmental policy in the field of innovations in industrialized and actively developing countries is aimed at providing the most favorable conditions to improve the performance of engineering/machine-building industry as the basis for increasing the innovative efficiency of the whole Economy.

One of the most important factors of the efficiency of innovative activity in engineering/machine-building sector is the cost component for Research and Advanced Development (R&D) issues. For example, in the United States, the share of machine-building sector is equal up to 20% of all R&D issues, and the annual investment amount in R&D in the field of engineering is 2,0 - 2,5% of GDP value. In EU countries, the average share of the annual expenditures on R&D in engineering/machine-building sector is up to 3% of GDP value [17, p. 94]. In Ukraine this factor (even at the mid-2000-ies, those years were the best period in terms of economic growth rates) was indicated at the level of 0,1% of GDP value [18].

The process of reforming the Ukrainian engineering/machine-building industry should be realized in several directions. Firstly, it's necessary to create the efficient mechanisms, in order to stimulate the investment promotion for modernization and technical re-equipment of the traditional branches of its industry (for example, introduction of special tax regime, when the tax rate for the enterprises of those branches should depend on the amount of investments into brand - new machinery and R&D issues). Secondly, it's necessary to develop strategies and mechanisms for managing the innovation processes at the machine-building enterprises, in order to ensure the priority development of knowledge-based and innovative technological processes for manufacturing the products with high value-added ratio.

Conclusions.

Nowadays, the priority focus has been set to the growth of agricultural industry, along with simultaneous neglecting of the problems related to the developing industrial potential; first of all, it will inevitably turn Ukraine into one of the countries of the world with an extremely low level and quality of life of its population, and, secondly, it is going to confirm the status of this country, being a "raw materials appendix" within the world economy.

The strategy for transition of Ukrainian economy from the raw-materials' export model to a competitive-and-innovative development model should be based upon acceptance of the need of thorough governmental intervention into the Economy. The governmental economic policy should be aimed at stimulating the development of high-tech industries that produce the competitive products with high value-added ratio, and first of all, we're speaking about the engineering/machine-building industry here.

The priority development of machine-building industry, which is the key source of the innovative development of Ukrainian Economy, should result in the revival of high-tech industrial facilities and stimulating the domestic consumer's demand.

The most important outcome of transition to a new economic development model will be the provision of sustainable economic growth in Ukraine, and, consequently, the positive dynamics in the level and quality of life of its population.

References

1. World Bank. Available at: http://pubdocs.worldbank.org/en/300571492021934576/data-ukr.pdf
2. National Bank of Ukraine. Inflation Report (January, 2019) . Available at: https://bank.gov.ua/control/uk/publish/article;jsessionid=D9941BD17ED1C1912CB72FA744C2DA84?art_id=63442015&cat_id=742185
3. "Europe Does Not Need Domestic Economy": Why is Ukrainian Economy is Increasingly Involved into Stagnation? Available at: https://nk.org.ua/ekonomika/evrope-promyishlennost-stranyi-ne-nujna-pochemu-ekonomika-ukrainyi-vse-bolshe-pogrujaetsya-v-stagnatsiyu-141690
4. P. Vernivskiy. New Time. - March, 26 (2018). - Available at: https://biz.nv.ua/experts/chto-ne-tak-s-rostom-vvp-2460126.html
5. World Bank. Available at: http://pubdocs.worldbank.org/en/426991542818421034/Global- Economic -Prospects-Jan-2019-Regional-Overview-ECA-RU.pdf
6. World Country Ranking in Terms of Gross National Income per Head of Population / Humanitarian Encyclopedia Available at: https://gtmarket.ru/ratings/rating-countries-gni/rating-countries-gni-info
7. World Data Atlas. Available at: https://knoema.ru/atlas
8. World Country Ranking in terms of GDR per head of population in 2017. Available at: http://investorschool.ru/rejting-stran-po-vvp-na-dushu-naseleniya-2017
9. State Statistics Committee of Ukraine. [Electronic resource]. - Available at: http://www.ukrstat.gov.ua/
10. State Tax Service of Ukraine. Available at: http://sfs.gov.ua/ms/f2
11. Structure and Features of French Economy / Economics, Belorussian State Economic University. Available at: https://www.economy-web.org/?p=382
12. Ukraine Was Classified the 123rd Position in Terms of Well-Being of its Population. Available at: https://news.finance.ua/ru/news/-/436881/ukraina-zanyala-123-mesto-v-mire-po-urovnyu-bogatstva-grazhdan
13. T. Unkovskaya. Mirror of the Week. 6, (2016). Available at: https://zn.ua/macrolevel/kak-ukraine-vyrvatsya-iz-lovushki-bednosti- .html
14. Reinert E. How Rich Countries Got Rich and Why Poor Countries Stay Poor Available at: http://uaterra.in.ua/wp-content/uploads/2017/11/Rajnert-Eryk.pdf
15. International Labour Organization. Available at: https://www.ilo.org/global/lang--en/index.htm
16. Polovinkin V.N., Fomichev A.B. Actual State and Problems of Development of Domestic Engineering Industry. Available at: http://www.proatom.ru/modules.php?name=News&file=print&sid=4639
17. Gerasimchuk V.I. Marketing and Management of Innovations. 3, 84-104 (2015). Available at: http://mmi.fem.sumdu.edu.ua
18. Danilishin B. New Time - April, 24 2015. - Available at: https://nv.ua/opinion/kak-ostanovit-agoniyu-ukrainskoy-ekonomiki-45647.html

Determinants of Social Entrepreneurship in Ukraine

Nataliia Gavkalova [1*], *Oleg* Amosov[2], *Alona* Zolenko[3] and *Iryna* Sierova [4]

[1]"Karazin Business School" V.N. Karazin Kharkiv National University
Management and Administration Department. Kharkiv, Myronosytska str. 1
[2]Kharkiv Regional Institute of Public Administration of the National Academy of Public Administration (Kharkiv)
Public Theory and Finances Department, Kharkiv, Moskovsky av. 75
[3]Simon Kuznets Kharkiv National University of Economics (Kharkiv)
Public Administration and Regional Economy Department, Kharkiv, Nauky av. 9-A
[4]Simon Kuznets Kharkiv National University of Economics (Kharkiv)
Economic Theory, Statistics and Forecasting Department, Kharkiv, Nauky av. 9-A

Abstract. The high importance of social entrepreneurship for the development of national and regional economies is widely recognized. This article contributes to the emerging theoretical discourse of social entrepreneurship by explicating the representatives of different schools. Social entrepreneurship occupies a third sector between the private sector of the economy and the non-profit sector as it aims to solve social problems. Social entrepreneurship functions as an organization that self-sustains at the expense of the innovative organization of its core business. The article suggests holistic approaches to entrepreneurship research based on entrepreneur-associated factors, characteristics of the political and legal environment. These determinants can have direct or indirect influences on entrepreneurial processes mandatory for the development of social entrepreneurship.

To determine the strengths and weaknesses of the development of social entrepreneurship in Ukraine, as well as its opportunities and threats, the SWOT analysis was carried out. The actions required to improve the current state of social entrepreneurship were justified. The detection of threats and opportunities has made it possible to identify priorities for the development of social entrepreneurship in Ukraine, taking into account existing risks.

Introduction

The transition of certain functions from the government to business plays a crucial role in the development of modern society under the current economic conditions and intense competition. This phenomenon could be due to a number of factors. The society in economically developed countries has begun to impose higher requirements for corporate social responsibility and many national and regional policy initiatives aim to promote

*Corresponding author: ngavl@ukr.net

 https://doi.org/10.1051/shsconf/20196706017

business start-ups and foster their growth as newly-founded firms contribute to the dynamism of the economy [1] whereas business representatives are getting aware of the need for a wider range of social functions and determinants. The combination of these factors has contributed to the creation of a social enterprise, that is, third sector organizations whose activities are aimed at solving or mitigating the problems of society.

Literature Review

Research on social entrepreneurship as a phenomenon has evolved and has been refined over the past few decades. However, before defining "social entrepreneurship|", it is crucial to determine the nature of the categories "entrepreneurship" and "entrepreneur" as well as its features and functions:

entrepreneurship is an interdisciplinary category and serves as an object of research in economics, management, sociology, psychology, and other socio-humanities;

entrepreneurship is an independent factor in production;

entrepreneurship has recently become a type of professional activity;

entrepreneurial skills are an integral part of innovation development. J. Schumpeter indicates that an individual can be referred to as an entrepreneur only if he or she utilizes a new mix of resources [2];

an entrepreneur is an innovator who organizes and coordinates economic resources.

Modern society faces a number of unsettled issues in the non-profit social sector. Third sector organizations in the form of public-private partnership and social entrepreneurship exercise socially oriented activities and launch various projects to meet the needs of the society.

In the study which set out to classify the third sector organizations activities, four main schools of thought have been distinguished – the English one, the European one, the socio-innovative one and the socio-entrepreneurial one. The representatives of the social innovation school of thought (S. Elward, J. Mayr, I. Marty, J. Thompson) consider social entrepreneurship as an activity of individuals who are responsive to social needs by innovative methods. The subject of the study of representatives of the socio-entrepreneurial school (R. Anderson, R. Dart, R. McDonald) is non-profit organizations that serve the non-profit social sector and receive additional income therewith. Within the framework of this school, scientists are engaged in the search for new ways of funding non-profit organizations, as well as implementing effective management methods in the activities of socially-oriented enterprises. The representatives of the European school (I. Vidal, R. Speer, U. Stefan, L. Ulaner), determine the purpose of socially oriented activities as the satisfaction of the society's interests and emphasize the impossibility of the entrepreneurship existence without the strong state support. According to the approach of the English school of thought of social entrepreneurship (L. Darby, S. Dixon, A. Clifford, K. Linging, D. Turner, H. Jenkins), solving social problems should rely entirely on business, with the proceeds being directed, first of all, for the satisfaction of social needs.

Factors that both enable and constrain innovation processes to meet the financial, social, cultural, and political expectations of stakeholders of social entrepreneurship ventures have been explored in several studies, such as additional sources of resistance that stem from its alternative, or multiple, institutional logics; greater dependence of social entrepreneurship on stakeholder support, as well as the measurement of social entrepreneurship success across the dual objectives of social impact and financial self-sufficiency; legitimacy of social entrepreneurship as a form of organizing and method of social change [3].

Although social entrepreneurship is not an entirely new phenomenon, there is very little scientific understanding of the mechanisms by which third sector organizations function and develop. However, as the report on the large-scale study of social enterprises in Europe, "A map of social enterprises and their eco-systems in Europe" points out, much of the research up to now has been unsatisfactory [4]. The main purpose of this mini research is to develop an understanding of the social entrepreneurship and determinants and analyze the dynamics of the social entrepreneurship development in Ukraine.

Social entrepreneurship is a new way of social and economic activity, which combines the social purpose of the organization with entrepreneurial innovation and a well-balanced structure for self-sustainability. According to K. Alter, the functioning of social entrepreneurship is based on social enterprises that are formed to address a particular social problem or problems and operate on the basis of innovation, financial discipline and business behavior adopted in the private sector [4, p. 12].

If we focus on the economic consequences of this phenomenon, social entrepreneurship increases overall economic efficiency, since it introduces into the economic turnover resources that were previously not used in this quality. This, first and foremost, concerns unused material and human resources - waste products, socially formed groups, solidarity and trust of people when they are united by a common goal.

Data from several foreign sources concerning the formation and development of a social enterprise have identified that the focus of social projects has changed significantly. Traditionally, many social projects were aimed at working with people with disabilities, nowadays more and more newly-formed businesses deal with issues of emigrants, national minorities, and ecology. These data confirm the state's interest in the formation of social institutions, which will contribute to the more effective implementation of the state policy.

Results

It is apparent that social entrepreneurship is a relatively new concept that combines many different determinants of the third organizations' activities. The primary mission of social entrepreneurship is the creation of social value and the pursuit of social change. Therefore, any forces that limit or constrain the social innovation processes are perceived by researchers as negative because they reduce the positive social change that could potentially be achieved [5]. Consequently, social entrepreneurship is an activity of enterprises that uses innovations and functions to solve any social problem or create benefits for society, rather than maximizing profits for the enterprise owners.

Conceptual provision for the formation and development of social entrepreneurship is presented in Fig. 1

The main task of social enterprises is to create positive changes that can be reflected in terms of measurable indicators. Social impact measurements serve as a tool for assessing, presenting, improving the activities of a social enterprise and attracting clients and investors.

Taken together, the following conclusions can be drawn from the present study, the main determinants of social entrepreneurship are the following factors, such as

favorable political climate;

adequate legal environment;

innovation and scientific infrastructure;

dynamic and active socio-cultural environment.

It should be emphasized that the political and legal environment provides opportunities for social entrepreneurship formation, while the socio-cultural environment creates favorable conditions for its existence, and government institutions stimulate social entrepreneurship growth and promote its further development.

https://doi.org/10.1051/shsconf/20196706017

The findings of this study suggest that social entrepreneurship can be characterized by the following features: the predominance of a social mission over a commercial component; the pursuit of long-term changes in the social sphere (its reformation); competitiveness, self-sustainability and cost-effectiveness; high responsibility of a social entrepreneur for the results of his/her activity; the use of innovations to solve social problems. Regarding the formation and development of social entrepreneurship in Ukraine, it should be recognized that the relevance of this study is confirmed by the lack of both legislative and methodological basis.

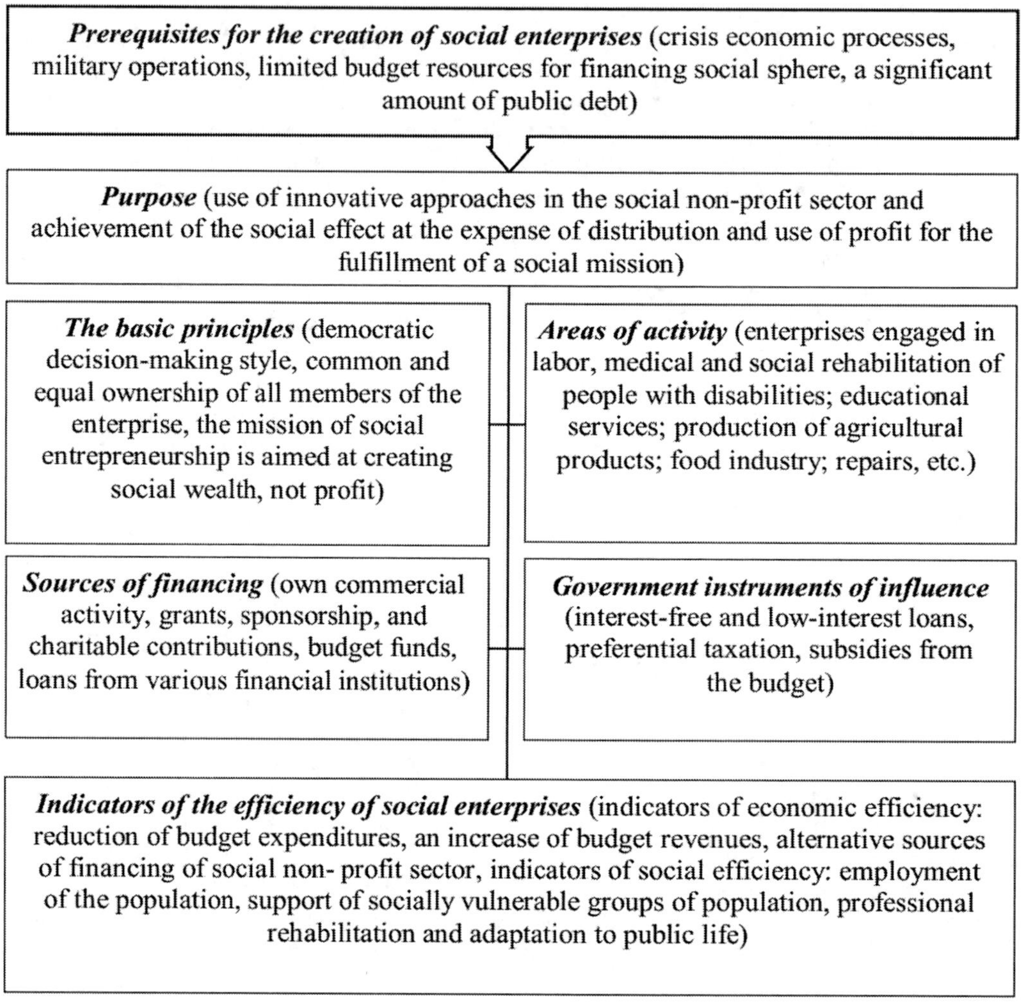

Fig. 1. Conceptual provision for the formation and development of social entrepreneurship.

Thus, according to most researchers [5-14], the proliferation of social enterprises in the country is hampered by the lack of legislation regulating their activities and the uncertainty of key concepts and criteria for identifying such enterprises.

To identify all external and internal aspects of the development of social entrepreneurship, it is expedient to conduct a SWOT analysis (Table 1). The implementation of such an analysis will allow identifying the current state and situation of the development of the social entrepreneurship sector in Ukraine. In addition, it will help to realize how to counteract threats and reduce risks, as well as how to effectively use the existing capabilities.

According to the SWOT analysis, it becomes clear that the obstacles to the development of social entrepreneurship are the following: the lack of specialized services and support; the impossibility to effectively compete in the field of public procurement; the limited access to investment financial resources; the insufficient qualification of employees, high rents and administrative expenses, etc.

Table 1. SWOT-analysis of the development of social entrepreneurship

Strengths	***Weaknesses***
- participation in continuous economic activity; - a clear social goal that benefits society; - employment of vulnerable groups of the population; - ensuring inclusive management and/or democratic decision-making processes; - provision of socially meaningful services; production of socially significant goods; - high responsibility for the results of their activities to the population.	- absence of regulatory and legal support; - poor awareness of citizens, enterprises, and organizations about social entrepreneurship; - the low motivation for the creation and operation of social enterprises; - low level of publicity and imperfect reporting.
Opportunities	***Threats***
- self-organization of social entrepreneurs for the purpose of exchange of experience; - influence on the formation of a favorable legal environment and effective public policy; - commercial efficiency, self-sustainability, and competitiveness; - innovation, innovation in the combination of social and economic resources; - founding business schools and conducting training for social entrepreneurship start-ups.	- lack of support from state and local authorities; - the ineffective policy of informing the population about social entrepreneurship; - low level of personnel qualification and lack of competent specialists for the enterprise organization; - A lack of management mechanisms that ensure the priority of the social goal including the interests of various stakeholders.

Therefore, the government support provided to social enterprises will ensure a systemic precondition for the development of the social entrepreneurship sector in Ukraine and public-private partnership in providing social services to the population. In addition, officially identifying and legitimizing the functioning of social entrepreneurship will help to define their activities, goals, tasks, etc.

Conclusions

This research offers some insight into the development of social entrepreneurship in Ukraine having applied modern analytical tools. Conceptual provision for the formation and development of social entrepreneurship has been presented. By means of voluminous analysis, it is possible not only to identify strengths and take into account weaknesses but also to substantiate all existing threats and opportunities for a social enterprise in Ukraine. To summarize, this article opens opportunities for further research using theoretical approaches of social entrepreneurship, determinants that can be used at the analysis of the influence of the different factors in the process of providing social entrepreneurship and the results of SWOT analysis to forecast development of the social entrepreneurship in Ukraine.

References

1. Christine Tamásy (2006) *Determinants of regional entrepreneurship dynamics in contemporary Germany: A conceptual and empirical analysis, Regional Studies*, 40:4, 365-384, DOI: 10.1080/00343400600612137

2. Schumpeter, J. A. The Theory of Economic Development. Cambridge, MA: Harvard University Press (1934)

3. How Context Shapes Innovation, Journal of Social Entrepreneurship, 5:2, 192-213, DOI: 10.1080/19420676.2014.889739

4. K. Alter, *Social enterprise typology* (VV LLC, Wilmington, 2007)

5. Elizabeth Chell, Katerina Nicolopoulou & Mine Karataş-Özkan (2010) Social entrepreneurship and enterprise: International and innovation perspectives, , 22:6, 485-493, DOI: 10.1080/08985626.2010.488396

6. 3 Z. I. Halushka, Fenomen sots. pidpr-tva: poniat.ta persp. rozv. v Ukr., Visnyk KNU im. T. Shevchenka, 148 (2013)

7. N. M. Horishna, Fenomen sots. pidpr-va: sutn. i kryt., Naukovyi visnyk Uzhh. univ, 2, 39 (2016)

8. Dzh. Kikal, T. Layons, Sots. predpr-vo: missiya – sdelat mir luchshe (Alpina pablisher, Moskva, 2014)

9. M. Naumova, Persp-vy i prob- rozv. sots. pidpr-va v Ukr., Visnyk KNU im. T. Shevch., 8, 185 (2016)

10. D. Bornstein, S. Davis, *Social Entrep-ship: What Everyone Needs to Know?* (Oxford University Press, 2010)

11. J. Mair, I. Marty *Social entrepreneurship research*, JWB, 41 (2006)

12. A. Nicholls *Playing the field: A new approach to the meaning of social entrepreneurship*, SEJ, 2.1 (2006)

The conditions of the balance of the national labor market in the process of institutional reform

Nataliia Gavkalova [1]*, *Volodimyr* Sybirtsev [2], and *Iryna* Sochynska-Sybirtseva [2]

[1]Department of Public Administration and Regional Economics, Simon Kuznets Kharkiv National University of Economics, av. Nauky 9-A, 61166 Kharkiv, Ukraine
[2]Department of Economics, Management and Business, Central Ukrainian National Technical University, Universytetskyy pt. 8, 25006 Kropyvnytskyi, Ukraine

Abstract. The aspects of implementation of the processes of functioning and development of the national labor market in the context of manifestation of its internal complexity and multifunctionality are determined. The regularity of the growth of needs in the development of the institutional structure of the national labor market in the conditions of transformation and restructuring of the system of socio-economic relations is proved. The existence of a multiplicative effect of distribution, dissemination of new institutional practices implemented on the national labor market, institutional regulation of a wide range of economic operations, phenomena and processes has been established. The substantive essence of the process of building the institutional space of the national labor market is determined, which consists in institutionalizing (institutionalizing) the processes of socialization of socio-economic relations related to productive employment, in the context of which there is an increase in the significance and role of human resources in the processes of creation of economic goods, expanded social reproduction, etc.

1 Introduction

Large-scale institutional reforms in various spheres of Ukrainian economy are conducted in a context of constant acceleration of changes in the technological mode of production, which are determined by the expansion of the creative content of labor and the increase in the level of freedom of relations between workers and means of labor, which in turn leads to an increase of needs for updating the entire system of industrial relations. One of the most important areas of reforms aimed at satisfying these needs is the national labor market, the state of development trends, the balance of which depends on the achievement of efficient allocation and use of labor resources, support for productive employment and welfare of population, solving the key issues to ensure the stability of economic development.

2 Analysis of recent research and publications

[1]Corresponding author: ngavl@ukr.net

Different conditions, principles, and aspects of the implementation of institutional reforms in the national economy were considered in the works of many domestic and foreign economists such as Weingast B. [12], Vellis D. [12], Winter C. [11], Heyets V. [1], Gritsenko A. [2], Greenberg R. [3], Zhalilo Y. [4], Zhukov L. [5], Kolesnichenko I. [6], Mazaraki A. [1], Nelson R. [11], North D. [12] etc. Problems and contradictions in the process of reforming the labor market in the conditions of market transformations of the national economy were studied by such researchers as Zvonar V. [7] Kurilo I. [7], Libanova E. [7], Lisogor L. [7], Makarova O. [7], Marchenko I. [8], Marshavin Yu [9], Pyshchulina O. [10] etc . However, the justification of a set of conditions for balancing the national labor market through institutional provision of its development on the basis of activation of reforming processes remains limited and requires further thorough research in this area.

3 Problem statement

The purpose of this study is to justify and develop a set of conditions for balancing the national labor market in the process of institutional reforms.

4 Research results

A comprehensive assessment of the principles of formation, state and current trends in the development of labor market as a component of national economy made it possible to distinguish the following key trends: deterioration of quality and stagnation of the process of development of human capital; deepening the imbalance between market demand and supply of labor; increasing the amount of hidden employment and increasing the scope of labor migration; reducing the size of real incomes of wage earners and correspondingly reducing the welfare of households; deterioration of the quality of life and safety of employees`life, etc.

According to the results of in-depth analysis of the dynamics of indicators of the state of national labor market in Ukraine for 2013-2017 (Table 1), an integrated and systematic nature of the impact on labor market, transformational changes associated with the implementation of the processes of market reform of the national economy was established.

In particular, the deterioration of economic conditions in the context of a significant aggravation of social contradictions and the growth of manifestations of crisis phenomena naturally conditioned the stagnation of economic activity of population. For the period of 2013-2017, the number of economically active population of working age has decreased from 19399.7 people to 17193.2 people (by 11.4%).

The number of employed population during 2013-2017 decreased by 16.3% (from 19314.2 to 16156.4 people). Another negative consequence of the manifestation of crisis phenomena in the socio-economic development of Ukraine should be the spread of the use of various forms of temporary or hidden (shadow) employment. Confirmation of the scale of scope of hidden employment for the studied period should be considered as a fairly high level of the proportion of economically inactive population in the total population of working age (the variation of values of this indicator ranged from 71.5% to 72.9%), of which only less than half falls on the number the unemployed population, which was taken into account by relevant state authorities (from 7.8% to 9.7%). The number of unemployed people of working age increased during the period under study from 1510.3 to 1697.3 people, i.e. 12.4%.

Thus, the emergence and spread of problems of functioning and development of national labor market is a direct consequence of the deployment of complex processes of social reforms, without which it seems almost impossible to successfully overcome a wide

https://doi.org/10.1051/shsconf/20196706018

and varied range of crisis phenomena, which, in turn, is a reflection of not only the current

Table 1. Indicators of the state of the national labor market in Ukraine in 2013-2017.

Indicator	Value of the indicator, over the years					Rate of change (%), over the years			
	2013	2014	2015	2016	2017	2014/ 2013	2015/ 2014	2016/ 2015	2017/ 2016
Economically active population (aged 15-70), thsd. persons	20824,6	19920,9	18097,9	17955,1	17854,4	-4,3	-9,2	-0,8	-0,6
Economically active population of able-bodied age, thsd.persons	19399,7	19035,2	17396,0	17303,6	17193,2	-1,9	-8,6	-0,5	-0,6
The employed population	19314,2	18073,3	16443,2	16276,9	16156,4	-6,4	-9,0	-1,0	-0,7
Economically inactive population, thsd. persons	11270,1	12023,0	10925,5	10934,1	10945,0	6,7	-9,1	0,1	0,1
Unemployed population of able-bodied age, thousand persons, from which	1510,3	1847,1	1654,0	1677,5	1697,3	22,3	-10,5	1,4	1,2
looking for work, trying to organize their own business	1537,2	1801,6	1617,4	1634,8	1662,5	17,2	-10,2	1,1	1,7
been employed during the year	541,9	494,4	444,7	409,0	398,6	-8,8	-10,1	-8,0	-2,5
Labor force demand, thousand people, incl	47,5	35,3	25,9	36	50,4	-25,7	-26,6	39,0	40,0
skilled	41,3	30,3	23,2	31,8	43,9	-26,6	-23,4	37,1	38,1
low-skilled	6,2	5	2,7	4,2	6,5	-19,4	-46,0	55,6	54,8
Load of registered unemployed at one vacant job (vacant post), persons / units	10	14	19	11	7	40,0	35,7	-42,1	-36,4

complications in business of domestic enterprises, but also a reflection of the accumulated imbalances and disproportions accumulated in previous years in the national labor market.

Modern conditions of functioning and development of national economy are marked by the emergence of a large number of new features of the organization of labor, management relations, the emergence of new non-standard forms of employment, dictated by the comprehensive development of productive forces, in particular, information and communication technologies, changes in industrial relations, increased competition in the world markets in the context of tendencies of globalization and postindustrialization of society. The structure of the regular trends in the development of forms of employment and employment should include the following: new forms of command organization of labor; projected labor organization of workers, in which the erosion of clear job descriptions and functional duties of employees takes place; the movement of production and management toward the formation of a cluster-network model; in this connection there are changes in the style of personnel management in the direction of autonomy of labor; the emergence and growth of the proportion of the remote type and new non-standard forms of employment. The spread of such innovations significantly affects the development of national labor market and leads to new requirements regarding the functioning of its system of regulation, which requires a detailed study of features and content and interconnection of the conditions of balance of national labor market in the process of its institutional reforming.

Condition 1. The national labor market is an extremely important, fundamental element of economic mechanism of market economy, which is in a complex relationship with all other markets, interconnected with a wide range of production-economic operations and processes, which in the complex constitutes a productive activity of society.

Condition 2. The national labor market due to the extremely high degree of complexity of its internal construction, the variety of internal and external factors that form the trajectory of its transformations, multifunctionality, as well as existing ties and relationships with other social entities, is characterized by a complex of processes of functioning and development that, first and foremost is reflected in the following aspects. Firstly, the national labor market should function as a mechanism for balancing (between employees and employers) the supply and demand supply parameters as the object of market exchange, a result of realization of which –the certain human potential (human capital, labor resources, labor potential, etc.) becomes a productive work, which, in combination with means of labor, creates a new value in the form of useful goods. Secondly, the national labor market should act as an important element of the system of production relations, influencing, at the same time, the techno-economic (the way of organizing the production forces, the state of development of factors of production, etc.) and socio-economic (methods and forms of ownership of the means of production, the combination of labor and means of production, the order of the factors of production and the appropriation of the results of productive use of these factors, etc.) are the components of these relationships. Thirdly, the national labor market, as the primary source for the formation of certain types of relations between members of society regarding the establishment of conditions for their participation in the process of creating economic wealth, should be a mechanism for the unification of socio-economic roles, functions, models and strategies of human behavior in economic sphere should be the mechanism of their goals and interests, motives and incentives, the basis for the establishment of a system of criteria and norms for assessing the utility and costs of goods as a whole.

Condition 3. In the system of national economy, the labor market occupies a special place and plays an important role, and the processes taking place in this market have a powerful (albeit indirectly) influence on socio-economic development of the entire society. In the context of determining the nature and extent of this impact, two basic provisions should be noted. Firstly, the agreements on hiring of employees concluded in the national labor market determine the size of a new value created in the process of productive employment, that is, directly determine the dynamics of changes in incomes and the welfare of the vast majority of households, which determines the volume and structure of domestic demand for goods and services, and, respectively, the pace of economic development of national economy. Secondly, the state's ability to fulfill a number of important social commitments to society is directly related to the economic conditions of productive employment of population (the cost of labor in the market). The need to ensure social protection of hired workers both in the process of productive employment and in the loss of them (for various reasons) of employment opportunities in the course of humanization and the socialization of industrial relations must become an integral part of the institutional framework for functioning and development of modern labor market.

Condition 4. The reform of national labor market on the basis of the development of mechanisms of institutional regulation primarily involves the development and improvement of the relevant institutional space within this market (a complex of institutional formations), adequate and appropriate to the requirements of adapting the system of industrial relations existing in society in the course of market transformation of national economy, to the conditions of the latest technological method of production, inherent in the modern stage of civilizational development. Such construction of the institutional space should consist of institutionalizing the processes of socialization of socio-economic relations connected with productive employment.

Condition 5. State support for the institutional reform of national labor market is to facilitate and participate in the creation and operation of a multi-agent information system for organizing the interaction and cooperation of labor market participants (a single, coherent, coordinated, open, self-regulated entity based on harmonizing and streamlining the balance of expectations and competitive -cooperative interaction of economic agents). The main element of such a system should be the mechanism of state regulation of institutional transformations of national labor market.

Condition 6. The mechanism of state regulation of institutional transformations of national labor market is an integrated complex of formal institutional norms and rules (a static aspect of institutional influence of the state) and institutional bodies and organizations (targeted activities to ensure the observance of formal institutions - the dynamic aspect of the institutional influence of the state). Functional filling of this mechanism is primarily connected with the implementation by the state, in the person of its authorized bodies and organizations of initiative, adaptation, distribution and control functions of institutional reforms.

Condition 7. The choice of the forms and methods of state incentives for institutional reforms in the national labor market must be based on the ordering of the institutional space of interaction of market participants using the architecture and infrastructure of information multiagent system that is focused on the search and selection, the intensive and effective use of innovations in the field of education (for example, the provision of continuous training and retraining) and the introduction of other forms of transitional measures to increase the human capital of the company.

Condition 8. Reform measures in the field of institutional transformation of national labor market must first of all take into account the state of socio-cultural norms, customs and traditions existing in society and be based on the influence (in relation to specific new institutional formations) to facilitate their adaptation to the changes of the corresponding institutional space and creation of a complex of mental norms and values, moral and ethical foundations of effective social and labor relations necessary for the formation of an institution of production relations, adequate to the requirements of the latest technological method of production.

5 Conclusion

The article describes a set of conditions for the balance of institutional reforms of national labor market, which is considered in the context of the manifestation of inherent complexity, multifunctionality, and also the diversity of its relations and relations with other social entities. The multidimensional implementation of the processes of functioning and development of national labor market must necessarily be taken into account when developing the system, mechanisms and tools for its regulation.

Institutional changes in the national labor market are characterized by the presence of a multiplicative effect of disseminating and popularizing new institutional formations to regulate a wide range of business operations, phenomena and processes that go well beyond the boundaries of this market. An important tool to overcome the problems, contradictions and limitations related to the need to adapt industrial relations to the requirements of the latest technological method of production is to improve the institutional basis of the relevant economic processes. Due to the importance of influence of this market on a complex system of socio-economic relations in society and through the manifestation of the multiple effect of the distribution of institutional practices described above, the success of such measures depends on the progressive institutional reform of national labor market.

Methodological aspects of choice of forms and methods of state incentives for institutional reforms in the national labor market require further research in the context of the implementation of the proposed set of conditions of balance.

References

1. V. Heiets *Priorytety natsionalnoho rozvytku v konteksti hlobalizatsiinykh vyklykiv* : monohrafiia (Kyiv : Kyiv. nats. torh.-ekon. un-t, 2010).
2. A. Hrytsenko *Ynstitutsionalnaia arkhitektonika i dinamika ekonomicheskikh preobrazovanyi* : monohrafiya (Kharkiv : Fort, 2008).
3. R. Hrinberh Ekon. Ukr. *Instytutsiini uroky rynkovykh transformatsii.* **1**, 27-38 (2012).
4. Ya. Zhalilo Teoriia ta praktyka formuvannia efektyvnoi ekonomichnoi stratehii derzhavy : monohrafiia (Kyiv : NISD, 2009).
5. L. Zhukova Probl. *Ekon. Sotsialni chynnyky evoliutsii derzhavnykh instytutiv*, **2**, 298-303 (2017).
6. I. Kolesnichenko Bizn. Inf. *Instytutsionalni zasady efektyvnoi ekonomichnoi polityky v Ukraini*, **3**, 8-15 (2017).
7. E. Libanova, V. Zvonar, I. Kurylo, L. Lisohor, O. Makarova *Liudskyi rozvytok rehioniv Ukrainy : analiz ta prohnoz : kolekt.* monohrafiia (Kyiv : In-t demohrafii ta sotsialnykh doslidzhen NAN Ukrainy, 2012).
8. I. Marchenko *Infrastrukturna pidtrymka rozvytku rynku pratsi Ukrainy* : monohrafiia (Kyiv : In-t demohrafii ta sotsialnykh doslidzhen imeni M. V. Ptukhy NAN Ukrainy, 2013).
9. R. Nelson, S. Winter *An evolutionary theory of economic change* (Harvard university press, 2009).
10. D. North, J. Wallis, B. Weingast *Violence and Social Order: A Conceptual Framework for Interpreting Recorded Human History* (New York : Cambridge University Press, 2009).

Transition to a New Economy: Transformation Trends in the Fields of Income and Salary Functions

Olena Grishnova[1] , *Andriy* Cherkasov[2], *Olena* Brintseva[3*], *Olena* Sakovska[4], *Anatoly* Ostapchuk[5]

[1]Taras Shevchenko National University of Kyiv, Ukraine
[2]Institute of Personnel Training at the State of Employment Service, Ukraine
[3]State Higher Educational Institution "Kyiv National Economic University named after Vadim Getman", Ukraine
[4] Uman National University of Horticulture, Ukraine
[5] National University of Life and Environmental Science, Ukraine

Abstract. The rapid spread of information technologies and other phenomena in the new economy causes significant changes in the social and employment spheres. The objective is mainly to analyze and systematize the peculiarities of labor and payment for it in the new economy. In this paper are used the analytical grouping method for characterizing salary functions, as well as changes in the social and labour sphere that take place in the new economy, correlation-regression analysis to determine the relationship between average income and factors affecting its level. The results of the paper show three important findings. Firstly, the gradual changes in the content of labour and forms of its organization that happen in modern conditions under the influence of dissemination of information and communication technologies require new approaches to income adjustment. Secondly, the authors also found that the fair employee compensation, ensuring productive employment in modern conditions, is a challenge not only of the national but also of international scale. Thirdly, the prospective directions of reducing inequality and social tension in society is the guarantee of the basic unconditional income; progressive tax on all types of income; reduction of the length of working time while maintaining payments and improving productivity.

INTRODUCTION

The rapid dissemination of information technologies and other phenomena in the new economy causes significant changes in the social and employment spheres. The noted events are primarily related to the changes in the content of work and forms of its organization, the emergence of new professions, in particular in the IT field, extensive development of high-

* Corresponding author: nadya-reznik@ukr.net

technology sectors of the economy, the expansion of opportunities for the formation and use of human capital, an increase in the options of human choice, but also the emergence of new risks. Gradual changes in the content of work and its organization forms also result in the transformation of the integrity and functions of remuneration. New criteria for assessing the personal labour input of employees lead to changes in compensation, forming different, rather diversified sources of income. A modern employee with the help of information and communication technologies can simultaneously participate in many projects. In the new economy, the diversity and flexibility of the employment conditions increases, an employee can enter into individual contracts. Furthermore, in the new conditions, the employee has the highest degree of freedom and independence, professional networks of specialists and freelances are becoming widespread, and the decentralization of business processes and management increases. However, changes in the social and employment sector, inherent in a new economy, can also result in a decrease of employment stability, less social guarantees for employees, an increase in the inequality of compensations and incomes in general.

During the 20th century, work from hard duty to satisfy the needs of survival gradually turned into a way of development and self-actualization. It happened for the following reasons: the majority of the population was engaged in hired labour; work and profession became the axis of human existence, determining the position in a society, income, lifestyle, social circle, etc.; the changes lead to the formation of a special employment democracy, employment values; people started to view employment as a source of property and well-being, it became the centre of economic progress.

At the same time, tireless work and its ever-increasing productivity leads to a steady decline in demand for human resources. Fewer people can produce more and more goods and services (just in the past 150 years, labour productivity has grown more than 50 times). Due to the automation of production and the development of robotics, deserted manufacturing sites are no longer surprising. According to forecasts, in the next decades, there will be an ongoing decline in manufacturing employment and it will approach the mark of 2-3%. The emergence of each new, more efficient workplace destroys dozens and hundreds of old ones. Therefore, modern economic growth, based on scientific and technological progress, is "growth without jobs." The overall objective of the paper is to analyze and systematize the peculiarities of labor and payment for it in the new economy, to investigate the gradual transformation of its functions, to identify the trends in changes of incomes of the population under the influence of the computerization of human life, and to substantiate the directions of agents' adaptation to these changes at different levels.

PRESENTATION OF THE MAIN MATERIAL

Salary and Wage Functions in Traditional and New Economy

At the same time, we cannot argue that, in modern conditions, there are revolutionary changes in the social and employment sphere. These, albeit very deep transformations that take place are, rather, *evolutionary,* as they are a natural consequence of social development – technological, economic, social progress, which has acquired special, inherent features with the advent of computer and the development of global Internet.

Considering the transformation of the classical functions of salaries and wages in the context of the transition to a new economy, it is necessary to distinguish the following distinctive features. Thus, society increasingly expects business to fulfill the principle of social responsibility and participation in solving important social issues. It is this principle that should become the basis for the image and mission of the company. 77% of the interviewed respondents called social responsibility an important or very important business component. Despite the connection between the social impact and financial performance of the company, only 18% of respondents regard social responsibility as the main priority of the corporate strategy. 34% of companies do not have enough social programs or such program's lack financing, and 22% do not pay attention to it at all.

Transformation Trends in the Field of Income and Salary in the Transition to a New Economy in the EU Member States and Ukraine

In order to identify and assess the transformation trends in the field of incomes and employee compensation in the transition to a new economy, we examined the interconnections and dynamics of the average income and the main factors influencing it (working time, average duration of training and social transfers), in the EU member states and in Ukraine. The distribution of annual income in the Eurozone is rather uneven: so the income in the "richest" and "poorest" countries can vary 10-20 times. In Ukraine, we determined this indicator, taking into account the average annual euro rate for the relevant year; the rapid drop in the course of the national currency in crisis years had a negative effect on the dynamics of average annual incomes.

Table 1. Change of Average Income in EU Member States and Ukraine in 2008-2016, Euro per year.

Country	Year								
	2008	2009	2010	2011	2012	2013	2014	2015	2016
Belgium	17,985	19,313	19,464	20,008	20,280	21,483	21,705	21,654	22,295
Bulgaria	2,171	2,828	3,017	2,914	2,860	2,924	3,311	3,332	3,151
Czech Republic	6,068	7,295	7,058	7,451	7,791	7,694	7,622	7,423	7,838
Denmark	24,161	25,029	25,668	26,944	27,184	27,444	27,861	28,364	28,665
Germany	18,309	18,586	18,797	19,043	19,595	19,582	19,733	20,668	21,275
Estonia	5,547	6,209	5,727	5,598	5,987	6,579	7,217	7,889	8,645
Ireland	22,995	22,445	20,512	19,726	19,748	19,856	20,169	21,688	22,407
Greece	10,800	11,496	11,963	10,985	9,513	8,371	7,680	7,520	7,500
Spain	13,966	14,795	14,605	13,929	13,868	13,524	13,269	13,352	13,681
France	18,899	19,644	19,960	19,995	20,603	20,924	21,199	21,415	21,713
Croatia	:	:	5,810	5,579	5,377	5,078	5,225	5,453	5,726
Italy	15,640	15,605	15,964	15,971	15,979	15,733	15,759	15,846	16,247
Cyprus	16,024	16,451	16,180	16,990	16,927	15,873	14,400	13,793	14,020
Latvia	4,740	5,355	4,488	4,195	4,450	4,666	5,203	5,828	6,365
Lithuania	4,111	4,715	4,030	3,857	4,337	4,698	4,823	5,180	5,645
Luxembourg	30,917	31,764	32,333	32,538	32,779	33,301	34,320	35,270	33,818
Hungary	4,400	4,739	4,241	4,493	4,696	4,449	4,512	4,556	4,768
Malta	10,009	10,503	10,435	10,862	11,449	12,093	12,787	13,493	13,572
Netherlands	19,522	20,156	20,292	20,310	20,562	20,839	20,891	21,292	22,733
Austria	19,413	20,469	21,058	21,463	21,807	22,073	23,211	23,260	23,694
Poland	4,155	5,097	4,405	5,025	5,060	5,164	5,336	5,556	5,884
Portugal	8,143	8,282	8,678	8,410	8,323	8,177	8,229	8,435	8,782
Romania	1,952	2,172	2,036	2,091	2,049	2,016	2,155	2,315	2,448

Slovenia	10,893	11,864	11,736	11,999	12,122	11,852	11,909	12,332	12,327
Slovakia	4,792	5,671	6,117	6,306	6,927	6,737	6,809	6,930	6,951
Finland	19,794	20,962	21,349	21,826	22,699	23,272	23,702	23,763	23,650
Sweden	20,217	20,477	18,897	21,584	23,852	25,401	25,838	25,306	25,164
UK	18,923	16,262	17,106	17,136	19,166	18,694	20,528	21,028	21,136
Iceland	32,206	22,362	18,320	18,973	19,361	21,234	22,487	24,554	28,379
Norway	31,798	33,737	32,399	36,453	40,106	42,909	43,775	41,483	39,573
Switzerland	26,987	28,958	30,666	33,936	39,407	40,791	38,162	39,586	44,253
Turkey	2,903	3,049	2,754	3,326	3,222	3,438	3,497	3,386	:
Ukraine	2,374	1,797	2,284	2,474	3,123	3,217	2,134	1,675	1,651

Source: Calculated and formulated by the authors basing on Eurostat (2018), State Statistics Service of Ukraine (2018), National Bank of Ukraine (2018).

The variation in the hours worked per week of full-time employment is significantly lower than among EU member states, but its dynamics in 2008-2016 indicates a gradual decrease in the number of hours worked against the rising incomes. The average length of training varies considerably more: from 10.9 years in Italy to 19.0 years in Iceland in 2016.

We determined the strength of a relationship between the average income and the factors affecting it in the EU and Ukraine on the basis of a correlation-regression analysis.

Table 2. Relationship between Average Income and the Factors Affecting It in Individual EU Member States and Ukraine.

Country	Hours Worked per Week of Full-Time Employment	Average Length of Training	Social Transfers
Belgium	0.804	0.948	0.888
Bulgaria	-0.762	0.817	0.608
Czech Republic	-0.775	-0.814	0.561
Denmark	0.134	-0.050	-0.370
Germany	-0.808	0.868	0.594
Estonia	-0.807	0.588	-0.538
Ireland	0.100	-0.664	-0.589
Greece	-0.876	-0.824	-0.654
Spain	0.421	-0.782	-0.722
France	-0.806	0.965	0.849
Croatia	0.383	-0.455	-0.075
Italy	-0.407	0.392	0.612
Cyprus	-0.511	-0.537	-0.555
Latvia	-0.068	-0.253	-0.195
Lithuania	-0.261	0.733	-0.783
Luxembourg	0.623	0.718	0.681
Hungary	0.017	-0.040	-0.253
Malta	0.258	0.191	0.498
Netherlands	0.205	0.637	0.536
Austria	-0.984	0.925	0.922
Portugal	-0.280	0.115	0.279

Romania	-0.652	-0.370	0.599
Slovenia	-0.919	0.287	0.759
Slovakia	0.458	0.661	0.759
Finland	-0.856	0.463	0.869
Sweden	-0.826	0.326	0.857
UK	0.500	0.341	0.618
Iceland	0.907	-0.190	-0.480
Norway	-0.384	0.746	0.804
Switzerland	0.426	0.935	0.955
Turkey	-0.804	0.856	0.172
Poland	-0.747	0.808	0.895
Ukraine	0.000	0.000	0.963

Relationship strength according to the correlation coefficient:

☐ - weak; (0.10—0.29) ☐ - moderate; (0.30—0.49) ☐ - significant; (0.50—0.69) ☐ - strong; (0.70—0.89) ☐ - very strong. (0.90—0.99)

"+" - the relationship is direct, "-" - the relationship is inverse.

Source: Calculated by the authors according to the data in Eurostat (2018), State Statistics Service of Ukraine (2017, 2018), United Nations Development Program (2018), National Bank of Ukraine (2018).

The results of the conducted analysis are the most important part; they indicate that under present conditions, the cumulative income does not depend on the time spent at the primary place of employment. Wages, salaries and traditional employment gradually lose their fundamental importance in the life of a modern person in the era of digital technology. The income sources are gradually being diversified. The paper revealed a rather interesting trend: the increase in the length of training in the system of classical formal education no longer guarantees high incomes throughout life (in some countries within the study, the relationship between the length of training and income is even inverse). We observed the strongest impact of social transfers on incomes of the population in Ukraine, Switzerland, Austria, Belgium, Finland and France.

CONCLUSIONS

In the paper, we obtained empirical results that confirm the abovementioned theoretical hypotheses about the dynamic changes in the social and employment sphere and society in general; and the world community as a whole, the public administration of each individual country and every real or potential employee in particular have to prepare for it carefully. The changes in the social and employment sphere, associated with the spread of processes and phenomena, briefly summarized by the phrase "new economy", require in-depth research from the side of science, as well as increased attention from the managers at all levels – from personal to global one.

The social and employment sphere undergo the following major changes:

✓ Rapid growth of productivity;

✓ A rapid change in the labour market demand with its general decline;

✓ Change in the forms and systems of labour organization towards increasing flexibility and instability;

✓ Corresponding change in the main functions and forms of remuneration towards individualization, contract basis, dependence on the outcome;

✓ Transformation of the value system of modern employees and the loss of salary's positions as the foremost motivator of hired employment;

✓ Differentiation of the income system, increasing the role of social transfers in the absence of jobs;

✓ Changing requirements for vocational education in terms of its specialization, intensification, accessibility and mobility as the main social elevator in an era of change;

✓ Further globalization and increasingly active redistribution of all resources around the world, including intellectual human capital.

Such radical transformations affect the lives of each person, opening up new opportunities and, at the same time, putting forward new threats. The global community as a whole, every single society, state, enterprise, every person must prepare themselves for such changes, because they simply cannot avoid them.

The study revealed a significant transformation of all wage functions in the transition to a new economy.

In such conditions, fair employee compensation and ensuring productive employment is a challenge not only of the national but also of international scale. Prospective directions of reducing inequality and social tension in society is the guarantee of the basic unconditional income; progressive tax on all types of income; reduction of the length of working time while maintaining payments and improving productivity. In addition to avoiding social dangers, it would be a proactive progressive function of expanding human development opportunities.

REFERENCES

1. O.G. Brintseva. Transformation of human capital in the context of changing the content of labour in the new economy. (Ukraine: aspects of labour, **3-4**, 46-49, 2016)
2. Deloitte. Human capital trends, https://hctrendsapp.deloitte.com/reports/2018/new-rewards.html (2018)
3. V. Dumanska. Remuneration of labor: Ukrainian problems against the background of world trends. (Ukraine: Aspects of Labor, **2**, 35-40, 2015)
4. Eurostat. Hours worked per week of full-time employment, http://ec.europa.eu/eurostat/tgm/table.do?tab=table&init=1&language=en&pcode=tps00071&plugin=1 (2018)
5. Eurostat. Mean and median income before social transfers (pensions included in social transfers) by household type, http://appsso.eurostat.ec.europa.eu/nui/show.do?dataset=ilc_di13b&lang=en (2018)
6. Eurostat. Mean and median income by age and sex - EU-SILC survey [ilc_di03], http://appsso.eurostat.ec.europa.eu/nui/show.do?dataset=ilc_di03&lang=en (2018)
7. M. Graham, V. Lehdonvirta, A. Wood, H. Barnard, I. Hjorth, D. P. Simon. *The Risks and Rewards of Online Gig Work At The Global Margins.* (Oxford: Oxford Internet Institute, 2017)
8. A. M. Kolot. Global social inequality of incomes: nature, trends, consequences. Social-Labour Relations: Theory and Practice, **2,** 8-47 (2017)
9. State Statistics Service of Ukraine. *Statistical Yearbook of Ukraine for 2016*, (Edited by Werner I.E., Kyiv, 2017)
10. State Statistics Service of Ukraine. People. Official web-site. http://www.ukrstat.gov.ua/operativ/operativ2007/ds/nas_rik/nas_u/nas_rik_u.html (2018)
11. The Future of Jobs Report. Insight Report. World Economic Forum, http://www3.weforum.org/docs/WEF_Future_of_Jobs_2018.pdf (2018)

Problems and prospects of development of intermediary business in Ukraine

Olena Gromova[1], *Alla* Derhousova[1], *Tetiana* Glushenko[1], and *Daria* Biletska[1*]

[1] Ukrainian State University of Railway Transport, Management and Administration Department, Feuerbach sq., 7, Kharkiv, 61050, Ukraine

Abstract. The article deals with the problem of identification and search of the future-oriented directions of intermediary business development in Ukraine in the conditions of transformation processes in the national economy. The article describes the essence of trade and intermediary activity, outlines the main components and represents the content of trade and intermediary transactions. On the basis of the general analysis of theoretical and practical works on the implementation of trade and intermediary activities, the main problems that impede the development of intermediary business in Ukraine are identified. It is noted that it is expedient to use the formulated signs and principles for the organization of trade and intermediary activities. The prospects of using the newest marketing tools (neuromarketing) in the trade-intermediary business are described. It is noted that it is necessary to improve the normative base, that now inhibits the development of trade and intermediary activities.

1 Problem statement

The economic stability of the organization, its survival and the effectiveness of its activities in the context of market relations are inextricably linked with its continuous improvement and development. At the same time, the organization's improvement should be based on the principle of adaptation to a changing environment.

This is especially true for enterprises operating in the trade and intermediary sector, which currently form an integral part of business and trade activities of the infrastructure entities in the product market. Therefore, the problem of identifying problems and looking for promising directions for the development of intermediaries in Ukraine are of particular relevance.

2 Analysis of research and publications

Issues of development of the mediation business are studied in the works of both domestic and foreign scientists. Among them are the works of Gaponchuk N.V., Gerchikova I.N., Didkovsky M.I., Dunskaya A.R., Tokmin L.N. and others.

[*]Corresponding author: d.beletskaya2015@ukr.net

Among the works devoted mainly to the foreign economic activity of the enterprise, scientists mostly study issues of the organization of intermediaries' work in the implementation of foreign economic activity of the enterprise.

Most sources compare the content of trade and intermediary activities with the essence of the concept of "trade" and pay more attention to the fact that it is an initiative, independent activity for profit. But this interpretation does not cover the broader nature of the activity, since its meaning reduces to the primitive processes of buying and selling.

3 Presentation of the main research

Sales transactions are a significant part of the trade and intermediary activity, and its content, as it has already been noted, is wider. This can be explained by the fact that trade mediation involves a large amount of work in the interests of the manufacturer, such as search for contractors for the sale of goods, preparation and execution of the transaction, lending for the parties, provision of guarantees for the sale of goods and receiving of appropriate means by the seller, insurance of goods and risks during transportation, carrying out promotional activities for the promotion of products to markets, provision of organizational, commercial, consulting and other services.

As a result, we can formulate the essence of the concept of "trade and mediation".

In general, trade and intermediary transactions refer to transactions involving the purchase and sale of goods, that are performed on behalf of the supplier (manufacturer or exporter / importer) by an independent commercial agent.

Such activities as procurement, distribution and sales also relate to the trade and intermediary activities, as shown by the study of world experience.

However, companies that perform these operations remain in an independent status. They do not act as producers or final customer, but are in the sales circle and carry out any transaction with the product at their own discretion. For the most part, they are engaged in commerce, while large intermediaries can sometimes perform production operations, mainly related to the processing of procured and sold goods, transportation, insurance, warehousing of goods, but these functions are subject to commercial operations.

Marketing activities and trade intermediation services are very closely linked. It is characterized by a modern marketing concept that is followed in most markets. Using the marketing toolkit, the staff of the company (both employees and managers) receives information about consumer wishes, pricing and purchasing power, the willingness to pay exactly this price, and in which specific regions the demand for certain products is higher, and therefore the sale of products can provide maximum profits.

Marketing allows you to determine which types of production are most promising for investing or establishing a new business. Marketing research also helps determine the direction of the organization of the sales process and how to build an advertising strategy. Marketing calculations make it possible to compare several variants of enterprise development on the efficiency of production and sales costs; determine which types of goods are sold to a particular consumer in a particular region, and which will subsequently bring the greatest return on each investment unit. Today, most intermediary companies are trying to enter the external market, starting to operate in a new economic environment, where market relations have reached a higher level of development.

However, if an enterprise does not follow and does not master advanced marketing methods, it will eventually become noncompetitive. Knowledge of marketing principles makes it possible to arrange work with the consumer, critically evaluate the company's potential and monitor competitors, and choose the right market segment and economic activity. Flexibility of intermediary activity should be manifested primarily in the operational account of market requirements.

For this purpose, a lot of attention and time should be devoted to the study and forecasting of commodity markets, the development of a sales advertising campaign, as well as the attraction of innovations in trade and intermediary activities, and, if necessary, to adjust the profile of activities, to make changes in organizational structures of commerce.

In the modern world, we have been witnessing the emersion of neuromarketing, which allows you to understand the consumer's response to marketing irritants better and to improve the effectiveness of marketing methods by studying the brain's response. Therefore, this aspect should be particularly focused on, especially when creating a trade and intermediary network and forming communication links.

The research of scientific and practical works has proved that trade and intermediary activities in conditions of essential market relations should be determined by the following features:

1. Motives and conditions of origin - the result of the objective laws of commodity production and circulation, the formation of real needs of economic entities.
2. The role and place of mediation - an organic part of the distribution and distribution system, which effectively promotes the promotion of goods from the manufacturer to the consumer.
3. Subject of trade mediation - any product is intended for professional use or resale.
4. Economic guarantees - the intermediary entrepreneur shares the risk with the manufacturers of goods related to the impossibility of ale or selling at prices that do not provide the necessary level of profit.
5. The main objectives of the intermediary - the intermediary operates in the interests of its counterparties and in their own interests in order to receive the corresponding profit.
6. Legal attitude to the subject of mediation – mediators mainly acquire the right to own goods.
7. Organizational conditions of operation – variety and flexibility of organizational forms of mediation, types of works and services.
8. The measure of freedom and the scope of activity is a free and appropriate specialization in the segment of the market where a qualifying advantage can be obtained.
9. Methods and techniques of work with consumers – a differentiated approach to various needs, focusing on immediate consumer's demands, carried out on the basis of market research in the segmented market.
10. Source of income – payment determined by socially necessary expenses for the implementation of these works within the wholesale price of the producer and the price of demand.
11. Payment of labor – based on the results of the activity and the real contribution of each employee.

Effective trading and intermediary activities should be based on a number of principles:

1. Increasing the role of strategic planning and management to ensure stable production and consumption of products in the national economy, creating conditions on this basis for a more stable and reliable functioning of the sphere of circulation.
2. Allocation of economic priorities in stimulation of material and financial flows in the most effective branches, based on economic-political and social problems, which are solved by the country at each stage.
3. Global mobilization of internal reserves, ensuring rational use and economics of financial and material resources in each link of the national economic complex.
4. Increase of responsibility for fulfillment of contractual obligations on deliveries and transportation of products by all participants of market turnover on the basis of mutual rights and obligations, stipulated by economic contracts, treaties, orders.
5. The most effective organization of the process of sales of products manufactured by enterprises, and procurement of material resources by consumers in the required range and

in the form prepared for industrial consumption at a minimum cost of treatment and aggregate stocks.

6. Creation of economic conditions and the use of marketing tools allow mediators to meet the demands of customers in a qualitative and efficient manner, flexibly maneuver with material resources and accelerate their reversibility in favor of stabilization and economic growth.

7. Creation of a modern highly equipped technical material base with key elements of the market infrastructure in the commercial chain, its continuous improvement for the proportional development of the sphere of circulation in accordance with the requirements of the sphere of production.

8. Use of scientific methods of management of material flows on the basis of modern logistic approaches and methods, wide use of opportunities of the electronic market, the Internet system in order to maximize and timely satisfaction of market requests of buyers.

9. Use of forms and methods of social and ethical marketing to establish mechanisms of mutually beneficial agreements not only with market partners, but also in society through the active use of communicative models, in particular public relations.

During trading and intermediary activities, participants face certain issues.

The most difficult one is the legislative framework, because it is weak and outdated.

It should be noted that the legal basis for regulation of trade and intermediary activities, as well as all economic turnover, consists of fragments of indirect laws.

Its basis consists of departmental regulations, instructions and regulations, and most often the development of the intermediaries themselves. Of course, such a base cannot be considered acceptable.

Another reason that slows progress in the area of development of trade and intermediary activities is organizational. The content of the problem is that in Ukraine organizational forms of trade mediation are not formed under control. For the most part, they have the appearance of somewhere else altered copies of foreign structures, which are important to adapt to the realities of national reality.

In the course of the study, it is proved that the structure of intermediary firms tends to simplify and is ineffective in functional terms. First of all it concerns exchanges, auctions, distributor, consignment firms, brokerage offices. They have a limited functional structure, but at the same time it is universal. This can be explained by two reasons: firstly, an attempt to reduce the cost of the operation of a brokerage firm; and secondly, the desire to universalize the activities of intermediaries.

There is a need to develop new, optimal organizational forms taking into account international experience and features of the domestic economy in the field of trade mediation.

The current issue is the territorial organization of intermediaries, their specialization. The reason is that a very important principle of the operation of most intermediaries is the division of territories. This means that the intermediary is territorially limited in its activities. The implementation of this principle requires the regulation of the territorial organization of intermediaries, their demarcation, specialization, revision of antimonopoly legislation.

The problem with personnel is an important issue to be addressed when it comes to organizing trade mediation.

Resellers carry out large-scale internal and external transactions, complex in commercial, financial and legal terms. But for some reason, they lack the qualifications.

In addition, higher education institutions in Ukraine do not train personnel for trade and intermediary activities.

In this regard, it is advisable to introduce the specialty "Trade and mediation" in colleges, trade and economic institutions of higher education in the country. Compliance with the principles of forming commercial relations will allow the trade and intermediary system to actively influence production, to identify discrepancies between supply and demand for certain types of goods and services, to form rational commercial relations between suppliers and consumers.

Multiple transactions in international trade are carried out with the help of intermediaries – individuals, organizations, etc., which have an intermediate position between producers and end users. From the economic point of view, it is entirely possible for them to apply the term "resellers", which is more common in international commercial matters.

The majority of commodity groups of machinery and equipment, raw materials, trade products, foodstuffs, consumer goods and basic semi-finished products are sold via resellers.

The feature of the modern business environment is that most of the international trade turnover is carried out with the participation of intermediaries. We mean trade and intermediary firms, organizations and individuals.

The importance of intermediaries is inherent both for the trade in mass demand and for trade in products for which demand is very limited due to their specific characteristics.

Without intermediaries, the sale of exclusive goods is not rarely done, in the event that there is only one monopoly producer and 2-3 end-users in the world market.

Companies use intermediaries to improve the efficiency and economic feasibility of foreign economic operations.

The economic efficiency of transactions performed by intermediaries, even taking into account the costs of remuneration, increases due to the following aspects:

- the involvement of intermediaries increases the speed of sales of goods, which leads to an increase in sales revenue due to accelerated capital turnover;
- intermediaries, more responsive to changing market conditions that allows you to sell goods on more favorable terms for the exporter;
- the involvement of intermediaries provides an opportunity to increase the competitiveness of goods at the expense of shorter delivery times and intermediate warehouses, better warehousing and storage of goods, pre-sale service and maintenance, special marking, additional equipment of products in the country of sale in accordance with local requirements;
- taking advantage of lending on various terms, some intermediaries finance exporters' transactions, invest their own capital in the creation and operation of a sales network, which gives the opportunity to benefit from the saving of money invested;
- intermediaries provide optimization of communication costs, accounting savings, help reduce credit risks, facilitate faster access to new markets;
- intermediaries are the source of valuable primary information about segmentation and market capacity, demand dynamics, margins of product price fluctuations, state of competitors, requirements for the quality of goods.

After analyzing the activity of international trade, we can conclude that, first and foremost, medium and small enterprises are turning to the resellers in the sphere of foreign trade in particular.

The involvement of a mediation tool is central to the marketing of their products. But large-scale industrial companies, including multinationals, are turning to independent intermediaries. Transnational corporations, meanwhile, involve intermediaries mainly for the implementation of secondary products, in some new or complex markets, in small-capacity markets, in the absence of their own sales network based on trade subsidiaries and representative offices, with relatively small volumes or with the occasional conclusion of an export-import operations.

In determining the importance of intermediaries, it should be noted that they have become virtually indispensable tools for the use of certain specific forms of sales by an exporter or importer, for example, stock trading, participation in international tenders for machinery, equipment and capital construction, patents trading and licenses. Referring to the involvement of intermediaries, it is necessary to remember the objective shortcomings, which entail the attraction of the trade and intermediary level to the sale of goods. The lack of a direct contact with the sales market is one of them.

Thus, when considering the issue of engaging a reseller, the manufacturer must clearly calculate and understand the economic efficiency of his activities.

In the event that the intermediary is not able to provide additional income compared with what the manufacturer receives in the case of self-distribution of goods on the market, its attraction is not economically meaningful.

Unfortunately, a thorough calculation of the possible economic effect of using an intermediary in practice is very difficult. It is impossible to estimate, for example, the lost profit from the use of an intermediary, which the manufacturer does not even suspect because of his ignorance.

4 Conclusion

Nowadays, Ukraine has no clear strategy for the development of intermediary business. Having pointed out the main shortcomings, we can state that now, in order to improve the activities of intermediaries in Ukraine as a whole, it is necessary first of all to pay attention to improving the legislative framework, to start actively using modern marketing tools (neuromarketing) and to pay attention to the training of specialized staff. Only a comprehensive approach will enable an accelerated development of intermediary business in Ukraine and increase the competitiveness of the country in the world market.

References

1. M. I. Belyavtsev, *Commodity market infrastructure*, 43-50 (2015)
2. N. Gaponchuk, Problems of the Sustainable Development of National Economies, 73-75 (2013)
3. I. Gerchikova, *Management* [3rd ed.] (1997)
4. I. Gerchikova, *International Commercial Business* [2nd ed.] (2001)
5. M. I. Didkivskyj, *Foreign economic activity of the enterprise* (2006)
6. A. R. Dunska, Economic Newsletter of NTU "KhPI", **9**, 89-95 (2012)

Economic development of industrial enterprises: motivation-oriented management methods

Oleksandr Husarov[1], *Olena* Antyptseva[2], *Kyrylo* Husarov[3], *Olga* Yurieva[4]

[1,2] Ukrainian Engineering Pedagogics Academy, Department of Management, 16, Universitetskaya str., Kharkiv, 61003 Ukraine
[3] National University of Civil Protection of Ukraine, Educational, Scientific and Production Center, 28, Lermontovskaya str., Kharkiv, 61024 Ukraine
[4] Chief State Auditor-Inspector of Internal Control Department of the Head Department of the State Tax Service in Kharkiv Region, 46, Pushkinskaya str., Kharkiv, 61057 Ukraine

Abstract. The article explores the main approaches to the concept of "motivation-oriented management". The importance of applying motivation-oriented management for the economic development of industrial enterprises is proved. It is determined that the economic development of enterprises is an extremely important factor for the development and competitiveness of the state as a whole. It is noted that in present unstable economic conditions, an innovative approach is necessary for the development of industry. The main and secondary processes that arise in the process of introducing motivational activities in the enterprise are described. It is noted that when managing the economic development of an enterprise, one should focus only on improving its condition, which should be confirmed by certain achieved results. It is indicated that the performance indicators of motivation affect all economic indicators of the enterprise, in particular the level of financial development of industrial enterprises. Conclusions are made regarding the feasibility of introducing motivation-oriented management at Ukrainian enterprises for their economic development.

Introduction

One of the leading sectors of the economy is industry, which is shaping the scientific and technological transformation, economic growth of the country and its social development. Mechanical engineering is the most important industry link in most developed countries of the world. Engineering covers many specialized industries and is the basis of heavy industry and plays a decisive role in developing the material and technical base, and produces tools for various sectors of the national economy. The main way to develop industrial enterprises is to ensure the production of competitive products.

The formation and strengthening of the competitive advantages of domestic industrial enterprises plays a significant role in increasing the competitiveness of Ukraine. However, there are many internal and external factors that impede this process.

These factors should not prevent increasing the competitiveness of the Ukrainian industry. Ways out of this situation should be constantly searched for, to develop, raise the level of workers, establish contacts with other countries, improve the level of product quality, and introduce new technologies etc.

The aim of the article is to describe motivation-oriented management as a method that plays a significant role in the economic development of Ukraine's industry.

Theory of the matter

New conditions of development require continuous adaptation of methods and approaches to the assessment of economic development of industrial enterprises, taking into account the rapid changes of the economic space. Theoretical and methodological aspects of managing the economical development of the enterprise were dealt with by such scientists as O. Arefyeva [1], V. Geyets [2], V. Dykan [3], M. Kizim [4], M. Porter [5], V. Prokhorova [6] and others. Problems of motivational-oriented management are considered in the works of such authors as O. Amosov [7], V. Kovalev [8], T. Kuzenko [9], V. Mazur [10], A. Maslow [11], M. Meskon [12] and others.

World experience indicates the transition from management technologies to taking into account the positions of the economy, which significantly affects the maximization of the financial results of enterprises. This necessitates focusing the key tasks of enterprise management on ensuring the competitiveness of the team on the basis of a motivational approach, which directly affects the level of economic development. The above determines the need for an integrated and fundamental approach to assessing the level of motivation-oriented financial management of engineering enterprises.

Despite the range of issues that have been studied in the works of foreign and domestic scientists, the problems of developing theoretical and methodological tools for motivationally oriented management of the economic development of industrial enterprises remain insufficiently investigated. The above confirms the relevance of the topic of our study.

Discussion of results

The functioning of modern enterprises is characterized by uncertainty and dynamism of the socio-economic environment. An effective confrontation of an enterprise with a change is possible only if it implements advanced measures aimed at maintaining viability and ensuring an appropriate level of competitiveness, but this is only possible under conditions of its constant development.

The development of the enterprise consists in making qualitative changes, updating the economic system and organizational structure, increasing the efficiency of functioning by improving technology, technology and labor organization in all structural divisions, improving the quality of products and services.

At the present stage of development of the domestic economy, a significant factor in reducing management efficiency is the lack of effective concepts and applied methods that have improved the process of enterprise development management and increased its profitability both in the short and long term. For the successful development of applied methods and models, it is appropriate to focus primarily on improving the categorical apparatus of the research subject area. Theoretical approaches to the definition of "development management" and "economic development" of an enterprise as structural elements of the research subject field are reflected in the works of domestic and foreign scientists and practitioners [6].

We emphasize that when it comes to the economic development of an enterprise, it is necessary to focus only on development in the management process in the understanding of improving its condition, which is characterized by certain achieved results. However, for the implementation of strategic management goals, it is appropriate to formulate tactical tasks that could focus on ensuring the process of determining and adjusting economic indicators and creating conditions for the course of planned processes. An effective solution to the problems of progressive economic development is impossible without awareness of the global organizational and managerial role of motivating factors.

In the organizational and managerial activities of the enterprise, the motivational aspects are becoming increasingly important, because the commercial success of any enterprise depends primarily on how employees realize their potential. The motivational aspect of labor management is widely used in countries with developed market economies.

The motivational mechanism at the enterprise is aimed at surviving and achieving success in a competitive environment [13].

In our opinion, special attention in this matter should be paid to studying the influence of directly motivating factors on the level of economic development of an enterprise, since it is motivation that is the tool that encourages a person to achieve high results in work, and therefore also has significant economic effect.

Studies on the problems of motivation and its methods are reflected in the works of scientific economists A. Maslow [11], A. Bogutsky [14], and others. In addition, out of the field of view of most researchers, there remained questions of the financial expression of the effect of motivational measures in the results of activities and development of the enterprise. Motivation-oriented management of economic development, as an independent direction of scientific research and a type of management activity, was formed as a result of the rapid development of entrepreneurship under the influence of European and world transformational changes, the rapid development of financial relations, the improvement of motivational technologies, and the growing risks of financial losses.

The work of V. Kapitonova [15] provides an interpretation of "economic motivation" as a process combining an employee and an enterprise, in contrast to classical interpretations in which economic motivation is seen as an incentive to action, as a way of rewarding employees, as a process of motivating people to activities at the employer level.

Motivation as an economic category expresses the economic relations between people who participate in the real process of reproduction regarding the realization of motivational potential. The managerial aspect of the motivation for the development of the enterprise lies in the emphasis on creating a certain culture with which the company can effectively solve the problems of external adaptation and internal integration.

The implementation of motivational measures is accompanied by a number of processes, both primary and secondary (aimed at providing) (Fig. 1). Fig. 1 shows that the main motivational process is the formation of a motivational idea and the study of the possibilities of its practical implementation. They are primarily focused on improving the management of enterprises, in particular, their economic development. A managerial decision on the implementation of the developed motivational idea in most cases is made on the basis of the economic efficiency of motivational measures. In the implementation of each of these processes, both primary and secondary, it is advisable to take into account the characteristics of motivational methods.

Considering the organization of the motivation system in enterprises, it is necessary to highlight the existing types of motivational factors of influence on the financial development of the enterprise. Motivational factors influencing financial development can be divided into endogenous and exogenous [16]. Endogenous factors of motivation are the motivation of the employee to work activity associated with the content of the work, the level of its significance. If work for an employee is important, useful, then it contributes to the realization of his abilities, leads to an increase in labor productivity.

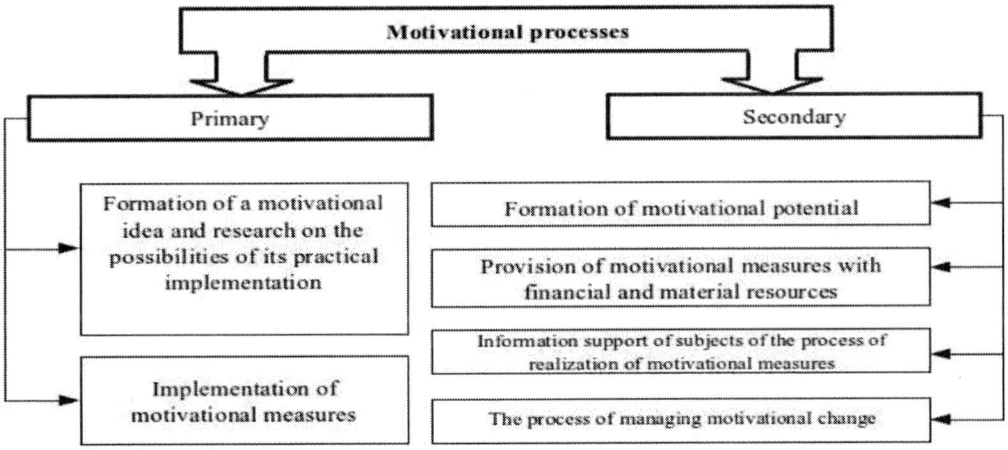

Fig. 1. - Systematization of motivational processes

Exogenous motivation is aimed at stimulating labor on the part of the socio-political system, the regulatory framework and the development of competing enterprises. Moreover, latent factors, intense and extensive, can be distinguished by the degree of influence on financial development; by the nature of the effect: stimulating, destimulating and supporting factors. It should be noted that the performance indicators of motivation affect all economic indicators of the enterprise, in particular the level of economical development of industrial enterprises.

To determine the effectiveness of the introduction of motivation-oriented management of the economic development of the enterprise, it is proposed to use the following criteria:

1. Accuracy of the display of goals. To implement this aspect, it is necessary to provide a quantitative expression of the qualitative content of the goals of the activity and development of the enterprise. With this expression, complex indicators of the economic development of the enterprise appear, and will be further compared with the results of operations. If the quantitative expression of goals is impossible, an assessment of the degree of their achievement in financial values is accepted using the expert method.

2. Compliance with organizational and managerial discipline, regulations and rules. The degree of fulfillment of this criterion characterizes the significance and cyclical nature of errors and failures in the process of performing organizational and managerial tasks, the key aspect of which is motivational support.

3. Accuracy of planning economic indicators. The essence of this criterion is that errors, cost and planned deviations should be monitored through the use of statistical methods of analysis of both indicators of economic activity and the implementation of motivational measures.

4. Effectiveness of economic models.

5. Payback measures of a motivational nature. The costs of organizing incentive events (staff costs, information collection) should be correlated with the economic effect received from them.

This set of criteria is universal, but it can be supplemented and detailed depending on the specifics of the enterprise.

Thus, from the general spectrum of vectors for managing the financial development of an enterprise, one should single out a motivation-oriented management of the financial development of enterprises to mobilize internal intellectual and competence resources for financial growth and activity in the external economic environment.

The internal result of motivation-oriented management is expressed in the growth of the potential of enterprises and its financial development by increasing the efficiency of using potential on the basis of the mechanism for introducing motivational measures.

Motivation-oriented enterprise management in the external economic environment is assessed by a combination of quantitative and qualitative parameters, including the company's market share, financial stability, business activity dynamics, image of the enterprise, its social activity, readiness to implement research and development, etc.

The structural-logical diagram of the relationship of private elements of the formalization of motivation-oriented management is shown in Fig. 2.

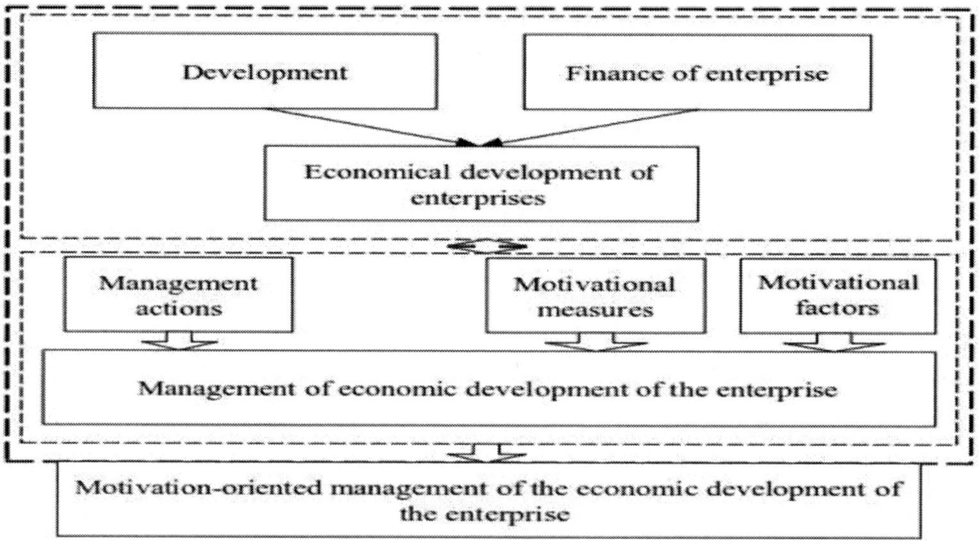

Fig. 2. - The structural-logical diagram of the relationship of the elements of the conceptual-categorical apparatus of the subject field of research of motivation-oriented management of the financial development of the enterprise

The authors determined the essence of motivation-oriented management of the economic development of the enterprise as a system of principles and methods for the development and implementation of managerial decisions, based on the diagnosis of the actual economic condition. Determining the factors that destabilize the sustainability of development and assessing indicators of financial development, allow us to bring the goals and objectives of the enterprise into line with the conditions of the internal and external environment, to ensure a balance between the financial interests of the enterprise and the need for our employees, through effective management actions, taking into account motivational factors in the process of transition to a new improved economic level.

The effectiveness of motivation-oriented management of the economic development of an enterprise is manifested in the dynamics of its development, namely: achievement of strategic goals that reflect natural and cost indicators, in the effective use of financial potential, expanding markets for products and replenishing the client base through effective motivational policies.

Conclusion

Thus, as a result of the our study, an analysis was made of the motivational factors of the influence of the external and internal environment on the economical development of the enterprise, which are important in constructing strategic plans for the development of business entities in modern conditions of the development of the knowledge economy, among the priority areas of which the motivational direction is distinguished.

After all, by taking into account the above factors to improve the financial effect of motivational measures, it is possible to adjust the influence of these factors and, thus, to improve the efficiency of activity and economical development of enterprises. Based on the results of the study, it can be argued that a well-designed organizational and management activities to implement motivational measures and taking into account the influence of motivational factors can create a sense of personal financial contribution to the activities of the entity.

In our opinion, effective motivational measures should ensure the successful transformation of a set of factors, principles, incentives, motives, value orientations, expectations, behavioral reactions from a linear discrete state into a closed, constantly repeating process with a tangible financial result in the strategic aspect of the economic development of the enterprise.

References

1. Arefyeva O. Upravlinnia rozvytkom ekonomichnykh system: teoriia, mekhanizmy rehuliuvannia ta upravlinnia: monohrafiia [Management of economic systems development: theory, mechanisms of regulation and management: monograph]. Kharkiv : UkrDAZT (In Ukrainian). (2010)
2. Geyets V. Suspilstvo, derzhava, ekonomika: fenomenolohiia vzaiemodii ta rozvytku [Society, state, economy: phenomenology of interaction and development]. Kyiv : NAS of Ukriane (In Ukrainian) (2009)
3. Dykan V. Ekonomyka predpryiatyia: uchebnoe posobye [Enterprise Economics: A Training Manual]. Kharkiv : UkrDAZT (In Russian). (2008)
4. Kizim M. Vysokotekhnolohichnyi sektor ekonomiky Ukrainy ta krain svitu: stan i tendentsii rozvytku [High-tech sector of economy of Ukraine and countries of the world: state and tendencies of development]. Problems of Economy №3 (In Ukrainian). (2009)
5. Porter M. Competitive strategy. Measuring business excellence. (In English). (1997)
6. Prokhorova V. Ekonomichnyi rozvytok pidpryiemstva: teoretyko – metodolohichnyi aspekt [Economic development of the enterprise: theoretical and methodological aspect]. Kharkiv: UkrDAZT (In Ukrainian). (2010)
7. Amosov O. Otsinka potentsialu konkrentospromozhnosti yak osnovnoho aspektu rozvytku pidpryiemstva [Assessment of competitive potential as a major aspect of enterprise development]. Problems of Economy (In Ukrainian). (2011)
8. Kovalev V. Rol truda v formirovanii obschestvennyh otnoscheniy. [The role of labor in the formation of social relations]. Kharkiv: NTMT (In Russian). (2015)
9. Kuzenko T. Optymizatsiia rivnia finansovoho rozvytku pidpryiemstva na osnovi taksonomichnoho analizu. [Optimization of the level of financial development of the enterprise on the basis of taxonomic analysis]. Kyiv : NAU (In Ukrainian). (2015)
10. Mazur V. Vdoskonalennia system motyvatsii pratsi. [Improvement of work motivation systems]. Ternopil: ANG (In Ukrainian). (2008)
11. Maslow A.A. Theory of Human Motivation. Psychological Review, vol. 50, pp. 370-396 (1943)

12. Meskon M. Management: Individual and Organizational Effectiveness. USA: Harper & Row. (1985)

13. Gergesha Y. Strukturno-funktsionalne modeliuvannia stratehichnoho upravlinnia finansovym rozvytkom pidpryiemstva [Structural and functional modeling of strategic management of financial development of the enterprise]. Development Management: A Collection of Scientific Papers (In Ukrainian). (2011)

14. Bogutsky A. Faktory rozvytku motyvatsii pratsi. [Factors of development of work motivation]. APK Economics (In Ukrainian). (2005)

15. Kapitonova V. Motivatsyonnaya sreda kak instrument aktivizatsii korporativnoy kultury na promyshlennom predpryiatii [Motivational environment as a tool to enhance corporate culture in an industrial enterprise]. Izhevsk: GTA. (In Russian). (2012).

16. Shakhovoy V. Motyvatsyia trudovoi deiatelnosty: Uchebno-metodycheskoe posobye [Motivation of work: educational-methodical manual]. Moscow: Alpha-press. (In Russian). (2006).

Logistic foundations of sustainable socio-economic development of the region in the context of decentralization

Alla Kasych [1], *Oksana* Onyshchenko[2*]

[1]Kyiv National University of Technologies and Design, Management Department, 01011, Kyiv, 2 Nemirovich-Danchenko Street, Ukraine
[2]Kremenchuk Mykhailo Ostrohradskyi National University, Department of Accounting and Finance, 39600, Kremenchuk, Pershotravneva str. 20, Ukraine

Abstract. Incompleteness of processes of structural reformation of Ukrainian economy reveals itself on the regional level. Argument towards the utility of decentralization reform was Ukraine-Europe integration. Decentralization should be viewed as a powerful tool of governmental policy for balanced development of territorial communities. In addition, it`s crucial to remember that the greatest effect will be achieved in a long term. Strategic development of united territorial communities (UTC) includes countless tasks that need to be solved as soon as possible. And organization of effective logistic activity in the region appears to be one of the ways to strengthen regional economy, to improve security level and development. The problem of ensuring stable socio-economic growth of a community is as well a logistic problem. It can be represented as a problem of optimization of flow processes of effective resource usage to achieve balanced processes of creation and consumption of resources in socio-economic system. Logistization of economic relationships on UTC level allows community to ensure better decision making of key tasks of regional management such as creation of optimal strategy to manage structural offers and material flows of social production; optimization of material, financial and informational flows of the region.

1 Introduction

Incompleteness of processes of structural reformation of the Ukrainian economy reveals itself on the regional level. The current administrative and territorial system in the country could not catch up with transformational processes and partially was an obstacle to the transformations in the country. It didn`t allow regions to effectively enforce it`s policies, as a result it hindered the growth of both regions and country in general. Argument towards the utility of decentralization reform was Ukraine-Europe integration. It required the appropriateness of administrative and territorial system of the country to European standards of managing regional development and local municipality. A number of studies

* Corresponding author: vikols75@ukr.net

on theoretical bases of impact of decentralization on effectiveness of provision of social services in the country (Musgrave, R.A. [1], Oates, W.E. [2], Tiebout, C.A. [3]); generalization of experience of different countries in decentralization (Baldersheim, H. [4], I.P. Kovacs [5], M. Kulesza [6], I. Otola [7], A. Feltenstein, S. Iwata [8]); empirical researches on the impact of decentralization on the economic growth dynamic (H. Baldersheim, [9], Ph. Bodman [10], J.K. Brueckner [11, 12]). Consideration of these researches is important for reformation processes in Ukraine.

Other country`s experience suggests that "increased economic growth and the increase in competitiveness of the region and welfare of its population relies on how much the government is paying attention to a balanced development of the country" [13, p.6]. Regional development is mostly viewed through lens of such economical aspects as economic growth, welfare and employment (level) [14], the number of number of workplaces, people's income and labour productivity [15], as well as the overall economic welfare of the region [16, p.5]. Accounting the following aspects like social inequality, ecological situation, effectiveness of local municipality bodies, cultural diversity [17] has lead to more profound understanding of decentralization and its effects not only in economical but in social and ecological spheres.

On the current stage of development of Ukraine, territorial organization of power, when local self-government stands as a starting point in organization of management system of social development, is believed to be most optimal.

2 Methods

As a main method of research, we use the critical analysis and comparison of analytical reports, region development strategies and scientific publications on the issues of sustainable social and economic development of regions in the context of decentralization and also conception of logistic.

To achieve the set goals, the following general scientific and specialized methods were used: theoretical generalization, comparison and systematization - to study the essence and features of logistics activity in united territorial communities; system analysis – for determining the state and level of development of the logistics system of Ukraine and the consequences of decentralization processes in Ukraine; abstract-logical – for theoretical summarization and conclusion; graphical method – for visualization the logistic system of the territorial community as an organizational and economic mechanism.

3 Results and Discussion

3.1 Decentralization reform and balanced regional growth

Reformation of administrative and territorial system in Ukraine began in 2014. On the governmental level it was announced that the new administrative and territorial system should become the foundation of the new model of territorial management, which will be based on the practices of decentralization, social solidarity, balance of the national and community`s interests, omnipresent bodies of local self-governing, ability and independency of territorial groups to manage matters of local importance [18].

Decentralization reform implied the creation of a new branch in the administrative system in Ukraine by means implementation of a new administrative and territorial unit – united territorial communities (UTC). According to the data [19] from 2014 to 2019, 3,443 territorial communities were united into 878 united territorial communities; they are home to 9.0 million people (25.2 % of the total population of Ukraine). Own revenues of local budgets has increased from 68.6 billion UAH in 2014 to 234.1 billion UAH in 2018. The decentralization reform changed the tax distribution formula, which affected the balance of interests between business and government. Communities are interested in attracting business to their territory, as a significant part of taxes remains in local budgets (Fig. 1).

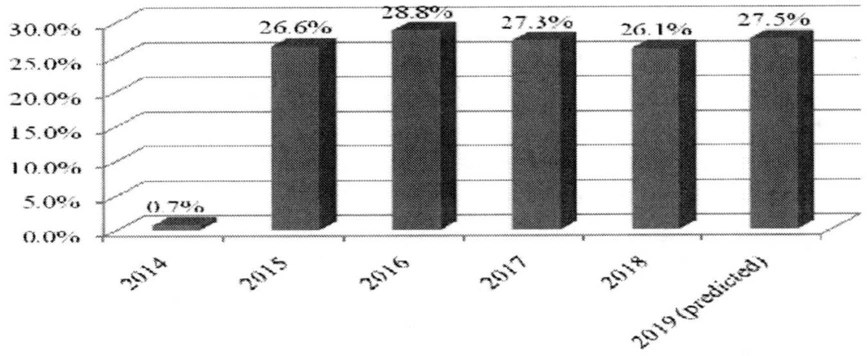

Fig. 1. Part of local taxes and fees in revenues of local budgets (2014-2019)
Source: According to the data [19].

The main purpose of the changes that are now taking place in most Ukrainian communities is to create conditions for effective development. Communities are not only united to gain more power and resources, they are consolidating their efforts to become stronger, to have resources and to be able to independently solve most of the issues that concern their residents. At the same time, the communities should also increase their level of responsibility to the business: the investor should not doubt that his funds can be used for other purposes [20]. In the 21st century understanding of the importance of balanced regional development emerges. Key priorities of European strategy 2020 are development of more effective, more ecological and more competitive economy [21, p. 8]. Decentralization should be viewed as a powerful tool of governmental policy for balanced development of territorial communities. According to A. Kasych, the mechanism of the impact of decentralization on the economic development of the country acts through the redistribution of state budget revenues in favor of the regions, the development of local infrastructure, ensuring efficient production of public goods, increasing the level of investment attractiveness and accelerating business development in the region [22].

3.2 Logistic approach to ensure stable growth of the territorial community

Strategic development of united territorial communities (UTC) includes countless tasks that need to be solved as soon as possible. In particular, key tasks are regarding the strengthening of economy bettering security level and balanced regional growth. However, while considering possibilities of usage of available potential, it's crucial to bear in mind the importance of logistic approach in choosing economic guides. In this aspect it is important to turn to planning, realizing and effective managing of logistic activity in the region, because many UTC territories provide many economic opportunities. Organization of effective logistic activity in the region appears to be one of the ways to strengthen regional economy, to improve security level and development. On the regional level, as the starting point of creation of new strategically guides of development of the region, it is important to consider the potential of logistic and, in particular, the ways to realize logistic activity. The main goal of logistic activity is to strengthen the regional economy, ensure its balanced growth and satisfy the needs of the customers. Logistic activity in region is there to solve such strategically problems as creation of environment to achieve competitive advantage and the formation of sufficient level of logistic potential.

Today conception of logistic became a foundation for innovative change in management of UTCs. So, the flows of various resources, owned by UTC (land, labor, capital, information, entrepreneurial abilities) are the object of logistic research.

While realizing logistic approach to ensure stable development of UTC it should be borne in mind that the community is an environment for people, foundation of management, means of consolidation of all spheres of social production, as well as the carrier of natural, economic, social and consumption potential. The problem of ensuring stable socio-economic growth of a community is as well logistic problem. It can be represented as a problem of optimization of flow processes of effective resource usage to achieve balanced processes of creation and consumption of resources in socio-economic system [23].

Logistization of economic relationships on UTC level allows community to ensure better decision making of key tasks of regional management such as creation of optimal strategy to manage structural offers and material flows of social production; optimization of material, financial and informational flows of the region. The execution of mechanisms of logistic management predetermines the need in creation of logistic system of UTC. Management of flow processes of resource usage is done on all stages of formation and operation of matter-energy flow, from provision of raw material resources till waste utilization. Thus, logistic system of UTC should comprise companies from production and service spheres, infrastructure objects, organizations and institutions interconnected in a process of decision making regarding the directions and ways of implementation of elements of resource flow. The logistic system of UTC should also have a system of provision of information. The logistic system of UTC is an organization mechanism of coordination of functions of resource flow management with certain characteristics that transpire in goal instructions, objects and tools of management (Fig.2).

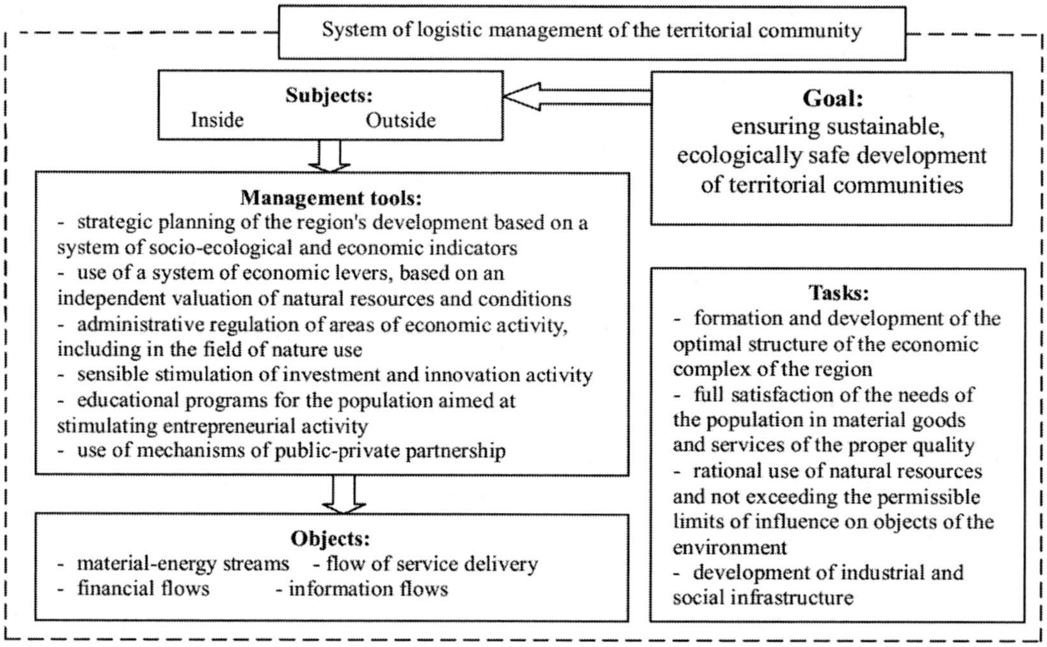

Fig. 2 Features of realization of logistic management in the regional aspect
Source: Authors

Within the framework of operational mechanism of logistic management of resource usage on UTC level, it is important to foresee, first of all, sufficient value assessment of a present natural resources potential, as well as adequate assessments of destructive economic activity, development of an effective tools to coordinate economic, social, and ecological interests of economic subjects and to solve existing contradictions between them.

4 Conclusion

Thus, implementation of logistic tool kit in regulating of resource usage processes on the level of UTC allows to create effective management approach to solve tasks regarding ensuring stable and balanced development of UTC – timely satisfaction of consumer needs in material and proper quality service, while maintaining appropriate level of quality of environmental and natural components, and minimizing social cost relatively to present organizational and economic, technical and technological conditions of management.

Further development of logistic concepts of stable socio-economic growth of regions in the context of decentralization requires the solution of the following tasks: deepening of structural and functional basis of creation of logistic systems of territorial communities; improvement of methodology used to analyze and optimize flow processes on the different hierarchy levels of management; development of structural and contextual foundations of an organizational and economic mechanism of logistic management of communities growth, as well as the development of a practical recommendations regarding its execution relatively to socio-economic conditions of a specific territorial community.

References

1. Musgrave, R.A.: The theory of public finance. McGraw-Hill, New York (1959)
2. Oates, W.E.: Fiscal Federalism. New York. http: www.jstor.org (1972)
3. Tiebout, C.A.: Pure Theory of Local Expenditures. The Journal of Political Economy. **64(5)**, 416–424 (1956)
4. Baldersheim, H.: Decentralisation in practice. European patterns and experiences. Publisher Ministry of Public Administration and Local Self – Government. http://www.undp.org (2011). Accesed 25 Mar 2019
5. Kovacs, I.P.: Europeanisation of Territorial Governance in Three Eastern/Central European Countries. Halduskultuur. **10**, 40–57 (2009)
6. Kulesza, M.: Methods and Techniques of Managing Decentralization Reforms in CEE Countries. The Polish Experience. Public Administration Department Faculty of Law and Administration Warsaw University, Poland. http: //www.oecd.org (2002)
7. Otola, I.: Decentralization Processes of Local Government Sector in Poland. The 16th NISPAcee Annual (2008)
8. Feltenstein, A., Iwata, S. Decentralization and macroeconomic performance in China: regional autonomy has its costs. Journal of Development Economics. **76**, 481–501 (2005)
9. Baldersheim, H.: Decentralisation in practice. European patterns and experiences. Publisher Ministry of Public Administration and Local Self – Government. http: //www.undp.org (2011). Accessed 04 Apr 2019
10. Bodman, Ph., Campbell, H., Heaton, K., Hodge, A.: Fiscal Decentralisation, Macroeconomic Conditions and Economic Growth in Australia. School of Economics, The University of Queensland, Macroeconomics Research Group (2009)
11. Brueckner, J.K.: Fiscal Decentralization with Distortionary Taxation: Tiebout vs. Tax Competition International Tax and Public Finance. **11**, 133–153 (2004)
12. Brueckner, J.K.: Fiscal federalism and economic growth. Journal of Public Economics. **90(10)**, 2107–2120 (2006)
13. Regionalnij rozvitok ta derzhavna regionalna politika v Ukrayini: stan i perspektivi zmin u konteksti globalnih viklikiv ta yevropejskih standartiv politiki.

https://bit.ly/2HIeWjF (2019). Accessed 14 Apr 2019
14. Armstrong, H., Taylor, J.: Regional Economics and Policy. Oxford, Blackwell (2000)
15. Storper, M.: The Regional World: Territorial Develop ment in a Global Economy. Guilford, London (1997)
16. Beer, A., Haughton, G., Maude, A.: Developing Locally: An International Comparison of Local and Regional Economic Development. Policy Press, Bristol (2003)
17. Haughton, G., Counsell, D., Haughton, G.: Regions, Spatial Strategies and Sustainable Development. Routledge and Regional Studies Association, London (2004)
18. Pavlyuk, A.P., Olijnik, D.I., Batalov, O.A., Dacko, O.I., Valyushko, I.V., Barvickij, S.Yu.: Teritorialna gromada yak bazova lanka administrativno-teritorialnogo ustroyu Ukrayini: problemi ta perspektivi reformuvannya. Analitichna dopovid. NISD, Kyiv (2016)
19. Decentralization of power. https://storage.decentralization.gov.ua/ uploads/ library/ file/ 378/10.02.2019_ukr.pdf (2018). Accessed 23 May 2019
20. Pochtovyuk, A., Semenikhina, V., Onyshchenko, O. and Ruban, B.: The formation and development of social responsibility of business: Ukrainian experience in a context of decentralization. SHS Web of Conferences **61**, 01018 (2019). doi: 10.1051/shsconf/20196101018
21. Europe 2020: A European strategy for smart, sustainable and inclusive growth. http://ec.europa.eu/eu2020/pdf (2010). Accessed 30 May 2019
22. Kasych, A.O.: Theoretical aspects of decentralization processes impact on economic development of the country. Actual Problems of Economics. **182 (8)**, 16-21 (2016)
23. Mishenin, Ye.V., Koblyanska, I.I: logistichni osnovi stalogo socialno-ekonomichnogo rozvitku region. Visnik Sumskogo nacionalnogo agrarnogo universitetu. **5(60)**, 3–7 (2014)

Social threats in the digitalization of economy and society

Olena Khandii*

Institute of Industrial Economics, National Academy of Sciences of Ukraine, 2 Zhelyabova Street, Kyiv, 03057, Ukraine

Abstract. As Ukraine slowly moves towards the digitalization of the economy and society, new trends in the social and labour sphere are revealed that have become challenges to the traditional system of social protection and regulation of labour relations. The impact of digitalization on the social sphere is ambiguous and creates both new opportunities and social threats that have become the purpose of the study. The social threats of digitalization for the nation include moral and spiritual degradation of society, reforming the economy without taking into account social priorities, the growth of structural unemployment, and cyberthreats in the national security system. It is substantiated that the main social threats posing a danger to every person in the process of digitalization are the Internet dependency, personality degradation, decline in mental health, further income differentiation, unemployment increase, increased state control over income and expenses of each person, decrease of personal life security, increased risk of loss of personal information, financial resources as a result of cybercrime and human negligence, computer illiteracy, transfer of decision-making powers to machines. Recommendations for improvement of institutional support of digitalization are proposed in order to minimize social threats in society.

1 Introduction

The digitalization processes taking place in the world are irreversible and provide a stable growth of economies at this stage of implementation. However, the scientific community is increasingly concerned about future changes in socioeconomic reality for the new generations of humankind growing in the era of high digital technology. Children of generation "Y" have become a new generation of the "digital person" – the generation "Z" which appeared in the conditions of globalization and digitalization, with its valuable ideas about the opportunities and challenges of the information era. The following basic components of social and labour relations are transformed: labour organization, payment and protection, guarantees of stable full-time employment, social protection of the unemployed and low-income groups of society. Generation "Z" does not idealize the approaches and traditions of the industrial age, freely possesses digital skills and values freedom in labour relations, looks for and follows its work and life balance. The conflict of

* Corresponding author: alkhandiy@ukr.net

priorities and values of managers and subordinates from different generations leads to the emergence of new methods of personnel management.

The works of supporters and active participants in the digitalization of the economy and society focus on the opportunities that will be facilitated by the intensification of this process. Digitalization is "the saturation of the physical world with electronic digital devices, systems and the establishment of electronic communication exchange between them, which actually enables the integral interaction between the virtual and physical, that is, creates a cyber-physical space" [1]. Among the main advantages of digitalization are the growth of the efficiency of the economy, the development of digital entrepreneurship, the improvement of the quality of services in education, medicine, transport, agriculture, increase in the environmental protection efficiency, reduction of the number and increase of the effectiveness of preventing disasters, public safety increase, reduction of corruption level in the system of public administration, increased participation of the population in social and political processes, creation of new jobs, at the micro-level – saving on the costs of organizing stationary working places. On the other hand, any changes in social and labour relations as a result of the digitalization of the economy, and, accordingly, new opportunities and threats in the social sphere, require the development and implementation of preventive measures that counteract the emergence and development of social threats and allow one to fully realize the potential for improving the quality of life of the population and the efficiency of the business entities.

The purpose of the work is to study the social threats that arise under the influence of the digitalization of the economy and society, in the social and labour areas.

2 Research results

Factors, conditions, phenomena, both existing and potential, which create or may create social danger are considered to be a social threat. Among the social threats posing a danger to every person when digitizing all spheres of his/her life should be singled out:

1. Internet addiction which is recognized by physicians more powerful and complex compared to other types of dependencies. The loss of the real world, the emergence of dependence on the virtual world, social networks negatively affects the mental and physical health of a person and allows manipulating the consciousness and actions of adults and children. Replacing real relationships with social life in social media creates an illusion of an intense active life, changes personality, spoils relationships with relatives and friends. The development of Internet addiction occurs in the form of participation in online financial transactions, game addiction, sexual communication, comments on materials in social media; permanent staying on dating websites, correspondence with strangers, browsing various websites without a specific purpose for hours. Internet addiction affects 2 to 10% of all network users [2].

2. Degradation of the population due to the preservation of necessary information on gadgets and the reduction of brain activity as a result of the use of online search systems to solve any problems, including in the learning process. The ability to get an answer to any question in a few minutes reduces the creativity of young people and children; ready-made solutions, examples and patterns form a one-track mind.

3. Decline in mental health of the population with easier access to violence, immoral videos depreciating the lives and health of other people and animals. The introduction of new technologies and the provision of new opportunities in the virtual world to all citizens must be accompanied by the establishment of new "rules of the game", norms of behaviour in the "new world", the definition of the rights, duties and responsibilities of each participant in the virtual world by all world society, since the virtual world does not have borders and does not recognize the boundaries of individual states, which complicates the

control of the content of Internet pages by government agencies. Therefore, there is an increasing need for strengthening and developing cultural values, moral guidelines, forming and maintaining an "acceptable" behaviour in society and condemning an "unacceptable" behaviour by public, popularizing socially meaningful content in the Internet.

4. Further differentiation of incomes and growth of the population, which belongs to the poor category, unemployment increase. According to various forecasts, the excess of the number of disappearing jobs over the number of created employment places in the near future may amount from 5.1 million to 85 million jobs. Carl Benedict Frey and Michael Osborne's study shows that 47% of jobs (in the United States) are at risk of automation [3]. Digital technology, machine learning, automation, previously used by large enterprises, are now being applied by medium and small businesses. Business executives have appreciated the benefits of digitalization of business processes that are associated with increased production efficiency and service delivery, increased productivity, and now jobs requiring both skilled and unskilled labour are being digitalized and automated.

Negative projections for job cuts do not concern IT professionals whose demand grows annually around the world. Creative, mobile employees with high digital skills are getting more valuable. "Social strata of "new rich" employees ("golden couples", "creative stars", etc.) and at the same time "new poor" employed in social production are appearing and growing... At one extreme of the employed in social production (top management, financial sector of international corporations, intermediary structures, investment funds, etc.) there is a concentration of income due to hypertrophied distribution both for property and for labor at the same time, which shows the asymmetry between the performance of these categories of the employed and received revenues. And at the other extreme, there is an expansion of the ghetto of non-standard employment and job market, requiring secondary and lower than the average qualification of the labour force" [4].

5. Increased state control over income and expenses of each person, decrease in protection of privacy, including freedom of speech, movement, rest. Increasing public safety by installing observation cameras in all public places, in the workplace, using face recognition technologies, on the one hand, provide protection against fraudsters and criminals, and on the other hand, allows controlling all the movements and conversations of employees and clients in the organization, the entire population on the whole territory where surveillance devices are installed.

There are different options for controlling income from fiscal services and the public, one of which is electronic declarations. The income declaration is either a voluntary declaration of the citizens, or a requirement of the legislation of the country, which enables income control. Since 2015, banking secrecy has ceased to operate in Ukraine due to amendments to the Law of Ukraine "On Banks and Banking" [5], according to which state bodies can make inquiries regarding the person's finances at the bank when making social payments, benefits and salaries at the expense of state funds. In compliance with the resolution № 148 of the National Bank of Ukraine [6], Ukrainians are obliged to make all purchases exceeding UAH 50,000 in a cashless form in order to allow fiscal services to control the movement of citizens' money. These legal documents and instruments allow controlling the income and expenses of each person, interfering with the people's personal life on the one hand, and on the other hand, preventing and counteracting the legalization of income and financing of terrorism.

6. Increased risk of loss of personal information, financial resources as a result of the growth of cybercrime and human negligence, computer illiteracy. The most vulnerable category of users is the elderly and children. "The entry point for the largest number of cyberattacks is personal, portable devices connected to the Web (smartphones, tablets, etc.) ... According to Cisco's data, today, up to 70% of cyber incidents start on personal mobile devices ... Cloud services, local corporate networks, Wi-Fi hotspots, industrial and home

appliances ... greatly increase the possibilities of cyberattacks for criminals" [7]. 65% of cyber incidents are due to deficiencies in modern security systems, up to 48% of incidents – to a low level of computer literacy among users, 55% of organizations fail to identify the cause of the incident [7].

7. Transfer of the right to make decisions to machinery (robots) following the prescribed rules and procedures, instructions may endanger the life and health of a person, unless a particular case is subject to general rules, but requires an individual approach and actions according to a particular situation. For example, the robot driving a car will follow the instructions given in accordance with the traffic regulations and may not stop in an unauthorized place if the passenger suddenly feels bad and needs to get out or get help. In turn, an automated recruitment system by certain criteria can leave talented and unique candidates for vacancies neglected if they do not, according to a certain criterion, score "the required number of points".

The main social threats of digitalization for a nation are moral and spiritual degradation of society, reforming the economy without taking into account social priorities, structural unemployment growth, and increased cyberthreats in the national security system.

The moral and spiritual degradation of society facilitates the manipulation of public opinion, formation of necessary moods in society, dissemination of false and half-truth information with a certain subtext, pushing for protest behavior, programming of the population for certain reactions and actions, including illegal. The author of articles [8-10] warns about the catastrophic impact of the digitization of the economy on the development of society while ignoring the cultural and moral values, norms and rules.

Reforming the economy without taking into account social priorities will exacerbate the social inequality and injustice existing in Ukrainian society; will lead to structural changes in industries, disappearance of uncompetitive enterprises, investment in which is ineffective, regardless of their social significance; the number of people at the poverty line will dramatically increase, and the need for social assistance from the state will increase from the part of the able-bodied population each year.

Changes in the sectoral structure under the influence of digitalization will lead to an increase in structural unemployment, which requires a rapid response from the education system and state support for training specialists in new professions and the creation of new jobs. Training specialists in new professions takes longer than the disappearance of outdated professions and the increase in the number of unemployed, therefore, without state regulation of the labor market, aimed at implementing social policy, it is impossible to overcome the negative phenomena in the social sphere that arise in the digitalization of the economy and society.

The growth of cyberthreats in the national security system requires active and preventive measures at all stages and levels of implementation of the Concept for the Development of the Digital Economy and Society of Ukraine. The number of cybersecurity specialists, both in Ukraine and around the world, is too small and their qualifications are insufficient. "The global need for cybersecurity specialists is now 12 times higher than other IT specialists need, ... 37% of the world's employers are not satisfied with the low level of training in this area" [7]. Traditional security technologies do not provide complete security, and the diversity of cyberthreats and the scale of their harm is constantly increasing, requiring new security technologies, advanced training for cybersecurity specialists and computer literacy users.

3 Conclusion

Thus, the digitalization of the economy and society requires simultaneous changes in the social policy of the state, including the labor legislation of Ukraine, in legislative acts concerning social protection of the population, national security of Ukraine, and the state budget of Ukraine. The absence of legally established rules, procedures and mechanisms for the consideration of social issues in the process of digitalization leads to discriminatory phenomena in the system of social and labour relations and the escalation of social threats in society.

Institutional support for digitalization should be accompanied by public support, which consists in the formation of "The Code of Ethics of the Internet User" and "The Code of Ethics of the Content Owner". "The proper use of the Internet possibilities and social networks" should be a separate issue of teaching digital disciplines from junior classes, which will teach "acceptable" and define "unacceptable" behaviour in the virtual world to preserve physical and mental health.

The level of freedom and non-interference in the life of every citizen must be balanced with the level of social security and ensured without violating the privacy of every person.

The digitalization of the economy should take place considering social priorities, adherence to the principles of social justice and development of the social and labor potential of the country, and the achievement of the economic effect should not lead to an increase in social threats and negative social effects.

References

1. Cabinet of Ministers of Ukraine, *The Concept of the Development of the Digital Economy and Society of Ukraine for 2018-2020 of 17 January 2018.* Retrieved from: https://zakon.rada.gov.ua/laws/show/67-2018-%D1%80 (2018)
2. S.M. Savenka, O.S. Yurtsenyuk, S.S. Rotar, *Online addiction is also a disease*. Retrieved from: https://www.bsmu.edu.ua/uk/news/digest/5440-internet-zalezhnist-%E2%80%93-tse-tezh-hvoroba (2017)
3. *Report of the Rome Club 2018, Chapter 1.11: Subversion Technologies and the Digital Revolution*. Retrieved from: https://habr.com/ru/company/philtech/blog/425401/ (2018)
4. A.M. Kolot, The social and labor sphere in the coordinates of the new economy: the expansion of opportunities and new threats. *Sotsialno-trudova sfera v umovakh stanovlennia novoi ekonomiky: hlobalni vyklyky ta dominanty rozvytku: zb. tez dop. uchasnykiv kruhloho stolu*. In A.M. Kolot (Ed.) (EDP KNEU, 2018).
5. The Verkhovna Rada of Ukraine, The Law of Ukraine On Banks and Banking № 2121-III of 7 December 2000. Vidomosti Verhovnoyi Rady, №. **5-6**, art. 30. Retrieved from: https://zakon.rada.gov.ua/laws/show/2121-14?lang=en (2000)
6. National Bank of Ukraine, Resolution On Approval of the Regulation on the Conduct of Cash Transactions in the National Currency in Ukraine № 148 of 29 December 2017, v0148500-17. Retrieved from: https://zakon.rada.gov.ua/laws/show/en/v0148500-17?lang=uk (2017)
7. S.L. Gnatyuk. National Institute for Strategic Studies. Retrieved from: http://www.niss.gov.ua/doslidzhennya/analitichni-materiali/informaciyni-strategii/kiberbezpeka-v-umovakh-rozgortannya (2019)
8. V.V. Kompaniets, Visnyk ekonomiky transportu i promyslovosti, **61**, 36-46. Retrieved from: http://nbuv.gov.ua/UJRN/Vetp_2018_61_6 (2018)
9. V.V. Kompaniets, Capitalism and post-capitalist future: socio-economic and moral assessment. *Sbornik materialov Pervogo Mezhdunarodnogo Kongressa pravoslavnyh uchenyh: Evangel'skie cennosti i budushchee pravoslavnogo mira: pravoslavie i nauka*, 147-151 (EDP Sciences, Voronezh: Istoki, 2017)
10. V.V. Kompaniets, Visnyk ekonomiky transportu i promyslovosti, **65**, 86-95 (2019).

The Influence of Mergers and Acquisitions on Competitiveness of Economies of Countries of the World

Viktoriia Khaustova[1]*, *Tetyana* Kolodyazhna[2] and *Hennadii* Kramarev[3]

[1]Research Centre for Industrial Problems of Development of NAS of Ukraine, Sector of Industrial Policy and Innovative Development, 1-a Inzhenernyi Ln., Kharkiv, 61166, Ukraine
[2]Kharkiv National University of Construction Engineering and Architecture, Department of Finance, 40 Sumska Str., Kharkiv, 61002, Ukraine
[3]Research Centre for Industrial Problems of Development of NAS of Ukraine, Applicant, 1-a Inzhenernyi Ln., Kharkiv, 61166, Ukraine

Abstract. Today, the most important factor which determines the development of the world economy is economic globalization. The article presents a scheme of the interdependence of the processes that determine competitiveness of the country in the context of globalization: CER.1 – the cause-effect relation reflecting the influence of M&A with the participation of national companies on concentration of a national economy; CER.2 – reflecting the impact of concentration of the national economy on its competitiveness; CER.3 – reflecting the impact of intensity of M&A with the participation of national companies on their capitalization; CER.4 – reflecting the impact of the level of national companies' capitalization on competitiveness of the country. The article puts forward and proves, based on the confirmation of the above mentioned cause-effect relations, the following hypotheses: 1. – merges and acquisitions with the participation of national companies influence concentration of a national economy which in turn determines the level of its competitiveness; 2. – integration with the participation of national companies through implementation of M&A increases their market capitalization, thereby affecting the level of competitiveness of the country. For simulating, the correlation and regression modeling is used. The statistical base of the study comprises a sample of 42 countries of the world.

1 Introduction

In a modern economy, competitiveness of the country determines the level of its socio-economic development. At the same time, today the most important factor determining the development of the world economy is economic globalization, the spread of which is the result of a whole range of processes taking place in society. Globalization has created new organizational forms of interaction among countries and individual companies, which allow overcoming obstacles created by state borders, and changed the conditions and the very principles of competition [1].

The basis of globalization is international economic integration, which is carried out both at the level of countries and at the level of companies. The activity of national companies in the

* Corresponding author: v.khaust@gmail.com

processes of mergers and acquisitions determines the degree of concentration of the national economy (economic concentration), affects market capitalization of companies and thereby determines competitiveness of the country.

The analysis of the scientific literature has shown that the influence of integration processes implemented through mergers and acquisitions of companies on competitiveness of an economy is insufficiently studied. The above indicates the relevance of further research on the outlined issue.

2 Theoretical basis and methodology

Within the framework of this research, it is proposed to understand competitiveness of the country as the ability of its scientific and technical resources and implemented economic policy to ensure the production of goods and services which are in demand in the market and create conditions for sustainable economic growth of the country and improvement of living standards of its population.

As mentioned earlier, the modern development of the world economy and competitive processes in it is determined by globalization of the world economy. The pace of globalization has intensified since the beginning of the 21st century and its processes are accompanied by amalgamation of economic entities, internationalization of national economies and, accordingly, growth in economic concentration [2]. Under these conditions, activities of transnational corporations (TNCs) have gained particular importance, and their influence on the economy of an individual country and development trends in the world economy increased significantly [3]. Thus, in the world, there observed a steady trend towards the spread of concentration processes both at the level of economies and that of individual industries and economic entities, the major concentration taking place within TNCs.

The analysis of the modern scientific literature has shown that the concept "concentration" is used by scientists and practitioners in different meanings and in relation to various levels of the economy and objects under consideration (economy, industry, market, etc.) [4]. In our opinion, concentration of capital should be investigated from two standpoints: 1 – as a result of the conversion of profits into capital by capitalizing surplus value; 2 – as strengthening capital through consolidating the capital of various owners [5]. Based a number of the most common definitions of the concept "concentration of an economy", in this study, it is proposed that the essence of the concept "concentration of an economy" be interpreted as degree of concentration of production volumes or provision of services by the largest national companies in relation to the national GDP.

In world economic practice, there are three main options for the development of companies which imply: 1 – independent internal development of tangible and intangible assets; 2 – gaining competitive advantages on a contractual basis, i.e., outsourcing; 3 – mergers and acquisitions (M&A). Despite the fact that many companies are successfully building their internal growth strategy, in recent decades, M&A are increasingly becoming the main competitive strategy, being the fastest way to acquire assets and expand into new types of sales markets, etc. According to experts, almost 80 % of all international investments are directed not to establishing new companies abroad but to acquiring assets of existing ones. Due to this, in the second half of the 20th century, the processes of concentration of production and centralization of capital manifested themselves in an avalanche-like increase in the number and value of M&A transactions in economies of different countries of the world. The intensity and focus of M&A processes are reflected in assessments of the level of concentration of the economy at all levels of its hierarchy. The state of development of the national economy and its competitiveness depend on the state of economic concentration. In this context, another manifestation of integration processes is reflected, in particular, that integration results in an increase in the market value (capitalization) of an integrated entity, making it more "strong" and competitive [6].

Thus, based on the above analysis, the following *research hypotheses* are put forward:

– *the hypothesis on the impact of M&A processes with the participation of national companies (as one of the most common forms of integration) on the economic concentration in a country and its level of competitiveness;*

– *the hypothesis on the impact of integration processes with the participation of national companies through implementation of M&A on the level of their market capitalization and the level of competitiveness of a country.*

Accordingly, it is proposed to analyze the impact of integration processes through M&A on competitiveness of economies of countries of the world by proving or refuting the hypotheses put forward with the help of identifying a number of cause-effect relations (CERs) (Fig. 2).

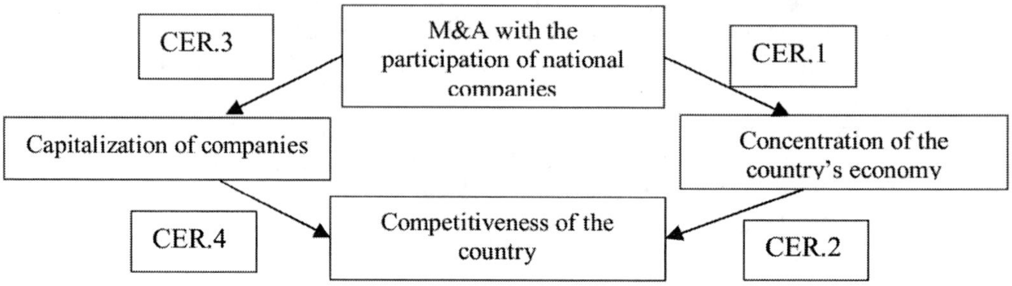

Fig. 2 Scheme of the interdependence of the processes that determine competitiveness of the country in the context of globalization, where: CER.1 is the cause-effect relation reflecting the influence of M&A with the participation of national companies on concentration of a national economy; CER.2 – reflecting the impact of concentration of a national economy on its competitiveness; CER.3 – reflecting the impact of intensity of M&A with the participation of national companies on their capitalization; CER.4 – reflecting the impact of the level of national companies' capitalization on competitiveness of the country.
Source: developed by the authors

Considering the objectives of this study – to prove or refute the hypotheses put forward by assessing the strength and direction of the relationship between certain factors and to establish the analytical expression of the dependence between them – the correlation and regression modeling is used. To solve the problem, a pair correlation, which demonstrates the strength and direction of relationship between the influencing and resultant factors, is applied. For a qualitative assessment of the relationship, the Cheddok scale is used. The statistical base of the study is compiled from a sample of 42 countries of the world. The calculations are carried out using the Statistica software package.

3 Results

According to the research objectives and scheme of the relationship of the processes determining competitiveness of the country shown in Fig. 2, there investigated:

1. The impact of integration processes of M&A with the participation of national companies on the level of their market capitalization. To determine the relationship between M&A of companies and their capitalization, the following indicators are used: the relative index of M&A volume (I_{MA}) and the relative index of market capitalization (I_{cap}). The relative index of M&A volume (I_{MA}) is calculated as the ratio of volume of transactions on M&A of companies (according to the World Investment Report published by UNCTAD [7]) to GDP of the countries in which they were created (according to the World Bank [9]). The relative index of market capitalization (I_{cap}) is calculated as the ratio of market capitalization of the largest companies (according to Forbes 2000 [8]) to GDP of the countries in which they were created (according to the World Bank [9]).

As a result of the calculations, the following models characterizing the impact of M&A processes on capitalization of companies involved are obtained: linear; exponential; 2nd -, 3[d] -; 4th -; 5th - and 6th degree polynomial. The 6th degree polynomial dependency, presented in Fig. 3, has the highest coefficient of determination between the indexes (R^2= 0.7091). Thus, the analysis of the relationship between M&A volume and market capitalization of companies involved has made it possible to substantiate the following statement: the countries in which M&A processes are more intensive have a higher level of capitalization of companies involved.

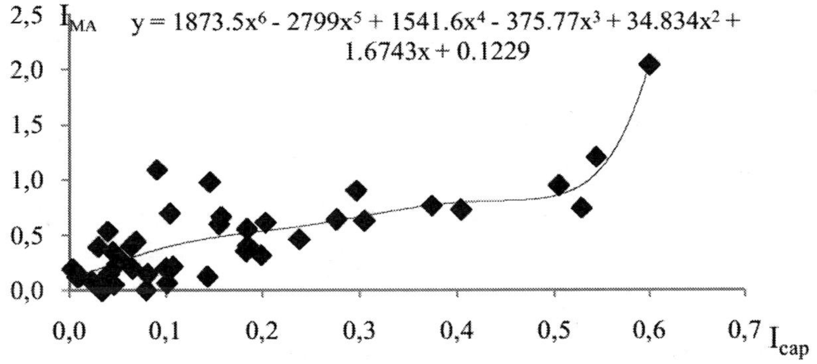

Fig. 3. The 6th degree polynomial correlation and regression model of the impact of M&A processes (I_{MA}) on capitalization of companies involved (I_{cap})
Source: developed based on the authors' calculations

2. The impact of capitalization of national companies on competitiveness of a country. To determine the relationship between market capitalization of national companies and competitiveness of the country in which they were created, the following indicators are used: the relative index of market capitalization (I_{cap}) and the relative index of competitiveness of the national economy (I_{com}). The relative index of market capitalization (I_{cap}) is calculated as the ratio of market capitalization of the largest companies (according to Forbes 2000 [8]) to GDP of the countries in which they were created (according to the World Bank [9]). The relative index of competitiveness of the national economy (I_{com}) is calculated as the Global Competitiveness Index (GCI) of the WEF [10].

As a result of the calculations, the models characterizing the impact of capitalization of national companies on competitiveness of economies of the countries are obtained. The 6th degree polynomial dependency (Fig. 4) has the highest coefficient of determination between the indexes (R^2 = 0.6729). Thus, the conducted study has allowed to confirm the existence of the cause-effect relation CER.4 and substantiate the statement: the countries of the world whose companies have higher capitalization have a higher competitiveness of the economy.

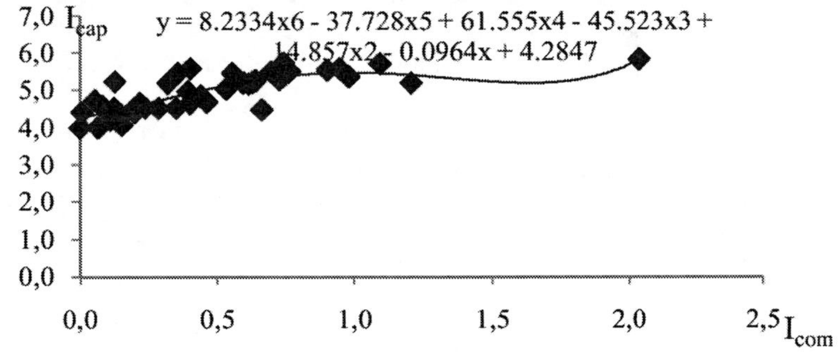

Fig. 4. The 6th degree polynomial correlation and regression model of the impact of capitalization of companies (I_{cap}) on competitiveness of the national economy (I_{com})
Source: developed based on the authors' calculations

3. The impact of M&A processes with the participation of national companies on economic concentration in a country. To determine the relationship between M&A with the participation of national companies and economic concentration in a country, the following indicators are used: the relative index of M&A volume (I_{MA}) and the relative index of concentration of an economy (I_{co}). As in previous models, the relative index of M&A volume (I_{MA}) is calculated as the ratio of volume of M&A transactions (according to the World Investment Report published by UNCTAD [7]) to GDP of the countries in which they were created (according to the World Bank [9]). The relative index of concentration of an economy (I_{co}) is calculated as the ratio of sales of the largest companies (according to Forbes 2000 [8]) to GDP of the countries in which they were created (according to the World Bank [9]).

As a result of the calculations, there obtained models which characterize the impact of M&A processes on concentration of the national economy. The 6th degree polynomial dependency (Fig. 5) has the highest coefficient of determination between the indexes ($R^2 = 0.6406$). Thus, the study has allowed to confirm the cause-effect relation CER.1 and substantiate the statement: the countries of the world whose companies are more active in M&A processes have a higher degree of economic concentration.

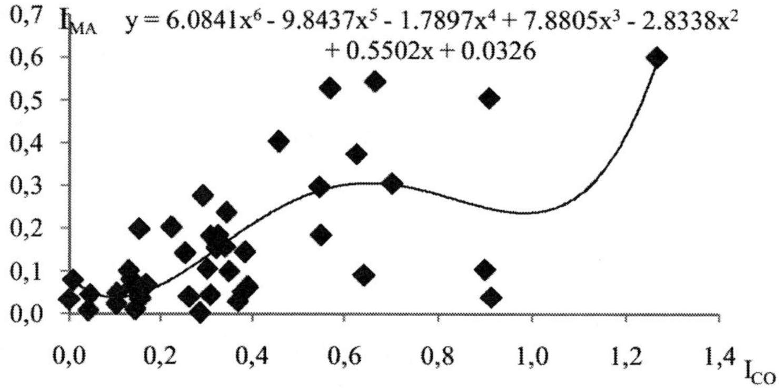

Fig. 5. The 6th degree polynomial correlation and regression model of the impact of M&A processes (I_{MA}) on concentration of economies of the countries of the world (I_{co})

Source: developed based on the authors' calculations

3. The impact of concentration of the national economy on competitiveness. To determine the relationship between the concentration of the national economy and its competitiveness, the following indexes are used: the relative index of concentration of economy (I_{co}) and the relative index of competitiveness of the national economy (I_{com}). As in previous models, the relative index of concentration of an economy (I_{co}) is calculated as the ratio of sales of the largest companies (according to Forbes 2000 [8]) to GDP of the countries in which they were created (according to the World Bank [9]). The relative index of competitiveness of the national economy (I_{com}) is calculated as the Global Competitiveness Index (GCI) of the WEF [10].

As a result of the calculations, the models characterizing the impact of concentration of the national economy on its competitiveness are obtained. The 6th degree polynomial dependency (Fig. 6) has the highest coefficient of determination between the indexes ($R^2 = 0.7137$). Thus, this has allowed to confirm the cause-effect relation CER.2 and substantiate the statement: countries of the world with a higher degree of economic concentration are more competitive.

4 Conclusions

Consequently, the confirmation of the above cause-effect relations (CER.1 - CER.4) has made it possible to prove the hypotheses put forward in this study (Fig. 1), namely: 1 – M&A

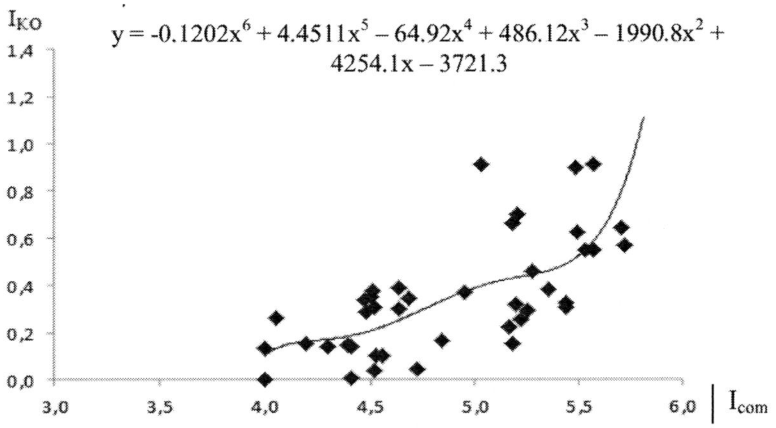

Fig. 6. The 6th degree polynomial correlation and regression model of the impact of concentration of the national economy (I_{co}) on competitiveness of the country (I_{com})

Source: developed based on the authors' calculations

with the participation of national companies influence concentration of the national economy which in turn determines the level of its competitiveness; 2 – Integration with the participation of national companies through implementation of M&A increases their market capitalization, thereby affecting the level of competitiveness of the country.

The carried out research suggests that national governments should pay attention to conditions for the implementation of integration processes in the countries, their stimulation and control (including antimonopoly control). Measures to support and stimulate the development of integration processes should be aimed at the formation of effective national integrated business structures with a sufficient amount of investment resources. The presence of such structures in key sectors of the national economy is the key to security of the national economy, the competitiveness of the country and the basis for the growth of the level and quality of life of its people.

References

1. Gelvanovsky, M. Competitiveness of the national economy: problems of statistical support. Economic issues, 7, 3–15 (2004)
2. Walter, J. Competitiveness: a common approach. Russian-European Center for Economic Policy, 52 p. (2005)
3. Limonova, E., Arkhipova, K. Transnational corporations: the main stages of development and regulation of their activities in the conditions of globalization. European vector of economic development, 1(16), 93–101 (2014)
4. Kyzym, M., Gorbatov, V. Concentration of economy and competitiveness of countries of the world. PH "INZHEK", 216 p. (2005)
5. Khaustova, V., Kolodyazhna, T. Influence of Mergers and Acquisitions on the Competitiveness of the Economy in the Conditions of Globalization, Publisher Liburkina L., 416 p. (2019)
6. Zagorsky, V., Khaustova, V. Theoretical Aspects of Capitalization of Entities. Innovations: Problems of Science and Practice. PH "INZHEK", 176–192 (2009)
7. UNCTAD. http://stats.unctad.org
8. Forbes Magazine. https://www.forbes.com
9. World Bank. https://data.worldbank.org/indicator/
10. World Economic Forum 2017. The Global Competitiveness Report 2016-2017. https://www.weforum.org/ reports/the-global-competitiveness-report-2017-2018

Contemporary economic development challenges of the industry in luhansk region

Yuliia Klius[1], *Iryna* Kushal[1] *Hanna* Pchelynska[1*], and *Eleanora* Chernodubova[2]

[1] Volodymyr Dahl East Ukrainian National University, Department of Accounting and Taxation, pr.Central 59-a, 93400, Severdonetsk, Ukraine
[2] Volodymyr Dahl East Ukrainian National University, Department of Finance and Banking, pr.Central 59-a, 93400, Severdonetsk, Ukraine

Abstract. The modern Ukrainian economy is characterized by significant transformational processes that are accompanied by a transition from heavy industry to the agro-industrial sector. In recent years, traditional key domestic industries with high added value in production have substantially lost their positions in the domestic and foreign markets. Especially acute the crisis phenomena are observed in the areas that are involved or are bordered by the area of the armed conflict. Before the beginning of hostilities in the East of Ukraine, Luhansk region was among the five most powerful industrial and economic regions of Ukraine. Now the region has turned from industrial to agrarian. Therefore, the matter of analysing the impact of post-conflict transformation processes in Ukraine on the region's economic development and identifying the reasons that hinder transformation in the areas of post-conflict transformation is relevant. The article assesses the impact of military actions in the east of Ukraine on the economic performance of the industry in the region. The modern problems of Luhansk region industrial enterprises development are determined. According to the results of the macro-environment factors PEST analysis of the industrial enterprises in the region the main measures to overcome the crisis phenomena in the region's economy are determined.

The modern economy of Ukraine is characterized by significant transformation processes that are accompanied by a transition from heavy industry to IT and agro-industrial complex (agriculture, food industry, equipment manufacturing, etc.) [1]. Over the past five years, traditional key domestic industries with high added value in production have substantially lost their positions in the domestic and foreign markets, in particular, engineering, chemical industry. The military actions in the east of the country and "opening up" of the European Union market have played a key role in transforming the Ukrainian economy. Especially acute crisis phenomena are observed in the territories involved or bordering the zone of military conflict. This article studies the influence of the post-conflict transformation

* Corresponding author: pchelinska2@gmail.com

processes on the economic performance of the Luhansk region industry, the definition of the main challenges of the regional economic development and ways to overcome them.

Prior to the start of the anti-terrorist operation in 2014, Luhansk region had a significant economic potential and was one of the five most powerful industrial and economic regions in Ukraine. Since the beginning of the military conflict, more than 2 thousand social infrastructure facilities, housing stock, life support systems have been destroyed; transport, energy and industrial infrastructures have been damaged in the region. The economic crisis of the region is confirmed by the regional statistics. The level of the volume indices of gross regional product in Luhansk region in 2014-2015, 2017 was the lowest among all regions of Ukraine. It decreased from 92.2% in 2013 to 47.7% in 2015, in 2017 it was 83.8%. The gross regional product per person indicator in 2014-2017 took the last place, despite the fact that in 2013 Luhansk region surpassed half of the regions of Ukraine in this indicator. For the analyzed period, the industrial production index decreased by 66.0% [2].

The extent of the impact of taxation on the development of the regional economy, the individual sectors of economic activity, territories or groups of economic actors allows us to estimate the level of tax burden. According to one approach, it is calculated as a relation of the total tax collections to the total gross regional product. In terms of tax burden, Luhansk region was ranked 4th in 2013, 5th in 2014 and 2015, 3rd in 2016, 12th in 2017 (calculations made according to [2]). A high level of tax burden on the economy of the region can be stated. This indicates the inefficiency of the system of redistribution and use of budgetary resources, forms and preserves the extensive type cycle of regional development, which adversely affects the competitiveness of the national economy as a whole and deepening the developmental imbalances.

During 2013-2017, in terms of the volume of products (services) sold in the region, industrial, commercial and agricultural enterprises as well as forestry and hunting enterprises occupy leading positions. The highest indicators are shown by the industrial enterprises, although until 2017 their sales volume decreased significantly due to the loss of a significant number of enterprises or a decrease in business volumes (Fig. 1, 2).

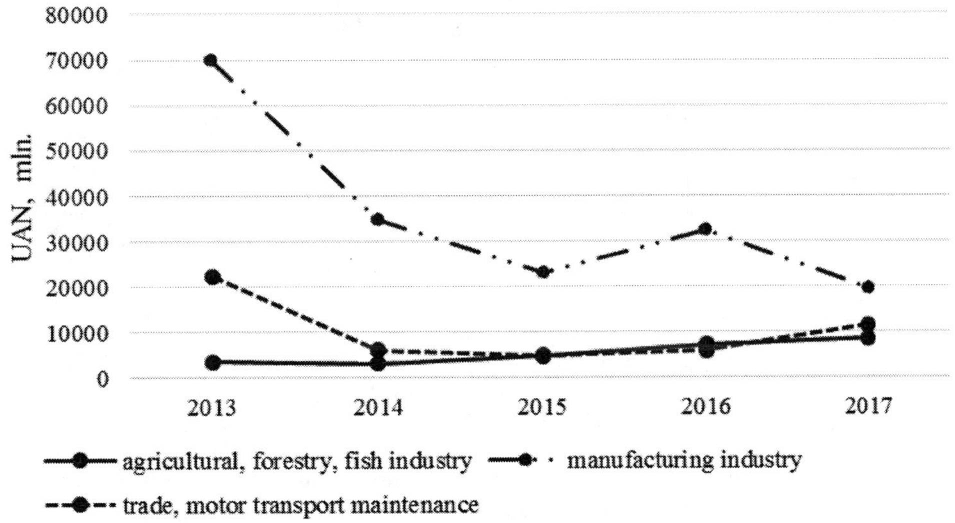

Fig. 1. Commerce dynamics of business entities of Luhansk region by the main types of economic activity [2]

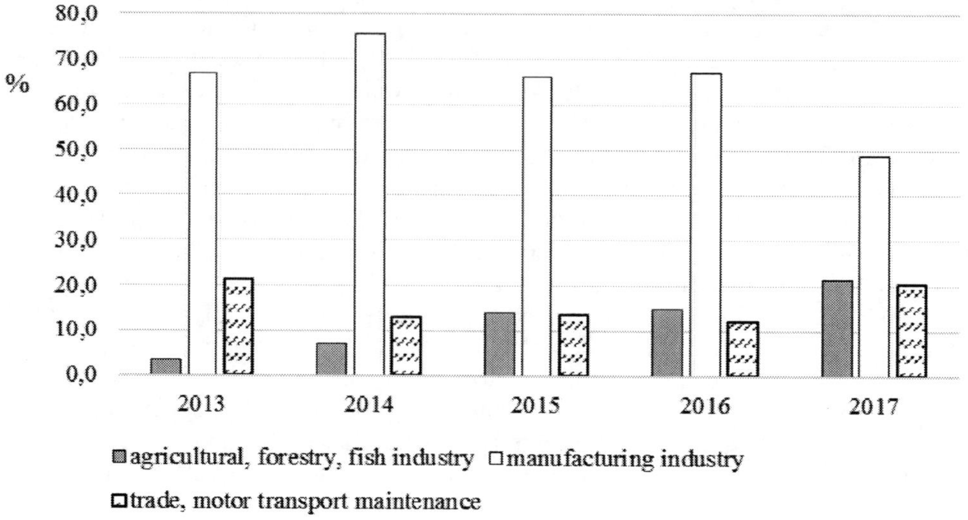

Fig. 2. Business Entities Commerce Volume in Luhansk Region Structured by the Main Types of Economic Activity [2]

The costs of the industry operating activities in the region during 2014-2017 exceed the sales volumes of enterprises (Table 1). The most critical situation is observed in 2015, 2017. In general, for the period under review, operating expenses and sales of the Luhansk region industrial enterprises products decreased by 40.8% and 43.9%, respectively. The increase in the fixed costs share (rent, utility bills, the share of labour costs, etc.) in their total with the increase in prices and tariffs (39.1% on average from 2014 to 2017) led to increasing unprofitableness in the operating activities of industrial enterprises from 22.9% in 2014 to 35.1% in 2017.

Table 1. The Dynamics of the Luhansk Region Industrial Enterprises Economic Indicators [2]*

Indicator	2014	2015	2016	2017
Sales Volume, UAH mln.	34681.8	22922.0	32441.5	19469.2
Operating Expenses, UAH mln.	59455.7	58484.1	49445.8	39023.2
Price Index for Hot Water, Heating,%	100.0	167.3	170.9	150.3
Electricity Price Index,%	111.3	166.7	160.0	128.1
Water Supply Price Index,%	100.0	168.7	151.4	116.2
Operating Profitability,%	-22.9	-24.2	-16.8	-35.1

*Data are given without taking into account the performance of banks and budget institutions. Excluding part of the temporarily occupied territory in Lugansk region.

After Luhansk region has lost the leading position in the key sectors of the regional economy - the chemical and mining industries, the mechanical engineering sector takes the first place in terms of financial performance; as well as manufacture of wood products, paper and printing activities (Table 2).

According to Table 2 we can draw the following conclusions. The activity of the mechanical engineering enterprises in the region, unlike the indicators of the industry in total, is profitable. A slight drop in operating profitability can be traced in 2015 - to 1.0%.

Table 2. Economic Indicators Dynamics of the Industrial Enterprises Financial Performance in Luhansk Region [2, 3]

Indicator	2014	2015	2016	2017
Share of Profitable Enterprises by Types of Industrial Activity in Total,%				
Industry:				
Average in Ukraine	63.3	72.9	72.8	71.6
Average in Luhansk Region	51.5	65.2	68.0	67.3
Engineering:				
Average in Ukraine	64.7	74.1	77.2	77.4
Average in Luhansk Region	59.5	61.9	72.1	77.3
Operational Cost Effectiveness,%				
Industry:				
Average in Ukraine	1.6	0.9	4.2	6.6
Average in Luhansk Region	-22.9	-24.2	-16.8	-35.1
Engineering:				
Average in Ukraine	-2.4	3.4	8.0	13.1
Average in Luhansk Region	3.1	1.0	14.2	8.6

In the second half of 2015, industrial production in Ukraine and the region gradually began to grow, which increased the operating profitability of the industry enterprises to 8.0% and 14.2% in 2016, respectively. However, the next increase in prices for inventories, tariffs for services, the minimum wage of workers increase in 2016-2017 led to a drop in the level of profitability of the enterprises of the mechanical engineering industry of the region to 8.6%. This is less than the average level of profitability in the industry by 4.5%. But the share of profitable enterprises increased from 59.5% in 2014 to 77.3% in 2017. Preservation of the overall profitability of the engineering industry in the region in 2015 was ensured by the activities of enterprises producing electrical equipment (operating margin of 104.5%). But in subsequent years, a decrease in the efficiency of this type of industrial activity to 9.1% was observed. The most stable in terms of operating profitability is the production of computers, electronic and optical products – 42-47%. This type of industrial activity is a leader in the engineering industry of Luhansk region in terms of profitable enterprises - in 2016-2017 it was 100%.

In order to study the macro environment factors of the industrial enterprises functioning in the region, the authors used PEST analysis. Its results allowed to identify the following groups of the major problems of economic development of industry [4-9]:

1) Political: continued joint forces operation in the country, campaigning for 2019, high level of privatization, excessive politicization of economic cooperation, mistrust of business to the authorities and its bodies, implementation of the Association Agreement with the EU, lack of a state program to support separate industries in the region, admission of Russian competing enterprises products to the Ukrainian market, etc.

2)) Economic: high inflation, devaluation of the hryvnia, instability and high exchange rates, commercial banks' high lending rates, low effective demand of the domestic market, unallocated losses of previous years business partners in related industries, limited own working capital, no real external investors , uncontrolled rise in prices for goods and services of natural monopolies (raw materials, energy), insufficient investment in industry, lack of government programs financing for the long-term development of industry for the next 5 years, increased competition in the domestic market from foreign producers, the dependence of the domestic enterprises on the supply of imported inventories, low profitability of production, loss of sales markets, reduction of exports, lack of alternative sales markets for primary products etc.

3) Social: increase in the outflow of specialists with work experience, reduce in the quality of specialist training, decrease of the working population, migration of the working

population, reducing the attractiveness of the labour in the industry, a significant gap between wages and employee needs, low prestige of labour in the manufacturing sector, shortage of labour occupations, etc.

4) Technological: a high level of material and energy intensity of the final products, a high level of depreciation of the fixed assets, an insignificant level of innovation, a mismatch between the innovative potential of the manufacturer and its customers-consumers, a slight level of production modernization, etc.

Thus, the impact study of the military operations in the east of Ukraine on the economic performance of the industry in Luhansk region, the identification of the main problems of the economic development of the region allows us to identify the following measures to overcome them [4-9]:

1) Political: public authorities support for the development of the State Program for the implementation of the EU technical standards and regulations, its funding; protection of the domestic producers from imported products, analogues of which are produced in Ukraine, by increasing the import duty rate; the state demand strategic planning for the infrastructure modernization; the free economic zones and territories of priority development update; simplification of the process and procedures for obtaining permits for the of innovative investment projects implementation in the industry; strengthening of the control and responsibility for the failure of the state governmental authorities to fulfill their obligations, enshrined in legislation, government programs, and etc.; business cooperation with legal institutions to protect the interests of the national manufacturer, etc.

2) Economic: timely adjustments of the business strategies in accordance with the market situations; production diversification; search for alternative suppliers, markets; establishment of the representative offices of the enterprises abroad; expanding the range of products; cost reduction by using information technology; expanding the range of the components of the own production; expansion of the cooperation with foreign partners; self-insurance at the expense of established reserves for future expenses and payments; land market liberalization; continue working on the energy independence.

3) Personnel: the preservation of highly qualified personnel by improving working conditions and its remuneration; the creation of local centres of the best practices and professional competence, and so on.

4) Technological: development of the high-tech products, creation of the new technology with exceptional capabilities; implementation of the best practices, methods and processes of economical production, investment in new technologies and modernization of production, etc.

The obtained research results constitute theoretical and practical basis for further development of the scientific base and improvement of the regional management theoretical and methodological foundations to ensure sustainable development in the territories of the post-conflict transformation on the basis of the formation of a comprehensive organizational and economic provision of the regional management system.

References

1. Struktura ekonomiky Ukrayiny. GTAI Germany Trade & Invest (2018). URL: http://ukraine.ahk.de/fileadmin/ahk_ukraine/GTAI/Struktura_ekonomiki__23-10-2018.pdf
2. Statystychna informatsiya. Derzhavna sluzhba statystyky Ukrayiny. URL: http://www.ukrstat.gov.ua/
3. Statystychna informatsiya. Holovne upravlinnya statystyky v Luhanskiy oblasti. URL: http://www.lg.ukrstat.gov.ua/statinform.php.htm
4. Bazy danykh pidpryyemstv. Ahent·stvo z rozvytku infrastruktury fondovoho rynku Ukrayiny. URL: https://smida.gov.ua/db

5. Y. A. Zhalilo, 2018: ochikuvannya ta ryzyky, Visnyk. Ofitsiyno pro podatky, No.1 (2018).
6. O. B. Mnykh, Kryza transportnoho mashynobuduvannya v Ukrayini ta marketynhovyy pidkhid do formuvannya rynkovoyi vartosti mashynobudivnykh pidpryyemstv. Aktualni problemy ekonomiky, No. 9, 77–86 (2016)
7. V. Sytnichenko, Stratehiya rozvytku mashynobuduvannya na osnovi systemnykh pokaziv
8. O. V. Shubina, A. S. Loz·hunova, Suchasnyy stan, problemy ta perspektyvy rozvytku mashynobudivnoho kompleksu Ukrayiny, Nauchnyy Vestnyk DHMA, No.3, 153–158 (2017)
9. L. Klishchuk, Ukrayinska promyslovist ta enerhetyka - trendi 2018, Na chasi (2018)

Approaches to formation of strategies for development of administrative territories

Oleksii Klok[1], *Olha* Loseva[1*], *Oleksandr* Ponomarenko[1]

[1] Simon Kuznets Kharkiv National University of Economics, Kharkiv, Ukraine

Abstract. The article studies theoretical and methodological bases of the strategic management of the development of administrative territories, considers the essence of strategic management and formulates the advantages of using it in management of administrative territory. Based on the analysis of the key provisions of the EU regional policy, the strategy of "smart specialization" is considered as the most common approach to territorial development. Using the experience of the countries of the European Union as a basis, a BPMN diagram, describing the conceptual bases for the formation of a competitive territory strategy, was built. Practical approaches to the formation of strategies for the development of administrative territories operating in Ukraine, regulatory acts, in particular, that had a direct impact on the formation of the existing model of strategic territorial management, were analyzed. The main requirements to the content of the strategic plan were considered and the list of key provisions and analytical methods (socio-economic analysis, comparative analysis, SWOT-analysis, PESTLE-analysis, sociological analysis) was formulated. Using the comparative legal analysis of the experience of the European Union as a basis, a number of features can be highlighted that must be taken into account in the process of forming the administrative territory development strategy.

1 Introduction

Methods of strategic management, which arose in view of the evolution of management systems under the influence of globalization, instability of environment and changes in the nature of competition, were first used by micro-level systems and only over time entered the practice of managing non-profit organizations, the state and its individual territories. Comparing previous years, more attention is paid to the strategic management of territorial development, as managers have become aware of the effectiveness of its use. At the present stage, the strategic management of administrative territories in Ukraine has a non-systemic nature and low efficiency in the use of tools, remains a poorly studied area, requires improvement of systems and approaches with the purpose of their further implementation.

Ukraine's integration into the European Union is impossible without an adequate practice of using strategic management in territorial development. Summarizing the current situation in Ukraine, it can be stated that all regions are characterized by the same problems

* Corresponding author: olha.loseva@hneu.net

of socio-economic development: uncertainty of strategic priorities, orientation of program documents to the use of extensive development factors, disproportionate nature of development.

2 Methodology

The theoretical and methodological foundations of strategic management of administrative territories development are presented in the studies of representatives of various scientific schools. The increasing degree of Ukraine's involvement in the world economy, in the processes of integration, in particular, internationalization and globalization, requires increasing attention to the world experience in territorial development. The implementation of the world experience allows you to act faster and increase effectiveness of managerial decisions. The information base for the study is legislative and other regulatory acts on the formation of strategies for the development of administrative territories, strategic planning, local self-government. The studies of Ukrainian and foreign authors, published in monographs and scientific journals, including the work papers of European Commission (EC), the Organization for Economic Co-operation and Development (OECD), Global Practitioners Network for Competitiveness, Clusters and Innovation (TCI) were analyzed.

The purpose of the study is to analyze the practical approaches to the formation of strategies for the development of administrative territories operating in the European Union and Ukraine on the basis of the theoretical and methodological support of the strategic management of administrative territories development.

A number of general scientific methods have been used to achieve this objective, such as: retrospective analysis; comparative-legal analysis; idealizations and analogies; content analysis; theoretical generalization.

3 Formation of Strategies for Development

3.1 Strategic Management

Strategy is more than objectives and actions. Strategy, in particular, is the choice of where and how to compete. Strategic management is a focused management activity in accordance with the chosen strategy, which allows us to function effectively in interaction with external environment using all possible potential, that is, a stable view of current activities, to achieve future objectives with constantly changing internal and external parameters [1].

The advantages of using the strategic management of the socio-economic development of the administrative territory include: raising the level of scientific and practical justification of the proposed activities; possibility to coordinate one's activities as efficiently as possible, the expanded list of participants involved in planning; prioritization and coordination of tasks between the performers, increasing the effectiveness of the managerial decisions made; decomposition into strategic and operational objectives; compilation of individual plans and programs into a single system for development of the territory; assessment and use of competitive advantages, potential of the territory; use of material and financial resources of subjects of various forms of ownership; development of territorial communities .

The development and implementation of an effective strategy for development of an administrative territory requires the experience and technical knowledge of managers of this level. The presence of professionals in narrow subject areas is not enough; there is a necessity of professionals who are able to look through the prism of politics, which greatly complicates the ability to create a quality strategy at the level of basic territorial units.

There is no consensus in the economic literature regarding an approach that would include the necessary elements of the system of administrative territory strategic development. However, the formation of the strategy involves a clear vision of the following provisions: identification of the main types of economic activities or scientific and technical areas in which the territory can compete; analysis of current or potential unique resources which may be used to strengthen competitive positions; identification of all possible participants and interested individuals who will be involved in strategic management; assessment of the level of interrelation of the region with external agents; understanding of one's own structural peculiarities, including the degree of concentration, polarization, specialization.

3.2 The Experience of the European Union

In order to overcome the systemic crisis phenomena in the EU economy, in 2010 the new European economic development strategy for the next 10 years was adopted - "EUROPE 2020: A strategy for smart, sustainable and inclusive growth". Based on the principles of reasonable, sustainable and comprehensive growth, five main areas of activity are assumed: employment; research and innovation; climate change and energy; education; fight against poverty.

In 2013, the parliament of the European Union defined the "Smart specialization strategy" as the national or regional innovation strategies which set priorities in order to build competitive advantage by developing and matching research and innovation own strengths to business needs in order to address emerging opportunities and market developments in a coherent manner, while avoiding duplication and fragmentation of efforts [2]. The idea of smart specialization is not new, but is only a modernized long-term practice of the European Union. Fig. 2 presents the process of developing a strategy for smart specialization by building a BPMN diagram.

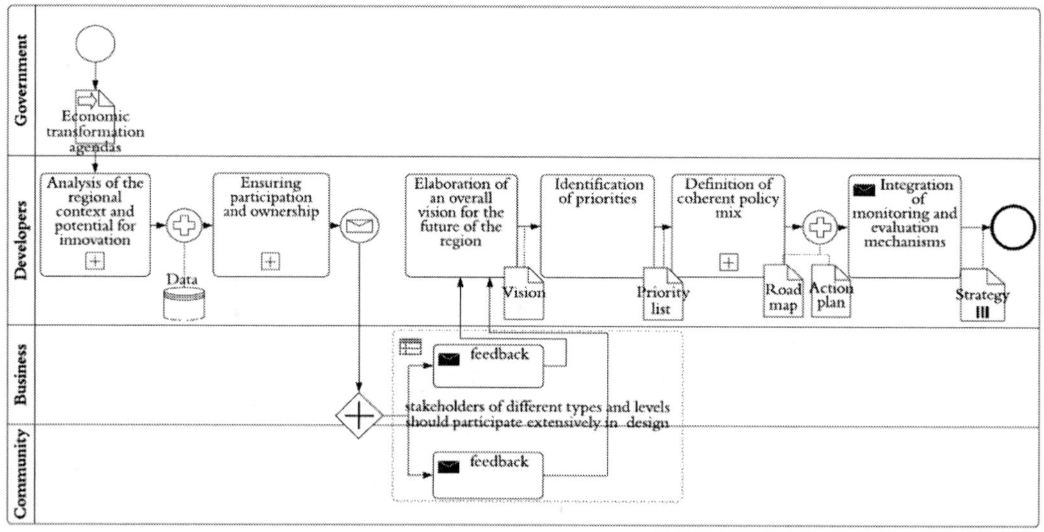

Fig. 2. Process of developing a strategy for smart specialization

In recent years, more than 100 smart specialization strategies (S3) have been developed in the EU, mainly at the regional level. Basically generation unique assets and capabilities based on the distinctive industry structures of regions and knowledge bases include: objectives ("why"), content ("what"), process ("how" and "by whom"), role ("where") [3].

The development of administrative territories is the process that requires efficient and effective institutions, as well as close cooperation between government, business and society at every level. Attaching due importance to this matter, the EU regional policy is an ongoing process of revision and adjustment, to ensure development in the dynamic environment. The experience of the European Union is the development of regions, which is based on three key assets: social capital, leadership, new management tools and three functions (learn and introduce innovations, build channels of interconnection, generate ideas).

3.3 Strategies of development of administrative territories in Ukraine

At the beginning of the 21st century, positive transformations, regarding the modernization of the management model, took place in Ukraine, which resulted in the formation of a qualitatively new state regional policy. The process of looking for relevant and effective ways of development of administrative territories in order to reduce uncertainty and timely response to the challenges of modernity is constant, regardless of the level of a administrative-territorial unit. The decisive role in shaping the strategies of economic development, including the system of strategic management of the territory development, belongs to the President of Ukraine, the Cabinet of Ministers of Ukraine, developers of national development programs, scientific, political and public institutions.

The formation of the foundations of the system of forecasting and program documents of the economic and social development of the country, individual sectors of the economy and individual administrative-territorial units by adopting the Law of Ukraine "About the State Forecasting and Development of Programs of Economic and Social Development of Ukraine" can be considered a significant event that had a direct impact on the development of strategic management in Ukraine [4].

Since 2001, Ukraine has the acting Concept of State Regional Policy as a component of the national strategy of socio-economic development of Ukraine. The most significant results for the development of conceptual approaches to strategic management can be considered: the formation of a network of regional development agencies; improving mechanisms of strategic state planning of regional development, the system of national and regional forecasting and planning of socio-economic development; creation of economic, organizational and legal prerequisites for implementation of the basic principles of sustainable, balanced development of the regions [5].

Establishing a single procedure for elaborating strategies of the development of basic administrative and territorial units of Ukraine (regions) with the adoption of the Order "On Adopting Methodological Recommendations for the Formation of Regional Development Strategies" (2002) can be considered the starting point for the introduction of a modern system of strategic management of administrative territories in Ukraine[6].

The strategic regional development plan should include: introduction with a short justification of the objective and prerequisites of strategy development; descriptive and analytical part, detailing the geographical, historical, demographic, economic and social characteristics, in which the analysis and assessment of the actual state of region's development is done; vision of the future – reflection (generalization) of the most common and long-lasting ideas about the prospects of the development of the region and the life of its population, the mission of the region; characteristics of competitive advantages and limitations of the region's development (SWOT-analysis); strategic objectives (directions) and step-by-step action plans with a detailed description of the events, deadlines, executives in charge, sources of funding; mechanisms for implementing the strategy; team of authors involved in the process of territory development strategy formation with indication of their contribution.

Taking into account the basic principles of the world practice of developing strategies, the content and objectives of the Strategy of Economic and Social Development of Ukraine "By Way of European Integration", the first strategies for the economic and social development of individual regions (Volyn, Odessa, Sumy, Ternopil, Khmelnitsky and the city of Kyiv) were adopted in early 2004 [5]. Awareness of the feasibility of using strategic planning and forecasting together with the adoption of the Law of Ukraine "On Stimulation of Regional Development" finds its place in the State Strategy for Regional Development of Ukraine for the period up to 2015 [7]. In connection with significant changes in the external and internal environment of the country, a new regional development strategy for the period up to 2020 has been operational since 2014. For the previous strategy, priority was given to the implementation of such tasks as: improving the competitiveness of regions and strengthening their resource potential; ensuring the development of human resources; development of interregional cooperation; creating institutional conditions for regional development [7-8].

The analytical methods, generally accepted in Ukrainian practice for formation of the development strategy of an administrative territory, are socio-economic analysis, comparative analysis, SWOT analysis, PESTLE analysis and sociological analysis. The territory profile is formed as a result of a socio-economic analysis. The territory profile is considered as one of the tools for looking for investors, sources of crediting and grants. The comparative analysis is carried out with the aim of positioning the territory in a competitive environment and identifying its own advantages, it helps to find out which of the development strategies against the background of existing strengths and weaknesses is favourable for an effective and efficient response to changes in the external environment and the use of opportunities and minimization of existing risks. The work [9], using the results obtained during the SWOT-analysis, proposes the following variants of strategies for the territories of Ukraine: based on the analysis of strengths and chances (strengthening / building strategy, opportunity realization strategy, binding / correlation strategy); based on the analysis of weaknesses and threats (elimination strategy, minimization strategy, transformation strategy). The use of the PESTLE model is justified by the presence of a significant number of factors affecting the development of the territory, in particular, political, economic, socio-cultural, technological, legal and environmental ones. Sociological analysis is most often carried out through a survey of the population and business representatives in the context of strategic development of the territory. For this, as a rule, a questionnaire is developed which consists of an introduction, a list of questions and personal data.

4 Conclusion

The obtained findings suggest significant differences in approaches to the formation of a territory development strategy in the EU and Ukraine. In the global world, 3S trends (Smart, Specialized, Strategic) dominate in developed countries. Smart specialization approach implies not only the identification of potential sectors of specialization, on the contrary, it is aimed at creating reliable and transparent means of identifying and developing new areas of activity at the regional level, allowing to explore and discover new technological and market opportunities [10]. Based on a high level of institutional development, focusing on improving competitiveness, such an approach is extremely difficult to implement into Ukrainian realities. It covers a broad view of innovation that goes beyond research and technology and is supported by effective monitoring mechanisms. Based on the comparative legal analysis of the experience of the European

Union, it is possible to single out a number of peculiarities that must be taken into account in the process of forming the development strategy of an administrative territory. These include: long-term strategic vision; objectivity of distribution and attraction of resources; strengthening ownership through co-financing and partnerships; institutional architecture of the legal field with a wide range of rules regulating the implementation of the strategy; development of integrated programs; grant programs; financial management and control, economic development planning, as well as identification and motivation of relevant stakeholders; monitoring and evaluation; transparency of the strategy implementation process.

Actual strategies of socio-economic development of the regions were developed up to 2020, thus, in the near future, work will begin on the formation of new strategies. Based on the results of monitoring and evaluation of current strategies, developers will not only have to eliminate current shortcomings, replacing tools that have demonstrated low efficiency, but also rethink approaches to the formation of development strategies. Strategic management requires a high degree of coordination between the government and other subjects involved in the strategy development process.

References

1. M. Avramenko, *Conceptual and methodological fundamentals for strategic management of territorial development in conditions of transformation processes: Scientific professional publication of state administration of Ukraine*, **1**, 214-224. (2015).
2. D. Foray, J. Goddard, I. Goenaga, M. Landabaso, K. Morgan, K, C. Nauwelaers, R. Ortega-Argilés, *Guide to Research and Innovation Strategies for Smart Specialisation. Brussels: European Commission*, 121. (2012).
3. C. Ketels, *What Is Regional Strategy? Lessons from Business Strategy*, **3**, 37–54. (2015).
4. *About the State Forecasting and Development of Programs of Economic and Social Development of Ukraine* (2000) [online]. Available at: http://zakon2.rada.gov.ua/laws/show/1602-14
5. *Strategy of economic and social development of Ukraine "By Way of European Integration"*. (2004) [online]. Available at: https://zakon.rada.gov.ua/laws/show/493/2004.
6. *On Adopting Methodological Recommendations for the Formation of Regional Development Strategies* (2002) [online]. Available at: http://zakon.rada.gov.ua/laws/show/z0632-16#n13.
7. *State Strategy of Regional Development of Ukraine for the Period up to 2015* (2006) Available at: https://zakon.rada.gov.ua/laws/show/1001-2006-п.
8. *State Strategy for Regional Development for the period up to 2020* (2014) Available at: https://zakon.rada.gov.ua/laws/show/385-2014-п.
9. O. Berdanova, V. Vakulenko. *Strategic planning of local development. Practical guide,* 88 p. (2012).
10. C. Nauwelaers, J. Blazek, E. Magro, P. McCann, K. Morgan, R. Ortega-Argiles, J. Wilson, *Smart Specialisation For Regional Innovation: European Union,* 145. (2015).

Paradigm of Development of Economic and Social Systems and Decent Work Provided in the Context of Systemic Changes

Victorya Kompaniets[1]*, *Vadim* Danilyan[2] and *Larisa* Shulgina[3], *Volodymyr* Pavlichenko[4] and *Boris* Ostapuk[1]

[1] Ukrainian State University of Railway Transport, Department of Economics and Management of Industrial and Commercial Business, 61000, Feuerbach Square, Kharkov, Ukraine
[2] Ukrainian State University of Railway Transport, Department of Philosophy and Sociology, 61000, Feuerbach Square, Kharkov, Ukraine
[3] "Voronezh State Technical University", Department of Economics and Fundamentals of Entrepreneurship, 394000, 20th Anniversary of October St., 84, Voronezh, Russian Federation
[4] Kharkiv Institute of Finance Kyiv National University of Trade and Economics , Pletnevsky lane 5, 610500 Kharkiv, Ukraine

Abstract. The article deals with the reasonable grounds for the idea of required transformations not only at social and economic level, but initially the changes should be realized at non-economic level, in order to provide the development of economic and social systems and decent work in the context of systemic changes. This approach involves the choice of a non-linear progress paradigm, transformation of the values for top-managers and entrepreneurs and society as a whole, formation of rules and incentives within the institutional system of national states and international community, being aimed at reorienting the economy towards human - oriented direction.

1 Introduction

At present, we face the profound transformation of worldwide and national economic and social systems under the impact of qualitative technological, cultural and demographic shifts.

Those factors involve such global risks as: altering the meaning and essence of Labour; increasing unemployment and migration, regional and gender difference in work environment; growing poverty and social and financial differentiation, social and economic and political instability; aggravated inability to control the economic and social systems at their critical level.

In order to reduce such risks and provide the qualitative development of economic and social systems, as well as to guarantee the decent work for all the employable members of society; only joint efforts of the entire world community would be able to realize it. The state of affairs becomes even more complicated due to the fact that such transformations are required not only at social and economic level, but initially the changes should be realized at non-economic level,

* Corresponding author: victorya.kom.ek@gmail.com

including the related changes in the paradigm of development of economic and social systems, along with the values for top managers and entrepreneurs and society as a whole. However, the necessity and priority of non-economic changes has been comprehended by few representatives of science, economics and politics only.

Therefore, in this publication we are going to provide the reasonable grounds for necessity and priority of such changes, as well as to demonstrate the interaction of non-economic and economic transformations at the stage of development of economic and social systems and the decent work provided herewith.

2 Research results

Unfortunately, the linear development paradigm dominates over the social theory and economic and social policy in contemporary society. Thus, the basis for further progress is being formed with the material component, i.e. the development in the fields of technology and economy goes on, while the cultural and moral standards of society are considered to be the "secondary" values, and the human being is considered to be not a goal but an instrument for transformation. However, such a policy that applies the economic or technological criteria in order to evaluate its success will result in fatal losses for environment, economy and humans themselves.

There is another type of a paradigm, i.e. a non - linear one, when the social culture is the basis for social development and its main estimation criterion (to wide extent, it comprises the level of morality, education, scientific development) and harmonious human development. Only a moral and educated person is able to determine the criteria related to the utility of engineering and technology and make it profitable for the whole society. Moreover, economic activity, which could be developed upon the profit-motivated basis without any moral restrictions, will result in further corruption and economic crimes, which could not be eliminated by any legal regulations.

This opinion related to the progress of society is based on ideas by N. Ya. Danilevskiy, P. A. Sorokin, A. Toynbee, F. List, as well as of the representatives of the German Historical School of Economics and the Eastern Slavic School of Economic Thought. The contemporary authors, who work in the framework of this paradigm, are J. K. Galbrait, I. Wallerstein, G.V. Zadorozhniy, V. Yu. Katasonov, A. S. Panarin, Yu. M. Osipov.

We should mention one of the works belonging to the historical heritage: it's one of the main concepts of the Eastern Slavic School related to the production factors, along with the similar idea by the researcher V. Sombart regarding the structure of economic and social system (the terms is called "economic system" according to the author's terminology).

The representatives of the Eastern Slavic School believed that the fourth (a non - economic) factor was the main production factor which used to determine the efficiency of application of another factors (as related to the work, nature and capital). They called it as "cultural and historical people's forces" or "spiritual and moral capital", as it united the spirit of the people, their culture and values [1, p.87].

According to V. Sombart's opinion, he defined the concept of economic system like "some spiritual unity that 1) penetrated with certain spirit or became involved into a certain meaning; 2) had a certain order and organization, and 3) applied a certain technique "[2]. That is, according to V. Zombart's opinion, the main element of the economic and social system is its spirit as a spiritual and cultural component which determines the national way of thinking, activity, motivation and psychology of economic activities of its subjects, in particular, organization of all the economic subsystems, regime of ownership and labor, application of engineering and technological approach. Such provisions require the circumstances for priority development of spiritual and moral capital of society, in comparison with its technical and economic development.

The non-linear paradigm approach became the background for the last published research work by J. K. Galbraith "The Economics of Innocent Fraud" [3], and the final report by the Global Commission on the Future of Work titled "Work for a Brighter Future" [4].

The authors of the Report (i.e. the experts of the International Labour Organization, ILO) underline that a human being and his work are the centers of economic and social policy and business practices [4, p. 27]. The authors draw our attention to the problem of destructive nature of technological changes and fundamental shifts in the field of labor, they invited all the parties involved to take responsibility in relation to the provision of economic and social development and decent work. Besides, they highlight the issue of new evaluation criteria for progress which reflects the human development, as well as guarantee the social justice, environmental protection, that is, the implementation of humanitarian, social and environmental goals of the economic and social system.

However, if the ILO experts only raised the problem related to criteria, yet the author J. K. Galbraith had already denied the basic concept of economic determinism in general, as it reflects the linear development paradigm related to the evaluation criteria for progress in the development of society. In compliance with contemporary economic theory and practice, the main criteria for progress are quantitative criteria of economic development. Nevertheless, J. K. Galbraith believed that "the practice to evaluate the progress of society with the indicators of production volume and GDP growth is not an innocent mistake" [3, p. 32]. The researcher underlined that the most significant achievements in the history of mankind took place in those societies whose criteria for success were the development of religion, culture and science. Unfortunately, according to J. K. Galbraith's opinion, the highest achievement of today's civilization is its total destruction.

We should realize that all human activities are regulated by two institutions: law (legislation) and moral standards. Basically, when we mean the rules and incentives in the field of Economics, then certain legislative changes are implied. However, they are not sufficient. The authors K. Homann and F. Blome-Drees reasonably argued that the acts of legislation can't be perfect enough, in terms of their ability to cover all the aspects of moral requirements. This process is caused by the impact of the factors of systemic arrangement, in particular: reactive character of institutional framework; unsatisfactory system of control over the implementation of legislative regulations; probability of conflict between legislation and moral standards; "increasing decline of competence in politics and the particular order in politics" [5, p. 276].

Therefore, those top - managers and entrepreneurs, who regulate the implementation of economic, technical or other policies should take moral responsibility for their decisions. The requirements for regulating the economic activities (both by the legal criteria and moral standards) involve the development of institutional and individual ethics.

In terms of material components of economic transformations related to the field of Labour and economy as a whole, we should discuss the fundamental changes in the investment policy of the state. However, we are going to demonstrate that all those transformations are also associated with the impact of non-economic factors.

The experts of the ILO identify three main investment trends: development of human abilities; institutions of labor market; decent and secure employment.

The investments in the field of decent and secure employment require the priority development of the real economy sector. This is a vital issue, because the contemporary economy, having become oriented at the speculative and debtor's approach, turned to be extremely inefficient, along with the actual risk of experiencing new crisis. At the same time the transition from financial speculations to development of the real sector is associated not only with economic incentives, but with the changing motivation and values for investors and entrepreneurs, i.e. it means the transition step from personal gain to the goals of creation and service to our society.

The proper examples can be the goals and motivation of business activities that had been realized by H. Ford, he was famous to have achieved tremendous success in his practices. He believed that the real objective of business activities should serve the people and society: "The ultimate well - being of a manufacturer also depends on the benefits to be brought to the people" [6]. He believed that human greed was some kind of "myopia" and declared the goal of his manufacture as "providing the maximum purchasing power" at the expense of cost savings, high payments of labour and marginal profits.

We should underline that the development of the real sector should be promoted on the basis of digital technologies as well. However, the implementation of new technologies in manufacturing sector (i.e. robotization, automation, artificial intelligence systems) involves significant risks. We'd better specify those risks: elimination of jobs; extremely inequitable distribution of technological dividends in favor of high-tech companies and highly skilled employees; significant gaps in training of the employees who should meet the requirements of new economy; decrease in degree of independence and control of the employees due to the automation, depreciation of their labor content, which might cause the loss of their skills and decrease in degree of their job satisfaction; exceeding exploitation of the employees who deal with digital platforms, greatly due to the absence of any international labour regulations for the employment practices to be realized on digital platforms; risks of disclosing the privacy of the employees due to the new technologies which generate large amount of data related to the employees [4,7]. Therefore, we should introduce and implement new technologies, taking into consideration the minimization of such risks and "adopting an approach to the management of artificial intelligence in compliance with the principle: "it's a human being who controls", it means that any final decisions, which might affect the labour model, shall be made by human beings only, not by the related algorithms" [4, p. 51].

The employment career of a person is characterized by some transition stages: starting from a certain education level and obtaining related skills up to his labour activities; transfer from one job to another, fundamentally new position; transition from employment to retirement. Their complex character is aggravated by continuous worldwide transformations. In order to make people strong enough to manage those changes, it's necessary to increase the investments in institutions, policies and strategies that could support people during such transition periods in future labour environment [4, p. 61].

The quality of education is a key factor in providing the employment opportunities and "smooth" transitional periods. The joint efforts taken by the state authorities, employers, educational institutions and society as a whole should be required, in order to create and support the development of the educational system throughout human life. However, the classical educational system should become its foundation, while the "new opportunities for broad participation in professional training, as well as opportunities to overcome time and resource limits" will introduce new digital technologies. Nevertheless, the high quality education can be provided with special quality programs, methods and training results, along with the universal access available for anyone. The properly trained teachers are highly demanded, because "whose skills, experience and tutorship could not be ever replaced by any technology", financial provision, creation of monitoring system and "a common system of recognition of qualifications and skills at the national and international level(s)" [4, p. 36].

Thus, one independent and complicated issue should be mentioned in relation to the sources of investment of the above - mentioned reforms in the field of Labour. In our opinion, the country should provide the primary mobilization of its internal resources. It's possible to establish the national development banks for strategic investments, in order to provide further accumulation of resources.

In this respect, a significant source for accumulation is prevention of the "drain" of resources in the form of illegal transfers of funds abroad. Indeed, one of the serious shortcomings of the existing economic system is the funds to have been stored on offshore accounts without any tax payments. The experts believe that the regular annual loses of the US budget is at least 100 billion USD, due to the offshore activities of the American companies. The biggest "tax evader" is the Company "Apple Inc."; it is estimated that at the beginning of year 2015, the amount equal to 181,1 billion USD was kept in its offshore assets. The amount of corporate tax savings (and, those are the losses of the US Treasury, accordingly) was estimated at 59,2 billion USD [8]. Moreover, according to the estimate presented by the International Monetary Fund (IMF), nearly 40% of all foreign direct investment in the amount of 12 trillion USD turns to be "artificial": those assets are the financial investments made through the "dummy corporations" that never deal with any actual activities [9].

An additional resource for investing in the economic development and providing the decent work is reduction of the level of corruption. According to the data presented by the United Nations Organization, the actual corruption (including the losses involved for the world economy in the annually amount equal to 2,6 trillion USD) is more than 5% of global GDP [10]. Corruption incurs real damage to social and economic development in all the countries all over the world. Thus, the annual corruption (including the losses involved for the European Union) amounts to 120 billion EUR, which could be compared to the amount of Ukraine's GDP [11].

The proper involvement of investment assets to the development of the national economy would be impossible without provision of the favorable domestic business climate and reorientation of business activities towards long - term investments in real economy and resolving the social problems. This approach requires both changes in psychology of a modern entrepreneur through special educational programs, as well as modifications of the state economic policy, it should direct the business activities towards long - term programs for development of the real economy sector and resolving the social problems through the reasonable tax policy.

Such a policy also implies reasonable taxation of economic subjects, in particular, that for the large corporations, medium and small entities; unfortunately, this approach has not been currently observed. Thus, over the past three decades in the countries of the Organization for Economic Cooperation and Development (OECD) the average corporate tax rates have been decreased from 45% in the middle 1980-ies up to 24% in 2018. This fact resulted in decrease in the share of corporate taxes in total government revenues and, according to the data presented by the Secretary General of OECD, raised "difficult issues" for those governments that used to seek the provisions for vital public services, social programs and infrastructure [quotation in 4, p. 74].

One of the most significant obstacles for the implementation of policies aimed at the development of a human being, provision of his decent work and environmental safety is the concentration of power by the authorities of contemporary technology companies which develop their policies based on intention to gain super-profits at any cost, even due to the violation of a human right to the decent work for all the citizens of the planet.

Nowadays the largest digital corporations operate according to the slogan: "the winner takes the greatest benefit'; the earnings appear due to the advantages of the "pioneers" and monitoring, analytical options for mass data and the "scale effect from the demand side" [12].

However, we're talking not about the high-tech companies only, but we mean the transnational companies (TNCs) as a whole. According to J.K. Galbraith, the corporate authorities, being motivated by their power and enrichment, actually impose such values upon the society that cause its gradual degradation. "Corporate authorities declare that the success of society means acquiring even more cars, TV sets, various clothes and other consumers' goods, along with buying even more and more lethal weapons. This is the measure of human achievements. Such negative consequences for society as environment pollution, destruction of natural landscape, unprotected health and civil rights, threating military acts and devastation are not taken into consideration" [3, p. 82]. Giving a moral assessment of the activities of TNCs, the well-known Western politician and researcher J.P. Buchanan called the transnational corporations (TNC) "Cain of our time" [13].

Thus, considering the directions of changes in investment policy to ensure qualitative transformations, again we face the question of the dramatic dependence of economic and social development on the moral qualities of those persons who deal with economic activities and actually need to change the situation.

3 Conclusion

By way of summing up, in this publication we have demonstrated that, under the circumstance of systemic changes, the provision of development of economic and social systems and decent work for all the members of society would be impossible without revising the progress paradigm which has been dominating over the theory, consciousness and politics, as well as without further transformation of the values for top-managers, entrepreneurs and society as a whole. It should involve the systemic changes in culture, politics and economics, both at the level of national states and international activities.

The cultural shifts should involve transformations in public opinion, when a human being becomes the main value and goal of the further transformations, while technological and economic development would become their instruments, and the economic activity would be evaluated in accordance with the moral standards and stipulated not only by the legislation, but also by those standards, namely. The alterations in goal setting and value system, the view of priority for non-economic factors within the development of economic and social systems are the key points to change the activities of top managers and entrepreneurs, and, therefore, the key to the existing model of economic and social development and provision of decent work.

It is necessary to set the rules and incentives that provide the economy reorientation towards the human - oriented direction within the entire institutional system of national states and international community.

References

1. V. V. Kompaniets, *Eastern Slavic Economic Thought in Modern Economic Science and Education*, Social Economy, Ed.1, 2(48), 84-95 (2014)

2. W. Sombart, *Selected Texts* (2005). Available at https://www.litres.ru/verner-zombart/izbrannye-raboty/chitat-onlayn/page-6/ (accessed January 23, 2019)

3. J. Kenneth Galbraith, *The Economics of Innocent Fraud: Truth for Our Time,* 88 (Publishing House "Europe", Moscow, 2009)

4. *Work for a Brighter Future*. Global Commission on the Future of Work. International Labour Organization (2019). Available at https://www.ilo.org/global/topics/future-of-work/publications/WCMS_662410/lang--en/index.htm (accessed January 23, 2019)

5. B. Sutor, K. Homann, F. Blome-Drees, *Political and Economic Ethics, transl. by S. Kurbatov, K. Kostyuk,* 368 (Grand-Fair, Moscow, 2001)

6. H. Ford, *My Life, My Achievements* (2013). Available at http://knigosite.org/library/read/7121 (accessed January 15, 2019)

7. *Issue Briefs No.6 The Impact of Technology on the Quality and Quantity of Jobs.* Global Commission on the Future of Work. International Labour Organization (2018). Available at https://www.ilo.org/wcmsp5/groups/public/---dgreports/---cabinet/documents/publication/wcms_618168.pdf (accessed February 22, 2018)

8. V. Yu. Katasonov, *Some Figures to Prove that the State Supports Business in the USA, not vice versa* (2017). Available at https://www.nakanune.ru/news/2017/4/14/ 22467004 (accessed April 14, 2018)

9. J. Damgaard, T. Elkjaer, N. Johannesen, *Piercing the Veil,* Finance & Development, Vol. 55, No. 2, 50-53 (2018).

10. *The Costs of Corruption: values, economic development under assault, trillions lost, says Guterres* (2018). Available at https://news.un.org/en/story/2018/12/1027971 (accessed December 9, 2018)

11. *Amount of Corruption in the EU is Equal to Ukraine's GDP* (2013). Available at http://podrobnosti.ua/892231-razmery-korruptsii-v-es-ravny-vvp-ukrainy.html (accessed March 7, 2018)

12. *Trade and Development*. Report 2018. Power, Platforms and Free Trade Delusion. Chapter III, Economic Development in a Digital World: Prospects, Pitfalls and Policy Options. New York and Geneva (2018). https://unctad.org/en/PublicationChapters/tdr2018ch3_en.pdf (accessed September 26, 2018)

13. Patrick J. Buchanan, *The Death of the West* (2003). Available at

https://royallib.com/book/byukenen_ patrik/smert_zapada.html (accessed October 23, 2018)

Readiness of workers for a new economy in condition of systemic changes (man - technologies – economy)

Viktoria Kompaniets[1*], *Viktoria* Polevaya[2], *Elena* Poliakova[1], and *Elena* Shramenko[1]

[1]Ukrainian State University of Railway Transport, Department of Economics and Management of Industrial and Commercial Business, Feuerbach Sq., 7, Kharkiv, 61050, Ukraine
[2]Kharkiv International Airport, Financial Analytics Service OOO «New Systems AM», Romashkin St. 1, Kharkiv, 61031, Ukraine

Abstract. The article analyses and systematizes requirements set to workers in the conditions of a new economy. It has been proven that a gap exists between those requirements and real skills that young people develop under the influence of information technologies and of modern society culture. A conceptual authors' vision of the educational system development is proposed. Principles of providing a lifetime economic support for the educational system have been singled out.

1 Introduction

The early 21st century heralded tectonic shifts in the culture, economy, technologies, politics, and, accordingly, in man himself. A new economy is dominated by intellectual resources, hi-tech and information – including digital – technologies. That economy imposes new requirements upon future workers [1-8].

But, as research demonstrates [9-16], today there is a problem of inconsistency between requirements set to a worker by the labour market and those values (motivation), skills and knowledge which develop in a modern man under the influence of culture, education and technologies. However, this problem has not been sufficiently recognized yet. At the same time, for those countries where the existing inconsistency is recognized, the lack of coordination between the state, employers, educational system and society as a whole with respect to requirements to the preparation of workers for the new economy and to adequate educational models remains a significant problem, the ILO (International Labour Organization) experts emphasize.

2 Basic material

By analyzing surveys and research [1-8] several trends may be identified in requirements that are imposed upon workers in the conditions of a new economy.

The first trend manifests itself in that employees shift their requirements towards such personal qualities of new workers which are defined by value-motivational personality

* Corresponding author: victorya.kom.ek@gmail.com

nucleus and by relevant skills (honesty, responsibility, empathy, communicative skills, etc.). In this complex world where technical development, including the one which produces enormous information flows, has begun to rapidly outpace human abilities and a humane, cultural development of the society, it is these values that ever more define the future of countries and civilizations, and of the world as a whole.

Thus, the founder and Executive Chairman of the World Economic Forum K. Schwab believes that one of the biggest threats of the fourth industrial revolution is robotization of humanity and destruction of values and traditional meanings of life [2, p. 136]. In his opinion, a key focus needs to be placed on values in order to direct benefits engendered by new technologies towards the development of the entire society and of human identity, as well as to minimize risks [2, p. 135]. Cooperation, responsibility, shared understanding of common fate, interdependency, traditional sources of meaning – such as labour, family and identity are those values that need to be preserved and cherished.

A study conducted by Deloitte LLP demonstrates that high-performing companies hire candidates according to such criteria as work ethics, values, potential, experience and skills [3]. At the same time, the representatives of the Business Education Association of the USA and the OECD place self-management skills, i.e., ethical behaviour, prominently among the skills that influence organizations' success [4].

Generally, the employers' focus has shifted towards the so-called soft, purely human, skills. One of the reasons behind that is that those skills cannot be replaced with robots.

Respondents in a survey undertaken within a research by Global Human Capital Trends forecast a vast future demand for such skills as complex problem solving (63%), cognitive abilities (55%), social skills (52%), as well as process skills (54%) [5].

The ten most popular skills by 2020 are as follows, according to the World Economic Forum research: complex problem solving, critical thinking, creativity, people management, coordinating with others, emotional intelligence, judgment and decision-making, service orientation; negotiation; cognitive flexibility [6].

The authors of report "Skills in future. What you need to know and be able to in a new complex world" believe that the following knowledge and skills will be the most sought-for in a new reality: emotional literacy, empathy; ability to co-create, to sincerely and dedicatedly serve the others; systemic thinking; ability to focus and maintain attention; ability to realize a set of future scenarios, to create individual and collective future strategies; ability to self-(educate); cross-culturalism; ecological thinking [7, p. 72-75].

Whilst the research listed above basically relied on expert – including employer – surveys, the following list of skill groups is compiled by the specialists of NTI University "20.35" based on an analysis of competence models of 50 leading companies from various economic spheres, taking into account that the results were processed by NLP (natural language processing) method. Based on the interviews with company employees and on the analysis of competence models 211 competences were identified and nine competence-based units were determined: value-based-regulatory; self-regulation; human interaction; thinking; work with systems and processes; work with innovation and changes, communication; work with information; entrepreneurship [8].

As we can see from the above – and this is the second trend – the complication of requirements to new-economy workers can be observed. If different research is supplemented, generalized and systematized [1-8], one may single out the following groups of qualities and skills relevant for new-economy workers:

1. Personal (in our opinion, they constitute a systemically important nucleus upon which the line of application of other qualities and skills is dependent – either for the benefit or to the detriment of the development of man, society and economy), etc.

1.1 value-based-conceptual (to understand the outside world; to find one's way around it, to understand the values of good and evil, to realize one's role and mission, to be able to choose target-based and concept-based objectives for one's actions, to make decisions);

1.2 value-based-regulatory (honesty, trustworthiness, work ethics, responsibility, reliability, discipline, decisiveness, self-management, adaptability, initiative, etc.);

1.3. those targeted at self-improvement (mastering methods of spiritual, intellectual, physical self-development, emotional self-regulation, self-support, etc.);

1.4 inter-cultural (ability to live with people from different cultures and religions);

2. Cognitive, such as:

2.1. memory, thinking agility, ability to concentrate and manage attention;

2.2. logical, analytical, critical thinking, research skills, including those in the field of inter-disciplinary research; cognitive flexibility;

2.3. ability to use information, ability to learn throughout life and to apply academic knowledge to achieve results in professional activities;

3. *Creative* (we have identified them as a separate group, apart from cognitive ones, as they are synthetic in nature and are significant), including, being inspired, able to generate ideas, being prognostic, critical, having one's own viewpoint, understanding inconsistencies; ability to unconventionally resolve complex and multi-faceted problems;

4. Socio-behavioural (action-related - emotional):

4.1 goal-setting; goal achievement, reflexive thinking, self-assessment;

4.2 empathy, ability to co-create and to serve the others with sincerity and dedication; communicative skills (methods of interaction with the others and with distant events and people; skills to work in a team, etc.)

5. *Digital:* 5.1 basic skills (ability to use a computer and Internet to resolve general tasks); 5.2 ability to use information technologies (understanding their application, strengths and weaknesses, ability to critically interpret information); ability to process large data arrays and to make judgments on that basis, to track trends and interrelations; 5.3 skills to learn online; 5.4 programming skills, IT literacy, ability to work with machine intelligence, algorithms (most often, those skills are professional).

6. *Professional* – skills relevant for a chosen professional sphere.

Therefore, a new-economy worker should have diversified qualities and skills. However, in practice we can observe an inconsistency between the actual and the requisite skills of potential workers that tells on a demand for them at the labor market. Some of employers and experts have begun to realize the problem. In one of the surveys 44% employers noted that the filling out of recruitment ad forms is hampered by the lack of such a communication skill. Thirty-three per cent told that the absence of personal skills (honesty, responsibility, teamwork, adaptability) at a workplace proved to be the most significant obstacle for meeting the employers' needs [9].

In a survey by LinkedIn, 58% recruitment managers say that lack of cognitive skills in employees restricts their company's efficiency, whereas 89% of the management surveyed by *Wall Street Journal* say that they experience «very serious » or «certain » difficulties trying to find candidates who possess such skills [9].

Other research and surveys have detected significant gaps in the critical thinking skills and in the skills to apply academic knowledge [9].

Of peculiar concern is the fact that a new generation, a product of the digital revolution, is joining the labour market. A rapid upsurge of information – including digital – technologies was seen in the late 1990s and continues to this day. That, alongside cultural changes in the society, globalization processes and a roller coaster of economic crises, has had a significant impact upon a socio-psychological portrait of a new generation born at the time. U. Strauss and N. Howe, the authors of the known generational theory, called it a digital generation. In year or two, the representatives of that generation will make up 20 per cent of the global workforce.

But what are the qualities and skills developed under the influence of information technologies, education and society's culture in the contemporary young people who are about to join the new economy and be competitive at the labour market?

As evidenced by many studies of the digital-generation, *human psyche is now undergoing dramatic changes as a result of the influence of digital technologies and contemporary culture*: a degradation of the natural intelligence is taking place, the sensuous and volitional sphere of human psyche is being damaged, and values have been changing.

At the level of personal qualities and skills this is manifested by the development of an inadequate perception of the world, the borrowing of values and needs from digital templates, a blurred system of values and lack of consistent preferences, moral relativism, egocentricity and individualism, hedonism, self-confidence and a strive for comfort [10,11].

At the level of *cognitive and creative* **qualities and skills this is manifested** in low and short-time focus of attention, intellectual dependence on technology (outsourcing the function of memory to gadgets), and, hence, shallow and short-time memory, reduction in analytical abilities and ability to think critically; development of "clip thinking" which is characterized by fragmentary and superficial nature [10, 12-14].

At the level of *socio-behavioural (action-related - emotional)* qualities and skills this is manifested in the elimination of a border between reality and illusion; an abrupt drop in an ability of social interaction, sympathy, self-control, including, of goal-setting, in the lack of tolerance (a strive for having it all and at once), in the perception of difficulties as a token of a wrongly chosen path [10, 11, 15].

So, the gaps in the requisite and real qualities of potential new-economy workers do exist. Overcoming them is a highly challenging task that requires, first and foremost, for the focus of the entire policy – governmental and corporate, technological and economic – to be shifted towards a man. It means that all innovations, governmental and corporate programs should be assessed in line with the humane and social utility criteria. Moreover, the educational system also requires to undergo essential changes.

In May 2017, Pew Research Center published the results of an expert survey "The Future of Jobs and Jobs Training", in which experts had assessed, among the other things, the readiness and perspectives of a contemporary system of education to those challenges which are connected with the changes in the labour sector triggered by the technical revolution [16]. Expert opinions over the future educational models were divided, of which two groups may be identified that embody contrary positions.

The first group is critical of the traditional educational system and comes forward with revolutionary changes involving a transition to digital technology and to online educational pattern. This approach relies on the opinion that knowledge and skill acquisition demands little human interaction and that, therefore, a half or more of the knowledge and skills may be developed through self-education with the help of digital technologies.

The second group – which is of the opinion that the authors of this publication share – believes that the classical system (with values, skills and knowledge being passed from man to man) which develops a nucleus of values and rules within a man, which teaches and develops personal, social, cognitive skills and gives fundamental knowledge and, in part, professional knowledge as well, should remain the pillar of the new educational model.

During the period of rapid technological development not only new opportunities arise for economy and society, but great risks also. That is why it is necessary to understand that the development and application of technologies is based on a particular system of values focusing either on a harmonious development of man and society, or on the realization of authoritative and economic interests of certain groups, contrary to and at the expense of the degradation and exploitation of the others. That is why a key issue of education is creation, support and development of the value- and sense-based personality nucleus, its spiritual and moral world, and, in terms of a country as a whole – formation and development of moral and intellectual potential of the society, of the society's moral and intellectual elite. These will form the basis upon which it may be possible to exercise consistent, innovative, socially beneficial, humane and safe development of the socio-economic system.

Digital technologies in education are supplementary to traditional ones; they need to be fully realized at workplaces during the process of the lifelong training of an adult (fully-formed) person. The sphere, volume, kind of digital technologies used in classical education needs, in our opinion, to be determined by psychophysiological norms, age and state of health of children and young people.

At the same time, soft skills, which are expected to be in deficit, are only acquired through practicing interpersonal communication, humanitarian disciplines and practices. And this is the experience that a person acquires in a family, educational system and in the process of their participation in communities that facilitate versatile development of personality (culture, sport, public service). This is a sphere of persistent priority for a man. That is why the classical educational model needs to remain basic.

In general, the ability to implement opportunities which the new economy will yield under the influence of technological, demographic and environmental changes will depend on the effectiveness of the organization and functioning of lifelong educational systems [17, p.34]. The state, employers and educational structures need to participate in the creation and operation of those systems, yet the state needs to retain the governing role.

The funding of expenditure on education, including, that in the sphere of professional training, re-training and advanced training, remains a key issue. Public funding may support and encourage educational programs with the help of such tools as guaranteed financing, payment of allowances, creation of guarantees of the quality of professional training, opening individual educational accounts, provision of subsidies, grants, credits and tax benefits [1, p.5]. This needs to take into account shrinking possibilities of public funding due to the build-up of debts, socio-demographic problems and deterioration of macroeconomic situation in general. This calls for diversification of the sources of funding, in particular, for active involvement of employers.

Governments need to create appropriate mechanisms of public funding and of encouraging entrepreneurs to increase investments into professional training, advanced training and re-training. They also need to enhance decent work opportunities for young people via programs of employment and support of young entrepreneurship. Private sector is required to play a special role by offering appropriate quality apprenticeship and first jobs to young people [17, p.37]. Among the other things, industry-specific or national tax preferences could become mechanisms for the engagement of private sector to the funding of professional training.

3 Conclusions

Therefore, the strengthening of the new economy is accompanied by the changes of requirements to workers. In parallel, their real skills developed under the influence of information technologies and contemporary society's culture contradict those requirements generating gaps.

To overcome the existing situation and to ensure the development of man, society and new information technology-based economy systemic changes need to be implemented:

- in the educational and socio-economic policy of nation-states towards strengthening the state's regulatory role in the management of the socio-economic system, including, of the educational one, and in the regulation of the labour market;
- in educational models and in the entire educational system, in order to create a different value-motivational portrait of new generations and to adapt these models and system to abrupt and in-depth changes of the external environment;
- in the construction and strengthening of a multilateral dialogue between the state,

employers (business sector), educational institutions and workers for the purposes of ensuring quality education and lifetime training.

References

1. Skills policies and systems for a future workforce. Issue Briefs No. 8. International Labour Organization, (2018), available online: https://www.ilo.org/global/topics /future-of-work/publications/issue-briefs/WCMS 618170/lang--en/index.htm
2. K. Schwab, *The Fourth industrial revolution*: transl. from Engl. ("E", Moscow.: 2017).
3. From careers to experiences: new pathways 2018 Global Human Capital Trends https://www2.deloitte.com/insights/us/en/focus/human-capital-trends/2018/building-21st-century-careers.html
4. National Business Education Association. This we believe about teaching the soft skills: human rlations, self-management, and workplace enhancement policy statement 67 https://www.nbea.org/newsite/curriculum/policy/no_67.pdf
5. AI, robotics, and automation: Put humans in the loop2018 Global Human Capital Trends https://www2.deloitte.com/insights/us/en/focus/human-capital-trends/2018/ai-robotics-intelligent-machines.html
6. A. Gray, *The 10 skills you need to thrive in the Fourth Industrial Revoluti*on. World Economic Forum, January 19, (2016), https://www.weforum.org/agenda/2016/01/the-10-skills-you-need-to-thrive-in-the-fourth-industrial-revolution/
7. Skills in future. What you need to know and be able to in a new complex world [Electronic resource]. – Access mode: https://worldskills.ru/media-czentr/dokladyi-i-issledovaniya.html
8. Digital competences basic model is included into a curriculum https://www.comnews.ru/print?nid=118097
9. R. Stephens, *Automate this: building the perfect 21st-century worker REPORT*. Published April 7, (2017), https://www.thirdway.org/report/automate-this-building-the-perfect-21st-century-worker
10. M. Spitzer, *Anti-Brain: Digital technologies and brain* [Electronic resource]. – Access mode: http://www.litres.ru/pages/ biblio_book/? art=6377279& lfrom=202213444
11. S. Turkle, *Life on the screen: identity in the age of the Internet* (Simon and Schuster, New York,1995)
12. Dr. N. Kardaras, *Screens In Schools Are a $60 Billion Hoax* [Electronic resource]. - Access mode: http://time.com/4474496/screens-schools-hoax/
13. V. L. Dunckley M.D., *Gray matters: too much screen time damages the brain* (2014) https://www.psychologytoday.com/us/blog/mental-wealth/201402/gray-matters-too-much-screen-time-damages-the-brain
14. Interview Clifford Nass // Digital Nation. Life on the digital frontier. – URL: http://www.pbs.org/wgbh/pages/frontline/digitalnation/interviews/nass.html
15. Stuart Wolpert, *In our digital world, are young people losing the ability to read emotions?* [Electronic resource]. - Access mode: https://www. universityofcalifornia.edu/news/are-young-people-losing-ability-read-emotions
16. R. Lee, J. Anderson, *The Future of Jobs and Jobs Training*. Pew Research Center, (2017), available online: http://www.pewinternet.org/2017/05/03/the-future-of-jobs-and-jobs-training/
17. Work for a brighter future – Global Commission on the Future of Work International Labour ORGANIZATION https://www.ilo.org/wcmsp5/groups/public/---dgreports/---cabinet/documents/publication/wcms_662410.pdf

Problems and prospects of management of investment attractiveness of subjects of business

Lidiya Kostyrko[1]*, *Ruslan* Kostyrko[2], *Olena* Sereda[1], and *Eleonora* Chernodubova[1]

[1]Volodymyr Dahl East-Ukrainian National University, Department of the Finance and Banking, Severodonetsk, Ukraine
[2]Banking University, Department of the Accounting and Taxation, Kyiv, Ukraine

Abstract. On the basis of the analysis of the scientific views of researchers, the essential characteristics of the category of "investment attractiveness" as an object of management are specified. According to the results of the study of macroeconomic indicators (GDP, gross fixed capital formation, financing of capital investments, financing of innovations, direct foreign investments), the current problems of investment attractiveness in the country are determined. The investment attractiveness of Ukraine is analysed in accordance with the international indices. The sequence of investment attractiveness management of business entities is proposed, where priority is given to the strategies of the financial regulation by the development of business entities. The urgency of the formation of a financial regulation strategy based on an estimation of investment attractiveness is substantiated. The priority directions of increase of investment attractiveness in the framework of realization of the strategy of financial regulation by development of the business entities, which stipulate the choice of sources of financing, optimization of the structure of capital, asset restructuring and financing of innovations are determined.

1 Introduction

Investment attractiveness of business entities plays a significant role in ensuring sustainable growth of the national economy. Attraction of investment resources creates opportunities for the strategic development of enterprises, financing innovative activity, and increasing the competitiveness of the business entities. Investment attractiveness is an indicator of the expediency of investing for the potential investors. Solving a complex of problems in the investment activity of the enterprise is connected with the substantiation of the perspective directions of development of enterprises and their financial support. Therefore, the investment attractiveness of business entities is an instrument for forming a financial regulation strategy for the development of business entities. The prerequisites that determine the relevance of building

* Corresponding author: lidiyakostyrko@gmail.com

a financial regulation strategy based on the assessment of investment attractiveness include increasing the impact of globalization on financial security; lack of reasonable methods for making strategic decisions on financial management of enterprises; necessity of analysis of factors influencing investment attractiveness of business entities.

Problems of improving the efficiency of investment attractiveness management were investigated by many scholars, among them T.V. Mayorova, L.M. Aleksieienko, I.A. Blank, N.Yu. Bryukhovetska, Yu.V. Budnikova, I.M. Vakhovich, I.V. Alexandrenko, M.S. Zabedyuk, A.O. Yepifanov, H.A. Dehtyar, T.M. Melnik, I.O. Shkolnik, Ya.Ye. Zadorozhna, L. P. Diadechko, G.V. Kozachenko, O.M. Antipov, O.M. Lyashenko, G.I. Dibnis, N.P. Makarii, O.G. Melnyk, Yu.L. Logvinenko, S.D. Suprun, S.V. Yuchymchuk, L.O. Chorna, L. Ye. Furdychko [1-13]. Main attention in the research papers is devoted to the problems of analysis and evaluation of investment attractiveness of enterprises. However, there are a number of unresolved issues related to the need to explore tools for managing investment attractiveness, discussion of approaches to the formation of an investment strategy. This necessitates the formation of an investment attractiveness management system in the context of the strategy of financial regulation of the development of the business entities.

The main purpose of the research is to determine the theoretical basis for managing investment attractiveness of business entities, analysis of external factors of the country on the investment attractiveness of enterprises and the justification of investment attractiveness management tools for business entities.

2 The Main Part

To organize the management of investment attractiveness, it is necessary to investigate the scientific approaches to the interpretation of the essence of the concept of "investment attractiveness". Investment attractiveness at different levels of the economy remains an actual object for scientific research. Investigating the genesis of the category of "investment attractiveness", we can conclude that there are various dominant approaches [1-13]. Most scholars consider investment attractiveness as a set of indicators of enterprises, which determine for the investor the most desired values of investment behaviour. However, it is not possible to ignore such important issues as the impact of external factors at the country and region level. The analysis of scientific approaches to the essence of the category "investment attractiveness " [1-13] allowed to single out the essential features inherent in the investment attractiveness of the business entities:

1. Set of factors of external and internal environment of the business entities, taking into account the investment attractiveness of the country and the region of operation, affecting the possibility of attracting investment resources. From the investor's point of view - these are the terms of investment that affect the choice of the investment object.
2. Integral characteristics of resource efficiency, asset liquidity, solvency and financial sustainability, as well as the value of a number of qualitative indicators.
3. Ability to develop on the basis of increase of profitability of invested capital and techno-economic level of production. That is a dynamic category that characterizes a certain vector (or system of vectors) of enterprise investment development.

The indicated intrinsic features are the basis for building an investment attractiveness management system for business entities.

Management of investment attractiveness of business entities is a system of instruments and measures of purposeful influence on the factors of enterprises activity in order to ensure investment attractiveness in the long-term.

Management of investment attractiveness of business entities as a system contains goals, objectives, criteria, constructed in such a way that factors of the macro level are taken into account.

The main internal factors influencing investment attractiveness are: corporate benefits; the availability of financial resources necessary for the development of production and increase sales; financial and economic condition; the pace of innovation development.

2.1 The External Factors of Investment Attractiveness Management of Business Entities

The external factors of investment attractiveness at the country level are: gross domestic product (GDP), gross fixed capital formation, level and sources of financing of capital investments and innovations, foreign direct investment.

An objective basis for the development of business entities and improvement of investment activity in the country should be an increase in GDP. According to the State Statistics Committee, a gradual decline in GDP in the period 2012-2015 and a slight increase in 2016 and 2017 are observed. Gross capital accumulation reflects investment in tangible and intangible assets of business entities and serves as the main source of implementation of the model of innovative economic development and should be implemented within certain limits at the level of 20-25% of GDP. In terms of accumulation rates below 10% of GDP, economic development is practically absent. The gross accumulation of fixed capital in the structure of GDP is low and there is a tendency towards a decline in the indicator in 2012-2015 from 18.9% to 13.6% [14]. In 2016-2017 there is a slight increase of 16% and 15.5%, respectively, which does not provide the need for modernization of the economy and leads to a reduction in the financial potential for future capital investments.

Investigating the dynamics of sources of financing for capital investment, it is appropriate to focus on financing the reproductive processes of the development of business entities. The main place in financing of capital investments during the years 2012-2017 is occupied by own funds of enterprises, the share of which varied in the range of 62.6-70.5%. The structure of capital investments by types of assets is characterized by a predominance of investments in tangible assets (93.3 to 96.7%); by types of economic activity - domination of investments in industry (35.3 to 36.3%); by types of industrial activity - in the processing industry (45.8 to 52.8%) [14]. Insignificant investments in intangible assets (3.3% to 6.7%) do not promote increase of business reputation and market capitalization of business entities. According to statistics, the main source of financing of innovation activity of business entities remains its own funds, the share of which in 2012 was 63.9%, in 2015 – 97.2% (maximum value), in 2017 – 84.5 % [14]. Financing of innovative projects at the expense of the state budget is insignificant and makes up 0.3-4.5% of the total investment, while foreign investors account for 0.1-13.1%.

In the economy of Ukraine, foreign investors from 76 countries of the world have invested $ 1630.4 million direct investment (equity). Significant inflows of foreign direct investment in 2017 were directed to financial and insurance activities - 26.4% and industrial enterprises - 27.3%. The main types of economic activity according to the criterion of mastered capital investments are: industry – 33.2%, agriculture, forestry and fisheries – 16.1%, construction – 13.9% [15].

The study of the international practice shows that there are different methods of assessing the investment attractiveness of countries that are dependent on macro-factors: the state and development of the country's financial and economic system, represented by the main macro indicators; state investment policy: privileges and priorities in industrial and financial markets; development of foreign economic relations; loyalty of legislation on foreign investment resources; and state support of innovation and investment spheres.

In the international practice, the assessment of investment attractiveness is based on the use of the business index, BDO investment attractiveness, UNCTAD investment attractiveness and EBA investment attractiveness [16] (Table 1).

The results of the analysis of the international indices of Ukraine's investment attractiveness show a slight improvement in the investment climate. The slow pace of socio-economic reforms, the unstable political situation in Ukraine, high interest rates on loans, and the corruption of power lead to risky investments. The development of business entities in Ukraine has the prospect of attracting foreign investment resources and improving investment attractiveness.

Table 1. Investment attractiveness of Ukraine in the international assessments

Indicator	Organization	Maximum value	Value				
			2013	2014	2015	2016	2017
Ease of Doing Business Index	World Bank	190 (rating)	137	112	96	83	80
International Business Compass index	International consulting company BDO	174 (rating)	99	109	89	130	134
EBA Investment Attractiveness Index	European Business Association	5	1.81	2.5	2.57	2.85	3.15

Source: [17-19]

2.2 Process of Management of Investment Attractiveness of Business Entities

Management of investment attractiveness of business entities is represented by the following stages (Fig. 1). The main objective of the management is to provide investment attractiveness, which will help to increase the market value of business entities.

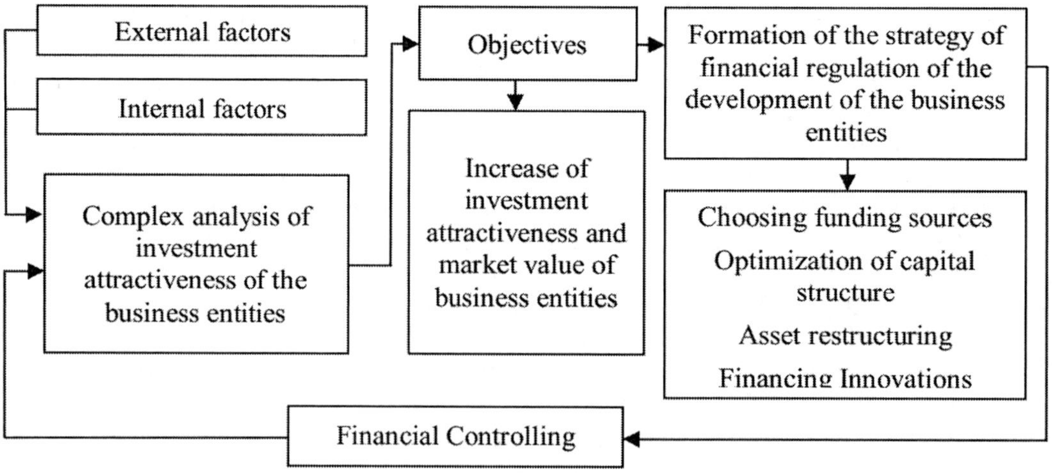

Fig. 1. The sequence of investment attractiveness management of business entities.
Source: compiled by the authors

The complex analysis of investment attractiveness is oriented on the formation of information for substantiation of the parameters of the strategy of financial regulation of the development of business entities. The basis for the formation of investment resources is the own and attracted capital of business entities. Potential opportunities to attract investment determine: gross capital accumulation, availability of free financial resources in the country and region, financial position of the enterprise. The enterprises are more attractive to the investor, provided they have a stable financial position, growth of the rates of basic financial indicators, and business reputation. Therefore, the basis for a comprehensive analysis of investment attractiveness is the methods of index analysis and integrated assessment. The aggregate estimation of investment attractiveness of business entities is calculated on the basis of integral indicators of investment attractiveness at the level of the country, region, and business entities.

Ensuring investment attractiveness is possible provided that the strategy of financial regulation of the development of business entities is effectively implemented. The strategy of financial regulation of the development of business entities is a strategic plan for financing the development of the enterprise, which provides an opportunity to assess the investment attractiveness of business entities in the long run.

When forming a strategy of financial regulation of the development of business entities, it is necessary to take into account the following features of investment activity: the formation of a long-term investment plan and a long payback period of the investment project; a significant amount of both own and attracted investments, includes an innovative component. Such a strategy should be aimed at an optimal structure of the financial potential, reduction of investment risks and determination of key areas of development of business entities. The choice of the financial regulation strategy for the development of the business entities is carried out according to the criteria of economic added value and risk.

Within the framework of the strategy of financial regulation of the development of business entities, the directions of increasing the investment attractiveness of the enterprise are determined, taking into account changes in the external and internal environment. The activities are carried out in the following areas.

1. Choosing funding sources. In the process of selecting sources of funding, consideration should be given to: methods for attracting investment resources; methodology of evaluation and selection of investment projects, investment activity tools of the enterprise. The choice of funding sources is preceded by the definition of the need for short- and long-term capital, taking into account new projects and identifying possible changes in the composition of assets and capital.

2. Optimization of the structure of capital. The optimal ratio of own and borrowed capital and the form of attraction of these resources taking into account the influence on the investment attractiveness of the enterprise are determined. Such measures are: a change in the volume of borrowed funds, replacement of short-term resources in the long-term and vice versa, increase (reduction of authorized capital), reduction (increase in the number of shares), replacement of borrowed funds with its own resources. The main purpose of these measures is to fulfil the conditions of financial equilibrium, reduce the cost of raising capital, and increase the manageability of the owners of the enterprise.

3. Restructuring of assets is aimed at asset renewal, cost optimization, and increased return on investment capital. The measures are carried out for all components of the assets structure: non-current assets, current assets, cash and financial investments. The measures include acquisition and renewal of fixed assets, write-off (sale) of unused assets, outdated assets, reduction (increase) of stocks, optimization of receivables. Under conditions of optimization of the financial investments, their term, volume, profitability change.

4. Financing Innovations. These measures are aimed at increasing competitiveness and expanding the range of products through the introduction of innovative technologies, which ultimately leads to an increase in the investment attractiveness of the enterprise.

Realization of the strategy of financial regulation of the development of business entities allows attracting investments in the necessary volume, improving the efficiency of management, increasing the market value of the enterprise.

3 Conclusions

Investment attractiveness of business entities is the integral characteristic of the investment object from the standpoint of development prospects. As a result of the conducted research, the sequence of management of investment attractiveness of business entities was substantiated, where the strategy of financial regulation was determined by the development of business entities. A comprehensive analysis of the investment attractiveness of business entities is based on the integral indicators that characterize the investment attractiveness of the country, region, and enterprises. It is proved that the strategy of the financial regulation by the development of business entities should be formed and implemented in accordance with the vector of enterprise development and the possibility of attracting investment resources.

References:

1. L.M. Aleksyeyenko, Economic forum, **3**, 94-102 (2009).
2. I.A. Blank, Investitsionnyy menedzhment (Investment management) (Elga-N, Nika-Tsentr, Kyiv 2001).
3. N.YU. Bryukhovetska, Economy of industry, **1**, 110-117 (2009).
4. YU.V. Budnikova, Innovative economy, **3**, 194-197 (2012).
5. I.M. Vakhovych, I.V. Oleksandrenko, M.S. Zabedyuk, Actual Problems of Economics, **11**, 154-160 (2014).
6. A.O. Yepifanov, H.A. Dekhtyar, T.M. Melnyk, I.O. Shkolnyk, Otsenka kredytospromozhnosti ta investytsiynoyi pryvablyvosti sub″yektiv hospodaryuvannya (Assessment of creditworthiness and investment attractiveness of business entities) (UAB NBU, Sumy, 2007).
7. YA.YE. Zadorozhna, L.P. Dyadechko, Investytsiyi: praktyka ta dosvid, **2**, 32-35 (2007).
8. H.V. Kozachenko, O.M. Antipov, O.M. Lyashenko, H.I. Dibnis, Upravlinnya investytsiyamy na pidpryyemstvi (Management of investments in the enterprise) (Libra, Kyiv, 2007).
9. N.P. Makariy, Economist, **10**, 52-60 (2010).
10. O.H. Melnyk, YU.L. Lohvynenko, Management and entrepreneurship in Ukraine: stages of formation and development problems. Bulletin of the National University "Lviv Polytechnic", **575**, 141-147 (2006).
11. S.D. Suprun, S.V. Yukhymchuk, Finance of Ukraine, **4**, 82-87 (2009).
12. L.O.Chorna, Investytsiyi: praktyka ta dosvid, **24**, 4-6 (2008).
13. H.V.Strokovych, O.P.Mykolenko, Financial and credit activity-problems of theory and practice **2**, (25), 246-252 (2018).
14. Official site State Statistics Service of Ukraine [Electronic resource]. http://www.ukrstat.gov.ua/.
15. Information and analytical materials on the investment climate in Ukraine in 2017, 2016, 2015 [Electronic resource]: http://www.me.gov.ua/?lang=uk_UA
16. Furdychko L. Ye., Financial and credit activity-problems of theory and practice, **3** (26), 281-291 (2018).
17. Official website of the World Bank Group [Electronic resource]. – www.doingbusiness.org.
18. Official website of BDO international business compass [Electronic resource]. – https://www.bdoibc.com.
19. Official website of the European Business Association [Electronic resource]. https://eba.com.ua.

Cryptocurrencies in the global space: factors and prospects of promotion

Anna Kramarenko*[1], *Anton* Kvitka[1], *Vitaliy* Diachek[1], and *Denis* Davydov[1]

[1]V. N. Karazin Kharkiv National University, 4 Svobody Sq., Kharkiv, 61022, Ukraine

Abstract. The article considers the place of cryptocurrency markets in the global economy. The key risks of active implementation of IT achievements in the financial sector were defined. It was shown that cryptocurrencies are associated with settlement operations, investment activities and speculation in the currency markets. It was argued that the availability of equipment for the production of digital currencies, the spread of trust on the part of users and business representatives contribute to the promotion of cryptocurrencies and the penetration of digital currencies in various spheres of economic activity. The attraction of additional resources in the economy in the conditions of insignificant costs for entering the cryptocurrency market that become a factor in the spread of cryptocurrency assets was considered.

1 Introduction

Cash flow management, effective fundraising, optimization of the tax burden, promotion of products by channels convenient for consumers are issues of concern to representatives of wide business circles. Besides, the creation and implementation of technological innovations at the present stage is the basis of the success of any company. The desire of enterprises to gain and maintain their market share requires significant efforts both in the promotion of existing products and in innovation, for example in the development of new products and updating existing products [1]. In this regard, the creation of technological innovations, the growth of business and the development of entrepreneurial activity are trends that mutually reinforce each other. The increase in the influence of information technology and entrepreneurial activity contributes to the modification of other areas of economic activity.

In particular, the rapid development of information technologies (IT) causes a radical transformation of the financial systems of most countries. At the same time, the most significant changes are observed at the level of financial infrastructure, primarily in the banking system and the foreign exchange market. One of the key risks of active implementation of IT achievements in the financial sector is the possible disappearance of the banking system in the "classical" form or its radical transformation due to the inability to compete with new technologies [2].

* Corresponding author: a.o.kramarenko@karazin.ua

At the same time, the most potentially dangerous for classical banks are two technologies – credit electronic platforms and blockchain. Even 3-4 years ago, the companies of credit electronic platforms looked like experimenters, but today the pioneers of this market "Landing Club", "Prosper" and others have a billion capitalization [3]. These companies independently determine the solvency of a person or enterprise using a huge database. After determining the level of solvency of a potential borrower, the electronic platform is looking for an investor. Thus, there is a partial replacement of the financial sector with resources of information technologies and services.

On the other hand, blockchain – a technology of storage and instant transmission of huge amounts of data in encrypted form - is becoming widely used. Blockchain allows anybody to easily manage personal accounts and transfers without the help of a bank. The system has multi-level protection, and most importantly – the corresponding code is seen by all participants in transactions, therefore potentially stolen funds can be observed in the network, so it is impossible to use them criminally. From a technical point of view, blockchain is a technology of distributed registers, which allows anybody to create a decentralized system of relations, in which each next transaction depends on the previous one. Therefore, the transaction data cannot be changed or deleted. For the first time, this term appeared to refer to a distributed database implemented in Bitcoin - the most popular cryptocurrency in the world [2]. It is the emergence of digital currencies, such as Bitcoin, which perform all the functions of money and which, thanks to the blockchain system, have a high degree of protection and security, contributed not only to the change in the functioning of the financial system but also to the modification of the principles of entrepreneurial activity.

2 Results

The user can buy Bitcoin or other types of cryptocurrencies, or generate them in the process of mining (the use of computing resources to perform a mathematical task to verify and implement cryptocurrency transactions with the help of special equipment - miners). The cost of the most popular miner produced by the Chinese company "Bitmain", which today occupies 75% of the world market, is about USD 2,000, and the payback period is about 12 months [4]. Bitcoin is not tied to defined national currencies or assets, and its value is not backed by anything other than the trust of its owners. Bitcoin exchange rate is set on virtual exchanges, where it is exchanged for real money. Digital exchange of cryptocurrencies is carried out on such world exchanges as "Bitfinex", "Poloniex", "Kraken", "Shapeshift", "Changelly", CHBTC, "Bitquare", "Bittrex", "BitMEX", BTER, "Yobit", "CoinExchange", "LiteBit", "Btc Markets" and others [5]. Online rates of different cryptocurrencies can be tracked [6]. The availability of equipment for the production of digital currencies, the spread of trust on the part of users and business representatives, the ability to exchange for real money without significant expenditure of time and resources, the possibility of operational monitoring and a high level of security – all this conditions increase the popularity of these types of assets.

Due to the demand for |fair money" in the world, complete anonymity and protection from counterfeiting, interest in cryptocurrencies is growing rapidly. The anonymity of payments makes the Bitcoin system attractive in a criminal environment [7]. Cryptocurrency is a convenient tool for money laundering, as this type of asset gives relative anonymity to the owner of the currency (the system does not require personal information about the user and his location). This feature minimizes the ability of law enforcement agencies to track criminal activity and identify the offender.

Also, foreign exchange transactions are out of state control, so any transactions with cryptocurrency do not need to be checked by third parties, such as a bank, government agency or non-governmental organization. In addition, the user can have more than one account and conduct transactions from different locations at the same time. Bitcoin allows for various money laundering schemes, which virtually blocks the possibility of identifying criminals. The mentioned characteristics of the cryptocurrency encourage legislators around the world to unify the legal system for the use of cryptocurrency as a means of payment [8]. The main attention is paid to the taxation of cryptocurrency turnover. Cryptocurrencies are associated with settlement operations, investment activities and speculation in the currency markets. All these operations have signs of objects of taxation. In this regard, the issues of monitoring the turnover of cryptocurrencies, the definition of the legal framework and economic instruments of taxation and tax control become relevant. To solve these issues, it is necessary to allocate groups of transactions with cryptocurrency, which can be an object of taxation. Attempt to create a similar classification was carried out by Volosych S. and Barannik Yu. It is presented in the figure below.

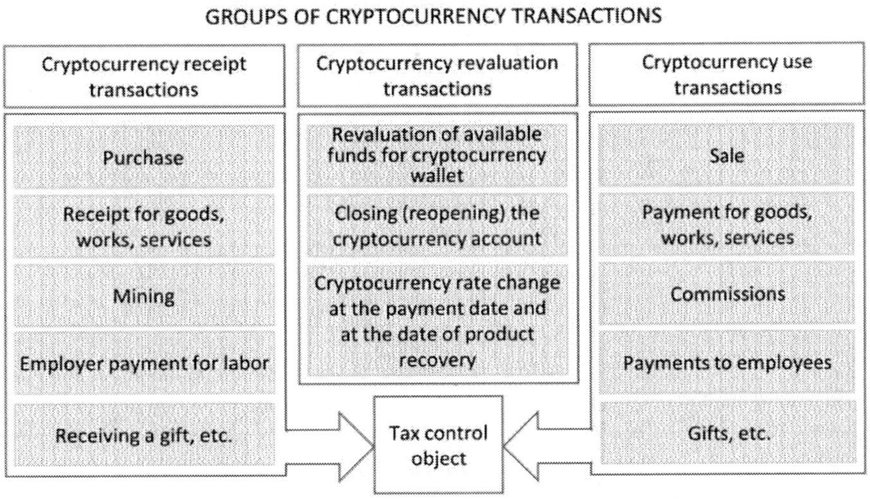

Fig. 1. Cryptocurrency transactions as a tax control object [9]

According to S. Volosych and Yu. Barannik, to prevent the use of cryptocurrencies in terrorist activities, it is advisable to carry out tax control of mining, exchange rate determination transactions, and transactions with cryptocurrencies.

For this purpose, at the macro level, the creation of legal documents regulating the cryptocurrency market should be ensured [9]. The need for structuring the institutional sectors of the economy and activities, which implies the need for a comprehensive study of the processes of differentiation, evolution and economic growth of the total financial potential of the country should be taken into account [10]. Understanding the total financial potential of the country and its timely and qualitative analysis, the choice of the necessary method of assessment and the system of indicators are fundamental for making effective management decisions [11]. At the same time, at the micro level, it is necessary to provide conditions for the identification of owners of electronic wallets, taking into account the importance of strategic management accounting information for strategic management purposes [12]. So, the issues of state regulation of the cryptocurrency market remain open. This situation, on the one hand, poses a threat to the centralized management of economic processes and reduces the number of effective instruments of influence on business entities. On the other hand, such difficulties create opportunities for small and medium-sized enterprises seeking to reduce the tax burden.

Cryptocurrency exists freely. All sorts of organizations, including those that provide related services, such as a guarantee of safety of coins in the wallet or ensure the security of transactions are trying and will try to adapt to cryptocurrency. Increased competition in the process of saturation of the market, along with increasing innovation and originality of the goods supplied, leads to the development of personal needs. Since entrepreneurs trying to win in tough competition, they are forced to create not only innovative products but also to identify new unique needs [1]. Increasing the level of consumerism against the background of growth in the number of small and medium-sized enterprises contributes to a significant adaptation of cryptocurrency technologies in global markets. Some payment systems accept Bitcoin, such as Visa, Paypal, and Bitpay. For example, the company Bitpay currently serves more than 15,000 business owners and is present in 200 countries [7].

New forms of investment, such as ICO (Initial Coin Offering), are associated with the formation of cryptocurrencies. In simple words, ICO can be compared to cryptocurrency crowdfunding. Most often projects involved in the ICO are somehow connected with the technology of the blockchain and cryptocurrency. By issuing its own "money" and exchanging it for one of the common cryptocurrencies (such as Bitcoin) or even for real currencies, the project can secure the funding needed to launch or develop. In addition, by issuing currency for the project, it is possible to accelerate its development and automatically solve the problem of future monetization [13]. Thus, the presence of significant investment opportunities, a significant level of decentralization, increased consumerism and innovations create favorable conditions for increased adaptation of cryptocurrencies in the business environment.

However, there are risks associated with failures in IT systems caused by various reasons. The need to deepen knowledge about the advantages and disadvantages of electronic payment systems, the need to increase the security of IT technologies and to increase IT literacy slow the development of non-cash payment instruments and, accordingly, limit the growth of cryptocurrency markets.

Representatives of the Central banks note significant risks of cryptocurrency development. Thus, the Head of the Reserve Bank of India said that the virtual currency is a potential financial, operational, legal threat to customers. Cryptocurrencies are vulnerable to hacker attacks, loss of passwords, unauthorized access, and lack of a system for resolving customer complaints [14].

M. Carney talks about the risk of fraud and theft of Bitcoins from the accounts of customers and exchanges. In all cases, the funds were stolen from “hot wallets” when the secret key was stored on a computer or device connected to the Internet [15]. The Head of the Central Bank of Australia believes that cryptocurrencies are of greater interest to criminals than to consumers: “Current fascination with these currencies is more like a speculative mania than using them as an effective and convenient form of electronic payment” [16]. The Vice-Chairman of the National Bank of the Czech Republic said that the rapid changes in purchasing power are a mortal enemy of any good currency [17]. The uniformity of business models of different cryptocurrencies creates the danger of simultaneous changes in the behavior of market players concerning these instruments. Players (miners) have branched virtual communities that use uncontrolled channels of moving money [18]. The high level of volatility in the cryptocurrency market, along with the creation of threats to the stability of the financial system, causes increased interest from investors. A small amount of initial investments and a high level of payback attract an increasing number of entrepreneurs, which in turn contributes to the growth of cryptocurrency markets.

According to Wang Chun Wei, increasing liquidity in the cryptocurrency market contributes to lower liquidity and increase the efficiency of the foreign exchange market [19]. So, the more widespread and developed cryptocurrency technologies are, the more stable their functioning is. The level of risk of investments in relevant instruments is reduced on the basis of regulation and increased transparency of market agents.

In this regard, cryptocurrencies are increasingly becoming the focus of investors' attention. Financial indicators in terms of risk and return are different from those of classical financial assets: investments in cryptocurrencies have relatively high volatility, that is, the risk, which, however, is compensated by the high expected return. Also, its low correlation with stock, bond, commodity, and foreign exchange rates creates diversification potential for investors. Investors also do not need to actively manage these assets [20].

Based on the above, it can be argued that cryptocurrencies are not only perceived by the main economic entities as alternative money but also recognized as an object of investment. Profitability expectations and trust positively affect the attitude to these assets. Currency characteristics associated with the expected profitability and compatibility of transactions have a positive impact on the attitude to cryptocurrency [21]. To replace the existing currencies cryptocurrencies lack full recognition from the largest trading platforms and an objective combination with the existing currencies. These factors become an obstacle to the spread of cryptocurrencies in the global space.

Nevertheless, the positive impact on the development of cryptocurrency markets has the fact that the profitability of the cryptocurrency can be predicted based on specific market impulses and the favor of investors [22]. The predictability of changes in market conditions and significant volatility is a significant advantage of cryptocurrencies, which can neutralize the shortcomings associated with the insignificant use of these assets in world trade.

Another opportunity for the growth of the cryptocurrency market is digital financial services (DFS), which can be defined as financial transactions using digital technologies, including electronic money, mobile financial services, online financial services, i-teller and banking without branches, whether banking or non-banking institutions. Digitalization of finance provides an opportunity to achieve broad financial integration by increasing the availability and penetration of digital financial services, benefiting from previously non-banking sectors of the population and offering digital alternatives to encourage consumers and entrepreneurs to leave the informal sector. DFS also opens up opportunities for the integration of low-income financially isolated groups [23]. The additional resources microentrepreneurs in conditions of the negligible cost of entry into the market of cryptocurrency become a factor of the promotion of cryptocurrencies.

3 Discussion

Thus, the spread of cryptocurrencies can affect the level of competition in the financial market and helps to reduce costs, provides an improved experience for entrepreneurs. Nevertheless, attention should be focused on the existence of new risks for consumers in the markets with DFS and cryptocurrencies. Risks such as inadequate disclosure and recovery mechanisms, diversity of suppliers, cross-border sales, arbitration, data protection, lack of coordination between authorities can pose a serious threat to the financial well-being of the businesses. These issues require a more detailed study in further studies.

4 Conclusion

Based on the presented analysis, the following conclusions and generalizations can be made. Active implementation of IT achievements in the financial sector creates conditions for the restructuring of business processes within the banking system. A factor in strengthening this trend is the emergence of digital currencies such as Bitcoin, which perform all the functions of money and which, thanks to the blockchain system, have a high degree of protection and security. The availability of equipment for the production of digital currencies, the spread of trust on the part of users and business representatives contribute to the promotion of these types of assets and the penetration of digital currencies in various spheres of economic activity. This situation, on the one hand, poses a threat to the centralized management of economic processes, and on the other hand, contributes to the possibility of reducing the tax burden on small and medium-sized enterprises. Significant investment opportunities, increased consumerism, and innovation create favorable conditions for increased adaptation of cryptocurrencies in the business environment. At the same time, attraction of additional resources in the economy in the conditions of insignificant costs for entering the cryptocurrency market becomes a factor in the spread of cryptocurrency assets.

References

1. A. Kvitka, A. Kramarenko, Problems and Perspectives in Management, **2**, 42-48 (2018)
2. D. Gladkyh, Bitcoin and other cryptocurrencies: the risks and prospects in Ukraine, available at: http://www.niss.gov.ua/content/articles/files/Glakih_bitcoin-34ba8.pdf (2018)
3. Gref predicted the future, available at: https://utro.ru/articles/2016/11/14/1304545.shtml
4. Are bitcoins legal in Ukraine, available at: https://www.epravda.com.ua/publications/2017/08/11/627993
5. O. Vlasuk, Strategichna panorama, **2**, 9-16 (2017)
6. CryptoCurrency Market Capitalizations, available at: https://coinmarketcap.com
7. A. Kvitka, E. Zaytseva, Ekonomichna teoriya ta pravo, **1**, 111-119 (2015)
8. V. Dyntu, O. Dykyi, Baltic Journal of Economic Studies, **5 (4)**, 75-81 (2018)
9. S. Volosovych, Yu. Baraniuk, Banks and Bank Systems, **2 (13)** (2018)
10. O. Portna, Actual Problems of Economics, **10 (172)**, 404–409 (2015)
11. O. Portna, Actual Problems of Economics, **3(165)**, 106–112 (2015)
12. M. Bondar, N. Iershova, Baltic Journal of Economic Studies, **1 (1)**, 47-54 (2015)
13. A. Kvitka, K. Balakleiets, Social economics, **55**, 107-114 (2018)
14. R. Gandhi, FinTechs and virtual currency, available at: https://www.bis.org/review/r170309d.htm (2017)
15. M. Carney, The Future of Money, available at: https://www.bankofengland. co.uk/-/media/boe/files/speech/2018/the-future-of-money-speech-by-mark-carney (2018)
16. E. Lam, What the World's Central Banks Are Saying About Bitcoin, available at: https://www.bloomberg.com/news/articles/2017-12-15/what-the-world-s-central-banks-are-saying-about-cryptocurrencies (2017)
17. M. Hampl, Central banks, digital currencies and monetary policy in times of elastic money, available at: https://www.bis.org/review/r170720b.htm (2017)
18. S. Kryvoruchko, Financial Journal, **5**, 120-129 (2018)
19. Chun Wei Wang, Economics Letters, **168**, 21–24 (2018)
20. T. Ankenbrand, D. Bieri, Investment Management and Financial Innovations, **15(3)**, 169-181 (2018)
21. Won-Jun Lee, Seong-Tae Hong, Taeki Min, Journal of Distribution Science, **16**, 47-56 (2018)
22. Yu. Liu, A. Tsyvinski, NBER Working Paper, **24877**, available at: http://www.nber.org/papers/w24877 (2018)
23. Digitalisation and Financial Literacy, available at: http://www.oecd.org/finance/G20-OECD-INFE-Policy-Guidance-Digitalisation-Financial-Literacy-2018.pdf (2018)

Organization and conduct of environmental and economic monitoring of the state of material flows in of thermal power facilities systems under the conditions of electromagnetic fields

Petro Kulikov[1], *Denis* Chernyshev[2], *Alla* Bielova[31], *Nataliia* Zhuravska[4]

[1] Kyiv National University of Construction and Architecture, Vozduhoflotsky Avenue 31, Kyiv, 03037, Ukraine
[2] Kyiv National University of Construction and Architecture, Vozduhoflotsky Avenue 31, Kyiv, 03037, Ukraine
[3]Department of Economics and Management of the GSP, "Institute of Innovation Education" of Kyiv, National University of Construction and Architecture, Vozduhoflotsky Avenue 31, Kyiv, 03037, Ukraine
[4] Kyiv National University of Construction and Architecture, Vozduhoflotsky Avenue 31, Kyiv, 03037, Ukraine

Abstract. The application of physicochemical analytical control for the determination of ecological compatibility of techno-natural systems as a factor mark of the state of material flows in the application of non-reagent water preparation in electromagnetic fields for heat-energy objects of the housing and communal sector and industrial enterprises with low production capacity is substantiated. Heat evaporation and hydrocarbons were selected as test gradients for the heating systems. In both cases, there was no excess emission limit value.

Introduction.

Problem statement

Managing enterprises that are geared towards obtaining stable profits and efficient work is a problem in improving the efficiency of the investment process with the optimal cost of all resources to achieve success in the modern development of Ukraine.

Taking into account the integral nature of the problem being solved, organizational and managerial decisions of the economy of nature management on heat energy objects should be considered everywhere in the prism of compatible interactions of engineering nanotechnologies and the ultimate technogenic impact of the impact on the environment (atmosphere).

Analysis of recent research and publications.

[1] Corresponding author: nzhur@ ua.fm

The best practices of existing problems are reflected in works by well-known domestic and foreign scientists: A.P. Algini, L.J. Savage, G.V. Chernova, M.V. Tersky, V.M. Lycha, O.A. Bondar, V.O. Pokolenko, S.P. Stetsenko, V.G. Fedorenko, K. Lancaster, D. Levy, L. Murphy, C. Lee, J. Pitirin and others, who made a significant contribution to the effectiveness of the organization and management of the construction industry.

At the same time, it should be noted that solving this problem for heat-energy facilities is due to the achievements of the development of such branches of science as ecology, energy, economics. It is also necessary to take into account that thermal energy objects are one of the most important energy consumers when their total thermal power is about 10% of the total capacity of the Ukrainian heat supply systems (E.S. Malkin, B.K. Draganov, B.V. Davydenko, A.O. Redko, O.T. Vozniak, V.M. Zhelich, Y.P. Morozov, I.E. Furtat, 1998-2010, 2014, 2016-2018) [1, 2].

That is, such a methodological tool for conducting experimental works allows to determine the innovative nature of the proposed non-reagent preparation of water in electromagnetic fields (EMF). Thus, the basis of the development of such a concept for the evaluation of the functional state of heat supply systems is the theoretical part of the function and control of production activity of heat and power facilities, and Applied part - environmental management, as a factor attribute (subordinated) of the system of passive monitoring of the state of production processes, which gives an economic effect.

Environmental analysis [3, 4], according to the passive monitoring of the state of the production processes of thermal power objects, provides for their mathematical formalization, the results of which determine the assessment of the effectiveness of the technology and the assessment of environmental impacts. The obtained results are used as a forecast for the study of non-reagent water preparation in heat supply systems with the aid of electromagnetic fields at present and in perspective [5].

The goal is to provide environmentally safe nature management [6, 7]. The peculiarity of this technology lies in the fact that the study uses artificial purposeful technogenesis to obtain magnetized water, which in the future is the basis of material flows of thermal energy objects. In connection with this, a special place is given to physicochemical analysis, as a driving and regulating tool for the functioning of the techno-natural systems of these research objects. The work was carried out within the framework of the tasks of the Law of Ukraine on the main principles (strategies) of the state policy for the period up to 2030 (№ 8328, dated 28.03.19). In accordance with these tasks (tactical intentions), expanded by us, the framework of ecological and economic control for heat-energy objects of the housing utility sector and industrial objects of the construction industry. It is the efficiency of the spatial management of thermal energy objects systems, on the basis of the interconnections and the interaction of the components of material flows, have allowed to expand the database of information control of technogenic-natural systems (the motive structure of passive monitoring) and thereby ensure the relevance of the topic.

Materials and methods.

The systematic approach to research using factor analysis of the systematization and formalization of the components of material flows of thermal energy objects (TEs) systems in electromagnetic fields (EMF) [8] was used in this work.

The research was carried out for five years in the laboratories of the Kiev National University of Construction and Architecture [5, 7].

When interpreting the results, the validity of the results was determined using modern methods of control [7].

The work is a continuation of the expansion, existing, established by us, of the two-level control system [9, 10], as one of the effective components of organizational and managerial decisions in the operation of the TEO systems in EMF.

The scheme of organization of environmental and economic analytical control is shown in Fig. 1.

Results and discussion

This publication is a continuation of previous studies and the final stage in relation to the scheme of experimental work and its evaluation (Fig. 1). In this work, a generalized indicator - criterion of environmental friendliness of non-reagent water preparation in the EMF of heat supply systems is defined, which characterizes the ecology of the results of the operation of the TEO systems, and thus evaluates the completeness and efficiency of the operation of the proposed nanotechnology. The definition of such an indicator is due to specific relationships in complex heterogeneous inorganic systems, namely, that in them (under the influence of increased catalytic activity - for EMF) there are intermolecular interactions regardless of the characteristics of micro particles, which reflect their physical and chemical properties.

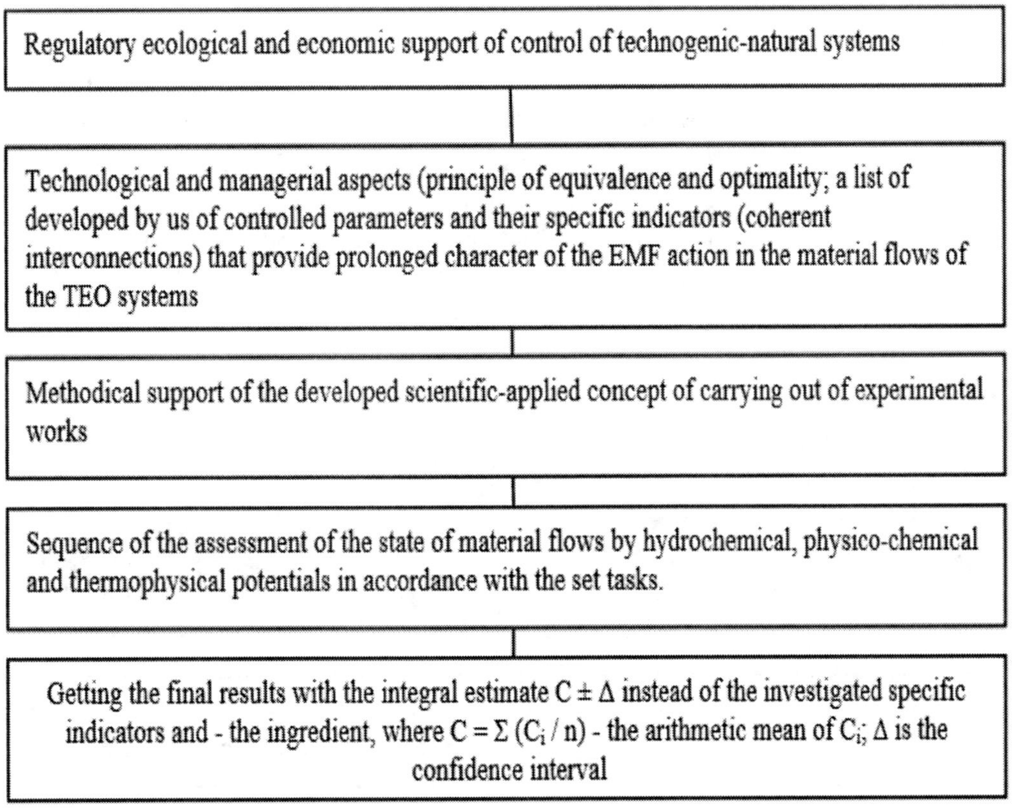

Fig. 1. Scheme of organization of ecological and economic (control) control

Such a course of physical and chemical reactions rules out the discrete nature of man-caused natural systems. Therefore, it is possible to determine the thermodynamic self-organization of them under the influence of artificial technogenesis (for EMF). In connection with this, there was a need to consider external costs (environmental protection) on the outside. In the process of various production processes in the industrial sphere there is a negative impact on the atmosphere - the ability to self-healing and self-organization of atmospheric processes [2] is partially lost.

Our studies have shown that volumes of evaporation heat decrease compared with the use of tap water by 20 % [5]. When the investigated nanotechnology is used, hydrocarbons in technological emissions do not exceed the normative data (Fig. 2).

It should be noted that in the city the X-ray diffraction systems of the TEO, where the pilot tests of nanotechnology took place, as shown in Fig. 2, negative environmental impacts were not determined.

Confirmation can be the calculations of the indicator-criterion of environmental friendliness (1):

$$J_{env} - \sum_{n=1}^{i}\left(\frac{C_1^h}{2DK_1}\right)K_r + \sum_{i=0}^{n}\left(\frac{C_2^{he}}{2DB_1}\right)K_r \quad , \qquad (1)$$

where J_{env}, indicator - criterion of efficiency of technological and economic results on the final effects on the atmosphere;
C_1^h – concentration of hydrocarbons in the atmospheric air;
C_2^{he} – volume, heat of evaporation;
$2DK_1$ - maximum permissible concentration of hydrocarbons in the atmosphere;
$2DB_1$ – maximum permissible emission according to technological norms (thermal balance);
K_r – constant rate of physicochemical reactions in material flows.

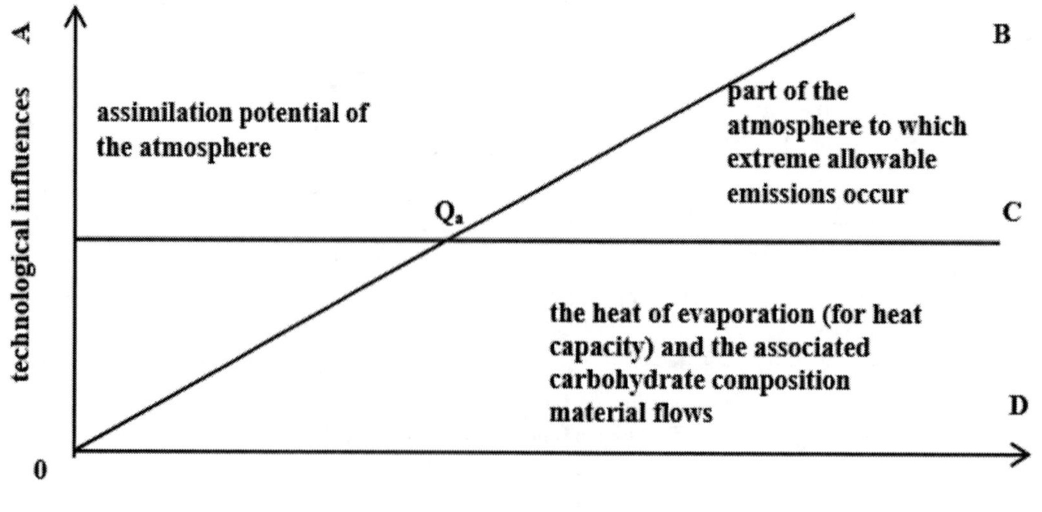

Fig. 2. Graphical conceptual model of external costs of thermal energy.
Q_a - volume of atmospheric emissions; BC – the content of heat of evaporation with hydrocarbon impurities not exceeding the regulated standards

According to practical results: the heat of evaporation is 0.01 cond. units (norm – 0.05) of the total thermal balance [regulation] – 0.901 cond. units in the case of hydrocarbons – 0.24, the indicator of relative aggressiveness (norm – 0.50).

Conclusions

In conclusion, it can be stated that:

1) we have made calculations for the most typical components entering the atmosphere with technological emissions for heat-energy facilities with a small amount of thermal energy production;

2) in both cases, normative indicators are identified.

Determining the ecological nature of man-caused processes by the final result of their implementation is an actual tactical tool for the economy of environmental protection measures.

References

1. A.O. Redko, A.V. Davydenko, S.V. Pavlovsky, N.V. Kulikova, V.Y. Kostiuk, O.I. Kirill. Bulletin of the National University "Lviv Polytechnic". Series: Theory and practice of construction, **844,** 180-187. (2016)
2. O.T. Vozniak, I.Y. Sukholova. Bulletin of the National University "Lviv Polytechnic". The theory and practice of construction, **844**, 26-33. (2016)
3. Determination of the influence of stakeholders on activity of construction enterprises / D.O. Prunenko, K.A. Mamonov. Economic results of 2013: problems, perspectives, efficiency: materials of the international scientific and practical conference. Odessa: "Helvetica", 78-79 (2013)
4. The Economics of Nature and the Nature of Economics (Advances in Ecological Economics Series): collection / J. Cleveland, I. Stern, R. Costanzo. – EE, 293 p. (2007)
5. E.S. Malkin, I.E. Furtat, N.E. Zhuravska. *Special issues of heat and mass exchange: a textbook*. Kyiv: Kyiv National University of Construction and Architecture, 288, (2017)
6. / M.A. Khvesik, A.V. Stepanenko. Economy of Ukraine, **1 (618),** .74-86, (2014)
7. Prospects for creating resource-saving technologies by magnetic processing of water and water solutions / Ventilation, lighting and heat and gas supply: NTZ, 17, E.A. Malkin, I.E. Furtat, V.P. Usachev, N.E. Zhuravska. - K.: Kyiv National University of Construction and Architecture, 120-127, (2014)
8. Ensuring technological reliability of energy-saving technologies with a reagent-free water treatment / USEFUL. SVP4U-KYIV-1-FUND LLC, 2, N. Zhuravska. – Miami, 1-8, (2018)
9. Determination of heat supply efficiency with external indicators for building industry / N. Zhuravska // International conference of European Academy of Science. Business, Economics & Management, 38 (2019)
10. Ecological scientific aspects of technogenic safety with nonchemical water treatment for a technical water-supply / N. Zhuravska, III International Scientific and Practical Conference "Underwater Technologies, 2017", K.: Kyiv National University of Construction and Architecture, 8, (2017)

The development and special aspects of Islamic finance in Europe at the present stage

Iryna Kurochkina[1] and *Ulija* Us[2]

[1] PHD (Econ), International marketing Department, Alfred Nobel University 49000 Naberezhna Sicheslavska str., 18, Dnipro, Ukraine,
[2] PHD (Econ), Department of Banking, Simon Kuznets Kharkiv National University of Economics, 61166, Nauka ave., 9-a Kharkiv Ukraine,

Abstract. The article is dedicated to the peculiarities and specificities of Islamic finance in Europe. We considered the prime tools of the Islamic economical system, historical aspects of the Islamic finance appearance in Europe. We also investigated the prospects of implementing and development of Islamic finance in Europe and in Ukraine.

The issues that concern the implementation of operating principles of Islamic financial institutions in the economic system of the country arise in the majority of the countries in these recent times. In contempt of the presence in more than 40 countries of the world, the greatest development of finance that was based on the principles of Islamic law was obtained in those countries where the legislative amendments were introduced to regulate new financial institutions and products, in a point of fact of Islamic banks and suitable Islamic financial instruments. The Industry leaders are Malaysia, UAE and Bahrain. Over recent years, considering a global migrant policies, the governments of Great Britain, France, Germany, along with CIS nations (Kazakhstan, Kyrgyzstan) have begun to develop Islamic finance in their countries [1].

Up to date, a clear definition of the concept of "Islamic finance" is almost absent in the economic and in legal sciences. It is linked with their key elements - the religious nature. In the constitutions of the countries in the East where Islamic finance is widespread foremost, it has been affirmed that religious sources of laws, inter alia the Quran and Sunnah, are the actions of a higher legal force.

The key and central principles of the Islamic economic doctrine, from the perspective of religious sources of law that properly distinguishes it from the traditional Western theory and practice, is the view of money.

The Sharia does not deny a consensual view of money as the means of circulation, a measure of value and a store of wealth. Islam shares the understanding of money as a human consciousness mechanism that was created to serve the economic needs of society, as an abstract category that is a social institution. The implication is that any product in certain circumstances may become an object of sale apart from money. They are considered only as a service, technical tool that does not have an intrinsic value.

Therefore, Islamic finance is a special type of money and debt relationship that do not contradict a traditional Muslim law (Sharia), between economic units that suffer a shortage of the financial resources, on the one hand, and they have spare cash, on the other hand.

Thus, the Islamic economy can be characterized as a system of management that is based on the norms and principles of Muslim law, that, however, does not mean that the activities of Islamic financial institutions are regulated solely by the rules of the Sharia. This is due to the fact that in the current context, geographically, the range zone of Islamic finance is not limited by the Arab world or Muslim countries. Islamic finance is used almost throughout the world as the local financial institutions, as well as internationally active banks and investment trusts. This type of "ethical finance" is the most widespread in Europe. With due time in Britain and France, where in early 2000s the issue about the development of Islamic finance was considered, some representatives of liberal parties and members of parliament sounded a note of warning that the development of Islamic banking facilities would mean the introduction of the Sharia in some European countries. However, after it became clear that the activities of Islamic banks and other Islamic financial institutions will be regulated by European law, all objections against the development of Islamic finance in Europe have been deposed.

The development of Islamic finance in Europe is influenced by at least two factors that at first sight, are not unrelated at all [2, 3]:

a) the commitment for an enlargement of European financial business to Muslim countries and the willingness of Muslim exporters of gas and oil resources to invest net worth into European countries, including using diversification of financial products and target segments of the market of the large banking capital;

b) the change in the structure in the population of many European countries and the rapid increase in the share of the Muslim population in them that is becoming an essential feature of purchasing power on similar financial services.

Respectively of the second factor it is noteworthy that although the proportion of Muslims in the population of Europe increased from 4% in 1990 to more than 7% in 2017, however, in the countries that received the largest number of migrants who profess Islam, their share in the population is still lower than in countries with native Muslim minorities. Thus, the share of Muslim population in France is less than 8%, in Germany it is about 6%, in the UK it is about 5%, while in Bulgaria, every seventh inhabitant professes Islam [4].

In recent years, Muslims have played an increasingly active role in the economy of the European Union, recruiting the businessmen and skilled wage workers. Simultaneously, the energizing of the Islamic banks activities that remained stability during the Great Recession and became the ownable sources of investment. According to Thomson Reuters [4], in 2014, there were 15 Islamic banks, 13 Islamic windows in conventional banks, 19 other Islamic financial institutions or Islamic windows common investment foundations, 3 Islamic insurance companies out of 1,143 Islamic financial organizations in the world, immediately in the European Union. The share of Islamic banking in universal European bank assets produced 0.15% at the beginning of 2015.

We should emphasize the following information in the capacity of elements in Islamic finance [2, 3, 7, 9]:

– system of prohibitions: Riba – prohibition of usury, Gharar – prohibition of uncertainty of contract terms, Maysir – prohibition of gambling and jobbing, Haram – prohibition of participation in illegal activities;

– a set of financial institutions which activities involve the fundraising and investment of funds (banks and non-bank financial institutions);

– principles of financial business that was established in compliance with the Sharia: Musharakah – partnership or ownership, Murabahah – trade financing, Mudarabah – trust management, Mufavada – mutual trust, Ijara – an analogue of finance lease, etc.;

– Muslim tax system: Zakat is a tax that was intended to ensure a fair distribution of benefits, and Jizyah is a tax on people of different faith.

The fundamental difference between Islamic finance from the finance of the West is the presence of two key elements – the institutes: Riba and Zakat. Negative attitude towards usury came to Islam from Judaism, and it is also contained in the Gospel Book. The prohibition of usury is not a solely particularity of Islam, but it received the greatest development in it. The idea of social justice, equal distribution of funds among members of the religious community, relief aid and the downtrodden is a key to understand the Islamic financial system. However, the biggest unicity of Islamic finance lies precisely in the tax system that came from the religious duty of every believer to pay a tax in support of the poor men to distribute equitable public goods. The European researchers do not pay attention at that light, connecting Islamic finance and Islamic banking. It is important that the funds from Zakat are always required to the public pursue of the head of the religious community and can be spent only on the goals that are set by the Quran, that are "the orphans, the men, wanderers, beggars, etc." [6].

As relating to the first Islamic financial institutions in Europe, it should be noted that they appeared only in the late 1970s. By and by with the change in national legislation, full-fledge Islamic banks or so-called "Islamic windows" in conventional banks began to be created. So, in 1978, the Islamic Banking System (later it was renamed on Islamic Finance House Universal Holding) was created in Luxemburg and in December, 1982 the first Islamic insurance company Takafol (both for investors from the Gulf countries) was created. Upon an initiative of the royal family in Saudi Arabia, Islamic Banking System International Holding was reorganized in 1978 in Geneva, the first Islamic financial structure in the form of a company that is limited in the West, and in 1981, the financial group Dar al-Mal al-Islami Trust with the authorized share capital of 1 billion dollars. In the 1980-s this financial group created several Islamic banks in Europe, but most of them turned out to be unsuccessful, while some turned into investment companies [4]. In Switzerland, because of the existing legal restrictions, the first full-fledge Islamic bank was opened only in 2006 (Faisal Private Bank) [4].

Another project of Islamic banking in Europe was the acquisition of Investment Group in the UK of Hargrave Securities in 1982 by the Saudi-based Al-Baraka, an organization that had a license to conduct bank deposits. In 1987, when British law was changed and a demand for Islamic financial products had grown sufficiently, it was converted into Islamic bank.

The question of legal relations of regulation of Islamic finance in European countries is quite relevant, in other words, can religious norms (Sharia) exist as regulators in European countries?

In practice, for example, in Turkey, Egypt or in some EU countries in financial and banking law, the Islamic finance is considered as a separate case of an innovative financial product, that is void of its beginning of religious content. The emphasis is made on its financial security and on the fact that it does not create additional destabilizing effects on the financial system of subordinated jurisdiction. However, in view of historical and cultural stereotypes and value-based approaches in Europe, the question is whether Islamic finance needs additional or special measures to regulate, control or liberalize the activities in comparison with monetary institutions.

A rule of thumb states that qualification of a financial institution as an Islamic per se can create some additional complexities and burdens for it. In Muslim countries, it has a manifestation in the form of additional regulatory requirements and more severe punishment for non-compliance, in non-Muslim countries – the alert and unacceptability of this type of institutions still continues.

All this reduces the competitiveness and maneuverability of Islamic financial institutions in comparison with usual credit establishments and it creates additional restrictions on the ways in which profits are received. To mitigate the effects of negative factors, some regulators refuse the legal use of the concept of "Islamic finance" in the money and credit systems that are under of their control [8, 10].

Speaking about the aspects of the development and existence of Islamic finance in Europe, in a point of fact the integration of Islamic banking into the economy of European countries, it should be reminded that it began its existence in its current form in the second half of the 20th century in Egypt as a single movement action of the Islamic idea of helping to the marginal population. Only one savings bank became both a kind of political demonstration and a reminder that faith in Islamic justice is alive and it requires action to improve the living conditions of the people and the return of religion as a powerful social societal glue.

The current conditions of social and economic evolution that are bitterly accelerating, continue to face the Arab countries with the challenges that threaten the loss of identity. To avoid this danger, they address insistently to their past in searching of roots that will keep them from being absorbed by the flow of mutual universalization. Other Arabic monarchies in Arabia succeeded in turning their faces to the national heritage; they found the mechanisms in the depths of their history that prevented them from getting lost in the traditions of the peoples and countries that followed the road of catching-up development and fell within the cultural and social pressure of globalization.

It is too early to talk about the future development and implementation of Islamic institutions in Ukraine, however, according to the authors, we shouldn't consider the aspects of the Islamic finance functioning from the point of view of demanding, that is widely seen, does not formulate offers in the information society, but rather on the contrary, follows it, and in the context of those economic benefits that will be able to obtain Ukraine from the development of Islamic financial products. One of the obvious benefits is the attraction of investments from the Middle East countries that is especially relevant in the current economic and financial crisis. Market appearance in Islamic financial institutions, at the core of which there is a lack of participation in distressed and speculative positions, can provide greater stability of the financial system of the country.

References

1. S. Krasnoliudska *The Subjectivity of Islam in World Political Processes.* Scientific notation. Pub. 37., Pg. 315 – 325 (2017)

2. R. Grassa, M. K. Hassan *Islamic Finance in France: Current State, Challenges and Opportunities*, Uluslararası İslam Ekonomisi ve Finansı Araştırmaları Dergisi, Pub. 1, P. 65 – 80 (2015)

3. Cl. *Domat What Is Islamic Finance And How Does It Work?* Available at: https://www.gfmag.com/topics/blogs/islamic-finance-faq-what-islamic-finance-and-how-does-it-work

4. C. Hackett *5 Facts about the Muslim Population in Europe.* Available at: http://www.pewrcsearch.org/fact-tank/2016/07/19/5- facts-about-the-muslim-population-in-europe/

5. *Islamic Finance Development Report 2015.* Global Transformation. 1CD - Thomson Reuters, 200 p. (2016)

6. I. Sobol *Islamic Banking in the European Union Countries.* European Integration Studies, no. 9, pp. 184-197 (2015).

7. A. Filonik *Islamskie banki: zadachi na blizhaishee budushchee.* Mirovaya ekonomika i mezhdunarodnye otnosheniya, vol. 61, no. 2, pp. 85-93. (2017).

8. A. Alharbi *Development of Islamic Finance in Europe and North America: Opportunities and Challenges.* International Journal of Islamic Economics and Finance Stuides, vol. 2, issue 3, pp. 109-136. (2016)

9. V. Malaev *Opportunities for adapting Islamic banking products to the Russian legislative framework.* Brics Law Journal Vol. IV, Issue 3, pp. 62-80 (2017)

10. P. Abedifar, Sh. Ebrahim, Ph. Molyneux, A. Tarazi. *Islamic Banking and Finance: Recent Empirical Literature and Directions for Future Research.* Available at: https://hal-unilim.archives-ouvertes.fr/hal-010731852014.

Actual Aspects of the Displaced Persons' Workforce Potential Realization

Marta Manukhina[1]*, *Marina* Melnik[2], *Alexander Rozmyslov*[3] *Inna* Tatciy[4]

[1,2,3,4]Volodymyr Dahl East Ukrainian National University, Department of Accounting and Taxation, pr. Central 59-a, 93400, Severodonetsk, Ukraine

Abstract. The article considers certain aspects of realizing the workforce potential of the internally displaced persons. The scientific approaches to the category of the "workforce potential" and the interpretation of the "workforce potential of the worker" are defined. The current state of the workforce potential utilization in the country is analysed, the conditions and possibilities of realizing the potential of the internally displaced persons are described. External and internal factors affecting the internally displaced persons adaptation are considered. By statistical evaluation of displaced persons and measures to solve their social problems, conditions and reasons that impede the full realization of workforce potential are considered. It was determined that the problem of employment and the ability to realize their workforce potential is crucial for the internally displaced persons successful adaptation to a new place. The recommendations to stimulate the employment of the displaced persons and their better adaptation in the places of resettlement were formulated.

Introduction

The world experience shows that the effective development and use of the workforce potential of the population is a decisive condition for the successful implementation of the socio-economic development strategy of the modern state. According to the World Bank's study of 192 countries, only 16% of the economic growth in the transitional economies countries are due to the production potential, 20% - to natural resources, and the remaining 64% are related to the human and social capital [1].

In this regard, it can be concluded that "the new development opportunities are revealed to those countries whose public improvement can ensure the realization of the workforce and creative potential of their citizens and satisfy their needs as much as possible" [2].

Research Analysis and Problem Statement

Scientific literature of the recent years shows a marked increase in interest in the problems of efficient use of workforce potential. These problems became the subject of research of such Ukrainian scientists as A. Amosha, S. Bandura, D. Boginia, Ye. Grishnova, A. Kolot,

* Corresponding author: marta.manukhina@gmail.com

E. Libanova, L. Lisogor, S.Paseka, M. Semykina, L. Shaulska, T. Bilorus and others. Despite the significance of the obtained scientific results, scientists adhere to the point of view that the problems of the workforce potential are still not fully investigated. The complexity of the problem is due to the instability of social and labour relations, a systemic crisis in all the spheres of the country's life, the impact of globalization and other external factors. In modern conditions, such factors include the annexation of the Crimea and the protracted conflict in the south-east of the country, which led to an increase in the number of internally displaced persons from the Crimea, Donetsk and Lugansk regions of Ukraine.

The article aims to analyze the current problems of the implementation of the workforce potential of the internally displaced persons (forced immigrants) in Ukraine.

Research Results

Note that the interpretation of the category "workforce potential" is still debatable. Despite the fact that there are various approaches in the literature, it is common for many authors to focus on the qualitative characteristics of the workforce potential - health, physical and psychological ability to work, education, qualifications, experience, morality, professionalism, labor, and creative activity. mobility and so on. Many authors (O. Grishnova, M. Semykina, L. Shualska [3; 4; 5]) consider the workforce potential multifaceted, at different economic levels. The authors distinguish the workforce potential of an individual employee, the workforce potential of an enterprise, the workforce potential of a region, a country as a whole with quantitatively inherent qualitative characteristics. Of course, such an integrated approach enriches the theoretical understanding of the workforce potential at the present stage of the development of science, and makes it possible to systematically evaluate the components of the workforce potential. [6].

Developing the authors' existing approaches to the definition of the concept "workforce potential" [3; 4, 5, 7], we propose to interpret the workforce potential of an employee - as a set of physical, mental, moral, educational, motivational capabilities of the individual and his\her abilities to work, which are formed, developed throughout life, realized in the process of work, being under the influence of subjective and "objective factors that determine the effectiveness of the workforce efforts and the degree of labour activity of an employee. It is through the development and realization of their workforce potential that each working, capable adult person feels as a full member of the society [8].

The workforce potential of Ukraine, which exists now, in its qualitative parameters does not correspond to the strategic objectives of economic and social development. Workforce potential is reduced quantitatively, deteriorating in its qualitative characteristics. These trends are explained by a combination of reasons – a grave demographic crisis, the lack of effective structural reforms in the economy, and ineffective socio-economic policies. As a result, there is a low birth rate, premature mortality, high morbidity, limited access of the population to quality medical services and quality education; ineffective employment, and low social standards. As a result of the imbalance in the demand and supply of the workforce in the labor market there is forced labor migration abroad, an outflow of the most active and educated young people.

In addition, today's realities have revealed that the society is not ready to solve new problems. First of all, it concerns the imperfection of the socio-economic mechanisms of workforce potential realization for the internally displaced persons in the context of the military conflict in the east of Ukraine. The acuteness of the problem dictates the need for a scientific study of the depth of the problems in the workforce potential realization of the internally displaced persons and finding ways to solve them. The Ukrainian society today find itself unprepared for the situation with the emergence of a large number of internally displaced persons inside the country and unable to coordinate actions to ensure the conditions for their workforce potential realization. The effectiveness of the state policy in solving the problems of the internally displaced persons is insufficient. Some achievements of legislative and regulatory support for internally displaced persons do not provide the expected results due to the low quality of the national and regional governance, inconsistency of the legislation and the possibilities for its implementation.

With regard to the current situation, we consider it is necessary to develop a clear program of measures for the effective use and development of the internally displaced persons workforce potential.

According to the UN Guiding Principles on Internal Displacement, internally displaced persons are "persons or groups of persons who have been forced or obliged to flee or to leave their homes or places of habitual residence, in particular as a result of or in order to avoid the effects of armed conflict, situations of generalized violence, violations of human rights or natural or human-made disasters, and who have not crossed an internationally recognized border" [9].

In international law, the term "internally displaced persons" is a category of forced migration, used along with the concepts of "refugees" and "displaced persons". As a rule, in case of forced migration, all the main indicators of the quality of life deteriorate: the well-being, living conditions, social status.

In Ukraine, the main reasons for such a move were the annexation of Crimea, the escalation of the military conflict in the Lugansk and Donetsk regions, the danger to the life and health of the population, civilian casualties, and the destruction of buildings. According to the United Nations Office for the Coordination of Humanitarian Affairs and the Ministry of Social Policy of Ukraine, as of the end of 2018, the number of internally displaced persons from Donbass and Crimea was 1,515,158 people. General information on displaced persons is presented in table 1.

Table 1. Information on the displaced persons as of the end of 2018 [13]

Regions of Ukraine	Number of the displaced from Donbass and Crimea (persons)	% of the total number of displaced persons	% of the Population of the region of arrival
Total	1,515,158	100.0	–
Vinnytsya	11,414	0.75	0.64
Volyn	2,949	0.19	0.28
Dnipropetrovsk	74,342	4.9	2.09
Donetsk	552,116	36.43	19.05
Zhytomyr	7,242	0.48	0.64
Zakarpattia	3,662	0.24	0.22
Zaporizhzhia	55,896	3.69	4.26
Ivano-Frankivsk	3,587	0.23	0.24
Kyiv	65,590	4.33	1.96
Kyiv (city)	164,813	10.88	2.7
Kirovograd	6,628	0.44	0.83
Luhansk	129,500	19.47	19.32
Lviv	10,947	0.72	0.32
Mykolaiv	8,556	0.56	0.7
Odesa	41,436	2.73	0.92
Poltava	25,413	1.68	1.7
Rivne	3,249	0.21	0.27
Sumy	11,248	0.74	1.15
Ternopil	2,121	0.14	0.22
Kharkiv	126,701	8.36	5.04
Kherson	14,199	0.94	0.95
Khmelnytsky	6,339	0.42	0.5
Cherkasy	11,344	0.75	0.9
Chernivtsi	2,449	0.16	0.24
Chernihiv	8,550	0.56	0.72

According to the table above, it can be concluded that most migrants are concentrated in the Donetsk, Kharkiv, Lugansk regions, Kyiv, Zaporizhzhia and Dnipropetrovsk regions, 53.3% of migrants are located in the first three regions, and 19.5% in the last three, that is, six regions received 72.8% of all IDPs.

Forced migrants face a whole range of problems, including:

housing problems (accommodation)
employment problems;
medical care problems;
problems with the provision of pre-school and school education services;
psychological problems;
lack of social protection.

A number of urgent issues regarding the registration and livelihoods of internally displaced persons and their family members in the new living conditions are considered by the Law of Ukraine "On Ensuring the Rights and Freedoms of Internally Displaced Persons" dated October 20, 2014 No. 1706– VII. However, it is impossible to solve the whole set of diverse problem situations in life support, employment, social security with one law. It is necessary to change the state and regional policies regarding migrants, to search for new mechanisms to stimulate the employment of internally displaced persons.

It should be noted that the resettlement of the internally displaced persons requires huge expenses, and therefore, together with the unresolved problems of the livelihood of a significant part of Ukrainian citizens already living in this territory, forced migration can aggravate the already tense situation in the labour market, housing, and exacerbate inter-ethnic conflicts. On the other hand, an increase in the able-bodied population could create favourable opportunities for improving the demographic situation, perhaps only with the implementation of an effective state policy in terms of preparation at all levels for receiving, accommodating and settling migrants [11].

Due to its social, economic and psychological consequences, forced relocation is a real life catastrophe for most adults, especially when it happens in a limited time interval, which makes it impossible at least to some extent to morally, financially, physically prepare for a change of residence. In the conditions of war, the overwhelming majority of immigrants found themselves in a situation of difficult life choices - to move, find a job and get used to living in a new place or to consider leave as temporal and be ready to return.

There is an urgent need to quickly adapt to new living conditions, and the success of such adaptation depends on external factors, relating to the resettlement region, and internal, characterizing the most forced migrants, their particular nature, mentality, readiness for any job, change of profession, retraining.

The external factors include such indicators as the density and demographic structure of the population of the region, the development of the economy, the need for workforce; availability of free housing; the development of social and industrial infrastructure; resource potential; climate and so on.

Internal factors include gender, age, education, specialty, qualification; marital status, level of material security, character trait.

The problem of employment and the ability to realize their workforce potential is, in fact, important for the successful adaptation of internally displaced persons to a new place. Having left as a result of involuntary resettlement without housing and a significant part of the acquired property, people retain practically their only capital - knowledge, skills, labor skills, certain abilities to display mobility, retraining, communications.

However, the issue of employment is very problematic due to the general economic crisis in the country, the ratio of prices in the housing and labor markets, forcing people to choose not the place where there is a job, but where they can find affordable housing. As long as such a discrepancy between the cost of housing and the existing level of salary will persist, there can be no rational use of the internally displaced persons workforce potential. [12].

In addition, it is necessary to note the internally displaced persons various abilities to adapt: unequal psychological readiness to find a job, retraining, get a new specialty. This is influenced by age, profession, level of education, skill level, and the like.

Conclusion

Considering the problems of comprehensive realization of the internally displaced persons' workforce potential, a number of new mechanisms to stimulate their employment in resettlement regions can be identified. The main ones include the following:

1. Providing employers who employ internally displaced persons with compensation for their wages.
2. Reimbursement to the employer of the costs for retraining and advanced training of employed forced migrant.
3. Expanded coverage of the internally displaced persons with vocational training, retraining and advanced training (by redistributing the licensed students recruitment for training, retraining and advanced training).
4. Expansion of the international cooperation (implementation of joint projects with international organizations that are aimed at developing employment).
5. Creation of new jobs with the help of technology parks and business incubators.

Thus, the internally displaced persons adaptation to the new conditions, creation of the possibility for the full realization of their workforce potential acts as a complex and multi-level process, affecting the legal, socio-economic, socio-cultural, socio-psychological aspects. A systematic approach to solving these complex tasks requires the state to develop a well-thought out concept of the state migration policy.

References

1. Human Development Report, 1994.-New York, Oxford: Oxford University Press, (1994).
2. Bilorus T. Visnyk Kyivskogo nacionalnogo universytetu imeni Tarasa Shevchenka. Seriya "Ekonomika", **137,** 42-45 (2012).
3. Grishnova O. A. *Lyudskiy rozvytok : Navch. Posib.* – K. : KNEU, 308 (2006).
4. Semykina M.V. Naukovi praci Kirovogradskogo nacionalnogo texnichnogo universytetu. Ekonomichni nauky. **16(1),** 3-10 (2009).
5. Shaulska L. V. *Strategiya rozvytku trudovogo potencialu Ukrayiny: monografiya.* Donetsk: NAN Ukrayiny, In-t ekonomiky prom-ti, 502 (2005).
6. Pasyeka S. R. Socialno-trudoviy potencial regionu: teoriya i praktyka rozvytku: Monografiya . Cherkasy: "IntroligaTOR", 608 (2012).
7. Ratnikov D.G. *Pidvy`shhennya konkurentospromozhnosti trudovogo potencialu naymanykh pracivnykiv v Ukrayini v umovax globalizaciyi.* In-t demografiyi ta socz.doslidzh. NAN Ukrayiny.- K., (2010).
8. Drozdova E. M. Dy`skussy`ya. **5**, 17-22 (2012).
9. Ofy`cy`al`nij sajt UVBK OON: [Elektronnuj resurs].URL: http://unhcr.org.ua/en/.
10. Libanova E.M. Vymushene pereselennya z Donbasu: masshtaby ta vyklyky dlya Ukrayiny. Vnutrishno peremishheni osoby v Ukrayini: realiyi ta mozhlyvosti /Materialy Mizhnarodnoyi naukovo-praktychnoyi konferenciyi (27 lyutogo 2015 r. m Kyiv).– Available at: http://ief.org.ua/wp-content/uploads/2015/02/Libanova2015.pdf.

11. Semykina M.V., Pustovit M.V. // Formuvannya ry`nkovoyi ekonomiky: zb. nauk. pr. – Specz. vyp. Pracya v XXI stolittya: novitni tendenciyi, socialnyj vymir, innovacijnyj rozvytok : u 2 t. –.2. – K.: KNEU, 2012. – S. 370 –379.
12. Pasyeka S.R., Semykina M.V., Fedunets A.D. Model ocinky socialno-trudovogo potencialu regionu v konteksti obgruntuvannya strategiyi modernizaciyi ekonomiky na zasadax innovacijnogo rozvytku / Innovacijna model rozvytku: tendenciyi ta perspektyvy ekonomichnoyi transformaciyi [Kolektyvna monografiya]: Uman, PP Zhovtyj O.O., 2014 r. – S. 202– 213.
13. www.kmu.gov.ua/ua/news/pyat-rokiv-peremishchennya-problemi-ta-uspihi-integraciyi-u-gromadi

Theoretical and methodological principles of renewable energy development based on the technologies' transfer

Marta Manukhina[1], *Nataliia* Riazanova[2]*

1Volodymyr Dahl East Ukrainian National University, Department of accounting and taxation, Severodonetsk, Ukraine
2State Establishment "Luhansk Taras Shevchenko National University", Department of Finance, Accounting and Banking, Starobelsk, Ukraine

Abstract. The article deals with the problems of economic provision of renewable energy development in interaction with innovation and technological elements: technology transfer, nanotechnology, intellectual environment for knowledge generation. The system of certain elements interaction in the form of innovation-technology consortium is analysed. The structure and composition of the engineering platform bases of the consortium are determined. Significant effects of nanotechnology projects in the field of renewable energy are formulated. The priority directions of scientific and technical policy in the relevant fields are proposed.

1 Introduction

Progressive growth rates of energy industry can be achieved both in extensive and intensive ways. The strategic aim of the innovation and science and technology policy in the energy sector is to create a sustainable national innovation system providing the fuel and energy complex with highly effective domestic technologies and equipment, scientific and technical and innovative solutions in the capacities necessary to support the country's energy security. However, it must be considered that the energy security of a certain state, based solely on its own innovative achievements, are more vulnerable than the energy security of countries that actively exchange technological benefits with partner countries on mutually beneficial terms. Hence, the interest in the development of renewable energy in the innovative-technological aspect significantly increase, which is confirmed by the highest level of growth in the use of non-traditional renewable energy sources among all energy sources in the world's balance.

1.1 Relevance and problem of research

1.1.1 The purpose and objectives of the study

* Corresponding author: natalirozez1975@gmail.com

Creation of an innovative and technological and efficient energy sector that is adequate to the needs of a growing economy in energy resources, as well as international interests of the country, providing the necessary contribution to the socially oriented innovative country's development, is the main aim of the country's strategic development. Achieving this goal requires a consequent solution of the complex set of tasks, including: increasing the efficiency of reproduction, extraction and processing of fuel and energy resources to meet internal and external demands; upgrade and creation of a new energy infrastructure in close interaction with the intellectual environment that generates knowledge.

The main purpose of the paper is to develop methodological recommendations on the economic provision of renewable energy development and to assess its impact on related economic spheres in close interaction with innovative technological elements, namely, technology transfer and nanotechnology.

1.1.2 The hypothesis of a scientific problem

The hypothesis of the scientific problem is to substantiate the position that the reliance on the innovation and technological development of renewable energy sources is a single-option choice for the countries aimed at global competitiveness increasing. The growing role of innovation, technology transfer, and nanotechnologies in the renewable energy sources development is revealed in conjunction with the peculiarities of the fossil energy sources replacement with renewable ones in the most energy-efficient energy markets of the industrialized countries and with certain new challenges for the energy policy of Ukraine.

1.2 Literature review

Problems of economic innovation in the renewable energy sector have been considered by foreign scholars such as Daniel H. Yergin, Simon Roberson, Hugh Sharman, Mark Z. Jacobson, Ernst Schwanhold, Mark A. Delucchi, S. Tugsy, Ilhan Ozturk, Aslan Alper. Key aspects of cutting-edge innovative development, theoretical aspects of innovation, innovative cycles were considered in the work of such scholars as Joseph A. Schumpeter, Igor Ansoff, Viktor I. Kushlin, Konstantin I. Pletnev, Aleksandr N. Folmyev.

In Ukrainian economic and technical papers attention to the problem of the RES was paid by such scientists as V. V. Kostetskyi, S. O. Kudria, L Y. Matviychuk, A. V. Prokip, B. P. Herasymchuk Certain researches on the renewable energy, energy-saving technologies transfer and commercialisation of results of intellectual works were described in the publications of P. A. Drohovoz, P. N. Tsybuliov, Y. Lashyn, S. N. Shykaev. Some attention to RES issues and its influence on energy safety was paid by such scientists as A. A. Arbatov, A. M. Mastepanov, A. A. Konopelnyk, A. A. Makarov, O. B. Brahynskyi, K. N. Mylovydov.

Theoretical issues of the RES commercial effectiveness assessment take significant place in all mentioned above papers. Nevertheless, issues on technology transfer in renewable energy a scientifically based promotion of the innovations in the RES application. Developing of qualitative tools and obtaining of the representational results of mentioned problems solving are the most important factors for increasing global technological effectiveness in the countries that are leaders in the RES consumption.

2 Results and discussion

2.1. Analysis of the situation

The main subjects of social and economic interaction in the processes of modernization based on the innovation, technology transfer and nanotechnology, are, on the one hand, creative environments that provide knowledge generation, and on the other - production structures that create high-tech products and services. Problematic aspects of such interaction are the instrumentation of the transfer area between these subjects, low research and innovation activity of enterprises, traditionally "closed" industrial culture, which restricted with localized technological partnership. However, the focus on innovation, elaborated and existing models of the external innovations implementation, as well as culture, which makes innovations an indispensable element of the production consciousness, based on the successful work of modern R & D departments.

It is possible to build effectively general interaction architecture including a system of direct relationships connecting education, science, industry and business in the innovation process. Open innovation model is one of the tools for organisation of such interaction [1] In this case company relies not only on its inner R&D products but actively involves innovations and competences from the outer environment. The understanding that "if we make the best use of internal and external ideas, we will win" is the basis of the open innovation paradigm is the [1; 2]. Industrial innovations' sphere is very sensitive for availability of its own R&D products and how fast technological innovations are commercialized. [3;4] Crowd-sourcing tools that allow to create innovation accepted ideas' management and perform open contests, can be very useful for overcoming problems caused by increased level of the clients comfortability [5;6]

Creation of consortium that consolidates environment of knowledge generation with environment where they are to be applied and provides deep engineering as result of interaction can be one of the way for innovation fracture overcoming. Within the frame of these consortia, the business groups of engineers, scientists and commercialization experts accompany and provide the production processes of energy saving technological innovations, and the research structures of universities and research institutes, together with R & D departments of high-tech enterprises, carry out the development of innovative challenges facing the energy sector. The ideology of open innovation, implemented with the help of such organizations, can change the production and technological culture and dramatically increase the speed of knowledge metabolism as complex exchange processes, allowing the industry to reach new knowledge, transform them into technology or product, to include in economic turnover [15]

In future, effectively composed interaction system can accumulate significant flow of the two-way transaction, performed in the innovation system of Ukraine. As a potential global player on the market, it attracts particular interest of domestic and foreign investors and partners. It is worth to note the importance of establishing a network in our country that is responsible for coordinating the technological demands of the energy sector and technological proposals from the scientific community.

The structure, which has the form of innovative technology consortium. can be the basic organization form for the interaction system. Its institutional core should consists of: high-tech enterprises and organizations, innovative and research universities, venture companies, funds and agency organizations.

The institutional basis of the engineering platform of the consortium consists of the knowledge generation environment, the environment of their technology and the catalytic environment for development (Fig. 1)

Disciplinary basis of the consortium engineering platform is represented by a complex of innovative knowledge systems that are significant for its activities. They include energy systems of the future and modern materials, nanotechnologies. The content of the disciplinary basis is determined by structuring its matrix, as the identification of specific relevant areas for one or another significant innovative knowledge system (Table 1). Based on the disciplinary matrix, it is possible to segment thoroughly the innovative interests of high-tech companies and organizations as components of the consortium, including for foreign partners. Simultaneously, the problem of using the most promising personnel is solved here.

The consortium is a structure that combines, on the one hand, the innovative interests of a group of leading high-tech enterprises, and on the other - a large number of universities and research institutes. It is capable to organize network projects stimulating the creation of horizontal and vertical connections important for technology transfer [7; 16]. Such formation can become a powerful tool for innovation development, both in the technological and in the investment aspects.

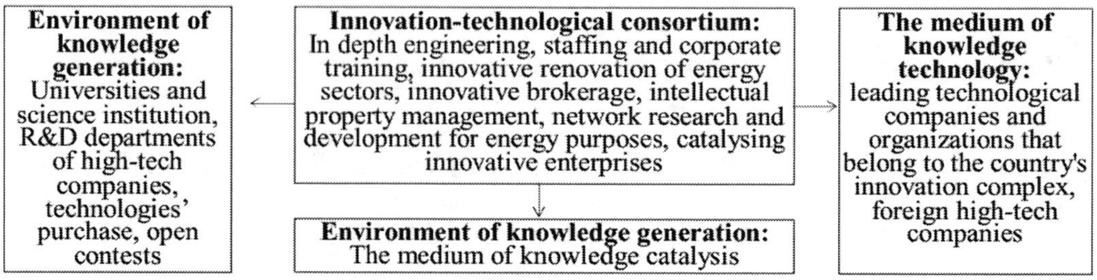

Fig. 1. Prototype of the institutional basis of the consortium engineering platform.

Table 1. Disciplinary basis of the consortium engineering platform.

Innovatively significant knowledge systems		Disciplinary matrix structuring
1	Energy systems of the future	Renewable energy: solar, geothermal, wind and tidal, biofuels; technologies of modern energy materials
		Traditional energy industry: creation of inexpensive hydroelectric power systems and thermal stations
		Nuclear power industry: creation of compact sources of "eternal" energy; power supply of space flights ; thermonuclear systems
2	Modern materials and nanotechnologies	Nanomaterials: structural and functional, including nanocomposites, which strengthen nanocoatings, conducting nanoelectronic materials (synthesis, diagnostics, use); high-purity substances
		Nanoengineering: micro and nanosystem technology, nanoelectronics, nanotechnology for energy production and management, nanometrology
		Nanorisks: safety of nanomaterials and nanoproducts (the impact of nanotechnologies and nanomaterials on human health during the development, production, consumption and utilization periods)
		The economy of the nanoscience: investments, high technology markets, competitiveness, industrial structure

2.2. Prospective of situation

Prospects for the modern energy development, primarily for the renewable one, are related to the transfer of technologies and nanotechnologies, nanomaterials, consider their strategic importance as catalysts of innovative upgrade of the economy, the creation of powerful internal and external impulses and the effects of development, stimulation of sustainable growth not only in the field of energy, but also national economy in general. The continuous development of technology transfer and nanotechnology defines innovation and economic growth in virtually all industries, including renewable energy [12; 15]. The massive introduction of nanotechnologies will bring large-scale economic effects to the energy sector.

The main areas of effective use of nanotechnologies in the power industry are [13; 14]: use of renewable sources; energy storage; reduction of materials' consumption; use of alternative materials. Particularly relevant nanotechnologies in the field of energy storage and solar energy [11; 17]. Significant losses of electricity up to 4-9% in the Europe, 7-10% in the US and Canada,

13-14% in Ukraine during its transportation to the consumer are still the most significant problems of the development of the energy sector of the world economy [9; 10].

The main economic benefits of using nanotechnology in the field of energy are to increase the efficiency of materials and reduce production costs [8].

In the future, nanomaterials will allow to reach a significant increase in the capacity of batteries, the growth of efficiency and cheaper production of solar panels, prolonging the useful life of photovoltaic cells and storage batteries for large energy storage systems.

Hence, the most significant effects of the implementation of nanotechnology projects in the renewable energy sphere in the short-term and medium-term prospective, which are significant for the country's economy in general, will be:

- strengthening the country's energy security on the basis of increasing the use of RES with the depletion of traditional;
- increasing the efficiency of generating electricity by improving the technology of production and conversion of energy received from RES, increasing the volume of converted into electricity energy, reducing the duration of the processes;
- reduction of the cost of production of electric energy received from RES, as a result of: reduction of materials consumption and increase their efficiency in case of nanomaterials use, reduction of production costs;
- development and application of innovative systems of electric energy accumulation and storage, as well as new technologies of energy saving, increase of economic efficiency of the renewable energy sector;
- price decreasing in the wholesale energy market as result of rising proposition of renewable energy prices, that fully corresponds with consumers' interests;
- reduction of the negative impact of traditional sources on the environment and the neutralization of such negative costs by the use of innovative renewable energy technologies;
- obtaining economic advantages of the leader in the field of development of the latest technologies in this field by the country, providing an equal position in the market of high-tech energy products and services with a limited range of participants, as well as participation in the creation of global mechanisms for obtaining and redistribution of rent from the monopoly to unique knowledge and high technology.

2.3 Recommendations

The successful transfer of technologies in the energy sector requires certain conditions, the main ones of which can be attributed: a common, coordinated energy policy; the presence of a competent scientific and innovative authority in the industry; the common strategic vision for the development of the industry and the ways of integration.

Priority should be given to the following areas of scientific and technical policy in the relevant fields. In power industry: development of new and upgrade of existing technologies and technical means in order to increase the efficiency; increase of reliability and efficiency of the electric power transportation system, including creation of highly integrated intelligent systems and distribution electric networks of the new generation (Smart Grids); guaranteed uninterrupted power supply to consumers.

In the field of RES and local fuels:

- development of technologies for the RES usage, as well as multifunctional energy complexes for autonomous energy supply of consumers in areas disconnected from the centralized energy supply networks;
- development of effective technologies of network supply on the basis of RES;
- development of technologies of combined use of RES, as well as technologies of power generation unevenness compensation by generating objects on the basis of wind and tidal energy;

- design and development of technologies for the modern materials use in the production of equipment and components for generating objects on the basis of RES in order to reduce the cost of their construction and increase the efficiency of functioning;
- expansion of production and use of new fuels derived from different types of biomass.

3 Conclusions

An important step in the implementation of the analysed problems is the creation of an innovation and technology consortium for the following tasks:
- organization of joint development and implementation of scientific and technical programs and innovative projects of consortium participants;
- assistance in determining of cooperation areas of mutual interest in the field of high technologies and innovations, based on national tasks and priority directions of scientific and technological, innovative and industrial development of participants;
- consolidation of effort of governmental authorities, public organizations expressing the interests of innovative business, entrepreneurs and commodity producers to increase the potential of high-tech branches of the economy as the basis of social and economic development;
- creation of conditions for attraction of investments to the innovative sphere.

Effective use of the capacities of fuel and energy complexes, as well as to provide national economies with the main types of energy resources, requires developing long-term mutually beneficial cooperation in the energy industry, technology transfer, and intellectual property, providing and pursuing coordinated energy policies, phasing the formation of common markets for energy resources based on the following principles: absence of technical, administrative and other obstacles to trade energy resources, equipment, technologies and related services; creation of favourable conditions for attraction of investments in the energy industry. The common market for energy resources creates an incentive for expert interest in the development of innovation, technology transfer, nanotechnology in the field of fuel and electricity transmission.

References

1. Chesbrough H. *Open Innovation. Moscow*: Pokolenie. (2007)
2. Denisov D. Business Journal. **6**. (2011)
3. Schumpeter J. A. Business cycles. A theoretical, historical and statistical analysis of the capitalist process. **1.** 220, (1939).
4. Andrew J. P., De Rocco E. S., Taylor A. *The Innovation Imperative in Manufacturing*:- How the United States Can Restore Its Edge. Boston: BCG, MI, NAM. (2009)
5. Yergin D. The Prize the Epic Quest for Oil, Money & Power. Simon & Schuster, New York, 945 (1991).
6. Sharman H. Renewables Won't Keep the Lights On. The Oil Drum.
7. Petrechenko V. (2010) Innovations: Switch on. Venture companies ecosystem of the harvesting cycle: Conditions and prospective. LLC "Nau-tech" M.: Business-journal; Computerra.
8. Building Connections, Fostering Solutions: 2012–2013ANSIAnnual Report. – New York, NY: ANSI, 40 (2013).
9. Russian nanotechnologies. **4, 11–12**, 20–22 (2009)
10. Interview with Academician V. E. Fortov IV St. Petersburg International Economic Forum "Smart Networks - Smart Energy - Smart Economy". St. Petersburg, 2010.
11. German Energy Turn - Arguments for a Renewable Energy Future. Energytransition. The German Energiewende.
12. Energy efficiency mechanisms: energy industry innovations. NanoNewsNet. 12Th of January, (2011).

13. Nanotechnology in the sectors of solar energy and energy storage. Technology Report / International Electrotechnical Commission. – Geneva, Switzerland : IEC, 101 (2013).
14. Make energy turning a successful project. Deutschland.de.1, 28-33 (2014).
15. Titov V. (2000) Transfer of Technology
16. Uskova O. (2011) Silicone Deadlock. Izvestiya. December 5
17. Tsybulov P. M, Intellectual Property Management. K.: "KIS". 448, (2005)

Problem aspects of use of the intellectual property of Ukrainian enterprises

Yurij Maslov[1], *Valentyna* Semenova[2], *Vadym* Luniachek[3], *Olga* Kerekesha[4]

[1] Odessa National Economic University, Department of Banking, 8, Preobrazhenskaya str., Odessa, 65082 Ukraine
[2] Odessa National Economic University, Department of the Economyof Enterprises and Entrepreneurship, 8, Preobrazhenska str., Odessa, 65082 Ukraine
[3] Ukrainian Engineering Pedagogics Academy, Department of Creative Pedagogy and Intellectual Property, 16 Universitetskaya st., Kharkiv, 61003 Ukraine
[4] Odessa National Economic University, Department of Foreign Languages, 26, Pushkinskaya str., Odessa, 65082 Ukraine

Abstract. The article is devoted to the study of the processes of formation and use of intellectual capital at Ukrainian enterprises and the analysis of the efficiency of innovative activity of domestic enterprises. The analysis conducted showed that domestic enterprises and organizations nowadays slow down the innovation activity and reduce the efficiency of the use of intellectual capital. Such tendencies are the consequence of the imperfection of the legislative framework for stimulating innovative activity of domestic enterprises; lack of incentive mechanism for developers and inventors, innovative enterprises and organizations. Problems of innovative development are inherent in both the state level of management and the management at the level of individual enterprises. It is necessary to form effective innovative state and regional development programs, to create modern innovative infrastructure, to develop at the legislative level the mechanism of preferential taxation and crediting of innovatively active enterprises, etc.

Introduction

The development of the world economy today is directly dependent on active technological changes. New branches of economy are being formed, technological processes are being transformed, and approaches to ensuring production processes and product promotion are changing qualitatively. The active development of the Enterprise 4.0 platform companies, the deepening of the digitalization of management and production processes in the global economic environment are challenges for domestic enterprises.

The effective activity of domestic enterprises at the Ukrainian sales markets and the ability to promote products to the international markets is conditioned by the ability to stimulate innovation through the introduction of new technologies, creation of innovative products and the formation of intellectual property. The conditions for taking leadership positions in the international community are created by stimulating the processes of intellectual capital formation, the creation of the objects of intellectual property and their further commercialization.

The purpose of the article is to study the processes of formation and use of intellectual capital at enterprises in Ukraine and to analyze the efficiency of innovative activity of domestic enterprises.

Theory of the matter

The problem of the study of intellectual capital remains relevant for a long time and takes on a new significance in the context of intensification of scientific and technological progress. Fundamental aspects of the formation of intellectual capital, its development and use have been reflected in the works of many domestic and foreign scholars. The studies of theoretical aspects of the essence of intellectual capital are devoted the works of such foreign scientists: A. Stuart Thomas [1], A. Brooking [2], L. Edwinson [3], V. Cohen [4], V. Landes [5], J. Roos [6], P. Sullivan [7] and others. Significant contribution to the consideration of methodological and practical aspects of the processes of formation of intellectual capital to make such domestic scholars as O.B. Butnik-Siverskiy [8], M.V. Vachevskiy [9], O.V.Kendyukhov [10], O. Ye. Kuzmin [11], I. Yu. Kuchumova [12] and others.

Despite the considerable number of publications, the issue of complex management of the intellectual capital at enterprises in the realities of the domestic economy needs further investigation. Today, at the level of individual enterprises, unfortunately, there is practically no effective system for managing intellectual capital formation aimed at stimulating innovation, creating intellectual property and technology transfer. This determines the relevance of the topic, and determines the direction of research in its scientific and practical aspects.

Discussion of results

Scientific and technological progress covers all spheres of human activity both in the domestic sense and in all sectors of the economy. In today's economic development, the intellectual activity of enterprises is crucially important. The processes of intellectual capital formation are the starting point for innovative development of production technologies, improvement of quality of products and services and creation of objects of intellectual property. The ability of economic systems and individual businesses and organizations to innovate and effectively use intellectual capital is a unique indicator of their success and competitiveness.

Global trends in the formation of individual economies as powerful technological systems and the transition to the creation of clusters of industry 4.0 drives enterprises to build an effective innovation strategy. Orientation of the domestic economy to the standards of the international market requires Ukrainian enterprises and organizations improve the development of innovative technologies and to integrate into the world tendencies of intellectualization of all management and production processes.

However, according to the Global Competitiveness Index, in 2018 Ukraine takes the 81st place in the ranking (index 4.11) [13]. Table 1 shows the indicators by which the Global Competitiveness Index of Ukraine in 2018 is formed.

Table 1. The system of indicators of global competitiveness index of Ukraine in 2018 [13]

Criteria	2018
1. Quality of institutions	3.2
2. Infrastructure	3.9
3. Macroeconomic environment	3.5
4. Health care and primary education	6.0
5. Higher education and training	5.1
6. Goods market efficiency	4.0
7. Labour market efficiency	4.0
8. Development of the financial market	3.1
9. Technological readiness	3.8
10. Market size	4.5
11. Competitiveness of enterprises	3.7
12. Innovation	3.4

The rating was led by Switzerland (index 5.86), the United States (index 5.85) and Singapore (index 5.71). The ten most developed and competitive countries include the Netherlands (index 5.66), Germany (index 5.65), Hong Kong (index 5.53), Sweden (index 5.52), the United Kingdom (index 5.51), Japan (index 5.49) and Finland (index 5.49).

Indicators of the Global Competitiveness Index include, in particular, "technological readiness" and "innovation", which characterize the efficiency of the use of intellectual capital and the innovative activity of enterprises and organizations. Let us analyze the results of the use of intellectual capital of domestic enterprises in the dynamics of the recent years. Figure 1 shows the dynamics of applications for inventions in Ukraine in the period from 2014 to 2018.

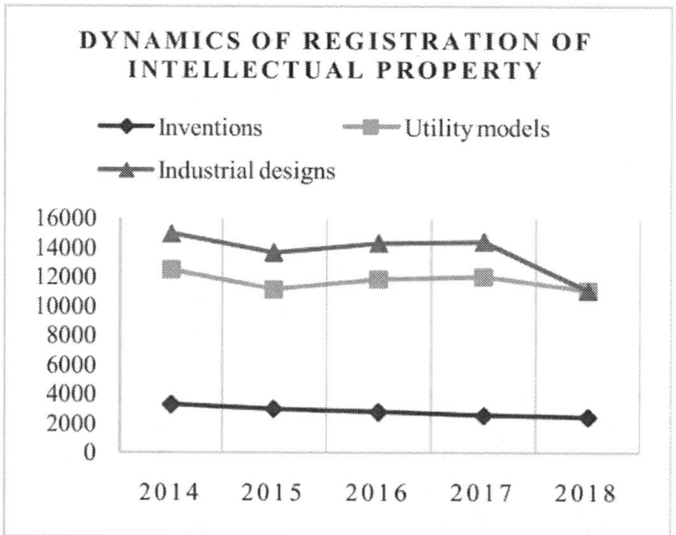

Fig.1. The dynamics of applications for inventions in Ukraine in the period from 2014 to 2018 [14].

Analysis of indicators in Fig. 1 shows that there is a clear downward trend in the amount of intellectual property in Ukraine during the study period. Table 2 shows the structure of applications for inventions and utility models from national applicants in Ukraine for 2014-2018.

Table 2. Structure of applications for inventions and utility models from domestic applicants in Ukraine [14].

Indexes	**2014**	**2015**	**2016**	**2017**	**2018**
Total amount of applications	100.0	100.0	100.0	100.0	100.0
Inventions	21.0	21.1	19.1	20.3	19.0
- entities	11.4	10.5	10.3	10.7	10.5
- individuals	9.6	10.6	8.8	9.6	8.5
Useful models	79.0	78.9	80.9	79.7	81.0
- entities	51.3	49.2	54.0	52.9	56.8
- individuals	27.7	29.7	26.9	26.8	24.2

The analysis of Table 2 shows that a significant part of inventions and utility models in Ukraine are registered by individuals. Taking into account the negative trends in the dynamics of registration of intellectual property objects (Fig. 1), it is possible to make a conclusion about the decline in innovation activity of domestic enterprises over a long time.

The detailed research on the formation and use of intellectual capital at Ukrainian enterprises that we conducted allowed making the following conclusions [15, p.194]:

- there is a tendency to decrease the total number of employees involved in the implementation of research and development;
- inventive activity is not developed at domestic enterprises, that are mainly focused on providing their own products with appropriate signs for goods and services;
- the number of specialists having a scientific degree and taking part in research and development decreased significantly;
- there is a decrease in the number of inventions and utility models in the domestic industry;
- there is a decrease in the innovation activity of domestic enterprises, which impedes the processes of commercialization of innovative development and technology transfer [15, p.194].

V.V. Prokhorova emphasizes that active innovative position of the enterprise is an important element of crisis management at enterprises [16]. Therefore, it is necessary to create an effective mechanism for enhancing the effective use of intellectual potential of domestic enterprises.

In our opinion, there are some obstacles to the effective utilization of enterprise intellectual capital, the processes of intellectual property creation and technology transfer, such as:

- imperfection of the legislative framework to the stimulation of innovative activity of domestic enterprises (contradictions in existing laws and regulations);
- underdevelopment of the incentive mechanism for developers and inventors;
- insufficient stimulation of innovation activity of enterprises and organizations;
- lack of effective innovative programs for development of individual regions;
- lack of effective national and regional innovative programs to support technology transfer;
- imperfection of mechanisms and criteria for selection of promising innovative projects;
- underdevelopment of innovative infrastructure (technology transfer centres, innovation incubators, etc.);
- lack of support at the national level of rationalization and invention;

- limited public funding of priority innovation programs;
- imperfect mechanism of formation of favourable investment climate in Ukraine, etc.

In our opinion, the priority tasks for activating the processes of formation of intellectual capital of domestic enterprises, creation of objects of intellectual property and promotion of technology transfer include:

- formation of the state tax and credit policy aimed at promoting the innovative development of enterprises;
- development of national and regional innovation programs;
- creation of infrastructure innovation system;
- building an integrated technology transfer network that integrates existing centres of innovation and technology transfer into a single network to formulate a common strategy for entering international markets and coordinate areas for international cooperation;
- formation and financing of innovative state programs aimed at activating priority directions of technological development;
- improvement of the system of protection of intellectual property rights;
- development of the state invention support programs and rationalization and promotion of participation of domestic inventors in international competitions and programs.

Conclusion

According to the Global Competitiveness Index, Ukraine took the 81st place in the ranking in 2018, indicating that there are systemic crises in all sectors of the economy, and these trends have been maintained over the past years. The domestic economic system is in the crisis, in comparison with the strong growth over the last decades of the world leaders on the basis of the activation and effective use of their intellectual capital.

The analysis of innovative processes and the use of intellectual capital at enterprises in Ukraine showed the presence of negative tendencies: the imperfection of the legislative framework to stimulate innovative activity of domestic enterprises; underdeveloped incentive mechanism for developers and inventors; insufficient stimulation of innovation activity of enterprises and organizations; the lack of effective innovative programs for the development of individual regions; lack of effective national and regional innovative technology transfer support programs; imperfection of mechanisms and criteria for selection of much promising innovative projects; the imperfection of the mechanism of formation of favourable investment climate in Ukraine, etc.

In our opinion, the formation of effective innovative state and regional development programs, comprehensive support for priority areas of economic development, creation of the necessary innovative infrastructure and the formation of a system of preferential taxation and crediting of enterprises-innovators will help to overcome crisis phenomena in the economy of Ukraine, to activate innovative activity of enterprises and to use their intellectual capital.

References

1. Stuart Tomas A. Intellectualnyy kapital. Novyy istochnik bogatstva organizatsii; per. s angl. V. Nozdrinoy. Moskva. Pokolenye. (In Russian). (2007).
2. Brooking A. Intellectual capital: Core Asset for the Third Millenium Enterprise / Annie Brooking. L.: International Thompson Business Press. (1998).
3. Edvinsson L., Malone M.S. Intellectual Capital: realizing your company's true value by finding its hidden brainpower (1st ed.). N.Y: Harper Business. (1997).

4. Cohen W.M., Goto A., Nagata A., Nelson R.R., Walsh J.P. R&D Spillovers, Patents and the Incentives to Innovate in Japan and the United States. Research Policy. (2002).

5. Landes W. M., Posner R. A. The economic structure of intellectual property law. The Belknap Press of Harvard University Press: Cambridge, Massachusetts and London, England. (2003).

6. Roos J., Roos G., Edvinsson L. Intellectual capital: navigating in the new business landscape. N.Y. : HarperBusiness. (1998).

7. Sullivan Patrick H. Value-driven Intellectual Capital; How to convert Intangible Corporate Assets into Market Value. Wiley. (2000).

8. Butnic-Siverskyy O. Transformatsiya intelektualnoyi vlasnosti v innovatsiynyy product. URL: http://iv.org.ua/index2.php?kontent/naykdial/articles/article-2.php. (2003).

9. Vachevskyy M.V. Dzherela patentnoyi documentatsyi ta patentnych opysiv do objectiv intelektualnoyi vlasnosti. Actualni problemy economiky. (In Ukrainian). (2004).

10. Kendjuhov O.V. Efektyvne upravlinnya intelektualnym kapitalom: Monohrafiya. NAS of Ukraine. Institute of Industrial Economics. Donetsk: DonIEE. (In Ukrainian). (2008).

11. Kuzmin O. E., Hiduha L. G., Lipych O. A. Upravlinnya intelektualnym kapitalom machynobudivnych pidpryemstv: teoretychnyi ta prykladnyi polozhennya: Monohrafiya. Luzk. Vezha-Druk. (In Ukrainian). (2014).

12. Kuchumova I.Y. Intelektyalnyi capital v systemy upravlinnya pydpryemstvom. Biznes Inform. (In Ukrainian). (2013).

13. The Global Competitiveness Index. URL: http://www3.weforum.org/docs/GCR2017-2018/05FullReport/TheGlobalCompetitivenessReport2017%E2%80%932018.pdf

14. The official website of the State Service of Intellectual Property of Ukraine URL: http://www.uipv.org/ [in Ukrainian].

15. Maslov Y., Semenova V. Finansovy aspekty formuvannja intelektualnogo kapitalu pidpryiemstv i zabezpechennja transferu technologij. Financial and credit activity: problems of theory and practice. (In Ukrainian). (2019).

16. Prokhorova V., Davydova O. Objectyvny umovy formuvannja innovaziynogo upravlinnja rozvytkom pidpryiemstv: naukovo-istorychne pidgrunttja podolannja naslidkiv krys. Bisnes Inform. (In Ukrainian). (2018).

Modification of the human development index in Ukraine in the regional aspect

Galyna Nazarova[1], *Volodymyr* Volikov[2] *and Alina* Demianenko[3]

[1] Simon Kuznets Kharkiv National University of Economics, Nauky ave. 9-a, 61166 Kharkiv, Ukraine
[2] North-East Scientific Center of the NAS of Ukraine and MES of Ukraine, Bagalíy str. 8, 61002, Kharkiv, Ukraine
[3] Simon Kuznets Kharkiv National University of Economics, Nauky ave. 9-a, 61166 Kharkiv, Ukraine

Abstract. The paper presents the results of research on the modification of the traditional index of human development to assess its level in Ukraine in the regional aspect. The difference between the proposed modified human development index from existing approaches is the inclusion of an additional component of human development: the development of information and communication technologies, the updating of the indicator system by the component "education" and "environmental welfare" in accordance with the domestic conditions of the development of society, as well as the method of determining weighting coefficients to calculate the integral index. In the process of research, the following methods have been used: the method of generalization, comparison, system analysis and hierarchy analysis method. The calculation of the modified index of human development was carried out on the example of statistical data of regions of Ukraine; the period of research was selected 2015-2017. The results of the measurement demonstrate the presence of the marked unevenness of Ukrainian regions by the level of human development.

1 Introduction

The study of key tendencies in human development is an important element of assessing the level of socio-economic development. Careful analysis, statistical data processing, comparison and confrontation will make it possible to draw appropriate conclusions regarding the decision to choose the future direction of human development, to work out or adjust the national development strategies of a country or region.

It should be noted that with the transition of humanity to a new stage of development and a shift in emphasis on the need to ensure the well-being of people and society, it is information and communication technologies (hereinafter – ICT) offer great opportunities for socio-economic development. As L. Kochubei notes, "information technologies very quickly became a vital stimulus for the development of not only the world economy, but also other spheres of human activity" [1]. Modern society is flooded with huge information flows, and the development of ICT penetrating practically all spheres of human activity became the driving force for change and accelerated development of society [2].

Giving statistical confirmation of this thought, as of December 31, 2017, more than 54.4% of the world's population has access to the Internet and this number increases every day [3]. Among the 20% of the poorest households only 7 out of 10 have a mobile telephone. Bringing an interesting statistical comparison by T. Pospelova, we agree that "the number of poorer households with a mobile phone is bigger than those who have access to the toilet and clean drinking water" [4]. So, it is expected that the dynamic development of ICT will increase the opportunities for human development, prosperity and social integration.

Information and knowledge today become the most important resource of humanity [5-9], determining not only the progress of the economy and culture in high-tech areas of activity, but also determine the speed and nature of the transformation of human development and influence on it from various institutions [9]. Therefore, the inclusion of this trend in the human development assessment system is considered to be well-founded both at the theoretical and practical levels.

2 Problem Statement

Among the most popular methods for evaluating complex, difficult formalized processes in world practice is using a single indicator - an integral indicator. So, for the purpose of determining the level of human development, UNDP specialists for the first time developed a methodology for measuring it, proposing the use of the integral index of human development [10]. The proposed methodological approach laid the foundation for further research in the direction of improving the methods for evaluating the process of expanding human potential.

The results of the authors' thorough study of peculiarities to the evaluation of human development in Ukraine and abroad [11-17], allowed to identify shortcomings. Thus, in determining the level of human development, the main shortcomings in our opinion are: the complication of a large number of indicators, not taking into account the quality of education components of the "education", not taking into account all levels of education: pre-school, full secondary and higher education, not taking into account the component of ICT, the same weight coefficients for components of human development generate the need for improvement the approach to human development assessment that takes into account the peculiarities of modern domestic human development conditions and focuses on the "innovation". The results of the authors` thorough study of peculiarities of the human development process, in addition to being simple, accessible and easy to apply.

3 Research Questions

In the process of research the authors faced the following questions:

What components of human development are the most important in determining its level in conditions of innovative society development?

What indicators should be used to measure ICT in the regions of Ukraine?

What requirements of a system of indicators and calculation algorithm should be followed in order to avoid contradictions in the evaluation of human development?

How to determine the weight coefficients for each component of human development?

4 Purpose of the Study

Improving the approach to assessing the level of human development, taking into account the development of ICT and the features of the domestic conditions.

5 Research Methods

To improve the approach to assessing the level of human development, the authors used the following methods: the method of theoretical generalization - to determine the list of major problems in determining the real level of acquired human potential; comparison and system analysis – to select the most important indicators of the main components of human development; the expert survey method and the hierarchy analysis method are used to determine the weight coefficients of the components of human development.

Thus, in order to calculate the modified index of human development, we propose the use of weighted coefficients calculated using the analytical hierarchical process method. After all, this method is a more reliable and consistent method of assigning estimates to weight factors by experts and is based on a pairwise comparison of each aspect of human development to determine the relative importance.

Table 1. The results of the analytical hierarchical process for the determination of weighting factors in the assessment of the modified index of human development

Components of human development	Values of weight coefficients	Measure of consistency
1. Longevity	0.379	5.5146
2.Education	0.108	5.4440
3. Economic welfare	0.254	5.4186
4. Environmental well-being	0.173	5.3071
5. ICT	0.086	5.3304
Consistency Index		0.101
Random Consistency Index		1.12
Consistency Ratio of expert opinions		0.090

As a result of calculations it is determined that Consistency Index is 0.101. Consequently, Consistency Ratio is equal to 0.090 (0.101 / 1.12). It should be noted that according to the results, the Consistency Ratio does not exceed the value of 0.10 according to the Saaty, it means that the experts were fairly objective in their assessments. Thus, the author proved the consistency of expert opinions, the calculated values are not random, which provides the basis for using the results of an expert survey in further research.

6 Findings

In the process of research it was clarified that the main three components of human development: longevity, education and economic welfare of the population and the indicators that determine their level, are in all methods considered by the authors. This suggests that these three components are the most important, and their use is justified and scientifically sound. Thus, the component of "longevity" is proposed to be evaluated according to the traditional indicator - the expected life expectancy at birth (formula 1). Where *actual value* is the actual value of the expected life expectancy at birth in the region, measured over the number of years. The standard formula used by the UNDP is used to rate the indicator [18].

$$L = \frac{actual\ value - 20}{85 - 20}, \qquad (1)$$

Measurement of the education component requires improvement and complementarity. In order to determine the level of educational potential in the country, according to the traditional model of human development measurement, it is proposed to use expected years of schooling and mean years of schooling. While higher and pre-school education is not taken into account, also there is no assessment of the quality of education.

In our opinion, the use of the above indicators to measure the educational situation in the regions of Ukraine with a view to comparing them is not feasible. After all, firstly, the term of studying for obtaining a complete general secondary education in institutions of general secondary education is the same in all regions of Ukraine and is eleven years [19]; and secondly, there is no consideration of the effective indicator of the quality of education. So, we agree with the opinion of O. V. Makarova and O. M. Hladun [20] about the necessity of attracting indicators that could reflect all levels of education: pre-school education, full secondary education and higher education, and we consider it expedient to overcome this idea when forming the composition of indicators of this the component. Thus, the list of indicators selected to assess the education is given below (formula 2).

$$Ed = \frac{\sum_{i=1}^{3} E_i}{3}, \tag{2}$$

where: E_1 is the provision of children by pre-school educational institutions in a year in the region; E_2 is the level of coverage of the population by secondary and higher education per year in the region; E_3 is the quality of education (the proportion of people who have an average and high score on the results of the external independent evaluation to the total number of external independent evaluation participants per year in the region).

Economic welfare component is determined by the indicator – GRP per capita (formula 3), where "actual value" is the actual value of GDP per capita, and "min" and "max" are respectively the minimum and maximum value of GRP per capita in Ukraine.

$$EW = \frac{ln(actual\ value) - ln(min)}{ln(max) - ln(min)} \tag{3}$$

The above-mentioned components are insufficient to determine the level of human development, because the classical system for calculating the level of human development does not take into account the ecological component, which contradicts the definition of the category of human development proposed by UNDP in 2010. Thus, "the formation of just and sustainable development on our planet" [21] is extremely important, which creates the need to supplement the component "environmental well-being" and indicators that determine the state of the environment in the region:

$$EN = \frac{\sum_{i=1}^{3} Eco_i}{3}, \tag{4}$$

where: Eco_1 is the level of waste utilization (the share of waste utilized to the total amount of waste generated, per year); Eco_2 - forest area (specific gravity of the area covered with forest vegetation of forest areas to the total area of forest areas, per year); Eco_3 is the level of water supply (percentage of water resources per inhabitant, per year).

The consideration of ICT in measuring human development was studied by Ukrainian scientists: O. Vlasiuk, S. Pyrozhkov [22]. The researchers drew attention to the need to evaluate ICT in human development. The list of indicators for assessing of ICT is marked by indicators: the number of TVs, radios and landlines. But we draw attention to the obsolete some indicators and consider expedient in the assessment of human development to take into account a comprehensive index (formula 5). The comprehensive ICT index would include: the level of use of stationary communication (I_1); level of mobile communication usage (I_2); and the level of Internet usage (I_3) that are calculated as follows:

$$ICT = \frac{\sum_{i=1}^{3} I_i}{3}, \tag{5}$$

A generalized system of indices of the modified index of human development is presented in Fig. 1.

The next important issue in determining the objective level of human development is the determination of weight coefficients for each of the components. The results of expert evaluation using the hierarchy analysis method allow solving this issue. The obtained results and confirmed by calculation of Consistency Ratio of expert opinions allow to be used for the calculation of the modified index of human development (MHDI), the formula for measuring which is as follows (formula 6):

$$MHDI = L \times 0.379 + Ed \times 0.108 + EW \times 0.254 + EN \times 0.173 + ICT \times 0.086, \tag{6}$$

where: 0.379, 0.254, 0.173, 0.108, 0.086 are weight coefficients for each of the components of the modified index of human development.

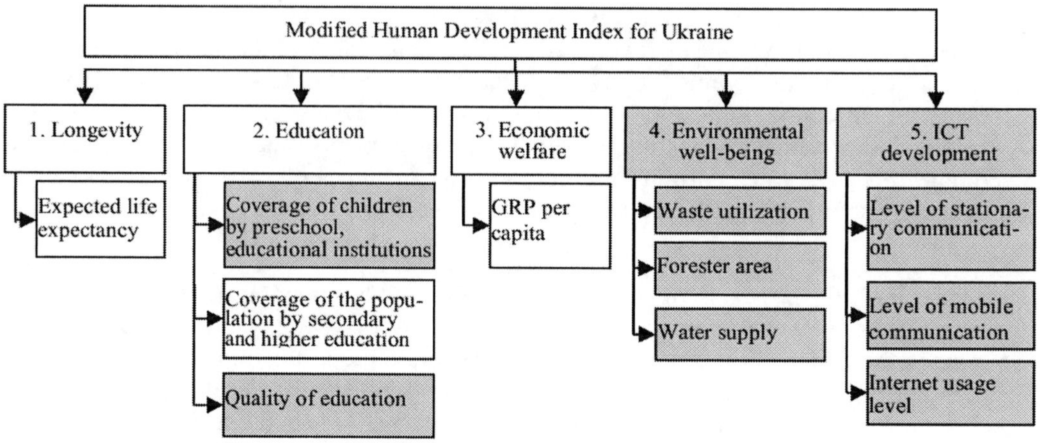

Fig. 1. Scheme of indices of the modified index of human development for Ukraine
Note: shaded squares are the author`s addition and clarification

The calculation of the modified index of human development was based on open access statistics [23] for 2015-2017 (Fig. 2).

As a result of the calculations revealed a pronounced unevenness in the level of human development of the regions of Ukraine. The determining factor is the low economic and environmental condition in the regions. The most favorable conditions for the development of human potential were recorded in Dnipro, Zaporizhzhia, Lviv, Odesa, Poltava, Cherkasy, Kharkiv and other regions. The lowest rates were recorded in Ternopil, Kirovograd, Mykolaiv, Kherson and others. Overall, there is a tendency to increase by 0.1 percentage points on average, indicating improvement in some regions. However, according to the human body's measurement of human slag, that proposed by UNDP, a high level of human development (which is defined in the range of 0.8 to 1) is absent. In Ukraine, the average and low level of development of human potential prevails. The obtained results will enable us to formulate the main recommendations for eliminating the sharply expressed unevenness of the development of the regions of Ukraine.

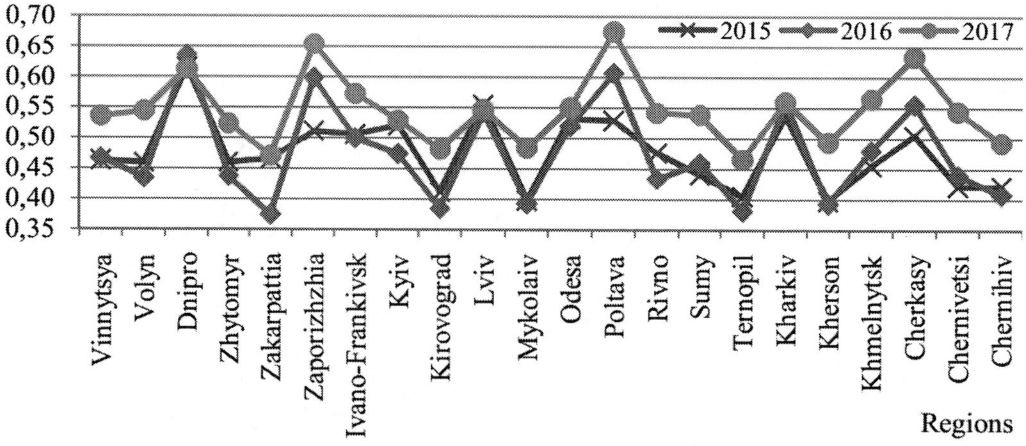

Fig. 2. Dynamics of change of the MHDI in the regional section in Ukraine Note: data exclude the temporarily occupied territory of the Autonomous Republic of Crimea and the part of temporarily occupied territories in Donetsk and Luhansk regions

7 Conclusion

The transition of humanity to a new stage of development determines the need to ensure the well-being of people and constant expansion of opportunities and freedoms. It is hard not to notice that modern society is full of huge flows of information, and the development of ICT penetrating practically in all spheres of human activity have become the driving force for change and accelerated development of society. That is why the methodology for determining the level of human development must meet the modern requirements, which creates an objective need for modification the index of human development.

The difference between the proposed modified index of human development, in comparison with the existing, is the inclusion in the system of assessment of the additional component of human development – ICT, updating indicators of the education and environmental welfare components and also a method for determining weight coefficients for calculating an integral index that involves the use of a hierarchy analysis method. Because this method is a more reliable and consistent for assigning estimates to weight coefficients by experts, and is based on a pairwise comparison with each aspect of human development to determine relative importance.

Approbation of the results on the example of statistical data of regions of Ukraine allows obtaining a modified model of integrated assessment of human development. It will allow formulating the main recommendations for elimination of sharply expressed unevenness of the development of regions of Ukraine and contributing to raising the level.

References

1. L. Kochubei, Scientific Papers Kuras IPES of NAS of Ukraine, **3(89),** 44-70 (2018)
2. Z. Yaremko, Economic Analysis, **15(1),** 216-227 (2014)
3. Internet World Stats. URL: https://www.internetworldstats.com/stats.htm (2019)
4. T. Pospelova, ESP, **3**, 366-376 (2017)
5. J. H. Nord, M. T. Riggio, J. Paliszkiewicz, JofCIS, **57(3),** 278-285 (2017)
6. O. Saebo, R. Gomez, L. F. Baron, HICSS Jan 05-08, 2697-2697 (2016)
7. A. X. Herrera Marquez, F. Cervantes, P. Cervantes, EDULEARN, 8062-8072 (2015)
8. M. D. Lytras, E. Kurilovas, Computers in human behavior, **30,** 361-361 (2014)
9. I. Bachylo, O. Sosnin, Yurydycheskyy visnyk Ukrayiny, **50** (2011)
10. UNDP, *Human Development Report* (New York: Oxford University Press, 1990)
11. UNDP, *National Human Development Report. Local and Regional Development* URL: http://hdr.undp.org/en/content/local-and-regionaldevelopment-poland-2012 (2012)
12. Lin. Yang, *An Inventory of Composite Measures of Human Progress* URL: http://hdr.undp.org/en/content/inventory-compositemeasures-human-progress (2014)
13. Z. Hermansons, *Measuring Human Development at the Local Level: A Case Study of Latvia* (Budapest, Hungary, 2015)
14. K. Mazumdar, Review of Social Economy, **61(4)**, 535-549 (2003)
15. M. A. Salas-Bourgoin Cepal Review, **112**, 29-44 (2014)
16. M. Klekotko, A. Jastrzębiec-Witowska, K.Gorlach, P. Nowak, EEC, **24** (2018)
17. R. P. Yakunina, G. A.Bychkov, Procedia Economics and Finance, **24** (2015)
18. Technical notes: Human development indices and indicators. URL: http://hdr.undp.org/sites/default/files/hdr2018_technical_notes.pdf (2018)
19. About general secondary education in Ukraine. URL: https://zakon.rada.gov.ua (2018)
20. O. V. Makarova, O. M. Hladun, Statistic of Ukraine, **1(56)**, 10-15 (2012)
21. UNDP, *Human Development Report. The Real Wealth of Nations: Pathways to Human Development.* (New York: Oxford University Press, 2010)
22. O. Vlasiuk, S. Pyrozhkov, *Human Development Index: Ukraine's Experience* (Kyiv: National Institute for Strategic Studies, 1995)
23. State Statistics Service of Ukraine. URL: http://www.ukrstat.gov.ua (2018)

Development of a strategy for improving organizational and production structures based on methods for non-convex programming

Igor Novakivskyi[1], *Vasyl* Kozyk[1], *Oleksandra* Mrykhina[1*] and *Orest* Koleshchuk[1]

[1]Lviv Polytechnic National University, Institute of Economics and Management, Bandery Str., 12, 79013 Lviv, Ukraine

Abstract. The scientific paper discloses problems of functioning of financial industrial groups, particularly their flexible organizational and production structures. The authors highlight that maintaining sustainability of development of financial industrial groups is oriented towards preserving integrity of their organizational and production structures and keeping a set of parameters within given limits, taking into account internal disturbances and external impacts. Generally, sustainable development of organizational and production structures contemplates concerted actions directed towards accomplishing the following two goals: maximizing indicators displaying results of activities of an organizational and production structure through determining optimal amounts of investments by different directions of activity of its modules, as well as minimizing a probable deviation from planned results by directions of activity. The authors have built a two criteria mathematical model for managing development of prospective directions of activities of organizational and production structures, which takes into account their riskiness and is based on retrospective data. To search the optimal value of a functional, the authors have developed a four-step algorithm for solving the indicated problem. To solve the indicated problem of constrained optimization, the authors have applied the method of steepest descent.

Introduction. The growing importance of organizational and production structures in the contemporary economy

Contemporary corporations implement elements of virtualization of their organizational and production structures to dynamically react to changes of a market under the scientific and technological progress. This fact is confirmed by a dramatic increase of a share of products having the five-year life cycle. Therefore, a standard of manufacturing such products should contemplate rapid dismantling and recycling both a product and production capacities. P. Drucker [2] asserts that corporations of the 21st century will be an organized set of production modules, which encompasses a production process stage or several

[*] Corresponding author: oleksandramrykhina@gmail.com

similar operations. Although general management of a corporation continues to perform its functions, an environment of an organizational and production structure will form own horizontal system of management for the interaction between its modules, as well as numerous stakeholders from an external environment. The problems of the strategy for improving organizational and production structures are described in the works [1-10].

Under the above-mentioned changes, the large-scale formation of modern viable organizational and production structures and developing effective mechanisms for managing them oriented towards the network market economy are important goals for development of financial industrial groups. In this case, a paramount aim consists in improving methods of management oriented towards the situational approach. This will enable to maintain sustainable development of organizational and production structures and to reduce negative risks of an external environment.

Methodical approaches to developing the strategy for improving organizational and production structures

Under traditional conditions, management of improving organizational and production structures is aimed at supporting sustainable development and protecting departments of an enterprise against the influence of external or internal threats. Maintaining sustainability of development of the potential of financial industrial groups is directed towards preserving integrity of their organizational and production structures and keeping a set of parameters within given limits, taking into account internal disturbances and external impacts, which dynamically grow under the increase of the informatization level of an external environment. Figure 1 shows a scheme of this process.

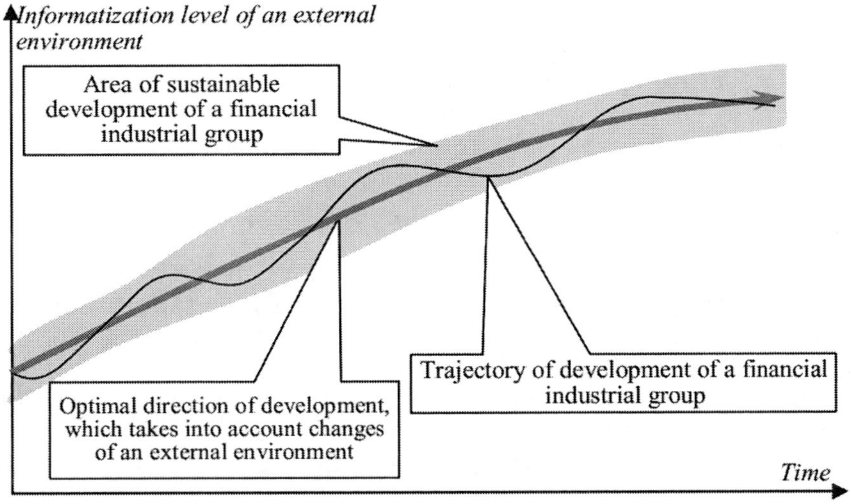

Fig. 1. Scheme of development of financial industrial groups over time.

Borders of the area of sustainable development change during evolvement of financial industrial groups while any change of their organizational and production structures influences the placement and size of the area of sustainability, which may be limited by the threshold value of selected indicators. Threshold values are marginal values the structures should not move beyond to reproduce internal elements and to avoid forming negative destructive tendencies. The approximation to their maximum permissible value indicates the increment of threats to socio-economic stability of organizational and production structures, i.e. a real threat to economic security of financial industrial groups. To ensure the high level of economic security, a structure should not exceed marginal costs.

Obviously, maintaining a higher level of economic security of an enterprise increases economic pressure on the potential of an organizational and production structure. This pressure is able to maintain a set of stabilization measures.

Economic security of functioning of financial industrial groups considerably depends on an accepted conception of risk management. Effective management should ensure sustainable development of an enterprise because its ability to accomplish determined goals may not only undermine competitive positions of financial industrial groups in a market but also lead to bankruptcy.

Mathematical model for management of improving organizational and production structures

Sustainable development of an enterprise implies achieving the following two objectives: maximizing indicators displaying results of activities of an organizational and production structure through determining optimal amounts of investments by different directions of activity of its modules; minimizing a probable deviation from planned results by directions of activity. The first objective can be achieved through maximizing the profitability index. In this case, it is expedient to use a quadratic form for determining dependence of the profitability index of an enterprise from amounts of investments by directions of activity of an organizational and production structure. This may be presented in the form of the following functional:

$$F = (1 \quad q_1 \quad \cdots \quad q_n) \times \begin{Vmatrix} \alpha_{00} & \alpha_{01} & \cdots & \alpha_{0n} \\ \alpha_{10} & \alpha_{11} & \cdots & \alpha_{1n} \\ \cdots & \cdots & \cdots & \cdots \\ \alpha_{n0} & \alpha_{n1} & \cdots & \alpha_{nn} \end{Vmatrix} \times \begin{pmatrix} 1 \\ q_1 \\ \vdots \\ q_n \end{pmatrix} \to max, \quad (1)$$

where F – criterion function for profitability of enterprise's investments, arb. units; α_{ij} – indicators presented in the quadratic form, which satisfy a condition $\alpha_{ij}=\alpha_{ij}$ and are determined owing to the correlation and regression analysis of statistical data describing activity of an enterprise, 1/hryvnia^2; vector $(1\ ;\ q_1;\ q_2;\ \ldots\ q_n)$ consists of an auxiliary unit of measurement and unknown amounts of investments $q_1;\ q_2;\ \ldots\ q_n$, hryvnias, n – number of directions of activity of the enterprise, units.

The criterion function (1) determines dependence of the profitability index of an enterprise from amounts of investments by directions of activity, i.e. the problem amounts to determining unknowns $\{q_i\}_{i=1}^{K}$. Affordable amounts of investments are typical additional conditions for this problem.

To ensure maximum sustainability by types of enterprise's activity, it is expedient to minimize a probable discrepancy of profitability indices by directions of enterprise's activity. In this case, it is necessary to use properties of the analysis of covariance applied in planning and statistical processing of experimental results as a method for reducing an experimental error, which cannot be directly measured. In general, the problem regarding minimizing uncertainty, which is stipulated by risks, in the integrated form by all declared directions of enterprise's activity may be amounted to ascertaining a minimum of a quadratic functional:

$$V = \frac{1}{\|q\|^2}(q_1 \quad \cdots \quad q_n) \times \begin{Vmatrix} v_{11} & v_{12} & \cdots & v_{1n} \\ v_{21} & v_{22} & \cdots & v_{2n} \\ \cdots & \cdots & \cdots & \cdots \\ v_{n1} & v_{n2} & \cdots & v_{nn} \end{Vmatrix} \times \begin{pmatrix} q_1 \\ q_2 \\ \cdots \\ q_n \end{pmatrix} \to min, \quad (2)$$

where V is functional, which determines the value of dispersion of profitability indices by directions of activity, v_{ij} is modified covariance coefficient calculated according to the formula:

$$v_{ij} = \frac{cov_{ij}}{\beta_i \times \beta_j}, \quad \beta_k = \begin{cases} \beta_k, & \beta_k \geq \beta_k^{(\text{crit})}; \\ \beta_k^{(\text{crit})}, & \beta_k < \beta_k^{(\text{crit})}, \end{cases} \tag{3}$$

$$cov_{ij} = \frac{\sum_{k=1}^{K}\left((\bar{\beta}_i - \beta_{i,k}) \times (\bar{\beta}_j - \beta_{j,k})\right)}{k-1},$$

$$\|q\|^2 = \left(\sum_{i=1}^{n} q_i^{\,2}\right)/n,$$

where $\beta_k^{(crit)}$ is value of profitability of investments given beforehand (if the value of the profitability index is less than the specified one, it is insignificant for an enterprise), hryvnias,

$$\bar{\beta}_i = {}^1\!/_n \sum_{k=1}^{n} \beta_{ik},$$

where $\bar{\beta}_i$ is average value of profitability of investments for the i direction of activity, hryvnias; $\{\beta_{ik}\}_{k=1}^{K}$ is data of statistical examination concerning the profitability ratio for the i direction of activity, arb. units; K is number of points of statistical observation, units.

To determine the optimal structure of investing in directions of activity of an organizational and production structure, the authors have proposed to apply the following functional:

$$W = {}^F\!/_V \rightarrow max, \tag{4}$$

where W – functional, which logically combines requirements of maximizing the functional F and minimizing the V. Formulas for calculation of the mentioned values are presented in the formulas (1) and (2). The functional W, which is presented in (4), reflects consolidated dependence of maximizing the generalized profitability index and minimizing the average dispersion of partial indices by directions of activity of an organizational and production structure. Thus, the two criteria problem has been amounted to a single-criterion problem. To solve the latter, the authors recommend using the method of nonlinear convex programming in a value domain. To ascertain optimal values of the functional F, which is disclosed in (4), the authors recommend using the following algorithm:

1. Collecting and processing of statistical data on a researched organizational and production structure regarding funds invested in certain directions of activity of an organizational and production structure and earned profits correspondingly.
2. Building the quadratic functional F, which is presented in (1), for calculation of the profitability index of an organizational and production structure according to amounts of investments in directions of its activity. To determine unknown coefficients of the required multi-factor equation, economists should apply classic methods of the correlation and regression analysis for collecting statistical data.
3. Building the quadratic functional V, which is presented in (2), for assessing the dispersion of profitability indices by directions of activity of an organizational and production structure. To achieve this goal, economists should calculate a covariance matrix according to the formulas (3) based on the collected statistical data.
4. Determining a maximum of the functional W, which is disclosed in (4), through applying an iterative optimization algorithm, namely the method of steepest descent.

Example of the calculation for the Energoresurs-Invest Corporation

To calculate indicators, the authors have used statistical data for 2002-2016 by two directions of activity of the corporation adjusted for inflation according to the last year of observation (Table 1).

Table 1. Results of activity of the Energoresurs-Invest Corporation

Years	Amount of investments for the first direction, thousand hryvnias	Amount of investments for the second direction, thousand hryvnias	Gross profit for the first direction, thousand hryvnias	Gross profit for the second direction, thousand hryvnias
2002	395.9	678.7	14.9	266.4
2003	508.2	564.7	193.2	27.1
2004	481.1	841.9	106.5	536.7
2005	963.8	783.1	766.8	413
2006	813.8	871.9	652	630.2
2007	987.4	603.4	681.8	140.4
2008	360	1020	-61.5	835
2009	720.9	961.2	509.9	752.7
2010	601.1	1322.4	309	824.2
2011	864.9	1383.9	701.3	513.6
2012	1012.3	1250.5	764.7	890.5
2013	1045.3	1045.3	627.9	852
2014	1322	1081.7	240.7	897
2015	1170.4	1114.7	405.5	865
2016	1204.9	1385.6	585.6	706.2

In case of considering only two directions of activity of the Energoresurs-Invest Corporation, the functional (4) may be presented by the following formula:

$$W = \frac{\alpha_{00}+2\alpha_{01}q_1+2\alpha_{02}q_2+\alpha_{11}q_1^2+2\alpha_{12}q_1q_2+\alpha_{22}q_2^2}{(v_{11}q_1^2+2v_{12}q_1q_2+v_{22}q_2^2)/(q_1^2+q_2^2)} \to max, \quad (5)$$

where W is calculated functional according to the statistical data shown in Table 1, q_1, q_2 are amounts of investments by two directions of activity of the corporation.

In consistency with paragraphs 1-3 of the suggested algorithm, the authors have computed coefficients of the functional W form the formula (5) according to the data shown in Table 1 and, therefore, obtained:

$$W = \frac{-1{,}979501+0{,}025607q_1+0{,}036732q_2-0{,}000138q_1^2-0{,}000045q_1q_2-0{,}000174q_2^2}{(q_1^2-0{,}1838_{12}q_1q_2+q_2^2)/(q_1^2+q_2^2)} \to max\,. \quad (6)$$

Additional conditions of the problem display requirements to amounts of investments:

$$125 \leq q_1 \leq 500; \quad 250 \leq q_2 \leq 625; \quad q_1 + q_2 \leq 1000. \tag{7}$$

Applying the method of steepest descent for solving the indicated problem of constrained optimization has enabled to obtain the following results: the maximum value of the functional W =1.82, the profitability index of the corporation equals 0.77, and recommended amount of investments for the first direction accounts for 427.9 thousand hryvnias and for the second direction accounts for 522.6 thousand hryvnias.

Conclusions

It is worth mentioning that the outlined approach may be exploited for solving a number of similar problems, e.g. regarding optimization of enterprise's profit. The authors have applied this approach for optimization of choosing the main directions at the Energoresurs-Invest Corporation. As a result, an amount of revenue of the corporation has increased, on average, by 3.2%.

Results of applying the suggested criterion of efficiency (4) for various goals related to choosing optimal solutions under conditions of uncertainty show that the proposed approach is effective.

In general, maintaining sustainability of development of financial industrial groups should primarily contemplate modernization of their organizational and production structure, exploiting new opportunities of the scientific and technological progress in economic activity, as well as in globalization of business interrelations.

References

1. I. Adizes, *Upravlenie zhiznennyim tsiklom korporatsii [Life Cycle Management Corporation]*, (Ed. by A. G. Seferyan), SPb.: Piter (2007). [In Russian]
2. P. F. Druker, *Ryinok: kak vyiyti v lideryi. Praktika i printsipyi [How to get to the leaders. Practice and principles]*, Moscow, Ekonomika (1992). [In Russian]
3. N. I. Chukhray, O. B. Mrykhina, *Theoretical and methodological basis for technology transfer from universities to the business environment*, Problems and Perspectives in Management, **16(1)** (2018).
4. O. I. Karyi, *Lohistychna infrastruktura v Ukraini: tsyfry i realnist [Logistics Infrastructure in Ukraine: Figures and Reality]*, Lohistyka, **762** (2013). [In Ukrainian]
5. O. Koleshchuk, V. Lesinskyi, O. Yemelyanov, O. Zarytska, A. Symak, *Substantiation of projects that account for risk in the resource-saving technological changes at enterprises*, Eastern-European Journal of Enterprise Technologies, **6(1-96)** (2018).
6. V. Kozyk, O. Mrykhina, T. Mirkunova, *Conceptual model for economic evaluation of innovative technologies*, Economics, Entrepreneurship, Management, **4(2)** (2017).
7. O. V. Moroz, N. P. Karachyna, T. M. Khalimon, *Korporatyvne upravlinnia na pidpryiemstvakh Ukrainy: postpryvatyzatsiinyi etap evoliutsii [Corporate Governance at Ukrainian Enterprises: Post-Privatization Stage of Evolution]*, Vinnytsia: UNIVERSUM (2008). [In Ukrainian]
8. J. M. Petrovich, I. I. Novakivskyi, *Modern concept of a model design of an organizational system of enterprise management*, Econtechmod, **1 (4),** Lublin – Lviv – Cracow (2012).
9. A. Toffler, *The adaptive corporation,* Aldershot: Gower (2005).
10. V. Trynchuk, P. Horyslavets, M. Plonka, *Experience marketing and its tools in promoting the insurance services,* Innovative Marketing, **14** (2018).

Conceptual vision of the policy of implementing social and labor potential of the population with decentralization

Olga Novikova [1], *Yaroslav* Ostafiichuk [1], *Olena* Khandii [1*], and *Marina* Deich [2]

[1]Institute of Industrial Economics, National Academy of Sciences of Ukraine, 2 Zhelyabova Street, Kyiv, 03057, Ukraine
[2]Vinnytsia Cooperative Institute, 59 Frunze St, Vinnytsya, 21009, Ukraine

Abstract. The article defines the conceptual vision of the policy of implementing social and labor potential of the population with decentralization, formulates its basic principles as guidelines for practical actions. Essence and role of inter-municipal cooperation for intensification of social potential of communities are revealed, possible ways of its institutional support are suggested. It is found that decentralization processes require effective instruments for coordinating interests, and social mobilization plays an important role in this, which is to involve a maximum number of people in solving community problems and participating in decision making. It is concluded that social potential may turn into a decisive factor in the development of a territorial community in the implementation of the following principles: comparativeness and competitiveness; motivation; social connections and inclusion; mutual responsibility; network availability; openness and transparency; value orientation; individual social interaction; limited solidarity.

1 Introduction

The decentralization reform in Ukraine is devoted to creating capable administrative-territorial units, improving the quality of social and public services throughout the country. The purpose of decentralization is to expand the potential of territorial communities, to strengthen their motivation to self-development on the basis of the formation of an adequate institutional environment, redistribution of powers and functions of the executive power, to increase the efficiency of use of local resources and own opportunities. The process of decentralization involves providing the community with available material and financial resources to form the material and financial basis for sustainable development of territories. The financial insolvency of territorial communities over the past decades was caused, in particular, by the specifics of the national budgetary system such as the limited revenue base of local budgets and their dependence on various state transfers. At the same time, financial decentralization should be considered not as an end in itself, but as a precondition

[*] Corresponding author: alkhandiy@ukr.net

and a mechanism for intensifying the social and economic potential of the community according to the "bottom-up" principle.

Focus on social potential is connected with the necessity of creating a new form of spatial organization of life – a capable united territorial community, in which the population is defined not only as inhabitants of a certain territory, but as an important social community capable, as a subject of management, of acting as a factor in social and economic development on the basis of common values and responsibility to the community [1]. Among other things, this needs to increase the level of professionalism, to overcome economic, entrepreneurial and social passivity, to develop social ties and networks. The management of sustainable development locally involves close interaction between the public, authorities, business entities, and other stakeholders, which requires an appropriate level of coordination, control and accountability between them. One of the results should be convergence of the local community, public consensus on the basis of involving all members of the community in the socio-economic processes. Ultimately, it is about building the social potential of sustainable development as a combination of opportunities, abilities, and willingness of a local community to make appropriate changes.

The aim of the study is to identify the conceptual framework and policy priorities for implementing social and labor potential with decentralization.

2 Research results

The content of social potential manifests itself in the level and nature of the harmonization of social relations in society as a whole, or within individual territorial communities on the basis of cooperation and partnership interactions. Realization of such relations ensures not only the efficient use of human potential, but also coordinates the social behavior of individuals within and beyond the economic activity, generating synergistic effects. The fundamental point is the recognition of the possibility of multi-directional action of social potential – it can provide development or disintegration. The quality of the formed social relationships (networks) and social institutions determines whether they are a factor of deterrence or, conversely, will contribute to the transition to a sustainable development model [2].

Creating conditions for sustainable development means overcoming barriers and accelerating engagement that extends real opportunities and freedoms for the development of an inclusive society in which diversity is a force, not a weakness. Within such a conceptual framework one can speak of processes of decentralization as processes of social attraction of communities as structured elements of society (social bodies) into a united territorial community (UTC).

The diversity of entities within the united communities increases opportunities, stimulates the development of social ties, the search for compromises, partnerships and, on this basis, the growth of the aggregate social potential. Decentralization can give the desired positive results in conditions of mobilizing social potential to address socially significant problems if it stimulates and enhances processes of participation of people in management at the regional or local levels. Otherwise, the central bureaucracy may be replaced by a local one, which, moreover, tends to be less professional. Decentralization, which is not accompanied by greater openness and the involvement of citizens in the management process, cannot be considered an effective institution for the formation of a capable territorial community.

Based on the statistically substantiated conclusion that larger communities generally have more financial resources [3-4], we assume that in addition to the positive financial effects and opportunities for developing social infrastructure, larger in size and quantity UTCs increasingly own other, not connected only with financial and technical preconditions, possibilities of building up, intensifying and realizing social potential, considered in the plane of the diversification of tools of interaction and constructive cooperation between subjects of UTC.

This statement follows from the concepts of social inclusion / engagement and social exclusion. Inclusion in the broad socio-philosophical sense is understood as a form of the common existence of many ordinary people, within which all interested parties should take an active part in obtaining the desired result [5]. As already mentioned, the field of social potential is heterogeneous. The difference in the potentials of different points in the social field generates tension. On the one hand, social tension of a certain degree and character intensifies the search for new variants and forms of involvement in social potential; on the other hand, an increase in social tension and its disorienting nature lead to the search for ways to destroy social potential or its localization. The diversity of subjects in the UTC (individuals, social groups, communities) expands opportunities, stimulates the development of social ties, the search for compromises, partnerships and, on this basis, the growth of the total social potential of UTC, while it increases the range of risks of decentralization that needs an adequate risk management system.

In this context, one of the areas of strengthening and implementing social potential is the cooperation of local communities (inter-municipal cooperation). According to the Law of Ukraine "On cooperation of local communities" it is a relationship between two or more territorial communities on a contractual basis in certain forms for the socio-economic and cultural development of the area, improvement of the quality of public services on the basis of common interests and goals, effective execution of the powers determined by the law by local self-government bodies. Taking into consideration this feature, one can conclude that cooperation as mutually beneficial relationships of interested parties has its logically justified ultimate meaning, that is the outcome, for which municipalities entered into a relationship on a contractual basis. This means that statutory mechanisms of combining existing resources of local communities should lead to more effective implementation of the powers defined by the law by local government, which in turn should provide socio-economic and cultural development of areas, increase the quality of public services on the basis of common interests and goals.

Accordingly, the most important in the cooperation of communities is a principle of mutual benefit, one of the most widely recognized principles of international law, which is closely linked with the principle of equality of the actors. After all, only within the framework of equality it is possible to take into account mutual interests in order to achieve mutual benefits. Therefore, in the Charter of Economic Rights and Duties of the States of 1974, these principles are regarded as complementary.

Another important principle is the principle of equality of actors, which means formal equality, i.e., the identity of the scope of rights and legally authorized opportunities of the parties to the legal relationship. The principle is expressed in the concepts of formal equality, prohibition of discrimination, lack of groundless, unfounded advantages or privileges for any of the actors of cooperation. Any of the actors has equal rights with regard to other actors during the organization and implementation of the cooperation itself, in spite of the size of the territory of the territorial community, the size of the budget, the number of inhabitants, the geographical location, the existing infrastructure, etc. This principle is relevant, given the need to protect the interests of peripheral rural communities within the framework of large-scale urban UTCs.

Based on the concept «rural-urban partnerships» OECD [6-7], Borodina O., A.Tkachuk and others [8-9] point to the priority for Ukraine considering the problems of cooperation of territorial communities in the format "city-village". "The essence of the idea is to organize cooperation between urban and rural areas on the basis of mutually beneficial cooperation, the division of powers, the use of existing advantages in order to bring rural areas closer to sources of concentration of innovation, social and financial capital, investments, social and transport infrastructure objects typical of urban settlements, together with more active involvement of rich resources of rural areas to processes of social production" [9].

Today domestic legislation on the cooperation of territorial communities regulates various types of cooperation that correspond to European practice and relate exclusively to voluntary cooperation. At the same time, it does not imply cases of compulsory cooperation between communities existing in different European countries. There are five forms in which communities can cooperate: 1) cooperation in the form of delegation to one of the actors of cooperation by other actors of cooperation to carry out one or more tasks with the transfer of appropriate resources to them; 2) cooperation in the form of joint projects implementation; 3) cooperation in the form of co-financing (maintenance) of enterprises, institutions and organizations of communal form of ownership – infrastructure objects by actors of cooperation; 4) cooperation in the form of the formation of joint communal enterprises, institutions and organizations – joint infrastructural objects by actors of cooperation; 5) cooperation in the form of the formation of a joint management body by actors of cooperation for joint implementation of the powers determined by the law.

As of March 20, 2019, 368 agreements were registered in the Register of Treaties on the Cooperation of Territorial Communities [10]. The analysis shows that most often problems of removal and disposal of solid household wastes, water supply and drainage in cities and suburban areas are solved in this way, but projects aimed at creating a complete cycle of collection, sorting, recycling and disposal of waste are only isolated cases. A practice of implementing projects aimed at developing social networks and forming a common social space in the united communities has not become widespread (conducting appropriate training activities, trainings, as well as joint activities, for example, joint tree planting in the forest or on the hills on the edge of communities improves mutual understanding). There are practically no projects of cooperation on creation of business incubators or other modern centers of entrepreneurial activity development, which is very important from the point of view of implementing the labor potential of communities on innovative grounds.

Thus, the cooperation of territorial communities is significant for intensifying the social potential of sustainable development at the local level through participatory democracy, implementation of social mobilization measures, more active involvement of citizens in solving local issues and strengthening interaction between the authorities and community. It is clear that the identification of possible topics for cooperation should begin with active consultations with citizens, studying the most urgent problems for inhabitants. Public discussions, hearings, consultations with community members are an essential element of the development of cooperation.

For Ukraine, the following three models of institutional support for the development of cooperation of territorial communities are the most feasible:

- creation of a resource center for the development of cooperation at the state authorities, in particular at the Ministry of Regional Development, Construction and Housing and Communal Services of Ukraine, which is the central executive body authorized to implement the reform of local self-government and development of cooperation of territorial communities;
- creation of a resource center of support at the associations of local self-government bodies. Such centers are established in France, Belgium and other countries;
- creation and development of a resource center as a non-governmental public organization [11].

Based on the experience of some European countries (in particular, Germany, Belgium), resource centers ("centers of competence") on cooperation of territorial communities at the regional and local levels may be useful. For instance, in the federal state of Hessen (Germany), a directive was adopted on promoting inter-municipal cooperation (Framework

Agreement), which sets out the goals and procedures for the facilitation. The Hessian Ministry of the Interior and Sports has established a Department for Inter-municipal Cooperation and an appropriate Center of Competence with the participation of the Association of Cities and Communities. The main areas of responsibility of the Center of Competence include strategy and definition of the content of inter-municipal cooperation, community counseling, preparation and implementation of information activities for the purpose of involving citizens and local authorities, development of appropriate methodological recommendations, brochures and publications, development and collection of information materials on opportunities for inter-municipal cooperation [11].

Social relations provide the interaction of all components of social potential. They help to identify and implement ways to systematically formulate and effectively use social potential. The development of social relations in this direction requires appropriate changes in social institutions. The growth of human consciousness in the system of social relations promotes the intensification of activity in this area, provides conditions for the transformation of opportunities and readiness of social potential for real actions.

The basic provision of the policy of realizing social potential is the provision of social mobilization of the population. Social mobilization is defined as the active involvement of all strata of the population (or of individual social groups) in addressing the priority community problems. It is aimed at increasing social activity and responsibility of the population, members of the territorial community, integrates its separate segments in the form of partnership and cooperation for the achievement of the common goal, leads to active participation of citizens in the processes of urban development. Social mobilization is possible under the condition of maximum engagement of citizens in decision-making. Involving the population in the management of the community allows one to increase the transparency of the management processes, including budget, and take into account the citizens' will. Thus, in general terms, social potential can turn into a determining factor in the development of a territorial community in the implementation of the following principles: comparability and competition; motivation; social connections and inclusion; mutual responsibility; network availability; openness and transparency; value orientation; individual social interaction; limited solidarity, i.e. compliance with the norms of support.

We believe that one of the conditions for the success of the decentralization reform in Ukraine is the strategic orientation of UTC to the development goals. Among other things, they include the financing and implementation of activities focused on business development and employment creation. As the analysis shows, most UTCs have significant employment constraints. This is largely due to such a challenge for the sustainable development of territorial communities as migration. In its turn, it raises a number of problems, including: reducing the activity of residents, included their participation in UTC issues, development planning, involvement in projects, as the most active members of a community with a high social potential often leave, while those who remain are passive and unprotected; lack of skilled personnel in the apparatus, which is a logical consequence of the decrease in the activity of local residents, significant demographic losses. The problem of migration beyond UTC is important for all types of communities, but it is most relevant for rural UTCs remote from cities.

3 Conclusion

Based on the assessment of current trends in the processes of decentralization of power, we consider it necessary to focus on such conceptual principles as guidelines for practical actions:

• mechanisms for stimulating decentralization processes should include both the creation of conditions for the formation of social potential, in particular through the development of social infrastructure and the expansion of the range of relevant services and conditions for its implementation, including in relation to the labor potential of the community. Providing benefits exclusively to projects with a purely social effect (currently prevalent) is, in the long run, associated with the risk of losing both social and labor potential, in particular in connection with the migration of young people, who will be deprived of opportunities for productive employment, their loss of professional skills and development on this basis, destructive changes in social potential;

• in addition to promoting individual entrepreneurial activity in the communities (in particular by facilitating business conditions), it is appropriate to focus on the opportunities and tools for the development of joint local entrepreneurship, the formation of a community as an active and interested subject of economic processes using innovative approaches to attracting labor potential and multiplying on this basis the constructive social potential of the community.

The development of new approaches to community self-development is one of the key tasks of decentralization policy, since, in our view, not only budgeting or business involvement, but first and foremost, the inclusion of the existing potential, its joint usage by the population is the basis of the community's economy and its sustainable development;

• the basis for the establishment of local labor markets is the development of cooperation of territorial communities (inter-municipal cooperation) as a tool for balancing labor demand and supply, as well as a social dialogue on a new institutional basis with the widest possible inclusion of various social actors in the process of achieving agreed goals in social and labor relations. This implies the transition from the formula "state – business – trade unions (employees)" to a wider institutional format of social interaction "state – business – society – community".

References

1. O. Novikova, O. Amosha, V. Antonyuk, V. Vishnevsky, *Social Resources for Decentralization of Management: Mechanisms for Mobilization and Effective Use* (2018)
2. O. Novikova, Ya. Ostafiichuk, O. Khandii, Mechanism of Strengthening Social and Labor Potential of Sustainable Development, Problemy Ekorozwoju, vol.**14**, **1**, 63-72 (2019)
3. Budget analysis of 665 AHs over 2018, URL: https://decentralization.gov.ua/news/10649 (2019)
4. Budget analysis of 665 AHs over 2018 in terms of oblasts, URL: https://decentralization.gov.ua/en/news/10674 (2019)
5. D. Acemoglu, J. A. Robinson, *Why Nations Fail: The Origins of Power, Prosperity, and Poverty Paperback* (Crown Business, New York, 2013).
6. Rural-Urban Partnerships: An Integrated Approach to Economic Development, URL: http://www.oecd.org/regional/regional-policy/rural-urban-partnerships-9789264204812-en.htm. (2013)
7. Governing rural-urban partnerships: lessons from the field, URL: http://www.oecd.org/cfe/regional-policy/Session%20III%20presentations.pdf (OECD, 2013)
8. O. Borodina, O. Shubravska, *Agrarian and rural development for the growth and renewal of the Ukrainian economy* (2018)
9. A. Tkachuk, O. Sergienko, O. Latsko, Overview of the situation regarding the implementation of partnership between urban and rural communities in conditions of decentralization and changes in the state regional policy in Ukraine: analytical note (2019)
10. Register of Territorial Communities Cooperation Agreements, URL: http://www.minregion.gov.ua/napryamki-diyalnosti/regional-dev/rozvytok-mistsevoho-samovryaduvannya/reyestr (2019)
11. *Cooperation of territorial communities (inter-municipal cooperation - MMC): Educational and practical manual*, V. Tolkovanova, T. Zhuravlya (ed.), 80–81 (2016)

Innovative activity of logistics enterprises in the modern world

Oksana Onyshchenko[1]*, *Yevhen* Yudenko[2]

[1]Kremenchuk Mykhailo Ostrohradskyi National University, Department of Accounting and Finance, 39600, Kremenchuk, Pershotravneva str. 20, Ukraine
[2] Kremenchuk Flight College of National Aviation University, Department of Logistics, 39600, Kremenchuk, Peremogy str. 17/6, Ukraine

Abstract. The peculiarities of application and development of innovations in logistic enterprises in modern conditions have been considered by the authors. The influence of innovations on the efficiency of logistics enterprises functioning was investigated. Innovations play a key role in providing quality services to enterprises and increasing their competitiveness. Finding and implementing innovations allows to ensure sustainable development of the enterprise and consolidate its competitive position in the market. The dependence of the Logistics Performance Index (LPI) on the general macroeconomic situation in the country is proved. Recommendations on the implementation of innovations in the process of logistics enterprises, which collectively solve an important scientific and applied task to identify ways to increase the efficiency of economic activity and the competitiveness of Ukrainian logistics enterprises have been provided. Directions of search of potential of innovations of the Ukrainian logistics enterprises have been offered. New logistic technologies reflecting the development of society, business and technological trends and innovations in the field of logistics have been considered. The necessity of introduction of innovations in the process of activity of logistics enterprises has been proved to increase the efficiency of their business activity and competitiveness.

1 Introduction

Logistics service innovation refers to a new, helpful idea, procedure, or practice in logistics operations that is different from a company's current practice [1]. Innovations are the basis for the development of enterprises of all types of activities, including logistics. As Chapman note, "Innovation is imperative for logistics firms serving the market in the new economy" [2, p. 641]. This sentiment is reinforced by Flint, "Innovation is critical to the success of many firms, including providers of logistics services" [3, p. 113]. The same is true for companies selling products: they may gain competitive advantage by focusing on logistics service innovation. Leveraging logistics service can help to build strong relationships with customers, generate barriers to competition, increase customer loyalty and switching costs, and make market activities more efficient [4]. In conditions of inconsistent environment and

* Corresponding author: vikols75@ukr.net

increasing competition among logistics enterprises their development relies on constant search and implementation of new technologies, delivery methods and service. In order to achieve competitive advantages in the cost, service time and added value (for the customer), logistic companies should use newest technologies and innovations, which are essential to cut costs of resources and to exclude additional, unproductive expenses. This is all because company`s innovative activity determines their competitiveness not only in the current moment but also in long run.

According to I. Afanasenko and V. Borisova the object of innovations in logistics is "trajectory optimization of circulating economic flows, in order to reduce the time of logistics cycle, to rationalize the total costs of commodities, to increase the effectiveness of development of logistic chain"[5]. Reserves for increasing the efficiency of logistics processes are mainly in the latest technologies and various innovations that use these technologies.

Opportunities for development are to be found mainly in new technologies and various innovations. Considering the underdevelopment of development in logistic infrastructure in Ukraine, the problem of innovation implementation in logistic activity of Ukrainian companies attracts significant interest.

2 Methods

As a main method of research, we use the critical analysis and comparison of analytical reports, innovative development strategies and scientific publications on the issues of innovative activity of logistics enterprises.

For the purpose of achieving the set goals, the following general scientific and specialized methods were used:

- theoretical generalization, comparison and systematization - to study the essence and features of innovative activity of logistics enterprises;
- system analysis – for determining the state and level of development of the logistics system of Ukraine;
- abstract-logical – for theoretical summarization and conclusion;
- graphical method – for visualization dynamics of the Logistics Performance Index (LPI) of Ukraine and others country.

3 Results and Discussion

Innovations are vital for the growth of all the enterprises, including those that specialize in logistics. Today, when opportunities for rapid increase in innovative activity of producers are restricted and the progress slowed down, companies are striving to create reserves to increase the effectiveness by means of logistics. The growing number of companies turning to logistic companies suggests that they see real opportunities to boost their competitiveness. When there`re couple suppliers of the same product on the market, the one that is be able to provide higher quality transportation and service will have an edge.

Namely for those who use the services of technologically advanced logistics enterprises, additional external competitive advantages are formed. On the other hand, logistics services related to the fast delivery of products to the consumer, wherever he was, require innovation in the field of vehicle production and in the infrastructure segment.

In countries with developed market economies logistics has already become the foundation of successful operating of market entities, and the formation of logistic system ensures the rise in effectiveness of economic processes and decline in company`s aggregate costs. Only a small proportion of Ukrainian enterprises work on bases of concepts of logistics.

Mostly it`s companies with foreign investment that have separate logistics departments and well-formed logistic system.

In order to objectively assess logistic system in Ukraine and compare it to logistic systems of other countries, we use the Logistics Performance Index (LPI) developed by World Bank (Fig.1). The Logistics Performance Index is an interactive benchmarking tool created to help countries identify the challenges and opportunities they face in their performance on trade logistics and what they can do to improve their performance. Construction of LPI following indicators were used, the efficiency of customs and border clearance, the quality of trade and transport infrastructure, the ease of arranging competitively priced shipment, the competence and quality of logistics services, the ability to track and trace consignment, the frequency with which shipments reach consignees within scheduled or expected delivery time [7].

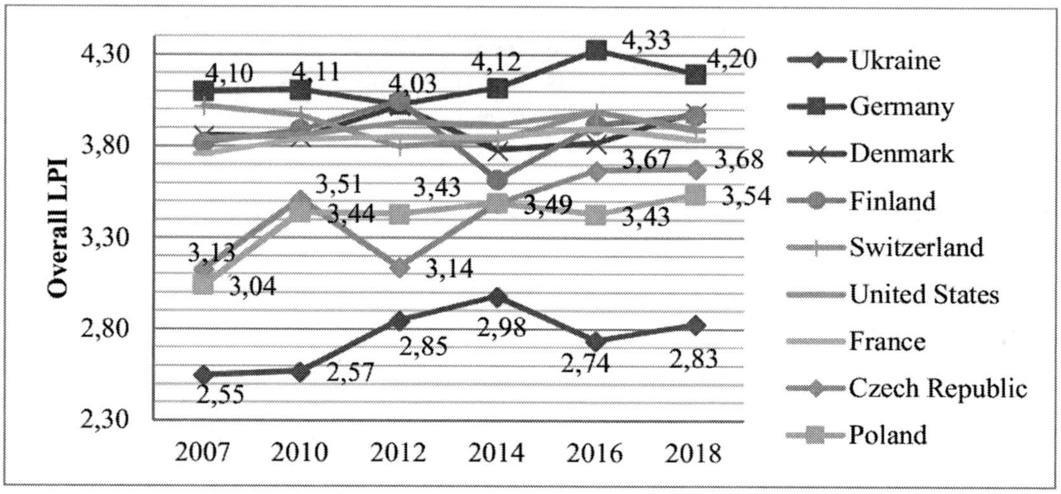

Fig. 1. Dynamics the Logistics Performance Index (LPI) of some of the worlds in 2007-2018
Source: According to the data [7].

According to LPI Ukraine is placed 66th out of 160 countries that got in World Bank`s rating. The world`s leading economies are found atop suggesting that there`s a high correlation between country's logistic performance and economic success. Ukraine was placed 73rdin 2007, 102thin 2010, 66thin 2012, 61thin 2014, 80thin 2016 and 66thin 2018. The decrease of LPI after 2014 can be attributed to worsening of political and economical situation in the country.

Macroeconomic situation in Ukraine was unstable throughout the last from 2010 till 2017. The factors of external impact influence the functioning of logistic companies, because their operating environment is fairly dynamic. Accounting the impact of external factors in practices of logistic companies provides gradual adaptation to modern demands: development of a network and the scale of activities, the increase in flexibility and mobility of the internal economy system, the decrease in costs. By cutting expenditures on separate stages of logistic`s chain total costs diminish, the labor organization improves, productivity and effectiveness of separate logistic processes increases. In order to secure a further growth in performance logistic companies are required to implement innovations. The implementation of innovations will facilitate optimization of logistic flows, which in turn will provide opportunities to gain economic effect and to boost competitiveness not only of the logistic that of other companies operating on the market. It will be possible to use the reserves of the company as a competitive advantage, the use of the latest technological programs that enable timely response to external and internal changes in order to achieve the desired result.

Ukrainian logistic companies are required to transition to an active phase of realization of system innovations. In modern circumstances the search of potential innovations is relevant in multiple directions:

1) Informatization and computerization of logistic processes allows to effectively operate various material flows. At the same time, there`s an opportunity to organize a complex transporting service based on the single agreement and to to formulate tasks for participants of the logistics chain on the basis of the order of the consumer of transport services [9, p.123]. Creation of database containing information about all logistic operations gives an ability for company to coordinate it`s shipping and transporting processes.

2) Transporting optimization, namely the cooperation of various transporting means on the same route in order to control transporting, loading, unloading, warehousing and delivering to the destination, etc., will assist in realization of integration of logistic chain and reduce the cost.

3) The execution of strategy to diminish costs by means of standardizing the products that require improvement of all inner functioning processes.

Innovations in logistics are realized on various stages of operating. Each operation in the company will be more effective if done using innovations. The usage of innovations in logistics makes customers more confident in company`s quality and timely service and significantly raises the competitiveness of company.

Richey [10] examined the relationship between reverse logistics innovation and a firm's performance. In doing so, they stated that logistics innovation should improve a firm's market effectiveness and internal cost efficiency. They also stated that logistics innovations can lead to increased revenues due to added services and improved customer satisfaction. The findings indicated a positive relationship between logistics innovation and strategic performance for large firms. Together, the findings highlighted above indicate a positive relationship between logistics innovation and the development of a competitive advantage. Persson [11] argues that logistics service innovation can provide firms with a competitive advantage. The author cites examples such as a firm's use of EDI to improve communications with customers, and the development of new services to open up new customer markets and add value to existing customers.

Attracting innovations, introducing and using innovative technologies allows enterprises not only to increase competitiveness, but also to minimize the impact on the environment, develop the social responsibility of their business [12].

Viewing the logistic trends, we can`t fail to notice the new emerging logistic technologies. DHL company regularly publishes key instruments for the global logistic community – DHL Trend Research, that reflects the progress in business and technological tendencies in the logistics field [13]. Logistics trends 2018/2019:

1) Social & Business Trends relevant in >5 years (Batch Size One, Grey Power Logistics, Servitization, Smart Containerization, Supergrid Logistics, Tube Logistics);

2) Social & Business Trends relevant < 5 years (Connected Life, Digital Work, Fair & Responsible Logistics, Fresh Chain, Green Energy Logistics, Logistics Marketplaces, Omni-channel Logistics, Sharing Economy);

3) Technology Trends relevant in >5 years (3D Printing, Artificial Intelligence, Bionic Enhancement, Blockchain, Next-generation Wireless, Self-driving Vehicles, Unmanned Aerial Vehicles, Virtual Reality & Digital Twins);

4) Technology Trends relevant in <5 years (Augmented Reality, Big Data Analytics, Cloud Logistics, Internet of Things, Low-cost Sensor Solutions, Robotics & Automation).

Just one in 10 firms is confident that its business has the right approach to breakthrough innovation, according to a new study from Arthur D. Little (ADL) [14]. In Breakthrough

Innovation, ADL found that 88% of business leaders were dissatisfied with their strategies for nurturing breakthrough – or radical – innovation. At the same time, companies expected the revenue contribution of breakthrough innovation to double over the next five years. The survey of leading European companies found that the time to deliver major revenue gains (three to 10 years) was a common barrier, due to short-termism and lack of project commitment. Companies with clearly defined innovation goals and dedicated breakthrough teams saw 15% more revenue gains and expressed more confidence in their breakthrough innovation activities.

Innovations is the direction in today`s entrepreneurship that requires significant financial resources. According to Forbes, the world`s most innovative companies spend close to 16 billion dollars a year [15]. On the other hand, innovations is the source of competitive advantages, that`s why recognized leaders in innovation always increase investment in this direction.

The level of innovative activity of enterprises in Ukraine remains relatively low. According to the National Statistic Association of Ukraine, only 16.2% of companies were innovatively active in 2017. This index has grown compared to the previous years, though it doesn`t correspond to the level of best economically developed countries (20-50%). The amount of aggregate expenses on innovations was 9117.5 thousand UAH, which it`s not enough to facilitate the required innovative activity [8].

The core to solving innovation progress problems in Ukrainian companies, including logistic companies, should become the complex government innovation-investment policy aimed at the raising the rate of innovation investment and the formation of an effective operating system to deal with innovations on all levels of economy [17].

The long-term strategy of innovative development by improving national innovation system can be considered the starting point of a state policy in the innovative sphere that will enable a country to develop adequate implementation toolkit and transform it into a legislative framework, targeted and comprehensive development programs [18].

3 Conclusion

Logistic companies require constant development and improvement as well as a profound study of all possible factors of influence and individualization of each innovation. Emergence of a new technical inventions successfully used by logistic enterprises considerably increases the effectiveness of the company. Realization of innovations in logistic companies allows to strengthen the competitive positions of stronger companies and to eliminate weaker ones from the market, which in turn will lead to the improvement in quality of service.

Companies successfully realizing innovations in their activities are constantly evolving and have a great demand on the services in the logistic market. Increase in quality of service and the ability to control the goods transporting increases the rating of a company, which influences the quantity of orders. Logistics are changing, new technologies and business models emerge. Even in crisis innovations shouldn`t be rejected. In today`s circumstances competitions increases considerably on the market, that`s we need to find and implement latent, not yet used reserves. Any changes that influence the development of the company in a positive way give it advantages.

References

1. Grawe, S.J.: Logistics innovation: a literature-based conceptual framework. International Journal of Logistics Management. **20**(3), 360–377 (2009)

2. Chapman, R.L., Soosay, C. and Kandampully, J.: Innovation in logistic services and the new business model: a conceptual framework. International Journal of Physical Distribution & Logistics Management. **33**(7), 630–650 (2003)
3. Flint, D.J., Larsson, E., Gammelgaard, B., Mentzer, J.T.: Logistics innovation: a customer value-oriented social process. Journal of Business Logistics. **26**(1), 113–147 (2005)
4. Bolton, R., Grewal, D. and Levy, M. Six strategies for competing through service: an agenda for future research. Journal of Retailing. **83**(1), 1–4 (2007)
5. Afanasenko, I.D., Borisova, V.V.: Logistika snabzheniya. Piter, SPb. (2010)
6. Chuxraj, N., Patora, R.: Innovaciyi ta logisty`ka tovariv: monografiya. Nacional`ny`j universy`tet L`vivs`ka politexnika, L`viv (2001)
7. The World Bank International LPI Data. https://lpi.worldbank.org (2019). Accessed 20 Mar 2019
8. State Statistics Service of Ukraine. htpp:/ukrstat.gov.ua (2019). Accessed 15 Mar 2019
9. Ry`bchuk, A.V.: Global`na vy`robny`cha infrastruktura svitovogo gospodarstva: teoriya i prakty`ka: monografiya. Vy`mir, Drogoby`ch (2009)
10. Richey, R.G., Genchev, S.E. and Daugherty, P.J.: The role of resource commitment and innovation in reverse logistics performance. International Journal of Physical Distribution & Logistics Management. **35** (3/4), 233–257 (2005)
11. Persson, G.: Achieving competitiveness through logistics. The International Journal of Logistics Management. **2**(1), 1–11 (1991)
12. Pochtovyuk, A., Semenikhina, V., Onyshchenko, O. and Ruban, B. : The formation and development of social responsibility of business: Ukrainian experience in a context of decentralization. SHS Web of Conferences **61**, 01018 (2019). doi.org/10.1051/shsconf/20196101018
13. DHL Group: Logistics Trend Radar. www.dhl.com/innovation (2019) Accessed 25 Mar 2019
14. Arthur D. Little: Innovation Excellence in Logistics – Value Creation by Innovation. Results of the European Logistics Association. http://www.adlittle.com/en/search/node/innovation (2019) Accessed 10 Mar 2019
15. The World's most innovative companies. https://www.forbes.com/ innovative-companies/list/tab:rank (2019) Accessed 05 Mar 2019
16. Kasy`ch, A.O., Vasen`ko, M.I.: Finansuvannya innovacijnoyi akty`vnosti pidpry`yemstv u suchasny`x umovax: teorety`chni ta prakty`chni aspekty`. Global`ni ta nacional`ni problemy` ekonomiky. **20**, 351–356 (2017)
17. Kasy`ch, A.O.: Dosvid formuvannya nacional`ny`x innovacijny`x sy`stem v krayinax, shho rozvy`vayut`sya. Aktual`ni problemy` ekonomiky. **5**(143), 46–49 (2013)
18. Kasych, A., Vochozka, M.: Conceptual provisions of development of Ukrainian national innovation system. Scientific bulletin of Polissia. **2**(10), 16–23 (2017)

Synthesis of the model of the socio-economic system

Oleksandr Poliarus[*], *Iryna* Nazarenko[2], *Yevhen* Poliakov[1], *Andrii* Lebedynskyi[1], and *Sergii* Ianushkevych[1]

[1]Kharkiv National Automobile and Highway University, Department of metrology and life safety, 61000 Kharkiv, Ukraine
[2]Ukrainian State University of Railway Transport, Department of economics and management of industrial and commercial business, 61050 Kharkiv, Ukraine

Abstract. The identification of the socio-economic system, which includes the scientific and scientific-technical activities of Ukraine, using the three input and output signals, is carried out in the article. The synthesis of the model is based on the minimization of the functional, which is the distance in the functional space with the quadratic metric between the real output signal and the theoretical signal containing the unknown parameters of the system model. These parameters include the system time constant and the coefficients that describe the power transformation function of the nonlinear block. A comparison of real output signals and signals calculated using the linear model of the system with memory and the Hammerstein model for a nonlinear inertial model is carried out. It is shown that the errors of the output signals representation for the quadratic model are less than for the linear one and are different for various output signals. If these errors exceed the allowable ones, it is recommended to change the parameters of the model or the number of members of the power function of the nonlinear block transformation.

1 Introduction

In the VUCA world, the requirements for the quality of socio-economic systems models increase. The socio-economic system is a complex stochastic dynamic system that covers the processes of production, exchange, distribution and consumption of material and other goods.

Despite the existence of many simulation methods (econometric, neural networks, etc.), there is still no reliable method for obtaining a model of a socio-economic system based on its input and output data, which change over time. This process is a system identification. In this case, the system is represented as a "black box", due to the impossibility of considering the main influences on it from both higher-level systems and other systems that are in contact with it.

We will identify the socio-economic system, which includes the scientific and technical activities of Ukraine. For this system, according to [1], the change of input parameters only

* Corresponding author: poliarus.kharkov@ukr.net

after a certain period of time affects the dynamics of the output parameters, i.e., the system is inertial. Taking into account the resource indicators and indicators of the efficiency of science proposed in [1, p. 11; 2, p. 385; 3, p. 23], as well as features of the statistics of science in Ukraine, we have constructed the general model of the studied socio-economic system (Fig. 1).

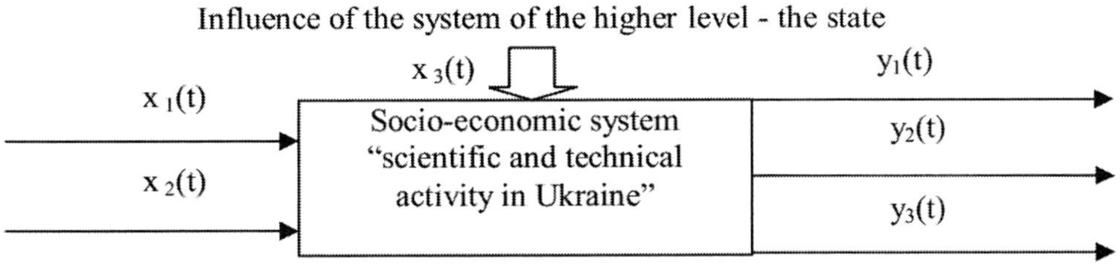

Fig. 1. The general model of the socio-economic system under study

The actions or input signals come to the input of the system:

$x_1(t)$ - the number of performers of scientific and technical works, thousand persons;

$x_2(t)$ - costs of carrying out research and development (in current prices), mln. UAH;

$x_3(t)$ - science-intensive gross domestic product (GDP), that is, the share of expenses for research and development in GDP,%.

The following signals are at the output of the system:

$y_1(t)$ - the volume of performed scientific and technical works, UAH million;

$y_2(t)$ - the share of the volume of performed scientific and scientific-technical works in GDP, %;

$y_3(t)$ - the number of patents received for inventions (in the name of the national applicants), unit.

The output data are given in Table 1.

Table 1. Output data (by [4] and earlier collections)

Years	**$x_1(t)$,** thousand people	**$x_2(t)$, mln.** UAH	**$x_3(t)$, %**	**$y_1(t)$,** mln. UAH	**$y_2(t)$, %**	**$y_3(t)$,** units
2010	115.6	8107.1	0.75	9867.1	0.90	2,034
2011	109.8	8513.4	0.65	10349.9	0.79	1,902
2012	105.9	9419.9	0.67	11252.7	0.80	1,557
2013	100.5	10248.5	0.70	11781.1	0.80	1,743
2014	89.5	9487.5	0.60	10950.7	0.69	1,701
2015	122.5	11003.6	0.55	12611.0	0.64	1,516
2016	97.9	11530.7	0.48	11516.5*	0.63*	1,277
2017	94.3	13379.3	0.45	11120.4*	0.6*	1,224

*Note: *forecast (in the absence of statistical data due to changes in statistical reporting forms)*

The graphs of time functions describing the behavior of input and output signals are shown in Fig. 2.

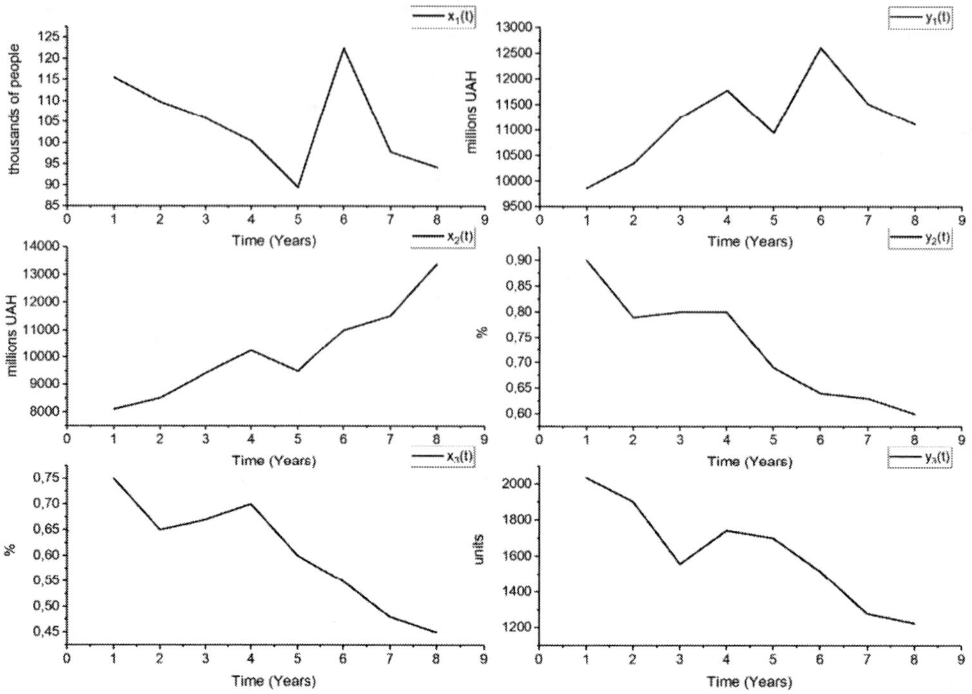

Fig. 2. The example of input and output signals of the socio-economic model

2 Construction of a model

It is convenient for the researcher to use the simplest models of socio-economic systems that adequately describe their behavior. First of all, we assume that the model of the socio-economic system is linear, and after its testing it is possible to switch to more complex models, if significant defects of the model are revealed. For a linear model, the relationship between the vector of input signals $\vec{x}(t)$ and the vector of output signals $\vec{y}(t)$ is determined by the dependence

$$\vec{y}(t) = \mathbf{H}(t)\vec{x}(t), \tag{1}$$

or in expanded form

$$\begin{pmatrix} y_1(t) \\ y_2(t) \\ y_3(t) \end{pmatrix} = \begin{pmatrix} h_{11}(t) & h_{12}(t) & h_{13}(t) \\ h_{21}(t) & h_{22}(t) & h_{23}(t) \\ h_{31}(t) & h_{32}(t) & h_{33}(t) \end{pmatrix} \begin{pmatrix} x_1(t) \\ x_2(t) \\ x_3(t) \end{pmatrix},$$

where $\mathbf{H}(t)$ is matrix operator whose elements are pulse characteristics or pulse responses $h_{ij}(t)$ $(i, j = 1, 2, 3)$, i.e., the reaction of the system on d-impulse, and for simulation conditions, this is the reaction of the system $y_i(t)$ at the i output on the pulse action at the j input. Consequently, at i output of the system using equation (1) we have:

$$y_i(t) = h_{i1}(t)x_1(t) + h_{i2}(t)x_2(t) + h_{i3}(t)x_3(t). \tag{2}$$

For the socio-economic system described by the differential equation of the first order, the pulse response of the system can be represented as

$$h_{ij}(t) = \frac{A_{ij}}{\tau_{ij}} e^{-\frac{t}{\tau_{ij}}}, \tag{3}$$

where A_{ij}, τ_{ij} are amplitude of the pulse characteristic and the system's time constant respectively.

Then expression (2), using the equation of convolution, is transformed to form

$$y_i(t) = h_{i1} \int_{-\infty}^{\infty} e^{-\frac{t-\tau}{\tau_{i1}}} x_1(\tau)d\tau + h_{i2} \int_{-\infty}^{\infty} e^{-\frac{t-\tau}{\tau_{i2}}} x_2(\tau)d\tau + h_{i3} \int_{-\infty}^{\infty} e^{-\frac{t-\tau}{\tau_{i3}}} x_3(\tau)d\tau, \tag{4}$$

where τ is a time shift in the convolution equation.

The solution of the equation (4) is carried out by the method of global random search (genetic algorithm), minimizing the functional

$$J_i = \int_{-\infty}^{\infty} [y_i(t) - m_i(t)]^2 dt. \tag{5}$$

where $m_i(t)$ is the right-hand side of equation (4) and depends on τ_{ij} and h_{ij} which are determined by global random search.

If the modulus of the average difference between $y_i(t)$ and $m_i(t)$ at the final stage of operation of the genetic algorithm exceeds the established threshold, then a linear model can be considered as an inadequate one, because it describes the connection between the input and the output of the system (Fig. 1) incorrectly. In this case, the Hammerstein model has the advantage and it virtually represents the system in the form of two consecutive blocks: nonlinear non-inertial and linear inertial. To evaluate the nonlinear properties of the system, we described a nonlinear block by a power (quadratic) function, for example, and then expression (2) has been transformed to the following form

$$\begin{aligned} y_i(t) = {} & a_{i10} \int_{-\infty}^{+\infty} h_{i1}(t-\tau)d\tau + a_{i11} \int_{-\infty}^{+\infty} x(t)h_{i1}(t-\tau)d\tau + a_{i12} \int_{-\infty}^{+\infty} x^2(t)h_{i1}(t-\tau)d\tau + \\ & a_{i20} \int_{-\infty}^{+\infty} h_{i2}(t-\tau)d\tau + a_{i21} \int_{-\infty}^{+\infty} x(t)h_{i2}(t-\tau)d\tau + a_{i22} \int_{-\infty}^{+\infty} x^2(t)h_{i2}(t-\tau)d\tau + \\ & a_{i30} \int_{-\infty}^{+\infty} h_{i3}(t-\tau)d\tau + a_{i31} \int_{-\infty}^{+\infty} x(t)h_{i3}(t-\tau)d\tau + a_{i32} \int_{-\infty}^{+\infty} x^2(t)h_{i3}(t-\tau)d\tau. \end{aligned} \tag{6}$$

In the equation (6) there are 9 unknown coefficients a_{imk} $(m, k = 0, 1, 2, 3)$ of power function, which are determined by minimizing the functional (5), but in this case $m_i(t)$ is the right-hand side of the relation (6). These coefficients can be further specified, as well as the parameters of the system pulse characteristics (Fig. 1), if they are considered unknown, but the average value is presented through previously calculated parameters and coefficients. We restrict ourselves to previously obtained results.

3 The results of modeling

Figure 3 shows the dependence $y_i(t)$ at the first input when $i = 1$. The calculation is carried out both for the linear model of the system, and for the nonlinear one. In Fig. 3 the real signal at the first input of the system is also given.

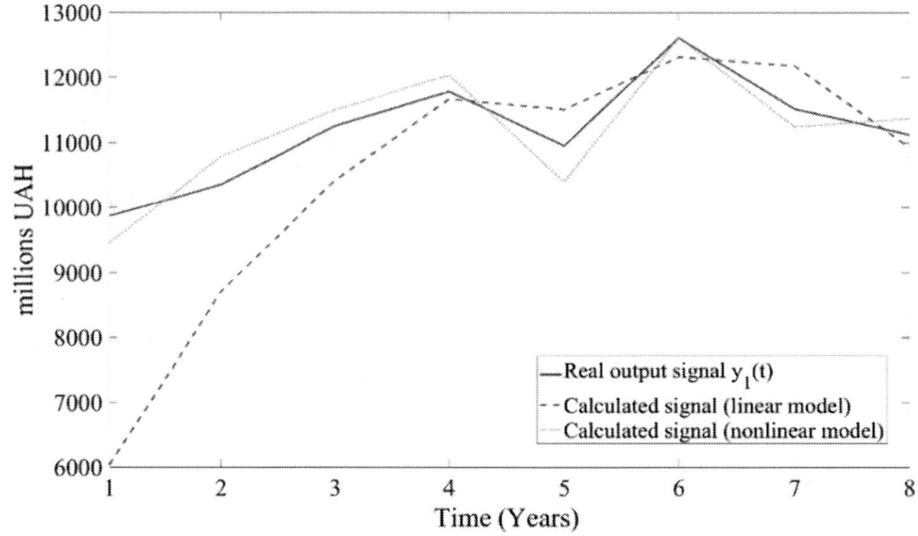

Fig. 3. Real and model signals on the first output of the system

The difference between the real signals and those calculated using linear and nonlinear models has been considered as the error. The graph of the relative error of output signal representation for two models is shown in Fig. 4.

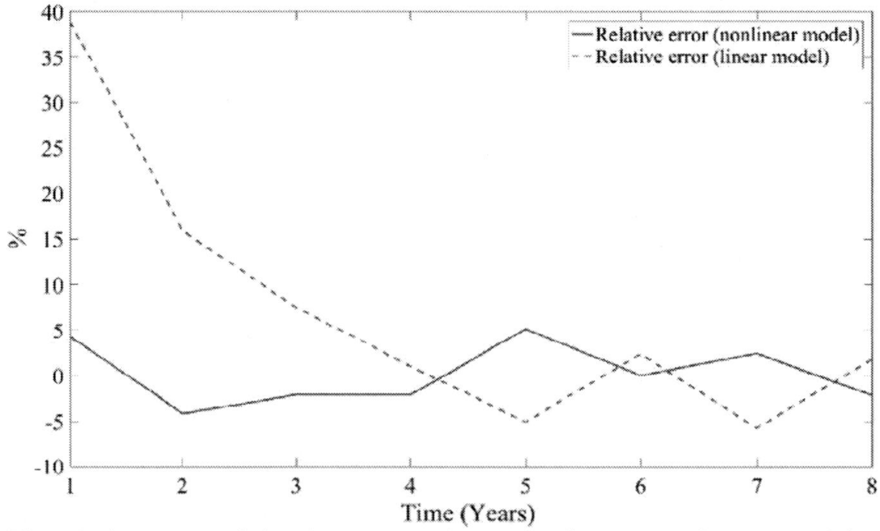

Fig. 4. The relative errors of signal representation at the first output for two models: dashed line – linear model, solid line – nonlinear one

Similar results are obtained at other outputs of the system. From Fig. 4 it follows that the application of the quadratic model of the socio-economic system provides the calculation of the first output signal with a relative error of not more than 5%. However, this model on other system outputs does not allow obtaining such an error. At different time intervals, it can exceed 10%. Thus, the developed method provides identification of the multi-channel socio-economic system, but for obtaining adequate results it is necessary to conduct research for many input and output signals with statistical processing. Unfortunately, for such a system there is only a limited set of data. In the presence of rapid changes in the parameters of input signals, it is expedient to use the method developed in [5]. The results obtained in mathematical modeling, are qualitatively similar to the results of the study of the measuring pressure channel [6], and the latter are confirmed experimentally.

4 Conclusions

1. The developed model of socio-economic system describes the nonlinear and inertial properties of the system and allows determining the output dependencies with known input ones.
2. Errors of model output signals compared to real ones are equal to several percent at all outputs, but at some moments of time they reach 20 - 30%.
3. Considering the nonlinear properties of the system leads to a reduction of these errors several times at the first output and 5-10% at the third output, but in the second output the error, on the contrary, increases. This indicates that it is not always necessary to use the Hammerstein model and its first nonlinear block.
4. The given errors can be reduced by repeated application of the genetic algorithm together with changes of the model parameters. To create a better model you need to have a wide range of experimental data.
5. The obtained model must be tested on other experimental samples. It should be expected that the model may be unique, that is, fair only for Ukraine. The method itself is suitable for any socio-economic systems that are similar to the system studied in the article.

References

1. A. V. Todosiichuk. *Science as a factor of social progress and economic growth.* 2nd edition, supplemented and revised. Moscow, 2005 [Electronic source]. - Available at: https://docplayer.ru/26309550-A-v-todosiychuk-nauka-kak-faktor-socialnogo-progressa-i-ekonomicheskogo-rosta.html
2. M. Arkhipova, V. Vlasova. *Monitoring and modeling of the main trends in the development of scientific activities in Russia.* Economics, statistics and informatics, 2014, 6(2), 384-389 [Electronic source]. - Available at: https://cyberleninka.ru/article/n/monitoring-i-modelirovanie-osnovnyh-tendentsiy-razvitiya-nauchnoy-deyatelnosti-v-rossii.
3. A. Burda. *Fundamentals of research (lecture course)*, Krasnodar, 2015 [Electronic source]. - Available at: https://kubsau.ru/upload/iblock/40f/40f35535e069a955ca6fc10d13d484b7.pdf
4. Scientific and innovation activity of Ukraine: Statistical collection, Kyiv, 2018 [Electronic source]. - Available at: http://www.ukrstat.gov.ua/druk/publicat/kat_u/2018/zb/09/zb_nauka_2017.pdf
5. O. V. Poliarus, Y. O. Poliakov, I. L. Nazarenko, Y. T. Borovyk, M. V. Kondratiuk. *Detection of Jumps Parameters in Economic Processes (on the example of modelling profitability).* International Journal of Engineering & Technology, **7**, No 4.3, p. 488-496, [Electronic source]. - Available at: https://docplayer.ru/26309550-A-v-todosiychuk-nauka-kak-faktor-socialnogo-progressa-i-ekonomicheskogo-rosta.html
6. O. Poliarus, O. Koval, Ya. Medvedovska, Ye. Poliakov, S. Ianushkevych. Identification of a nonlinear inertial measuring pressure channel. - Ukrainian Metrological Journal, 2019, №1, p. 63-70.

Social responsibility development in the conditions of globalization: financial flows focus

Oksana Portna*[1], *Svitlana* Kachula[2] *and Liubov* Lysiak[2]

[1]V. N. Karazin Kharkiv National University, 4 Svobody Sq., Kharkiv, 61022, Ukraine

[2] University of Customs and Finance, 4/2 Vernadskogo av., Dnipro, 49000, Ukraine

Abstract. The article deals with social responsibility system development of the state, business, society. It is shown that the characteristic of development and the result of social responsibility of the state, business, society is the high quality of socio-economic development, balancing the interests of all participants in social relations. Modern complex social relations actualize social issues both within individual countries and in the world dimension. The problems of social responsibility of the state, business, and society concern a wide range of economic, social, financial, environmental and political aspects. The article presents generalized standards and conditions for the development of socially responsible state, business, and society, which are mediated by financial flows. A meaningful characteristic of state, business and society social responsibility development is the social relations institutional participants financial flows focus, which should be considered in unity with the global system and relevant world processes during globalization. The problems of financial flows direction in the conditions of financial and economic globalization are considered. In Ukraine, social responsibility of the state, business, society is non-systemic, fragmentary, situational.

1 Introduction

Problems of social responsibility not only of business but of the state, the population both within the borders of any country and in the world remain relevant because of the deepening of socio-economic asymmetries in the context of the development of globalization processes and the modern complex configuration of the world financial and economic order and countries development. his gives an additional impetus to deepen the research of this problem, especially in terms of the direction of financial flows in a situation where institutional actors are trying to accumulate considerable amounts of money resources. This provides an additional impetus for deepening the research of this issue, especially as regards the orientation of financial flows in a situation when institutional actors are trying to concentrate considerable amounts of money resources.

* Corresponding author: portna@karazin.ua

The world scientific community representatives, M. Porter [1], T. Wilson [2] and others, made a significant contribution to the study of a wide range of corporate social responsibility issues. The thorough processing of the globalization processes was carried out by M. Waters [3], T. Levit [4], and others. Fundamental researches of financial imbalances in the conditions of globalization were carried out by P. Krugman [5], J. Stiglitz [6].

Currently, "social responsibility" category theoretical aspects are analysed in separate sections of the state, business, society. However, the aspect of the development of social responsibility of the state, business, society as a system, especially in the context of globalization processes, characterized by a certain orientation of financial flows, remains unexplored. Insufficient attention is paid to the definition of social relations subjects socially irresponsible behaviour.

The purpose of this article is to consider social responsibility of the state, business, society as a systemic phenomenon, to determine the quantitative and qualitative limits of social responsibility and/or irresponsibility of the authorities, business, and society of the country, the basis of which is the focus of financial flows in the globalization dimension.

In today's conditions of global and national financial and economic crises, the spread of social tension, development rates inhibition, etc. one of the important concepts of the world order is the states, business and society social responsibility development, which is consistent with the concept of sustainable development. Different manifestations of social responsibility of the state, business, society are characterized by different characteristics, influence of various factors actions, formation and implementation of the system of social responsibility principles. A significant characteristic of development and the result of social responsibility of the state, business, society is the high quality of socio-economic development, balancing the interests of all participants in social relations.

Modern complex social relations actualize social issues both within individual countries and in the world dimension, reflected in the global growth gap between income and wealth of the poor and the rich, the economic inequality of the development of countries and their population, large differences in the efficiency of sectoral economic activity, etc.

Existing theoretical and practical developments of social responsibility various aspects, mostly in form and content, relate to the formulation of individual social actions of the state, business [1; 2], society, etc., and the declaration of the influence of institutions (official and informal, state and non-state) on the rules and norms of behaviour and obligatory observance by members of social relations.

The problems of social responsibility of the state, business, and society concern a wide range of economic, social, financial, environmental, political, and other aspects. But these aspects are mostly considered within individual states, individual businesses. In our opinion, this is a simplified approach to the problem of development of social responsibility, especially in the context of the globalization processes strengthening. That is, it is about a system of social responsibility, in which all participants in social relations, namely the state, business and society, mutually influence and interact.

2 Discussion

All members of public relations (state, business, society) in order to realize their interests declaratively define the elements of social responsibility. The state social responsibility is determined by the very nature of the state and is embodied in ensuring each person's conditions for an acceptable standard of living, establishing sound social standards and ensuring their observance, preventing anti-social behaviour caused by poverty, chronic diseases, inability to receive education, etc. At the same time, the state aims to ensure equity, reduce income inequality by redistributing it in favour of those who are objectively unable to work and the poor. An important tool for achieving the stated goal is the state financial policy and its main directions: fiscal, monetary, investment, social and others. Using the financial policy tool, the state makes indexation of incomes, price regulation, labour market, establishment of the minimum wage, state social transfers, regulates the labour market and others. The targeted and effective financial flows direction, considering the harmonization of the interests of the state, business and society, involves the comprehensive and responsible use of the instruments of state financial policy.

According to the world practice, the criteria for assigning a business to a socially responsible (full payment of taxes in accordance with the legislation, production and sales of quality products, staff development, social investment, charitable and sponsorship projects implementation, etc.), more and more Ukrainian companies can be called socially responsible by certain criteria. Major Ukrainian business mainly implements charity and sponsorship projects (accompanied by their aggressive advertising in the mass media), provides social investments, both internal (training of personnel directly in the workplace) and external ones (development of local communities, realization of certain public initiatives, protection of the environment, landscaping, etc.). Business is responsible for the organization's culture, compliance with business and corporate ethics, and the socio-economic implications of managerial decisions for its own staff and society. At the same time, it should be noted that the principles of social responsibility of business in Ukraine are being implemented slowly and implemented in the most part fragmentary.

If we consider the standards and conditions of development of social responsibility of the state, business and society, then we can cite such as: development of economy, business, creation of workplaces, life quality improvement, increase of incomes, increase of solvency of the population, growth of consumer demand and commodity circulation, expansion of the base taxation, payment of taxes in full, formation of expenditures on budgets on the basis of socially oriented state, etc.

In Ukraine, social responsibility of the state, business and society is non-systemic, fragmentary, situational. According to the study by O. Moroz [7, p. 142], an efficient and effective system of realization of harmonizing the interests of the state, business, society and ensuring at all levels mutually beneficial cooperation between bodies of all levels of the state (regional, municipal) authorities, business owners (employers) and employees' representatives (trade union organizations), which will ensure the prevention and/or mitigation of contradictions that arise in the process of social dialogue is currently practically absent. An optimal system of reconciliation of interests can only be created through the achievement by the parties of the social dialogue of a fundamental consensus on the issues: preservation and maintaining the stability of the development of the national economy; promotion of socio-economic growth and competitiveness of business; joint regulation and control in the field of economic and social policy by the parties of social partnership (dialogue); support of social peace.

In our opinion, the substantive characteristic of the development of social responsibility of the state, business, and society is the focus of financial flows of all institutional participants in social relations on social development policy implementation. This is especially significant in the context of globalization. In the context of globalization processes, financial determinants of deepening the asymmetries of global development [8], asymmetry and paradoxes of financial globalization [9] the problems of social responsibility of the state, business, society should be considered in unity with the global system and relevant global processes.

According to N. Kravchuk [8, p. 346-347], the transformational processes in the world economy have significantly accelerated and become systematic, creating the preconditions for the formation of a global model of economic development. The profound qualitative changes in the character and driving forces of the economic system development, with the prevalence of global forms of its functioning on the basis of the internationalization of production and exchange, the growth of the level of trans nationalization and the national economies integration, lead to the modernization of a modern market economy and the adoption of universal life standards.

However, on the other hand, globalization processes are challenging, especially with regard to the accumulation of global financial resources and financial flows direction. According to Z. Lutsyshyn [9, p. 111], globalization cannot progress without the financial resources of world markets, which evolve under the direct influence of globalization trends. Globalization of finance as an important component of the globalization of the world economy is based on a powerful over-scale transfer of monetary capital between countries - in international financial flows. The negative effect of financial globalization is, among other things, chronic debt crisis that involves dozens of countries, mainly developing countries. The economic backwardness of developing countries is increasing from the advanced ones. This gap often contributes to fundamental differences in the political and social structures of developed and backwards countries. And as a result of these differences, the dependence of the backward countries on the advanced ones increases.

That is, in the conditions of deepening of globalization processes, strengthening of financial globalization, the direction of financial flows of all subjects of national economies are indicators of social responsibility. Thus, the volume of public debt of Ukraine is characterized by significant amounts of consolidated budget expenditures for debt servicing [10, p. 226]. In 2013, debt servicing amounted to UAH 33,172.1 million; in 2014 - UAH 49,434.6 million; in 2015 - UAH 86,172.7 million; in 2016 - UAH 96,105.2 million; and in 2017 - UAH 110,578.2 million. If we consider the dynamics of the quantitative and qualitative indicators of the consolidated budget of Ukraine 2016-2017, then the dynamics of consolidated budget revenues is 129.9%, while the state debt service is 115.1%; the expenditure dynamics is 126.5%; the dynamics of expenditures on social protection and social security – 110.6%; the dynamics of expenditures on environmental protection – 117.5%. The share of expenditures of the Consolidated Budget of Ukraine in 2017 for servicing the state debt is 10.5%; and spiritual and physical development - 2.3%; for the protection of the environment - 0.7%. The situation when the services of interests of external borrowers spend more resources than national development (with the tendency of debt total amount growth and its maintenance), testifies to the socially irresponsible activity of the state, business, population, that is, the entire system of national social and economic relations.

As I. Chugunov notes, the problem of fiscal financial flows, the development of financial relations in Ukraine is always particularly acute [11]. Chugunov also argues that in order to increase the efficiency and balance of budget flows, it is necessary to implement balanced budgetary regulation during a certain period in order to positively influence its economic growth, and budget policy should be determined on the basis of the cyclical nature of economic development, as well as the state of public finances, stock stability of the budgetary system in the light of the existing institutional environment [12].

When analysing direct investment (share capital) from Ukraine in the economies of the world, it should be noted that more than 90% of resources are directed to offshore (Table 1).

Table 1. Direct investment (share capital) from Ukraine (USD million) [10]

Indexes	Direct investment on January 1						December 31, 2017
	2012	2013	2014	2015	2016	2017	
Total	6456.2	6588.7	6702.9	6456.2	6315.2	6346.3	6339.8
Cyprus	5899.7	5917.6	5925.1	5926.1	5923.7	5930.5	5932.5
Proportion: Cyprus / Total%	91.4	89.8	88.4	91.8	93.8	93.4	93.6

If we consider the factor of the constant decrease of the UAH to the US dollar, in the analysed period [10], we can conclude that in the UAH equivalent, the volume of direct investment in offshore is significantly increasing.

3 Results

Such an orientation of the national economy financial flows to investment activities in the conditions of financial globalization actually demonstrates both the social irresponsibility of the national business and the irresponsible behaviour of the state, society as a system, after all, businesses withdraw resources in the offshore, and the state at the legislative and executive levels does not restrict leakage of resources from the national economy.

That is, for example, indicators of social irresponsibility of the state are the reduction of social expenditures, low incomes, low domestic demand due to the low purchasing power of the national currency, the growth of external public debt, etc. Indicators of social irresponsibility of the business are shadow activity, tax evasion, unofficial income, low wages of hired workers, withdrawal from the offshore, etc. For example, one of the indicators of corporate social responsibility is conscientious payment of taxes in accordance with the current legislation. At the same time, the results of tax planning and unplanned controls carried out by the State Fiscal Service of Ukraine during 2016-2018 show that taxpayers (large, medium and small businesses) every year violate tax laws and do not fully pay taxes. Thus, a significant amount of underpayment to the budget of value added tax, income tax, excise tax and land tax was revealed. In the sectoral section in 2018: 140 subjects of economy with revenues of over 200 billion UAH were inspected from oil and gas industry and 5.8 billion UAH were added to the budget; in the construction industry - 372 enterprises with incomes of more than 33 billion UAH and paid to the budget by 850 million UAH, in the processing and scrap metal trade - 21 business entities with revenues of more than 70 billion UAH, taxes were added up to 570 million UAH [13]. The percentage of tax control of enterprises of large, medium and small businesses during the period 2016-2018 is decreasing and averaging over the period of about 60% [13]. The control over 2016-2018 covers every fifteenth largest and mid-sized business entity that, judging by the data presented, is socially irresponsible.

Indicators of social irresponsibility of the population are evasion from work and duties, passivity to increase their own professional qualification level, careless attitude towards their health and health of others, shadow/illegal activity and obtaining shadow/illegal income, therefore tax evasion, and so on.

According to the Corporate Social Responsibility Development Promoting Strategy in Ukraine for the Period up to 2020, the necessity of developing social responsibility of business in Ukraine is determined by the following factors: low level of corporate culture; legal nihilism; low level of public awareness of the role of business; Ukrainian companies secrecy from the public; uneven consideration of interests of stakeholders [14]. Unfortunately, the Strategy does not take into account the need of the state, society and each of its members social responsibility development, which is a prerequisite for the harmonization of social and economic social relations.

4 Conclusion

Thus, a number of aspects of the state, business and society social responsibility development are updated, namely, the clear delimitation of quantitative and qualitative national and global indicators, as well as the quality of information support, information base [15], for monitoring of state, business, population social responsibility and/or irresponsibility.

For the development of the system of social responsibility of the state, business, society, especially in the context of deepening globalization processes, it is important:

- to create the appropriate institutional mechanisms that can neutralize the influence of negative factors on the development of social responsibility of business, state, society, promote integrated and responsible use of instruments of state financial policy regarding purposeful and effective direction of financial flows for taking into account the state, business, society interests harmonization;
- to create a modern high-quality information base for monitoring the state, business, and population social responsibility system;
- to introduce public control over observance of the state, business, society socially responsible behaviour;
- align the interests of social dialogue participants based on consideration of their real needs and capabilities based on responsible government financial policy;
- to define methodological and methodical principles and recommendations concerning indicators of the state, business, society social responsibility system; and
- to develop criteria for assessing the degree of social responsibility of the state, business, society.

References

1. M. Porter. Harvard Business Review, **84 (12)**, 78–94. (2009)
2. Th. Wilson. Brotherhood of St. Laurence's Research & Policy Centre. *CSR and banks: the role that banks could and should play in addressing financial exclusion*. Fitzroy Vic. (2008)
3. M. Waters. London; N.Y.: Routledge. *Globalization (Key Ideas)*. 272. (1995)
4. T. Levitt. Harvard Business Review. *The Globalization of Markets*. № 61 (3). P. 92 102. (1983)
5. P. Krugman. The New York Times. *Deficits saved the world*, available at: https://krugman.blogs. nytimes.com/2009/07/15/deficits-saved-the-world/
6. J. Stiglitz. New York: WW Norton & Co. *Freefall: America, Free Markets and the Sinking of the World Economy*. 361 (2010)
7. O. S. Moroz. *Ekonomika promyslovosti - Economy of The Industry,* **4 (65)**, 135 – 143. (2013).
8. N. Ya. Kravchuk. *Naukovi zapysky. Seriya Ekonomika.* **15**, 346 – 356. (2010).
9. Z. Lutsyshyn. *Mizhnarodna ekonomichna polityka - International Economic Policy,* **1-2**, 88–121. (2008).
10. Werner I. Ye. (Eds.). *Statystychnyi chorichnyk Ukrainy za 2017 rik.* Kiyv. (2018).
11. I.Ya. Chugunov. *Visnyk Kyivskoho natsionalnoho torgovelno-ekonomichnogo universytetu*, **4**, 13-22. (2011)
12. I.Ya. Chugunov. *Visnyk Kyivskoho natsionalnoho torgovelno-ekonomichnogo universytetu*, **5**, 64-77. (2014)
13. Zvit Derzavnoij fiskalnoi sluzhby Ukrainy za 2016-2018 roky. fvailable at: http://sfs.gov.ua/data/files/240396.pdf
14. Strategiya spriyannya rozvytku sotsialnoi vidpovidalnosty biznesu v Ukraini na period do 2020 roku, available at: http://www.svb.org.ua/sites/default/files/201309_strategiya_spriyannya_rozvitku_svb_v_ukrayini.pdf.
15. N. U. Ershova. *Aktualni problemy ekonomiky*, **8 (158)**, 368 – 374. (2014)

Management of financial risks in the security system of enterprise development: economic and legal aspects

Viktoriya Prokhorova[1]*, *Svitlana* Mushnykova[2], *Oleksiy* Bytiak[3], and *Krystyna* Slastianykova[3]

[1] Ukrainian Engineering and Pedagogical Academy, Department of Economics and Organization of the Activity of Business Entities, 61003, 16 Universitetskaya st., Kharkiv, Ukraine
[2]National Metallurgical Academy of Ukraine, Department of Finance, 4 Gagarin ave., 49005 Dnipro, Ukraine
[3]Yaroslav Mudryi National Law University, Department of Economic Law, Pushkinskaya str., 77, 61024 Kharkiv, Ukraine

Abstract. The state of market relations which is determined by the high level of competition, instability and environmental variability, requires from each subject of the weighted actions to manage the safe development. The risks accompanying the activities of each enterprise are the components of its management. The necessity of considering financial risks as a generalizing factor of the security system of enterprise development is substantiated. The research proposes a methodological approach within the framework of economic and legal aspects to the management of financial risks in the security system of the enterprise development, with the definition of the main theoretical positions, namely: goals, objectives, principles, legal entities and methods of managing financial risks in accordance with the stage of the life cycle enterprises The necessity of improvement of legal support of legal relations with the enterprise in the system of security of enterprise development is determined.

1. Introduction

The current state of market relations is determined by the high level of competition, instability and volatility of the environment. These are also the main barriers, and in some cases, and stimulants on the way to the development of industrial enterprises, which can only be overcome if the development of a system for development security within an enterprise management system is created. One of the areas of the development security system is risk management at the enterprise.

*Corresponding author: *vkprohkorova@gmail.com*

The activity of any entity is associated with the emergence of a large number of risks and threats to their implementation throughout its life cycle. Over the past decades, many scholars and practitioners have focused on defining the concept of risk, evaluating its occurrence and managing it using various methods and techniques [1, 9, 11, 12]. Based on the experience gained, it was concluded that risks are not a sentence but an existing reality. In general, the risk can be represented as the probability of occurrence of an event and / or a certain result. Moreover, if the risks were predictable (expected) and timely measures were taken to avoid or minimize, the result of the activity can be extremely positive.

On the other hand, the security of enterprise development should also take into account the legal support of the subjects of relations in the management of risks. However, legal support has not received enough attention at any level of the national economy. Although, according to Article 42 of the Commercial Code of Ukraine, entrepreneurship is an independent, initiative, systematic, at its own risk, economic activity carried out by economic entities (entrepreneurs) in order to achieve economic and social results and profit [16], there is no integrated regulatory framework for risk management regulation. All these conditioned the relevance of this study.

2. Main body

One of the main groups of risks associated with the economic activity of industrial enterprises are financial risks, which, in they turn, can be considered as a generalization indicator. They accompany enterprises at all stages of the production and financial cycles, and reflect the positive or negative impact of existing risks on it. Moreover, depending on the stage of the life cycle, starting from the stage of entering the market, passing the stage of growth or expansion of activity, maturity, saturation, even before the exit from the market, there may be various threats and risks. Therefore, it should be considered necessary to determine the methodological approach to managing financial risks in the system of security of enterprises development, taking into account both the economic and legal aspects of enterprises.

2.1 Methodological Approach to Financial Risk Management in the Enterprise Security System

Each industrial enterprise wants to develop with a certain level of protection against possible threats and risks accompanying their activities at every stage of the life cycle, not only in economic activity, but also in legal relationships with different actors. Therefore, according to the authors, the main objective of managing financial risks in the system of enterprise development security in both the economic and legal aspects is to achieve a safe development of an enterprise with the optimal level of financial risk (Fig.1).

Achievement of the set goal is possible only under conditions of thorough and reliable analysis of existing and/or possible threats and risks of both internal and external origin, assessment of the current status of the results of the enterprise and forecasting them with the possibility of preventing possible future risks.

The management of financial risks within the framework of the security system of enterprise development should be based on a set of economic and legal principles. Within the framework of the imperative principles [16]:

1) protection by the state of all subjects of ownership and management;

2) ensuring equal conditions for their rights and guarantees of equality before the law;

3) independence of the development of business entities action program - in the management of financial risks within the security system of the enterprise must adhere to the following basic principles of economic activity, namely:

1) awareness, manageability, risk management efficiency, which involve taking risks only those that are subject to neutralization, and the costs of their neutralization should not exceed the sum of possible financial losses on it, even with the highest degree of likelihood of occurrence of a risk case;

2) comparison of the level of risks with the level of profitability of financial transactions and the level of financial capacity of the enterprise;

3) independence of management of individual risks of the enterprise, that is, financial losses do not depend on each other on individual risks.

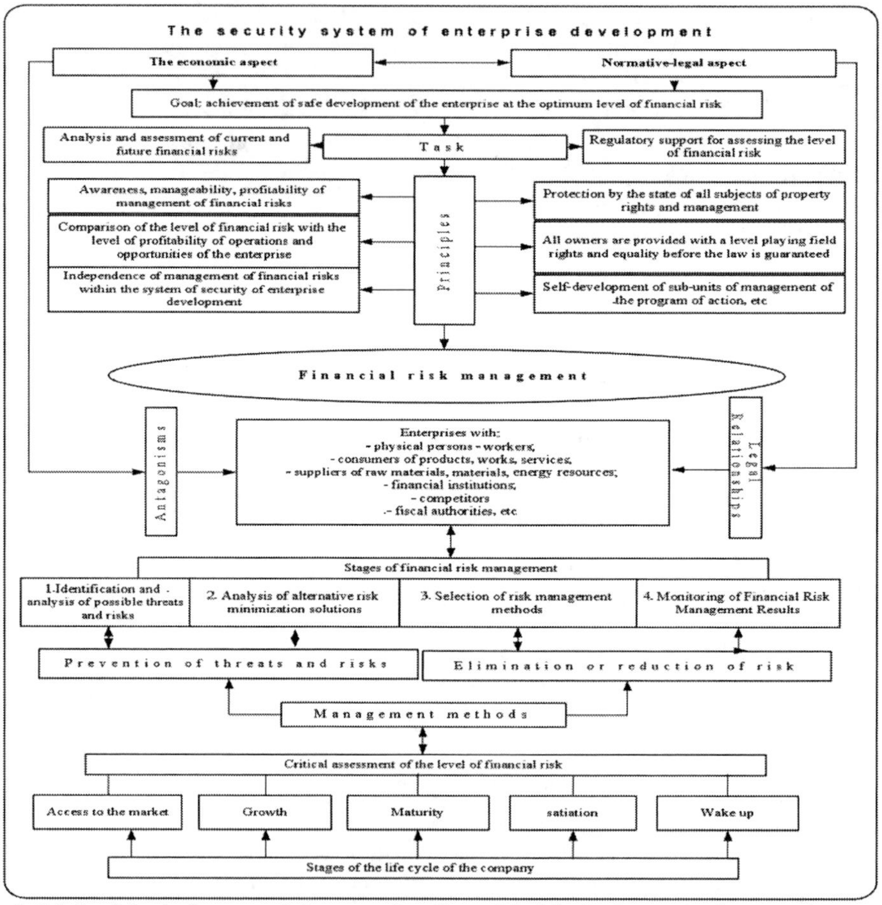

Fig 1. Methodological approach to managing financial risks in the enterprise security system

Adhering to the proposed principles for financial risk management, the company forms legal relationships with various entities with pronounced antagonisms between them, which, in turn, affect the emergence and/or aggravation of existing threats and risks in the relationship. However, today in Ukraine, the state authorities, only legal documents that partially regulate the production risks of certain sectors of the national economy (customs, building legislation, etc.) are developed, and the risks of the financial sector, which is mediating for enterprises (insurance activities [2] , banking [3], securities market activity, audit), which exacerbates the contradictions between market participants as a result of the unilateral protection of their activities, thus restricting the rights of economic entities. Therefore, the real conditions of existence require from business entities sound decision-making related to financial risks and support of a comprehensive regulatory framework for regulating their management.

2.2 Management of financial risks in the stages of the life cycle of the enterprise

Each stage in the life cycle of an industrial enterprise has certain causes of the risks and must have effective measures to minimize or overcome them. However, there are risks that accompany the company at all stages of the life cycle (table). One of them is the risk of quality and incomplete information. This risk is due to uncertainty or asymmetry of information, since in a competitive environment, industrial enterprises have a desire not to disclose all available information. This leads to multicriteria and conflict in the assessment of events, when one has to consciously make compromise solutions.

Table 1 Financial risks in the life cycle of the enterprise

Stage of Life Cycle	Probability of occurrence of risks	Specific financial risks	Economic methods of managing financial risks	Regulatory and legal support for financial risk management
Access to the market	Emergence of threats	Investment; credit; tax; inflationary	Prevention of threats and risks: statistical; expert assessments; analogies; "tree of solutions"; insurance, etc.	Constitution of Ukraine The Commercial Code (Chapter 4, Articles 42-51; Part 1, Article 142) [16] Art. 8 of the Law of Ukraine "On Employment of the population" [5] Art. 17 of the Law of Ukraine "On the basis of social protection of the disabled in Ukraine" [8] Art. 1 of the Law of Ukraine "On the Procedure for Settlements in Foreign Currency" [9] Civil Code [11] "On Fundamentals of National Security of Ukraine" [7], "On Protection of Economic Competition" [6] financial and legal regulation of legal relations
	The onset of a risky situation		Risk Remedy: Sensitivity analysis; script analysis; "Tree of failure"; expediency; utility function, etc.	
Growth (expansion of activity)	Emergence of threats	Percentage monetary; credit; loss of solvency; reduction of financial stability; innovative; tax; inflationary	Prevention of threats and risks: analytical; "tree of solutions"; game theory	
	The onset of a risky situation		Elimination of risks: "tree of failures"; securitization; insurance, etc.	
Maturity	Emergence of threats		Prevention of threats and risks: analytical; "tree of solutions"; game theory	
	The onset of a risky situation		Risk Remedy: Kornai Software Modeling; marketing matrices; scoring programs	
Saturation	Emergence of threats		Prevention of Threats and Risks: Simulation of Monte Carlo; marketing matrices, etc.	
	The onset of a risky situation		Elimination of risks: "tree of failures"; insurance	
Leaving the market	Emergence of threats	Investment; tax; credit; loss of solvency; reduction of financial stability; price	Prevention of Threats and Risks: Risk Map; analogies; game theory	
	The onset of a risky situation		Elimination of risks: "tree of failures"; insurance	

Along with the generally accepted types of threats and risks of an enterprise, there are some and specific ones. The main specific types of systematic (market) financial risks at the stage of entry of the company into the market can be attributed:

- investment risk, characterized by the possibility of financial losses in the process of investment activity of the enterprise;

- the credit risk that occurs in the financial activity of the enterprise in providing it with a commodity loan to buyers for further establishment of relations with them;

- tax risk arises as a result of changes in legislative acts regarding the probability of introducing new taxes and fees, changes in the procedure for charging and timing of payment of existing obligatory payments, cancellation of tax benefits, etc .;

- inflationary risk is characterized by the ability to depreciate the real value of capital and reduce revenues from the results of financial activity of the enterprise in the conditions of inflation.

At the stages of growth, maturity, saturation, there may be specific financial risks such as:

- interest rate risk arises as a result of an unexpected change in the interest rate on the financial market;

- currency risk arises in the course of foreign economic activity as a result of changes in the exchange rate of foreign currency;

- the risk of reducing the structure of capital, which leads to imbalances in the volume of positive and negative cash flows of the enterprise;

- the risk of insolvency of an enterprise is characterized by a decrease in the level of liquidity of circulating assets;

- innovative risk associated with the introduction of new financial technologies, the use of new financial instruments, etc.

Possessing reliable information about the legal relationship between different market players, it is possible to predict more or less probable threats and risks. Estimation of the probability of occurrence of risk is possible with the use of qualitative and quantitative methods and techniques, and depending on the stage of the life cycle of the enterprise, they may differ. Considering that the company's financial risk is an indicator of complexity that combines a large number of external and internal factors, or simple risks [1, 13], we propose a model for an integrated assessment of a company's financial risk in the enterprise security system.

First of all, it is necessary to determine the frequency of occurrence of some level of losses:

$$p = n / n_{zg} \quad (1)$$

where p is the frequency of occurrence of some level of losses;

p - number of cases of occurrence of a specific level of losses;

p_{zg} - the total number of cases in the statistical sample, which covers all, in that account, and successfully carried out operations.

It is expedient to determine the degree of risk in absolute terms for the characteristics of certain types of losses, and for relative - when comparing the projected level of losses with real, medium-sized, average in the market segment, etc.

In absolute terms, the degree of risk (the measure of expected failure in achieving the goal) can be defined as the product of the probability of failure (undesirable consequences) by the magnitude of these undesirable effects (losses, payments, etc.) [1]:

$$Rj = Bj \times pj \quad (2)$$

where Rj is the magnitude of the risk;

Bj is amount of losses (losses);

pj is probability of unwanted risks.

The probability (p) with a sufficient degree of accuracy is calculated on the basis of statistical data. To evaluate the actual data, all probabilities take the same and are determined as follows:

$$P = 1 / n \quad (3)$$

The average expected value is associated with the uncertainty of the situation, it is expressed in the form of the weighted average of all possible results where the probability of each result (p) is used as the frequency or weight of the corresponding value (B).

Then, hopefully, the most probable value of losses (losses, income, profit) - will be:

$$B_u = \sum_{i=1}^{n} B_i \times p_i \quad (4)$$

where n - number of cases;

B_i - the magnitude of losses (losses, income, profit) in the second case;

p_i -is the probability of occurrence of the th case.

The average expected value is a generalized quantitative characteristic and does not allow a decision to be made in favor of any option. For the final decision it is necessary to measure the fluctuations of the indicators, that is, to determine the extent of the fluctuations of the possible result. Fluctuations of the possible result are the degree of deviation of the expected value from the mean value. For its determination, usually the variance is calculated, or the mean square deviation.

The dispersion is the weighted average of the squares of deviations of actual results from the mean value:

$$\sigma^2 = \sum_{i=1}^{n} (B_i - B_u)^2 \times p_i \quad (5)$$

The mean square deviation is calculated by performing statistical tests of various hypotheses, as well as for identifying the relationships between random variables. This statistic is the most common type of deviations used in calculating quantitative risk indicators and is determined by the formula:

$$\sigma = \sqrt{\sigma^2} = \sqrt{\sum_{i=1}^{n} (B_i - B_u) \times p_i} \quad (6)$$

The coefficient of variation (g) that allows you to compare the fluctuations of the signs having different units of measurement. The higher the coefficient of variation, the stronger the oscillation of the sign.

$$r = \sigma / B_u \quad (7)$$

Based on the results of calculating the value of the coefficient of variation, the following scale can be used to assess financial risk:

0,0-0,1 - minimal risk;

0,1-0,25 - low risk;

0,25-0,50 - acceptable risk;

0,50-0,75 - critical risk;

0,75-1,0 - catastrophic risk.

Thus, real business conditions require industry to make informed decisions about financial risks. Therefore, before taking a managerial decision regarding the further development of the enterprise, it is necessary to establish which risk the group, type and type will have to deal with, as well as quantify the degree of potential hazard from the occurrence of a risk event.

References

1. V.V. Vitlinskyi, H.I. Velykoivanenko, *Rizikologsya v ekonomitsi ta pidpriemnitstvi: monogr*, Kiyv, 480 (2004).

2. *Zakon Ukrayini «Pro strahuvannya»* [Tekst] : za stanom na 10 bereznya 2005 roku / Verhovna Rada Ukrayini. – Ofitsiyne vidavnitstvo. – K. : Parlamentske vidavnitstvo, 2005. – 42 s. – (SerIya «Zakoni Ukrayini»). – ISBN 966-611-356-2.

3. *Pro banki i bankivsku diyalnist (iz zminami ta dopovnennyami)*: Zakon Ukrayini vid 07.12.2000 №2121 – III. [Elektronniy resurs], rezhim dostupu http://zakon2.rada.gov.ua/laws/show/2121-14

4. *Pro zaynyatist naselennya* : zakon Ukrayini vid 1 berez. 1991 r № 803-XII // Vidomosti Verhovnoyi Radi URSR. – 1991. – №14. – st. 170.

5. *Pro zahist ekonomichnoyi konkurentsiyi* : zakon Ukrayini vid 11 sIch. 2001 r. №2210-III // Vidomosti Verhovnoyi Radi Ukrayini. – 2001. – №12. – st. 64.

6. *Pro osnovi natsIonalnoyi bezpeki Ukrayini*: zakon Ukrayini vid 19 cherv. 2003 r. №964-IV // Vidomosti Verhovnoyi Radi Ukrayini. – 2003. – № 39. – st. 351.

7. *Pro osnovi sotsialnoyi zahischenosti invalidiv v Ukrayini* : zakon Ukrayini vid 21 berez. 1991 r. №875-XII // Vidomosti Verhovnoyi Radi URSR. – 1991. – № 21. – st. 252.

8. *Pro poryadok zdiysnennya rozrahunkiv v inozemniy valyuti* : zakon Ukrayini vid 23 veres. 1994 r. № 185/94-VR // VIdomostI Verhovnoyi Radi Ukrayini. – 1994. – №40. – st. 364.

9. K.D. Semenova, *Riziki dIyalnostI promislovih pidpriemstv: integralne otsinyuvannya. Monografiya,* Odesa, 234 (2017).

10. *Tsivilniy kodeks Ukrayini* : zakon Ukrayini vid 16 sich. 2003 r. №435-IV // Vidomosti Verhovnoyi Radi Ukrayini. – 2003. – №40. – st. 356.

11. Yu.H. Kim, *Finansovi ryzyky v systemi finansovo-ekonomichnoi bezpeky pidpryiemstva,* Finance of Ukraine, 6, 13-18 (2005).

12. V.V. Vitlinskyi, *Kontseptualni zasady ryzykolohii u finansovii diialnosti.* Finance of Ukraine, 3, 4-9 (2003).

13. V.Z. Kuzminskyi, O. Chynarov, *Finansovi ryzyky subiektiv hospodariuvannia.* Mizhnarodnyi naukovyi zhurnal «Internauka», 2 (24), 2 t., 97-101 (2017).

14. A. Kovalchuk, *Finansove pravo: problemy utverdzhennia ta efektyvnoho zastosuvannia,* Publichne pravo, 1 (25), 10-17 (2017).

15. V.V. Kovalenko, *Finansovi ryzyky ta shliakhy yikh minimizatsii (pravovyi aspekt)*, Kyiv, 319 (2011).

16. *Gospodarskiy kodeks Ukrayini vid 16.01.2003 № 436-15,* (2003): http://zakon4.rada.gov.ua/laws/show/436-15 . – Nazva z ekranu.

Economic and mathematical models of development of the railway tourism

Oleksandr Pshinko[1], *Larysa* Martseniuk[1]1, *Volodymyr* Bobyl[1], *Olena* Kakhovska[2]

[1]National University of Railway Transport named after Academician V. Lazaryan, Dnipro, Ukraine
[2]Prydniprovs'ka State Academy of Civil Engineering and Architecture, Dnipro, Ukraine

Abstract. The authors used a complex of mathematical models meant for the optimal investment of the processes of creation and use of the railway transport infrastructure. The specific features of these models are as follows: long-term period of realization of the infrastructure projects, many participants with special interests, which do not always coincide, a lot of factors of uncertainty connected with the choice of quantities of the calculation parameters, etc. The authors created economic-mathematical models of the discrete optimal planning of the railway tourism operations. This takes into account conditions of risks and cooperation, and allows to determine sets of effective routes which are most profitable ones, provided they meet the established demands. The latter includes routes which meet maximum criteria of the net costs under the conditions of the established system of demands for the activity of the tourism operators.

Introduction

Nowadays tourism is one of the most fast-developing, dynamic and manifold branches of business all over the world. The profit of enterprises with tourist services is been fast growing [1]. Rail freightage is tightly connected with tourism, and in many cases rail freightage is multimodal. This article mainly concerns the problems of the railway usage as a basis for the development of the tourism potential in the regions.

Brief Literature Review

A number of contemporary scientific publications are dedicated to the problems of studying and planning factors that influence tourists' preferences and their trips [2-4]. Methods and means of providing tourism business growth in different regions, tourist's safety, ecological problems, etc, are also widely discussed [5-7]. At the present moment, a lot of railways are being closed because of less volume of passenger transportation and, consequently, unprofitability [8].

*1 Corresponding author: rwinform1@ukr.net

The analysis of the literature on the theme allows us to state that interaction of railway and automobile transport is rational both from the point of view of safety and ecology of multimodal travelling. On the continent these kinds of transport are rather convenient and accessible as for the price parameters. It should be mentioned that in the contemporary scientific literature the problems of formation and development of the tourist infrastructure have not been completely investigated yet. At the same time formation of the new tourist routes which can be attractive for all the stakeholders is a topical issue nowadays.

Results

Models and algorithms of the multi-staged optimal planning of the investment to the tourist activity.

The set of economic-mathematical models of investment and planning of railway tourism is developed. These models focus on the possibilities of the changes of project realizations, restrictions of the uncertainty factors as risks and investors cooperation as well.

We are to give an analytical representation of the objective function (index NPV) and the criterion of the choice touristic routes.

$$NPV = \sum_{t=0}^{T} \frac{\left(Д_t - В_t\right)\left(1 - \frac{\gamma}{100}\right) + A_t \frac{\gamma}{100} - K_t}{\left(1 + E_m\right)^t} + R \rightarrow \max, \qquad (1)$$

In this NPV is given net cost, (hryvnia.); D – annual profit that tourist companies can obtain from all kinds of activities in different scenarios, hryvnia.; B_t – total annual expenses in all kinds of activities in different scenarios, hrv.; before the amortization deductions, hryvnia.; Y – the value of the income tax, %; A_t – amortization deductions; K_t – annual investments for railway tourist transportation in different scenarios, hryvnia; T – the number of the accounted year, t=0,1,2 … (T is a term of the realization of the railway tourist transportation); R – reversion, or investment that gives profit after it has been justified; E_m – real, or modified discount stake.

In the model the general criterion of the optimum planning takes the form (1); the calculated parameters and restrictions are given below.

The data structure for separate routes is as follows:

$$St_i\left[M_i, Rw_i, l_i, T_i, SP_i(t), Z_i, S_i, RTur_i\right], \qquad (2)$$

In this M_i – tourist route identificator; RW_i – list of the railway hubs belonging to M1: L_i – length of the route; T_i – duration of the route; $SP\,i$ (t) – evaluation of the necessary expenses for M introduction before functioning for t period; Z_1 – variants of realization; $S\,i$ – variants of calculated costs for tourism service; RTur – the planned number of tourists on the route M;

We are to study an introduction of every railway route as a project Pr_i (t).

In order to establish the optimal succession of the tourist routes introduction into the development project Pr_i (t), we are to form the following matrix of the economic expediency of the investments and operations in the stages.

$$E(\Pr, t) = \left|e(i, t)\right|. \qquad (3)$$

In this, elements e (i, t) determine efficiency for (2) project concerning routes M_i, if they start in the period t. It is necessary to stress that in the matrix all the elements e (i, t) are calculated due to the number of realization variants Z+K according to the matrixes Zt, Zik E Zt.

The general scheme of the algorithm of the model realization for the variant B consists of the following:

1. To determine the initial values of the variables and the evaluation of the objective function.

$$\bar{X}(0)=(x_1=0, x_2=0, \ldots, x_m=0);\ NPV_{1*}(\bar{X})=NPV_{1\min};$$
$$\bar{X}_{opt}=\bar{X}(0).$$

2. To determine the initial values of the calculator of the variants CN_x =0. It is important that the binary representation of the number CN_x gives a variant of the tourist activity realization x; E {0, 1}, (I = 1, m).

3. To generate the number of the order of the variant CNx+1; in case CN_x +1 is more than N_x = M to finish the optimization procedure (p. 9).

4. To form a current vector of activity variants on the basis of CN_x . In it the value x_1=1 testifies to the introduction of the route "i" into the current plan (for

B1 the variants with one chosen route are used, ($\bar{x}$ (i)=1).

5. To determine a calculator of the number of the realization variants for every x1=1, according to the matrixes Zt, Zik E Zt.

6. To calculate the indices of the model.

7. To revise the restriction system. In case of non-correspondence to the requirements, move to p. 3.

8. To compare the previous value of the objective function $NPV_1(\bar{x})$ with the current $NPV_1(\bar{x})$. With $NPV_{1*}(\bar{X})<NPV_1(\bar{X})$ to change $NPV_1(\bar{x})$ to remember $x_{opt}=\bar{x}$. Move to p.3.

9. Delivery of the results of the optimal planning

$$\{\bar{X}_{opt};\ NPV_{1*}(\bar{X}_{opt})\}.$$

The given algorithm scheme is used for calculating models of all the representations B1–B6.

The models of the route choice as for the conditions of risks and cooperation. We are to analyze the task of planning and the choice of the succession of the route introduction in calculating stochastic factors $\theta=\theta_1, \theta_2, \ldots, \theta_s$. That is the plurality of the sporadic states, which determine a certain foreseen scenario of the denial realization (disturbances in the processes of gauge development, purchasing rolling stock, lack of timely investment, etc.) Such planning models can occur under the risk conditions when probabilities are known concerning possible state disturbances or system parameters; representation variants B2, B4 [9; 10]. In forming models in this case for every possible disturbance the denial scenarios V

$_k$ are determined, which will be described with the separate template, that is Hk (V_k, H_n, P_k).

The latter is necessary for the compensation of disturbances and their probability P_k. The value $\{P(\theta_i)\}$ established by the expert way is considered known. While states θ of disturbances will be determined as ranges $[d_i^1, d_i^2]$ of the values of some deviations of the totality of plan indices,

$$\theta_i = \left\langle \left[d_i^1, d_i^2\right], h_i(\theta_i), p(\theta_i) \right\rangle; \; \sum_i p(\theta_i) = 1, \tag{4}$$

In this $h_1(\theta_i)$ we ask evaluations or the additive expenses for plan correction under the conditions θ_i, $\theta = \theta_1, \theta_2 \ldots\ldots \theta_3$ and functions of the additional expenses formalize representations of the two-stage tasks of railway tourism planning. The two-stage planning model can be generally presented in the way of

$$\left\{\Phi\left(\bar{X}\right) = NPV_1\left(\bar{X}\right) + M\left[f_h\left(\bar{X}, Y\left(\bar{X}, \theta\right), \theta\right)\right]\right\} \Rightarrow \max_{X \in G_X}. \tag{5}$$

$NPV_1(\bar{x})$ – the determined function – cost evaluation of the planning vector;

$f_h(*)$ – the function of the additional expenses, which are necessary for plan realization under conditions of θ_i $Y(X, \theta)$;

$M[*]$ – the sign of the mathematical expectation while realizing (5) by means of the stochastic programming [11] for any $X^1 \varepsilon Y_x$ and for every $\theta_1 \varepsilon \theta$ we should calculate and generalize every $P(\theta)$ value $f_h(X_1^1 Y_1^1 \theta)$, which together with NPV(x) gives $X^1 : \Phi(x^1)$, that calculate the quality $X^1 \varepsilon Y_x$ in the two-stage planning (5).

Correspondingly (5), the optimal development $\bar{x}_{opt}$ ensures maximum index of the given net value under the conditions of the expectations of the additional expenses in the disturbances (4) it the processes of railway tourism realization. In the above given algorithm the model B1 realization can be also used for the numerical realization of representations B2–B6.

Because all the representations, analysis models and successions of the railway routes are constructed by means of the simple connection of all the decision variants and resources restrictions, the forms of criterion in the determined case BC are as follows:

$$\left\{\Phi\left(\bar{X}\right) = \sum_k NPV_k\left(\bar{X}\right)\right\} \Rightarrow \max_{X \in G_X}, \tag{6}$$

Under the conditions of risks B4 are as follows:

$$\left.\begin{array}{l} \left\{\Phi\left(\bar{X}\right) = \sum_k v_k\left(NPV_k\left(\bar{X}\right) + M\left[f_{kh}\left(\bar{X}, Y\left(\bar{X}, \theta\right), \theta\right)\right]\right)\right\} \Rightarrow \max_{X \in G_X}; \\ \sum_i v(\theta_i) = 1. \end{array}\right\}$$

(7)

Route numbers are represented with the index K in (6), (7). Those route members are included into the group of the optimal route numbers. All the other planned models (6), (7), which are indicated, save the values, which were established before. The model CR differs in the frequencies of the trip numbers per year in every route that is included into the optimal group. The frequencies of the trip realization are calculated for the tourism routes in which the maximum general index level is saved NPV (x). In (7) the condition of norming $\sum_i v(\theta_i) = 1$ meets those route variants from the variable vector which are included into the optimal plan X opt. The task of the linear programming, as for determination of the optimal number of the separate routes realization, is formed for calculating optimal values of the frequencies $\sum_i v(\theta_i) = 1$ It is necessary to choose separate roots in the optimal cooperative plan X opt with the optimal values mentioned above.

Multi-criteria multi-stage planning.

The given planning models of the railway tourism regional development are formed provided there are definite conditions of interests of the united investment center. That is why the maximum values of the total income are represented as optimum criteria. The models mentioned above do not envisage participation in the project and competition with several investors (UH(K)) who are interested in the urgent development of the different tourism routes. We are to consider briefly some problems of formalization of the competition mechanisms and their efficiency analysis. First of all, the given criteria are based on the rule of the external solution of the conflicts among UH (K). In this case, the mechanism of the redistribution of the obtained maximum income is established outside the planning model. The creation and realization of such a mechanism is connected with the solution of many non-formal problems, which are not discussed in this paper. The game principle of the guaranteed result [11] is offered to be used as a method of conflict solution UH (K), which realize the demand of the equality of the infrastructure investors

$$F_C(V) = \max_V (\theta = \min_k (V_k / D_k)) = \theta_C , \quad (8)$$

where V – is a value of K-investor's income, and Dn- the volume of their investment. The realization of the planning model with the criteria (8) permits to choose tourists routes and the trip frequencies which provide income UH (K) corresponding to their contribution in the project of the railway tourism development.

The results of our planning models realization include the task of the succession of the railway tourism routes due to the present and future infrastructure, availability of the rolling stock etc, which provide maximum planned income for every participant in the tourism operation for a definite period. The calculation samples, which are represented here, show certain possibilities of the proposed models for the choice of the succession of the railway tourism routes as a factor of the development of the regional tourism potential.

References

1. Orçun AYDINa. Tourism Income of Turkey: A Panel Data Approach. Procedia Economics and Finance 38 (2016), pp. 245–256. Available online at www.sciencedirect.com doi: 10.1016/S2212-5671(16)30196-4.

2. Md. Samsul Alam, Sudharshan Reddy Paramati. The impact of tourism on income inequality in developing economies: Does Kuznets curve hypothesis exist? Annals of

Tourism Research. Volume 61, November 2016, pp. 111-126. https://doi.org/10.1016/j.annals.2016.09.008

3. Peter Josef Stauvermann, Ronald Ravinesh Kumar. Productivity growth and income in the tourism sector: Role of tourism demand and human capital investment. Tourism Management. Volume 61, August 2017, pp. 426-433. https://doi.org/10.1016/j.tourman.2017.03.006

4. Kevin X. Li, Mengjie Jin, Wenming Shi. Tourism as an important impetus to promoting economic growth: A critical review. Tourism Management Perspectives. Volume 26, April 2018, Pages 135-142. https://doi.org/10.1016/j.tmp.2017.10.002

5. Chiang-Ming Chen, Yen-ChienChen. Income source effect on retiree's tourism behavior. Annals of Tourism Research Volume 68, January 2018, Pages 4-6. https://doi.org/10.1016/j.annals.2017.10.002

6. Amin Sokhanvar, Serhan Çiftçioğlu, Elyeh Javid. Another look at tourism- economic development nexus. Tourism Management Perspectives. Volume 26, April 2018, Pages 97-106. https://doi.org/10.1016/j.tmp.2018.03.002

7. LiangWang, BinFang, RobLaw. Effect of air quality in the place of origin on outbound tourism demand: Disposable income as a moderator. Tourism Management Volume 68, October 2018, pp. 152-161. https://doi.org/10.1016/j.tourman.2018.03.007

8. Guerrieri, M., Ticali, D., 2012. Design standards for converting unused railway lines into greenways. ICSDC 2011, 654–660. http://dx.doi.org/10.1061/41204(426)80

9. O.M. Pshinko, U.S. Barash, V.V. Skalozub, L.V. Martseniuk. Economic-mathematical model of formation of the sphere of railway tourism in Ukraine.Collection of scientific papers Dnipropetrovsk. National University of Rail Transport named after Acad. V. Lazaryan «Problems of Transport Economics». 2017, (13): pp. 83-96. (In Ukr.).

10. U.S. Barash, V.V. Skalozub, L.V. Martseniuk. Multistage investment model for the development of rail tourism in Transcarpathia. Investments: practice and experience. 2017, (14): pp. 5-13. (In Ukr.).

11. Zirka V.V. Manipulative games in advertising: Moscow, 2014, 256 p. (In Russ.).

Improving the methodology for assessing labour potential in socio-economic systems based on the organizational and methodological model

Maryna Savchenko[1*], *Larysa* Shaulska[2], *Olga* Shkurenko[3*] *and* *Larysa* Sarkisyan[1]

[1]Vasyl Stus Donetsk National University, International Economic Relations Department, Akademik Yangel st. 4, 21007, Vinnytsa, Ukraine

[2]Vasyl Stus Donetsk National University, Management and Behavioural Economics Department Akademik Yangel st. 4, 21007, Vinnytsa, Ukraine

[3]State University of infrastructure and technology, Business Logistics and Transport Technologies Department, Ogienko st. 19, 03049, Kyiv, Ukraine

Abstract. The article examines the methodological approaches to the assessment of the labour potential of the socio-economic system. On the basis of the research, it has been established that the proposed algorithm for assessing the labour potential of the socio-economic system at the micro level. The approbation of the methodology for assessing the labour potential of the socio-economic system was carried out on the basis of the performance of an industrial enterprise. To improve the methodology for assessing the labour potential of socio-economic systems, the use of an organizational and methodological model for assessing the labour potential of the system has been proposed. The use of the organizational and methodological model is carried out in several stages: the construction of the Ishikawa diagram, the rating score, the construction of the Pareto diagram and the conduct of the ABC analysis. The implementation of the organizational and methodological model for assessing the labour potential of the socio-economic will allow for the effective management of the labour potential of the system and, as a result, will increase its competitiveness in the information economy.

1 Introduction

The effective functioning of any socio-economic system depends on the human resources at its disposal. Describing human resources as a human potential for labour, most scientists associate this category with the concept of "labour potential" - the possible quantity and quality of labour that society (labour collective) has at a given level of science and technology development [1, p. 324].

The economic category "labour potential" was formed in the 1970s to express a generalizing assessment of personnel, taking into account not only its actual characteristics but also reserves and potential.

* Corresponding author: m_savchenko@donnu.edu.ua, dondyy@ukr.net

The most developed under the labor potential of the socio-economic system is based on the professional knowledge, practical skills, personal and psychophysiological characteristics of workers, the ability to make the required contribution to the realization of the enterprise's goals, with proper material, technical, informational and organizational support of the labor process [2, p. 20].

Since its inception, this category has undergone an evolutionary path of development from its complete identification with labour resources to its presentation as a set of abilities to work. However, the question of the development of a practically reasonable methodology for the effective assessment of the labour potential of a socio-economic system at any level remains important.

2 Degree of Problem Development

The development of methods for analyzing the labour potential of the enterprise was carried out by many leading scientists. In particular, scientists A. Ledneva [3] dealt with the analysis and evaluation of the labor potential of industrial enterprises, E. Lipatova, Y. Yukhtanova [4], V. Potudanskaya, L. Trunking [5], A. Fleenko [6], N. Hadasevich [7].

Despite the development of methods for analyzing the labour potential of the socio-economic system, many issues remain unresolved. This fact encourages us to conduct further theoretical, methodological and practical research in this scientific direction.

The purpose of the study is to develop the methodology, theoretical, methodological and practical recommendations and approaches to the assessment of the labour potential of the socio-economic system.

3 Research results

The labour potential of the socio-economic system should be considered as the synergistic ability of its labour resources to the effective functioning of activities and the achievement of certain productivity of them, which depends on the complex of social, communication, organizational, economic, logistical and other capabilities of the enterprise [8].

A generalized assessment of the labour potential of the socio-economic system should be carried out using the integral indicator ($\boldsymbol{I_{LP}}$), which is determined by the formula:

$$I_{LP} = f\,(T_p;\ T_s;\ T_o;\ T_m;\ T_c) \tag{1}$$

where $\boldsymbol{T_p}$ – the complex indicator of the personnel component of the labour potential;

$\boldsymbol{T_s}$ – complex indicator of the component of social security and security of labour potential;

$\boldsymbol{T_o}$ – the complex indicator of the organizational component of the labour potential;

$\boldsymbol{T_m}$ – complex indicator of the motivational component of the labour potential;

$\boldsymbol{T_c}$ – complex indicator of the creative and intellectual component of the labour potential.

To improve the methodology for assessing the labour potential of socio-economic systems on the basis of the proposed use of the organizational and methodological model for assessing the labour potential of the socio-economic system (Fig. 1).

For practical testing, methodological tools have been proposed and the rationale for each stage of the assessment of the labour potential of the socio-economic system has been carried out.

Stage 1. Construction Ishikawa diagrams.

Ishikawa diagram is a graphical way of researching and determining the most significant causal relationships between factors and consequences in a situation or problem under investigation. This diagram allows you to identify key relationships between various factors and more accurately understand the process being studied. The diagram contributes to the determination of the main factors causing the most significant contribution to the one under consideration and to the prevention or elimination of their actions.

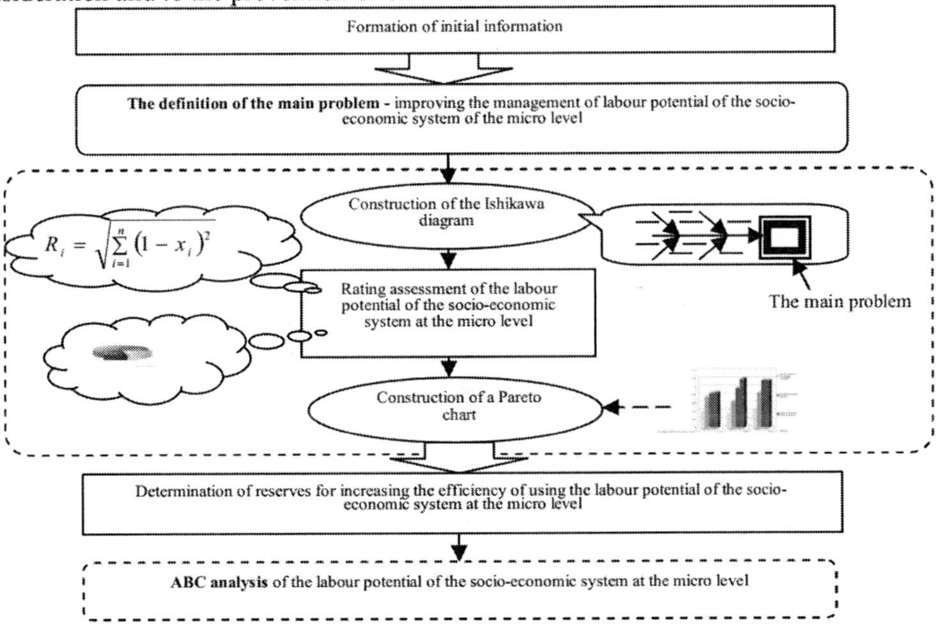

Fig. 1. The organizational model for assessing the labour potential of the socio-economic system

The Ishikawa diagram is used as an analytical tool to view the effects of possible factors and to highlight the most important causes, the effects of which give rise to specific consequences and are manageable.

Note that the Ishikawa diagram is a unique tool that allows you to create a system of indicators to assess the labour potential of the enterprise. Work with the Ishikawa diagram is carried out in several stages:

1. Identification and collection of all factors and causes, in any way, affect the studied result.
2. Grouping of factors by semantic and causal blocks.
3. Ranking of these factors within each block.
4. Analysis of the resulting picture.
5. "Liberation" of factors that cannot be affected.
6. Ignoring insignificant and unprincipled factors.

The study of the policy of management and positioning of the socio-economic system shows that the effectiveness of the system's activity is determined by the level of its employment potential. This, in turn, made it possible to single out the main problem of the system, which is to increase the level of the labour potential of the socio-economic system. The use of the Ishikawa diagram (Fig. 2) allows us to identify the main problem and detailing factors that influence this problem.

Stage 2. Rating evaluation of the labour potential of the socio-economic system.

At this stage, an effective tool for assessing the management of labour potential of the socio-economic system is the "radar of labour potential". "Labour potential radar" is a method of assessing the state of labour potential management, aimed at identifying weak points in the entire system of relations inherent in the management of labour potential.

The "labour potential radar", namely, the graphics-analytical method of analysis, makes it possible to systematically establish quantitative and qualitative links between individual elements of labour potential, its level of development and competitiveness, and on the basis of this, timely justify and implement management decisions to improve the functioning of the socio-economic system. It has several specific theoretical aspects that must be considered when justifying management decisions.

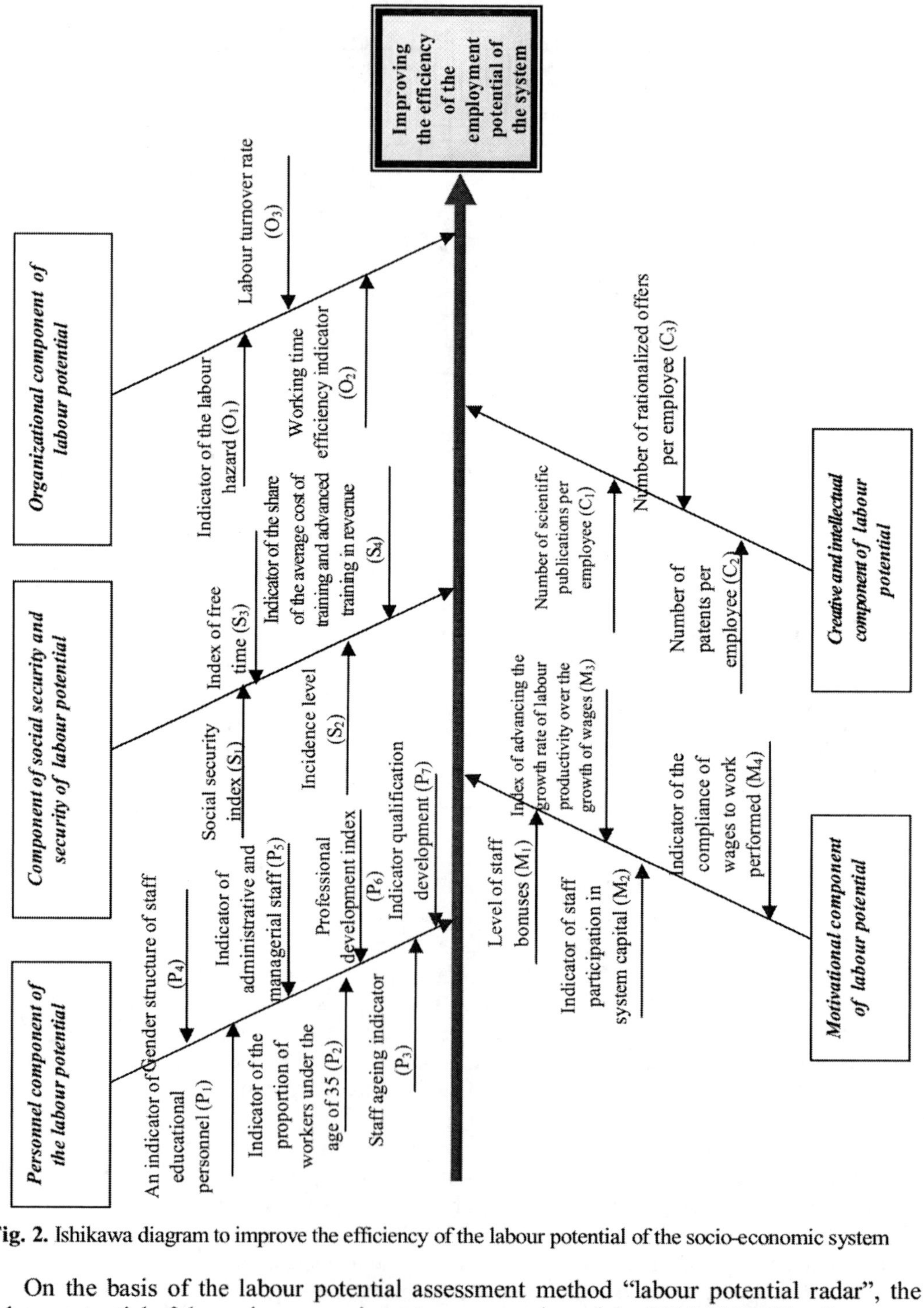

Fig. 2. Ishikawa diagram to improve the efficiency of the labour potential of the socio-economic system

On the basis of the labour potential assessment method “labour potential radar”, the labour potential of the socio-economic system was evaluated for 2015-2018 (Fig. 3).

It should be noted that the first two stages characterize the current state of labour potential, that is, the above economic tools present a general picture of labour potential, respectively, they do not enable to rank the factors in order of importance (priority).

Stage 3. Construction of a Pareto chart.

The search for solutions to these problems begins with their classification by individual factors in order to clarify the main ones, that is, those that are associated, for example, with the greatest costs. To identify the main factors build Pareto charts and then make their analysis. The Pareto diagram makes it possible to distinguish an «important minority» compared with the "unimportant majority".

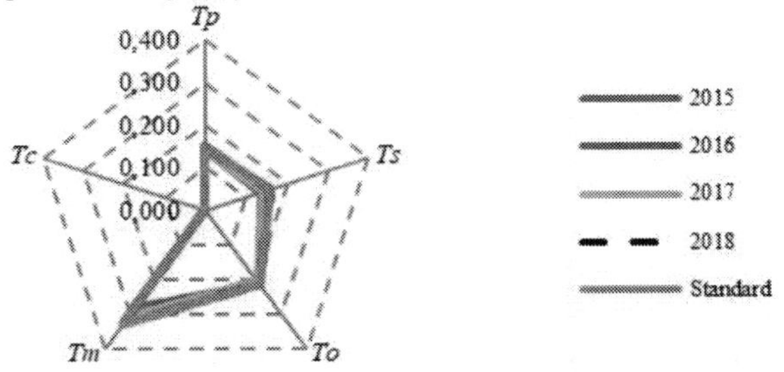

Fig. 3. Radar of labour potential of the socio-economic system in 2015-2018

Thus, the use of the Pareto chart allows modern enterprises to identify 20% of the factors that determine 80% of the total result of socio-economic systems. Based on the Ishikawa chart data, a system of indicators has been defined that characterizes our main problem and calculated the rate of change, and according to the developed scale for evaluating the results obtained, the dynamics of factor changes (Table 1), the points are distributed among the indicators.

Table 1. Scale score for the indicator

Stimulator scale (C), points		Scale de-stimulators (D), points	
0	<100	0	> 150
1	100-110	1	133-150
2	111-121	2	122-132
3	122-132	3	111-121
4	133-150	4	100-110
5	> 150	5	<100

The ranking of indicators is carried out in accordance with the assigned points. At the next stage, the Pareto data for the socio-economic system is calculated by a cumulative total.

According to the principles proposed by Pareto, a graph is plotted: to the left of the intersection point, there are factors that provide 80% of the result.

The results show that the effectiveness of the labour potential of the socio-economic system for 2015-2018 in general, it is determined by an indicator of «ageing» of personnel, an indicator of an occupational hazard, an indicator of labour turnover, an indicator of a business career.

Stage 4. ABC analysis. After identifying the problem by constructing a Pareto chart from the results, it is important to determine the reasons for its solution. It is necessary to solve it. When using the Pareto chart to identify performance and causes, the most common method is ABC analysis.

The results of the ABC analysis are shown in fig. 4.

The calculations made it possible to assert that the indicators of group A are 18,2%, which are managed by the greatest influence on the state of the labour potential of the socio-economic system for 2015-2018, namely: the indicator of «ageing» of personnel, the indicator of occupational hazard, the indicator labour turnover and a business career indicator.

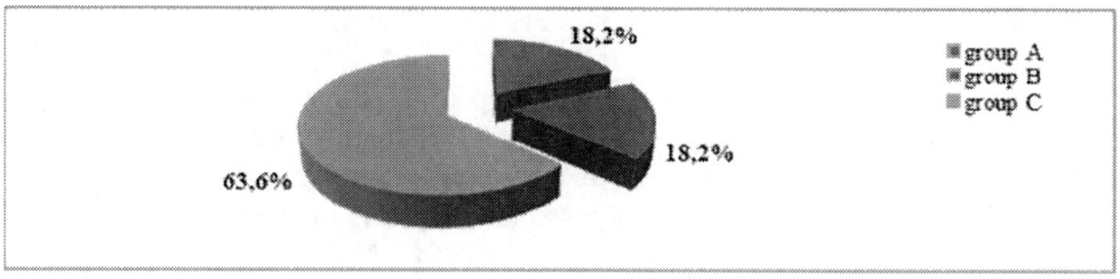

Fig. 4. Summarizing the results of the ABC analysis of the socio-economic system for 2015-2018

Conclusion

Thus, in accordance with the stated goal of the study, the authors of the article developed recommendations for solving theoretical and practical issues and problems related to the formation and development of the labour potential of the system. As a result of the study of a set of methodological tools and methodological tools that will be used in the course of reproduction and adjustment of the state of the labour potential of the socio-economic system, a methodology is proposed for monitoring the labour potential of the system according to a certain system of indicators. The article tested the proposed methodology based on the results of the socio-economic system of the micro level. The formation and development of the labour potential of the socio-economic system involves the creation of conditions for the full use and realization of the labour potential at all levels of management. For this, it is necessary to ensure the improvement of the governance mechanism itself in the socio-economic system through the use of an organizational-methodological model for assessing the labour potential of the socio-economic system.

References

1. Yu. G. Odegov, G. G. Rudenko, *Labour Economics: textbook*, Moscow (2011).
2. S. G. Radko, *Labor potential as a socio-economic category*, St. Petersburg (2010).
3. A. A. Ledneva, *Methodology for assessing the efficiency of using human resources*. Cybernetics, 4-8 (2010).
4. E. D. Lipatova, Yu. A. Yukhtanova, *Study of methods for assessing labor potential and labour resources*. Student Forum: electron. scientific journals, **21 (21)**, (2017).
5. V. F. Potudanskaya, L. V. Trunkina, *Evaluation of the labour potential of the personnel of the enterprise*, Business. Education. Right, Bulletin of the Volgograd Institute of Business, **4 (17)**, 96-101 (2011).
6. A. N. Fleenko, M. O. Barancheev, *Methods for assessing the labour potential of an organization*. Scientific world, 8-14 (2015).
7. N. R. Hadasevich. *Evaluation of labor potential: approaches and methods*, Cybernetics, **6**, 8-16 (2014).
8. O. V. Posilkina, O. A. Yaremchuk, Yu. S. Bratishko, *Modern approaches to the assessment and management of the labour potential of pharmaceutical companies*, Pharmaceutical Journal. K. 5, September-October, 3-9 (2006).

STRATEGY OF DEVELOPMENT OF LOGISTIC INFRASTRUCTURE OF UKRAINE IN THE CONDITIONS OF EUROPEAN INTEGRATION

Olga Shkurenko [1*] and *Maryna* Savchenko [2*]

[1] State University of Infrastructure and Technology, Business Logistics and Transport Technologies Department, Ogienko st. 19, 03049 Kyiv, Ukraine
[2] Vasyl' Stus Donetsk National University, International Economic Relations Department, Akademika Yangel st. 4, 21007, Vinnytsa, Ukraine

Abstract. The article defines the role and place of logistics infrastructure in the development of the country's economy. The division of the transport and logistics centers of the European Union into five groups is substantiated: the main general, the main parts of European importance, secondary regional significance, auxiliary national and auxiliary regional centers. The problems of formation and development of logistic infrastructure of Ukraine are investigated. The state of development of logistic infrastructure of European countries and Ukraine is investigated, and their competitive advantages are determined. A comparative analysis of the transport infrastructure of Ukraine and Poland is carried out. The authors define the principles of forming logistics centers: security, competitiveness, openness, interaction and organizational management, implementation of which will enable the formation of an effective logistic infrastructure of Ukraine and, as a result, will increase its competitiveness in and positions in international ratings.

1 Introduction

The current conditions of Eurointegration, which Ukraine aspires to, demands increasing improvement, growth and attention in the field of freight transportation logistics. The quality and efficiency of freight transport depends on optimization of coordination processes of different types of transport, distribution of volumes of transportation, design of logistics freight transportation systems, and the clear formation of necessary management decisions [1-4]. Today, one of the most important issues for Ukraine is the formation of an infrastructure that, according to international standards, meets economic principles and forms different spheres of activity and could sufficiently satisfy market demand on the basis of reliability, quality, timeliness and minimum costs. That is why studies related to comparing the effectiveness of the Ukrainian logistics infrastructure with international ratings able to provide a continuous flow of all necessary logistics operations are relevant.

* Corresponding author: dondyy@ukr.net, m_savchenko@donnu.edu.ua,

2 Degree of Problem Development

Thorough studies of the logistics infrastructure are generally undertaken by such foreign and domestic scientists as A. Harrison, D. Lambert, E. Krikavsky, N. Chornopyska, M. Dovba, O. Sumets, T. Babenkova, M. Grigorak, V. Kolodin, Y. Kuzmenko, V. Kopytko, N. Savina [1-8]. In their publications, the content of the definition of "logistics infrastructure" is deduced, as well as its components and operating conditions are considered. Despite the significant scientific achievements in the study of infrastructure development, in particular logistics, there are still insufficient research issues regarding the implementation of foreign experience in infrastructure management in domestic practice.

3 Research results

One of the most important meanings for efficient functioning of the economy is the highly developed logistics infrastructure, which would provide a sufficiently low cost of transportation to markets of other regions and countries, reduce the impact on the distance between different regions, and ensure the integration of the national market. A prerequisite for access to high-quality and advanced economic services and processes of the least developed settlements is highly developed transport infrastructure. Highways and railways, airports and high quality ports, i.e., efficient modes of transport, guarantee for entrepreneurs systematic and absolutely timely delivery of their services and products to the market. The highly developed telecommunications network provides a fast and uninterrupted information flow. This allows market players to make optimal decisions for the full amount of available efficiency and improve overall economy performance.

In Western European countries, one of the most significant sphere, is the share of logistics services in the economy. The European countries in which the number of employees employed in the enterprise leads the provision of auxiliary transport services, such as Germany, France, Italy, Spain and the United Kingdom are worth highlighting. In Italy, one of the largest transport and logistics centers in the European Union, called Bologna Freight Village, is located at the intersection of four motorways and five major railways. This transport and logistics center serves about 16% of international cargo and 35% of freight Italian traffic. Bologna Freight Village includes almost 100 logistic and transport companies from all over the world, and includes customs, mail, banking institutions, restaurants and hotels.

As a result of the rich historical practice regarding the introduction of transport and logistics centers and their research within the European Union, the classification of transport and logistics centers was carried out into five categories: main transport and logistics centers of the general European value; main transport and logistics centers of partial European significance; secondary transport and logistics centers of regional importance; auxiliary national transport and logistics centers; auxiliary regional transport and logistics centers. Most of the major transport and logistics centers of the European Union are created with the support from the government. All European transport and logistics centers are subject to a generally accepted structure. In recent years, one of the main pillars of the European transport policy has been the development of piggyback cargo transport, which predicts the operation of transport chains as a single and continuous transport container, based on the principles of logistics. Almost one-third of all international door-to-door freight traffic is carried out thanks to container trains, as evidenced by international practice [8, p.21]. Since the goods are provided with the door-to-door principle, road and rail transport, in particular container transportation, turns from partners into competitors. Ukraine in its development of logistics infrastructure is not in the best position, it is on the 137th place out of 144 in the rating of the international non-governmental organization The World Economic Forum on Road Quality [9]. Rather a big problem of Ukraine is an inadequate quality of most of the transport, and first of all, roads. Even the most developed and modern logistics center cannot provide high-quality and fast delivery of goods, if there is no participation of the organization responsible for the state of the regional highways [2]. According to estimates of transport infrastructure, Ukraine is significantly inferior to highly developed infrastructure leaders, as shown by a detailed analysis of the LPI index. But it is interesting to note that surveys of logistics operators have showed that Ukraine has no problems finding logisticians with high qualifications, while in Germany, about 34-37% have difficulty in finding candidates for managerial positions and almost 41% are in the search for skilled managers in the field of logistics. In comparison with Germany, Ukraine, of course, lags behind in the development of logistics infrastructure, and also shows a negative dynamics, if it is compared even to countries whose incomes are lower than the average (Table 1).

Table 1. – The value of LPI in countries whose incomes are below the average for 2015 and 2017

Country	Place in rating for 2015	Result for 2015	Place in rating for 2017	Result for 2017
India	54	3.08	35	3.42
Kenya	74	2.81	42	3.33
Egypt	62	2.97	49	3.18
Vietnam	48	3.15	64	2.98
Pakistan	72	2.83	68	2.92
Ukraine	61	2.98	80	2.74
El Salvador	64	2.96	83	2.71
Guyana	124	2.46	85	2.67

Source: composed by the authors on the basis of [5]

For a more detailed consideration of the development of the market for logistic infrastructure services, the state of development of the logistic infrastructure of the countries of Europe and Ukraine should be investigated (Fig. 1). In the context of deepening European integration, the role of transport and transport infrastructure is increasing substantially, as labor mobility and the intensification of both international and domestic commodity circulation are increasing. This forces national governments to seek additional resources for the development of transport and the modernization of transport infrastructure. The transport sector and transport infrastructure of Poland are almost the most developed in the whole of Eastern Europe and serve as an example for many states, in particular for Ukraine. Existing potential for development of transport infrastructure objects was effectively used and developed due to its transition to market principles of functioning. Some of the state-owned companies, including the railways of the Polish National Rail Company "Polskie Koleje Panstwowe" (PKP SA), were stocked, others were included in the field of open market competition. The railway industry, which is one of the main carriers of resources, goods and people, both within the country and on an intergovernmental scale, has developed especially successfully. The most loaded are the railways of the Upper Silesian district, which are transported by coal. One third of the railways are electrified, in particular, the most important highways from Silesia and Krakow through Warsaw to Trumuesti (Gdansk, Gdynia, Sopot); from Silesia to Szczecin, as well as along the entire Polish part of the trans-European highway Hamburg-Berlin-Warsaw-Minsk [6].

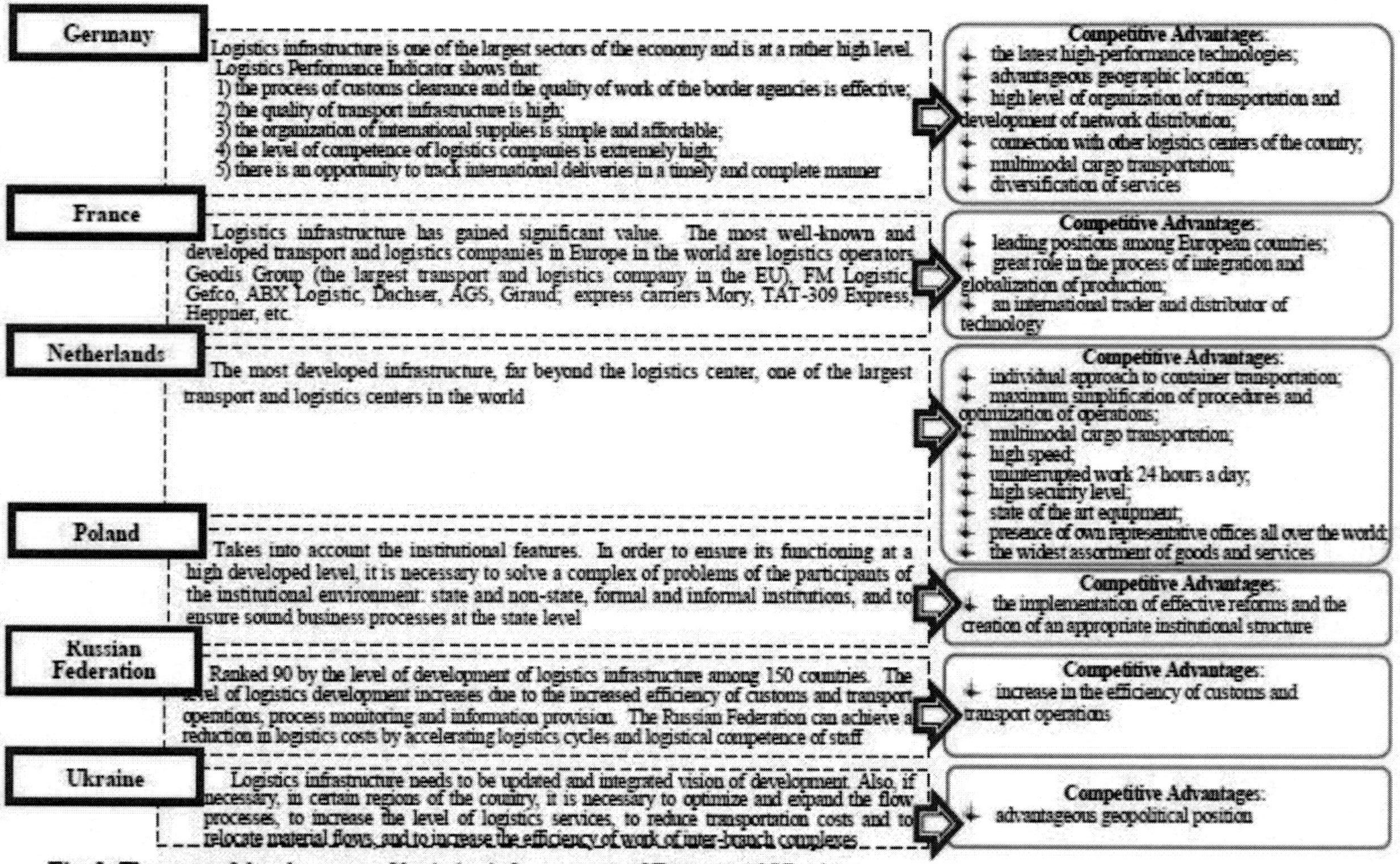

Fig. 1. The state of development of logistics infrastructure of Europe and Ukraine

Thanks to the reforms undertaken in preparation for Poland's accession to the EU, the transport infrastructure of the country has undergone fundamental quality as well as significant quantitative changes. As a result of this, in comparison with Ukraine, modern Poland significantly exceeds our country by most indicators (Table 2).

Table 2. Comparison of certain elements of the transport infrastructure of Ukraine and Poland

Indexes	Ukraine	Poland
Area, thousand km^2	603.7	312.6
Population, million people	42	38
Road length, km	169.4	424.0
Density of roads, km / thousand km^2	280.6	1,355.9
Length of railways, thousand km	21.7	22.3
Electrified railway tracks, thousand km	10.2	11.7
Cargo transportation in 2017, billion tkm	191	31
Number of locomotives (work park)	2,195	1,250
Number of freight cars (working park)	67,000	60,000
Number of employees of state railway companies, thousand people	274	81
Density of railways, km/thousand km ^{2}	35.9	71.4

Source: composed by the authors on the basis of [6-7]

The main problem of Ukraine today is not that the logistics of the infrastructure is deteriorating, but that the provision of general logistics activities in Ukraine is developing much slower than in other countries, this provision of our state significantly reduces the place and rating in the rating.

One can conclude that, if we look at the experience of foreign countries in the creation of transport and logistics centers, for the same quality of construction of centers in Ukraine, the following principles have to be taken into account: cooperation and internal cooperation with transport and logistics centers located on the border with other states or in other regions and districts is required; the proper placement of transport and logistics centers should be observed, so that they are located near airports, railway stations and highways; also transport and logistics centers should be located in those places where high rates of activity of import and export can be clearly observed; transport and logistics centers should be located near the centers of highly developed infrastructure.

Also, in addition to the principles mentioned above, the following principles can be developed for the operation of transport and logistics centers: 1) security - all transport and logistics centers should have in their possession the equipment necessary for transportation of cargo (loading and unloading, tracking of cargo and its control, forwarding, storage and handling of goods, transportation, research, insurance, customs clearance, gas stations, car wash etc.); 2) competitiveness – a large number of logistic, transport and forwarding competing companies in their structure is required; 3) openness – all transport and logistics centers, regardless of their legal and administrative forms, should be accessible to any other companies; 4) interaction - in all transport and logistics centers in order to optimize freight flows, it is necessary to carry out activities related to logistics operations and transport and logistics services in accordance with national and international cooperation; 5) organizational management – transport and logistics centers should be managed by only one governing body – an independent law firm, this company should be located within the transport and logistics center, act on its behalf and protect the interests of the company

Conclusion

If we take into account the tendencies of European countries in the sector of logistic infrastructure, it is immediately clear that Ukraine is significantly inferior in its development, because it only passes through the stage of formation. Nevertheless, despite all the above, one of the most attractive sectors in the Ukrainian investment sector is the transport logistics.

In order to form an effective strategy for the development of Ukrainian logistic infrastructure in the context of European integration, certain principles and methods of making changes should be applied, namely: accounting for a number of territorial specifications, such as climatic and socio-demographic conditions, geographic location; proper formulation of the goals of functioning and development of logistics infrastructure, as well as accounting for the social and economic development of the country; accounting for all features of management, state of development and provision of governmental resources through its specialization; applying the principle of co-operation to regulate the supply of raw materials between regions; using the principle of complexity of development of the subjects of logistic infrastructure, because they should holistically promote prospective directions and development of the state; accounting for non-partial shifts, and general indicators of activity of logistic infrastructure on the basis of realization of such measures for formation of the national logistic infrastructure, which bring the greatest effect.

Therefore, logistics infrastructure has been one of the most important factors for Ukrainian as well as for the world economy as a whole; therefore, the problems of its development and modernization should be approached in an effective and rational way, as well as to investigate the problems of introducing innovations into the logistic infrastructure.

References

1. Harrison Alan, Van Hoek Remko. *Upravlinnia lohistykoiu: Rozrobka stratehii lohistychnykh operatsii*. Per. z anhl.; Za nauk. red. O.Ie. Mikheitseva. – Dnipropetrovsk: Balans Biznes Buks, 368 (2007).
2. Pepich T.A., Velykyi D.Yu. *Optymizicija logistychnoi' infpastpuktupy mizhnapodnyh vantazhnyh perevezen'*. Efektyvna ekonomika. **1.** (2017).
3. Bondapenko O.S. *Obgpuntuvannja sutnosti logistychnoi' infpastpuktupy u finansovyh potokah pidppyjemstv*, Investytsii: praktyka ta dosvid, **8**, 51-55 (2015).
4. Falovych V.A. *Vykopystannia zasobiv lohistychnoi infpastpuktupy v lantsiuhu postavok*, Ekonomika i suspilstvo, **10**, 389-395 (2017).
5. Arvis J.-F. Connecting to Compete 2016. *Trade Logistics in the Global Economy. The Logistics Performance Index and Its Indicators*, Communications Development Incorporated, Washington, 62 (2016).
6. *Transport v Polshchi* [Elektronnyi resurs]. URL: http://www.geograf.com.ua/poland/695-poland-transport/.
7. Kopytko V. *Napriamy rozvytku transportnoi infrastruktury Ukrainy*, Visn. Dnipropetr. nac. un-tu zalizn. transp. im. akad. V. Lazarjana, Dnipro, **41**, 261-265 (2012).
8. Kotenko A.M. *Tekhnolohii i tekhnichni zasoby kombinovanykh perevezen vantazhiv za chynnykamy hlobalnoi*, **1/2(6),** 21-25 (2015).
9. *Ukrainski dorohy u desiattsi naihirshykh u sviti*. URL: http://zik.ua/news/2017/06/06/ukrainski_dorogy_u_desyattsi_naygirshyh_u_sviti_1109301.

Adaptation of Enterprises to Continuous Socio-Economic Changes in a Region

Valentina Smachylo[1], *Veronika* Khalina[1], *Olena* Kolmakova[1*], and *Anastasiya* Ustilovska[1]

[1]National University of Civil Engineering and Architecture, Department of Economics, 61001 Kharkiv Sumskaya street 40, Ukraine

Abstract. The article proposes a mathematical model for assessing development of regional resources. The general model includes components which take into account the onset of crises. This model is the basis of the mechanism for assessing socio-economic asymmetry in development of regions.

Asymmetry in socio-economic development of regions is inherent to any country of the world. It leads to social tension and political destabilization in regions with low socio-economic development, causing an outflow of capital and population from depressed regions to prosperous ones or even other countries. Accordingly, territories which have advantages are increasing their potential, while the development of lagging ones is becoming more and more problematic. The main objectives of the national regional policy of Ukraine include welfare of its citizens, social justice, integrity and sustainable development of the country. The implementation of these goals largely depends on the overcoming of the existing asymmetry in socio-economic development of regions of Ukraine, formation and development of its modern regional structure, and, consequently, preservation of the national unity. It is asymmetry that intensifies resource flows and influences their redistribution among the regions. The issue of reducing the asymmetry in the development of Ukraine's regions is becoming relevant, and this can be done only through inflow of resources to depressed regions. But here, too, certain contradictions arise: with the advent of new "players", enterprises working in the region are threatened; new "players" have certain difficulties when entering a new market; at the country level, an inflow to one region causes an outflow from another.

Therefore, there is an urgent need to overcome the asymmetry not only in the regional development, but also in the socio-economic system for redistributing resources among the regions. Harmonious development of a region directly depends on success in the development of its enterprises. If we take, e.g., labour resources of the region, the majority of its population (about 44% of the total population and 90% of the economically active population) are employed in the regional enterprises. This means that the qualitative and quantitative characteristics of labour resources of the region are shaped by stimulating and motivation mechanisms of its enterprises. It should be noted that the formation of labour resources in a region depends on the policy of forming labour potential of enterprises in the region. Positive trends in development of the economy, increase in the number of enterprises and expansion of the existing ones create additional jobs, and, accordingly, contribute to an inflow of labour resources into the region (Fig. 1).

* Corresponding author: elenkolmakova78@gmail.com

Whatever the resources of the region are, their formation is directly affected by enterprises working in it. To connect two socio-economic systems, the authors propose to build a model to forecast and assess development of resources of an individual region.

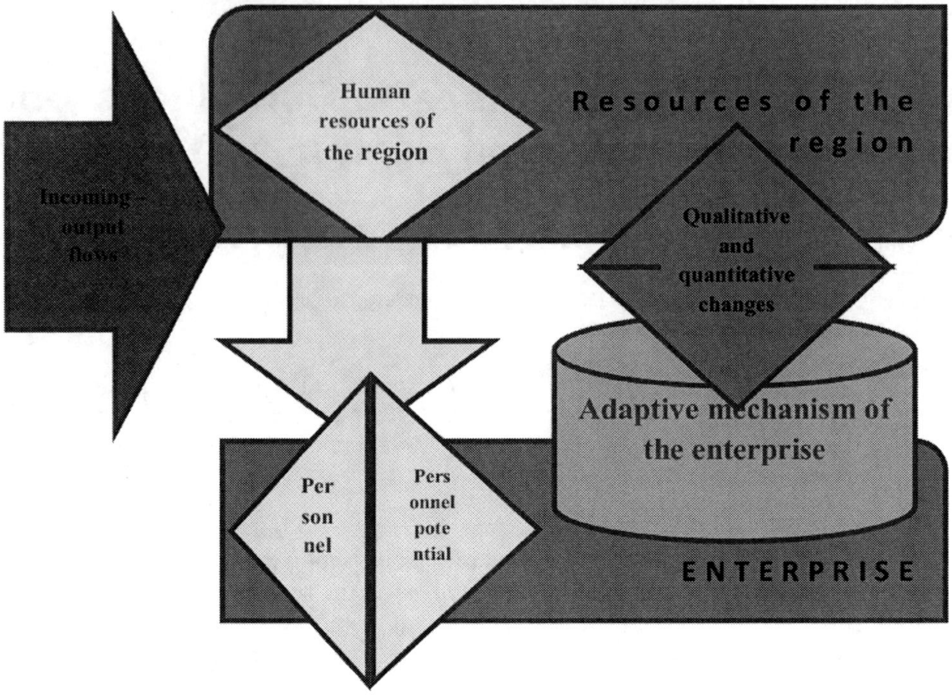

Fig. 1. Impact of enterprise development on the development of regional resources

Before building the forecast and assessment model, it is necessary to investigate factors affecting formation of enterprise environment. The analysis of researches in this area [1-8] has made it possible to summarize and determine the prerequisites for emergence of asymmetry, which can be broken down as follows (Fig. 2):

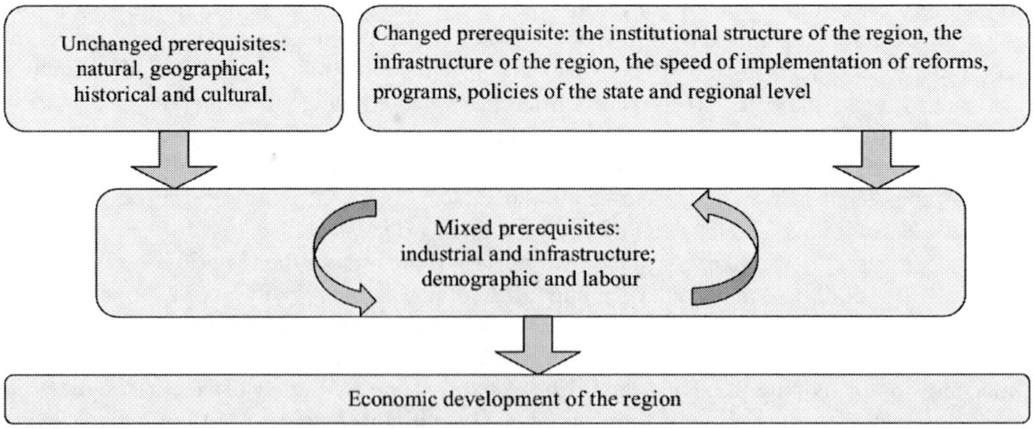

Fig. 2. Prerequisites for the emergence of regional asymmetry

- prerequisites which are almost constant (basic, objective) – those that are formed under the influence of time and are characterized by natural conditions, mentality of the population living in the region, etc. Among these prerequisites scientists single out the following: natural and geographical conditions; historical and cultural factors, in this case natural and geographical prerequisites being the basic ones.

- variable prerequisites (secondary, subjective) – those that arose under the influence of reforms, programs, policies at the country and regional level, as well as under the influence of business entities. Among these prerequisites scientists single out the following ones: institutional structure of a region, infrastructure of a region, speed of implementation of reforms, programs, policies at the country and regional levels.

- mixed prerequisites (indirect ones) – those that are formed under the influence of the previous two groups. Researchers recommend including in this group of prerequisites production and infrastructure, demographic and labour ones, which are interconnected. Previously, the location of a production complex depended on the availability of production resources in the region. Overcoming asymmetry in development of regions and optimizing redistribution of regional resources is within the competence of authorities, but it directly depends on enterprises working in the regions. This means that actions of authorities without coordination with the commercial sector will not have a great effect either at the regional or at the country level. That is why the authors propose a model to forecast and assess development of a region. Evaluation of the future state of resources of a region should be the basis for two mechanisms (Fig. 3): 1) a mechanism for managing resources of a region; 2) a mechanism for adapting enterprise to changes in the surrounding socio-economic environment.

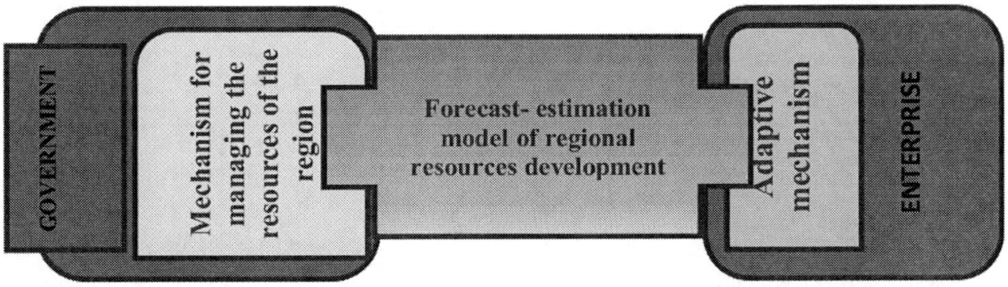

Fig. 3. Coordination of actions of the power of the region and enterprises of the region to overcome the asymmetry of the region

It should be noted that these mechanisms complement each other and the proposed model can become their common component. To obtain an objective assessment of regional asymmetry, a comprehensive approach is needed. It requires consideration of all prerequisites defined in Fig. 1, among which the authors single out a number of characteristics of the level of regional development (and, accordingly, asymmetry): ecological and territorial characteristics – features of the natural landscape, availability of natural resources, environmental conditions, etc. (ET); economic ones – economic development of the region, its production structure, level of income of its enterprises and citizens, profitability of business, increase in GNP, etc. (E); political – political stability, state of the legislative and executive systems, level of the judicial system, institutional structure of a region, infrastructure of a region, speed of implementation of reforms, programs, policies at the country and regional level, existence of corruption in a region, ease of doing business, etc. (P); socio-demographic – social security of citizens, their standard of living, population growth including natural and migratory increase (SD), others – level of innovation in the region, technological development (others).

The authors conducted a study of the available methodologies for assessing the level of development of a region using various characteristics [1–8]. For the research, the methodologies which are used in international practice and have confirmed their feasibility were selected.

As a result of the research, the author broke down the existing methods and methodologies into 5 groups, based on the criterion of comprehensiveness. Groups were formed depending on the number of characteristics used to assess the state of the region and its development.

Table. 1. Calculation of indicators characterizing asymmetry [9]

Asymmetry indicators	Formulas of calculation and normative values	Components	Number of employees		Volume of sold products		Number of enterprises in the region	
			2016	2018	2016	2017	2016	2017
Atkinson index	$A = 1 - \frac{\sqrt[n]{x_1 \dots x_n}}{\bar{x}}$	x_1,x_n - basic distribution in a separate measurement; $\bar{x}$ - average of aggregate x; n - sample size; P – share of the unit in the total sample size; S – standard deviation.	0,09	0,08	0,50	0,18	0,28	0,16
Gini coefficient	$G = 1 + \frac{1}{n} - \frac{2}{n^2\bar{x}}(x_1 + 2x_2 + 3x_3 + \dots + nx_n)$ A value from 0 to 1. A zero value characterizes absolute equality, and 1 is a complete inequality.		0,24	0,056	0,20	0,13	0,44	0,07
Theil index	$T_1 = \frac{1}{n}\sum_{i=n}^{n}\left(\frac{x_i}{\bar{x}} \times \ln\frac{x_i}{\bar{x}}\right)$ Full symmetry - Teil's index is equal to 0.		0,10	0,09	0,85	0,21	0,4	0,18
Gerfindal coefficient	$H = P_1^2 + P_2^2 + \dots + P_n^2$ The closer to 10,000 the greater the asymmetry, the smaller the coefficient, the less the asymmetry		491	501,1	1893	617,5	407	585,4
Kurtosis formula	$Es = \frac{n(n+1)\sum_{i=1}^{n}(x_i - \bar{x})^4}{(n-1)(n-2)(n-3)S^4} - \frac{3(n-1)}{(n-2)(n-3)}$ Less than 2 - insignificant level of asymmetry, 2-3 - moderate level of asymmetry, more than 3 high level of asymmetry		0,85	4.93	20	6.65	16	4.02
Level of asymmetry			Low High		High Moderate		Moderate High	
Integral coefficient			0,90	0.98	3218	19.60	320,91	4.87
Level of asymmetry			Low		High		Moderate	

Group 1. Narrow assessment of regional asymmetry, including the calculation of: Living Planet Index (ET), Environmental Performance Index (ET), Environmental Vulnerability Index for the Black Sea and Azov regions (ET), Quality of Life Index (SD), Index of Knowledge Societies (SD). Group 2. Assessment of asymmetry using two characteristics, including the calculation of: Index of Sustainable Economic Welfare (SD, E), Human Development Index (E, P), Environmental Sustainability Index (ET, P). Group 3. Assessment of asymmetry with the help of three characteristics, including the calculation of Indicators of Development (Economic, Social & Environmental) (ET, E, DM). Group 4. Assessment of asymmetry using four characteristics, including the calculation of: Index of Economic Freedom (P, E, SD, others), Competitiveness Index (P, E, SD, others). Group 5. Comprehensive assessment of asymmetry, including the calculation of Indicators of

Sustainable Development of the UN Commission on Sustainable Development (ET, P, E, DM, others).

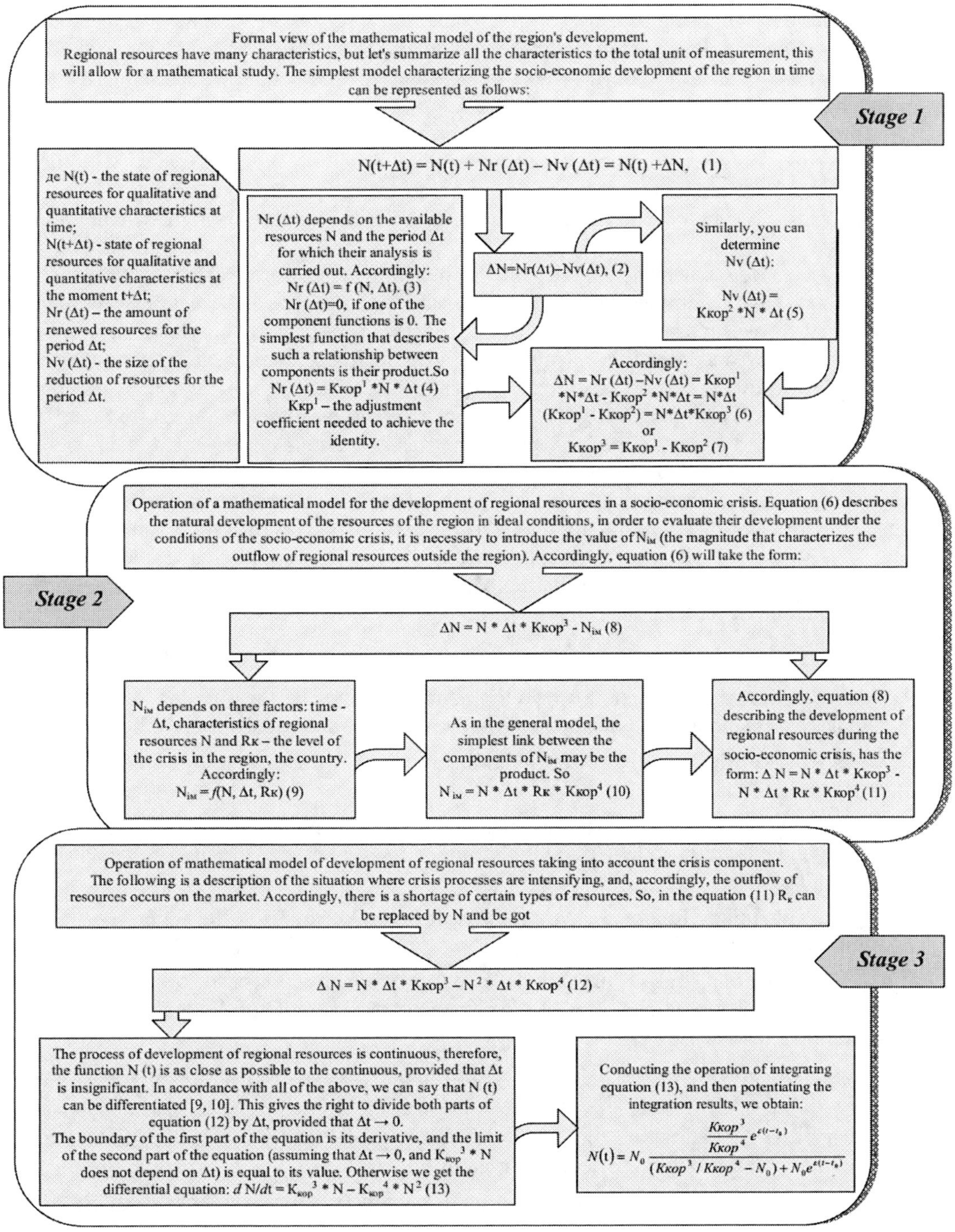

Fig. 4. Stages of development of the forecast-estimation model of regional resources development

Whatever happens in a region, it either affects a person's life, or is the result of their activities. Group 3 and Group 4 contain methodologies which are intended for comprehensive assessment of the level of regional development but do not take into account some characteristics. Calculation of Indicators of Sustainable Development proposed by the UN Commission on Sustainable Development is currently the only generally known official methodology for assessing the level of development of a region. In Ukraine, this methodology was adapted and implemented to evaluate regional asymmetry [9]. However, this methodology allows assessing the level of development of a region but does not allow identifying the factors which have conditioned it, only stating the facts.

To eliminate these contradictions, the authors propose a simple but reliable mathematical model for assessing socio-economic development of a region over time, which will allow to evaluate and then model (mitigate, reduce the gap between regions) asymmetry in regional development of a country. The stages of the formation of the mathematical model are presented in Fig. 4.

The developed mathematical model allows predicting the value of N (t) at any time moment. The authors have determined and summarized the prerequisites for emergence of regional asymmetry, for which a statistical study was conducted. The statistical analysis and grouping of the regions allowed to reveal the asymmetry in their socio-economic development, which was also confirmed by the calculation of other indicators characterizing asymmetry. A grouping of methodologies for comprehensive assessment of regional asymmetry was carried out, which made it possible to propose a model for assessing socio-economic development of a region over time, its formal description being presented. For a detailed development of the model, it is necessary to determine the qualitative and quantitative characteristics of regional resources, develop a mechanism for evaluating these characteristics and constructing a system of integral indicators which will allow to represent the value of N (t). It would be advisable to develop a document flow, report forms, software to speed up the processing of information and implementation of forecasts.

References

1. A. Yu. Fedorov Asymmetry of development of regions of the Russian Federation and tools for its smoothing. Questions of modern science and practice. 2, 4 (2006)
2. B. Lavrovsky Measurement of regional asymmetry on the example of Russia. Voprosy ekonomiki. 3 (1999)
3. V. G. Ignatov Asymmetry of the socio-economic development of the regions of the Russian Federation and the main directions of its weakening. TERRA ECONOMICUS. 7, 2 (2009)
4. O. Yu. Krasilnikov Regional asymmetry of structural changes in the economy // Society and the economy, 2 (2001)
5. N. G. Duplenko, L.V. Puryzhova Asymmetry of the potential of socio-economic development of municipalities of the region on the example of the Kaliningrad region. Young Scientist. 19.1 (2014)
6. N. V. Bozhidarnik, V.I. Chuzhikov European model of the identity of regional asymmetries (social aspect). Ukraine: aspects of practice. 6 (2005)
7. Zhovtyak G.A. Estimation of Spatial Asymmetry of Placing Productive Forces in the Economy of the Region. Effective Economics. 9 (2015)
8. I. M. Vakhovich, O.E. Tabalova Regional asymmetry of sustainable development of Ukraine: diagnostics and alignment mechanisms (2012)
9. The State Statistics Service of Ukraine
10. A. D. Pilko, Models of Assessment and Analysis of Asymmetry of Regional Development. Development Economics. 2, 86 (2018)

Assortment Portfolio Price Positioning of a Transnational Corporation in the Ukrainian Market

Serhii Smerichevskyi[1*], *Tetiana* Kniazieva[1], *Svitlana* Smerichevska[2], *Anna* Shevchenko[1]

[1]Marketing Department, National Aviation University, Ukraine
[2] Logistics Department, National Aviation University, Ukraine

Abstract. The paper deals with the critical stages of the evolution of theoretical approaches to the positioning strategies content. On this basis, the key approaches to the market positioning processes management formulated. The correlation between the concept of creating consumer value and the price positioning strategy proved. The algorithm of the process of forming and implementing the strategy of assortment portfolio price positioning of a transnational corporation on the Ukrainian market proposed.

1. Introduction.

The development of management strategies under different market conditions and the choice of optimal positioning methods, including a wide range of producing products price positions, is an essential and complicated task for the companies' management under current unstable market conditions, with a decrease in people's consumer activity. The search for sound management decisions is complicated due to many factors of the internal and external environment and requires a fully justified algorithm of actions. The management process is a whole series of interconnected analytical procedures and decisions that allow the company to move consistently towards achieving its intended strategic goal. These procedures and decisions are also typical for the process of price positioning formation, the importance of which is exceptionally high, given the fact that the price as a monetary category concentrates the expectations of financial results on which the company focused. In this respect, pricing becomes the basis of all other product management decisions.

In current conditions, when economic instability contributes to the growth of the price sensitivity of people and reducing their consumer activity, transnational corporations are facing another round of crisis behaviour in most categories of convenience goods. The need for the systematic development of management concepts and methods in this area is determined by the relevance of the topic of this article.

In foreign and domestic scientific literature, some problems of price positioning of transnational corporations considered many authors [1-10]. In these works, the definitions of the key concepts of strategic management that form the basis of the positioning concept are

* Corresponding author: s_f_smerichevsky@ukr.net

given: 1) positioning related to the long-term strategy of the company since it takes time to create the desired position; 2) positioning occurring in the minds of consumers; 3) positioning based on obtaining the benefit; 4) the product's position in the mind of one buyer may differ from the same product's position in the mind of another buyer. Therefore, it is crucial to understand the position the product occupies in the minds of consumers of all market segments which are essential for a company; 5) positioning is a relative concept in a competitive market environment because the products occupy the positions relative to competing ones. Positioning is an essential means of ensuring the competitiveness of national and international companies.

Despite numerous works in the field of market strategies development, most of them leave rather unaddressed issues related to the development and implementation of product portfolio management strategies of the company. The factors arising within the product portfolio and directly influencing the development of the positioning strategy for the company's aggregative portfolio are not sufficiently studied. Many companies do not control positioning in distribution channels, which leads to significant problems in the perception of the companies' products and financial losses. A real challenge for any company operating on the consumer market is to turn the pricing process into a strategic management competence within the company.

The objective of the article is theoretical justification and development of a management algorithm of forming the strategy of a broad assortment portfolio price positioning based on systematisation the transnational corporations' strategic management in the Ukrainian market.

2. The genesis of the approaches to the market positioning strategies development.

The market strategy of a company is a complex, multi-aspect category. It should set a specific direction of development and coordinate the efforts within the company on its basis to achieve the desired position in the future, reflect the nature of the organisation and demonstrate its distinctive features, as well as contain a set of rules that will guide the company in making a management decision. The market strategy is designed to provide sustainable competitive advantages in the present and in the medium-term perspective to achieve the desired strategic position of the company in the future.

In current market conditions, the statement that two key factors determine the choice of competitive strategy of the company remains relevant: the profitability of the industry and the positioning of the company within one or another industry [3, 5]. Depending on the company's positioning relative to other market participants, its revenues will be higher or lower than the average level in the industry. These competitive advantages, combined with the competitive environment of the market segment in which the company seeks to achieve the desired position, form the key management parameters and allow developing types of competitive strategies: leadership in minimising costs, differentiation or focusing.

The management practice of transnational corporations (such as Coca-Cola, Samsung, Toyota, Nestle, Nokia, Metro Cash&Carry, Hewlett-Packard, British American Tobacco, etc.) shows the features of its application:

1) when developing a strategy of the market segment, the company's management uses specific attributes that meet the identified customers' needs and position products or brands uniquely;

2) the differentiation of goods or brands is based on the unique properties of the product itself, the peculiarities of implementation or particular marketing approach in the field of merchandising and product placement within retail facilities;

3) several companies in the Ukrainian consumer market at the same time successfully implement the strategy of differentiation;

4) several parameters appreciated by buyers characterise products and brands in the consumer market;

5) key market players seek to bring their costs closer to their target competitors by managing costs, reducing costs in areas that are not directly related to the chosen direction of differentiation;

6) companies' management invests significant financial and human resources in support and development of market expertise, analysis and forecasting of market trends, the so-called macro trends, as well as in the technologies development.

Within the globalising process of the transnational corporations brand portfolio, it is also necessary to decide which brands have the potential to become global and which will remain national trademarks.

The study of the evolution of the theoretical views within the positioning concept allows comprehending the changes which occurred over the past few decades and transformed strategic management into its current state, analysing the accumulated the theory and practical application of fundamental theoretical approaches in the positioning strategy development. Based on the analysis of the formation and development of the positioning concept, it is possible to distinguish four main stages of theoretical thought evolution within the price positioning concept.

It can be said that the conceptual framework of strategic positioning (the first stage of the theoretical thought development) are presented in the works of John Francis Trout and Al Ries [9]. The key idea of the positioning concept is that positioning is the management of the consumer's opinion about the position of a particular product among many other brands of a given or related product group. The positioning aims to create such an impression among consumers that they see a unique, one-of-a-kind product for which there is no equivalent replacement. At the same time, the position of a brand in the consumer's mind is a relative concept since based on the consumer's comparative assessment of this product with competing analogues. Strategic positioning is the result of related management decisions, which create a set of ideas about the product. The main types of characteristics used by the company's management to develop a positioning strategy are product properties; advantages, problem-solving and essential needs; price and quality; features of use; product consumer.

At the second stage of the positioning concept evolution, theorists and practitioners of strategic management developed methodical management approaches to the effective positioning strategy formation and described the key stages of the strategic positioning process. For example, Jean-Jacques Lambin systematised quantitative methods used to develop an active positioning strategy, methods of assessing the current perception of products based on a multi-attribute product model.

At the third stage of the positioning concept evolution, key aspects are built in the framework of the brand management concept [1, 4]. Unlike John Francis Trout, representatives of this stage focus on emotional benefits as one of the most powerful bases for product differentiation, and within this direction, the key components of brand positioning, such as identity elements, value proposition, brand-communications, and advantages highlighted and developed. The concept of portfolio approach was also introduced, which is extremely important for modern management, because, in this approach, manufactured goods are not considered separately from each other, but as an interconnected aggregation of brands, therefore, for maximising brand portfolio effective management, a clear understanding of the roles of each of them in the portfolio is needed.

The fourth stage of the management approaches to the positioning concept evolution is interesting because it focuses on the interrelated managerial stages of the company's actions in creating and launching products into the market. Within this direction, the following most well-known models should be mentioned: 4D branding (a model, which includes functional, social, mental and spiritual aspects), Bates Worldwide Brand Wheel model (the model allows consistently linking the brand's essence with goods or services specific properties), the Thompson Total Branding model, the Brand Key model of Unilever, the Brand Platform model of Young & Rubicam, the Censydiam System of the Synovate company, the Brand Pyramid model of Mars and others. Within these models' framework, the positioning strategy management based on various key aspects that are most important for creating and launching products onto the market in multiple industries and competition conditions.

Analysis of various theories shows that the development directed to the understanding of multiple technologies of the positioning processes management. A systemic analysis of the positioning concept transformation leads to understanding that creating customer value is the key result of the positioning strategy.

3. The algorithm of a process of price positioning development and implementation in strategic management of a transnational corporation.

The consumer value concept focuses on the satisfaction of basic needs and involves the process of creating values that are difficult to copy for a transnational corporation's competitors. Having analysed the modern development of conceptual approaches to the process of value production management, it can be concluded that the scheme of the central management relationships between the market segment strategy and the positioning strategy should be supplemented by management blocks of controlling the positioning strategy implementation, and monitoring the chain of creation and proposition of value to the target market segment [10]. It is this algorithm of management decisions that should be the basis for the management of positioning processes.

The role of price as a significant attribute of the product for the consumer is typical for many consumer markets. On this basis, companies sometimes build the entire strategy of the product or brand positioning on the price. Thus, from the management position, it is necessary to run not a technical analysis, but the analysis of quality and price perception.

It is important to note several characteristics of the price as the market strategy element [2, 6-8]: 1) the price affects the level of demand and determines the level of buyers' activity depending on the sensitivity to the price; 2) the price directly affects the profitability of the company's commercial activity and the achievement of the company's operational goals (quantitative terms of the company's strategic goals); 3) the price has a direct effect on the overall product or brand perception and is one of the essential factors determining the positioning for the target audience; 4) the price, unlike other product attributes, allows a direct comparison of competing goods; 5) the price must comply with other elements of the company's market strategy, particularly taking into account the distribution strategy and providing the necessary trade margin.

Thus, price positioning is the most critical element of the product positioning strategy and should aim at strengthening the proposition of consumer value. Based on the analysis of the relationship between the basic positioning strategy and the price strategy, it is possible to formulate an algorithm for the price positioning developing process and the pricing strategy (Fig. 1) and describe the critical stages of the process.

The stage of the positioning strategy implementation and delivering value to the target segment is critical to minimise the differences between the conceived strategy and the implemented one but does not guarantee its full implementation.

On the basis of this algorithm, the pricing strategy can be determine das a reasonable choice of several price options aimed at achieving the maximum (normative) profit for a transnational corporation on the market within the planned period, and to analyse three possible directions in the pricing strategy development: 1) cost-oriented pricing solutions (methods and strategies), 2) demand-oriented pricing solutions, 3) competition-oriented pricing solutions. From this statement, to conclude that demand-oriented pricing strategies are the most characteristic of the daily demand products categories.

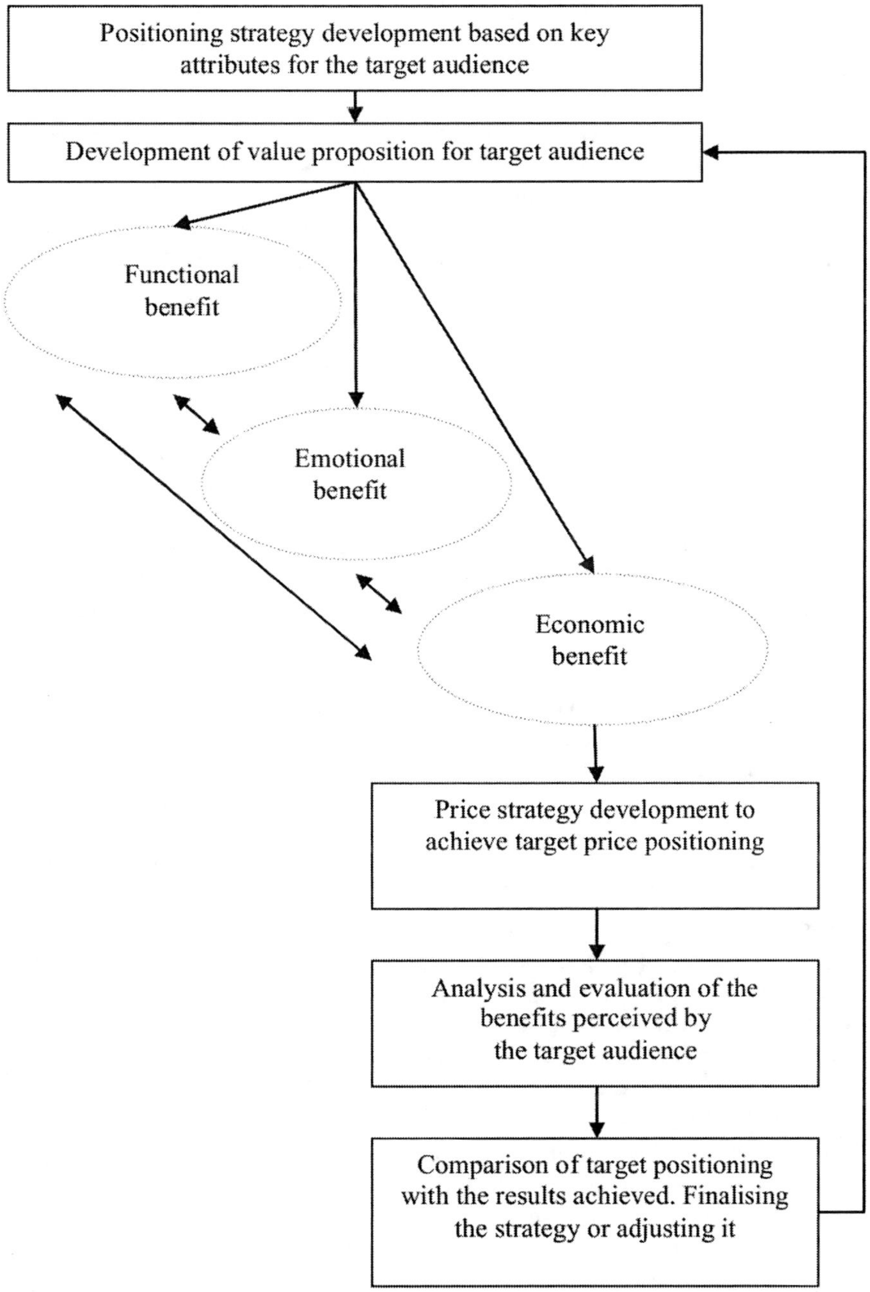

Fig. 1. The algorithm of price positioning development and implementation in strategic management of a transnational corporation

Thus, the paper specifies the vital management principle concerning the current market situation: the strategy is not static and can be subject to corrective changes, both at the stage of its implementation and at the monitoring stage to strengthen the brand's position on the market. Besides, it is necessary to keep in mind that the presence of significant pricing factors differences in the international activity of a company stimulates management to adapt not only the pricing strategy but also the brand positioning strategy.

The conducted research confirmed that the scheme of management of logical relationships between the market segment strategy and the price positioning strategy, proposed in the paper, should include blocks for monitoring the positioning strategy implementation, and monitoring the chain of creating and offering value for the target market segment. This approach should be the basis of the company's management algorithm. In this algorithm, it is necessary to pay special attention to the price positioning strategy implementation in key distribution channels. The lack of an adequate level of control can damage the identity of the forming brand. It is also important to predict the key competitors' potential reaction and to develop a plan in advance for maintaining and strengthening the critical brand's positions in a changing competitive environment.

4. Conclusions.

The determined necessity of transition from a traditional organisation and management models of a company to more complex models with enhanced external company's orientation, which require a quick management response within the company to the company's external environment changes at the structure, systems, and business processes levels. The systematisation of theoretical approaches to positioning as a method of strategic management is presented, which showed the current key factors and strategies of market positioning. Four stages of the positioning theory evolution are identified: the first stage – highlighting the positioning as a functional management area, the second stage – developing key elements of positioning, the third stage – the positioning concept transformation within the brand management, the fourth stage – developing a range of differentiated positioning models.

In the context of the relationship between consumer value and price positioning, the logical chain of management relations and management decisions is built, which should accompany the development of the element-functional strategy of a company.

It is proved in the paper that demand-oriented pricing strategies are the most typical for the transnational corporations' activities on the Ukrainian market in the daily demand goods categories, which allowed substantiating the most effective management method – the value-oriented method of positioning development. The management algorithm of forming and implementing the strategy of a company's assortment portfolio price positioning on the consumer market developed and the corrective measures proposed of positioning strategy management under the features of price-forming factors on the regional markets: 1) to improve distribution and representation in retail facilities: increasing a new product's quantitative distribution in key regional cities and implementing the MSL (Must Selling List) tool for key channels of modern retail to increase the average number of items; developing key principles of the critical brand categories placement in a retail facility within the relevant categories and close to key players in the category; 2) to run an additional product test to quantify the key product characteristics of the brand in the category; 3) to provide sales promotion: developing a promotional calendar of activities for the calendar year.

References

1. Becker-Ritterspach, F., Lange, K., & Becker-Ritterspach, J. Critical perspectives on international business, **13(3),** 186-203 (2017). *https://doi.org/10.1108/cpoib-06-2016-0024.*
2. Blajut, L. P. SEA – Practical Application of Science, **1(7),** 97-102 (2015)...
3. Calabrese, A. International Journal of Production Economics, **135**, 800-812 (2012)..
4. Ganushchak-Efimenko, L., Shcherbak, V., Nifatova, O., Kolodiziev, O. & Rębilas, R. Innovative Marketing, 15(2), 14-29 (2019).. *doi:10.21511/im.15(2).2019.02.*
5. Iazzolino, G. & Migliano, G. Investment Management and Financial Innovations, **13(4),** 106-119 (2016).. *doi:10.21511/imfi.13(4).2016.09.*
6. Kandrashina, E. A. & Zotova, A. S. Problems and Perspectives in Management, **16(1),** 437-444 (2018).. *doi:10.21511/ppm.16(1).2018.41.*
7. Melnychenko, R., Pugachevska, K., & Kasianok, K. Investment Management and Financial Innovations, **14(4),** 40-49 (2017).. *doi:10.21511/imfi.14(4).2017.05.*
8. Romaniuk, M. Transfer pricing in the tax system of Ukraine. Proceedings of the International Economic Conference on World Trends and Prospects for the Development of the Financial System of Ukraine. Taras Shevchenko Kyiv National University, Kyiv, Ukraine. (2013).
9. Trout, J. & Ries, A. (1981) *Positioning: The Battle for Your Mind.* New York: McGraw-Hill Education.
10. Wolff, M. Problems and perspectives in management, **5(2),** 66-74 (2007).

Institutional System of Human Development and Institutional Environment: Relationship Features

Olena Stryzhak [1*]

[1] Simon Kuznets Kharkiv National University of Economics, Tourism Department, 9-A, Nauky Ave., Kharkiv, 61166, Ukraine

Abstract. The article considers the features of human development in the context of the concept of sustainable development. The need to study the institutional system of human development as regulated multilevel system of interconnected formal and informal institutions is justified. The relationship between the level of human development and the quality of the institutional environment is determined using the methods of correlation analysis. Analyse covers 214 countries and territories for 2017. The results of the correlation analysis show that there is a strong direct interconnection between HDI and WGI. The correlation ranges from significant to very strong one. This gives grounds for conclusion that there is influence of the quality of institutional environment on the level of human development.

Introduction

In modern conditions of the overpopulation of the planet, exhaustion of natural resources, global warming, the attention of scientists and public figures is attracted not only to the problems of achieving economic prosperity and ensuring the rates of economic growth, but also the problems of implementing the economic interests of society together with the development of other spheres, including social, political, moral, cultural, ethical, ecological and other aspects of development. In 1987, the Bruntland Commission in its report defined sustainable development as "development that meets the needs of the present without compromising the ability of future generations to meet their own needs" [5]. The implementation of the ideas of the sustainable development concept directs the state to the organic development of all sectors of society, focusing on the long-term perspective and creating a growth potential for future generations.

The human development is an important component and one of the prerequisites for sustainable development. UN Development Program defines human development as a process of enlarging people's choices. But human development is also the objective, so it is both a process and an outcome. Human development implies that people must influence the processes that shape their lives. In all this, economic growth is an important means to human development, but not the end. Human development is the development of the people

* Corresponding author: sssselllennnn@gmail.com

through building human capabilities, by the people through active participation in the processes that shape their lives and for the people by improving their lives. It is broader than other approaches, such as the human resource approach; the basic needs approach and the human welfare approach [3, p. 2]. The implementation of the main principles embodied in the concept of human development allows us to create conditions for the full development of a person, observance of their rights, providing opportunities for a long, healthy, full life in harmony with nature and society.

It should be noted that the economic component is rather important, but not the only condition for ensuring effective human development. It is also important to determine how income is distributed in society, what needs it is directed at, what the distribution inequality is, and so on. Accordingly, in formation of human development opportunities, priority is given not only to economic but also to its institutional support, that is, how the institutions, that provide human development, are organized, what their qualitative composition is, how they are combined within a single system. From these positions, the institutional system of human development is a regulated multilevel system of interconnected formal and informal institutions directed at ensuring the conditions of human development.

Both formal and informal institutions are important for insuring human development opportunities. And if the formal ones can be established by a directive, by adopting one or another law or by-law, then informal ones are evolving in a society in evolutionary way, and therefore are more complex for regulation, management and change. For example, the institution of remuneration of labour includes both a formal institution of minimum subsistence level, which is established by law, and informal one - the market rate of wages by spheres and sectors of economy.

Research methodology

The level of human development is assessed by the appropriate Human Development Index (HDI). The methodology of calculating the index includes the following indicators: Life expectancy at birth; Expected years of schooling; Mean years of schooling; Gross national income (GNI) per capita.

The state of the institutional environment in the country can be assessed on the basis of Worldwide Governance Indicators (WGI, which is calculated by the World Bank according to the methodology of D. Kaufmann, A. Kraay, M. Mastruzzi [1, p. 4].

The indicators cover three areas of management and include six aspects:

(a) The process by which governments are selected, monitored, and replaced:

1. Voice and Accountability (VA);
2. Political Stability and Absence of Violence/Terrorism (PV);

(b) The capacity of the government to effectively formulate and implement sound policies:

3. Government Effectiveness (GE);
4. Regulatory Quality (RQ);

(c) The respect of citizens and the state for the institutions that govern economic and social interactions among them:

5. Rule of Law (RL);
6. Control of Corruption (CC).

The indicators take values in the range from -2.5 to 2.5, where the bigger value of the indicator reflects the more effective management. Accordingly, the percentage value among all countries is determined, ranging from the lowest (0) to the highest (100) rank. The indicators are based on several hundred individual variables of the perception of management, received from 35 separate data sources from 33 different organizations from around the world. They cover 214 countries and territories for 1996, 1998, 2000, and annually for 2002-2018.

In order to assess the relationship between the level of human development and the quality of institutional environment, the study used the methods of correlation analysis.

The Pearson correlation coefficient was used in this research since the analyzed parameters are measured in the rank scales (scales of order). However, as the Pearson correlation coefficient is predominantly intended for use in metric spaces of indicators, the Spearman and Tau Kendall rank correlation was also used in the analysis, which is specially designed for the objects of non-numeric statistics. The Spearman rank correlation coefficient more accurately takes into account the quantitative degree of correlation between the variables and is a nonparametric analogue of the classical Pearson correlation coefficient, but when it is calculated, the ranks are taken into account and not the indicators of comparable variables (arithmetic mean and dispersion) related to the distribution. Unlike the Spearman coefficient, Tau Kendall analyzes the interrelations between variables in more detail, sorting all possible matches between pairs of values of variables.

This study makes calculations using the software Statistica. Obtaining unambiguous results with the use of all these coefficients of correlation provides grounds for recognition of the obtained findings as objective ones.

Research findings

The institutional system of human development consists of basic and derivative institutions, which can be both formal and informal.

The basic institutions of human development include:

- institution of education (pre-school, school, higher, industrial, self-study, advanced training, retraining);
- health care institution (state, private, insurance, non-traditional medicine, self-medication, etc.);
- social security institution (social transfers, compensations, subsidies, benefits, established minimum subsistence level ;
- labor pay institution (minimum wage, fair remuneration for labor, market rate of wages, labor legislation);
- institution of protection of human rights and freedoms (guaranteeing the right to work, religion, education, freedom of speech, distribution of income from labor participation in the production process, etc.).

The derivative institutions are created on the basis of basic institutions, which complement them and ensure the sustainability of functioning of the institutional system of human development.

The institutional system of human development is part of the institutional environment of the country. The assessment of the relationship between the quality of the institutional environment and the level of human development can be obtained by identifying a correlation between these indicators. The Pearson (product-moment) correlation, the Spearman rank correlation, and Kendall's Tau correlation are most often used to identify correlation dependence. The calculations of the coefficients were made on the basis of a sample for 187 countries and territories for 2017, for which data were available by all the indicators under analysis.

The relationship between HDI and WGI is presented in Table 1.

Table 1. The relationship between HDI and WGI

Variable	Marked correlations are significant at p < ,05000 N=187 (Casewise deletion of missing data)		
	Pearson Correlations	Spearman Rank Order Correlations	Kendall Tau Correlations
VA	0.591586	0.597818	0.431848
PS	0.628240	0.632721	0.454088
GE	0.854106	0.869506	0.692737
RQ	0.805191	0.827067	0.650074
RL	0.770315	0.788669	0.603344
CC	0.715771	0.734156	0.548212

The results of the correlation analysis show that there is a strong direct interconnection between HDI and WGI. The correlation ranges from significant to very strong one. This gives grounds for conclusion that there is influence of the institutional environment on the level of human development. In this case, the highest positive interdependence is observed between HDI and GE and RQ. This confirms the necessity for direct state institutional regulation of human development, especially in the direction of supporting institutions of education and health care and ensuring an adequate level of funding.

Graphical representation of the correlation relationship between indicators of HDI and WGI is presented in Fig. 1.

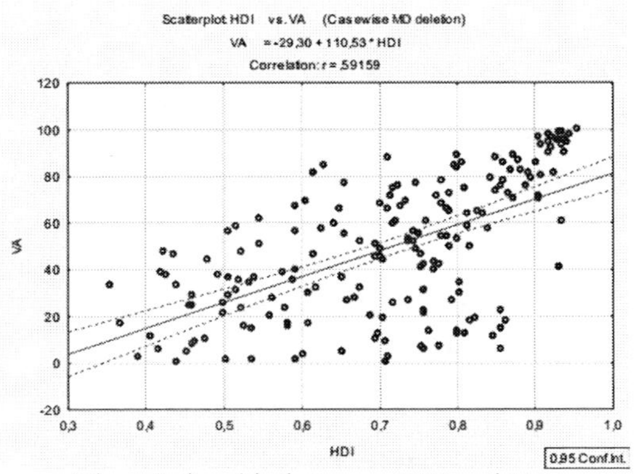

The relationship between HDI and VA

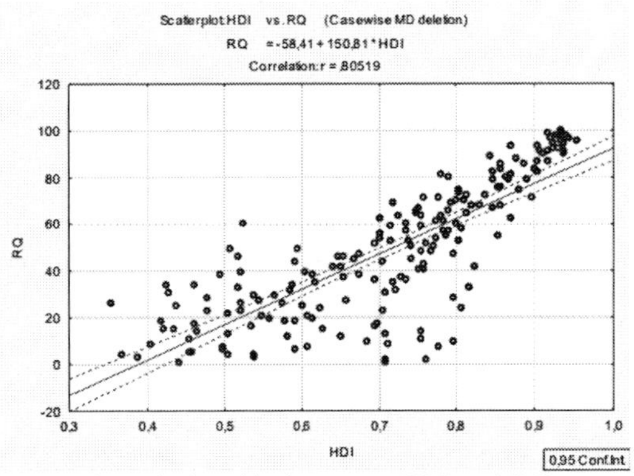

The relationship between HDI and RQ

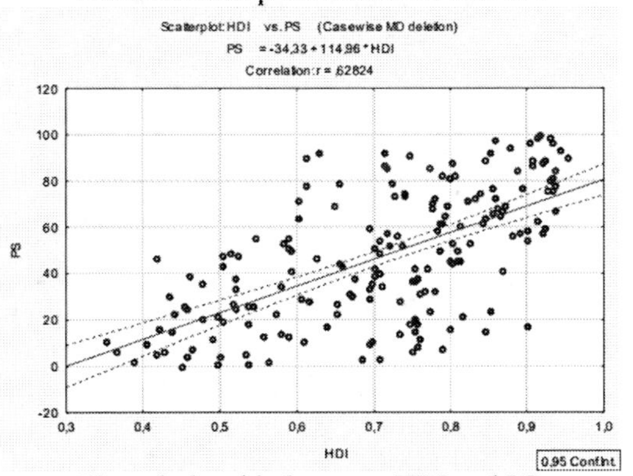

The relationship between HDI and PS

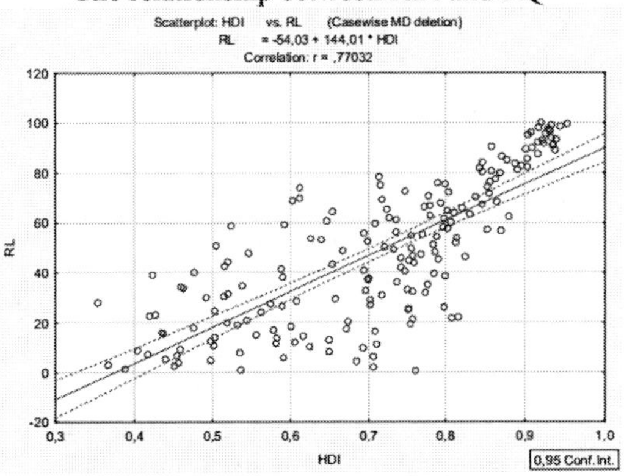

The relationship between HDI and RL

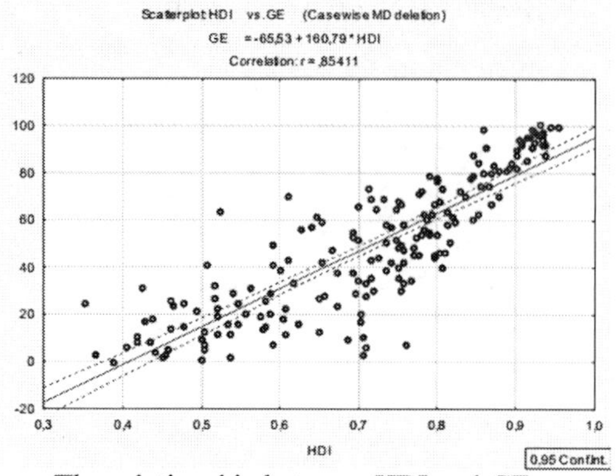

The relationship between HDI and GE

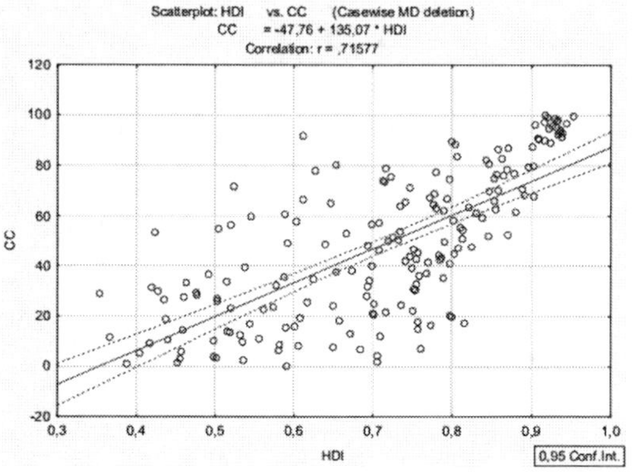

The relationship between HDI and CC

Fig. 1. The relationship between HDI and WGI

Based on the results of the calculations, the following conclusions can be drawn. First, the level of human development depends on the quality of institutional environment. Correlation dependence is observed between all the analyzed variables. However, the relationship between HDI and WGI is different for different aspects of management and varies from average to very strong. Secondly, as it can be seen from the graphical interpretation of the research results, there is an uneven spread of the values of the analyzed indicators. Thus, the scatter of RQ and GE indicators in relation to HDI is much less than when determining the level of HDI connection with other indicators.

Hence, the effectiveness of the institutional system of human development can be ensured under conditions of the presence of a favorable institutional environment. The unfavorable institutional environment with the state's support of ineffective formal institutions or the absence of appropriate sanctions for their removal can lead to deformations of the institutional system and the widespread use of destructive informal institutions (corruption, bribery, nepotism, etc.) to destabilize the institutional system and slow down development processes. For example, a significant increase in wages by the state in certain sectors of economy leads to an imbalance in the labor market, and a corresponding drop in demand for educational services for the training of professionals in those specialties that are paid lower. The increase of tax pressure, the lack of effective control and sanctions on the part of the state in relation to economic entities generates more serious problems for the national economy, starting from shadow employment, hiding of income, the transfer of the part of economic activity "to the shadow" and financial resources to other countries or offshore zones, ending with "brain drain" and growth of criminal activity.

In our view, state regulation of all aspects of development, including human, must be expressed not only in the provision of a sufficient number of formal regulatory institutions, but also in the creation of conditions for the effective use of informal institutions, which, together with the formal ones, form an institutional system. As K. Charakhchyan rightly notes, the problem is that an increase in the number of formal institutions does not automatically lead to positive changes in the institutional environment of the business entities. Moreover, a set of simulacrum institutions (dummy institutions), which drive out really useful and effective institutions for firms and households, can be formed [6, p. 69]. Thus, for the implementation of the basic principles and ideas of human development as a component of the concept of sustainable development, the construction of an effective institutional system, the formal and informal institutions of which complement each other, is of great importance.

Conclusions

The contradictions between formal and informal norms can lead to an imbalance in the institutional system of human development, hampering its development, or even blocking its functioning. Consequently, the harmonization of the formal rules that are introduced, with those informal institutions that have been in society for many years, that are generally recognized and are often more consistent and effective than the new formal ones, should take place. The rejection of the latter may lead to an institutional conflict, which may negatively affect the activities of some spheres of society and sectors of the national economy, as well as the economic system as a whole. Therefore, the realization, formation and adoption of such formal norms, which together with the existing informal ones will increase the efficiency of the functioning of institutional system, is an important task of the state. An ineffective formal institution, introduced from outside, may subsequently stimulate the emergence of a simulacrum institution - an institution that, formally existing and occupying its place in the institutional hierarchy of society, in reality is only a visibility, an "empty form", the presence of which conceals the fact of its real absence. In the general sense, the simulacrum is a representation (a copy) of an object whose original does not exist in reality, which occupies the place of this object and is subsequently perceived by others as the primary object. This is a copy that does not have the original and which replaces the really existing objects and relationships.

The practice of state support of inefficient formal institutions leads to incorporation of simulacrum institutions, which subsequently oppose to the introduction of more effective institutions, into the institutional system. Since the simulacrum institutions are closely interconnected, they either "intertwine" with the existing institutional system or form their alternative, in parallel acting simulacrum institutional system. Research and assessment of the impact of simulacrum institutions on functioning of the institutional system as a whole, and the human development system, in particular, is, in our opinion, a promising direction for further research.

References

1. D. Kaufmann, A. Kraay, M. Mastruzzi, *The Worldwide Governance Indicators: Methodology and Analytical Issues.* World Bank Policy Research Working Paper No. 5430. Available at SSRN: https://ssrn.com/abstract=1682130 (2010).
2. *Human Development Indices and Indicators: 2018 Statistical update.* [Electronic resource]. URL: http://hdr.undp.org/en/2018-update
3. *Human Development Report 2016: Human Development for Everyone*, Publisher: United Nations. – 286 p. [Electronic resource]. URL: http://hdr.undp.org/sites/default/files/2016_human_development_report.pdf
4. *The Worldwide Governance Indicators.* [Electronic resource]. URL: https://info.worldbank.org/governance/wgi/
5. *United Nations General Assembly. Report of the world commission on environment and development: Our common future.* Oslo, Norway: United Nations General Assembly, Development and International Co-operation: Environment. (1987).
6. К. Чарахчян, *Перспективы трансформации институциональной политики государства.* Научные труды ДонНТУ. Серия: экономическая. Выпуск 38-1, С. 69-73. (2010).

3R philosophy for sustaining recreational attractiveness and development of hotel business in Ukraine

Olena Sushchenko[1], *Olena* Prokopishyna[2] and *Natalia* Kozubova [11]

[1] Simon Kuznets Kharkiv National University of Economics, Department of tourism, 61001 Kharkiv, Ukraine

[2] Simon Kuznets Kharkiv National University of Economics, Department of accounting and business consulting, 61001 Kharkiv, Ukraine

Abstract. The paper highlights the problem of waste accumulation and ways of its solution. It was established that landfills occupy 7% of the total territory of Ukraine and their area is larger than the area of natural reserve funds. Based on the dynamics of garbage accumulation the paper forecasts increase solid household waste in geometric progression. The article shares experience in field of utilization and waste management and grounds the necessity of separating rubbish into fractions and processing each component that is sold separately on the market. It is proposed to use a systematic approach to waste management in hotel business. The paper justifies the conclusion that only the proliferation of 3R philosophy – Reuse-Reduce-Recycle – in behaviour of tourists and the local population in everyday life will preserve the recreational attractiveness of Ukraine and provide the basis for the development of hotel business.

1 Introduction

Economic growth affects all aspects of human life, including production and consumption, which, in turn, leads to an increase of waste. This problem is especially vivid for developing countries, where environmental laws are not yet fully formed, and acceleration of economic development is being accompanied by the deterioration of the environment. Experts of the Organization for Economic Cooperation and Development (OECD) have calculated that a 1% growth in the national income increases the solid domestic waste by 0.69% [1]. In addition, the composition of waste is changing: the number of urban food waste is increasing (by 44% between 1995 and 2013), the share of complex waste (electronic, nanomaterials, household goods) also increases. According to World Bank forecasts, in 2025, city dwellers will produce an average of 1.42 kg / person of solid household waste per day compared with the current 0.64 kg. The total amount of waste generated by residents of cities will increase from 0.68 to 2.2 billion tons per year. As for most developing countries, the problems of waste management appeal to be quite acute for Ukraine.

[1] Corresponding author: kozubova_natali@ukr.net

https://doi.org/10.1051/shsconf/20196706050

Despite economic fluctuations in Ukraine, in recent years the process of waste accumulation has continued as a result the lack of a concept for efficient utilization at the state level. The environmental situation around the cities is worsening (especially with a population exceeding 1 million inhabitants), hazardous landfills emerge throughout the country, hurt ecology situation and tourism attractiveness, hamper development of hotel business.

1.1 The relevance and the problem of the research

As the cost of raw materials is high and increases it is irrational, both from an economic and an environmental point of view, to underground raw materials which are suitable for reuse. The main problem with the use of raw material is its removal from the funny mass, because it is almost unusable for recycling. Therefore, with the improvement of the system's technical sorting, deductions are worked out. For this economic policy, legal measures, agitation and propaganda among the population should be carried systematically.

The purpose of this research is to analyze existing waste problems and develop methodical approaches for waste management for preserving recreational suitability and ensuring the sustainable development of the hotel business.

1.2 Literature Review

Different countries apply many approaches to implementing the concept of waste disposal. In developed countries, garbage is recycled and reused. Waste is returned to the state of raw materials and is being processed by enterprises. Michael Hiete, Julian Stengel, Jens Ludwig and Frank Schultmann highlighted due to economic development demand increases as well as consumption and therefore the accumulation of waste accelerates, so only re-use of materials can reduce the burden on the environment [2].

Under the concept of full waste utilization, the multi-stage process of sorting and recycling of garbage, which minimizes the amount of waste at disposal, maximizes economic efficiency and minimizes the impact on the environment. As Nadim K. Copty, Didar Ergene and Turgut T. Onay note, the consumer pays for the disposal of garbage [3]. The cost can be paid separately (for example, utilities) or included in the price of the goods. Paul Isely and Aaron Lowen in their work emphasize that a very important factor in the development of the waste recycling industry are pricing and tariffs for the export and sorting of rubbish [4]. In many countries, citizens pay a certain amount of money for the collection and utilization of rubbish, which is included in the payment for utilities. Another approach is that garbage recycling is handled by the company's producers, but the cost of recycling is also translated into the consumer and is reflected in the price of the product.

2 Results and discussion

Most countries of the world have been confidently reducing the amount of solid domestic waste for several decades, while Ukraine, despite a number of laws, such as “On Waste” and “On Environmental Protection”, continues to bury them at landfills and burn them, which has a negative impact. to the environment. Handling solid domestic waste threatens Ukraine with an environmental problem due to their complex environmental impact and environmental safety. In 2017, it was formed 366,054.0 thousand tons. domestic waste and waste from economic activity, and this indicator has significantly increased after a few years of decline. Today, 12,442.2 million tons are in places of waste disposal [5].

According to the statistics of Ukraine, there is a tendency to decrease the amount of waste in recent periods (Table 1), but the reason for this is the difficult political situation in the regions, which at the moment are not reflected in the reports, therefore for the forecast the data from 2010 to 2013 should be adjusted by corresponding amount of generated waste in these regions.

Table 1. Dynamic of the wastes generation, thousand tons

Year	Official statistic values	Adjusted values
2010	419191.8	344083.1
2011	447641.2	363539.8
2012	450726.8	373360
2013	448117.6	374128.3
2014	354803	354803
2015	312267.6	312267.6
2016	295870.1	295870.1
2017	366054	366054.0

Polynomial model (Figure 1) has the best approximation value, the rate of determination ensure that forecast is quite accurate. Predictive values demonstrate that official statistic tends to conceal waste problems while adjusted data reveal inevitable increase in the amount of solid household waste in a geometric progression in future.

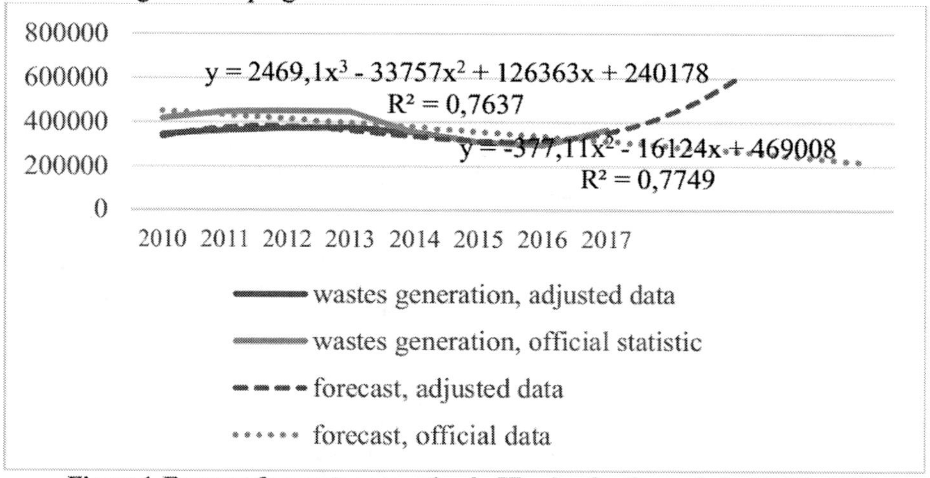

Figure 1. Forecast for waste generation in Ukraine for the period of 2019 - 2022

The constant increase in the amount of generated waste leads to an increase in the seizure of valuable, recreationally, land for burial grounds and loss of resource and value components, while Ukraine increases the import of re-raw materials. In 2017, the amount of imported waste paper increased by 27% compared with 2016 (346.4 thousand tons versus 271.9 thousand tons), during the 4 months of 2018 by 30% (130 thousand tons versus 99.9 thousand tons tone). Imports of secondary polymer materials in 2017 grew by 2.6 times compared to 2015 (38.6 thousand tons vs. 15.1 thousand tons in 2015), and in 2018, the import of these wastes is expected in volume of 48,0 thousand tons, which is 1,5 times more than in 2017 [5].

Excessive import of waste has led to a significant collapse of domestic specialized enterprises, since the recycling of waste paper and secondary polymer materials from domestic suppliers to processing enterprises is limited, and for certain types - discontinued. The warehouses of the enterprises are crowded with sorted waste paper and secondary polymer materials. Collection and harvesting of these types of secondary raw materials is limited, and some enterprises are discontinued.

Unnecessary domestic waste paper and waste from plastics continue to replenish unauthorized dumps, clog up forest parks, forest plantations, yards of cities and villages of Ukraine.

An increase in the import of secondary raw materials into the territory of Ukraine jeopardizes the fulfillment of the tasks set by the National Waste Management Strategy in Ukraine, namely, the introduction by 2023 of 2,500 settlements for the separate collection of waste, which are suitable for reuse and recycling [6].

In accordance with the Law of Ukraine "On Waste" a mandatory separate collection of household waste is established. General provisions for separate collection household waste are specified in the Method separate collection household waste, approved by the order of the Ministry of Regional Development as of 01.08.2012 number 133, which defines the general aspects of the implementation separate collection system. Separate collection household waste, including hazardous waste in their composition, is carried out by waste holders in accordance with waste legislation. At the same time, the selection of recycled raw materials from household waste collected in containers or loaded with garbage can only be allowed at specialized waste sorting and processing enterprises in accordance with waste legislation. At present, there is no normative act that clearly and transparently defines the legal relationship (rights, responsibilities, responsibilities) of all participants in the process of implementation of separate collection of household waste, including its sorting.

In 2017, a small number (822) of settlements in Ukraine introduced a separate collection household waste. There are 25 garbage collection lines in 20 settlements, including 6 in Kyiv. However, it's still too early to state the success of sorting. For example, we can consider the proportion of settlements where separate collection of solid waste is implemented, in relation to the total number of settlements in this region in 2017: in Vinnytsya oblast - 2.9%; in Zhytomyr - 1%; in Odesa - 0.3%; in Kharkiv - 10.9%; in Kyiv - 2.8%.

3 Recommendations

Only the proliferation of the 3R philosophy – Reuse, Reduce, Recycle – in everyday life and at business will preserve the recreational appeal of Ukraine.

Reuse means the opportunity to use things again, or to find a new consumer. Organic waste can be turned into composted fertilizer, old clothes may be taken to a shelter. Reduce means lower consumption. This is a decrease in the number of things used in everyday life, the first and most important step on the way to saving Ukraine from garbage. Unnecessary packaging of consumer goods should be avoided. Recycle means reprocess materials, for instance, recycled aluminium bottle need 80% less electricity than new ones.

Consequently, the existing waste management practices do not allow for the economic potential reuse of resources and reduce the burden on the environment. Most of the existing landfills are morally and physically obsolete and will not be able to absorb growing waste volumes in the long run and ultimately lead to serious environmental impacts.

In Ukraine, there are 6,148 landfills, of which only 2,600 are certified. The area of landfills is 7% of the total territory of our country and is larger than the area of the natural reserve fund; 32,984 dumps are unauthorized.

The efficiency of the end result of disposal depends to a large extent on the implementation of the principles of a systematic approach, which should include: a gradual reduction the total amount of domestic waste generation in human settlements; a sharp decrease in the amount of landfill disposal at landfills and the transition to waste recycling technologies; coverage 100% of settlements by collection and removal household waste; introduction of waste sorting in the settlements of the region; liquidation, sanation, reclamation of unauthorized landfills and closed landfills; consolidation of existing polygons and their transformation into regional ones in accordance with current environmental and sanitary requirements; creation of new production facilities for the sorting and processing household wastes and selected secondary raw materials; creation in the regions network of sorting and processing enterprises for the processing of all components household waste by modern advanced technologies; creation of preconditions for the complete collection, removal and recycling household waste with the placement land only on the landfills that are not processed, inert to the environment.

The social and responsible attitude of the hotel and tourist business to the problems of domestic waste management is very important, as tourist enterprises, on the one hand, are "producers" of sufficient quantity of resource-valuable waste, and on the other hand their activity depends on the environment state. Today, the concept sustainable development of tourism involves focusing on the understanding its environmental impacts and the development more sustainable forms its development to preserve the basic natural and socio-cultural resources of the environment.

Among the tourism sub-sectors, the hotel business has a significant impact on the environment. This sphere consumes a significant amount of energy, water, chemicals, produces a significant amount household waste. The adverse impact on the environment increases and deepens. Such enterprises partially or completely do not utilize waste, use wood for fuel from the surrounding area, dispose untreated sewage into the rivers, develop additional services that are oriented towards the active use the environment.

Substantial damage causes pollution by household waste, the source of which is the place of the longest tourists stay – accommodation establishments [7]. In addition to the deteriorating aesthetic appeal of the territory, household rubbish threatens the existence many representatives of the fauna. Animals undergo changes in behavioural reflexes, chemical additives and gene modulators in food waste cause impairment of immunity and genetics, poisoning and death. The problem of environmental pollution with household rubbish is especially relevant for hotels, separated in hard-to-reach places, which causes uncontrolled violation the waste management rules. Solid household waste is spontaneously stored in unsuitable places for this purpose, with non-compliance with environmental standards of utilization. Local landfills polarize the scale of the problem of pollution by dispersing garbage by the wind, falling into surface and groundwater. Unique natural conservation and recreational areas are particularly exacerbated by this problem [8].

The nature and consequences of the impact of the hospitality infrastructure are due to the object location of natural or socio-cultural interest. Subjects with a socio-cultural organization higher level are significantly less negatively affected. Such objects provide an effective stay of visitors, maintain an appropriate comfort level, and at the same time organize structures for limiting negative environmental impacts. In addition to utilizing the products of the activity, limiting the impact on the surrounding area by limiting recreational and tourist activities (buffer forest park areas, protected areas, monitoring the status of individual natural objects, etc.). In addition to the spatial sign of the hospitality impact on the environment, it is necessary to take into account the temporal phenomenon identified in the peculiarities of tourism development - seasonality. Most hotels, especially with the active use natural resources of the surrounding area (resort hotels, tourist hotels, tourist shelters), have a profound effect on ecosystems during a certain part of the year, in other seasons it is possible to restore it. Thus, the effect is temporary and coincides with the tourist season (skiing, beach, mountaineering) or long-term, with irreversible ecological changes - the disappearance certain species of the organic world or generally biocenoses. An important feature of hospitality is the nature of the relationship with the environment - technological. Hotel establishments for the most part provide a continuous mode of operation with an average load of 60-80%. Despite the application of the latest technologies, special equipment for saving energy, water, chemicals, hygiene products, food and other resources, their use remains significant.

4. Conclusions

On the basis of the conducted research it can be concluded that there is regularity in the rate of growth of waste accumulation over time. Since the total amount of waste is not utilized, the amount of waste steadily increases. Together with the increase in the amount of garbage, the area at which landfills are located, are increasing by 160 hectares per year. By 2022, the projected square of landfills will exceed 1 million hectares.

According to the results of the study, it was found that the lack of developed infrastructure for the collection, removal, recycling of household waste, lack of awareness of society about the need to sort household waste by their separate components, which are secondary raw materials, is hindered by the implementation of separate collection of household waste. The main factors contributing to this situation are non-fulfillment by local self-government bodies of the requirements defined by the Laws of Ukraine "On Housing and Communal Services" and "On Waste" as part of the insufficient organization of separate collection of useful components of household waste.

The use of 3R philosophy in the hotel business is especially relevant for Ukraine, as it contributes not only to the acquisition of additional customers advocating for ecology. Such objects provide an effective stay of visitors, maintain an appropriate level of comfort, and at the same time organize structures for limiting negative environmental impacts. In addition to the utilization of products, 3R philosophy decrease the impact on the surrounding area by limiting recreational and tourist activities due to establishment of buffer forest park areas, protected areas and monitoring the status of individual natural objects.

References

1. Organisation for Economic Co-operationand Development, OECD http://www.oecd.org (2019). Accessed 30 Mar 2019
2. Hiete, Michael, et al. "Matching construction and demolition waste supply to recycling demand: a regional management chain model." Building Research & Information **39.4** (2011): 333-351.
3. Copty, Nadim K., Didar Ergene, and Turgut T. Onay. "Stochastic model for landfill gas transport and energy recovery." Journal of Environmental Engineering 130.9 (2004): 1042-1049.
4. Isely, Paul, and Aaron Lowen. "Price and substitution in residential solid waste." Contemporary Economic Policy **25.3** (2007): 433-443.
5. State Statistics Service of Ukraine. http://www.ukrstat.gov.ua (2019). Accessed 22 Mar 2019
6. Solid domestic waste in Ukraine: Potential for developing scenarios for the development of the solid waste management industry. URL:https://www.ifc.org/wps/w952.S (2019). Accessed 22 Mar 2019
7. Sushchenko, O., Prokopishyna, O., & Kozubova, N. Eco-Friendly Behaviour of Local Population, Tourists and Companies as a Factor of Sustainable Tourism Development. IJE & T, **7(4.3),** 514-518 (2018). doi:http://dx.doi.org/10.14419/ijet.v7i4.3.19926
8. O. Prokopishyna, N. Kozubova Organizational support for internal tourism based on rational use of Ukraine tourist and recreational resources, NVIU **13**, 54 (2015)

The company human capital basis in the context of career component

Olena Sushchenko[1 1], *Karyna* Tymoshenko[1], *Volodymyr* Yermachenko[1], *Serhii* Sushchenko[1]

[1]Department of Tourism, Simon Kuznets Kharkiv National University of Economics, Kharkiv, Ukraine

Abstract. An industrial enterprise is a complex mechanism, the stability and efficiency of which depends on the influence of many factors, the main among which is the provision of an enterprise with production personnel with appropriate education, qualification level, skills and experience. Consequently, investing in personnel development creates conditions for expanding their competencies in order to ensure the competitiveness of personnel and a high level of enterprise performance. It is precisely the inadequate professional skills of the staff of most enterprises that cause low productivity, lead to significant shortcomings in the system of labour organization and management, and as a result – inefficient work of the enterprise. On the information analytical basis, the concept of a new approach to professional development is formulated as the mutual responsibility of the employee and the organization.

Introduction

In the current economic conditions, a new generation of workers are formed and begins to dominate, so called knowledge workers. "As evidenced by the data of the Organization for Economic Cooperation and Development (OECD), in the member countries of this organization, since the 1960s, they have steadily increased (on average, 3% of the annual investment in knowledge (science, education, public and private education and software). So, between 1985 and 1992 in the OECD member countries spent on knowledge-related investments on average from 8 to 11% of their GDP. In 1998, their total investments in knowledge reached 8.8%" [1].
In other words, the transformation of the economic system and society as a whole under the influence of scientific, technical and informational revolutions leads to the need to form a completely different type of worker. Under these conditions, an employee's characteristics such as creativity, ability to innovate, a high degree of adaptation to production technologies, ability to make quick decisions, etc., come to the fore. That is, employees are not only highly qualified professionals, but also have developed creative intelligence any working conditions and their changes, as well as a rapid change of profession, cultural and social environment, self-knowledge and self-improvement.

[1] Corresponding author: helen.sushchenko@gmail.com

Thus, the process of human capital development involves qualitative changes in it, which are provided by professional (professional development, retraining, etc.), social (improvement of working conditions, team relations, organizational culture, etc.) and career growth. Since the proportion of the last component in practical activities is insignificant compared to others, attention should be focused on the first two.

The main components of human capital development

It should be noted that with the change in technological modes, the very content of labour also changes, it is increasingly becoming intellectual, innovative, and creative. E. Toffler in the mid-80s noted that most "third wave" companies no longer increase their income by "squeezing sweat" from their employees; the main value acquired information and creativity [2]. According to V. Inozemtsev, at the beginning of the 21st century, the proportion of intellectual workers in the US was 70% of the total workforce, with knowledge itself becoming the object of knowledge [3].

Because of the effective use of productive abilities of the subjects of the formation of human capital, the maximum development of the latter occurs. In view of the above, we can assert that a person is able to manage his development in the appropriate limits, applying precisely the abilities that are most needed in this case. It is important not only the volume of knowledge and abilities, but also the ability to use them.

Consequently, the achievement of a high level of efficiency of production and economic activity is possible only with the condition of ensuring a continuous and intensive process of qualitative accumulation, increase and preservation of human capital in the form of professional knowledge, skills and abilities of employees of the enterprise.

In the study of human capital development it is expedient to consider three of its components, namely: professional development; social development; career development. Professional development of personnel is a system of interrelated activities, which include the development of strategy, forecasting and planning of staffing requirements, career management and professional growth, the organization of the process of adaptation, training, and the formation of organizational culture. Professional development also influences the employees themselves. By developing their skills and acquiring new skills and knowledge, they become more competitive in the labour market and receive additional opportunities for professional growth both inside and outside of their organization. This is especially important in today's conditions of rapid aging of professional knowledge.

Social development involves the improvement of forms, methods and conditions of life of workers based on changes in their development, the growth of social activity of staff, the growth of the effectiveness of the enterprise, the growth of material well-being of staff. The conditions that ensure the social development of workers include the compliance of each employee's personal capabilities with the structure and content of his activities, the similarity of employees 'moral positions, the homogeneity of the main motives of activities and individual goals of team members, the possibility of real complementarity and the organic combination of employees' abilities collective [4].

Career growth of human capital is the personal contribution of each employee to the overall workforce planning process in an enterprise in order to improve the skills and qualification of employees.

Career growth is necessary to create a successful career program, which includes the desire to develop and promote employees within the organization, the formation and support of a positive image of the employer. To do this, it is necessary to create a special personnel service in the enterprise the specialists of which will be responsible for providing tools and information that can help in making career decisions. They should also help managers create a climate and culture that should foster growth and stimulate professional development. Employees, in turn, should take advantage of this climate and gain knowledge about the important components of career planning.

So, career development of personnel is a process that requires employees and organizations to partner, which expands the knowledge, skills and abilities needed to solve current and strategic tasks and, as a result, to improve the abilities of employees and at the same time update the organization and improve the efficiency of its activities [5].

Career planning by providing labour resource development services (LRDS)

So how to plan a career? LRDS specialists are responsible for providing tools and information that can help in making career decisions. They should also help managers create a climate and culture that is conducive to growth and stimulate professional development. They need to develop career realization plans, analyze potential career paths, and determine whether employees have the skills, abilities, and knowledge necessary to be placed in appropriate positions. Just as in career planning, LRDS specialists are also responsible for developing and providing tools for acquaintance with a career – choosing career analysis tools, assisting in transitioning from performance evaluations to assessing development, helping managers and employees in developing growth and development plans and managing reorientation of the organization to the culture of development. As a result, employees, managers and specialists of LRDS become equal partners in career awareness activities.

Career planning is the personal contribution of the employee to the overall process of planning the company's workforce. Thus, career planning is a procedure for identifying individual career goals and developing measures for their implementation. It can also be regarded as a personal process consisting of three criteria:

1) planning of life is carried out to unite career and life;
2) development planning is used to improve the skills and qualifications of employees;
3) performance planning is used to increase productivity and organizational efficiency.

Target career planning is essential for creating a successful career program. The organization, through the LRDS, should stimulate employees and assist them in their own career planning.

There are several reasons to use career planning programs:

- The desire to develop and promote employees within the organization;
- Lack of potential talent;
- Desire to contribute to career planning;
- Strong interest of employees;
- Desire to increase productivity;
- Personal interest of department heads;
- Formation and maintenance of a positive image of the employer.

Any of the above points is an important reason for organizations to support and stimulate career-planning efforts. Moreover, LRDS specialists must understand that any or all of these reasons may be present in their organization. Regardless of the number of reasons for career planning, the main task of LRDS specialists remains unchanged: the development of labour resources within the organization [6]. This includes developing as well as moving up the career ladder using career planning. All this, of course, implies the growth and development of employees in the interests of the organization. However, in the practice of career planning programs, barriers may arise:

1) insufficiently clear distribution of roles in the organization in the field of career planning responsibilities;

2) lack of qualified LRDS specialists involved in career planning and similar processes;

3) the organization's underestimation of the importance of career planning.

The organization needs to introduce some kind of career planning philosophy, consider organizational development procedures, including the duties of LRDS specialists and managers, additional training and a long-term commitment.

Career planning tasks is to provide employees necessary knowledge about themselves and the surrounding reality. Career planning techniques include counselling, seminars, self-help materials, professional information and assessment programs. Although employees are fully responsible for their careers, the organization is their equal partner in career planning. Therefore, the main goal of career planning activities is to encourage employees to be accountable in this matter [7].

Informing staff about the company's long-term plans and strategies creates the prerequisites for employee career planning, since it gives an idea of new career opportunities, the needs of the organization in new competencies.

Today, many factors contribute to the introduction of career planning programs that are supported by the organization:

1) innovative processes in technique, technology, management;

2) talented employees who are waiting for opportunities for career advancement;

3) corporate growth and development leading to the need for more skilled labour;

4) a global competitive environment that requires more and more knowledge from employees;

5) outsourcing, staff reduction and optimization of organization size;

6) recognizing that employees are the most valuable asset of an organization.

Professional development programs can be significantly expanded if LRDS specialists use the five-step approach (Table 1).

Table 1. A five-step approach to professional development

Step	**Employee actions**	**Organization actions**
1. Orientation	Career responsibility	Individual conversation or seminar for new employees
2. Identification of his position	Analysis of own career	Manager as a career adviser
3. How am I treated?	Self-examination	Performance evaluation
4. What are my options?	Acquaintance with the organization. Using Career Resource Centres	Vacancy announcements. Career Resource Centres. Workforce planning
5. How will I achieve my goal?	Career planning	Manager in the role of career adviser. Career planning

The process begins with the identification of employee orientation, during which the focus is on individual career responsibility. To do this, it is advisable to hold a seminar – a business game, during which the participants' abilities, interest in career growth and the interest of participants in achieving the goals of the organization are well manifested. This allows you to identify ways to achieve a balance of personal and corporate interests. Another way to achieve greater unity involves the use of a manager – a direct manager (as a career adviser) to express and convey the idea that professional development is primarily an employee's responsibility [8].

The second stage is calling individual self-analysis. At this stage, the employee answers the question "Who am I?" The main goal here is to get familiar with the career. A manager as a career adviser and using surveys can help identify interests, values, and abilities related to professional development.

Stage three is authentication. The employee defines "How am I treated?" in the organization. Development assessments are key events that organizations use to provide feedback to staff on these issues. At this stage, it is necessary to determine the opportunities available to the employee. Employees gain knowledge about the organization and seek alternatives inside or outside the organization by participating in growth and development activities.

At the fourth stage, LRDS specialists reveal the organization's resource planning needs so that employees can determine how well they fit their organization. LRDS specialists should also encourage employees to study announced vacancies. In addition, mentors and managers (as career advisers) should also express their views on career opportunities.

Finally, at the fifth stage, employees should turn their attention to developing an action plan to achieve career goals. Career planning is the main activity at this final stage. At the same time, LRDS specialists should be involved in the training and engagement of managers as career advisors to help employees draw up career plans.

To ensure the success of joint efforts of employees and organizations, an important role is played by the corporate culture, the tasks of the management and the team is to form it as a culture of development. In practice, a culture of development cannot succeed unless leaders, managers and employees collectively use their abilities to achieve strategic business goals and objectives. Improper handling of labour resources impedes obtaining the desired results.

Organizational performance and effectiveness cannot be achieved without improving the skills, qualifications, talents and intellectual abilities of all participants in the organization. It is important to organize personnel work in such a way as to ensure the solution of the tasks of the innovative development of enterprises. The problem of career growth of employees should be addressed in the aspect of the formation of innovative personnel to ensure sustainable competitiveness [9].

Results and Discussion

Staff development is a process that requires from employees and the organization of such a partnership that broadens the knowledge, skills, abilities and settings needed to solve current and strategic tasks. Professional development is the quintessence of the whole progress of the organization – the improvement of employees' abilities, while at the same time updating the organization and increasing the efficiency of its activities. Human capital must fully comply with the organization's strategy. Such an approach fundamentally changes the function of labour resources and focuses on the work of HR managers on the development of the profile of competencies of employees, assessment of the readiness of human capital to implement the strategy of the enterprise. At the same time, the existence of such an approach is the basis for the development of human capital development programs designed to increase the degree of strategic readiness for implementation of the strategy.

References

1. D.A. Ashirov, O.YU. Leonova *Upravleniye chelovecheskimi resursami* (Moscow, MMIEIFP, 2003)
2. Ye. Toffler *Tretya khvilya* (Kyiv, VD "Vsesvít", 2000)
3. V. Inozemtsev V poiskakh razumnoy strategii Inozemtsev, Econ. i MO, **7**, 124-128 (2006)

4. O. A. Sushchenko *Development of employment mediation services and labour market in the tourism industry in Ukraine* (Riga: "Landmark" SIA, 104-114, 2016)
5. N. Nenkov, O. Sushchenko, Y. Dyachenko Role of chief information officer within the system of human resource development in the service organizations (tourism). Economic Annals-XXI, **165**, 97-103 (2017)
6. I.A. Esaulova, Perm: Izd-vo Permskogo gos. tekhnicheskogo un-ta, 11, (2009)
7. M. Petrova, M. Tepavicharova, L. Boykova Possibilities for human capital development in the mining and quarrying sector in Bulgaria. E3S Web of Conferences, **41**. Retrieved from: https://doi.org/10.1051/e3sconf/20184104017 [available on 6 May 2019] (2018)
8. Jerry W.Gilley, A. Steven, Eggland and Ann Maycunich Jilley. *Principles of Human Resource Development.* Second edition. (Cambridge, Perseus Publishing, 2002)
9. K.V. Timoshenko Globalní ta natsíonalní problemi ekonomíki, **19**, 271-275 (2017)
10. I.M. Trunina Development of entrepreneurship entity competitive strategy using competence-based approach. Actual Problems of Economics, **3(177)**, 191-198 (2015)
11. C. Koch The big uneasy, CIO Magazine, Vol. **11**, 40-47, (1997)

SHS Web of Conferences **67**, 06052 (2019) https://doi.org/10.1051/shsconf/20196706052
NTI-UkrSURT 2019

The specifics of formation of the loyalty strategy in the field of alternative energy

Victoriia Tretyak[1], *Karina* Moskalenko[*1], *Tetiana* Chelombitko[1], *Nataliia* Vasylechko[1]

[1]V. N. Karazin Kharkiv National University, 4 Svobody Sq., Kharkiv, 61022, Ukraine

Abstract. The essence of the definitions of "strategy", "loyalty" and "loyalty strategy" is considered, and the peculiar vision of the definition "loyalty strategy in the field of alternative energy" is proposed. The specifics of formation of the loyalty strategy in the field of alternative energy are determined. The stages of formation of the loyalty strategy for stakeholders in the field of alternative energy are grouped. It is proved that the branch of alternative energy has a significant impact on territorial development, in particular on the transformation of socio-economic processes in the region.

1 Introduction

Formation of effective governmental long-term policy requires the socio-economic development of the territories. The concept of the development of a separate region is aimed at defining tasks and tools for solving social problems, productivity of the economy, and profitability of business and income of the population [1].

Transformational processes of decentralization of the economy will help to increase the role of regions in organizing and ensuring real processes of sustainable development, mobilizing resources and funds for certain goals. Therefore, the use of alternative energy is highly relevant.

Under today's conditions of economic competition, the formation of the loyalty strategy in the field of alternative energy appears as a way of representing market participants, their values, goals and socio-ethical commitments.

The peculiarities of the development of communication management through the formation of the loyalty strategy are reflected in the publications of such scholars as I. Kliachenko [2], Yu. Kniazyk [3], K. Gurdjyian [4], D. Aaker [5], and others. However, under modern conditions of economy management, it is necessary that one should further study and improve the issue of the formation of the loyalty strategy for stakeholders in the field of alternative energy.

2 Main Results

[*] Corresponding author: kmoskalenko@karazin.ua

At the core of any research lies the definition of the essence of economic categories. Therefore, within the framework of this research, we consider that special attention should be paid to studying the scientists' views on the interpretation of such concepts as: "strategy", "loyalty" and "loyalty strategy". The definitions of the term "strategy" are given in Table 1.

Table 1: The definitions of the term "strategy" [6-9].

No.	Author	Definition
1	2	3
1.	I. Ansoff	Strategy is a systematic approach that ensures balance and overall management of the organization [6].
2.	M. H. Meskon M. Albert F. Khedouri	Strategy is a detailed and complex plan which has to ensure accomplishing of the mission of the organization and achieving its goals [7].
3.	B. Carlof	Strategy is a generalized model of action that is needed to coordinate the company's activities and allocate resources [8].
4.	T. Kalinescu Yu. A. Romanovskaya, O. D. Kirilov	Strategy is a set of measures (goals, tasks and ways of their accomplishment) to ensure the profitability of the enterprise [9].

At the time of this research, market conditions in the field of alternative energy in Ukraine are characterized by high competition among producers of goods and services in the sphere of generating energy from alternative sources, the emergence of new channels of communication between stakeholders, such as: sale of goods and services through the Internet sites and social networks of generating companies and the marketing management system of production.

Thus, in the field of alternative energy, the "buyer's market" prevails when the volume of goods or services exceeds the demand of the buyers. Therefore, the formation of loyalty of the stakeholders takes the leading place among the strategic objectives of the enterprise.

The term "loyalty" encompasses various aspects in relations between stakeholders. The definitions of the term "loyalty" are given in Table 2.

Table 2: The definitions of the term "loyalty" [5, 10, 11, 12].

No.	Author	Definition
1	2	3
1.	D. Aaker	Loyalty as measure of a customer's commitment to the brand. It shows the level of probability of switching to another brand, especially when changes in price or any other indicators take place [5].
2.	J. Rossiter and L. Percy	Loyalty as a regular (recurring) purchase of a branded product based on familiarity with and favorable attitude to the brand [10].
3.	F. Reichhold	Loyalty as property inherent to a user of value who each time returns to his/her source and leaves this source as inheritance. That means that loyalty is commitment to his/her source of value [11].

4.	F. Kotler	Loyalty as advantage of the customer formed as a result of the personification of feelings, emotions, thoughts regarding an organization and services [12].

At the present stage of development of economic thought, the definition of the term "loyalty strategy" has also acquired its further development. The definitions of the term "loyalty strategy" are given in Table 3.

Table 3: The definitions of the term "loyalty strategy" [2, 13, 14, 15].

No.	Author	Definition
1	2	3
1.	E. Taylor	Loyalty strategy is a course of action on the basis of which a company will create an advertising product or service, that means that loyalty strategy under today's market conditions is a guiding approach to communicating with stakeholders [13].
2.	H. Mintzberg	Loyalty strategy is a method of management to increase the competitiveness of an enterprise [14].
3.	Ye. Romat	Loyalty strategy as program of integrated marketing communications used to cover goods and services [15].
4.	I. Kliachenko	Loyalty strategy is a long-term program of interaction between the company and its stakeholders [2].

The consideration of the theoretical material enabled us to develop our own vision of the definition of "the loyalty strategy in the field of alternative energy" as program of long-term mutually beneficial relations between stakeholders, in which they can not only choose from a variety of goods or services, but also aim to expand the policy of market actors in such a way that it corresponds their interests as much as possible. The following features have to be taken into account when developing the loyalty strategy in the field of alternative energy:

- establishment of long-term relations between stakeholders which require analysis of all stages of purchase, supply, transportation, storage and consumption of energy;
- state policy in the field of alternative energy, the work of regulatory authorities;
- interests of generating companies, energy companies, end users;
- a complicated process of purchasing energy resources, which involves a wide range of individuals.

Taking into account the 0peculiarities of the alternative energy sector, we have grouped the stages of the formation of loyalty strategy for stakeholders in this area (Fig. 1).

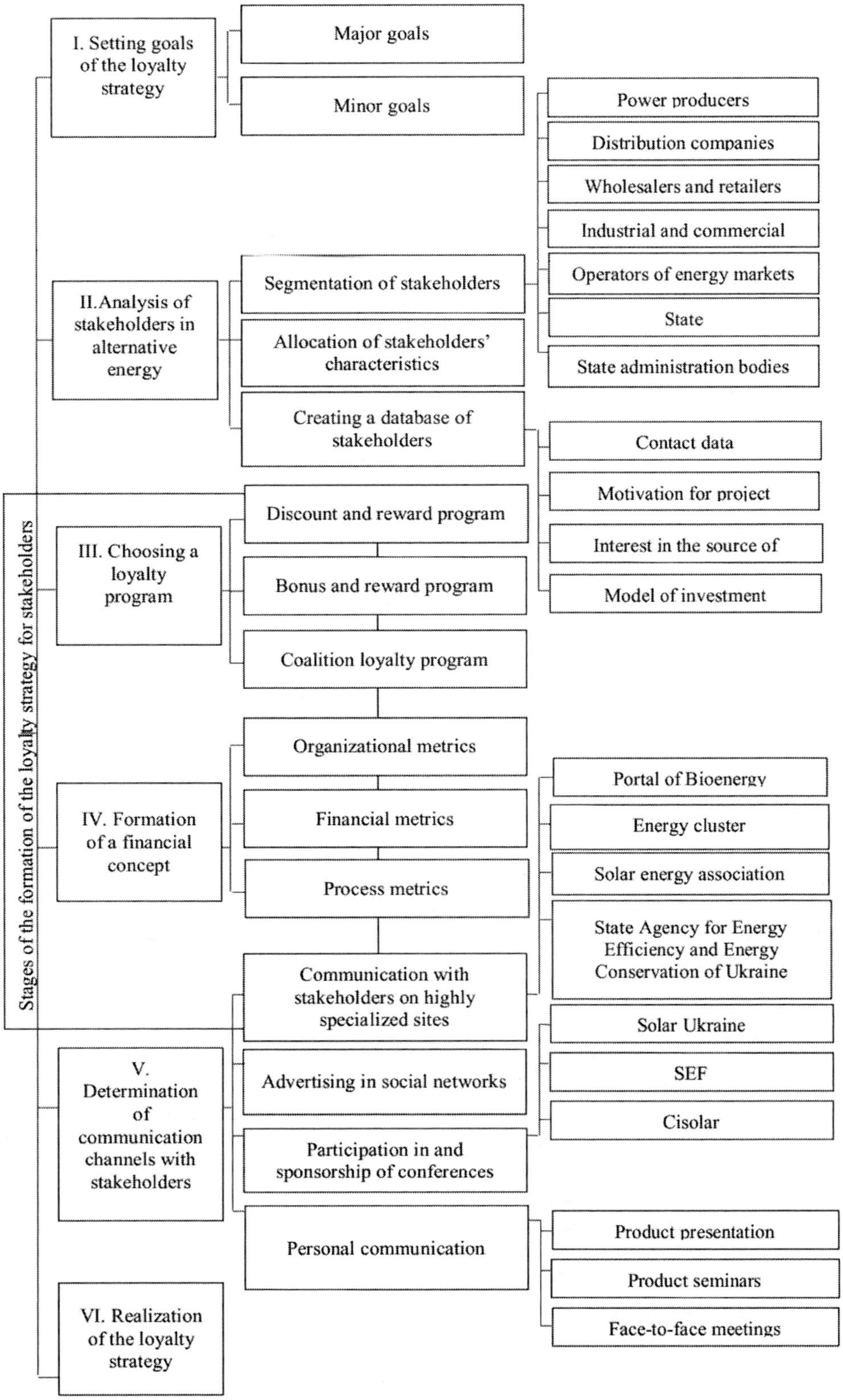

Fig. 1. Stages of the formation of loyalty strategy for stakeholders in the field of alternative energy.

The formation of the loyalty strategy for stakeholders in the field of alternative energy is a crucial factor in territorial development, as the interest in the development of energy resources arises if stakeholders express their loyalty to the energy market. Characteristics of the impact of alternative energy introduction on the territorial development, taking into account the legislative norms on alternative energy sources [16], are presented in Figure 2.

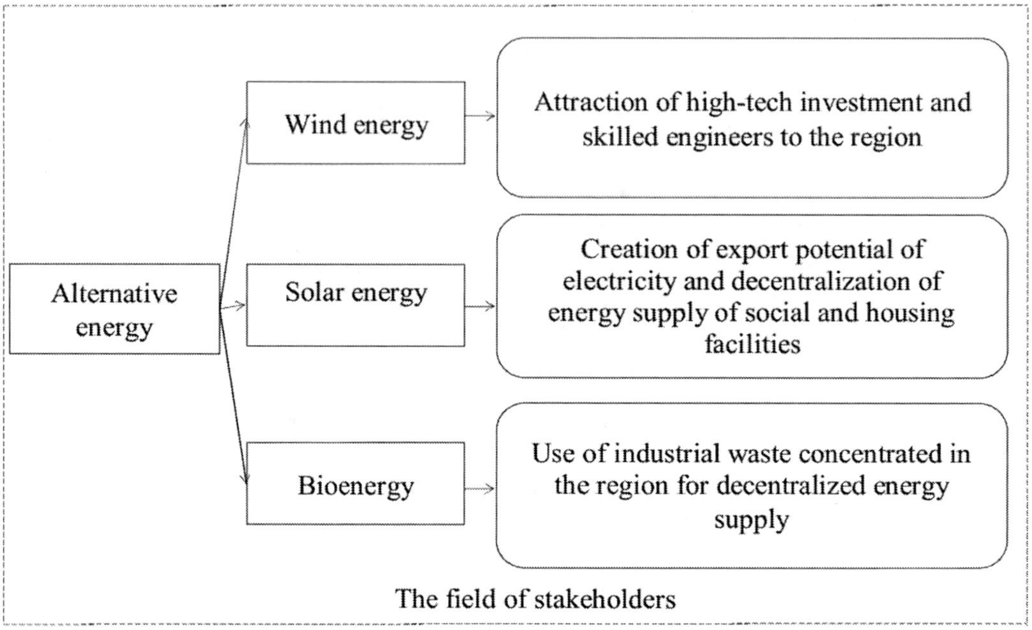

Fig. 2. The structure and characteristics of the impact of alternative energy introduction on the territorial development.

Thus, energy sources have a fairly even distribution and can be used to decentralize energy supply. This reduces the cost of energy transportation, diversifies the risks of the destruction of energy networks and if using alternative sources of energy, it brings down the use of fossil fuels, which is the basis of energy security both for separate territories and for the state as a whole.

3 Conclusion

The field of alternative energy has a significant impact on territorial development, in particular on the transformation of socio-economic processes in the region, i.e.: if free energy resources are available on a certain territory, potential investors express their loyalty to the implementation of new projects, thereby developing the territory (creating new generating capacities and jobs, consuming agricultural production waste, etc.). The introduction of alternative energy will promote the development of small and medium-sized businesses, as well as improve living standards.

The loyalty strategy in the field of alternative energy is a program of long-term mutually beneficial relations between stakeholders, which includes communication support with a set of incentive benefits.

References

1. Official bulletin of Ukraine, **70**, 34 (2014)
2. I. O. Klyachenko, O. V. Zozulov, *Consumer Loyalty to the Brand Program*, 28 (2012)
3. Y. M. Knyazyk, T. V. Lebid, *Formation of consumer loyalty as one of the directions of increasing profitability of the enterprise*, 86-90 (2006)
4. K. V. Gurdjian, *Essence and types of loyalty*, 78-86 (2015)
5. D. Aaker, *Creation of strong brands*, 320 (2008)
6. I. Ansoff, *New corporate strategy*, 416 (1999)
7. M. Meskon, M. Albert, F. Hedouri, *Fundamentals of Management*, 702 (1992)
8. B. Carloff, *Business Strategy: Concept, Content, Symbols*, 239 (1991)
9. T. V. Kalinescu, U. A. Romanovskaya, O. D Kirilov, *Strategic potential of the enterprise: formation and development*, 272 (2007)
10. J. Rossiter, L. Percy, *Advertising and promotion of goods*, 136-143 (2000)
11. Y. Skobeleva, Measuring *loyalty to the brand according to the "residual promoter index" method by Frederick F. Reichheld*, available at: http://www.satio.by/publications/researches/76.html
12. P. Kotler, *Marketing 3.0: from products to consumers and further - to the human soul*, 240 (2011)
13. R. E. Taylor, *A Six-Segment Message Strategy Wheel*, 7-12 (1999)
14. G. Minzberg, D. Quinn, S. Goshall, *Strategic Process: Concepts, Problems, Solutions*, 684 (2000)
15. E. V. Romat, *Advertising*, 496 (2001)
16. Data of the Verkhovna Rada of Ukraine, **24**, 155 (2003)

The role of stakeholders' interests management mechanism in the transport services market development

V.P. Tretyak[1], *G.P.* Rekun[1], *K.O.* Andreieva[1*], *and M.S.* Bieliai[1]

[1]V.N. Karazin Kharkiv National University (Kharkiv, Ukraine)

Abstract. Article deals with stakeholder's interests analysing in transport services market. The main goal of the paper is to propose the best solution for current stakeholders' conflicts in transport market. Common groups of stakeholders in transport services market are identified. Stakeholder's interests are highlighted and analysed. Social, corporate, business, ecological, and political interests are compared. One-level and inter-level conflicts of stakeholders' interests are figured out. Analysis according to Importance and Influence Stakeholder Analysis framework has been made. In particular, it is determined that the stakeholder's conflicts are complex and entailed the emergence of a conflict web. A set of tools for most effective solving current problem and achieving existence opportunities are selected. Searching for a "win-win" solutions is proposed as the basic strategy for resolving stakeholders' conflicts of the transport services market, achieving the goals of sustainable development, and acceleration of European integration processes.

Transformation processes in modern economic systems contribute to the creation of disparities and discontinuities in the chains between key players in any market - consumers and producers of goods and services. The imbalance of interests and tools for their harmonization is the main reason of it. This problem becomes of particular importance for transport enterprises taking into account the role of the transport system and its participants in ensuring the development of the state.

The purpose of the article is identifying the key stakeholders of the transport services market in Ukraine and the best tools of forming a mechanism for resolving conflicts between them.

1 Main part

An effective and flexible system for coordinating stakeholders' interests should be created for the successful operation and development of the transport services market. For this purpose, it is advisable to classify the main stakeholders by the importance of their participation in the process and to identify their priority groups, interests and possible contradictions. A mechanism for coordinating stakeholders' interests is developed on the basis of the above [1].

* Corresponding author: andreeva@karazin.ua

Consider the stakeholder classification on the transport services market example (Fig. 1). The key stakeholders of this market include the state (the main regulator), transport enterprises (transport corporations and small transport companies), states with transport links (main partners during the carriage), employees of transport companies, suppliers (vehicles, components, energy carriers, fuel, related products) and consumers of transport services. The effective interaction between them and the management of their interests are the key to the sustainable development of the transport services market in Ukraine.

Economic, social, ecological, and political interests can be identified as the most important for the activities of transport companies. [2]. Also, they are divided into: interconnected, coherent, or conflicting. That depends on the synergistic effect for each group.

The existing in our country model of finding a consensus and of developing a transport system (tools, set of approaches) does not gives a positive result. This requires a change in the approach to concentrating on subjects (stakeholder groups) for effective managing conflicts of interest between them.

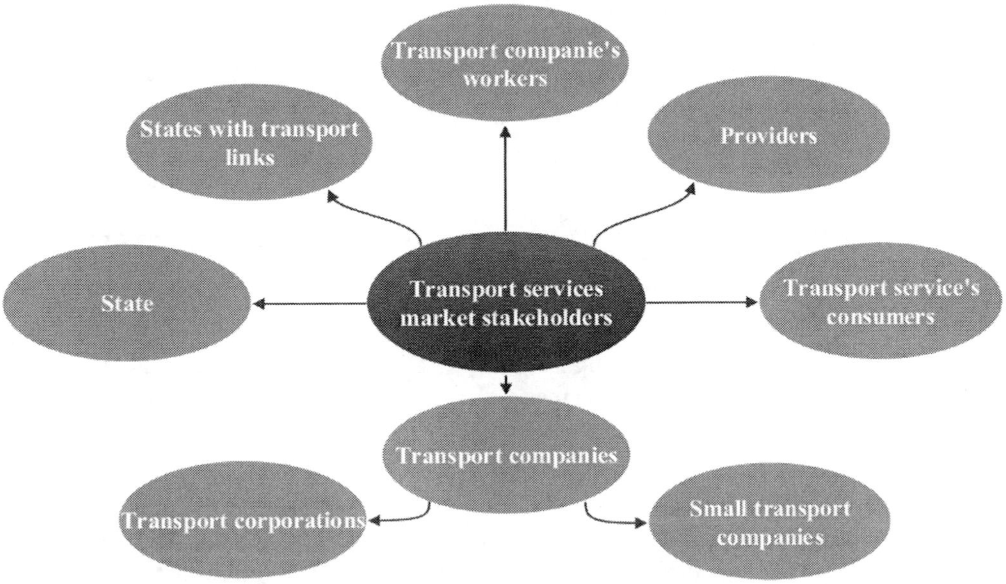

Fig. 1. Classification of The transport market stakeholders [1]

In general, the interests of certain stakeholders can be grouped in the following box:

- Economic;
- Social
- Ecological;
- Political;

We considered the thesis that characterizes the interests of each stakeholders' group by the proposed box.

Table 1. The transport market stakeholders' interests

	Economic	Social	Ecological	Political
State	Reduce the cost of maintaining and refurbishing transport	Ensuring social standards and fulfilling their social function.	Ensuring environmental norms, environmental	Preservation of state sovereignty, border protection, political and

	infrastructure, and increasing the amount of collected taxes.		protection, reducing CO_2 emissions	information security
State with the transport links	Reduce the cost of transport systems integrating.	Protecting own citizens while using Ukrainian transport.		
Transport corporations	Providing stable profits and company development, reducing costs (taxes, % on loans, salaries, etc.)	Compliance with legislation and social standards.	—	Strengthening influence on society, getting of monopoly positions.
Small transport companies			—	Saving of positions, protection from monopoly, dumping and admin impact.
Transport companies' workers	High salary.	Good working conditions, high social standards, corporate culture.	Safety of life and health while performing their duties.	
Providers	Greater price for their product, reducing costs.	—	—	Levels of rights for suppliers' lack lobbying.
Consumers	Reduce the price of transport services.	High quality of services	Safety of life and health during using transport services.	

Source: [1,3]

A detailed study of each stakeholder's interests does not reveal a synergistic effect on the level of stakeholders. It is understood that only interests of one direction can be conflicted, such as the economic interests of users with the economic interests of transport corporations. However, there is a situation where the interests of different levels conflict through the tools, which these interests are achieved. So, the ecological interests of the state conflict with the economic interests of transport companies through instruments (ecological requirements and taxes). At the same time, it exacerbates the conflict of economic interests (one level) of the company, user and company's employee.

As Figure 2 shows, a significant proportion of stakeholders is important for the transport services market in Ukraine, but only three categories differ by grate influence: the state, transport corporations and suppliers. It is these groups of stakeholders who can independently defend their interests, which forms two additional threats to conflicts.

First, the conflicts that arise between these stakeholders considerably exhaust the resource that could be used in a positive way. Secondly, interested parties that do not have such influence automatically fall into the dependence on "strong" players. A similar situation of power balance in a conflict of interest exists in corporate governance. In particular, conflict between minority and majority shareholder. Therefore, it is proposed to use a special toolkit for corporate governance to manage conflicts of this level [5].

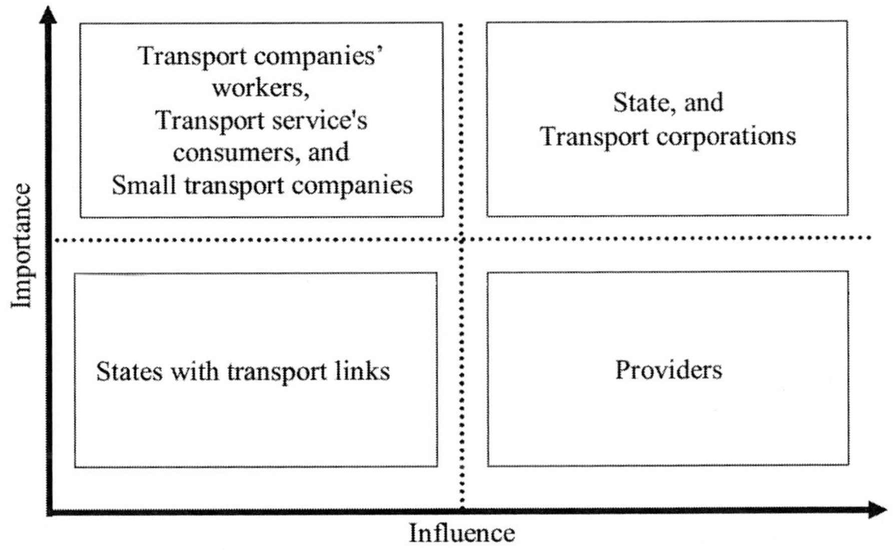

Fig. 2. Classification of stakeholders by Influence and Importance [1,4]

To achieve the goals of sustainable development of the state, including in the transport sector, and acceleration of European integration processes it is proposed not to seek compromise solutions but to meet the needs of each group using current world experience and corporate governance practices. It must be understood that a comprehensive system will not only overcome the gaps that arise from conflicts, but also strengthen the already existing positive synergistic effect. Therefore, we recommend the use of the following tools to resolve the problem [2]:

- eco-lending (can simultaneously meet the economic interests of corporations and the state's ecological interests and, as a result, reduce the intensity of the conflict between business, users and workers).
- the use of international standards, not only in technological aspects, but also in management (for solving the conflict between corporations and small businesses, meeting the social and environmental needs of the state and the interests of the state on the combination).
- "Smartization" of the transport services market (at the same time, it satisfies the needs of the state, economic needs of business, social and political needs of users and workers).
- public-private partnership instruments (while satisfying the economic, political, social and ecological interests of the state, the economic needs of corporations, the political needs of small businesses, and the political, economic and social needs of users and workers).
- ESCO-mechanism (allows satisfying simultaneously the economic interests of corporations and the state's ecological interests, economic interests of users and the political and ecological interests of workers).
- SROI tool and social investment (can balance the social and economic interests of all stakeholders);
- marketing and business tools.

2 Conclusions

In conclusion, we can say that in modern conditions, ensuring the efficient functioning of the transport services market and the stable competitive advantages of its participants are ensured by the coordination (management) of their key interests. However, today in Ukraine it is difficult to achieve the required balance, because the previous instruments are not effective. Accordingly, there is a need for the formation of a new effective mechanism for management conflicts between the stakeholders' interests and the implementation of its instruments at different levels of management.

References

1. Development strategy of PJSC "Ukrzaliznytsya" for 2017 - 2021 years. Retrieved from: http://publications.chamber.ua/2017/Infrastructure/Session_1_Kravtsov.pdf [in Ukrainian]
2. Babina, O.E., Strelkov, O.V. "Investments: practice and experience". *Interests of transport service providers* **16**, 15-21 (2017)
3. Schaar, David, and Lance Sherry. "Analysis of airport stakeholders." *Integrated Communications, Navigation, and Surveillance Conference Proceedings*. IEEE, (2010)
4. McCormick, D., Stakeholder Analysis, Paper Presented in REME Training Workshop (1997)
5. Landini, Fabio, and Ugo Pagano. "Stakeholders' conflicts and corporate assets: an institutional meta-complementarities approach." Socio-Economic Review (2018).

Increasing social responsibility in tourism based on volunteer tourism

Iryna Trunina[1], *Inna* Khovrak[2*] and *Maryna* Bilyk[1]
[1]Kremenchuk Mykhailo Ostrohradskyi National University, Tourism Department, 39600 Kremenchuk, Ukraine
[2]Kremenchuk Mykhailo Ostrohradskyi National University, Accounting and Finance Department, 39600 Kremenchuk, Ukraine

Abstract. The aim of the paper is to determine the impact of volunteer tourism on the level of social responsibility in the tourism industry of Ukraine. The online survey included 440 respondents (77.3% women; 22.7% men). Physicians were asked about the importance of volunteering and their participation in the volunteer movement. According to the results of the survey, 59.1% of respondents do not have volunteering experience, 38.6% of respondents have episodic experience and only 2.3% of respondents constantly participate in volunteer activity. Although a quarter of respondents who do not have volunteering experience do not consider it appropriate to have such experience. This allowed us to identify the motives for engagement in volunteering and the factors hindering such activities. Participants were also asked about the impact of volunteer tourism on the prosperity of regional communities, building a democratic society, education of socially responsible citizens. The research has shown that for the development of volunteer tourism the most important is the promotion of volunteerism in society (61.4% of respondents) and cooperation with international organizations (50.0%). This allowed the authors to suggest directions and forms of international cooperation for the development of volunteer activities in the tourism.

1 Introduction

In recent years, the ideas and principles of socially responsible tourism as a world-wide conceptual tendency of mass recreation of citizens have been becoming popular in the world. This combination of pleasure travels to any unique place with the opportunity to visit its attractions and provide socially useful assistance to the benefit of the region. To streamline these processes, the Cape Town Declaration on Responsible Tourism was adopted in 2002 and its principles were formally approved. In June 2016, the concept of socially responsible tourism was presented at the 12th International Conference on Responsible Tourism in Finland.

*Corresponding author: inna.khovrak@gmail.com

There are the following types of responsible tourism: (1) Ecological (contributes to the protection of the environment and the maintenance of the welfare of the local population, different from natural tourism with the presence of commitments in relation to environmental protection and the development of regions). (2) Volunteer (carries out volunteer missions in support of responsible tourism, especially in the objects and territories studied by scientists or other specialists). (3) Religious (combines religious values with the natural, cultural, historical heritage of the region and contributes to making a work contributes to improving the lives of churches and monasteries). It is worth pointing out that the ecological and religious aspects of responsible tourism are well-known and common in Ukraine. However, the phenomenon of volunteer tourism in Ukraine is rather new and little known. At the same time, at the present stage of development of the society, the volunteer institute is a widespread phenomenon in many countries. Moreover, the work of volunteers with each passing year becomes an increasingly significant resource for the development of the world economy, in particular, the tourism industry. Participation in the volunteer movement benefits both society as a whole and the volunteer itself. Attracting people into volunteer activity is one of the important indicators of the level of self-organization of society. Volunteerism creates opportunities for socially active young people to realize their potential, to help others, to find support and to feel the social value of their work, to obtain certain life skills, and so on.

That is why the purpose of the study is to determine the impact of volunteer tourism on the level of social responsibility in the tourism industry in Ukraine. The research was carried out in the following sequence: the essence and characteristics of voluntarism were determined, an evaluation of the potential of voluntarism in Ukraine was outlined, prospects of international cooperation for the development of volunteering activities in the tourism sector were outlined.

2 Data and Methods

The online survey among 440 respondents (77.3% or 340 people are women; 22.7% or 100 people are men). Distribution of respondents by age under 18 years old (2.3%), 18-25 years (63.6%), 26-35 years (22.7%), 36-50 years (9.1%), over 50 years (2.3%). The survey was opened online on May 1, 2019, and closed on May 19, 2019. Each participant could complete the questionnaire only once. All of the participants were asked about their age and gender. Physicians were asked about the importance of volunteering and their participation in the volunteer movement. Participants were also asked about the impact of volunteer tourism on the prosperity of regional communities, building a democratic society, education of socially responsible citizens. This allowed us to identify the motives for engagement in volunteering and the factors hindering such activities. The authors proposed directions and forms of international cooperation for the development of volunteering in the tourism.

3 Results

3.1 The Essence and Characteristics of Volunteer Tourism

Volunteer tourism is based on the sustainable development of the tourism industry in the long run and the support of the socially weak sections of society in countries where tourism is extremely productive. One of the priorities of the concept of volunteering is to support the cultural exchange between tourists and the local population, taking into account the identity of each nationality, and a reasonable redistribution of income from tourism. A significant innovation in the strategy of volunteer tourism is the reduction of costs and excessive consumption, which will positively affect ecological equilibrium. In volunteer tourism, the desire for profit is inferior to the more altruistic desire to travel in order to help communities. So the predominance of altruistic motives over commercial volunteer tourism is primarily an alternative to mass tourism. Table 1 shows the main characteristics of volunteer tourism.

Table 1. The main characteristics of volunteer tourism

	[6]	[7]	[8]	[9]	[10]	[11]	[12]
help for certain social groups	+	+	+	+			
individuals travelling on a nonpaid working holiday for worthy causes		+					
participating in a variety of social, educational and environmental projects			+	+	+		
does not focus on profits				+			
a short-term trip, less than three months					+		
yet another form of post-colonialism and globalization						+	
responsible tourism				+			
counteraction to corporate burnout							+

Volunteer tourism is a form of tourism that involves voluntary assistance for certain social groups, participation in social projects of different orientations, not focusing on profit, but on sustainable development of regions. Volunteer tourism has some advantages over other types of tourism: (1) Attracting a significant number of tourists by creating volunteer tours in the region; (2) Volunteer tourism does not limit volunteers to the choice of season; (3) If the event is long-term, the time of stay of volunteer tourists is increasing; (4) Volunteering tours are usually for different age groups; (5) This type of tourism does not require presence on the territory of natural resources. One of the most learned components of volunteer activity is motivation. The volunteer activity represents a unique opportunity for man to combine the satisfaction of his personal needs with the needs of society. World experience proves that the formation of the motivation plays an important role in the environment, especially educational [13-18]. In addition, there are a number of specific barriers caused by the natural fears and risks of the individual, in connection with the unrelatedness and danger of volunteer tourism. That is why it is relevant to identify the most promising directions of cooperation between Ukraine and other countries in terms of volunteer activity. In addition to countries, international cooperation actors also include domestic and foreign volunteer organizations, non-governmental organizations, higher educational institutions, etc.

3.2 Estimation of Volunteer Tourism in Ukraine

According to the results of the survey, 59.1% of respondents do not have volunteering experience, 38.6% of respondents have episodic experience and only 2.3% of respondents constantly participate in volunteer activity. It is important that none of the respondents chose the answer "do not consider it necessary to have such experience". However, answering questions about the frequency of volunteering, 13.6% of respondents indicated that they did not consider it appropriate to engage in volunteering. That is, from 59.1% of respondents who do not have experience in such activity, practically a quarter does not consider it appropriate to have such experience. In this case, 13.6% of respondents consider it necessary to engage in volunteering once a week and more often, 50% - 1-2 times a month, 13.6% - 1-2 times a half year, 9.1% - 1-2 times a year and less often.

Among the reasons that most impede engagement with volunteering, the respondents preferred the lack of information about volunteer organizations and initiatives (34.1% of respondents), lack of time (31.8%), and lack of confidence in volunteer organizations (18.2%). Also, the reasons for low activity in volunteering are the presence of own problems (9.1% of respondents), lack of self-confidence / knowledge / experience (4.5%) and lack of willingness to engage in volunteer activities (2.3%).

Mostly, the volunteer activity of respondents was participation in community projects and fundraising for patients, helping vulnerable people and animals, distributing information and more. The survey found that respondents considered the following areas of volunteering to be the most necessary in Ukraine: assistance to children (63.6% of respondents chose this sphere as one of the 5 most important ones), assistance to people with special needs (59.1%), homeless animals (56.8%), urban improvement and community development (50.0%), assistance to the elderly (50.0%). For the development of the volunteer movement, it is necessary to understand the motives that can bring the population to volunteer activity. Respondents were asked to select a maximum of five out of fourteen options. The most popular were motives like helping others (56.8% of respondents), feeling their own need (40.9%), acquiring professional skills (40.9%) and gaining new knowledge (40.9%). The least motivation for volunteering is to follow the example of friends / relatives / colleagues (only 6.8% of respondents chose this answer). According to respondents, the main qualities of volunteers are the desire to help solve the problem (72.7% of respondents believe), responsibility (70.5%), selflessness (61.4%), kindness (47.7%) and communication skills (45.5%).

Absolutely all respondents indicated that they love to travel. At the same time, 40.9% of respondents indicated that they would like to try themselves volunteers abroad in their lifetime. In this case, 9.1% of respondents do not plan to receive such experience, and 50% have not yet been determined. One of the reasons for this situation is the lack of awareness of the population about the opportunities and benefits of volunteering, as well as the lack of knowledge about the existence of volunteer programs, projects or organizations. Thus, 65.9% of respondents did not hear about the existence of international volunteer programs, projects or organizations, 13.6% - have the positive experience of cooperation, 13.6% plan to join international programs. However, 6.9% of respondents know, but do not use the possibilities of international cooperation in the field of volunteering. It was found that volunteer activity has the greatest impact on the education of socially responsible citizens (38.6% of respondents recognized the maximum impact of volunteering on this factor). According to respondents, the significant influence of volunteer activity is possible on the construction of a democratic society, as well as on the well-being and prosperity of regional communities. In order to accelerate the development of voluntarism, the most important is the promotion of volunteerism in society (61.4% of respondents) and cooperation with international organizations (50.0%). According to respondents, the least impact is the consolidation of the efforts of civic organizations (2.3%).

3.3 The prospects of international cooperation for the development of voluntary activities in the tourism sector

As the research showed, one of the most effective areas of cooperation today is the development of joint international social projects, with the involvement of volunteers. The second area of cooperation is the organization of cooperation at the expert level. The third area of cooperation is raising awareness of the Ukrainian population about the development of volunteer activities, volunteer projects and programs (Fig. 1).

4 Conclusion

Socially responsible tourism is a set of measures, recommendations and rules for travelers and providers of tourist services that promotes sustainable development of the regions. It applies to any type of tourism and encourages all tourists to contribute to the development of territories, to help local communities, to have an active civic position. Accordingly, tourists are able to travel consciously and be attracted to the life of that small, original world, where they travel. Socially responsible tourism can be internal (when representatives of the tourism industry, in addition to profit, are aimed at socio-cultural, ecological and economic transformation of the society within the region of residence) and external (when representatives of the tourism industry, in addition to profit, are aimed at socio-cultural, ecological and economic transformation of the society within the regions of temporary stay).

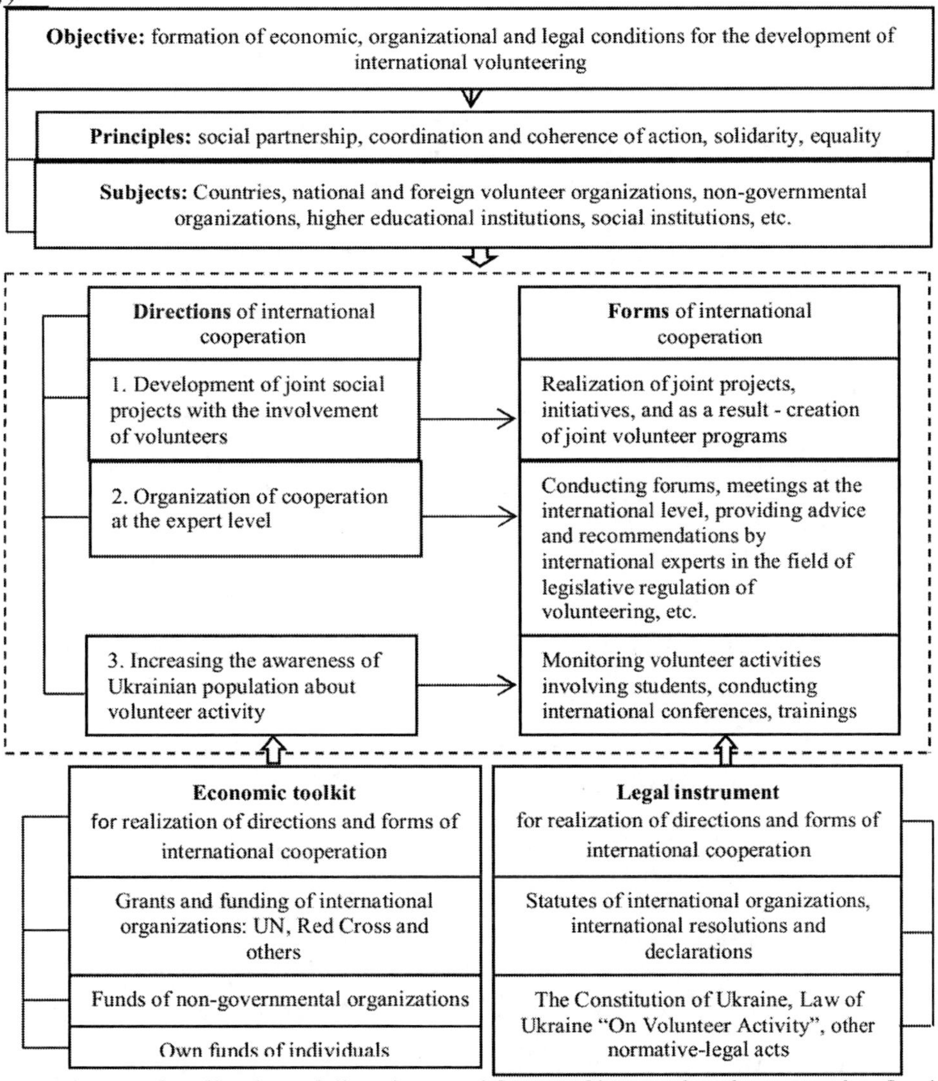

Fig. 1. Scheme of realization of directions and forms of international cooperation for the development of volunteer activity* source: own research

A socially responsible tourist is a person who supports objects and initiatives of socially responsible tourism, adheres to the rules and principles of responsible tourism in travel and in life. The tourist voluntarily participates in projects of socially responsible tourism, independently determines its conditions, or accepts the terms of the project. Participation of a tourist may be non-material (socially useful labour) or material (fully or partially covers the cost of the tour, taking part in volunteer activities based on charitable or membership fees). Stakeholders of responsible tourism are (1) authorities (state, regional and municipal levels) in the field of tourism and in other related fields (sports and youth policy, ecology, industry, economy, culture, investments); (2) business (all interested representatives carrying out socially responsible actions and supporting projects of responsible tourism);

(3) non-profit sector and especially funds that lead the direction of responsible tourism; (4) higher educational establishments (especially those who specialize in tourism); (5) tourists who travel consciously. The study found that 59.1% of respondents do not have volunteering experience, 38.6% of respondents have episodic experience and only 2.3% of respondents constantly participate in volunteer activity. Although, a quarter of respondents who do not have experience in volunteering does not consider it appropriate to have such experience. However, respondents believe that volunteer tourism has a significant impact on the prosperity of regional communities, building a democratic society, education of socially responsible citizens. Therefore, in order to make volunteer tourism a priority for tourism development, it is necessary to outline the benefits of volunteer tourism, as well as actualize the motives of people in the voluntary sphere. Understanding the motivations of volunteers is a key skill for good leaders of volunteer programs in tourism. This is a prerequisite for helping tourists to realize their motivation and prevent cases where motivation can cause certain complications. An important aspect of volunteering development for Ukraine is the formation of international cooperation in this field.

References

1. Wearing S. *Volunteer Tourism: Experiences That Make a Difference*. New York: CABI (2001)

2. Cheung S., Michel M., Miller D. Voluntourism: Give a Little, Gain a Lot. Ontario: Georgian College Research Analyst Program. [online]. Available at: https://voluntourismgal.files.wordpress.com/ (2010)

3. Benson, A. (ed.) *Volunteer Tourism: Theory Framework to Practical Applications*. New York: Routledge (2011)

4. Carter K. Volunteer Tourism: An Exploration of the Perceptions and Experiences of Volunteer Tourists and the Role of Authenticity in Those Experiences. [online]. Available at: https://researcharchive.lincoln.ac.nz/ (2008)

5. Daldeniz, Bilge and Hampton, Mark P. VOLUNtourists versus volunTOURISTS: a true dichotomy or merely a differing perception? In: Benson, Angela M., ed. Volunteer Tourism: Theory Framework to Practical Applications. Routledge (2010)

6. McGehee, N. G., & Andereck, K. Volunteer tourism and the "voluntoured": The case of Tijuana, Mexico. Journal of Sustainable Tourism, **17(1)**, 39-51 (2009)

7. Brown, Sally A. *Voluntourism – Traveling with a Purpose: Understanding the Motives and Benefits*. Ph.D. diss., Purdue University, West Lafayette, Ind. (2005)

8. Kratt O. Trunina I., Zagirniak D. The assessment of the provision of the industry by electrical engineers. *International Conference on Modern Electrical and Energy System (MEES)*. P. 356-359 (2017)

9. Buzko I., Vartanova O., Trunina I., Khovrak I. Theoretical aspects of regional sustainable development in the EU and Ukraine. SHS Web of Conferences. Volume **61**, 01001 (2019)

10. Sushchenko O., Trunina I., Klok O., Loseva O. Management technologies of ensuring environmental protection as the territory development strategic priority. SHS Web of Conferences. Volume **61**, 01026 (2019)

11. Trunina I., Kushal I., Zagirniak D. An imitation model of financial-economic mechanism of taking strategic decisions at the enterprise. SHS Web of Conferences. Volume **61**, 01027 (2019)

12. Zagirniak D., Khovrak I., Perevozniuk V. The Convergence of the Systems of Education at Scottish and Ukrainian Universities in the Context of Ethic Leadership. International Journal of Engineering & Technology, Vol **7**, № 4.3, Special Issue 3. P.: 550-556 (2018)

13. Kratt O., Pryakhina K., Bilyk M. Ukrainian-Chinese collaboration: Prospects of development. SHS Web of Conferences. EDP Sciences, **39**, 01014 (2017)

Complex synthesis of factors for modern development of welfare of territorial communities

Valentyna Varuk[1,1], and *Olena* Parkhomenko

[1] Education and research institute "Karazin business school" V.N. Karazin Kharkiv National University, 1, Myronosytska St., Kharkiv, 61002, Ukraine

Abstract. It is substantiated that providing citizens' welfare, which occurs through a complex synthesis of factors, is one of those that is in the constant field of view of the socio-economic sphere. Analyzing, systematizing and summarizing the scientific works of many scientists, welfare factors are identified that are important for prospects' understanding of territorial communities' functioning, their role and place in the national economy, conditions that ensure competitiveness, and directions for improving efficiency of activities. At the same time, factors of welfare of territorial communities can be individual and subjective, with different significance for different people, may vary throughout a lifetime, and may also have a direct or indirect impact on each member. So, some people pay more attention to financial wealth, others – to their physical and/or mental health. The following factors of welfare of territorial communities are proposed and described: socio-economic, demographic, legal, political, cultural and psychological. It is concluded that to date, in the context of intensive transformations and reforms, communities' welfare is a product of many and often interconnected factors, understanding the impact of which is important for determining competitiveness of the communities and directions to improve efficiency of activities.

1 Introduction

Determining prospects of functioning of territorial communities, their role, place, conditions of activity is important in the modern economy of the country. There is a need to study welfare factors to ensure the above. Analyzing and systematizing the scientific works of many scientists [1, 2, 3], it was established that the question of which factors can affect the social life's satisfaction remain relevant forever. Therefore this study is aimed at researching and describing welfare factors, as there is no generalization of this issue.

2 Summarize factors that affect welfare of territorial communities

In conditions of intensive transformation and reforms, welfare factors of territorial communities need improvement and development. It is necessary to identify several important groups of welfare factors of territorial communities, namely (Fig. 1):

[1] Corresponding author: varuk@karazin.ua

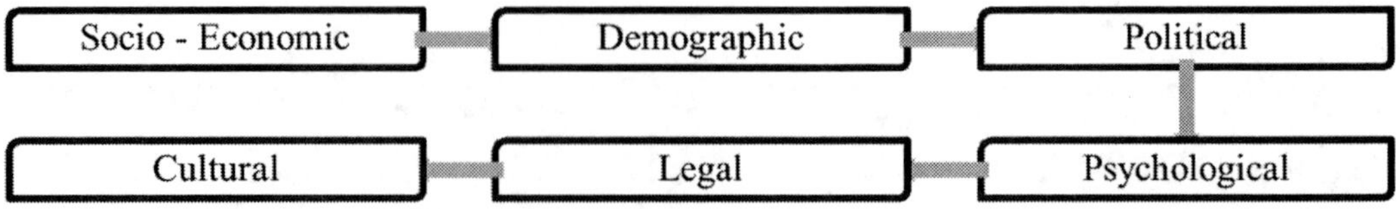

Fig. 1. Factors of growth of welfare of territorial communities (developed by the author)

The first factor is **social and economic welfare** of members of territorial communities, namely the level of prosperity and life's quality of people. Scientists determine social and economic welfare of territorial communities' members as a process of improving the life's quality of people, which living in relatively isolated and sparsely populated areas. Economic welfare of community members can be measured by various indicators such as GDP and other indicators that reflect population welfare (for example, literacy, doctors' number, pollution levels, etc.). Economic welfare of community members has a positive impact on the lives of the population and society as a whole.

Regarding social welfare of community members, it manifests itself in a sufficient human satisfaction with a dynamic system of social ties, with a presence of positive interpersonal relationships that manifested in the form of friendship, love and support. A level of social welfare of community members can be high when a person has the guarantees of good living conditions, decent education, medical care, etc. A career growth can also have a positive impact on human welfare. A work that brings pleasure allows a person to self-fulfill, provides social guarantees and stable income.

The social and economic factors of community members' welfare characterize a level of creating conditions for the fullest realization of labor and intellectual potential of population, providing them with material wealth, decent living standards, etc [3].

Workforce welfare is largely determined by a situation on the labor market and an employment level. Therefore, an effective regulation of the labor market and employment is one of the main social and economic factors of welfare. A special place among the social and economic indicators of community welfare is investment in human capital, that is, an asset of all qualities and abilities of a person, able to positively influence a development of all spheres of society life. Labor employment is not an absolute and only condition for ensuring material welfare of a person and his family. An important role is played by the wages size. One of the most significant factors that determining the size payments in the modern state is a level of minimum wage, which is related to the basic social guarantees.

The fundamental social and economic factor of disabled population welfare is formation of social funds (the Pension Fund, the Social Insurance Fund and the Compulsory Health Insurance Fund) to provide, assist and support socially vulnerable groups of the population. A main source of formation of such funds is deduction of employers and self-employed citizens. These contributions are calculated as a percentage of the workers' total income. Contribution collection and control over the correctness of their accrual are carried out by tax authorities. Funds are spent on provision of guaranteed health insurance services, pension payment, benefits, compensations to certain social categories [4].

A financial state and unemployment are considered as factors which are part of the overall political and economic environment that affects territorial communities' welfare. Today, there is an impact of a society general state on their welfare, including the growth of aggression in society, false reports and fake news in the media. Money provides access to other factors that affect community members' welfare. Meeting basic needs (such as housing, food, warmth, clothing, etc.) is a goal in itself, but it also gives a chance to have a steady life. At the basic level, employment is related to community members' welfare, because satisfaction from work (desired work satisfaction) is an important factor of welfare.

Educational skills concern not only knowledge acquisition in a formal sense, but also a general desire to expand their knowledge and to be exposed to new ideas. For many, educational opportunities are not only related to their own professional development, but also with a financial ability to send their children to the educational institutions they would like to attend [3, 5]. Health is seen as a very important factor affecting community members' welfare and is understood as absence of illness, suffering or any weakening or disturbance of public life. First of all, it is physical preparedness and physical exercise that affect an energy level, the cardiac performance and physical form. In a similar way, nutrition affects human health, and therefore welfare.

Innovative factors of community members' welfare include: achievements, a level of scientific and technical development, volumes of research results, degree of introduction of innovations and information technologies, a level of result accessibility of scientific and technological activities, innovations, information technologies, etc. A level of environmental pollution, mechanisms for regulating nature use, ensuring protection, an impact of weather are factors of community members' welfare. These factors, as well as the territory features and an availability of natural resources, may have a positive impact on welfare.

Therefore, the social and economic factors of community members are the following elements: income, employment, housing, education, work, duration and quality of life, leisure, an environment, the happiness levels.

The next factor of welfare of territorial communities is a connection with demographic characteristics. The **demographic factors** of territorial communities' welfare include a population size, density, a structure by age and sex, a migration level and its geographical distribution, as well as these parameters' dynamics.

Moreover, the predicted link between demographic parameters and territorial communities' welfare is considered from two points of view. According to the first perspective, changes in a nature of population growth create changes in welfare. The second perspective is based on the fact that changes in welfare lead to changes in the population characteristics. Despite the fact that the basic hypothesis about the existence of the relationship between welfare and the population characteristics lies at the heart of the demographic system and comes to ahead, these relationships have not yet been studied.

According to the research results of the International Bank for Reconstruction and Development, the greatest impact on community members' welfare provides a subject belonging to different gender groups: women are less than men satisfied with life spheres, and at an emotional level they experience tension and distrust regarding their health. Subjects' belonging to different ages and educational groups also affects welfare level, but in somewhat lesser degrees (satisfaction with age of life spheres decreases and tensions increase). By a influence's degree on a level of person welfare, an environment of human habitation has no less significant effect (city or village): rural inhabitants, in comparison with town inhabitants, are more satisfied with a social and economic state of the country, their material provision (a material situation, living conditions, leisure, an opportunity to spend holiday), an education and relations with different generations of relatives [6].

That is, we can conclude that, according to a person belonging to one or another social and demographic group, a large difference is demonstrated by a cognitive component of welfare, which includes an assessment of satisfaction or dissatisfaction with life aspects.

Legal factors of territorial communities' welfare are the third factor, which are oriented on a formation and development of regulatory and legislative bases that ensures guarantees' observance of human rights and freedoms in all spheres of its life, as well as an organization of a society life in accordance with a law in force in the state. The legal factors of community members' welfare are based on legal norms which represent an official expression of values and norms of morality that are embodied in legal laws. Laws are a basic foundation for social stability, security and welfare of social life.

Scientists acknowledged an influence of current governments and politicians on the daily lives of territorial communities' members through decisions and laws, which implemented by politicians and government officials, and also by creating economic and political stability, providing services (including social and medical services), jobs and investments in cultural objects (museums, theaters, cultural centers, etc.) [6, 7]. So, political institutions with a high degree of trust can convert this trust into an influence on community members' welfare, while ensuring respect for human rights and a law rule. Freedom to express their opinions without sanctions from the authorities is seen as a key factor of welfare. A feeling of general confidence that human rights are respected and that they are protected by the authorities gives a certain understanding and, thus, positively influences on territorial communities' welfare.

A main mechanism that motivates a person to recognize and follow to legal norms is formal social control, the organization and implementation of which is an indispensable condition for an existence of society and the state as a whole. It includes two interconnected subsystems [8]:

- Subsystem tracking is a timely detection and elimination of undesirable behavior, maintenance and promotion of the desired;
- Subsystem of influence is to encourage people to comply with the procedure established by law.

That is, the legal factors, in aggregate, namely norms, laws, freedom, contribute to the creation and maintenance of stable, safe, comfortable conditions of social life, which are the foundation of community welfare.

The fourth factor of community members' welfare is **a political** one, which are associated with the main mechanisms for a formation and implementation of state policy, which are an integral part of the state's domestic policy and are focused on solving current problems. Its tasks are to harmonize social relations, create conditions for a progressive development of a social sphere, improve conditions, methods and quality of life, provide necessary assistance to the population, support and protect socially vulnerable groups and groups of territorial communities' members.

National projects, social programs and state standards promote to increase territorial communities' welfare. These policy tools most fully reflect the priorities, directions, goals and objectives of social policy, identify social problems and groups of people who need special attention, as well as specify methods for solving existing social contradictions. Social programs are created and operated in order to consolidate valuable, highly skilled professionals, increase their level of material satisfaction, and facilitate a creation of additional motivation for career growth in labor activity. State standards, which are established by the state norms and regulations with an aim [2, 4]:

- To determine a guaranteed structure and norms of consumption of material goods and services by citizens;
- Carrying out a single social policy in different regions of the country;
- Assessing the living standards of the population and the implementation of social policy, etc.

That is, these instruments determine a lower line of welfare and are set in a form of natural, cost, time and financial norms and standards, rules and regulations, which governing a procedure for their establishment and application.

Cultural activity is the fifth factor, which contributing to welfare of individuals, families and communities on time. Examples of recreational indicators for territorial communities' members can be the following: visits to theaters, concerts, museums, that is, possession of freedom and resources for participation in cultural programs. Formation of cultural factors for territorial communities' welfare provides for the development of personality. Leisure contributes to goals 'number that improve community members' welfare: providing relaxation, entertainment, escape from reality and from everyday work and realization, the opportunity to be creative and practice your hobbies.

Cultural factors of community members' welfare (moral values and norms, spirituality, religion, social ideals) are associated with a formation and development of social culture, which is understood as a culture of social life, communication, social relations and interactions that provide a social system's «grip», a social life's stability and a progressive development of social actors at different levels [9].

Thus, moral values and norms form a basis of moral culture and play an important role in regulating social relations and interactions, regulation or ensuring social life and prosperity, stability and order in society. A development of an individual spiritual world through an art's attraction is a mechanism for a social culture's formation. And also religion, as spirituality and religion contribute to welfare sense of every person. But despite this statement, these two concepts are not considered the same. The population speaks of itself more about spiritual rather than religious. Social ideals are interpreted as higher social goals and models, and also form a foundation of social culture. They are formed on a basis of universally recognized values, and also structure and organize a world picture. In a way, they organize an activity of social subjects and communities in general, and define their strategic target points.

That is, together cultural factors will promote an individual socialization in the community through a formation of social qualities, knowledge, etc.

The sixth psychological (subjective) factor can be attributed to a person's mood that positively or negatively affects their welfare. In turn, welfare sense can affect a person's mood. It largely depends on a person's nature and a general views on life. Subjective welfare of a person also affects a number of uncontrolled factors, such as luck, fate, genetics and personality, in which people can have very little influence. Satisfaction, which follows from a feeling of a good attitude towards yourself or belief in yourself, as well as values, self-realization, recognition, success, can be a great positive factor in person's welfare. A personal growth and development have a positive impact on welfare, which contributes to both stability and predictability in life and work.

The psychological factor of welfare affects a people understanding of a nature and purpose, the awareness and experience of a life meaning, and is one of the most important factors of human welfare. Not perceiving itself as the universe's part without thinking about its purpose, without feeling responsible for welfare of other people - a person can't be safely herself. So, Carole D. Riff, author of well-known welfare concepts, based on an integration of various theories, proposed a generalized model of psychological welfare [10].

Psychological welfare is achieved through two instrumental goals: stimulation (activation) and comfort. Stimulation refers to activity that causes excitement, including mental and sensory stimulation and physical effort. Comfort is a somatic and psychological state, based on a lack of thirst, hunger, pain, fatigue, fear, extreme unpredictability, etc. Activating within a pleasant range and comfort is a positive connection to physical welfare.

Psychological welfare is a complex and multifaceted construct that is directly related to its own personal experience and consistent living activity, which each considers to be adequate to its needs and desire. This type consists of [11]:

- Life satisfaction: what everyone thinks about their lives;
- Particular satisfaction: a determining factor is a subject's life such as work, health, family, friendship and economics;
- Positive symptoms: an experience of pleasant positive emotions;
- Negative symptoms: feeling of negative emotions.

In this case, according to Abraham Maslow, centers of psychological welfare are in a realization and development of a genuine individual potential [1]. Psychological welfare is achieved when a person lives by agreement with oneself and reaches the delivered. That is, when life's activities are consistent with personal values. In this case, the components of psychological welfare are: autonomy, personal growth, positive relationships, happiness (which can be explained by 50% genetic factors), competition and life purpose.

3 Conclusions

Community members' welfare depends on the formation of the necessary conditions for the socio-economic growth of the territory. Understanding of welfare factors is an important for the determination of the communities' competitiveness and the areas of improving the activities' efficiency.

Thus, the socio-economic factors of communities' welfare are related to optimization of a labor market and employment, ensuring incomes' decent level, protecting the purchasing power of wages, investing in human capital and creating funds for the provision of socially vulnerable groups. The demographic factors, namely: population size, density, structure by age and gender, migration level and geographical distribution, have a significant impact on welfare of everyone. Legal factors of community members' welfare are represented by social legislation, which ensures observance and protection of human rights and freedoms, the spread of legal awareness, legal culture and formal social control.

National projects, social programs and state standards are instruments of political factors that help harmonize social relations, create conditions for social development, improve conditions, way and quality of life, provide necessary assistance to the population, etc. The main cultural factor of welfare of man and the community as a whole is a full socialization, no matter what the role of others was. That is, it is a process of mastering the socio-cultural experience of previous generations, the formation of social qualities and properties, knowledge and skills, values, ideals and norms of social behavior, through which the individual becomes a full member of society. Psychological factors of welfare are connected with self-confidence, adequate self-evaluate, positive attitude to life, goodwill, sociability, emotional stability, etc.

So, territorial communities' welfare is carried out through a comprehensive synthesis of factors that influence on our lives' satisfaction.

References

1. Yu.I. Kashliuk, The main factors influencing the individual's psychological welfare, Problems of modern psychology, **34**, 170 - 186 (2016)
2. O.V. Maistrenko, O.V. Mazorenko, Social programs: concept and structure, Sustainable development of the economy, **28**, 44 - 49 (2015)
3. A.S. Mahdich, The social factors' influence on economic growth: The national aspect, Nobel Economic Journal, **1 (8)**, 68 – 78 (2015)
4. N.O. Stefanopulos, Redistrations and Territorial Communities, University of Pennsylvania, **160**, 1379 - 1477 (2012)
5. The European Commission, Well-being, Eurobarometer Qualitative Studies, **1**, 24 – 50 (2011)
6. International Bank for Reconstruction and development, Gender issues' evaluation in Ukraine, Public Disclosure Authorized, **1**, 28 - 74 (2016)
7. O.O. Karmazina, Social indicators of Ukraine population's living standards, State service of Ukraine statistics, **1**, 15 - 167 (2018)
8. V.M. Palchenkova, Review of foreign concepts of social control, Theory and history of state and law, **1 (18)**, 3 – 7 (2015)

9. A.V. Furman, O. Morshchakova, Social culture, Psychology and Society, **1**, 26 – 36 (2015)
10. K.D. Riff, C.L. Keies, The rebuilt structure of psychological welfare, Journal of personality and social psychology, **69 (4),** 719 - 727 (1995)
11. H. Shamborovskyi, Welfare as a socio-psychological category in economic science, Economy, **2 (122)**, 78 – 82 (2013)

Specialization issues of production of agrarian enterprises

Tatiana Vlasenko[1,*], *Vitaliy* Vlasovets[1], *Sergiy* Timofeev[2] and *Anna* Kravets[2]

[1]Kharkiv Petro Vasylenko National Technical University of Agriculture, 61002 Alchevski Str, Kharkiv, Ukraine
[2]Ukrainian State University of Railway Transport, 61000 Feierbakh Square 7, Kharkiv, Ukraine

Abstract. The relevance of the study is explained by the fact that the combination of branches of livestock farming and crop production within agrarian enterprises creates a certain complexity in the use of material, labour and land resources and makes it necessary to create an optimal ratio on the basis of the formation of the most perspective and efficient industries. However, the contradictory and controversial nature of the data on the impact of various interconnected factors, the presence of primarily theoretical modelling while determining the optimal level of specialization of agrarian enterprises does not allow to achieve a given level of efficiency and efficiency of functioning performance Therefore, the purpose of the research is to provide scientific substantiation of the efficiency improving directions of agrarian enterprises through their specialization improvement. They are the empirical basis of the structured choice of directions to increase the functioning efficiency of agrarian enterprises. For the crop production industry, the main direction that will increase the management efficiency is the crops productivity increase due to the production profitability increase for the most management conditions. It was determined that regional features are important for plant cultivation and animal husbandry sectors while determining the influence factors on specialization. It was found out that for both industries, the negative impact of labour security and ineffective management practices are typical, and state support through subsidies for agricultural enterprises does not guarantee the increase in the production efficiency.

1 Introduction

According to modern studies searching for tendency of the increase of agrarian enterprises efficiency, there is a sectoral structure that is the ratio of interdependence of certain leading industries (crop production, livestock farming) and sub-sectors, which reflect the integral system the agrarian enterprise production [1]. The functioning effectiveness of such a system at each level of specialization causes the necessity to create an optimal ratio on the basis of the creation of the most perspective and efficient industries [2].

2 Problem Statement

* Corresponding author: vlasovez@ukr.net

However, the contradictory and controversial nature of the data on the impact of various interconnected factors, the presence of primarily theoretical modelling while determining the optimal level of specialization of agrarian enterprises does not allow to achieve a given level of efficiency and efficiency of functioning performance. Despite the wide range of applied theoretical models in the studied works, there are no practical recommendations for determining the impact of enterprises specialization on their functioning efficiency in dynamics.

3 Research Questions

The theoretical and methodological basis of the study is a systematic approach to studying the laws of development of agrarian enterprises in dynamics, the basic postulates of economic theory, macro- and microeconomics. The main tasks of the research are to determine the scientific principles of the sectoral structure formation of agrarian production and the peculiarities of its specialization levels formation, as well as the methodological principles substantiation for determining the optimality of the sectoral structure of agrarian enterprises in the dynamics.

4 Purpose of The Study

The purpose of the research is to provide scientific substantiation of the efficiency increasing directions of agrarian enterprises through their specialization improvement.

5 Research Methods

The analysis of the density of the scientific research, cluster and temporal analyses were used to estimate the scientometric data for the period of 1990-2019 to establish the time-division of the connections between the clusters of the scientific directions of the most cited publications. This approach made it possible to establish the scientific priorities of the research on the specialization of agrarian enterprises.

6 Findings

The sectoral structure of agrarian enterprises is not sustainable. The ratio of industries is constantly changing under the influence of natural, economic and social conditions [3]. The analysis of the scientific research density aimed at studying the specialization of agricultural production in the world has allowed to outline the main factors that influence the formation of specialization levels of agrarian enterprises in the dynamics, namely: natural and climatic conditions, market conditions, labour supply level, development of agrotechnologies, material and technical support Regression and correlation analyzes are often used to determine the influence of factors on the resulting feature. Our research has established [4] that the use of regression and correlation analysis allows us to reliably describe processes regularities only in the presence of a normal population distribution, and the other factors immutability, which gives the stability of the basic dependence. However, when studying economical phenomena in dynamics, the population with normal distribution is rare, and at the feature compatibility - is practically absent. Therefore, for a further studies, a relatively new method, the statistical equations of dependencies [5], has been used.

Using temporal analysis, clusters of scientific research on the specialization of agrarian enterprises (Fig. 1 a) and their main connections are established on the basis of the analysis of the most cited research (Fig. 1 b).

The first cluster of research, based on the work of R. Pieri, D. Rama, L. Venturini [6] aimed at the studying of the impact of agrarian enterprises specialization on the industries competitiveness. The relationship between agrarian production factors and vertical specialization is studied in I. Ferto works. [3] Scientists R. Sarker and Y. Surry [2] on the base of I. Ferto research substantiated the need to take into account not only the influence of production factors, but also their intensity and product differentiation on the agrarian enterprises specialization. Studies by Bojnec S. made an assessment of the relationship of value added products and the enterprises specialization [7]. Despite the wide range of applied theoretical models in the works, there are no practical recommendations for determining the impact of enterprises specialization on their functioning efficiency in dynamics and determining the optimality of the sectoral structure.

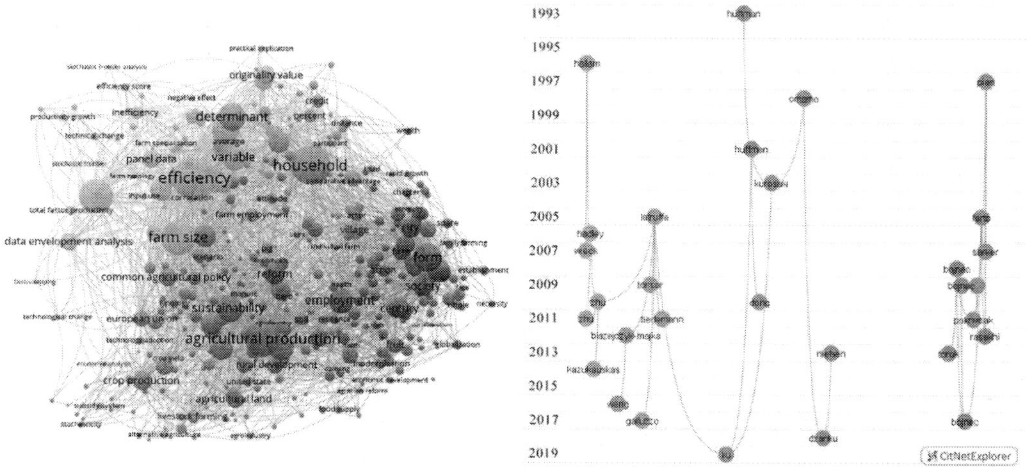

a) b)

Fig. 1. Clusters of scientific research (a) on the agricultural production specialization in the world (red and yellow - the most intense development, green and blue - less dynamics) and links between the scientific directions of the most cited authors (b) for the period of 1990-2019. Source: Web of Science Core Collection, Clarivate Analytics (Data Processed: April 10, 2019)

The second cluster of research is aimed at studying the influence of various factors on the specialization of the primarily crop production industry. It is based on the works of W. E. Huffman and R. E. Evenson focusing on the study of the factors influence (agrochemical prices, salaries, exploitation costs of agricultural machinery, farm sizes) and state programs on specialization and production efficiency [8]. They emphasize the special role of market factors in the sectoral structure formation. S. W. Omamo offers developing countries to focus primarily on the rational specialization of industries that have high transport costs when importing products into the country [9]. Increasing the efficiency of transportation by rural road networks is one of the effective ways of increasing small farms specialization. Empirical studies by T. Kurosaki aimed at the impact determining of the concentration of crop cultivation area on the specialization and diversification of the enterprises sectoral structure allowed to establish that the increase in the crops productivity mainly increases the production profitability [10]. Studies of this group of scientists covered, on the one hand, a significant historical interval of the various factors influence on specialization, but on the other hand, they did not take into account their specific effect in the context of agrarian production activity in Ukraine, as well as the formation of the sectoral structure of farms.

The third cluster of research is devoted to the study of technical efficiency and specialization mainly of livestock enterprises on their financial figures of efficiency. Thus, on the basis of studies of the impact of a significant number of production efficiency factors D. Hallam and F. Machado have established that for Portugal, there is a stable correlation between the size of the dairy farm and the degree of specialization [11]. The average production efficiency varies within 60-70% and increases with the enterprises consolidation. D. Hadley studied the impact of technical efficiency and specialization on livestock farming and crop production in the UK [12]. He outlined two main groups. The first one concerns farms operating on the upper limit of profitability for which the technical efficiency increase quickly leads to the production efficiency increase. To the second - the enterprises for which the increase of technical efficiency leads to less intense dynamics. For both groups, it has been established that the herd size has primarily effect on the development of the livestock sector, and for both industries - the volume of credit commitments, the average age of workers, the level of specialization and ownership. For Polish farms, the influence of the technical effect and scale of production was studied by L. Latruffe, K. Balcombe, S. Davidova and K. Zawalinska [13]. They confirmed that the livestock sector is on average more sensitive to technical and scale effects than crop production. But for both industries, the negative impact of ineffective management practices is characteristic. It is established that the low level of education of workers employed in farms is one of the most important reasons of inefficiency of technical modernization method of agrarian production. The research was further conducted by L. Blazejczyk-Majka, R. Kala and K. Maciejewski, who considered the effect of technical efficiency on the farms for traditional EU countries and its new members [14]. It has been determined that the greatest efficiency is achieved at the expense of large farms, but those of the new EU member states at the same time had low scale efficiency, while those belonging to the traditional EU countries acted on a scale close to the optimal. In addition, it has been confirmed that longer period of agriculture under relatively stable conditions contributes to higher efficiency, regardless of the agricultural production type. On the other hand, in contrast to the approach where higher specialization contributes to greater efficiency, it has been found out that farms with crop production specialization are on average less effective than mixed farms, although the difference between efficiency decreases with the size increase.

The comparison of L. Latruffe and L. Blazejczyk-Majka studies shows that the data of the technical efficiency calculation are significantly dependent on the scale of the research carried out and have significant differences for the interstate, state and regional levels. In addition, the absolute value of maximum technical efficiency is not comparable. Some authors do not directly link technical and economic efficiency. A closer connection exists for dairy farms, and for the crop production it is less obvious. In addition, S.C. Kumbhakar, E.G. Tsionas and T. Sipilainen's studies in the Finnish dairy industry have shown that traditional production technologies are more efficient, as well as technically more than organic, which usually require state support. State support through subsidies for agrarian enterprises also does not guarantee efficiency increase. Thus, the share of subsidies in the total amount of factors has a negative impact on technical efficiency in Germany (the farms are expanding), but has little positive impact in Sweden (the specialization of farms is increasing) and practically does not affect the Netherlands.

7 CONCLUSION

The temporal analysis of research on management of agrarian enterprises specialization made it possible to reveal a number of tendencies that are characteristic for its various levels. It was established that while studying the problems of the impact of agrarian enterprises specialization on the competitiveness of industries, it is necessary to take into account not only the influence of production factors, but also their dynamics intensity and product differentiation. For the crop production industry, the main factor influencing the specialization level is the increase in crop yields, which increases the production profitability for the most farming conditions. Besides such factors as prices on agrochemicals, salary, agricultural machinery exploitation prices, the size of farms and state programs of technical support are secondary. It has been established that specialization increase for dairy farms (technical effect), along with their size increase (scale effect), positively influences the increase of production profitability, especially for regions with low prime cost. It is confirmed that the livestock farming industry is on average more sensitive to technical and scale effects than crop production. It has been determined that regional features are the most important for determining the influence factors both for the plant and livestock farming industries of the enterprise. Both industries are characterized by the negative impact of labour supply and ineffective management practices, and the state support through subsidies for agrarian enterprises does not guarantee increased production efficiency.

References

1. P. Sckokai, D. Moro, Modeling the reforms of the common agricultural policy for arable crops under uncertainty. *AJAE*, **88**(1), 43-56 (2006)
2. R. Sarker, Y. Surry, Product differentiation and trade in agri-food products: Taking stock and looking forward. *International Agricultural Trade and Development,* (Hauppauge: Nova Science Publishers Inc., 39-80, 2007)
3. I. Ferto, Vertically differentiated trade and differences in factor endowment: The case of agri-food products between Hungary and the EU. *JAE*, **56**(1), 117-134 (2005)
4. T.S. Skoblo, V.M. Vlasovets, Specific Features of the Formation of Structures in 60Kh2N4GDeF Precipitation-Hardening Steel. *Materials Science*, **12**, 26-29 (2001)
5. T.V. Vlasenko, Modeliuvannia dynamiky rozvytku dlia udoskonalennia upravlinnia spetsializatsiieiu pidpryiemstva. *Tekhnolohycheskyi audyt y rezerv proyzvodstva*, **5/4**(31), 9-15 (2016)
6. R. Pieri, D. Rama, L. Venturini, Intra-industry trade in the European dairy industry. *European Review of Agricultural Economics*, **24**(3-4), 411-425 (1997)
7. S. Bojnec, I. Ferto, Agro-food trade competitiveness of Central European and Balkan countries. *Food Policy*, **34**(5), 417-425 (2009)
8. W.E. Huffman, R.E. Evenson, Structural and productivity change in US agriculture, 1950-1982. *Agricultural Economics,* **24**(2), 127-147 (2001)
9. A. Naseem, D.J. Spielman, S.W. Omamo, Private-Sector Investment in R&D: A Review of Policy Options to Promote its Growth in Developing-Country Agriculture. *Agribusiness*, **26**(1), 143-173 (2010)
10. T. Kurosaki Specialization and diversification in agricultural transformation: The case of West Punjab,1903-92. *AJAE,* **85**(2), 372-386 (2003)
11. D. Hallam, F. Machado, Efficiency analysis with panel data: A study of Portuguese dairy farms. *European Review of Agricultural Economics,* **23**(1), 79-93 (1996)
12. D. Hadley, Patterns in technical efficiency and technical change at the farm-level in England and Wales, 1982-2002. *JAE,* **57**(1), 81-100 (2006)
13. L. Latruffe, K. Balcombe, S. Davidova, K. Zawalinska, Technical and scale efficiency of crop and livestock farms in Poland: does specialization matter? *Agricultural Economics.* **32**(3), 281-296 (2005)

14. L. Blazejczyk-Majka, R. Kala, K. Maciejewski, Productivity and efficiency of large and small field crop farms and mixed farms of the old and new EU regions. *Agricultural Economics-Zemedelska Ekonomika*, **58**(2), 61-71 (2012)

AUTHOR INDEX

AUTHOR INDEX

AUTHOR INDEX

AUTHOR INDEX